5.2 FUNDAMENTAL IDENTITIES

$\sin \theta \csc \theta = 1$ $\sin^2 \theta + \cos^2 \theta = 1$

$\cos \theta \sec \theta = 1$ $\tan^2 \theta + 1 = \sec^2 \theta$

$\tan \theta \cot \theta = 1$ $\cot^2 \theta + 1 = \csc^2 \theta$

$\tan \theta = \dfrac{\sin \theta}{\cos \theta}$ $\cot \theta = \dfrac{\cos \theta}{\sin \theta}$

$\sin(-\theta) = -\sin \theta$ $\csc(-\theta) = -\csc \theta$

$\cos(-\theta) = \cos \theta$ $\sec(-\theta) = \sec \theta$

$\tan(-\theta) = -\tan \theta$ $\cot(-\theta) = -\cot \theta$

5.3 FUNCTIONS OF IMPORTANT ANGLES AND COFUNCTION RELATIONSHIPS

θ	$\sin \theta$	$\cos \theta$	$\tan \theta$
$0°$	0	1	0
$30°$	$\frac{1}{2}$	$\frac{\sqrt{3}}{2}$	$\frac{\sqrt{3}}{3}$
$45°$	$\frac{\sqrt{2}}{2}$	$\frac{\sqrt{2}}{2}$	1
$60°$	$\frac{\sqrt{3}}{2}$	$\frac{1}{2}$	$\sqrt{3}$
$90°$	1	0	undefined
$180°$	0	-1	0
$270°$	-1	0	undefined

$\sin \theta = \cos(90° - \theta)$ $\cos \theta = \sin(90° - \theta)$

$\tan \theta = \cot(90° - \theta)$ $\cot \theta = \tan(90° - \theta)$

$\csc \theta = \sec(90° - \theta)$ $\sec \theta = \csc(90° - \theta)$

5.4 RADIAN MEASURE

π radians = 180 degrees

1 degree $= \dfrac{\pi}{180}$ radians   1 radian $= \dfrac{180}{\pi}$ degrees

$s = r\theta$ (θ in radians)   Formula for arc length

$A = \frac{1}{2}r^2\theta$ (θ in radians)   Formula for the area of a sector

$v = r\omega$ Formula for linear velocity

5.6, 5.7 GRAPHS OF THE TRIGONOMETRIC FUNCTIONS

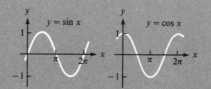

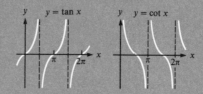

5.6–5.8 AMPLITUDE, PERIOD, AND PHASE SHIFT

Function	Amplitude	Period	Phase Shift
$y = a \sin b(x + c)$	$|a|$	$\frac{2\pi}{b}$	$|c|$
$y = a \cos b(x + c)$	$|a|$	$\frac{2\pi}{b}$	$|c|$
$y = \tan b(x + c)$	—	$\frac{\pi}{b}$	$|c|$

6.1 RIGHT TRIANGLE TRIGONOMETRY

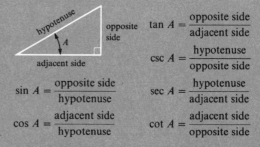

$\tan A = \dfrac{\text{opposite side}}{\text{adjacent side}}$

$\csc A = \dfrac{\text{hypotenuse}}{\text{opposite side}}$

$\sin A = \dfrac{\text{opposite side}}{\text{hypotenuse}}$

$\sec A = \dfrac{\text{hypotenuse}}{\text{adjacent side}}$

$\cos A = \dfrac{\text{adjacent side}}{\text{hypotenuse}}$

$\cot A = \dfrac{\text{adjacent side}}{\text{opposite side}}$

6.3 THE LAW OF COSINES

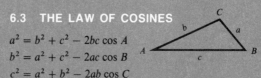

$a^2 = b^2 + c^2 - 2bc \cos A$

$b^2 = a^2 + c^2 - 2ac \cos B$

$c^2 = a^2 + b^2 - 2ab \cos C$

COLLEGE ALGEBRA
AND TRIGONOMETRY Second Edition

Books in the Gustafson and Frisk Series

COLLEGE ALGEBRA AND TRIGONOMETRY

Second Edition

R. DAVID GUSTAFSON
PETER D. FRISK

Rock Valley College

Brooks/Cole Publishing Company
Monterey, California

To our wives, Carol and Martha,
and our children, Kristy and Steven;
Sarah, Heidi, and David

Consulting Editor: *Robert J. Wisner*

Brooks/Cole Publishing Company
A Division of Wadsworth, Inc.

Printed in the United States of America
10 9 8 7 6 5 4 3 2 1

Library of Congress Cataloging-in-Publication Data

Gustafson, R. David (Roy David), [date]
  College algebra and trigonometry.

  Includes index.
  1. Algebra. 2. Trigonometry. I. Frisk, Peter D.,
[date]   . II. Title.
QA154.2.G873 1986   512'.13   85-19558

ISBN 0-534-06480-9

Sponsoring Editor: *Jeremy Hayhurst*
Editorial Assistant: *Eileen Galligan*
Production Editor: *Candyce Cameron*
Production Assistant: *Dorothy Bell*
Manuscript Editors: *Charles Hibbard, Paul Monsour*
Permissions Editor: *Carline Haga*
Interior Design: *Sharon L. Kinghan*
Cover Design: *Sharon L. Kinghan, David Aguero*
Cover Illustration: *David Aguero*
Art Coordinator: *Michele Judge*
Interior Illustration: *Lori Heckelman*
Typesetting: *Syntax International, Singapore*
Cover Printing: *The Lehigh Press Co., Pennsauken, New Jersey*
Printing and Binding: *R. R. Donnelley & Sons Co., Crawfordsville, Indiana*

PREFACE

TO THE INSTRUCTOR

A wave of educational reform is sweeping the country. Educators and legislators, alarmed by what they perceive to be an erosion in the quality of education, are pressing for pervasive changes in the educational system. At the heart of the matter is the desire to promote greater student achievement through higher academic standards.

Implementing these reforms at the college level will challenge even the best of teachers, especially in these transition years when many students have had marginal academic preparation. To help students meet increased expectations, mathematics teachers will need a battery of teaching materials, beginning with a text that fulfills students' needs without compromising instructors' standards.

This new edition of College Algebra and Trigonometry, invited by the widespread success of the previous edition, answers the challenge of the next decade. College Algebra and Trigonometry, Second Edition, provides a thorough, no-nonsense approach to the topics of college algebra and trigonometry. It is designed to be easy for students to use and understand and to show them that these topics are useful and interesting.

Thanks to helpful suggestions from many instructors who used and reviewed this text, we have been able not only to maintain the precision and accessibility of previous editions, but also to refine many of the following special features:

Solid Mathematics
The treatment of college algebra and trigonometry is direct and straightforward. Although the treatment is mathematically sound, it is not so rigorous that it will confuse students. Every effort has been made to ensure the accuracy of the mathematics and of the answers to the exercises. The book has been critiqued by dozens of reviewers, and each author and several problem checkers have worked each exercise. Although the exercise sets are designed primarily to provide practice and drill, they also contain problems that will challenge even the best of your students. The text contains over 4600 exercises.

Accessibility to Students
The text is written in a way that students can read and understand. On the Frey readability test, the writing is at the tenth-grade level. The numerous problems in each exercise set are carefully keyed to over 500 worked examples in which author's notes explain many of the steps used in the problem-solving process. Students will

v

like the review exercises with all answers provided, the functional use of a second color, and the endpapers that list, in order of presentation, the important formulas that are developed in the text. They will also appreciate the *Student Solutions Manual,* which contains solutions to the even-numbered exercises.

Emphasis on Applications   To show that mathematics is useful, we have included a large number of word problems and applications throughout the text.

Built-in Redundancy   Because skills taught in the early chapters are used throughout the text, students have several opportunities to review or even relearn material. This constant review helps to improve retention.

In response to the suggestions we have received, we have made the following major changes in this edition:

1.   Many sections have been rewritten using simpler and more concise language. The writing is now even more accessible to students.
2.   The chapter on functions has been rewritten. The material on the coordinate system, graphing linear equations, and writing equations of lines now precedes the discussion of functions. This organization better prepares the student for the more formal aspects of functions.

      The horizontal line test is now included as a test for one-to-one functions.
3.   The treatment of trigonometry has been expanded to four chapters. Chapter 5 introduces the trigonometric functions as functions of angles with both degree and radian measure, along with their graphs. Chapter 6 is devoted entirely to applications. Chapter 7 thoroughly treats identities and equations, and Chapter 10 covers complex numbers and polar coordinates.
4.   Gaussian elimination is now emphasized as a method of solving simultaneous equations.
5.   The chapter on exponential functions and logarithms has been expanded to give more emphasis to base-*e* exponential expressions and natural logarithms. More applications are included that show how the exponential and logarithmic functions model certain events in nature.

Supplementary Materials   For your convenience, there is a *Test Bank of Items*, including three forms of chapter test for each chapter, available in hardcopy manual format.

ORGANIZATION AND COVERAGE

The text can be used in a variety of ways. Several chapters are sufficiently independent to allow you to pick and choose topics that are relevant to student needs. The following diagram shows how the chapters are interrelated:

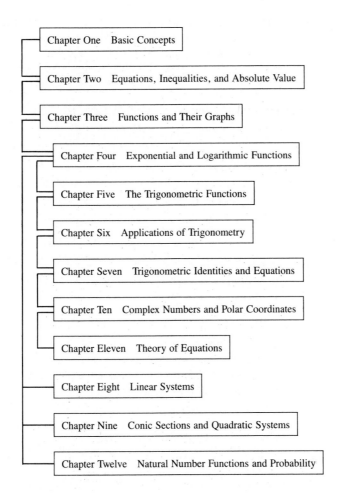

Chapter One   Basic Concepts

Chapter Two   Equations, Inequalities, and Absolute Value

Chapter Three   Functions and Their Graphs

Chapter Four   Exponential and Logarithmic Functions

Chapter Five   The Trigonometric Functions

Chapter Six   Applications of Trigonometry

Chapter Seven   Trigonometric Identities and Equations

Chapter Ten   Complex Numbers and Polar Coordinates

Chapter Eleven   Theory of Equations

Chapter Eight   Linear Systems

Chapter Nine   Conic Sections and Quadratic Systems

Chapter Twelve   Natural Number Functions and Probability

CALCULATORS

The use of calculators is assumed throughout the text. We believe that students should learn calculator skills in the mathematics classroom. They will then be prepared to use a calculator in science and business classes and for non-academic purposes. The directions within the exercise sets clearly indicate which exercises require the use of a calculator.

TOPICS COVERED

REVIEW

Chapters 1 and 2 primarily review basic algebra—the real number system and its properties, exponents and radicals, polynomial arithmetic, solutions of equations and inequalities, and absolute value.

FUNCTIONS

The concept of function is introduced in Chapter 3, with emphasis on function notation and graphing polynomial functions. Also included is a section on variation and a thorough treatment of inverse functions.

EXPONENTIAL AND LOGARITHMIC FUNCTIONS

Chapter 4 includes exponential functions, logarithms, and many of their applications. Base-e and natural logarithms are treated in detail. The calculator is emphasized in this chapter.

TRIGONOMETRY

Trigonometry is introduced by considering the trigonometric functions of angles. This is the way trigonometry developed historically, and we are convinced that this approach makes sense to students. However, we make the transition from angle domains to real number domains very early in the discussion. Calculators are emphasized throughout.

SYSTEMS OF EQUATIONS AND INEQUALITIES

Chapter 8 includes techniques for solving systems of linear equations. Matrix methods are developed and some matrix algebra is presented. Linear programming in two variables is discussed. The topic of partial fractions is introduced as an application of systems of linear equations.

CONIC SECTIONS

Chapter 9 develops the basic forms of the equations for the conic sections and provides opportunity for graphing these equations. Solutions of simultaneous second-degree equations are obtained both graphically and algebraically.

COMPLEX NUMBERS

Complex numbers are treated both algebraically and trigonometrically in Chapter 10. Complex roots of polynomial equations are considered in Chapter 11.

THEORY OF EQUATIONS

Chapter 11 discusses methods for finding rational roots of polynomial equations. Several examples illustrate the interplay between the fundamental theorem of algebra, Descartes' rule of signs, the remainder and factor theorems, the rational root theorem, and the conjugate pairs result. Binary chopping is used to approximate irrational roots.

NATURAL NUMBER FUNCTIONS AND PROBABILITY

Mathematical induction is introduced in Chapter 12. Proof by induction is used in the discussion of sequences, series, and progressions. The binomial theorem, permutations, and combinations lead into a presentation of simple and compound probability, odds, and mathematical expectation. The proof of the binomial theorem for natural number exponents is given in Appendix I.

TO THE STUDENT

We have tried to write a text that you can read and understand. We have provided an extensive number of worked examples and have keyed them to the exercises within each exercise set. Be sure to read the explanations carefully since they contain much of the value of the text. Answers to all the even-numbered exercises are available in the *Student Solutions Manual*.

The material presented here will be of value to you in later years. Therefore, we suggest that you keep your text after completing this course. It will be a good source of reference and will keep at your fingertips the material that you have learned here.

We wish you well.

ACKNOWLEDGMENTS

We are grateful to the following people who reviewed the manuscript in its various stages. All of them had valuable suggestions that have been incorporated into this text.

John Adam
Old Dominion University

Daniel Anderson
University of Iowa

James Arnold
University of Wisconsin

Wilson Banks
Illinois State University

Jerry Bloomberg
Essex Community College

Dale Boye
Schoolcraft College

James Choike
Oklahoma State University

Lee R. Clancy
Golden West College

Romae J. Cormier
Northern Illinois University

John S. Cross
University of Northern Iowa

Lena Dexter
Faulkner State Junior College

Emily Dickinson
University of Arkansas

Robert E. Eicken
Illinois Central College

Mary Jane Gates
University of Arkansas at Little Rock

Jerry Gustafson
Beloit College

Jerome Hahn
Bradley University

Douglas Hall
Michigan State University

Robert Hall
University of Wisconsin

David Hansen
Monterey Peninsula College

William Hinrichs
Rock Valley College

Arthur M. Hobbs
Texas A & M University

Jack E. Hofer
California Polytechnic State University

Warren Jaech
Tacoma Community College

William B. Jones
University of Colorado

William Lakin
Old Dominion University

Marcus McWaters
University of Southern Florida

James W. Mettler
Pennsylvania State University

Eldon L. Miller
University of Mississippi

Stuart E. Mills
Louisiana State University, Shreveport

Gilbert W. Nelson
North Dakota State

Marvin Papenfuss
Lora's College

Anthony Peressini
University of Illinois

Janet P. Ray
Seattle Central Community College

Paul Schaefer
SUNY, Geneseo

L. Thomas Shiflett
Southwest Missouri State University

Richard Slinkman
Bemidji State University

John Snyder
Sinclair Community College

Warren Strickland
Del Mar College

Ray Tebbetts
San Antonio College

Carroll G. Wells
Western Kentucky University

Charles R. Williams
Midwestern State University

Albert Zechmann
University of Nebraska

We wish to thank Jeanne Wyatt and Daniel Ropp for their help in preparing the manuscript. We also thank the staff at Brooks/Cole, especially Craig Barth, Jeremy Hayhurst, Candy Cameron, Dorothy Bell, Sharon Kinghan, Michele Judge, and Lori Heckelman for all their help.

R. David Gustafson
Peter D. Frisk

CONTENTS

1 BASIC CONCEPTS

The concept of number is fundamental to mathematics. For this reason, we begin by discussing various sets of numbers and their properties.

1.1 THE SET OF RATIONAL NUMBERS AND ITS SUBSETS

A **set** is any collection of objects. Each object is called a **member** or an **element** of the set. To denote a set, we use braces to enclose the list of its elements. For example, the notation $\{a, b, c\}$ means the set whose elements are the letters a, b, and c. We read the statement

$$b \in \{a, b, c\}$$

as "b is an element of the set containing a, b, and c."

In mathematics, the most basic set of numbers is the set of **natural numbers**. These are the numbers that are used for counting.

$$\mathbf{N} = \{1, 2, 3, 4, 5, 6, 7, 8, 9, 10, 11, \ldots\}$$

The three dots, called the **ellipsis**, following the natural number 11 indicate that the list of natural numbers continues on forever. There is no largest natural number.

Certain natural numbers are called **prime numbers**. The prime numbers are those natural numbers greater than 1 that are divisible without remainder only by 1 and by the number itself:

$$\mathbf{P} = \{2, 3, 5, 7, 11, 13, 17, 19, \ldots\}$$

Because every prime number is also a natural number, we say that the set of prime numbers is a **subset** of the set of natural numbers, and we write

$$\mathbf{P} \subset \mathbf{N} \qquad \text{Read as "}\mathbf{P}\text{ is a subset of }\mathbf{N}\text{."}$$

If a natural number is greater than 1 and can be divided without remainder by some number other than itself and 1, the number is called a **composite number**. The set of composite numbers is the set

$$\mathbf{C} = \{4, 6, 8, 9, 10, 12, 14, 15, 16, \ldots\}$$

Because every composite number is also a natural number, we have

$$\mathbf{C} \subset \mathbf{N} \qquad \text{Read as "}\mathbf{C}\text{ is a subset of }\mathbf{N}\text{."}$$

If we include 0 with the set of natural numbers, we have the set of **whole numbers**:

W = {0, 1, 2, 3, 4, 5, 6, 7, 8, 9, . . .}

Sets of numbers can be represented geometrically as points on a **number line**. To construct a number line, we draw a line, choose some arbitrary point on it, and label the point 0. We then mark off equal distances to the right and to the left of 0, labeling the points as in Figure 1-1. The line and the number labels on it continue forever in both directions.

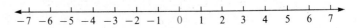

Figure 1-1

The numbers that represent points to the right of 0 are **positive numbers**, and the numbers that represent points to the left of 0 are **negative numbers**. The number 0 is neither a positive number nor a negative number.

To graph the set of natural numbers from 1 to 10 on the number line, we place large dots on the points represented by 1, 2, 3, 4, 5, 6, 7, 8, 9, and 10, as shown in Figure 1-2**a**. The point associated with each number is called the **graph** of that number. The number is called the **coordinate** of its corresponding point. The graph of the prime numbers less than 12 is shown in Figure 1-2**b**, and the graph of the whole numbers less than 10 is shown in Figure 1-2**c**.

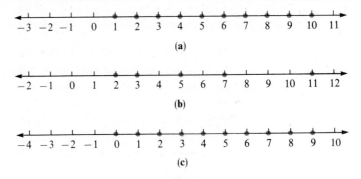

Figure 1-2

If the numbers −1, −2, −3, . . . are joined with the whole numbers, we have the set of **integers**:

J = {. . . , −6, −5, −4, −3, −2, −1, 0, 1, 2, 3, 4, 5, 6, . . .}

The graph of the integers is shown in Figure 1-3. Both the line and the points marked on it continue forever in both directions.

Figure 1-3

Those integers that are divisible by 2 without a remainder are called **even integers**, and those that are not are called **odd integers**. If **E** is the set of even integers and **O** is the set of odd integers, then

$$\mathbf{E} = \{\ldots, -8, -6, -4, -2, 0, 2, 4, 6, 8, \ldots\}$$
$$\mathbf{O} = \{\ldots, -7, -5, -3, -1, 1, 3, 5, 7, \ldots\}$$

If a number can be written as a fraction with an integer for its numerator and a nonzero integer for its denominator, it is called a **rational number**. To denote the set **R** of rational numbers, we use **set-builder notation**, in which a **variable** is used to represent an element of the set.

$$\mathbf{R} = \{x \mid x \text{ is a number that can be written in the form } \frac{a}{b}$$

where a and b are integers, and $b \neq 0.\}$

The previous statement is read as "**R** is the set of all real numbers x such that x is a number that can be written in the form $\frac{a}{b}$, where a and b are integers and b is not equal to 0." The rational numbers include numbers such as

$$\frac{3}{4}, \quad \frac{-1}{3}, \quad \frac{5}{1}, \quad \frac{-8}{4}, \quad \frac{0}{5}, \quad \frac{99}{113}$$

Numbers such as 3, 0, -0.25, and $0.333\ldots$ are rational numbers also because each one can be written in the required form:

$$3 = \frac{3}{1}, \quad 0 = \frac{0}{7}, \quad -0.25 = \frac{-1}{4}, \quad 0.333\ldots = \frac{1}{3}$$

It is also possible to graph the rational numbers on a number line. For example, the point with a coordinate of $\frac{3}{2}$ is at a distance midway between the points with coordinates of 1 and 2. The point with a coordinate of $\frac{1}{3}$ is at a distance one-third of the way from 0 to 1. These points and others are shown in Figure 1-4.

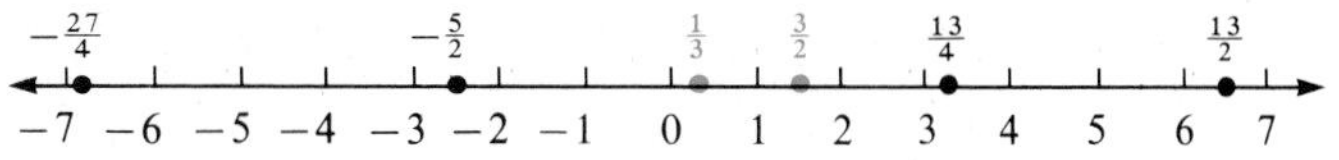

Figure 1-4

Every rational number in fractional form can be changed to decimal form. For example, to change $\frac{3}{4}$ to a decimal fraction, we divide 3 by 4 to obtain 0.75:

$$
\begin{array}{r}
0.75 \\
4{\overline{\smash{\big)}\,3.00}} \\
\underline{2\,8} \\
20 \\
\underline{20} \\
0
\end{array}
$$

Because the division leaves no remainder, the quotient is a **terminating decimal**. If we change a fraction such as $\frac{4}{15}$ to a decimal fraction, we obtain a **repeating decimal**:

$$
\begin{array}{r}
0.266\ldots \\
15\,\overline{)\,4.000} \\
3\,0 \\
\overline{1\,00} \\
90 \\
\overline{100} \\
90 \\
\overline{10}
\end{array}
$$

It can be shown that the decimal forms of all rational numbers are either terminating decimals or repeating decimals. It is also true that any decimal fraction that is a terminating decimal or a repeating decimal can be changed to fractional form.

Example 1   Change the *terminating decimal* 0.75 to fractional form.

Solution   Because the decimal fraction 0.75 means 75 hundredths, write 0.75 as $\frac{75}{100}$ and then reduce the fraction $\frac{75}{100}$ to obtain $\frac{3}{4}$. ■

Example 2   Change the *repeating decimal* 0.3 876 876 876 . . . to fractional form.

Solution   Note that the decimal has a repeating block of three digits, 876. Form an equation by setting x equal to the decimal.

   1. $x = 0.3\,876\,876\,876\ldots$

Then form another equation by multiplying both sides of Equation 1 by 10^3, which is 1000.

   2. $1000x = 387.6\,876\,876\,876\ldots$

Subtract each side of Equation 1 from the corresponding side of Equation 2 to obtain $999x = 387.3$.

$$
\begin{array}{r}
1000x = 387.6\,876\,876\,876\ldots \\
x = 0.3\,876\,876\,876\ldots \\
\hline
999x = 387.3\,000\,000\,000\ldots
\end{array}
$$

Finally, solve this equation for x and simplify the fraction.

$$999x = 387.3$$

$$x = \frac{387.3}{999} \qquad \text{Divide both sides by 999.}$$

$$= \frac{3873}{9990} \qquad \text{Multiply both the numerator and the denominator by 10.}$$

$$= \frac{1291}{3330} \qquad \text{Reduce.}$$

Use a calculator to show that the decimal representation of $\frac{1291}{3330}$ is

0.3 876 876 876 . . .

The key step in this example was multiplying both sides of Equation 1 by 10^3. If there had been n digits in the repeating block of the decimal, you would have multiplied both sides of Equation 1 by 10^n.   ■

Because it is always possible both to write rational numbers in fractional form as terminating or repeating decimals and to write terminating or repeating decimals as rational numbers in fractional form, these two sets of numbers are one and the same. Thus, the set of rational numbers can be described as the set of all decimals that either terminate or repeat.

$$\mathbf{R} = \{x \mid x \text{ is a terminating or a repeating decimal.}\}$$

Exercise 1.1

1. Find the first prime number larger than 50.
2. Find the largest prime number less than 100.
3. Draw a number line and graph the even integers between 19 and 31.
4. Draw a number line and graph the integers from -5 to -1.

In Exercises 5–28, indicate whether each statement is true. If a statement is false, explain why.

5. The number 27 is a prime number.
6. The number 2 is a prime number.
7. The set of composite numbers is a subset of the set of integers.
8. The set of prime numbers is a subset of the set of odd integers.
9. The product of two prime numbers is a prime number.
10. The sum of two prime numbers is a prime number.
11. The square of a prime number is a prime number.
12. The number 1 is a prime number.
13. The sum of two prime numbers can be a composite number.
14. The sum of two composite numbers is a prime number.
15. All natural numbers are integers.
16. The number 0 is neither even nor odd.
17. No even integers are prime numbers.
18. All odd integers are prime numbers.
19. The sum of an even integer and an odd integer is an odd integer.
20. The sum of three odd integers must be another odd integer.
21. If the product of several integers is an even integer, then at least one of the integers is an even integer.
22. If the product of several integers is an odd integer, then each of the integers must be an odd integer.
23. If the sum of three integers is an even integer, then each of the integers must be an even integer.
24. The set of even integers combined with the set of odd integers forms the set of integers.

25. The number 0 is not a rational number.

26. The symbol $\frac{16}{0}$ represents a rational number.

27. All integers are rational numbers.

28. Some integers are greater than -1.

In Exercises 29–34, change each fraction into an equivalent decimal fraction.

29. $\dfrac{1}{4}$ **30.** $\dfrac{7}{2}$ **31.** $\dfrac{2}{9}$ **32.** $\dfrac{3}{11}$ **33.** $-\dfrac{5}{12}$ **34.** $-\dfrac{1}{7}$

In Exercises 35–46, change each decimal fraction into an equivalent common fraction.

35. 0.3   **36.** 0.375   **37.** $-0.7575\ldots$ **38.** $-0.2828\ldots$
39. 0.123123 ...   **40.** 0.841841 ...   **41.** 0.3456456 ...   **42.** 0.9245245 ...
43. $-8.6161\ldots$ **44.** $-4.321321\ldots$ **45.** 1.61717 ...   **46.** 2.35151 ...

47. Change the repeating decimal 1.999 . . . to a common fraction and thereby show that 1.999 . . . = 2.

48. Change the repeating decimal 4.999 . . . to a common fraction and thereby show that 4.999 . . . = 5.

49. Does 0.999 = 1? Explain.

50. Determine the pattern in the decimal

0.1 21 221 2221 . . .

Does this decimal represent a rational number? Explain.

1.2 THE REAL NUMBERS AND THEIR PROPERTIES

Numbers whose decimal representations neither terminate nor repeat are called **irrational numbers**. Numbers such as

$$\sqrt{2} = 1.414213562\ldots$$

and

$$\pi = 3.141592653\ldots$$

are examples of irrational numbers because their decimal representations are nonterminating, nonrepeating decimals. To express the set of irrational numbers, we write

H = $\{x \,|\, x$ is a nonterminating, nonrepeating decimal.$\}$

If the elements of some set A are joined with the elements of some set B, the **union** of set A and set B is formed. The union of set A and set B is denoted as

$A \cup B$ Read as "the union of set A and set B."

The union of the set of rational numbers (the set of terminating or repeating decimals) and the set of irrational numbers (the set of nonterminating, nonrepeating decimals) is the set of all decimals. This set is called the set of **real numbers**:

$\mathscr{R} = \{x \,|\, x$ is a decimal number.$\}$

In symbols,

$$\mathscr{R} = \mathbf{R} \cup \mathbf{H}$$

Note that both the set of rational numbers and the set of irrational numbers are subsets of the set of real numbers.

The set of elements that are common to two sets A and B is called the **intersection** of set A and set B. The intersection of set A and set B is denoted as

$A \cap B$ Read as "the intersection of set A and set B."

For example, if $A = \{0, 1, 2, 3\}$ and $B = \{2, 3, 4, 5\}$, then

$$A \cap B = \{2, 3\}$$

If a set has no elements, it is called the **empty set**, which is denoted by the symbol $\emptyset$. The intersection of the set of rational numbers and the set of irrational numbers is the empty set:

$$\mathbf{R} \cap \mathbf{H} = \emptyset$$

The graphs of many subsets of the real numbers are portions of the number line called **intervals**. Figure 1-5 shows the graph of the real numbers x that are between -2 and 4. We describe this set with the expressions $\{x \mid -2 < x < 4\}$ or $\{x \mid -2 < x \text{ and } x < 4\}$, or more simply by

$$-2 < x < 4 \qquad \text{or} \qquad -2 < x \quad \text{and} \quad x < 4$$

where the symbol $<$ is read as "is less than."

Figure 1-5

The open circles at points -2 and 4 indicate that the endpoints are not included in the graph. When neither endpoint is included, the interval is called an **open interval**. The open interval between -2 and 4 is often denoted as $(-2, 4)$. The parentheses in this notation indicate that the endpoints are not included.

The graph of the set of all real numbers from 1 to 6 is shown in Figure 1-6. We describe this set with the expressions

$$1 \leq x \leq 6 \qquad \text{or} \qquad 1 \leq x \quad \text{and} \quad x \leq 6$$

where the symbol $\leq$ is read as "is less than or equal to." The closed circles at points 1 and 6 indicate that these points are included in the graph. When both endpoints are included, the interval is called a **closed interval**. The closed interval from 1 to 6 can be denoted as $[1, 6]$. The brackets in this notation indicate that the endpoints are included.

Figure 1-6

Sometimes an interval is half-open. For example, Figure 1-7 shows the graph of the half-open interval $(-6, -1]$. This interval can also be described with the expressions

$$-6 < x \leq -1 \qquad \text{or} \qquad -6 < x \quad \text{and} \quad x \leq -1$$

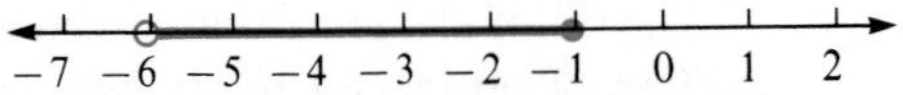

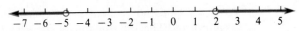

Figure 1-7

As a final example, the graph of all real numbers x such that $x < -5$ or $x > 2$ (read the symbol $>$ as "is greater than") is shown in Figure 1-8.

Figure 1-8

There is a distinction between the words *and* and *or* as used in previous examples. The statement $-2 < x$ *and* $x < 4$ means that both conditions on x must be true simultaneously. The statement $x < -5$ *or* $x > 2$ means that only one of the conditions on x needs to be true.

We note that if a and b are real numbers and $a < b$, then the graph of a is to the left of the graph of b. If $a > b$, then the graph of a is to the right of the graph of b.

Two real numbers are **negatives** of each other if their graphs on a number line lie at equal distances from 0 on opposite sides of 0. For example, the negative of 2 is negative 2, denoted as -2, and the negative of -2, denoted as $-(-2)$, is 2. See Figure 1-9. In general, for any real number x,

$$-(-x) = x$$

This property of negatives and others are summarized as follows:

Properties of Negatives.

1. $-(-x) = x$

2. $(-x) + (-y) = -(x + y)$

3. $x - y = x + (-y)$

4. $(-1)x = -x$
    $(-x)y = x(-y) = -xy$
    $(-x)(-y) = xy$

5. $\left. \begin{array}{l} \dfrac{-x}{y} = \dfrac{x}{-y} = -\dfrac{x}{y} \\[2ex] \dfrac{-x}{-y} = \dfrac{x}{y} \end{array} \right\}$ provided $y \neq 0$

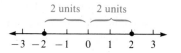

Figure 1-9

The **absolute value** of any real number a, denoted as $|a|$, is the distance on a number line between 0 and the point with a coordinate of a. Because points with coordinates of 4 and -4 both lie 4 units from 0, $|4| = |-4| = 4$. Similarly, because points with coordinates of -17 and 17 both lie the same distance from 0, $|-17| = |17| = 17$. In general,

$$|a| = |-a|$$

for any real number a.

The absolute value of a number can be defined more formally. In the following definition, read the symbol $\geq$ as "is greater than or equal to."

Definition.   If $x \geq 0$, then $|x| = x$.
If $x < 0$, then $|x| = -x$.

The definition of the absolute value of x indicates that if x is a positive number or 0, then x is its own absolute value. However, if x is a negative number, then $-x$ (which is a positive number) is the absolute value of x. Thus, $|x|$ always represents a nonnegative number. In other words, $|x| \geq 0$ for all x.

Example 1   Find **a.** $|3|$, **b.** $|-4|$, **c.** $|0|$, and **d.** $|1 - \sqrt{2}|$.

Solution   **a.** $|3| = 3$

b. $|-4| = -(-4)$
$ = 4$

c. $|0| = 0$

d. $|1 - \sqrt{2}| = -(1 - \sqrt{2})$ Because $1 < \sqrt{2}$, the number $1 - \sqrt{2}$ is negative.
$\phantom{|1 - \sqrt{2}|} = \sqrt{2} - 1$ ∎

Consider the number line in Figure 1-10. The distance between the points with coordinates of 1 and 4 is $4 - 1$, or 3 units. However, if the subtraction were done in the other order, the distance between these points would be $1 - 4$,

$$d = |4 - 1| = 3$$

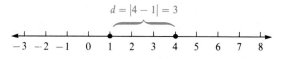

Figure 1-10

or -3 units. To guarantee that the distance between two points is always positive, we use absolute value symbols. Thus, the distance d between points 1 and 4 is $|4 - 1| = |1 - 4| = 3$.

The previous discussion suggests that we can define the distance d between two points with coordinates of a and b on the number line by using the concept of absolute value.

Definition. If a and b are the coordinates of two points on a number line, then the distance d between those points is given by the formula

$$d = |a - b|$$

Example 2   Find the distance d on a number line between points with coordinates of **a.** 3 and 5, **b.** -2 and 3, and **c.** -5 and -1.

Solution   **a.** $d = |5 - 3| = |3 - 5| = 2$

b. $d = |3 - (-2)| = |-2 - 3| = 5$

c. $d = |-5 - (-1)| = |-1 - (-5)| = 4$ ■

Any two numbers a and b can be compared. Either a and b are equal, or they are not. And if they are not equal, then one or the other must be the larger. The possibilities are summed up in the **trichotomy property**.

The Trichotomy Property.   For any two real numbers a and b, *exactly one* of the following relationships must hold:

1. a and b are equal $(a = b)$
2. a is less than b $(a < b)$
3. a is greater than b $(a > b)$

To indicate that x is not equal to y, we use the notation $x \neq y$. If x is not less than y, then x must be greater than or equal to y, which is denoted as $x \geq y$. Similarly, $x \leq y$ denotes that x is less than or equal to y.

Three properties of equality are basic in algebra.

Properties of Equality.   If a, b, and c are real numbers, then

$a = a$ (Reflexive property)

If $a = b$, then $b = a$.   (Symmetric property)

If $a = b$ and $b = c$, then $a = c$.   (Transitive property)

Example 3 Determine if the relation $<$ is **a.** reflexive, **b.** symmetric, and **c.** transitive.

Solution **a.** To determine if the relation $<$ is reflexive, ask the question "Is a real number less than itself?" Because the answer is no, the relation is *not* reflexive.

b. To determine if the relation $<$ is symmetric, ask the question "If one real number is less than a second, is it also true that the second number is less than the first?" Because the answer is no, the relation is *not* symmetric.

c. To determine if the relation $<$ is transitive, ask the question "If one real number is less than a second, and if that second real number is less than a third, does this mean that the first is less than the third?" Because the answer is yes, the relation *is* transitive. ∎

There are many assumptions, called the **field properties**, that we must use when working with real numbers. If a, b, and c are real numbers, then we have

---

The Closure Properties of Real Numbers.

$a + b$ is a real number.

$a - b$ is a real number.

ab is a real number.

$\dfrac{a}{b}$ is a real number   (provided $b \neq 0$).

The Associative Properties of Addition and Multiplication.

$$(a + b) + c = a + (b + c) \qquad (ab)c = a(bc)$$

The Commutative Properties of Addition and Multiplication.

$$a + b = b + a \qquad ab = ba$$

The Distributive Property of Multiplication over Addition.

$$a(b + c) = ab + ac$$

The Identity Elements.   There exists a real number 0 such that

$$a + 0 = 0 + a = a$$

There exists a real number 1 such that

$$a \cdot 1 = 1 \cdot a = a$$

The number 0 is called the **additive identity element** and the number 1 is called the **multiplicative identity element**.

The Inverse Elements.   For every real number a, there exists a real number $-a$ such that

$$a + (-a) = -a + a = 0$$

There also exists a real number $\dfrac{1}{a}$ $(a \neq 0)$ such that

$$\frac{1}{a} \cdot a = a \cdot \frac{1}{a} = 1$$

The number $-a$ is called the **additive inverse** or the **negative** of a. The number $\dfrac{1}{a}$ is called the **multiplicative inverse** or the **reciprocal** of a.

Example 4   The statements in the left column are true because of the property of equality or the property of real numbers listed in the right column.

$a < b$ or $a = b$ or $a > b$.	trichotomy property
$2 + 7$ is a real number.	closure property for addition.
$2(7)$ is a real number.	closure property for multiplication
$9 + 3 = 3 + 9$	commutative property for addition
$8 \cdot 3 = 3 \cdot 8$	commutative property for multiplication
$9 + (2 + 3) = (9 + 2) + 3$	associative property for addition
$2(xy) = (2x)y$	associative property for multiplication
$2(x + 3) = 2x + 2 \cdot 3$	distributive property
$(a + b) + c = c + (a + b)$	commutative property for addition
$37 + 0 = 37$	identity property for addition
If $a = 37$ and $37 = b$, then $a = b$.	transitive property of equality
$\frac{3}{7} + (-\frac{3}{7}) = 0$	additive inverse property

If $a = b + c$, then $b + c = a$.   symmetric property of equality

$\frac{3}{7} \cdot \frac{7}{3} = 1$ multiplicative inverse property

$32 = 32$ reflexive property of equality   ∎

Exercise 1.2

In Exercises 1–8, indicate whether each statement is true. If a statement is false, explain why.

1. The rational numbers are a subset of the real numbers.
2. Every real number is an irrational number.
3. Some irrational numbers are integers.
4. Some real numbers are integers.
5. There are real numbers that are neither rational nor irrational.
6. The smallest irrational number is $\sqrt{2}$.
7. The intersection of the set of real numbers and the set of rational numbers is the set of decimals that either terminate or repeat.
8. The union of the set of rational numbers and the set of irrational numbers is the set of real numbers.

In Exercises 9–26, graph each subset of the real numbers on the number line, if possible.

9. $3 < x < 10$
10. $-2 \le x \le 3$
11. $-4 \le x \le 0$
12. $5 < x < 15$
13. $3 < x$ and $x \le 7$
14. $x \le -4$ or $x > 8$
15. $x < -2$ or $x \ge 10$
16. $x > 0$ and $x > 5$
17. $x > 0$ and $x < 0$
18. $x > 0$ or $x < 0$
19. $(-2, 3)$
20. $[-5, 5]$
21. $[4, 8)$
22. $(-10, -2]$
23. $[-2, 0] \cup [1, 2]$
24. $(-5, -1) \cap (-3, 0)$
25. $x|x \ne 0 \cap (-3, 2]$
26. $x|x > 4 \cup [0, 4)$

In Exercises 27–32, find the value of each expression.

27. $|-7|$
28. $-|-6|$
29. $|7| - |-6|$
30. $|8 - 5| + |5 - 8|$
31. $|\sqrt{2} - 1|$
32. $|1 - \pi|$

In Exercises 33–38, assume that $x = -5$, $y = -2$, and $z = 3$. Find the value of each expression.

33. $|xy|$
34. $|x + y| + z$
35. $x|yz|$
36. $x|y - z|$
37. $|x - y|z$
38. $|y| + |xz|$

In Exercises 39–40, assume that $x = -5$, $y = -2$, and $z = -3$.

39. Find the distance on the number line between two points with coordinates of x and y.
40. Find the distance on the number line between two points with coordinates of x and z.

In Exercises 41–48, tell if the given relation is reflexive, if it is symmetric, and if it is transitive.

41. $>$

42. $\geq$

43. $\leq$

44. $\neq$

45. $\cong$ ("is congruent to," from geometry)

46. "is divisible by" on the set of natural numbers

47. "is the same color as"

48. "is taller than"

In Exercises 49–64, indicate which field property or property of equality justifies each given expression.

49. $(8 + a) + b = b + (8 + a)$

50. $3(x + y) = 3x + 3y$

51. $\dfrac{2}{3}\left(-\dfrac{3}{4}\right)$ is a real number.

52. $2 + 0 = 2$

53. If $3 = a$ and $a = c$, then $3 = c$.

54. $\dfrac{2}{5}\left(\dfrac{5}{2}\right) = 1$

55. $5 = 5$

56. $(a + b)c = c(a + b)$

57. $-5 + (-8)$ is a real number.

58. $3(4 + t) = 3 \cdot 4 + 3t$

59. $a + (b + c) = a + (c + b)$

60. $a(bc) = (bc)a$

61. If $a = b$, then $b = a$.

62. $3(1) = 3$

63. $[a(b + c)]d = a[(b + c)d]$

64. $5 + 4 + (2 + 1) = 5 + (4 + 2) + 1$

1.3   INTEGER EXPONENTS AND SCIENTIFIC NOTATION

When two or more quantities are multiplied, each is called a **factor** of the product. The expression x^4 indicates that x is to be used as a factor four times. Thus,

$$x^4 = x \cdot x \cdot x \cdot x$$

In general,

Definition.   For any natural number n,

$$x^n = \underbrace{x \cdot x \cdot x \cdot \cdots \cdot x}_{n \textbf{ factors of } x}$$

The number x in the **exponential expression** x^n is called the **base.** The number n is called the **exponent** or the **power** to which the base is raised. The expression x^n is called a **power of x.**

We begin to develop the rules of exponents by considering the product $x^m \cdot x^n$. Because x^m indicates that x is to be used as a factor m times, and because x^n indicates that x is to be used as a factor n times, there are $m + n$ factors of x in the product $x^m \cdot x^n$.

$$x^m \cdot x^n = \underbrace{\underbrace{x \cdot x \cdot x \cdot \cdots \cdot x}_{m \text{ factors}} \cdot \underbrace{x \cdot x \cdot x \cdot x \cdot \cdots \cdot x}_{n \text{ factors}}}_{m + n \text{ factors of } x} = x^{m+n}$$

This argument suggests that when exponential expressions with the same base are multiplied, their exponents are added.

Theorem (The Product Rule for Exponential Expressions).   If m and n are natural numbers, then

$$x^m x^n = x^{m+n}$$

Note that the product rule applies to exponential expressions with the same base only. A product of two powers with different bases, such as $x^4 y^3$, cannot be simplified.

Example 1   **a.** $4^2 \cdot 4^3 = 4^{2+3} = 4^5$ Read 4^2 as "four squared," 4^3 as "four cubed," and 4^5 as "four to the fifth power."

b. $x^5 \cdot x^7 = x^{5+7} = x^{12}$

c. $x^5 y^7$ does not simplify because the bases are not the same.

d. $x^2 y^3 x^5 y^1 = x^2 x^5 y^3 y^1 = x^{2+5} y^{3+1} = x^7 y^4$

e. $y^{p+1} y^{1-p} = y^{p+1+1-p} = y^2$

f. $x^3 x^3 = x^{3+3} = x^6$

g. $x^3 + x^3 = 2x^3$

Note that $x^3 + x^3$ is *not* equal to x^{3+3}. In general, $x^m + x^n$ is *not* equal to x^{m+n}.   ■

A second property of exponents is found by considering the expression $(x^m)^n$. The exponent n indicates that x^m is to be used as a factor n times:

$$(x^m)^n = \underbrace{\underbrace{(x^m)(x^m)(x^m) \cdot \cdots \cdot (x^m)}_{n \text{ factors of } x^m}}_{m \cdot n \text{ factors of } x} = x^{mn}$$

Thus, when an exponential expression is raised to a power, the exponents are multiplied.

Theorem (The Power Rule for Exponential Expressions).   If m and n are natural numbers, then

$$(x^m)^n = (x^n)^m = x^{mn}$$

Example 2   **a.** $(3^5)^2 = 3^{5 \cdot 2} = 3^{10}$

b. $(3^2)^5 = 3^{2 \cdot 5} = 3^{10}$

c. $(x^5)^9 = x^{5 \cdot 9} = x^{45}$

d. $(x^p)^{p+1} = x^{p(p+1)} = x^{p^2+p}$ ■

Another property of exponents can be found by considering the expression $(xy)^n$. The exponent n indicates that the product xy is to be used as a factor n times:

$$(xy)^n = \underbrace{(xy)(xy)(xy) \cdot \cdots \cdot (xy)}_{n \text{ factors of } xy}$$

We can rearrange the factors on the right side of the previous equation and express the result as follows.

$$(xy)^n = \underbrace{(x \cdot x \cdot x \cdot \cdots \cdot x)}_{n \text{ factors of } x}\underbrace{(y \cdot y \cdot y \cdot \cdots \cdot y)}_{n \text{ factors of } y}$$

$$= x^n y^n$$

A similar argument shows that

$$\left(\frac{x}{y}\right)^n = \frac{x^n}{y^n}$$

provided $y \neq 0$. These two properties are summarized in the following theorem.

Theorem.   If n is a natural number, then

$$(xy)^n = x^n y^n$$

and if $y \neq 0$, then

$$\left(\frac{x}{y}\right)^n = \frac{x^n}{y^n}$$

Example 3   **a.** $(3xy)^3 = 3^3 x^3 y^3 = 27x^3 y^3$

b. $(x^2 y)^4 = (x^2)^4 y^4 = x^8 y^4$

c. $\left(\dfrac{5x^2 y}{z^3}\right)^2 = \dfrac{5^2(x^2)^2 y^2}{(z^3)^2} = \dfrac{25x^4 y^2}{z^6}$ provided $z \neq 0$

d. $(3 \cdot 4)^2 = 3^2 \cdot 4^2 = 9 \cdot 16 = 144$

e. $(3 + 4)^2 = 7^2 = 49$

Note that $(3 + 4)^2$ is *not* equal to $3^2 + 4^2$.   ■

The quotient rule for exponential expressions can be developed by removing the common factors in both the numerator and the denominator of the fraction $\frac{x^m}{x^n}$, where m and n are natural numbers and $x \neq 0$. We must consider the cases where $m > n$, where $m = n$, and where $m < n$.

Case 1

Assume that m and n are natural numbers, $x \neq 0$, and $m > n$.

$$\frac{x^m}{x^n} = \frac{\overbrace{x \cdot x \cdot x \cdot \cdots \cdot x}^{m\text{ factors}}}{\underbrace{x \cdot x \cdot \cdots \cdot x}_{n\text{ factors}}}$$

To simplify this fraction, n factors of x can be removed from both the numerator and denominator of the fraction, leaving a surplus of $m - n$ factors of x in the numerator. Hence, we have the following rule when $m > n$ and $x \neq 0$.

$$\frac{x^m}{x^n} = x^{m-n}$$

Case 2

Assume that m and n are natural numbers, $x \neq 0$, and $m = n$.

$$\frac{x^m}{x^n} = \frac{\overbrace{x \cdot x \cdot x \cdot \cdots \cdot x}^{m\text{ factors}}}{\underbrace{x \cdot x \cdot x \cdot \cdots \cdot x}_{n = m\text{ factors}}}$$

Because both the numerator and denominator of this fraction are the same, the value of the fraction is 1. Hence,

$$\frac{x^m}{x^n} = \frac{x^m}{x^m} = 1 \quad \text{provided } m = n \text{ and } x \neq 0$$

If we use the rule developed in Case 1 and subtract exponents, we have

$$\frac{x^m}{x^n} = \frac{x^m}{x^m} = x^{m-m} = x^0 \qquad \text{Remember that } m = n.$$

We will define x^0 to be 1 so that the formula

$$\frac{x^m}{x^n} = x^{m-n}$$

holds true when $m = n$.

Definition.   If $x \neq 0$, then $x^0 = 1$.

Case 3

Assume that m and n are natural numbers, $x \neq 0$, and $m < n$.

$$\frac{x^m}{x^n} = \frac{\overbrace{x \cdot x \cdot \cdots \cdot x}^{m \text{ factors}}}{\underbrace{x \cdot x \cdot x \cdot \cdots \cdot x}_{n \text{ factors}}}$$

To simplify this fraction, m factors can be removed from both the numerator and denominator, leaving a surplus of $n - m$ factors of x in the denominator. Hence,

$$\frac{x^m}{x^n} = \frac{1}{x^{n-m}} \quad \text{provided } m < n \text{ and } x \neq 0$$

If we use the rule developed in Case 1 and subtract exponents, we have

$$\frac{x^m}{x^n} = x^{m-n}$$

The expressions x^{m-n} and $\frac{1}{x^{n-m}}$ are equal if we define a negative exponent in the following way.

Definition.   If $x \neq 0$ and n is a natural number, then

$$x^{-n} = \frac{1}{x^n} \quad \text{and} \quad x^n = \frac{1}{x^{-n}}$$

Because of the previous definition it follows that

$$x^{m-n} = x^{-1(n-m)} = x^{-(n-m)} = \frac{1}{x^{n-m}}$$

The results of the previous three cases justify the quotient rule for exponential expressions with natural number exponents.

Theorem (The Quotient Rule for Exponential Expressions).   If m and n are natural numbers and $x \neq 0$, then

$$\frac{x^m}{x^n} = x^{m-n}$$

Because of the definition of a negative exponent, the Product Rule, the Power Rule, and the Quotient Rule also hold true for integral exponents.

Example 4 Assume that all variables represent nonzero numbers.

a. $\dfrac{x^9}{x^5} = x^{9-5} = x^4$

b. $\dfrac{x^5}{x^9} = x^{5-9} = x^{-4} = \dfrac{1}{x^4}$

c. $3^{-1} + 5^{-1} = \dfrac{1}{3} + \dfrac{1}{5} = \dfrac{5}{15} + \dfrac{3}{15} = \dfrac{8}{15}$

d. $(3+5)^{-1} = 8^{-1} = \dfrac{1}{8}$

Note that $(3+5)^{-1}$ does *not* equal $3^{-1} + 5^{-1}$.

e. $3x^0 = 3(x^0) = 3(1) = 3$

f. $(3x)^0 = 1$

g. $-(xy)^2 = -x^2y^2$

h. $(-xy)^2 = (-1)^2 x^2 y^2 = x^2 y^2$

i. $\dfrac{x^{m+3}}{x^m} = x^{m+3-m} = x^3$

j. $\dfrac{x^{m-2}}{x^m} = x^{m-2-m} = x^{-2} = \dfrac{1}{x^2}$

k. $\left(\dfrac{x^3 y^{-2}}{x^{-2} y^3}\right)^{-2} = (x^{3-(-2)} y^{-2-3})^{-2} = (x^5 y^{-5})^{-2} = x^{-10} y^{10} = \dfrac{y^{10}}{x^{10}}$

l. $\left(\dfrac{x}{y}\right)^{-n} = \dfrac{x^{-n}}{y^{-n}} = \dfrac{x^{-n} x^n y^n}{y^{-n} x^n y^n} = \dfrac{y^n}{x^n} = \left(\dfrac{y}{x}\right)^n$ ∎

Part l of Example 4 establishes this theorem.

> **Theorem.** If n is a natural number and neither x nor y is 0, then
>
> $$\left(\dfrac{x}{y}\right)^{-n} = \left(\dfrac{y}{x}\right)^n$$

Reflect for a moment on the following equations and settle in your mind the difference between ax^n and $(ax)^n$, and between $-x^n$ and $(-x)^n$:

$$ax^n = a \cdot \overbrace{x \cdot x \cdot x \cdots x}^{n \text{ factors of } x}$$

$$\overbrace{}^{n \text{ factors of } ax}$$
$$(ax)^n = (ax)(ax)(ax) \cdots (ax)$$

$$\overbrace{}^{n \text{ factors of } x}$$
$$-x^n = -x \cdot x \cdot x \cdots \cdot x$$

$$\overbrace{}^{n \text{ factors of } -x}$$
$$(-x)^n = (-x)(-x)(-x) \cdots (-x)$$

Example 5 Given $x = -2$, evaluate **a.** $3x^3$, **b.** $(3x)^3$, **c.** $-x^4$, and **d.** $(-x)^4$.

Solution **a.** $3x^3 = 3(-2)^3 = 3(-8) = -24$

b. $(3x)^3 = [3(-2)]^3 = (-6)^3 = -216$

c. $-x^4 = -(-2)^4 = -16$

d. $(-x)^4 = [-(-2)]^4 = 2^4 = 16$ ■

Scientific Notation

Scientists and engineers often work with numbers that are either very large or very small. These numbers can be written compactly by expressing them in **scientific notation**.

Definition. A number is written in **scientific notation** if it is written as the product of a number between 1 (including 1) and 10, and an appropriate power of 10.

Example 6 Light travels 29,980,000,000 centimeters every second. Express this number in scientific notation.

Solution To write the number 29,980,000,000 in scientific notation, you must express it as a product of a number between 1 and 10 and some power of 10. The number 2.998 lies between 1 and 10. To get the number 29,980,000,000, the decimal point in 2.998 must be moved ten places to the right. This is accomplished by multiplying 2.998 by 10^{10}. The number 29,980,000,000 written in scientific notation is 2.998×10^{10}.   ■

Example 7 One meter is approximately 0.00062 miles. Express this number in scientific notation.

Solution To write the number 0.00062 in scientific notation, express it as a product of a number between 1 and 10, and some power of 10. The number 0.00062 may be obtained by moving the decimal point in 6.2 four places to the left. This is accomplished by multiplying 6.2 by 10^{-4} because

$$6.2 \times 10^{-4} = 6.2 \times \frac{1}{10^4} = \frac{6.2}{10,000} = 0.00062$$

The number 0.00062 written in scientific notation is 6.2×10^{-4}. ■

Example 8   Express 3.27×10^{-5} in standard notation.

Solution   The factor of 10^{-5} indicates that 3.27 is divided by 5 factors of 10. Because each division by 10 moves the decimal point one place to the left, you must move the decimal point in the number 3.27 five places to the left. In standard notation, 3.27×10^{-5} is 0.0000327. ■

Study each of the following numbers written in both scientific and standard notation. In each case, note that the exponent gives the number of places the decimal point moves, and the sign of the exponent indicates the direction in which it moves.

a. $3.72 \times 10^5 = 372{,}000$ 5 places to the right

b. $9.93 \times 10^9 = 9{,}930{,}000{,}000$ 9 places to the right

c. $5.37 \times 10^{-4} = 0.000537$ 4 places to the left

d. $1.529 \times 10^{-1} = 0.1529$ 1 place to the left

e. $7.36 \times 10^0 = 7.36$ No movement of the decimal point

Example 9   Use scientific notation to calculate $\dfrac{(3{,}400{,}000)(0.00002)}{170{,}000{,}000}$.

Solution

$$\frac{(3{,}400{,}000)(0.00002)}{170{,}000{,}000} = \frac{(3.4 \times 10^6)(2.0 \times 10^{-5})}{1.7 \times 10^8}$$

$$= \frac{6.8}{1.7} \times 10^{6+(-5)-8}$$

$$= 4.0 \times 10^{-7}$$

$$= 0.0000004$$
■

Exercise 1.3

In Exercises 1–44, simplify each expression. Express all answers with positive exponents only. Assume that all variables are restricted to those numbers for which the expression is defined.

1. $2^2 + 3^2$
2. $(3+5)^3$
3. $(2+3)^2$
4. $3^3 + 5^3$
5. 0^5
6. 5^0
7. $2^{-1} - 3^{-1}$
8. $(2-3)^{-1}$
9. $x^2 x^3$
10. $y^3 y^4$
11. $y^{-2} y^{-3}$
12. $x^2 x^{-5}$
13. $(z^2)^3$
14. $(t^3)^2$
15. $(3x)^3$
16. $(2y)^4$

17. $(x^2y)^3$ **18.** $(x^3z)^2$ **19.** $\left(\dfrac{x^2}{y}\right)^3$ **20.** $\left(\dfrac{x}{y^2}\right)^4$

21. $(-3xy^2)^5$ **22.** $(-2x^2yz)^6$ **23.** $\left(\dfrac{3x^2z^{-1}}{y^3}\right)^3$ **24.** $\left(\dfrac{2x^2y^3}{xy^{-2}z^{-1}}\right)^2$

25. $\left(\dfrac{x^5y^{-2}}{x^{-3}y^2}\right)^2$ **26.** $\left(\dfrac{2x^{-7}y^5}{x^7y^{-4}}\right)^3$ **27.** $\left(\dfrac{5x^{-3}y^{-2}}{3x^2y^{-2}}\right)^{-2}$ **28.** $\left(\dfrac{3x^{-2}y^{-5}}{2x^{-2}y^{-5}}\right)^{-3}$

29. $\left(\dfrac{x^2+3x+y}{2x^5-y}\right)^0$ **30.** $\left(\dfrac{3x^5y^{-3}}{3x^5y^3}\right)^{-2}$ **31.** $(-x^2y^x)^5$ **32.** $(-x^2y^y)^6$

33. $-(x^2y^3)^{4xy}$ **34.** $-(x^2y)^{3xy}$ **35.** $x^{m+1}x^{3-m}$ **36.** $(y^{m+1})^m$

37. $(x^{2n+1})^n$ **38.** $y^{2n+1}y^4y^{-2n}$ **39.** $\dfrac{x^{3m+5}}{x^{3m}}$ **40.** $\dfrac{12x^{m+3}}{3x^m}$

41. $\dfrac{(8^{-2}z^{-3}y)^{-1}}{(5y^2z^{-2})^3(5yz^{-2})^{-1}}$ **42.** $\dfrac{(m^{-2}n^3p^4)^{-2}(mn^{-2}p^3)^4}{(mn^{-2}p^3)^{-4}(mn^2p)^{-1}}$

43. $\left[\dfrac{(m^{-2}n^{-1}p^3)^{-2}}{(mn^2)^{-3}(p^{-3})^4}\right]^{-2}$ **44.** $\left[\dfrac{(3x^2)^{-4}(3p)^2}{(x^{-2}p^4)^{-3}(3x^{-3})^{-2}}\right]^{-3}$

In Exercises 45–62, let $x = -2$, $y = 0$, and $z = 3$, and evaluate each expression.

45. x^2 **46.** $(-x)^2$ **47.** $-x^2$ **48.** x^3

49. $(-x)^3$ **50.** $-x^3$ **51.** $2z^2$ **52.** $(2z)^2$

53. $5x^2 - 3y^3z$ **54.** $3(x-z)^2 + 2(y-z)^3$

55. $3x^3y^7z^{15}$ **56.** $5x^{-2}z$ **57.** $-3x^{-3}z^{-2}$ **58.** $(-5x^2z^3)^y$

59. $x^zz^xy^z$ **60.** x^yz^y **61.** $x^xz^zx^y$ **62.** $x^yz^xx^{x+z}$

In Exercises 63–76, express each number in scientific notation.

63. 372,000   **64.** 89,500   **65.** 177,000,000   **66.** 23,470,000,000

67. 0.007   **68.** 0.00052   **69.** 0.000000693   **70.** 0.000000089

71. one trillion   **72.** sixty-three billion

73. one trillionth   **74.** forty-three billionths

75. 99.7×10^{-4} **76.** 0.0085×10^5

In Exercises 77–82, express each number in standard notation.

77. 9.37×10^5 **78.** 4.26×10^9 **79.** 2.21×10^{-5} **80.** 2.774×10^{-2}

81. 0.00032×10^4 **82.** 9300.0×10^{-4}

In Exercises 83–86, use the method of Example 9 to perform each calculation. Write all answers in scientific notation.

83. $\dfrac{(65,000)(45,000)}{250,000}$ **84.** $\dfrac{(0.000000045)(0.00000012)}{45,000,000}$

85. $\dfrac{(0.00000035)(170,000)}{0.00000085}$ **86.** $\dfrac{(0.0000000144)(12,000)}{600,000}$

87. The speed of sound (in air) is 3.31×10^4 centimeters per second. Use scientific notation to compute the speed of sound in meters per minute.

88. Calculate the volume of a box that has dimensions of 6000 by 9700 by 4700 millimeters. Use scientific notation to perform the calculation, and express the answer in scientific notation.

89. The mass of one proton is 0.00000000000000000000000167248 grams. Find the mass of one billion protons.

90. The speed of light (in a vacuum) is approximately 30,000,000,000 centimeters per second. Find the speed of light in miles per hour, and express your answer in scientific notation. There are 160,934.4 centimeters in one mile.

In Exercises 91–96, use a calculator to work each problem.

91. Show that $4.57^0 = 1$.

92. Show that $(1.2)^3(3.2)^3 = [(1.2)(3.2)]^3$.

93. Show that $(4.1)^2 + (5.2)^2 \neq (4.1 + 5.2)^2$.

94. Show that $[(3.7)^2]^3 = (3.7)^6$.

95. Show that $(3.2)^{1 \cdot 2}(3.2)^{2 \cdot 1} = (3.2)^{3 \cdot 3}$.

96. Show that $4.75^{-0.2} = \dfrac{1}{4.75^{0.2}}$.

1.4   RADICALS AND FRACTIONAL EXPONENTS

The numbers 8 and -8 are called **square roots of 64** because $8^2 = 64$ and $(-8)^2 = 64$. The two square roots of 25 are 5 and -5 because $5^2 = 25$ and $(-5)^2 = 25$. In general,

---

Definition.   If a and b are two numbers and $b^2 = a$, then b is a **square root of a**.

If a is a positive real number, then a has two square roots—one positive and one negative. The positive square root of a is called its **principal square root**.

Definition.   If a is a positive real number, then the **principal square root** of a, denoted as $\sqrt{a}$, is the positive square root of a.
The principal square root of 0 is 0: $\sqrt{0} = 0$.

The above definition implies that $\sqrt{x} \geq 0$ for all nonnegative numbers x.

Expressions such as $\sqrt{x}$ are called **radical expressions**. The symbol $\sqrt{}$ is called the **radical sign**, and the number under the radical sign is called the **radicand**.

$$\sqrt{81} = 9 \text{ because } 9^2 = 81 \text{ and } 9 \text{ is positive}$$
$$\sqrt{225} = 15 \text{ because } 15^2 = 225 \text{ and } 15 \text{ is positive}$$

The negative square root of a positive number x is denoted by the symbol $-\sqrt{x}$. Thus,

$$-\sqrt{81} = -9 \text{ because } (-9)^2 = 81 \text{ and } -9 \text{ is negative}$$
$$-\sqrt{225} = -15 \text{ because } (-15)^2 = 225 \text{ and } -15 \text{ is negative}$$

The definition of a square root of x implies the truth of the following statement.

For all nonnegative real numbers x,

$$\sqrt{x}\,\sqrt{x} = x \qquad \text{and} \qquad (-\sqrt{x})(-\sqrt{x}) = x$$

Example 1   If x is a positive number, find $\sqrt{25x^2}$.

Solution   Note that $25x^2$ is a positive number and that $\sqrt{25x^2}$ represents the principal, or positive, square root of $25x^2$. Thus,

$$\sqrt{25x^2} = 5x$$

because $(5x)^2 = 25x^2$, and because x is positive, so is $5x$.   ∎

Example 2   If x is a negative number, find $\sqrt{36x^2}$.

Solution   Note that $36x^2$ is a positive number and that $\sqrt{36x^2}$ represents the principal, or positive, square root of $36x^2$. Thus,

$$\sqrt{36x^2} = -6x$$

because $(-6x)^2 = 36x^2$, and because x is negative, $-6x$ is positive.   ∎

If a number x is unrestricted, we must use absolute value symbols to guarantee that the principal square root of x^2 is a nonnegative number:

$$\sqrt{x^2} = |x|$$

Example 3   If x can be any real number, find $\sqrt{49x^2}$.

Solution   Because $7x$ might be a negative number, $\sqrt{49x^2}$ is not simply $7x$. To force $7x$ to be nonnegative, use the absolute value of x:

$$\sqrt{49x^2} = 7|x|$$

because $(7|x|)^2 = 49x^2$ and $7|x|$ is nonnegative.   ∎

Numbers such as 4, 9, 16, 25, and 36, which have integers for their square roots, are called **perfect integer squares**. Other positive integers, such as 2 and 3, do not have integer square roots; their square roots are irrational numbers. To find approximations to the square roots of such integers, you can use a calculator or Table A in Appendix III.

We can extend the concept of square root to define the nth root of a number.

Definition.   If a and b are two numbers and n is a positive integer such that $b^n = a$, then b is an **nth root of a**.

If $n = 3$ in the above definition, then b is called a **cube root** of a.

Example 4   **a.**   5 is a cube root of 125 because $5^3 = 125$.

b. -5 is a cube root of -125 because $(-5)^3 = -125$.

c.   Both 3 and -3 are fourth roots of 81 because $3^4 = 81$ and $(-3)^4 = 81$.

d. -81 has no real number for a fourth root because no real number's fourth power can equal -81.   ∎

In general, we have the following:

If n is a positive odd integer and a is a real number, then a has exactly one real nth root called the **principal nth root of a**. It is denoted as $\sqrt[n]{a}$.

If n is a positive even integer and a is a positive number, then a has two nth roots—one positive and one negative. The positive nth root is called the **principal nth root of a**. It is denoted as $\sqrt[n]{a}$.

If n is a positive even integer and a is a negative number, then a has no real nth roots.

The nth root of 0 is 0: $\sqrt[n]{0} = 0$.

In the notation $\sqrt[n]{a}$, n is called the **index** of the radical, and the radical is said to be of **order n**. If n is not written, it is understood to be 2.

Example 5   **a.**   The principal fifth root of 32 is 2 because $2^5 = 32$, and we write $\sqrt[5]{32} = 2$.

b.   The principal fifth root of -32 is -2 because $(-2)^5 = -32$, and we write $\sqrt[5]{-32} = -2$.

c.   Both 5 and -5 are fourth roots of 625 because $5^4 = 625$ and $(-5)^4 = 625$. The principal fourth root of 625 is 5, and we write $\sqrt[4]{625} = 5$.   ∎

Example 6   If x is any real number, simplify **a.** $\sqrt[4]{x^4}$, **b.** $\sqrt[3]{x^3}$, and **c.** $\sqrt{x^8}$.

Solution   **a.** $\sqrt[4]{x^4} = |x|$ The symbol $\sqrt[4]{x^4}$ represents the principal fourth root of x^4. Because the principal fourth root of x^4 cannot be negative, absolute value symbols are needed.

b. $\sqrt[3]{x^3} = x$ This is true whether x is positive, negative, or 0.

c. $\sqrt{x^8} = x^4$ No absolute value symbol is necessary because x^4 cannot be a negative number. ■

Rational Exponents

It is possible to raise numbers to fractional powers. For example, we consider the symbol $3^{\frac{1}{2}}$. If fractional exponents are to obey the same rules as integral exponents, then $(3^{\frac{1}{2}})^2$ can be evaluated by multiplying exponents:

$$(3^{\frac{1}{2}})^2 = 3^{(\frac{1}{2})2} = 3^1 = 3$$

However, $\sqrt{3}$ is a number whose square is 3:

$$(\sqrt{3})^2 = 3$$

Because $(\sqrt{3})^2$ and $(3^{\frac{1}{2}})^2$ are both equal to 3, we define $3^{\frac{1}{2}}$ to be $\sqrt{3}$. In a similar manner, $3^{\frac{1}{3}} = \sqrt[3]{3}$, and so on.

Definition. *If n is any positive integer, and $x \geq 0$ when n is even, then*

$$x^{\frac{1}{n}} = \sqrt[n]{x}$$

and if $x \neq 0$, then

$$x^{-\frac{1}{n}} = \frac{1}{\sqrt[n]{x}}$$

The definition above can be extended to cover exponential expressions such as $3^{\frac{5}{4}}$ that have fractional exponents with numerators different from 1. Again, we want the definition to be consistent with the rules for integral exponents. To be consistent with the Power Rule, the expression $3^{\frac{5}{4}}$ should mean either $(3^5)^{\frac{1}{4}}$ or $(3^{\frac{1}{4}})^5$. Hence,

$$3^{\frac{5}{4}} = (3^5)^{\frac{1}{4}} = \sqrt[4]{3^5}$$

or

$$3^{\frac{5}{4}} = (3^{\frac{1}{4}})^5 = (\sqrt[4]{3})^5$$

The generalization of this example leads to the following definition.

Definition. If m and n are positive integers, and x is a real number, then

$$x^{\frac{m}{n}} = \sqrt[n]{x^m} = (\sqrt[n]{x})^m$$

and if $x \neq 0$,

$$x^{-\frac{m}{n}} = \frac{1}{x^{\frac{m}{n}}}$$

provided $\sqrt[n]{x}$ is a real number.

Example 7 The left column shows how to simplify an expression containing fractional exponents by using radicals. The right column shows how to simplify the same expression by using the rules of exponents.

a. $32^{\frac{1}{5}} = \sqrt[5]{32} = 2$ $32^{\frac{1}{5}} = (2^5)^{\frac{1}{5}} = 2^{5\left(\frac{1}{5}\right)} = 2^1 = 2$

b. $(0.01)^{-\frac{1}{2}} = \dfrac{1}{\sqrt{0.01}} = \dfrac{1}{0.1} = 10$ $(0.01)^{-\frac{1}{2}} = (10^{-2})^{-\frac{1}{2}} = 10^1 = 10$

c. $(-8)^{\frac{1}{3}} = \sqrt[3]{-8} = -2$ $(-8)^{\frac{1}{3}} = (-1 \cdot 2^3)^{\frac{1}{3}}$

$\qquad\qquad\qquad\qquad\qquad\qquad\qquad\qquad = (-1)^{\frac{1}{3}}(2^3)^{\frac{1}{3}} = -2$

d. $25^{\frac{3}{2}} = (\sqrt{25})^3 = 5^3 = 125$ $25^{\frac{3}{2}} = (5^2)^{\frac{3}{2}} = 5^{2\left(\frac{3}{2}\right)} = 5^3 = 125$

e. $1000^{\frac{2}{3}} = (\sqrt[3]{1000})^2 = 10^2 = 100$ $1000^{\frac{2}{3}} = (10^3)^{\frac{2}{3}} = 10^2 = 100$ ∎

Simplifying Radicals

The properties of exponents can be used to discover many of the corresponding properties of radicals. If $\sqrt[n]{x}$ and $\sqrt[n]{y}$ are real numbers, we can write $\sqrt[n]{xy}$ in exponential form and use the properties of exponents to transform the term into a product of two nth roots:

$$\sqrt[n]{xy} = (xy)^{\frac{1}{n}} = x^{\frac{1}{n}}y^{\frac{1}{n}} = \sqrt[n]{x}\,\sqrt[n]{y}$$

In the exercises, you will be asked to show that

$$\sqrt[n]{\frac{x}{y}} = \frac{\sqrt[n]{x}}{\sqrt[n]{y}} \quad \text{provided } y \neq 0$$

> **Theorem.** If all of the radicals involved represent real numbers, then
>
> $$\sqrt[n]{xy} = \sqrt[n]{x}\,\sqrt[n]{y}$$
>
> Also, if $y \neq 0$, then
>
> $$\sqrt[n]{\frac{x}{y}} = \frac{\sqrt[n]{x}}{\sqrt[n]{y}}$$

A generalization of the first part of the previous theorem asserts that the nth root of a product of several factors is the product of the nth roots of those factors. The second part asserts that the nth root of a quotient is the quotient of the nth roots.

Example 8   Simplify **a.** $\sqrt{49x^4}$, **b.** $\sqrt[3]{216x^9y}$, **c.** $\sqrt{50} + \sqrt{200}$, and **d.** $3\sqrt[5]{64} - 2\sqrt[5]{2}$.

Solution   **a.** $\begin{aligned} \sqrt{49x^4} &= \sqrt{49}\,\sqrt{x^4} \\ &= 7x^2 \end{aligned}$

b. $\begin{aligned} \sqrt[3]{216x^9y} &= \sqrt[3]{216}\,\sqrt[3]{x^9}\,\sqrt[3]{y} \\ &= 6x^3\sqrt[3]{y} \end{aligned}$

c. $\begin{aligned} \sqrt{50} + \sqrt{200} &= \sqrt{25 \cdot 2} + \sqrt{100 \cdot 2} \\ &= \sqrt{25}\,\sqrt{2} + \sqrt{100}\,\sqrt{2} \\ &= 5\sqrt{2} + 10\sqrt{2} \\ &= (5 + 10)\sqrt{2} \\ &= 15\sqrt{2} \end{aligned}$

d. $\begin{aligned} 3\sqrt[5]{64} - 2\sqrt[5]{2} &= 3\sqrt[5]{32 \cdot 2} - 2\sqrt[5]{2} \\ &= 3\sqrt[5]{32}\,\sqrt[5]{2} - 2\sqrt[5]{2} \\ &= 3 \cdot 2\sqrt[5]{2} - 2\sqrt[5]{2} \\ &= 6\sqrt[5]{2} - 2\sqrt[5]{2} \\ &= (6 - 2)\sqrt[5]{2} \\ &= 4\sqrt[5]{2} \end{aligned}$ ∎

We now consider radicals such as $\sqrt{\frac{5}{3}}$, whose radicands are fractions. Because of the previous theorem, this radical can be written in the form $\sqrt{5}/\sqrt{3}$. There is a process, called **rationalizing the denominator**, that enables us to write the fraction $\sqrt{5}/\sqrt{3}$ as a fraction with a rational number in the denominator. All that we must do is to multiply both the numerator and the denominator of the fraction $\sqrt{5}/\sqrt{3}$ by $\sqrt{3}$.

$$\frac{\sqrt{5}}{\sqrt{3}} = \frac{\sqrt{5}}{\sqrt{3}} \cdot \frac{\sqrt{3}}{\sqrt{3}} = \frac{\sqrt{15}}{3}$$

Example 9   Rationalize each denominator and simplify, if possible, **a.** $\dfrac{1}{\sqrt{7}}$, **b.** $\sqrt{\dfrac{3}{x}}$, **c.** $\sqrt{\dfrac{3a^2}{5x^5}}$,

d. $\sqrt{\dfrac{1}{2}} + \sqrt{\dfrac{1}{8}}$, and **e.** $\sqrt[3]{\dfrac{1}{y^2}} - \sqrt[3]{\dfrac{16}{x^4}}$. Assume that x, y, and a are positive numbers.

Solution   **a.** $\dfrac{1}{\sqrt{7}} = \dfrac{1}{\sqrt{7}} \cdot \dfrac{\sqrt{7}}{\sqrt{7}} = \dfrac{\sqrt{7}}{7}$

b. $\sqrt{\dfrac{3}{x}} = \dfrac{\sqrt{3}}{\sqrt{x}} = \dfrac{\sqrt{3}}{\sqrt{x}} \cdot \dfrac{\sqrt{x}}{\sqrt{x}} = \dfrac{\sqrt{3x}}{x}$

c. $\sqrt{\dfrac{3a^2}{5x^5}} = \dfrac{\sqrt{3a^2}}{\sqrt{5x^5}} = \dfrac{a\sqrt{3}}{\sqrt{5x^5}} \cdot \dfrac{\sqrt{5x}}{\sqrt{5x}} = \dfrac{a\sqrt{15x}}{5x^3}$

d. $\sqrt{\dfrac{1}{2}} + \sqrt{\dfrac{1}{8}} = \dfrac{1}{\sqrt{2}} + \dfrac{1}{\sqrt{8}}$

$\qquad = \dfrac{1}{\sqrt{2}} \cdot \dfrac{\sqrt{2}}{\sqrt{2}} + \dfrac{1}{\sqrt{8}} \cdot \dfrac{\sqrt{2}}{\sqrt{2}}$

$\qquad = \dfrac{\sqrt{2}}{2} + \dfrac{\sqrt{2}}{\sqrt{16}}$

$\qquad = \dfrac{\sqrt{2}}{2} + \dfrac{\sqrt{2}}{4}$

$\qquad = \dfrac{3\sqrt{2}}{4}$

e. $\sqrt[3]{\dfrac{1}{y^2}} - \sqrt[3]{\dfrac{16}{x^4}} = \dfrac{\sqrt[3]{1}}{\sqrt[3]{y^2}} - \dfrac{\sqrt[3]{16}}{\sqrt[3]{x^4}}$

$\qquad = \dfrac{1}{\sqrt[3]{y^2}} - \dfrac{2\sqrt[3]{2}}{\sqrt[3]{x^4}}$ $\qquad \sqrt[3]{16} = \sqrt[3]{8}\sqrt[3]{2} = 2\sqrt[3]{2}$

$\qquad = \dfrac{1}{\sqrt[3]{y^2}} \cdot \dfrac{\sqrt[3]{y}}{\sqrt[3]{y}} - \dfrac{2\sqrt[3]{2}}{\sqrt[3]{x^4}} \cdot \dfrac{\sqrt[3]{x^2}}{\sqrt[3]{x^2}}$

$\qquad = \dfrac{\sqrt[3]{y}}{\sqrt[3]{y^3}} - \dfrac{2\sqrt[3]{2x^2}}{\sqrt[3]{x^6}}$

$\qquad = \dfrac{\sqrt[3]{y}}{y} - \dfrac{2\sqrt[3]{2x^2}}{x^2}$ ∎

Another property of radicals can be derived from the properties of exponents. If all of the expressions involved represent real numbers, then

$$\sqrt[n]{\sqrt[m]{x}} = \sqrt[n]{x^{\frac{1}{m}}} = (x^{\frac{1}{m}})^{\frac{1}{n}} = x^{\frac{1}{mn}} = \sqrt[mn]{x}$$

Similarly,

$$\sqrt[m]{\sqrt[n]{x}} = \sqrt[mn]{x}$$

> **Theorem.**   If all of the expressions involved represent real numbers, then
>
> $$\sqrt[m]{\sqrt[n]{x}} = \sqrt[n]{\sqrt[m]{x}} = \sqrt[mn]{x}$$

Example 10   Calculate $\sqrt[3]{\sqrt{8}}$.

Solution   By the previous theorem, it follows that

$$\sqrt[3]{\sqrt{8}} = \sqrt{\sqrt[3]{8}} = \sqrt{2}$$ ■

Fractional exponents can often be used to simplify radical expressions. The following examples show how.

Example 11   Assume that x and y are positive numbers and simplify **a.** $\sqrt[6]{4}$, **b.** $\sqrt[12]{x^3}$, and **c.** $\sqrt[9]{8y^3}$.

Solution   **a.** $\sqrt[6]{4} = 4^{\frac{1}{6}} = (2^2)^{\frac{1}{6}} = 2^{\frac{2}{6}} = 2^{\frac{1}{3}} = \sqrt[3]{2}$

b. $\sqrt[12]{x^3} = x^{\frac{3}{12}} = x^{\frac{1}{4}} = \sqrt[4]{x}$

c. $\sqrt[9]{8y^3} = (2^3 y^3)^{\frac{1}{9}} = (2y)^{\frac{3}{9}} = (2y)^{\frac{1}{3}} = \sqrt[3]{2y}$ ■

Example 12   Simplify **a.** $\sqrt{2}\sqrt[3]{4}$, **b.** $\dfrac{\sqrt{3}}{\sqrt[3]{3}}$, and **c.** $\dfrac{\sqrt[3]{3}}{\sqrt{5}}$.

Solution   **a.** $\sqrt{2}\sqrt[3]{4} = 2^{\frac{1}{2}} 4^{\frac{1}{3}} = 2^{\frac{1}{2}}(2^2)^{\frac{1}{3}} = 2^{\frac{1}{2}} 2^{\frac{2}{3}} = 2^{\frac{1}{2}+\frac{2}{3}} = 2^{\frac{7}{6}}$

$$= 2^{\frac{6}{6}+\frac{1}{6}} = 2^{\frac{6}{6}} 2^{\frac{1}{6}} = 2\sqrt[6]{2}$$

b. $\dfrac{\sqrt{3}}{\sqrt[3]{3}} = \dfrac{3^{\frac{1}{2}}}{3^{\frac{1}{3}}} = 3^{\frac{1}{2}-\frac{1}{3}} = 3^{\frac{1}{6}} = \sqrt[6]{3}$

c. $\dfrac{\sqrt[3]{3}}{\sqrt{5}} = \dfrac{\sqrt[3]{3}}{\sqrt{5}} \cdot \dfrac{\sqrt{5}}{\sqrt{5}} = \dfrac{\sqrt[3]{3}\sqrt{5}}{5}.$

Note that $\sqrt[3]{3}\sqrt{5}$ cannot be simplified because $3^{\frac{1}{3}} 5^{\frac{1}{2}}$ is the product of two exponential expressions with unlike bases.   ■

▓▓▓▓   **Exercise 1.4**   ▓▓▓▓▓▓▓▓

In Exercises 1–28, simplify each expression. Write all answers without using negative exponents.

1. $\sqrt{49}$ **2.** $\sqrt{81}$ **3.** $-\sqrt{64}$ **4.** $-\sqrt{36}$

5. $\sqrt[3]{8}$ **6.** $\sqrt[3]{-27}$ **7.** $-\sqrt[3]{64}$ **8.** $-\sqrt[3]{-125}$

9. $\sqrt[4]{81}$ **10.** $\sqrt[5]{-243}$ **11.** $\sqrt[5]{\dfrac{32}{243}}$ **12.** $\sqrt[4]{\dfrac{256}{625}}$

13. $9^{\frac{1}{2}}$ **14.** $-16^{\frac{1}{2}}$ **15.** $4^{\frac{3}{2}}$ **16.** $(-8)^{\frac{2}{3}}$

17. $-1000^{\frac{2}{3}}$ **18.** $100^{\frac{3}{2}}$ **19.** $64^{-\frac{1}{2}}$ **20.** $25^{-\frac{1}{2}}$

21. $64^{-\frac{3}{2}}$ **22.** $32^{-\frac{3}{5}}$ **23.** $-9^{-\frac{3}{2}}$ **24.** $-(27^{-\frac{2}{3}})$

25. $\left(\dfrac{4}{9}\right)^{\frac{5}{2}}$ **26.** $\left(\dfrac{25}{81}\right)^{\frac{3}{2}}$ **27.** $\left(\dfrac{27}{64}\right)^{\frac{2}{3}}$ **28.** $\left(-\dfrac{125}{8}\right)^{\frac{4}{3}}$

In Exercises 29–40, assume that all variables represent positive real numbers. Simplify each expression. Write answers without using negative exponents.

29. $\sqrt{49x^2}$ **30.** $-\sqrt{25y^6}$ **31.** $\sqrt{x^2y^4}$ **32.** $\sqrt{a^4b^8}$

33. $\sqrt[3]{8y^3}$ **34.** $\sqrt[3]{-27z^9}$ **35.** $(x^{10}y^5)^{\frac{3}{5}}$ **36.** $(a^6b^{12})^{\frac{5}{6}}$

37. $(x^9y^{12})^{-\frac{2}{3}}$ **38.** $(r^8s^{16})^{-\frac{3}{4}}$ **39.** $\left(\dfrac{a^8c^{24}}{b^{20}}\right)^{0.25}$ **40.** $\left(\dfrac{u^{10}v^0}{z^5}\right)^{0.2}$

In Exercises 41–48, assume that all variables are unrestricted. Simplify each expression and use absolute value symbols if necessary. Write answers without using negative exponents.

41. $\sqrt{4a^2}$ **42.** $\sqrt{16b^4}$ **43.** $(r^2s^4)^{\frac{1}{2}}$ **44.** $(u^2v^2)^{-\frac{1}{2}}$

45. $\sqrt[3]{27x^3}$ **46.** $\sqrt[4]{16x^4}$ **47.** $(32y^{10}z^5)^{-\frac{2}{5}}$ **48.** $(625a^4b^8)^{-\frac{3}{4}}$

In Exercises 49–60, assume that all variables represent positive real numbers. Simplify each expression.

49. $\sqrt{8} - \sqrt{2}$

50. $\sqrt{75} + 2\sqrt{27}$

51. $\sqrt{200x^2} + \sqrt{98x^2}$

52. $\sqrt{128a^3} - a\sqrt{162a}$

53. $2\sqrt{48y^5} - 3y\sqrt{12y^3}$

54. $y\sqrt{112y} + 4\sqrt{175y^3}$

55. $2\sqrt[3]{81} + 3\sqrt[3]{24}$

56. $3\sqrt[4]{32} - 2\sqrt[4]{162}$

57. $\sqrt[4]{768z^5} + \sqrt[4]{48z^5}$

58. $-2\sqrt[5]{64y^2} + 3\sqrt[5]{96y^2}$

59. $(\sqrt{20} + \sqrt{5})^2$

60. $(\sqrt[3]{54} - \sqrt[3]{2})^3$

In Exercises 61–78, assume that all variables are positive real numbers. Rationalize each denominator and simplify.

61. $\dfrac{3}{\sqrt{3}}$ **62.** $\dfrac{10}{\sqrt{5}}$ **63.** $\dfrac{2}{\sqrt{x}}$ **64.** $\dfrac{3}{\sqrt{y}}$

65. $\dfrac{2}{\sqrt[3]{2}}$ **66.** $\dfrac{3}{\sqrt[3]{9}}$ **67.** $\dfrac{5a}{\sqrt[3]{25a}}$ **68.** $\dfrac{16}{\sqrt[3]{16b^2}}$

69. $\dfrac{2b}{\sqrt[4]{3a^2}}$ **70.** $\dfrac{64a}{\sqrt[5]{16b^3}}$ **71.** $\sqrt{\dfrac{x}{2y}}$ **72.** $\sqrt{\dfrac{5y}{3x}}$

73. $\sqrt[3]{\dfrac{2u^4}{9v}}$ **74.** $\sqrt[3]{-\dfrac{3s^5}{4r^2}}$ **75.** $\sqrt{\dfrac{1}{3}} - \sqrt{\dfrac{1}{27}}$ **76.** $\sqrt{\dfrac{1}{5}} + \sqrt{\dfrac{2}{5}}$

77. $\sqrt{\dfrac{1}{8x}} - \sqrt{\dfrac{1}{2y^3}}$

78. $\sqrt[3]{\dfrac{4}{x}} + \sqrt[3]{\dfrac{9y}{x^4}}$

In Exercises 79–82, simplify each expression. Assume that all variables represent positive real numbers.

79. $\sqrt[3]{\sqrt{27}}$

80. $\sqrt{\sqrt[3]{100}}$

81. $\sqrt{\sqrt[3]{625x^8}}$

82. $\sqrt[3]{\sqrt{512y^{12}}}$

In Exercises 83–94, assume that all variables represent positive real numbers and simplify. Write answers without using negative exponents.

83. $\dfrac{\sqrt[3]{x^6 y^3}}{\sqrt{9x^2 y}}$

84. $\dfrac{\sqrt[6]{128x^{-9} y^8 z^7}}{\sqrt[6]{x^6 y z^4}}$

85. $\dfrac{\sqrt[4]{81x^{-6} y^8}}{\sqrt[4]{x^{-2} y^4}}$

86. $\dfrac{\sqrt[4]{625x^8 y^4}}{\sqrt{16x^3 y^2}}$

87. $\dfrac{\sqrt[5]{32x^6 y^{11}}}{\sqrt[10]{x^{12} y^{12}}}$

88. $\dfrac{\sqrt[3]{-27x^{10} y^{14}}}{\sqrt[3]{64x^7 y^2}}$

89. $\sqrt{8x^2 y} - x\sqrt{2y}$

90. $3x\sqrt{18x} + 2\sqrt{2x^3}$

91. $\sqrt[3]{16xy^4} - y\sqrt[3]{2xy}$

92. $\sqrt[4]{512x^5} - \sqrt[4]{32x^5}$

93. $\dfrac{3x\sqrt{x} - 2\sqrt{x^3}}{\sqrt{18x} - \sqrt{2x}}$

94. $\dfrac{4x\sqrt{3x} + \sqrt{12x^3}}{\sqrt{27x} + \sqrt{12x}}$

In Exercises 95–98, simplify each radical. See Example 11.

95. $\sqrt[4]{9}$

96. $\sqrt[6]{27}$

97. $\sqrt[10]{16x^6}$

98. $\sqrt[6]{27x^9}$

In Exercises 99–102, write each expression as a single radical. See Example 12.

99. $\sqrt{2}\,\sqrt[3]{2}$

100. $\sqrt{3}\,\sqrt[3]{3}$

101. $\dfrac{\sqrt{3}}{\sqrt[4]{3}}$

102. $\dfrac{\sqrt{5}}{\sqrt[3]{5}}$

103. For what values of x does $\sqrt{x^2} = x$?

104. For what values of x does $\sqrt[3]{x^3} = x$?

105. For what values of x does $\sqrt[4]{x^4} = -x$?

106. If all of the radicals involved represent real numbers, and $y \neq 0$, prove that $\sqrt[n]{\dfrac{x}{y}} = \dfrac{\sqrt[n]{x}}{\sqrt[n]{y}}$.

107. If all of the radicals involved represent real numbers, and there is no division by 0, prove that
$$\left(\dfrac{x}{y}\right)^{-\frac{m}{n}} = \sqrt[n]{\dfrac{y^m}{x^m}}.$$

108. The definition of $x^{\frac{m}{n}}$ requires that $\sqrt[n]{x}$ be a real number. Why is this important? (*Hint:* Consider what happens when n is even, m is odd, and x is negative.)

In Exercises 109–114, let $x = 3.5$, $y = 1.2$, and $z = 1.4$. Use a calculator to work each problem.

109. Show that $(xy)^z = x^z y^z$.

110. Show that $\left(\dfrac{x}{y}\right)^z = \dfrac{x^z}{y^z}$.

111. Show that $(x^y)^z = (x^z)^y$.

112. Show that $(x + y)^z \neq x^z + y^z$.

113. Show that $\sqrt[z]{\sqrt[y]{x}} = x^{\frac{1}{zy}}$.

114. Show that $(x^y)^z \neq x^{(y^z)}$.

1.5 ARITHMETIC OF POLYNOMIALS

A **monomial** is either a number, or the product of a number and one or more variables with whole-number exponents. The number is called the **numerical coefficient**, or simply the **coefficient**, of the monomial. For example, x, $-5ab^2c^4$, and -12 are monomials with coefficients of 1, -5, and -12, respectively.

The **degree** of a monomial is the sum of the exponents of its variables. The degree of x is 1, the degree of $-5ab^2c^4$ is 7, and the degree of -12 is 0 (note that $-12 = -12x^0$). All nonzero constants have a degree of 0. The constant 0 does not have a defined degree.

A single monomial or a finite sum of monomials is called a **polynomial**. Each of the monomials in that sum is called a **term** of the polynomial. A polynomial with two terms is called a **binomial**, and one with three terms is called a **trinomial**. The **degree of a polynomial** is the degree of the term in that polynomial with highest degree. The only polynomial with no defined degree is the **zero polynomial**, 0.

Example 1

a. $3x^2y^3 + 5xy^2 + 7$ is a trinomial. It is of degree 5 because its term of highest degree (the first term) is of degree 5.

b. $3xy + 5x^2y$ is a binomial of degree 3.

c. $5x + 3y + \sqrt[3]{3}\,z - \sqrt{7}$ is a polynomial because its variables have whole-number exponents. It is of first degree.

d. $5x^{\frac{1}{2}} + 3\sqrt[5]{y}$ is not a polynomial because its variables do not have whole-number exponents. ∎

If two terms of a polynomial have the same variables and the same exponents, they are called **similar** or **like terms**. The distributive property enables us to combine such terms.

Example 2 Combine the terms in the binomial $3x^2y + 5x^2y$.

Solution The terms of this binomial are similar terms. By the distributive property, the common factor of x^2y can be written outside of a set of parentheses, and the resulting expression can be simplified as follows:

$$3\,x^2y + 5\,x^2y = (3 + 5)\,x^2y$$
$$= 8x^2y$$ ∎

Polynomials are added by combining any similar terms that are contained within those polynomials.

Example 3   Perform the addition $(3x^3y + 5x^2 - 2y) + (2x^3y - 5x^2 + 3x)$ by combining similar terms.

Solution   Because the terms in this expression represent real numbers, they can be re-arranged, regrouped, and combined:

$$(3x^3y + 5x^2 - 2y) + (2x^3y - 5x^2 + 3x)$$
$$= (3x^3y + 2x^3y) + (5x^2 - 5x^2) - 2y + 3x$$
$$= 5x^3y + 0x^2 - 2y + 3x$$
$$= 5x^3y - 2y + 3x \qquad \blacksquare$$

Example 4   Perform the subtraction $(2x^2 + 3y^2) - (x^2 - 2y^2 + 7)$ by combining similar terms.

Solution   This expression indicates that the second polynomial is to be subtracted from the first. To do so, you must subtract each term of the second polynomial. Then, combine similar terms:

$$(2x^2 + 3y^2) - (x^2 - 2y^2 + 7) = 2x^2 + 3y^2 - x^2 - (-2y^2) - 7$$
$$= 2x^2 + 3y^2 - x^2 + 2y^2 - 7$$
$$= (2x^2 - x^2) + (3y^2 + 2y^2) - 7$$
$$= x^2 + 5y^2 - 7 \qquad \blacksquare$$

Example 5   Simplify the expression $(2y^2 + 13x^2) - 5(y^2 - 3x^2)$.

Solution   Use the distributive property to remove parentheses and combine similar terms.

$$(2y^2 + 13x^2) - 5(y^2 - 3x^2) = 2y^2 + 13x^2 - 5y^2 - 5(-3x^2)$$
$$= 2y^2 - 5y^2 + 13x^2 + 15x^2$$
$$= -3y^2 + 28x^2 \qquad \blacksquare$$

Polynomials can be multiplied, as the following examples indicate.

Example 6   Find the product of $3x^2y^3z$ and $5xyz^2$.

Solution   This type of product was discussed in Section 1.3. The commutative and associative properties of multiplication allow the regrouping and rearranging of the factors, and the properties of integral exponents can be used to simplify the resulting expression:

$$(3x^2y^3z)(5xyz^2) = 3 \cdot 5 \cdot x^2 \cdot x \cdot y^3 \cdot y \cdot z \cdot z^2$$
$$= 15x^3y^4z^3 \qquad \blacksquare$$

Example 7   Find the product of $3xy^2$ and $2xy + x^2$.

Solution   Use the distributive property to multiply the binomial by the monomial, and simplify:

$$3xy^2(2xy + x^2) = 3xy^2 \cdot 2xy + 3xy^2 \cdot x^2$$
$$= 6x^2y^3 + 3x^3y^2$$ ■

Example 8   Find the product of $x + 3$ and $y + 2$.

Solution   Use the distributive property to multiply each term of $y + 2$ by the quantity $x + 3$, and proceed as follows:

$$(x + 3)(y + 2) = (x + 3)y + (x + 3)2$$
$$= xy + 3y + 2x + 6$$

Observe that the right side of this equation is simply the result of multiplying each term in the second binomial by each term in the first binomial.   ■

Example 9   Find the product of $\sqrt{3} + x$ and $2 - \sqrt{3}\,x$.

Solution   Multiply each term in the second binomial by each term in the first binomial and simplify:

$$(\sqrt{3} + x)(2 - \sqrt{3}\,x) = 2\sqrt{3} - \sqrt{3}\sqrt{3}\,x + 2x - x\sqrt{3}\,x$$
$$= 2\sqrt{3} - 3x + 2x - \sqrt{3}\,x^2$$
$$= 2\sqrt{3} - x - \sqrt{3}\,x^2$$ ■

Example 10   Expand $(x + 3)^2$.

Solution   The expression $(x + 3)^2$ indicates the product of two binomials: $(x + 3)(x + 3)$. Proceed as follows:

$$(x + 3)(x + 3) = x \cdot x + 3x + 3x + 3(3)$$
$$= x^2 + 6x + 9$$

Students often make the mistake of interpreting $(x + 3)^2$ as $x^2 + 9$. This is not correct. When squaring binomials, always remember to include the middle term:

$$(x + 3)^2 = x^2 + 6x + 9$$ ■

The discussion in Example 10 suggests the following two product formulas:

$$(x + y)^2 = x^2 + 2xy + y^2$$
$$(x - y)^2 = x^2 - 2xy + y^2$$

Example 11   Find the product of $x + y$ and $x^2 - xy + y^2$.

Solution   Multiply each term in the second expression by each term in the first expression and simplify:

$$(x + y)(x^2 - xy + y^2) = x(x^2) + x(-xy) + x(y^2) + y(x^2) + y(-xy) + y(y^2)$$
$$= x^3 - x^2y + xy^2 + x^2y - xy^2 + y^3$$
$$= x^3 + y^3$$ ■

Example 12 Expand $(x + 3)^3$.

Solution Write the expression as a product and multiply:

$$(x + 3)^3 = (x + 3)(x + 3)^2$$
$$= (x + 3)(x^2 + 6x + 9)$$
$$= x^3 + 6x^2 + 9x + 3x^2 + 18x + 27$$
$$= x^3 + 9x^2 + 27x + 27$$ ■

Example 13 Find the product of $\sqrt{7}x + 2$ and $\sqrt{7}x - 2$.

Solution Multiply each term of the second binomial by each term of the first binomial and simplify:

$$(\sqrt{7}x + 2)(\sqrt{7}x - 2) = (\sqrt{7}x)(\sqrt{7}x) - 2\sqrt{7}x + 2\sqrt{7}x - 4$$
$$= 7x^2 - 4$$

Note that the middle term of this product—the term that would involve x—drops out. ■

The discussion in Example 13 suggests another product formula.

$$(x + y)(x - y) = x^2 - y^2$$

We can use the preceding product formula to write fractions whose denominators contain radicals as equivalent fractions whose denominators are free of radicals.

Example 14 Simplify the fraction $\dfrac{6}{\sqrt{7} + 2}$ by rationalizing its denominator.

Solution Multiply both the numerator and the denominator of the given fraction by $\sqrt{7} - 2$. Because $(\sqrt{7} + 2)(\sqrt{7} - 2) = 7 - 4 = 3$, this operation will clear the denominator of radicals:

$$\frac{6}{\sqrt{7} + 2} = \frac{6(\sqrt{7} - 2)}{(\sqrt{7} + 2)(\sqrt{7} - 2)}$$
$$= \frac{6(\sqrt{7} - 2)}{3}$$
$$= 2(\sqrt{7} - 2)$$ ■

Example 15   Simplify $\dfrac{\sqrt{3} - \sqrt{2}}{\sqrt{3} + \sqrt{2}}$ by rationalizing the denominator.

Solution   Multiply the numerator and the denominator by $\sqrt{3} - \sqrt{2}$ and simplify:

$$\frac{\sqrt{3} - \sqrt{2}}{\sqrt{3} + \sqrt{2}} = \frac{(\sqrt{3} - \sqrt{2})(\sqrt{3} - \sqrt{2})}{(\sqrt{3} + \sqrt{2})(\sqrt{3} - \sqrt{2})}$$

$$= \frac{(\sqrt{3} - \sqrt{2})(\sqrt{3} - \sqrt{2})}{(\sqrt{3})^2 - (\sqrt{2})^2}$$

$$= \frac{\sqrt{3}\sqrt{3} - \sqrt{3}\sqrt{2} - \sqrt{2}\sqrt{3} + \sqrt{2}\sqrt{2}}{3 - 2}$$

$$= \frac{3 - 2\sqrt{6} + 2}{1}$$

$$= 5 - 2\sqrt{6} \qquad\blacksquare$$

Gottfried Wilhelm Leibniz (1646–1716)

In calculus, it is sometimes necessary to rationalize a numerator.

Example 16   Rationalize the numerator of the fraction $\dfrac{\sqrt{x + h} - \sqrt{x}}{h}$.

Solution

$$\frac{\sqrt{x + h} - \sqrt{x}}{h} = \frac{(\sqrt{x + h} - \sqrt{x})(\sqrt{x + h} + \sqrt{x})}{h(\sqrt{x + h} + \sqrt{x})}$$

$$= \frac{x + h - x}{h(\sqrt{x + h} + \sqrt{x})}$$

$$= \frac{h}{h(\sqrt{x + h} + \sqrt{x})}$$

$$= \frac{1}{\sqrt{x + h} + \sqrt{x}} \qquad \text{Divide out the common factor of } h. \qquad\blacksquare$$

Division of polynomials is accomplished by using a process similar to that used for the long division of whole numbers. Consider this division:

$$\frac{2x^2 + 11x - 30}{x + 7}$$

This division can be rewritten in long division form as

$$x + 7 \,\overline{)\, 2x^2 + 11x - 30}$$

The binomial $x + 7$ is called the **divisor**; the trinomial $2x^2 + 11x - 30$ is called the **dividend**. The final answer, called the **quotient**, will appear above the long division symbol.

We begin the division process by asking "What expression when multiplied by x gives $2x^2$?" The answer to this question is $2x$ because $x \cdot 2x = 2x^2$. We

Sir Isaac Newton (1642–1727) In Newton's lifetime, there was a bitter controversy over who first invented calculus. It is now believed that Newton and Leibniz invented the calculus independently.

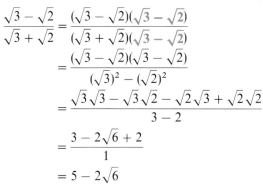

place $2x$ as the first term in the quotient, multiply each term of the divisor by $2x$, subtract, and bring down the -30:

$$
\begin{array}{r}
2x \\
x + 7 \overline{)\, 2x^2 + 11x - 30} \\
2x^2 + 14x \\
\hline
-\ 3x - 30
\end{array}
$$

We continue the division by asking the question "What expression when multiplied by x gives $-3x$?" We place the answer, -3, as the second term in the quotient, multiply each term of the divisor by -3, and subtract:

$$
\begin{array}{r}
2x \quad -3 \\
x + 7 \overline{)\, 2x^2 + 11x - 30} \\
2x^2 + 14x \\
\hline
-\ 3x - 30 \\
-\ 3x - 21 \\
\hline
-\ 9
\end{array}
$$

Because the degree of the remainder, -9, is less than the degree of the divisor, the division process is completed. You can express the result in *quotient* $+$ $\dfrac{remainder}{divisor}$ form as follows:

$$
2x - 3 + \frac{-9}{x + 7}
$$

Note that you have subtracted from $2x^2 + 11x - 30$ first $2x$ times $x + 7$, then -3 times $x + 7$, leaving a remainder of -9.

Example 17   Divide $6x^3 - 11$ by $2x + 2$.

Solution   Set up the division, leaving space for the missing powers of x in the dividend:

$$
2x + 2 \overline{)\, 6x^3 + 0x^2 + 0x - 11}
$$

The division process continues as usual: "What expression when multiplied by $2x$ gives $6x^3$?" The answer, $3x^2$, is the first term of the quotient. The division proceeds as follows:

$$
\begin{array}{r}
3x^2 \\
2x + 2 \overline{)\, 6x^3 -\ 11} \\
6x^3 + 6x^2 \\
\hline
-\ 6x^2
\end{array}
$$

Because $2x(-3x) = -6x^2$, the second term of the quotient is $-3x$:

$$
\begin{array}{r}
3x^2 - 3x \\
2x + 2 \overline{)\, 6x^3 - 11} \\
\underline{6x^3 + 6x^2 } \\
-6x^2 \\
\underline{-6x^2 - 6x } \\
+6x - 11
\end{array}
$$

Because $2x(3) = 6x$, the third term of the quotient is 3:

$$
\begin{array}{r}
3x^2 - 3x \; + 3 \\
2x + 2 \overline{)\, 6x^3 - 11} \\
\underline{6x^3 + 6x^2 } \\
-6x^2 \\
\underline{-6x^2 - 6x } \\
+6x - 11 \\
\underline{+6x + 6} \\
-17
\end{array}
$$

Hence,

$$
\frac{6x^3 - 11}{2x + 2} = 3x^2 - 3x + 3 + \frac{-17}{2x + 2}
$$

■

Example 18   Divide $-3x^3 - 3 + x^5 + 4x^2 - x^4$ by $x^2 - 3$.

Solution   The division process works most efficiently when the terms in both the divisor and dividend are written with their exponents in descending order. The long division is set up and accomplished as follows:

$$
\begin{array}{r}
x^3 - x^2 + 1 \\
x^2 - 3 \overline{)\, x^5 - x^4 - 3x^3 + 4x^2 - 3} \\
\underline{x^5 - 3x^3 } \\
-x^4 + 4x^2 \\
\underline{-x^4 + 3x^2 } \\
x^2 - 3 \\
\underline{x^2 - 3} \\
0
\end{array}
$$

Hence,

$$
\frac{x^5 - x^4 - 3x^3 + 4x^2 - 3}{x^2 - 3} = x^3 - x^2 + 1
$$

■

Exercise 1.5

In Exercises 1–12, indicate whether the given expression is a polynomial. If it is a polynomial, give its degree, if possible, and tell if it is a monomial, binomial, or trinomial.

1. $x^2 + 3x + 4$ **2.** $x^3 - 5xy$ **3.** $x^3 + y^{\frac{1}{2}}$ **4.** $x^{\frac{3}{2}} - 5y$

5. $\sqrt{5}x^3 + 4x^2$ **6.** x^2y^3 **7.** $\sqrt{15}$ **8.** $\dfrac{5}{x} + \dfrac{x}{5} + 5$

9. 0 **10.** $\sqrt{x}$

11. $4x^{-3} - 3x^{-2} + 2x + 1$ **12.** $3x^3 + 4x^2 + 2x + 2$

In Exercises 13–52, perform the indicated operations and simplify.

13. $x^3 - 3x^2 + 5x^3 - 8x$ **14.** $2x^4 - 5x^3 + 7x^3 - x^4$

15. $(x^5 + 2x^3 + 7) - (x^5 - 2x^3 - 7)$ **16.** $(3x^7 - 7x^3 + 3) - (7x^7 - 3x^3 + 7)$

17. $2(x^2 + 3x - 1) - 3(x^2 + 2x - 4) + 4$ **18.** $5(x^3 - 8x + 3) + 2(3x^2 + 5x) - 7$

19. $8(x^2 - 2x + 5) + 4(x^2 - 3x + 2) - 6(2x^2 - 8)$

20. $3x(x - 5) - x(2 + 3x) + 3(x + 2)$

21. $x(x^2 - 1) - x^2(x + 2) - x(2x - 2)$

22. $x(x - 4) - (x^2 + 3) + x(2x + 3)$

23. $(2x^2y^3)(4xy^4)$ **24.** $(-15a^3b)\left(\dfrac{1}{3}ab^4\right)$ **25.** $-4rs(r^2 + s^2)$ **26.** $6u^2v(2uv^3 - y)$

27. $(x - 5)(x + 5)$ **28.** $(x - 6)(x + 5)$ **29.** $(3x - 2)(x + 2)$ **30.** $(2x - 3)(4x + 1)$

31. $(2x + 3)(x - 4)$ **32.** $(3 - x)(2 + 3x)$ **33.** $(5x - 1)(2x + 3)$ **34.** $(-x + 1)(2x - 3)$

35. $(2 - 4x)(3 - 2x)$ **36.** $(3x - 4)(4x + 3)$ **37.** $(-2x + 3)(3x + 1)$ **38.** $(4x + 5)(3x - 4)$

39. $(9x - 1)(x^2 - 3)$ **40.** $(8x^2 + 1)(x + 2)$ **41.** $(5x + 2)(x^2 - 1)$ **42.** $(1 - 2x^2)(x^2 + 3)$

43. $(3x + 9)^2$ **44.** $(5x^2 - 1)^2$ **45.** $(3x - 1)^3$ **46.** $(2x - 3)^3$

47. $(xy^2 - 1)(x^2y + 2)$ **48.** $(xy - z^2)(xy + z^2)$

49. $(3x + 1)(2x^2 + 4x - 3)$ **50.** $(2x - 5)(x^2 - 3x + 2)$

51. $(x^2 + x + 1)(x^2 + x - 1)$ **52.** $(x^2 - x + 1)(x^2 + x + 1)$

In Exercises 53–56, multiply the expressions as you would multiply polynomials.

53. $x^{\frac{1}{2}}(x^{\frac{1}{2}}y + xy^{\frac{1}{2}})$ **54.** $ab^{\frac{1}{2}}(a^{\frac{1}{2}}b^{\frac{1}{2}} + b^{\frac{1}{2}})$ **55.** $(a^{\frac{1}{2}} + b^{\frac{1}{2}})(a^{\frac{1}{2}} - b^{\frac{1}{2}})$ **56.** $(x^{\frac{3}{2}} + y^{\frac{1}{2}})^2$

In Exercises 57–70, rationalize each denominator.

57. $\dfrac{2}{\sqrt{3} - 1}$ **58.** $\dfrac{1}{\sqrt{5} + 2}$ **59.** $\dfrac{3x}{\sqrt{7} + 2}$ **60.** $\dfrac{14y}{\sqrt{2} - 3}$

61. $\dfrac{6}{\sqrt{5} - \sqrt{2}}$ **62.** $\dfrac{15}{\sqrt{7} + \sqrt{2}}$ **63.** $\dfrac{x}{x - \sqrt{3}}$ **64.** $\dfrac{y}{2y + \sqrt{7}}$

65. $\dfrac{y + \sqrt{2}}{y - \sqrt{2}}$

66. $\dfrac{x - \sqrt{3}}{x + \sqrt{3}}$

67. $\dfrac{\sqrt{5} - 2}{\sqrt{5} + 2}$

68. $\dfrac{\sqrt{7} + 4}{\sqrt{7} - 4}$

69. $\dfrac{\sqrt{2} - \sqrt{3}}{1 - \sqrt{3}}$

70. $\dfrac{\sqrt{3} - \sqrt{2}}{1 + \sqrt{2}}$

In Exercises 71–72, rationalize each numerator.

71. $\dfrac{\sqrt{x + 3} - \sqrt{x}}{3}$

72. $\dfrac{\sqrt{2 + h} - \sqrt{2}}{h}$

In Exercises 73–86, simplify each fraction by performing a long division. If there is a nonzero remainder, write the answer in quotient $+ \dfrac{remainder}{divisor}$ *form.*

73. $\dfrac{3x^2 + 11x + 9}{x + 3}$

74. $\dfrac{3x^2 + 11x + 6}{3x + 2}$

75. $\dfrac{2x^2 - 19x + 35}{2x - 5}$

76. $\dfrac{2x^2 - 19x + 35}{x - 7}$

77. $\dfrac{x^3 - 2x^2 - 4x + 3}{x^2 + x - 1}$

78. $\dfrac{x^3 - 2x^2 - 4x + 5}{x - 3}$

79. $\dfrac{x^5 - 2x^3 - 3x^2 + 9}{x^2 - 2}$

80. $\dfrac{x^5 - 2x^3 - 3x^2 + 9}{x^3 - 3}$

81. $\dfrac{x^5 - 32}{x - 2}$

82. $\dfrac{x^4 - 1}{x + 1}$

83. $\dfrac{36x^4 - 121x^2 + 120 + 72x^3 - 142x}{11x - 10 + 6x^2}$

84. $\dfrac{-121x^2 + 72x^3 - 142x + 120 + 36x^4}{x + 6x^2 - 12}$

85. $\dfrac{11x^5 - 9x^2 + 12x^6 + 3x^4 + 3x + 10x^3 - 6}{4x^4 - 3 + 5x^3}$

86. $\dfrac{3x^4 + 10x^3 + 3x - 6 - 9x^2 + 12x^6 + 11x^5}{2 + 3x^2 - x}$

In Exercises 87–92, simplify each expression.

87. $x^{3a+2} x^{2a-3}$

88. $x^{3 - y^2} x^{3y^2 + y}$

89. $(b^{3a})^{a+1}$

90. $(a^{2b^2 + 3})^{2b^2 - 3}$

91. $\dfrac{a^{2(x+1)}}{a^{3 - 2x}}$

92. $\dfrac{b^{x^2 + 1}}{b^{3 - 2x^2}}$

93. Show that any trinomial can be squared according to the formula $(a + b + c)^2 = a^2 + b^2 + c^2 + 2ab + 2bc + 2ca$.

94. Show that $(a + b + c + d)^2 = a^2 + b^2 + c^2 + d^2 + 2ab + 2ac + 2ad + 2bc + 2bd + 2cd$.

1.6 FACTORING POLYNOMIALS

When two or more polynomials are multiplied together, each of those polynomials is called a **factor** of the resulting product. For example, the indicated product $7(x + 2)(x + 3)$ has polynomial factors of 7, $x + 2$, and $x + 3$. The process of writing a polynomial as the product of several factors is called **factoring** that polynomial.

In this section, we will discuss factoring over the set of integers. If a polynomial cannot be factored by using integers only, we will call the polynomial **prime** or **irreducible** over the set of integers.

Example 1   Factor the binomial $3xy^2 + 6x$.

Solution   Note that each term contains a factor of $3x$:

$$3xy^2 + 6x = 3x(y^2) + 3x(2)$$

Use the distributive property to factor out the common factor of $3x$:

$$3xy^2 + 6x = 3x(y^2 + 2)$$

The factored form of $3xy^2 + 6x$ is $3x(y^2 + 2)$. This type of factoring is called **factoring out a common** (monomial) **factor**.   ∎

Example 2   Factor the binomial $x^2y^2z^2 + xyz$.

Solution   The binomial can be rewritten, and the common factor xyz on the right side of the following equation can be factored out:

$$x^2y^2z^2 + xyz = xyz \cdot xyz + xyz \cdot 1$$
$$= xyz(xyz + 1)$$

It is important to understand where the $+1$ comes from. The last term in the expression $x^2y^2z^2 + xyz$ has an understood coefficient of 1. When the xyz is factored out, that understood 1 must be made explicit.   ∎

Example 3   Factor the expression $ax + bx + a + b$.

Solution   There is no factor that is common to all four terms, but the factor x is common to the first two terms. Factor out that x and enclose the last two terms with parentheses:

$$ax + bx + a + b = x(a + b) + (a + b)$$

The common factor $(a + b)$ is present in each term on the right side and can be factored out:

$$ax + bx + a + b = x(a + b) + (a + b)$$
$$= x(a + b) + 1(a + b)$$
$$= (a + b)(x + 1)$$

The technique used in this example is called **factoring by grouping**.   ∎

Example 4   Factor the binomial $49x^2 - 4$.

Solution   Observe that each term is a perfect square:

$$49x^2 - 4 = (7x)^2 - (2)^2$$

The difference of the squares of two quantities is the product of two factors. One factor is the sum of those quantities, and the other is the difference of those quantities. Thus, $49x^2 - 4$ factors as follows:

$$49x^2 - 4 = (7x)^2 - (2)^2$$
$$= (7x + 2)(7x - 2)$$

Verify that $(7x + 2)(7x - 2) = 49x^2 - 4$ by multiplying $(7x - 2)$ by $(7x + 2)$:

$$(7x + 2)(7x - 2) = 49x^2 - 14x + 14x - 4$$
$$= 49x^2 - 4$$

This type of problem is called factoring the **difference of two squares.**   ∎

Example 4 suggests the following formula for factoring the difference of two squares.

$$x^2 - y^2 = (x + y)(x - y)$$

Example 5   Factor the binomial $16x^4 - y^4$.

Solution   The binomial $16x^4 - y^4$ can be factored as the difference of two squares:

$$16x^4 - y^4 = (4x^2)^2 - (y^2)^2$$
$$= (4x^2 + y^2)(4x^2 - y^2)$$

However, the second factor, $4x^2 - y^2$, is also a difference of two squares and can be factored:

$$16x^4 - y^4 = (4x^2 + y^2)(4x^2 - y^2)$$
$$= (4x^2 + y^2)(2x + y)(2x - y)$$

With only integers to work with, $4x^2 + y^2$ cannot be factored.   ∎

Example 6   Factor $18x^2 - 32$.

Solution   Begin by factoring out a 2.

$$18x^2 - 32 = 2(9x^2 - 16)$$

Because $9x^2 - 16$ is the difference of two squares, it can be factored:

$$18x^2 - 32 = 2(9x^2 - 16)$$
$$= 2(3x + 4)(3x - 4)$$

   ∎

Certain quadratic trinomials are squares of binomials. Such trinomials can be factored according to the following formulas.

$$x^2 + 2xy + y^2 = (x + y)^2$$
$$x^2 - 2xy + y^2 = (x - y)^2$$

Example 7   Factor the quadratic trinomial $a^2 - 6a + 9$.

Solution   Note that $a^2 - 6a + 9$ can be written in the form

$$a^2 - 2(3a) + 3^2$$

and this form matches the left side of the formula $x^2 - 2xy + y^2 = (x - y)^2$. Hence,

$$a^2 - 6a + 9 = (a - 3)^2 \qquad \blacksquare$$

Factoring quadratic trinomials that are not squares of binomials requires more thought. Suppose that a quadratic trinomial has no common monomial factors. If this trinomial is factorable, it must factor into the product of two binomials. The process to follow when factoring quadratic trinomials is illustrated in the next three examples.

Example 8   Factor the quadratic trinomial $x^2 + 3x - 10$.

Solution   Note that the terms of the trinomial $x^2 + 3x - 10$ have no common monomial factors. So, if it is factorable, it will factor into the product of two binomials of the form $(ax + b)$ and $(cx + d)$. Because $x \cdot x$ gives the first term (x^2) of the given trinomial, start by considering this partial factorization.

$$x^2 + 3x - 10 = (x + ?)(x + ?)$$

Because the last term of the trinomial is -10, the product of the second terms of each binomial factor must be -10, and they must give a middle term of $3x$. There are four possibilities:

$$(x + 1)(x - 10)$$
$$(x - 1)(x + 10)$$
$$(x + 2)(x - 5)$$
$$(x - 2)(x + 5)$$

Of these possibilities only the last one provides the correct middle term of $3x$. Thus,

$$x^2 + 3x - 10 = (x - 2)(x + 5)$$

This type of problem is called **factoring the general quadratic trinomial.** $\blacksquare$

Example 9 Factor the quadratic trinomial $2x^2 - x - 6$.

Solution The $2x^2$ can be factored as $2x \cdot x$.

$$2x^2 - x - 6 = (2x \quad)(x \quad)$$

The number -6 in the trinomial can be factored as $3(-2)$, $(-3)(2)$, $1(-6)$, or $(-1)(6)$. The only possibility that provides a middle term of $-x$ is

$$(2x + 3)(x - 2)$$

Hence,

$$2x^2 - x - 6 = (2x + 3)(x - 2) \qquad \blacksquare$$

Example 10 Factor the quadratic trinomial $12 + 5x - 3x^2$.

Solution Factor $-3x^2$ as $-3x \cdot x$.

$$12 + 5x - 3x^2 = (\quad - 3x)(\quad + x)$$

The first term of the trinomial, 12, can be factored in many ways. The only factors that give a middle term of $5x$, however, are -4 and -3. Hence,

$$12 + 5x - 3x^2 = (-4 - 3x)(-3 + x)$$

This factorization could have been handled differently. The $-3x^2$ could have been split apart into factors of $3x$ and $-x$. In this case, the factorization of the given trinomial would be

$$12 + 5x - 3x^2 = (4 + 3x)(3 - x)$$

Although these two factorizations look different, they are both correct. $\qquad \blacksquare$

Two other factoring types are the **sum of two cubes** and the **difference of two cubes.** Like the difference of two squares, they follow a definite pattern.

$$x^3 + y^3 = (x + y)(x^2 - xy + y^2)$$
$$x^3 - y^3 = (x - y)(x^2 + xy + y^2)$$

Example 11 Factor the binomial $x^3 - 8$.

Solution This binomial can be written in the form $x^3 - 2^3$, which is the difference of two cubes. Substituting into the formula for the difference of two cubes gives

$$x^3 - 2^3 = (x - 2)(x^2 + 2x + 2^2)$$
$$= (x - 2)(x^2 + 2x + 4)$$

Hence,

$$x^3 - 8 = (x - 2)(x^2 + 2x + 4) \qquad \blacksquare$$

Example 12   Factor the binomial $27x^6 + 64y^3$.

Solution   This expression can be rewritten in the form of the sum of two cubes and factored as follows:

$$27x^6 + 64y^3 = (3x^2)^3 + (4y)^3$$
$$= (3x^2 + 4y)[(3x^2)^2 - (3x^2)(4y) + (4y)^2]$$
$$= (3x^2 + 4y)(9x^4 - 12x^2y + 16y^2)$$ ∎

Example 13   Factor $x^2 - y^2 + 6x + 9$.

Solution   This expression can be rewritten and factored as follows:

$$x^2 - y^2 + 6x + 9 = x^2 + 6x + 9 - y^2$$
$$= (x + 3)^2 - y^2 \qquad \text{Factor } x^2 + 6x + 9.$$
$$= (x + 3 + y)(x + 3 - y) \qquad \text{Factor the difference of two squares.}$$

Note that this is a factoring-by-grouping problem.   ∎

Example 14   Factor the trinomial $z^4 - 3z^2 + 1$.

Solution   An attempt to factor $z^4 - 3z^2 + 1$ as the product of two binomials will fail. One reasonable attempt is $(z^2 - 1)(z^2 - 1)$, but this product is $z^4 - 2z^2 + 1$. However,

$$z^4 - 3z^2 + 1 = z^4 - 3z^2 + z^2 + 1 - z^2 \qquad \text{Add and subtract } z^2.$$
$$= (z^4 - 2z^2 + 1) - z^2$$
$$= (z^2 - 1)(z^2 - 1) - z^2$$
$$= (z^2 - 1)^2 - z^2$$

The expression $(z^2 - 1)^2 - z^2$ is the difference of two squares and can be factored as

$$z^4 - 3z^2 + 1 = (z^2 - 1 + z)(z^2 - 1 - z)$$

or

$$= (z^2 + z - 1)(z^2 - z - 1)$$

Hence, the factorization of the given trinomial is

$$z^4 - 3z^2 + 1 = (z^2 + z - 1)(z^2 - z - 1)$$ ∎

Example 15   Factor the trinomial $x^4 - 6x^2 + 1$.

Solution   Again, factorization as the product of two binomials fails. The only reasonable possibility is $(x^2 - 1)(x^2 - 1)$, but that product is $x^4 - 2x^2 + 1$. However,

$$x^4 - 6x^2 + 1 = x^4 - 6x^2 + 4x^2 + 1 - 4x^2$$
$$= (x^4 - 2x^2 + 1) - 4x^2$$
$$= (x^2 - 1)^2 - (2x)^2$$
$$= (x^2 - 1 + 2x)(x^2 - 1 - 2x)$$
$$= (x^2 + 2x - 1)(x^2 - 2x - 1)$$ ■

Exercise 1.6

In Exercises 1–100, completely factor each algebraic expression over the set of integers, if possible.

1. $3x - 6$

2. $5y - 15$

3. $8x^2 + 4x^3$

4. $9y^3 + 6y^2$

5. $7xy^2 + 14x^2y$

6. $25y^2z - 15yz^2$

7. $3a^2bc + 6ab^2c + 9abc^2$

8. $5x^3y^3z^3 + 25x^2y^2z^2 - 125xyz$

9. $b(x + y) - a(x + y)$

10. $b(x - y) + a(x - y)$

11. $4a + b - 12a^2 - 3ab$

12. $x^2 + 4x + xy + 4y$

13. $3x^3 + 3x^2 - x - 1$

14. $4x + 6xy - 9y - 6$

15. $2txy + 2ctx - 3ty - 3ct$

16. $2ax + 4ay - bx - 2by$

17. $ax + bx + ay + by + az + bz$

18. $6x^2y^3 + 18xy + 3x^2y^2 + 9x$

19. $6xc + yd + 2dx + 3cy$

20. $ax + ay + az + bx + by + bz$

21. $4x^2 - 9$

22. $36z^2 - 49$

23. $4 - 9r^2$

24. $(x - y)^2 - 9$

25. $(x + z)^2 - 25$

26. $16 - 49x^2$

27. $1 + 25x^4$

28. $36x^4 + 121$

29. $x^2 - (y - z)^2$

30. $z^2 - (y + 3)^2$

31. $(x - y)^2 - (x + y)^2$

32. $(2a + 3) - (2a + 3)^2$

33. $x^4 - y^4$

34. $z^4 - 81$

35. $x^8 - 64z^4$

36. $1 - y^8$

37. $3x^2 - 12$

38. $3x^3y - 3xy$

39. $18xy^2 - 8x$

40. $27x^2 - 12$

41. $x^2 + 10x + 21$

42. $x^2 + 7x + 10$

43. $x^2 - 4x - 12$

44. $x^2 - 2x - 63$

45. $x^2 - 2x + 15$

46. $x^2 + x + 2$

47. $12x^2 - x - 6$

48. $8x^2 - 10x - 3$

49. $24y^2 + 15 - 38y$

50. $10x^2 - 18 + 3x$

51. $-15 + 2a + 24a^2$

52. $-32 - 68x + 9x^2$

53. $6x^2 + 29x + 35$

54. $10x^2 - 17x + 6$

55. $-35 - x + 6x^2$

56. $-5x - 6 + 6x^2$

57. $12y^2 - 58y - 70$

58. $3x^2 - 6x - 9$

59. $6x^3 - 23x^2 - 35x$

60. $y^2 + y - 90$

61. $6x^2 - 11x - 35$

62. $12 + 17x - 7x^2$

63. $35 - 47x + 6x^2$

64. $14x^2 + 11x - 15$

65. $x^4 + 2x^2 - 15$

66. $x^4 - x^2 - 6$

67. $2x^2y^2z^2 - 16xyz + 80$

68. $3x^2 + 30x + 75$

69. $8z^3 - 27$ **70.** $125a^3 - 64$ **71.** $2x^3 + 2000$ **72.** $3y^3 + 648$

73. $(x + y)^3 - 64$ **74.** $(x - y)^3 + 27$ **75.** $1 - (x + 1)^3$ **76.** $1 + (x - 1)^3$

77. $64a^6 - y^6$ **78.** $a^6 + b^6$

79. $4x^2 + 6x^2y + 6x + 9xy$ **80.** $(a^2 - y^2) - 5(a + y)$

81. $64x^6 + y^6$ **82.** $z^2 + 6z + 9 - 225y^2$

83. $x^2 - 6x + 9 - 144y^2$ **84.** $x^2 + 2x - 9y^2 + 1$

85. $(a + b)^2 - 3(a + b) - 10$ **86.** $2(a + b)^2 - 5(a + b) - 3$

87. $6(u + v)^2 + 11u + 4 + 11v$ **88.** $8(r + s)^2 - 10(r + s) - 3$

89. $x^6 + 7x^3 - 8$ **90.** $x^6 - 13x^4 + 36x^2$

91. $a(c + d) + a + b(c + d) + b$ **92.** $c(a + b) + 3c + 3d + d(a + b)$

93. $x^4 + 3x^2 + 4$ **94.** $x^4 + x^2 + 1$

95. $x^4 + 7x^2 + 16$ **96.** $y^4 + 2y^2 + 9$

97. $4a^4 + 1 + 3a^2$ **98.** $x^4 + 25 + 6x^2$

99. $2x^4 + 8$ **100.** $3x^4 - 21x^2 + 27$

In Exercises 101–106, factor the indicated monomial from the given expression.

101. $3x + 2; 2$ **102.** $a + b; a$ **103.** $x + x^{\frac{1}{2}}; x^{\frac{1}{2}}$ **104.** $2x + \sqrt{2}y; \sqrt{2}$

105. $ab^{\frac{3}{2}} - a^{\frac{3}{2}}b; ab$ **106.** $ab^2 + b; b^{-1}$

In Exercises 107–108, simplify each expression.

107. $\dfrac{\sqrt{5} - 1}{\sqrt{5} + 1}$ **108.** $\dfrac{\sqrt{7} + 3}{\sqrt{7} - 3}$

1.7   ALGEBRAIC FRACTIONS

We begin the discussion of algebraic fractions with a review of the basic properties of fractions.

Always remember that *the denominator of a fraction can never be 0.* This fact is implicit whenever we write a fraction.

Property 1 of Fractions. $\dfrac{a}{b} = \dfrac{c}{d}$ if and only if $ad = bc$

Property 1 enables us to decide whether two fractions are equal. If the same product occurs when "cross-multiplying," the fractions are equal. For example, $\frac{2}{3} = \frac{4}{6}$ because $2 \cdot 6 = 3 \cdot 4$. Note that $\frac{3}{4} \neq \frac{5}{7}$ because $3 \cdot 7 \neq 4 \cdot 5$.

Property 2 of Fractions. $a \cdot 1 = a, \quad \dfrac{a}{1} = a, \quad$ and $\quad \dfrac{a}{a} = 1$

Property 2 asserts that any number multiplied by 1 is left unchanged, that any number divided by 1 is left unchanged, and that any nonzero number divided by itself is equal to 1. For example, $5 \cdot 1 = 5$, $\frac{5}{1} = 5$, and $\frac{5}{5} = 1$.

Property 3 of Fractions. Fractions can be multiplied and divided according to the following rules:

$$\frac{a}{b} \cdot \frac{c}{d} = \frac{a \cdot c}{b \cdot d} = \frac{ac}{bd}$$

$$\frac{a}{b} \div \frac{c}{d} = \frac{a}{b} \cdot \frac{d}{c} = \frac{ad}{bc}$$

The rule for the division of fractions is the familiar "invert and multiply" idea. In the exercises, you will be asked to prove this rule.

It is often necessary to change the form of a fraction without altering its value. Properties 2 and 3 of fractions imply that both the numerator and denominator of a fraction can be multiplied by the same nonzero expression without changing the value of the fraction. For example,

$$\frac{3}{5} = \frac{3}{5} \cdot 1 = \frac{3}{5} \cdot \frac{2}{2} = \frac{6}{10}$$

The generalization of this idea gives another property of fractions.

Property 4 of Fractions. $\dfrac{a}{b} = \dfrac{ax}{bx} \quad$ provided $x \neq 0$

It is Property 4 that we use to reduce fractions because it allows common factors that appear in both the numerator and denominator of a fraction to be removed, or canceled. For example,

$$\frac{4}{10} = \frac{2 \cdot 2}{5 \cdot 2} = \frac{2 \cdot 2}{5 \cdot 2} = \frac{2}{5}$$

To *simplify a fraction* means to reduce it to lowest terms.

Remember that only factors of the entire numerator may cancel factors of the entire denominator. Do not make the mistake of canceling terms or factors of a single term within a many-termed algebraic expression. The following false proof illustrates this error:

$$\frac{7}{10} = \frac{1+6}{1+9} = \frac{1+2\cdot3}{1+3\cdot3} = \frac{1+2\cdot3}{1+3\cdot3} = \frac{1+2}{1+3} = \frac{3}{4}$$

The conclusion that $\frac{7}{10} = \frac{3}{4}$ is not true because $7\cdot4 \neq 10\cdot3$.

Property 4 of fractions can also be used to insert factors. Suppose we want to write the fraction $\frac{2}{5}$ as a fraction with a denominator of 15. Multiplying both the numerator and the denominator by 3 gives

$$\frac{2}{5} = \frac{2\cdot3}{5\cdot3} = \frac{6}{15}$$

Similarly, to express $-\frac{3}{7}$ as a fraction with a denominator of -14, we multiply the numerator and denominator by -2:

$$-\frac{3}{7} = \frac{(-3)(-2)}{7(-2)} = \frac{6}{-14}$$

The following examples illustrate the multiplication and division of some simple fractions.

Example 1
$$\frac{2}{7}\cdot\frac{3}{5} = \frac{2\cdot3}{7\cdot5} = \frac{6}{35}$$
∎

Example 2
$$\frac{5}{9}\cdot\frac{3}{10} = \frac{5\cdot3}{9\cdot10} = \frac{5\cdot3}{3\cdot3\cdot2\cdot5} = \frac{1}{6}$$
∎

Example 3
$$\frac{3}{4}\div\frac{9}{12} = \frac{3}{4}\cdot\frac{12}{9} = \frac{3\cdot4\cdot3}{4\cdot3\cdot3} = 1$$
∎

Property 5 of Fractions. There are three signs associated with every fraction: the sign of the fraction, the sign of the numerator, and the sign of the denominator. Any two of these signs may be changed without altering the value of the fraction. (If no sign is indicated, a + sign is understood.)

$$\frac{a}{b} = \frac{-a}{-b} = -\frac{a}{-b} = -\frac{-a}{b}$$

$$-\frac{a}{b} = \frac{a}{-b} = \frac{-a}{b} = -\frac{-a}{-b}$$

Example 4   The fractions $\frac{4}{2}$ and $\frac{-4}{-2}$ are equal because each reduces to $+2$. Note that the signs of the numerator and denominator have changed, but the understood positive sign in front of the fraction remains unaltered.   ∎

> **Property 6 of Fractions.** Fractions may be added and subtracted according to the following rules:
>
> $$\frac{a}{b} + \frac{c}{b} = \frac{a+c}{b}$$
>
> $$\frac{a}{b} - \frac{c}{b} = \frac{a-c}{b}$$

To add fractions with different denominators, we must find a common denominator, preferably the least common denominator. To find this common denominator, we use Property 4 of fractions and insert whatever factors are needed to make the denominators equal:

$$\frac{3}{5} + \frac{2}{7} = \frac{3 \cdot 7}{5 \cdot 7} + \frac{2 \cdot 5}{7 \cdot 5} = \frac{21}{35} + \frac{10}{35} = \frac{21 + 10}{35} = \frac{31}{35}$$

Example 5 Add the fractions $\dfrac{3}{5}$ and $\dfrac{7}{10}$.

Solution In this case, 10 is a common denominator, so only the fraction $\dfrac{3}{5}$ needs to be altered:

$$\frac{3}{5} + \frac{7}{10} = \frac{3(2)}{5(2)} + \frac{7}{10} = \frac{6}{10} + \frac{7}{10} = \frac{13}{10}$$

∎

Example 6 Subtract $\dfrac{3}{10}$ from $\dfrac{4}{15}$.

Solution The common denominator could be $(15)(10) = 150$, but a smaller number will work and is more convenient: 30 is divisible by both 10 and 15:

$$\frac{4}{15} - \frac{3}{10} = \frac{4(2)}{15(2)} - \frac{3(3)}{10(3)} = \frac{8}{30} - \frac{9}{30} = \frac{-1}{30} = -\frac{1}{30}$$

∎

There is a method for finding the least common denominator. In the example above, note that the factors of the first denominator, 15, are 3 and 5. The factors of the second denominator, 10, are 5 and 2. Because the common denominator must be divisible by both 10 and 15, it must have factors of 3, 5, and 2. The product of 3, 5, and 2 is the least common denominator, 30.

Fractions that involve algebraic expressions behave exactly as arithmetic fractions do. However, because division by 0 is not defined, fractions such as

$$\frac{2x}{(x-3)(x-4)}$$

are not defined for those values of x that make the denominator 0. Thus, in this fraction, x cannot be either 3 or 4. Such restrictions always will be understood as we work with algebraic fractions.

Example 7   Simplify the fraction $\dfrac{x^2 - 9}{x - 3}$.

Solution   Factor both the numerator and the denominator and simplify:

$$\frac{x^2 - 9}{x - 3} = \frac{(x + 3)(x - 3)}{x - 3} = \frac{x + 3}{1} = x + 3$$

■

Example 8   Simplify the fraction $\dfrac{x - y}{y - x}$.

Solution   Note that -1 can be factored out of the expression in the denominator, that two signs can be changed, and that the fraction can then be simplified.

$$\frac{x - y}{y - x} = \frac{(x - y)}{-(x - y)} = -\frac{x - y}{x - y} = -1$$

In the solution, -1 was factored from the denominator. It could have been factored from the numerator with the same result. A similar argument shows that

$$\frac{x - 2}{2 - x} = -1 \quad \text{and} \quad \frac{a - x}{x - a} = -1$$

■

Example 8 indicates that the quotient of a nonzero quantity and its negative is -1. We use this fact again in the next example.

Example 9   Simplify the fraction $\dfrac{x^2 - y^2}{y - x}$.

Solution   Factor the numerator as follows:

$$\frac{x^2 - y^2}{y - x} = \frac{(x + y)(x - y)}{(y - x)}$$

and recognize that

$$\frac{(x + y)(x - y)}{(y - x)} = (x + y) \cdot \frac{x - y}{y - x}$$

Because of the result of Example 8, $\dfrac{x - y}{y - x} = -1$. Hence,

$$\frac{x^2 - y^2}{y - x} = (x + y)(-1) = -(x + y)$$

■

Example 10 Simplify the fraction $\dfrac{x^2 - 3x + 2}{x^2 - x - 2}$.

Solution Factor both the numerator and the denominator and simplify:

$$\frac{x^2 - 3x + 2}{x^2 - x - 2} = \frac{(x - 1)(x - 2)}{(x + 1)(x - 2)} = \frac{(x - 1)\cancel{(x - 2)}}{(x + 1)\cancel{(x - 2)}} = \frac{x - 1}{x + 1}$$

∎

Example 11 Simplify $\dfrac{x^2 + 2x - 3}{x^2 - x - 2}$.

Solution After factoring, note that there are no factors common to both the numerator and the denominator:

$$\frac{x^2 + 2x - 3}{x^2 - x - 2} = \frac{(x - 1)(x + 3)}{(x + 1)(x - 2)}$$

Because there are no common factors, this algebraic fraction is already in lowest terms and cannot be simplified.

∎

Example 12 Find the product of $\dfrac{x^2 - x - 2}{x^2 - 1}$ and $\dfrac{x^2 + 2x - 3}{x - 2}$.

Solution Before multiplying the fractions together, factor and then remove any factors common to the numerators and denominators of the fractions:

$$\begin{aligned}
\frac{x^2 - x - 2}{x^2 - 1} \cdot \frac{x^2 + 2x - 3}{x - 2} &= \frac{(x - 2)(x + 1)}{(x + 1)(x - 1)} \cdot \frac{(x - 1)(x + 3)}{(x - 2)} \\
&= \frac{\cancel{(x - 2)}\cancel{(x + 1)}}{\cancel{(x + 1)}\cancel{(x - 1)}} \cdot \frac{\cancel{(x - 1)}(x + 3)}{\cancel{(x - 2)}} \\
&= \frac{(x + 3)}{1} \\
&= x + 3
\end{aligned}$$

∎

Example 13 Simplify $\dfrac{x^2 - 2x - 3}{x^2 - 4} \div \dfrac{x^2 + 2x - 15}{x^2 + 3x - 10}$.

Solution Change the division to a multiplication, factor everything you can and simplify:

$$\begin{aligned}
\frac{x^2 - 2x - 3}{x^2 - 4} \div \frac{x^2 + 2x - 15}{x^2 + 3x - 10} &= \frac{x^2 - 2x - 3}{x^2 - 4} \cdot \frac{x^2 + 3x - 10}{x^2 + 2x - 15} \\
&= \frac{(x - 3)(x + 1)}{(x + 2)(x - 2)} \cdot \frac{(x - 2)(x + 5)}{(x + 5)(x - 3)} \\
&= \frac{x + 1}{x + 2}
\end{aligned}$$

∎

Example 14   Simplify $\dfrac{2x^2 - 5x - 3}{3x - 1} \cdot \dfrac{3x^2 + 2x - 1}{x^2 - 2x - 3} \div \dfrac{2x^2 + x}{3x}$.

Solution   Again, factor everything you can, change the division to a multiplication and simplify:

$$\frac{2x^2 - 5x - 3}{3x - 1} \cdot \frac{3x^2 + 2x - 1}{x^2 - 2x - 3} \div \frac{2x^2 + x}{3x}$$

$$= \frac{(x - 3)(2x + 1)}{3x - 1} \cdot \frac{(3x - 1)(x + 1)}{(x + 1)(x - 3)} \cdot \frac{3x}{x(2x + 1)}$$

$$= \frac{\cancel{(x - 3)}\cancel{(2x + 1)}}{\cancel{(3x - 1)}} \cdot \frac{\cancel{(3x - 1)}\cancel{(x + 1)}}{\cancel{(x + 1)}\cancel{(x - 3)}} \cdot \frac{3\cancel{x}}{\cancel{x}\cancel{(2x + 1)}}$$

$$= 3 \qquad\qquad \blacksquare$$

Example 15   Add the fractions $\dfrac{2}{x + 5}$ and $\dfrac{3x}{x + 5}$.

Solution   Because these fractions have the same denominator, add the numerators and keep the common denominator.

$$\frac{2}{x + 5} + \frac{3x}{x + 5} = \frac{2 + 3x}{x + 5} \qquad\qquad \blacksquare$$

Example 16   Add the fractions $\dfrac{1}{x - 2}$ and $\dfrac{2}{x + 1}$.

Solution   To add fractions, a common denominator is required, which in this case is the product of $x - 2$ and $x + 1$. Multiply the numerators and denominators of each fraction by an expression that will result in this common denominator:

$$\frac{1}{x - 2} + \frac{2}{x + 1} = \frac{1(x + 1)}{(x - 2)(x + 1)} + \frac{2(x - 2)}{(x + 1)(x - 2)}$$

Because the fractions now have the same denominator, add them by finding the sum of their numerators and keeping their common denominator:

$$\frac{1}{x - 2} + \frac{2}{x + 1} = \frac{(x + 1) + 2(x - 2)}{(x - 2)(x + 1)}$$

$$= \frac{3x - 3}{(x - 2)(x + 1)}$$

Always attempt to factor the numerator on the chance that the answer can be simplified:

$$= \frac{3(x - 1)}{(x - 2)(x + 1)}$$

In this case, the fraction does not simplify. $\qquad\qquad \blacksquare$

Example 17 Combine and simplify $\dfrac{x-2}{x^2-1} - \dfrac{x+3}{x^2+3x+2} + \dfrac{3}{x^2+x-2}$.

Solution Factor the denominators before finding the least common denominator:

$$\frac{x-2}{x^2-1} - \frac{x+3}{x^2+3x+2} + \frac{3}{x^2+x-2}$$

$$= \frac{x-2}{(x-1)(x+1)} - \frac{x+3}{(x+1)(x+2)} + \frac{3}{(x-1)(x+2)}$$

When the denominators are in factored form, it is easy to see that the common denominator must have factors of $x-1$, $x+1$, and $x+2$. Change these fractions so that each has this denominator:

$$= \frac{(x-2)(x+2)}{(x-1)(x+1)(x+2)} - \frac{(x+3)(x-1)}{(x+1)(x+2)(x-1)} + \frac{3(x+1)}{(x-1)(x+2)(x+1)}$$

Now, combine the fractions and simplify to get

$$= \frac{(x^2-4) - (x^2+2x-3) + (3x+3)}{(x-1)(x+2)(x+1)}$$

$$= \frac{x^2-4-x^2-2x+3+3x+3}{(x-1)(x+2)(x+1)}$$

$$= \frac{x+2}{(x-1)(x+2)(x+1)}$$

$$= \frac{1}{(x-1)(x+1)}$$

It is acceptable to leave the denominator in factored form. ∎

A fraction that has a fraction in its numerator or its denominator is called a **complex fraction**.

Example 18 Simplify the complex fraction $\dfrac{\dfrac{1}{x}+\dfrac{1}{y}}{\dfrac{x}{y}}$.

Solution 1 Determine that the lowest common denominator of the three fractions in the given complex fraction is xy. Multiply both the numerator and denominator of the given fraction by xy and simplify:

$$\frac{\dfrac{1}{x}+\dfrac{1}{y}}{\dfrac{x}{y}} = \frac{\left(\dfrac{1}{x}+\dfrac{1}{y}\right)xy}{\left(\dfrac{x}{y}\right)xy} = \frac{\dfrac{xy}{x}+\dfrac{xy}{y}}{\dfrac{xxy}{y}} = \frac{y+x}{x^2}$$

Solution 2 Combine the fractions in the numerator of the complex fraction to obtain a single fraction over a single fraction. Then use the fact that any fraction indicates a division; that is, the numerator is divided by the denominator:

$$\frac{\dfrac{1}{x} + \dfrac{1}{y}}{\dfrac{x}{y}} = \frac{\dfrac{1(y)}{x(y)} + \dfrac{1(x)}{y(x)}}{\dfrac{x}{y}} = \frac{\dfrac{y + x}{xy}}{\dfrac{x}{y}}$$

Change this result to a multiplication problem and simplify.

$$\frac{\dfrac{y + x}{xy}}{\dfrac{x}{y}} = \frac{y + x}{xy} \div \frac{x}{y} = \frac{y + x}{xy} \cdot \frac{y}{x} = \frac{y + x}{x^2}$$

Exercise 1.7

In Exercises 1–4, simplify each fraction.

1. $\dfrac{x^2 - 16}{x^2 - 8x + 16}$

2. $\dfrac{4 - x^2}{x^2 - 5x + 6}$

3. $\dfrac{6x^2 + x - 12}{4x^2 + 4x - 3}$

4. $\dfrac{x^3 - 8}{x^2 + ax - 2x - 2a}$

In Exercises 5–38, perform the indicated operations and simplify.

5. $\dfrac{x^2 - 1}{x} \cdot \dfrac{x^2}{x^2 + 2x + 1}$

6. $\dfrac{y^2 - 2y + 1}{y} \cdot \dfrac{y + 2}{y^2 + y - 2}$

7. $\dfrac{2x^2 + 32}{8} \div \dfrac{x^2 + 16}{2}$

8. $\dfrac{x^2 + x - 6}{x^2 - 6x + 9} \div \dfrac{x^2 - 4}{x^2 - 9}$

9. $\dfrac{z^2 + z - 20}{z^2 - 4} \div \dfrac{z^2 - 25}{z - 5}$

10. $\dfrac{ax + bx + a + b}{a^2 + 2ab + b^2} \div \dfrac{x^2 - 1}{x^2 - 2x + 1}$

11. $\dfrac{3x^2 + 7x + 2}{x^2 + 2x} \cdot \dfrac{x^2 - x}{3x^2 + x}$

12. $\dfrac{x^2 + x}{2x^2 + 3x} \cdot \dfrac{2x^2 + x - 3}{x^2 - 1}$

13. $\dfrac{x^2 + x}{x - 1} \cdot \dfrac{x^2 - 1}{x + 2}$

14. $\dfrac{x^2 + 5x + 6}{x^2 + 6x + 9} \cdot \dfrac{x + 2}{x^2 - 4}$

15. $\dfrac{3x^2 + 5x - 2}{x^2 + 2x} \cdot \dfrac{2x^2 + 5x}{6x^2 + 13x - 5}$

16. $\dfrac{x^2 + 7x + 12}{x^2 - x - 6} \cdot \dfrac{x^2 - 3x - 10}{x^2 + 2x - 3} \cdot \dfrac{x^2 - 4x + 3}{x^2 - x - 20}$

17. $\dfrac{x^2 + 13x + 12}{8x^2 - 6x - 5} \div \dfrac{2x^2 - x - 3}{8x^2 - 14x + 5}$

18. $\dfrac{x^2 - 2x - 3}{21x^2 - 50x - 16} \cdot \dfrac{3x - 8}{x - 3} \div \dfrac{x^2 + 6x + 5}{7x^2 - 33x - 10}$

19. $\dfrac{x^3 + 27}{x^2 - 4} \div \left(\dfrac{x^2 + 4x + 3}{x^2 + 2x} \div \dfrac{x^2 + x - 6}{x^2 - 3x + 9} \right)$

20. $\dfrac{x(x - 2) - 3}{x(x + 7) - 3(x - 1)} \cdot \dfrac{x(x + 1) - 2}{x(x - 7) + 3(x + 1)}$

21. $\dfrac{3}{x + 3} + \dfrac{x + 2}{x + 3}$

22. $\dfrac{3}{x + 1} + \dfrac{x + 2}{x + 1}$

23. $\dfrac{4}{x - 1} - \dfrac{3x}{x + 1}$

24. $\dfrac{2}{5 - x} + \dfrac{1}{x - 5}$

25. $\dfrac{a+3}{a^2+7a+12}+\dfrac{a}{a^2-16}$

26. $\dfrac{x}{x^2-4}-\dfrac{1}{x+2}$

27. $\dfrac{1}{3a+4}-\dfrac{a-7}{3a^2+13a+12}-\dfrac{4}{a^2+4a+3}$

28. $\dfrac{x+1}{6x^2+x-1}+\dfrac{x}{4x^2-1}-\dfrac{x-1}{6x^2-5x+1}$

29. $\dfrac{1}{x-2}+\dfrac{3}{x+2}-\dfrac{3x-2}{x^2-4}$

30. $\dfrac{x}{x-3}-\dfrac{5}{x+3}+\dfrac{3(3x-1)}{x^2-9}$

31. $\left(\dfrac{1}{x-2}+\dfrac{1}{x-3}\right)\cdot\dfrac{x-3}{2x}$

32. $\left(\dfrac{1}{x+1}-\dfrac{1}{x-2}\right)\div\dfrac{1}{x-2}$

33. $\dfrac{3x}{x-4}-\dfrac{x}{x+4}+\dfrac{3x+1}{x^2-16}$

34. $\dfrac{7x}{x-5}-\dfrac{3x}{x-5}+\dfrac{3x-1}{x^2-25}$

35. $\dfrac{1}{x^2+3x+2}-\dfrac{2}{x^2+4x+3}+\dfrac{1}{x^2+5x+6}$

36. $\dfrac{-2}{x-y}+\dfrac{2}{x-z}-\dfrac{2z-2y}{(y-x)(z-x)}$

37. $\dfrac{3x-2}{x^2+x-20}-\dfrac{4x^2+2}{x^2-25}+\dfrac{3x^2}{x^2-16}$

38. $\dfrac{3x+2}{8x^2-10x-3}+\dfrac{x+4}{6x^2-11x+3}-\dfrac{1}{4x+1}$

In Exercises 39–64, simplify each complex fraction.

39. $\dfrac{\dfrac{3a}{b}}{\dfrac{6ac}{b^2}}$

40. $\dfrac{\dfrac{3t^2}{9x}}{\dfrac{t}{18x}}$

41. $\dfrac{\dfrac{3a^2b}{ab}}{27}$

42. $\dfrac{\dfrac{3u^2v}{4t}}{3uv}$

43. $\dfrac{\dfrac{x-y}{ab}}{\dfrac{y-x}{ab}}$

44. $\dfrac{\dfrac{x^2-5x+6}{2x^2y}}{\dfrac{x^2-9}{2x^2y}}$

45. $\dfrac{\dfrac{1}{x}+\dfrac{1}{y}}{xy}$

46. $\dfrac{xy}{\dfrac{11}{x}-\dfrac{11}{y}}$

47. $\dfrac{\dfrac{1}{x}+\dfrac{1}{y}}{\dfrac{1}{x}-\dfrac{1}{y}}$

48. $\dfrac{\dfrac{1}{x}-\dfrac{1}{y}}{\dfrac{1}{x}+\dfrac{1}{y}}$

49. $\dfrac{\dfrac{3a}{b}-\dfrac{4a^2}{x}}{\dfrac{1}{b}+\dfrac{1}{ax}}$

50. $\dfrac{1-\dfrac{x}{y}}{\dfrac{x^2}{y^2}-1}$

51. $\dfrac{x+1-\dfrac{6}{x}}{x+5+\dfrac{6}{x}}$

52. $\dfrac{2z}{1-\dfrac{3}{z}}$

53. $\dfrac{3xy}{1-\dfrac{1}{xy}}$

54. $\dfrac{x-3+\dfrac{1}{x}}{\dfrac{1}{x}-x+3}$

55. $\dfrac{3x}{x+\dfrac{1}{x}}$

56. $\dfrac{2x^2+4}{2+\dfrac{4x}{5}}$

57. $\dfrac{\dfrac{x}{x+2}-\dfrac{2}{x-1}}{\dfrac{3}{x+2}+\dfrac{x}{x-1}}$

58. $\dfrac{\dfrac{2x}{x-3}+\dfrac{1}{x-2}}{\dfrac{3}{x-3}-\dfrac{x}{x-2}}$

59. $\dfrac{1}{1+x^{-1}}$

60. $\dfrac{y^{-1}}{x^{-1}+y^{-1}}$

61. $\dfrac{3(x+2)^{-1}+2(x-1)^{-1}}{(x+2)^{-1}}$

62. $\dfrac{2x(x-3)^{-1}-3(x+2)^{-1}}{(x-3)^{-1}(x+2)^{-1}}$

63. $\dfrac{1}{1 + \dfrac{1}{x^{-1}}}$

64. $\dfrac{ab}{2 + \dfrac{3}{2a^{-1}}}$

65. Use Properties 2 and 3 of fractions to prove Property 4.

66. Prove that $\dfrac{a}{b} \div \dfrac{c}{d}$ is equivalent to $\dfrac{a}{b} \cdot \dfrac{d}{c}$.

REVIEW EXERCISES

1. Draw a number line and graph the prime numbers between 10 and 20.

2. Change $\frac{25}{27}$ to decimal form.

3. Change $0.861\ 61\ 61 \ldots$ to a common fraction.

4. Does the decimal 2.3773 represent an irrational number? Explain.

In Review Exercises 5–8, graph each subset of the real numbers on the number line, if possible.

5. $-3 < x \le \dfrac{1}{3}$

6. $\dfrac{2}{3} \le x < \dfrac{5}{3}$

7. $x \ge 0 \quad$ or $\quad x < -1$

8. $[-2, 4) \cap (0, 5]$

In Review Exercises 9–10, assume that $x = -2$, $y = 3$, and $z = -1$. Find the value of each expression.

9. $|xyz|$

10. $|xz| - |y|$

In Review Exercises 11–16, indicate which field property justifies each given expression.

11. $(a + b) + 2 = a + (b + 2)$

12. $a + 7 = 7 + a$

13. $(6 + a) + 0 = 6 + a$

14. $4(2x) = (4 \cdot 2)x$

15. $(a + b) \cdot \dfrac{1}{a + b} = 1$

16. $1(a - b) = a - b$

In Review Exercises 17–22, simplify each expression. Express all answers with positive exponents only. Assume that all variables represent positive numbers.

17. $\left(\dfrac{x^3}{y^2}\right)^3$

18. $-(x^3 y^{-2})^2$

19. $\left(\dfrac{x^{-2} y^2}{2}\right)^{-3}$

20. $\left(\dfrac{x^2 y}{3x}\right)^{-3}$

21. $\left(\dfrac{3x^2 y^{-2}}{x^2 y^2}\right)^{-2}$

22. $\left(\dfrac{-3x^3 y}{xy^3}\right)^{-2}$

In Review Exercises 23–28, evaluate each expression if $x = 0$, $y = -1$, and $z = 2$.

23. $x^2(3^x)$

24. $y^3(3^y)$

25. $z^{yz}(z^{xy})$

26. $(z^z y^z)^y$

27. zy^x

28. $(y^z y^{yz})^x$

In Review Exercises 29–34, simplify each expression.

29. $\sqrt{36}$ **30.** $-\sqrt{49}$ **31.** $-4^{\frac{1}{2}}$ **32.** $8^{\frac{2}{3}}$

33. $-8^{\frac{2}{3}}$ **34.** $(-8)^{\frac{5}{3}}$

In Review Exercises 35–38, assume that x and y are positive. Simplify each expression.

35. $\sqrt{x^2 y^4}$ **36.** $\sqrt[3]{x^3}$ **37.** $(x^{12} y^2)^{\frac{1}{2}}$ **38.** $\left(\dfrac{x^{14}}{y^4}\right)^{-\frac{1}{2}}$

In Review Exercises 39–42, assume that x and y are unrestricted numbers. Simplify each expression.

39. $\sqrt{x^2 y^4}$ **40.** $\sqrt[3]{x^3}$ **41.** $(x^{12} y^2)^{\frac{1}{2}}$ **42.** $\left(\dfrac{x^{14}}{y^4}\right)^{-\frac{1}{2}}$

In Review Exercises 43–48, rationalize each denominator and simplify.

43. $\dfrac{2}{\sqrt{5}}$ **44.** $\dfrac{8}{\sqrt{8}}$ **45.** $\dfrac{1}{\sqrt[3]{2}}$ **46.** $\dfrac{2}{\sqrt[3]{25}}$

47. $\dfrac{2}{\sqrt{3}-1}$ **48.** $\dfrac{2}{\sqrt{3}-\sqrt{2}}$

In Review Exercises 49–54, simplify each expression and combine terms.

49. $\sqrt{50}+\sqrt{8}$ **50.** $\sqrt{12}+\sqrt{3}-\sqrt{27}$

51. $(\sqrt{2}+\sqrt{3})^2$ **52.** $(2+\sqrt{3})(\sqrt{3}-2)$

53. $(\sqrt{2}+1)(\sqrt{3}+1)$ **54.** $(\sqrt[3]{3}-2)(\sqrt[3]{9}+2\sqrt[3]{3}+4)$

In Review Exercises 55–58, indicate whether the given expression is a polynomial, and if so, give its degree and then indicate whether it is a monomial, binomial, or trinomial.

55. x^3-8 **56.** $8x^2-8x-8$

57. $\sqrt{3}x^2$ **58.** $3x^2-\sqrt{x}$

In Review Exercises 59–62, simplify each fraction by performing a long division. If there is a nonzero remainder, write the answer in $quotient+\dfrac{remainder}{divisor}$ form.

59. $\dfrac{2x^3+13x^2+17x-14}{2x+7}$ **60.** $\dfrac{x^4+x^3+2x^2+x+1}{x^2+1}$

61. $\dfrac{3x^2+1}{x+1}$ **62.** $\dfrac{x^3-x+5}{x^2+3x}$

In Review Exercises 63–76, completely factor each polynomial, if possible.

63. $3x^3-3x$ **64.** $5x^3-5$

65. $6x^2+7x-24$ **66.** $3a^2+ax-3a-x$

67. $8x^3 - 125$

68. $6x^2 - 20x - 16$

69. $x^2 + 6x + 9 - 4t^2$

70. $3x^2 - 1 + 5x$

71. $8z^3 + 343$

72. $1 + 14b + 49b^2$

73. $121z^2 + 4 - 44z$

74. $64y^3 - 1000$

75. $2xy - 4zx - wy + 2zw$

76. $x^8 + x^4 + 1$

In Review Exercises 77–92, perform the indicated operations and simplify.

77. $\dfrac{x^2 - 4x + 4}{x + 2} \cdot \dfrac{x^2 + 5x + 6}{x - 2}$

78. $\dfrac{2x^2 - 11x + 15}{x^2 - 6x + 8} \cdot \dfrac{x^2 - 2x - 8}{x^2 - x - 6}$

79. $\dfrac{2x^2 + x - 3}{3x^2 - 7x + 4} \div \dfrac{10x + 15}{3x^2 - x - 4}$

80. $\dfrac{x^2 + 7x + 12}{x^3 + 8x^2 + 4x} \div \dfrac{x^2 - 9}{x^2}$

81. $\dfrac{x^2 + x - 6}{x^2 - x - 6} \cdot \dfrac{x^2 - x - 6}{x^2 + x - 2} \div \dfrac{x^2 - 4}{x^2 - 5x + 6}$

82. $\left(\dfrac{2x + 6}{x + 5} \div \dfrac{2x^2 - 2x - 4}{x^2 - 25} \right) \dfrac{x^2 - x - 2}{x^2 - 2x - 15}$

83. $\dfrac{2}{x - 4} + \dfrac{3x}{x + 5}$

84. $\dfrac{5x}{x - 2} - \dfrac{3x + 1}{x + 3}$

85. $\dfrac{x}{x - 1} + \dfrac{x}{x - 2} + \dfrac{x}{x - 3}$

86. $\dfrac{x}{x + 1} - \dfrac{3x + 7}{x + 2} + \dfrac{2x + 1}{x + 2}$

87. $\dfrac{3(x + 1)}{x} - \dfrac{5(x^2 + 3)}{x^2} + \dfrac{x}{x + 1}$

88. $\dfrac{3x}{x + 1} + \dfrac{x^2 + 4x + 3}{x^2 + 3x + 2} - \dfrac{x^2 + x - 6}{x^2 - 4}$

89. $\dfrac{\dfrac{5x}{2}}{\dfrac{3x^2}{8}}$

90. $\dfrac{\dfrac{3x}{y}}{\dfrac{6x}{y^2}}$

91. $\dfrac{\dfrac{1}{x} + \dfrac{1}{y}}{x - y}$

92. $\dfrac{\dfrac{1}{x} + \dfrac{1}{y}}{\dfrac{1}{y} - \dfrac{1}{x}}$

2

EQUATIONS, INEQUALITIES, AND ABSOLUTE VALUE

The topic of this chapter is equations, one of the most fundamental concepts in mathematics. Equations are used in almost every academic discipline and vocational area, especially chemistry, physics, medicine, economics, electronics, and business.

2.1 LINEAR EQUATIONS AND THEIR SOLUTIONS

An **equation** is a statement indicating that two mathematical expressions are equal. An equation can be true or it can be false. For example, the equation $1 + 3 = 4$ is a true equation, and the equation $2 + 3 = 6$ is a false equation. An equation such as $3x - 2 = 10$ could be true or false depending on the value of x. If $x = 4$, then the equation is true:

$$3x - 2 = 10$$
$$3 \cdot 4 - 2 = 10$$
$$12 - 2 = 10$$
$$10 = 10$$

However, the equation is false for all other numbers x.

The set of all permissible replacements for a variable such as x is called the **domain of the variable**. In the equation $3x - 2 = 10$, the domain of x consists of all real numbers. Of all the numbers in the domain of x, only the number 4 makes the equation true. Thus, 4 is called a **solution** or a **root** of the equation. The set containing all of the solutions of an equation is called the **solution set** of the equation. The solution set for the equation $3x - 2 = 10$ is $\{4\}$. To **solve an equation** is to find all of its solutions.

Francois Vieta (Viète) (1540–1603) Vieta simplified the subject of algebra by developing the symbolic notation that we use today.

Example 1   Find the domain of x in the equation $\sqrt{x} = \dfrac{2}{x - 1}$.

Solution   The domain of x consists of all real numbers that are permissible replacements for x. Note that x must be a nonnegative number for $\sqrt{x}$ to be a real number. Note also that x cannot be 1 because that would cause $x - 1$ to be 0, and the denominator of a fraction cannot be 0. Thus, the domain of x consists of all nonnegative real numbers except 1.   ■

For some equations, every member of the domain of the variable is a solution of the equation. Such equations are called **identities**. For others, no member of the domain of the variable is a solution. For example, if x is a real number, then

$$x^2 - 9 = (x + 3)(x - 3)$$

is an identity because *every* number in the domain of x is a solution. The equation

$$3x + 3 = 3x + 2$$

has no solutions.

An equation whose solution set contains some, but not all, of the members of the domain of the variable is called a **conditional equation**. We now discuss how to solve many of these equations.

Definition. If two equations, each in one variable, have the same solution set, they are called **equivalent equations**.

There are three properties that we can use to transform equations into equivalent, but less complicated, equations. If we use only these three properties, the resulting equation will be equivalent to the original equation, and the two equations will have the same solution set.

Property 1. Either side of an equation can be simplified, provided the value represented by that side remains the same. That is, a quantity can be substituted for its equal.

Property 2. Any algebraic expression that represents a number can be added to or subtracted from both sides of an equation: If $a = b$ and c is a number, then

$$a + c = b + c \qquad \text{and} \qquad a - c = b - c$$

Property 3. Both sides of an equation can be multiplied or divided by any algebraic expression that represents a nonzero number. If $a = b$ and $c \neq 0$, then

$$ac = bc \qquad \text{and} \qquad \frac{a}{c} = \frac{b}{c}$$

We will solve equations by using these three properties to replace one equation with a simpler, but equivalent, one. We will do this until an equation is obtained whose solutions are obvious. We begin by considering equations that are equivalent to **linear equations**.

Definition. An equation that is equivalent to an equation of the form

$$ax + c = 0$$

where $a \neq 0$, is called a **linear** or a **first-degree equation**.

Example 2 Solve the linear equation $2x + 3 = 0$.

Solution

$$2x + 3 = 0$$

$$2x + 3 - 3 = 0 - 3 \qquad \text{Use Property 2 and subtract 3 from both sides.}$$

$$2x = -3 \qquad \text{Use Property 1 by simplifying.}$$

$$\frac{2x}{2} = \frac{-3}{2} \qquad \text{Use Property 3 and divide both sides by 2.}$$

$$x = -\frac{3}{2} \qquad \text{Use Property 1 by simplifying.} \qquad \blacksquare$$

As the previous example suggests, every linear equation has exactly one solution.

Theorem. If $a \neq 0$, the solution of the equation $ax + c = 0$ is $x = -\dfrac{c}{a}$.

In the exercises, you will be asked to verify that $-\frac{c}{a}$ is a root of the equation $ax + c = 0$.

Example 3 Find the solution set for the equation $3(x + 2) = 5x + 2$.

Solution Proceed as follows:

$$3(x + 2) = 5x + 2$$

$$3x + 6 = 5x + 2 \qquad \text{Use Property 1 by removing parentheses.}$$

$$3x + 6 - 3x = 5x + 2 - 3x \qquad \text{Use Property 2 and subtract } 3x \text{ from both sides.}$$

$$6 = 2x + 2 \qquad \text{Use Property 1 by simplifying.}$$

$$6 - 2 = 2x + 2 - 2 \qquad \text{Use Property 2 and subtract 2 from both sides.}$$

$$4 = 2x \qquad \text{Use Property 1 by simplifying.}$$

$$2 = x \qquad \text{Use Property 3 and divide both sides by 2.}$$

All of the equations above are equivalent, and the solution set of the final equation is $\{2\}$. This is also the solution set of the original equation.   ∎

Example 4   Solve the equation $3(x + 5) = 3(1 + x)$.

Solution   Proceed as follows:

$$3(x + 5) = 3(1 + x)$$

$$3x + 15 = 3 + 3x \qquad \text{Use Property 1 by removing parentheses.}$$

$$3x + 15 - 3x = 3 + 3x - 3x \qquad \text{Use Property 2 and subtract } 3x \text{ from both sides.}$$

$$15 = 3 \qquad \text{Use Property 1 by simplifying.}$$

Because the final equation is false and is equivalent to the original equation, the original equation has no solutions. Its solution set is $\varnothing$.   ∎

Example 5   Solve the equation $5 + 5(x + 2) - 2x = 3x + 15$.

Solution   Proceed as follows:

$$5 + 5(x + 2) - 2x = 3x + 15$$

$$5 + 5x + 10 - 2x = 3x + 15 \qquad \text{Use Property 1 by removing parentheses.}$$

$$3x + 15 = 3x + 15 \qquad \text{Use Property 1 by simplifying.}$$

Both sides of the final equation are identical, so any value of x will make the equation a true equation. The solution set of the simplified equation is the same as the domain of its variable. Because this simplified equation is an identity and is equivalent to the original equation, the equation $5 + 5(x + 2) - 2x = 3x + 15$ is an identity also. Thus, its solution set is the set of all real numbers.   ∎

Example 6   Solve the equation $\dfrac{3}{2} x + \dfrac{2}{3} = \dfrac{1}{5} x$.

Solution   Multiply both sides of this equation by the least common denominator of its three fractions to clear the equation of fractions. Then, proceed as follows:

$$\frac{3}{2} x + \frac{2}{3} = \frac{1}{5} x$$

$$30 \left(\frac{3}{2} x + \frac{2}{3} \right) = 30 \left(\frac{1}{5} x \right) \qquad \text{Use Property 3 and multiply both sides by 30.}$$

$$45x + 20 = 6x \qquad \text{Use Property 1 by removing parentheses and simplifying fractions.}$$

$$45x = 6x - 20 \qquad \text{Use Property 2 and add } -20 \text{ to both sides.}$$

$$39x = -20 \qquad \text{Use Property 2 and add } -6x \text{ to both sides.}$$

$$x = -\frac{20}{39} \qquad \text{Use Property 3 and divide both sides by 39.}$$

The root of the given equation is $-\frac{20}{39}$.   ∎

Equations often contain fractions that have numerators and denominators that are polynomials. Such equations are called **rational equations**. The following two examples illustrate one rational equation that has a root and one that does not.

Example 7   Solve the equation $\dfrac{2}{3-x} = 1$.

Solution   Note that x cannot be 3 because $3 - 3$ is 0, and the denominator of a fraction can never be 0. Thus, if 3 should appear as a suspected solution, you must discard it. Solve the equation as follows:

$$\frac{2}{3-x} = 1$$

$(3 - x)\dfrac{2}{3-x} = 1(3-x)$	Use Property 3 and multiply both sides by $3 - x$.
$2 = 3 - x$	Use Property 1 by simplifying.
$2 + (-3) = 3 - x + (-3)$	Use Property 2 and add -3 to both sides.
$-1 = -x$	Use Property 1 by simplifying.
$-1(-1) = (-1)(-x)$	Use Property 3 and multiply both sides by -1.
$1 = x$	Use Property 1 by simplifying.

If you solve a rational equation in one variable, say x, and obtain a number that is in the domain of x, that number will be a root of the original equation. To make certain that you made no errors, however, it is a good idea to check all suspected roots. To verify that 1 is a solution of the original equation, substitute 1 for x and simplify:

$$\frac{2}{3-x} \overset{?}{=} 1$$

$$\frac{2}{3-1} \overset{?}{=} 1$$

$$\frac{2}{2} \overset{?}{=} 1$$

$$1 = 1$$

This verifies that the solution set for the given equation is $\{1\}$.   ■

Example 8   Solve the equation $\dfrac{x+1}{x-2} = \dfrac{3}{x-2}$.

Solution   Note that x cannot be 2 because $2 - 2$ is 0, and the denominator of a fraction can never be 0. Thus, if 2 should appear as a suspected solution, you must discard it. Solve the equation as follows:

$$\frac{x+1}{x-2} = \frac{3}{x-2}$$

$$(x-2)\frac{x+1}{x-2} = \frac{3}{x-2}(x-2) \qquad \text{Use Property 3 and multiply both sides by } x-2.$$

$$x+1 = 3 \qquad \text{Use Property 1 by simplifying the fractions.}$$

$$x = 2 \qquad \text{Use Property 2 and add } -1 \text{ to both sides.}$$

Because x cannot be 2, this false root, called an **extraneous root**, must be discarded. This also becomes apparent if you attempt to check the result $x = 2$ by substituting **2** for x in the original equation:

$$\frac{x+1}{x-2} \overset{?}{=} \frac{3}{x-2}$$

$$\frac{2+1}{2-2} \overset{?}{=} \frac{3}{2-2}$$

$$\cancel{\frac{3}{0} \overset{?}{=} \frac{3}{0}}$$

Because neither side of the preceding equation is defined, the number 2 cannot be a root of the given equation. The equation $\dfrac{x+1}{x-2} = \dfrac{3}{x-2}$ has no roots. Thus, its solution set is $\varnothing$. ∎

Example 9   Solve the equation $\dfrac{x+2}{x+3} + \dfrac{1}{x^2 + 2x - 3} = 1$.

Solution   Note that x cannot be -3, because that would cause the denominator of the first fraction to be 0. To find any other restricted values, factor the trinomial in the denominator of the second fraction:

$$x^2 + 2x - 3 = (x+3)(x-1)$$

This denominator will be 0 if $x = -3$ or $x = 1$. Thus, x cannot be either -3 or 1. Solve the equation as follows:

$$\frac{x+2}{x+3} + \frac{1}{x^2 + 2x - 3} = 1$$

$$\frac{x+2}{x+3} + \frac{1}{(x+3)(x-1)} = 1 \qquad \text{Factor the second denominator.}$$

$$(x+3)(x-1)\left[\frac{x+2}{x+3} + \frac{1}{(x+3)(x-1)}\right] = (x+3)(x-1)1 \qquad \text{Use Property 3 to multiply both sides by } (x+3)(x-1).$$

$$(x+3)(x-1)\frac{x+2}{x+3} + (x+3)(x-1)\frac{1}{(x+3)(x-1)} = (x+3)(x-1)1 \qquad \text{Remove brackets.}$$

$$(x-1)(x+2) + 1 = (x+3)(x-1) \qquad \text{Simplify each fraction.}$$

$$x^2 + x - 2 + 1 = x^2 + 2x - 3 \qquad \text{Multiply the binomials.}$$

$$x - 1 = 2x - 3 \qquad \text{Use Property 2 to add } -x^2 \text{ to both sides and combine terms.}$$

$$2 = x \qquad \text{Use Property 2 to add } 3 - x \text{ to both sides.}$$

Because 2 is in the domain of x, it is a solution. However, it is always a good idea to check the solution. Show that 2 is a root by verifying that it satisfies the given equation by substituting 2 for x in the original equation and simplifying:

$$\frac{x + 2}{x + 3} + \frac{1}{x^2 + 2x - 3} \overset{?}{=} 1$$

$$\frac{2 + 2}{2 + 3} + \frac{1}{2^2 + 2(2) - 3} \overset{?}{=} 1$$

$$\frac{4}{5} + \frac{1}{5} \overset{?}{=} 1$$

$$1 = 1$$

Because 2 satisfies the original equation, it is a root. ∎

Exercise 2.1

In Exercises 1–8, each quantity represents a real number. Find the domain of x.

1. $x + 3 = 1$

2. $\dfrac{1}{x} = 0$

3. $\dfrac{1}{x^2 - 9} = 5$

4. $\sqrt{x} = 36$

5. $\sqrt{x + 5} = 17$

6. $\dfrac{24}{\sqrt{x - 3}} = 15$

7. $\dfrac{1}{x - 3} = \dfrac{4x^2}{x^2 + 7x + 12}$

8. $\dfrac{1}{x^2 - 7x + 10} = 39$

In Exercises 9–22, solve each equation, if possible, and classify it as an identity, a conditional equation, or an equation with no solution.

9. $2x + 5 = 15$

10. $3x + 2 = x + 8$

11. $2(x + 2) = 2x + 5$

12. $3(x + 2) - x = 2(x + 3)$

13. $\dfrac{x + 7}{2} = 7$

14. $\dfrac{x}{2} - 7 = 14$

15. $2(a + 1) = 3(a - 2) - a$

16. $x^2 = (x + 4)(x - 4) + 16$

17. $3(x - 3) = \dfrac{6x - 18}{2}$

18. $x(x + 2) = (x + 1)^2$

19. $\dfrac{3}{b - 3} = 1$

20. $x^2 - 8x + 15 = (x - 3)(x + 5)$

21. $2x^2 + 5x - 3 = (2x - 1)(x + 3)$

22. $2x^2 + 5x - 3 = 2x(x + 2.5 + 7)$

In Exercises 23–61, solve each equation, if possible.

23. $2x + 7 = 10 - x$ **24.** $9x - 3 = 15 + 3x$ **25.** $\dfrac{5}{3}z - 8 = 7$ **26.** $\dfrac{4}{3}y + 12 = -4$

27. $\dfrac{z}{5} + 2 = 4$ **28.** $\dfrac{3p}{7} - p = -4$ **29.** $\dfrac{3x - 2}{3} = 2x + \dfrac{7}{3}$ **30.** $\dfrac{7}{2}x + 5 = x + \dfrac{15}{2}$

31. $5(x - 2) = 2x + 8$ **32.** $5(r - 4) = -5(r - 4)$

33. $2(2x + 1) - \dfrac{3x}{2} = \dfrac{-3(4 + x)}{2}$ **34.** $(x - 2)(x - 3) = (x + 3)(x + 4)$

35. $7(2x + 5) - 6(x + 8) = 7$ **36.** $(t + 1)(t - 1) = (t + 2)(t - 3) + 4$

37. $(x - 2)(x + 5) = (x - 3)(x + 2)$ **38.** $\dfrac{3x + 1}{20} = \dfrac{1}{2}$

39. $\dfrac{3}{2}(3x - 2) - 10x - 4 = 0$ **40.** $a(a - 3) + 5 = (a - 1)^2$

41. $x(x + 2) = (x + 1)^2 - 1$ **42.** $\dfrac{2 + x}{3} + \dfrac{x + 7}{2} = 4x + 1$

43. $\dfrac{(y + 2)^2}{3} = y + 2 + \dfrac{y^2}{3}$ **44.** $2x - \dfrac{7}{6} + \dfrac{x}{6} = \dfrac{4x + 2}{6}$

45. $2(s + 2) + (s + 3)^2 = s(s + 5) + 2\left(\dfrac{17}{2} + s\right)$ **46.** $\dfrac{3}{x} + \dfrac{1}{2} = \dfrac{4}{x}$

47. $\dfrac{2}{x + 1} + \dfrac{1}{3} = \dfrac{1}{x + 1}$ **48.** $\dfrac{3}{x - 2} + \dfrac{1}{x} = \dfrac{3}{x - 2}$

49. $\dfrac{9x + 6}{x(x + 3)} = \dfrac{7}{x + 3}$ **50.** $x + \dfrac{2(-2x + 1)}{3x + 5} = \dfrac{3x^2}{3x + 5}$

51. $\dfrac{2}{(n - 7)(n + 2)} = \dfrac{4}{(n + 3)(n + 2)}$ **52.** $\dfrac{2}{a - 2} + \dfrac{1}{a + 1} = \dfrac{1}{a^2 - a - 2}$

53. $\dfrac{2x + 3}{x^2 + 5x + 6} + \dfrac{3x - 2}{x^2 + x - 6} = \dfrac{5x - 2}{x^2 - 4}$ **54.** $\dfrac{3x}{x^2 + x} - \dfrac{2x}{x^2 + 5x} = \dfrac{x + 2}{x^2 + 6x + 5}$

55. $\dfrac{3x + 5}{x^3 + 8} + \dfrac{3}{x^2 - 4} = \dfrac{2(3x - 2)}{(x - 2)(x^2 - 2x + 4)}$ **56.** $\dfrac{1}{n + 8} - \dfrac{3n - 4}{5n^2 + 42n + 16} = \dfrac{1}{5n + 2}$

57. $\dfrac{1}{11 - n} - \dfrac{2(3n - 1)}{-7n^2 + 74n + 33} = \dfrac{1}{7n + 3}$ **58.** $\dfrac{4}{a^2 - 13a - 48} - \dfrac{2}{a^2 - 18a + 32} = \dfrac{1}{a^2 + a - 6}$

59. $\dfrac{5}{y + 4} + \dfrac{2}{y + 2} = \dfrac{6}{y + 2} - \dfrac{1}{y^2 + 6y + 8}$ **60.** $\dfrac{6}{2a - 6} + \dfrac{3}{3a - 3} = \dfrac{1}{a^2 - 4a + 3}$

61. $\dfrac{3y}{3y - 6} - \dfrac{2y}{2y + 4} = \dfrac{8}{y^2 - 4}$

62. Verify that $-\dfrac{c}{a}$ is a root of the equation $ax + c = 0$.

2.2   APPLICATIONS OF LINEAR EQUATIONS

In this section, we will apply equation-solving techniques to word problems. The list of steps below provides a strategy to follow.

1. Read the problem several times until you understand the facts that are given. What information is given? What are you asked to find? Draw a sketch or diagram, if possible, to help you visualize the facts of the problem.
2. Pick a variable to represent the quantity that is to be found, and write a sentence telling what that variable represents. Express all of the other quantities mentioned in the problem as expressions involving this single variable.
3. Organize the data and find a way of expressing the same quantity in two different ways. This may involve a formula from geometry, finance, or physics.
4. Set up an equation indicating that the two quantities found in Step 3 are equal.
5. Solve the equation.
6. Answer the questions posed by the problem. Has all the requested information been found?
7. Check the answers in the words of the problem.

This list does not apply to all situations, but it can be applied to a wide range of problems with only slight modifications. The following examples use these steps to solve a variety of word problems.

Example 1   The denominator of a fraction exceeds its numerator by 3. If 5 is added to the numerator and 3 is subtracted from the denominator, the resulting fraction equals 2. Find the original fraction.

Solution   Use n to represent the number in the numerator of the fraction. The denominator exceeds the numerator by 3, so the denominator must be $n + 3$. The fraction can now be expressed as $\frac{n}{n+3}$. If you add 5 to the numerator and subtract 3 from the denominator, the result is 2. Set up the following equation and solve for n:

$$\frac{n+5}{n+3-3} = 2$$

$$\frac{n+5}{n} = 2$$

$$n + 5 = 2n$$

$$5 = n$$

Because the numerator of the original fraction is n, and $n = 5$, the denominator is $5 + 3$ or 8. The desired fraction is $\frac{5}{8}$. To check this result, note that if 5 is added to the numerator of the fraction $\frac{5}{8}$, and 3 is subtracted from the denominator, the result is $\frac{10}{5}$, which is equal to 2.   ■

Example 2   A woman invests $10,000, part at 9% annual interest and the rest at 14%. In each case, the interest is compounded annually. The total annual income is $1275. How much is invested at each rate?

Solution Let x represent the amount invested at 9% interest. Then, $10{,}000 - x$ (the rest of \$10,000) represents the amount invested at 14% interest. The annual income from any investment is the product of the interest rate and the amount invested. The total income from these two investments can be expressed in two different ways: as \$1275 (which is given) and as the sum of the incomes of the two investments:

<table>
<tr><td>The income from the
9% investment</td><td>+</td><td>the income from the
14% investment</td><td>=</td><td>the total
income.</td></tr>
</table>

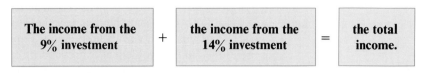

$$9\% \text{ of} \begin{pmatrix} \text{amount invested} \\ \text{at } 9\% \end{pmatrix} + 14\% \text{ of} \begin{pmatrix} \text{amount invested} \\ \text{at } 14\% \end{pmatrix} = \begin{pmatrix} \text{total} \\ \text{income} \end{pmatrix}$$

$$0.09(x) \qquad + \qquad 0.14(10{,}000 - x) \qquad = \qquad 1275$$

Multiply both sides of this equation by 100 to clear it of decimal fractions and solve for x:

$$9x + 14(10{,}000 - x) = 127{,}500$$
$$9x + 140{,}000 - 14x = 127{,}500$$
$$-5x = -12{,}500$$
$$x = 2500$$

The investment at 9% was \$2500, while \$10,000 − \$2500, or \$7500, was invested at 14%. These amounts are correct because 9% of \$2500 is \$225, and 14% of \$7500 is \$1050. The total income from both investments is \$225 + \$1050, or \$1275. ∎

Example 3 If a man can paint a house in six days and his daughter can paint the same house in eight days, how long will it take them to paint the house working together?

Solution Let n represent the number of days it would take them to paint the house if they work together. In one day, the two working together can paint $\frac{1}{n}$ of the house. In one day, the father working alone can paint $\frac{1}{6}$ of the house. In one day, the daughter working alone can paint $\frac{1}{8}$ of the house. The work that they can do together in one day is the sum of what each can do in one day. This gives the equation

<table>
<tr><td>The part of the house
the father can do in
one day</td><td>+</td><td>the part of the house
the daughter can do
in one day</td><td>=</td><td>the part of the house
they can do together
in one day.</td></tr>
</table>

$$\frac{1}{6} + \frac{1}{8} = \frac{1}{n}$$

Multiply both sides of this equation by $24n$ to clear the fractions, and solve for n:

$$24n\left(\frac{1}{6} + \frac{1}{8}\right) = 24n\left(\frac{1}{n}\right)$$

$$4n + 3n = 24$$

$$7n = 24$$

$$n = \frac{24}{7}$$

The paint job will take $\frac{24}{7}$, or $3\frac{3}{7}$, days. To check this answer, verify that if the father paints $\frac{1}{6}$ of the house each day and the daughter paints $\frac{1}{8}$ of the house each day, and if they both work together for $\frac{24}{7}$ days, then one complete house is painted:

$$\frac{1}{6} \cdot \frac{24}{7} + \frac{1}{8} \cdot \frac{24}{7} \overset{?}{=} 1$$

$$\frac{4}{7} + \frac{3}{7} \overset{?}{=} 1$$

$$1 = 1 \qquad \blacksquare$$

Example 4   If a bottle holding 3 liters of milk contains $3\frac{1}{2}\%$ butterfat, how much skimmed milk must be added to dilute the milk to 2% butterfat?

Solution   Let L represent the number of liters of skimmed milk that must be added to the milk. Then, $3 + L$ represents the number of liters in the final mixture. The butterfat in the final mixture is the sum of the butterfat in the original 3 liters of milk and the butterfat in the skimmed milk added. This leads to the following equation:

The butterfat in the milk	$+$	the butterfat in the skimmed milk	$=$	the butterfat in the total mixture.

$$3\tfrac{1}{2}\% \text{ of 3 liters} + 0\% \text{ of } L \text{ liters} = 2\% \text{ of } (3 + L) \text{ liters}$$

$$0.035(3) + 0 = 0.02(3 + L)$$

Multiply both sides by 1000 to remove the decimal fractions, and solve the resulting equation for L:

$$35(3) = 20(3 + L)$$

$$105 = 60 + 20L$$

$$45 = 20L$$

$$\frac{9}{4} = L$$

To dilute the milk to a 2% mixture, $2\frac{1}{4}$ liters of skimmed milk must be added. To check this answer, note that the final mixture contains $0.02(5.25) = 0.105$ liters of pure butterfat, and that this is equal to the amount of pure butterfat, $0.035(3) = 0.105$ liters, in the original solution.   ■

Example 5   A mathematics student has scores of 74%, 78%, and 65% on three examinations. What score is needed on a fourth examination for the student to earn an average grade of 80%?

Solution   Let x be the required grade on the fourth examination. On the one hand, the average grade is one-fourth of the sum of the four grades; on the other hand, the average is to be 80.

The average of the four grades	=	the required average grade.

$$\frac{74 + 78 + 65 + x}{4} = 80$$

Simplify and solve this equation for x:

$$\frac{217 + x}{4} = 80$$

$$217 + x = 320$$

$$x = 103$$

To score an average of 80% on all four examinations, this unfortunate student must score 103% on the fourth exam. Of course, this is impossible.   ■

Example 6   A man leaves home driving at the rate of 50 miles per hour. After discovering that he forgot his wallet, his daughter drives at the rate of 65 miles per hour to catch up to him. How long will it take her to overtake him and return the wallet if the man had a 15-minute head start?

Solution   Uniform-motion problems are based on the formula $d = rt$, where d is distance, r is rate, and t is time. You can organize the information of this problem in a chart such as Figure 2-1. Let t represent the number of hours the daughter

	d	$=$	r	$\cdot$	t
man	$50(t + \frac{1}{4})$		50		$t + \frac{1}{4}$
daughter	$65t$		65		t

Figure 2-1

must drive to overtake her father. Because the father had a 15-minute, or $\frac{1}{4}$ hour, head start, he was on the road for $t + \frac{1}{4}$ hours.

Set up the following equation and solve for t:

The distance the man drove	=	the distance the daughter drove.

$$50\left(t + \frac{1}{4}\right) = 65t$$

$$50t + \frac{25}{2} = 65t$$

$$\frac{25}{2} = 15t$$

$$\frac{5}{6} = t$$

It will take the daughter $\frac{5}{6}$ hours, or 50 minutes, to overtake her father. ∎

Exercise 2.2

Solve each word problem.

1. If a certain number is added to both the numerator and the denominator of the fraction $\frac{3}{5}$, the resulting fraction is $\frac{5}{6}$. Find the number.

2. The denominator of a fraction exceeds the numerator by 1. If 4 is added to the numerator of the fraction, and 3 is subtracted from the denominator, the result is 7. Find the original fraction.

3. One number is 3 more than twice another. Their sum is 54. Find the numbers.

4. The sum of three consecutive odd integers is 69. Find the integers.

5. An executive invests some money at 7% and some money at 6% annual interest. If a total of $22,000 is invested and the annual return is $1420, how much is invested at 7%?

6. A student invests some money at 8% and twice as much at 9% annual interest. The total income from these two investments is $2080. How much was invested at each rate?

7. An adult ticket for a college basketball game costs $2.50, and a student ticket costs $1.75. The total receipts from the game were $1217.25 and 585 tickets were sold. How many of these were student tickets?

8. Of the 800 tickets sold to a movie, 480 were adult tickets. The gate receipts totaled $2080. What was the cost of a student ticket if an adult ticket cost $3?

9. A child has equal numbers of nickels, dimes, and quarters. The coins are worth $3.20. How many of each type are there?

10. Maria has twice as many quarters as dimes. If all the dimes were quarters and all the quarters were dimes, she would have 60¢ less. How much money does she have?

11. If a woman can mow a yard with a lawn tractor in 2 hours, and another woman can mow the same lawn with a push mower in 4 hours, how long will it take them if they work together?

12. A garden hose can fill a swimming pool in 3 days, and a larger hose can fill the pool in 2 days. How long will it take to fill the pool if both hoses are used?

13. An empty swimming pool fills in 10 hours, and when full, this same swimming pool drains in 19 hours. How long will it take to fill this pool if the drain is accidentally left open?

14. Sam stuffs shrimp as his part-time job as a seafood chef. Sam can stuff 1000 shrimp in 6 hours. If his sister Sally helps him, they can stuff 1000 shrimp in 4 hours. If Sam is sent home sick, how long will it take Sally to stuff 500 shrimp?

15. A small car radiator has a 6-liter capacity. If the liquid in the radiator is 40% antifreeze, how much liquid must be replaced with pure antifreeze to bring the mixture up to a 50% solution?

16. A nurse has 1 liter of a solution that is 20% alcohol. How much pure alcohol must she add to bring it to a solution that contains 25% alcohol?

17. If there are 400 cubic centimeters of a chemical in 1 liter of solution, how many cubic centimeters of water must be added to dilute it to a 25% solution? (There are 1000 cubic centimeters in 1 liter.)

18. A swimming pool contains 15,000 gallons of water. How many gallons of chlorine must be added to "shock the pool" and bring the water to a $\frac{3}{100}$% solution?

19. Assume that an automobile engine will run on a mixture of gasoline and a substitute fuel. Also assume that gasoline costs $1.50 per gallon and that the substitute fuel costs 40¢ per gallon. What percent of the mixture must be substitute fuel to bring the total cost down to $1.00 per gallon?

20. How many liters of water must evaporate to make 12 liters of a 24% salt solution into a 36% salt solution?

21. John scored 5 points higher on his midterm, and 13 points higher on his final, than on his first exam. What did he score on that first exam if his mean (average) score was 90?

22. Sally took four tests in science class. On each successive test, her score improved by three points. If her mean score is 69.5%, what did she get on the first test?

23. A motorboat goes 5 miles upstream in the same time that it requires to go 7 miles downstream. The river flows at 2 miles per hour. What is the speed of the boat in still water?

24. John drove to his uncle's house in a distant city in 5 hours. When he returned home, there was less traffic and the trip took only 3 hours. If John drove 26 miles per hour faster on the return trip, how fast did he drive each way?

25. Suzi drove home for spring vacation at 60 miles per hour, but her brother Jim, who left at the same time, could only drive at 48 miles per hour. When Suzi got home, Jim still had 60 miles to go. How far did Suzi drive?

26. One morning John drove 5 hours before stopping to eat. After lunch, he increased his speed by 10 miles per hour in order to complete a 430-mile trip in 8 hours of driving time. How fast did he drive in the morning?

27. A farmer wishes to mix 2400 pounds of cattle feed that is to be 14% crude protein. Barley, which is 11.7% crude protein, represents 25% of the mixture. The remaining 75% is made up of oats (11.8% crude protein) and soybean oil meal (44.5% crude protein). How many pounds of barley, oats, and soybean oil meal ought to be used?

28. If the farmer in Exercise 27 wants only 20% of the mixture to be barley, how many pounds of barley, oats, and soybean oil meal ought to be used?

2.3   QUADRATIC EQUATIONS

In this section we discuss methods for solving equations such as $2x^2 - 11x - 21 = 0$. This type of equation is called a **quadratic equation**.

> **Definition.**   An equation of the form $ax^2 + bx + c = 0$, where $a \neq 0$, is called a **quadratic**, or **second-degree**, **equation**.

If x and y are two real numbers and $x = 0$ or $y = 0$, then $xy = 0$. The converse of this property of real numbers makes it possible to solve quadratic equations.

> **The Zero Factor Theorem.**   If x and y represent two real numbers and $xy = 0$, then either $x = 0$ or $y = 0$.

Proof   Suppose that $xy = 0$. There are two possibilities: either $x = 0$ or $x \neq 0$. If $x = 0$, then we are finished because at least one of x or y is 0. If $x \neq 0$, then x has a multiplicative inverse $\frac{1}{x}$. We can multiply both sides of the equation $xy = 0$ by $\frac{1}{x}$ and simplify to obtain

$$xy = 0$$
$$\frac{1}{x}(xy) = \frac{1}{x}(0)$$
$$\left(\frac{1}{x}x\right)y = 0$$
$$1y = 0$$
$$y = 0$$

Hence, if $x \neq 0$, then y must be 0, and the proof is complete. $\square$

Example 1   Solve the quadratic equation $2x^2 - 11x - 21 = 0$.

Solution   The left side of the equation can be factored to get

$$(2x + 3)(x - 7) = 0$$

If either factor is 0, then the product will be 0. So, use the zero factor theorem and set each factor equal to 0 and solve for x:

$$
\begin{array}{ll}
2x + 3 = 0 & \text{or} \quad x - 7 = 0 \\
2x = -3 & \qquad x = 7 \\
x = -\dfrac{3}{2} &
\end{array}
$$

Because the product $(2x + 3)(x - 7)$ can be 0 only if one of its factors is 0, the numbers $-\frac{3}{2}$ and 7 are the only roots of the equation $2x^2 - 11x - 21 = 0$. Verify that each of these roots does satisfy the original equation.   ∎

Example 2   Solve the equation $\dfrac{2}{x} + 1 = 3x$.

Solution   When both sides of the equation are multiplied by x, the result is a quadratic equation, which can be solved by factoring:

$$\frac{2}{x} + 1 = 3x$$

$$x\left(\frac{2}{x} + 1\right) = x(3x) \qquad \text{Multiply both sides by } x.$$

$$2 + x = 3x^2 \qquad \text{Remove parentheses and simplify.}$$

$$0 = 3x^2 - x - 2 \qquad \text{Add } -2 - x \text{ to both sides to get the equation in quadratic form.}$$

$$0 = (3x + 2)(x - 1) \qquad \text{Factor the trinomial.}$$

$$3x + 2 = 0 \quad \text{or} \quad x - 1 = 0 \qquad \text{Set each factor equal to 0.}$$

$$3x = -2 \qquad\qquad x = 1$$

$$x = -\frac{2}{3}$$

Because both $-\frac{2}{3}$ and 1 are in the domain of x, they are both solutions. Verify this by showing that both roots satisfy the original equation.   ■

To develop other methods for solving quadratic equations, we consider the equation $x^2 = c$. If c is positive, the two real roots of $x^2 = c$ can be found by adding $-c$ to both sides, factoring the binomial $x^2 - c$, setting each factor equal to 0, and solving for x:

$$x^2 = c$$
$$x^2 - c = 0$$
$$x^2 - (\sqrt{c})^2 = 0$$
$$(x - \sqrt{c})(x + \sqrt{c}) = 0$$
$$x - \sqrt{c} = 0 \quad \text{or} \quad x + \sqrt{c} = 0$$
$$x = \sqrt{c} \qquad\qquad x = -\sqrt{c}$$

Hence, the roots of the equation $x^2 = c$ are $x = \sqrt{c}$ and $x = -\sqrt{c}$. This fact is often called the **square root property**.

The Square Root Property.   If $c > 0$, then the equation $x^2 = c$ has two real roots:

$$x = \sqrt{c} \qquad \text{or} \qquad x = -\sqrt{c}$$

Example 3   Solve the equation $x^2 - 8 = 0$.

Solution   Solve the equation for x^2 and apply the square root property:

$$x^2 - 8 = 0$$
$$x^2 = 8$$
$$x = \sqrt{8} \quad \text{or} \quad x = -\sqrt{8}$$
$$x = 2\sqrt{2} \quad \Big| \quad x = -2\sqrt{2} \qquad \sqrt{8} = \sqrt{4}\sqrt{2} = 2\sqrt{2}.$$

Verify that each root satisfies the equation $x^2 - 8 = 0$.   ■

Example 4   Solve the equation $(x + 4)^2 = 1$.

Solution   Apply the square root property to $(x + 4)^2 = 1$ and solve for x:

$$(x + 4)^2 = 1$$
$$x + 4 = \sqrt{1} \quad \text{or} \quad x + 4 = -\sqrt{1}$$
$$x + 4 = 1 \quad \Big| \quad x + 4 = -1$$
$$x = -3 \quad \Big| \quad x = -5$$

Verify that each root satisfies the original equation.   ■

Completing the Square

We now discuss another method, called **completing the square**, that can be used to solve quadratic equations.

Example 5   Use the method of completing the square to solve the equation $x^2 - 10x + 24 = 0$.

Solution   Begin by adding -24 to both sides:

$$x^2 - 10x \qquad = -24$$

Then add 25 to both sides of the equation and combine terms:

$$x^2 - 10x + 25 = -24 + 25$$
$$x^2 - 10x + 25 = 1$$

Note that adding 25 makes the left side of the equation a perfect trinomial square. When the left side is factored, the equation becomes

$$(x - 5)^2 = 1$$

You can solve this equation by using the square root property:

$$(x - 5)^2 = 1$$
$$x - 5 = 1 \quad \text{or} \quad x - 5 = -1$$
$$x = 6 \quad \Big| \quad x = 4$$

Verify that each root satisfies the original equation.   ■

Example 6   Use the method of completing the square to solve the equation $x^2 + 4x - 5 = 0$.

Solution   Rewrite this equation in the form

$$x^2 + 4x \qquad = 5$$

and add 4 to both sides to get

$$x^2 + 4x + 4 = 5 + 4$$
$$x^2 + 4x + 4 = 9$$

Adding 4 to both sides makes the trinomial on the left a perfect trinomial square. Now solve for x by using the square root property:

$$x^2 + 4x + 4 = 9$$
$$(x + 2)^2 = 9$$
$$x + 2 = 3 \quad \text{or} \quad x + 2 = -3$$
$$x = 1 \qquad \qquad x = -5$$

Verify that each root satisfies the original equation.   ■

A question is raised by the previous two examples. What number must be added to $x^2 + bx$ to make a perfect trinomial square? There is a pattern in the following list of perfect trinomial squares. We look at the right side of each equation for a pattern relating the constant term and the coefficient of x.

$$(x + 1)^2 = x^2 + 2x + 1$$
$$(x + 3)^2 = x^2 + 6x + 9$$
$$(x + 5)^2 = x^2 + 10x + 25$$
$$(x - 7)^2 = x^2 - 14x + 49$$
$$(x - 8)^2 = x^2 - 16x + 64$$
$$(x - a)^2 = x^2 - 2ax + a^2$$

In each case, the constant term that completes the square is the square of one-half of the coefficient of x. This fact is applied in the next example.

Example 7   Solve the equation $x(x + 3) = 2$.

Solution   Remove parentheses to obtain

$$x^2 + 3x = 2$$

To solve this equation, begin by calculating the number to be added to both sides of the equation to complete the square. One-half of 3 (the coefficient of

x) is $\frac{3}{2}$. The square of $\frac{3}{2}$ is $\frac{9}{4}$, so add $\frac{9}{4}$ to both sides of the above equation and solve for x:

$$x^2 + 3x + \frac{9}{4} = 2 + \frac{9}{4}$$

$$x^2 + 3x + \left(\frac{3}{2}\right)^2 = \frac{8}{4} + \frac{9}{4}$$

$$\left(x + \frac{3}{2}\right)^2 = \frac{17}{4}$$

$$x + \frac{3}{2} = \frac{\sqrt{17}}{2} \qquad \text{or} \quad x + \frac{3}{2} = -\frac{\sqrt{17}}{2} \qquad \text{Use the square root property.}$$

$$x = \frac{-3 + \sqrt{17}}{2} \qquad\qquad x = \frac{-3 - \sqrt{17}}{2}$$

Verify that each of these roots satisfies the original equation.   ■

In Example 7, the coefficient of the x^2 term is 1. This is a necessary condition before the square can be completed. To solve a quadratic equation involving the variable x by completing the square, follow these steps.

1. Make sure that the coefficient of x^2 is 1. If it is not, make it 1 by dividing both sides of the equation by the coefficient of x^2.
2. If necessary, add a number to both sides of the equation to get the constant on the right-hand side of the equation.
3. Complete the square.
   a. Identify the coefficient of x.
   b. Take half the coefficient of x.
   c. Square half the coefficient of x.
   d. Add that square to both sides of the equation.
4. Factor the trinomial square and combine terms.
5. Solve the resulting quadratic equation by applying the square root property.

Example 8   Solve $6x^2 + 5x - 6 = 0$.

Solution   Begin by dividing both sides by 6 to make the coefficient of x^2 equal to 1. Then, proceed as follows.

$$6x^2 + 5x - 6 = 0$$

$$x^2 + \frac{5}{6}x - 1 = 0 \qquad \text{Divide both sides by 6.}$$

$$x^2 + \frac{5}{6}x = 1 \qquad \text{Add 1 to both sides.}$$

$$x^2 + \frac{5}{6}x + \frac{25}{144} = \frac{25}{144} + 1$$

Take $\frac{1}{2}$ of $\frac{5}{6}$ to get $\frac{5}{12}$. Then square $\frac{5}{12}$ to get $\frac{25}{144}$. Add this to both sides.

$$\left(x + \frac{5}{12}\right)^2 = \frac{169}{144}$$

Factor and combine terms.

Now apply the square root property.

$$x + \frac{5}{12} = \sqrt{\frac{169}{144}} \qquad \text{or} \qquad x + \frac{5}{12} = -\sqrt{\frac{169}{144}}$$

$$x + \frac{5}{12} = \frac{13}{12} \qquad\qquad x + \frac{5}{12} = -\frac{13}{12}$$

$$x = -\frac{5}{12} + \frac{13}{12} \qquad\qquad x = -\frac{5}{12} - \frac{13}{12}$$

$$x = \frac{8}{12} \qquad\qquad x = -\frac{18}{12}$$

$$x = \frac{2}{3} \qquad\qquad x = -\frac{3}{2}$$

Verify that both roots check.   ■

The Quadratic Formula

The method of completing the square can be used to solve the **general quadratic equation** $ax^2 + bx + c = 0$, where $a \neq 0$. The solution is as follows:

$$ax^2 + bx + c = 0$$

$$x^2 + \frac{bx}{a} + \frac{c}{a} = 0$$

Divide both sides by a.

$$x^2 + \frac{b}{a}x = -\frac{c}{a}$$

Add $-\frac{c}{a}$ to both sides.

$$x^2 + \frac{b}{a}x + \frac{b^2}{4a^2} = \frac{b^2}{4a^2} - \frac{c}{a}$$

Add $\frac{b^2}{4a^2}$ to both sides to complete the square.

$$x^2 + \frac{b}{a}x + \left(\frac{b}{2a}\right)^2 = \frac{b^2}{4a^2} - \frac{4ac}{4aa}$$

Rewrite the left side, and get a common denominator on the right side.

$$\left(x + \frac{b}{2a}\right)^2 = \frac{b^2 - 4ac}{4a^2}$$

Factor the left side and add the fractions on the right side.

$$x + \frac{b}{2a} = \frac{\sqrt{b^2 - 4ac}}{2a} \qquad \text{or} \qquad x + \frac{b}{2a} = -\frac{\sqrt{b^2 - 4ac}}{2a}$$

Apply the square root property.

$$x = \frac{-b}{2a} + \frac{\sqrt{b^2 - 4ac}}{2a} \qquad\qquad x = \frac{-b}{2a} - \frac{\sqrt{b^2 - 4ac}}{2a}$$

Add $-\frac{b}{2a}$ to both sides of each equation.

$$x = \frac{-b + \sqrt{b^2 - 4ac}}{2a} \qquad\qquad x = \frac{-b - \sqrt{b^2 - 4ac}}{2a}$$

Add the fractions on the right of each equation.

These two values of x are the roots of the equation $ax^2 + bx + c = 0$. They are usually combined into a single expression, called the **quadratic formula**.

The Quadratic Formula. The solutions of the general quadratic equation $ax^2 + bx + c = 0$, where $a \neq 0$, are

$$x = \frac{-b \pm \sqrt{b^2 - 4ac}}{2a}$$

The above expression should be read twice, once using the plus sign, and once using the minus sign. The quadratic formula implies that

$$x = \frac{-b + \sqrt{b^2 - 4ac}}{2a} \qquad \text{or} \qquad x = \frac{-b - \sqrt{b^2 - 4ac}}{2a}$$

Example 9 Use the quadratic formula to solve the equation $2x^2 + 8x + 7 = 0$.

Solution In this equation, $a = 2$, $b = 8$, and $c = 7$. Substitute these numbers into the quadratic formula and simplify:

$$x = \frac{-b \pm \sqrt{b^2 - 4ac}}{2a}$$

$$= \frac{-8 \pm \sqrt{8^2 - 4(2)(7)}}{2(2)}$$

$$= \frac{-8 \pm \sqrt{8}}{4}$$

$$= \frac{-8 \pm 2\sqrt{2}}{4}$$

$$x = -2 + \frac{\sqrt{2}}{2} \qquad \text{or} \qquad x = -2 - \frac{\sqrt{2}}{2}$$

Both values satisfy the original equation. ■

Exercise 2.3

In Exercises 1–12, solve each equation by factoring.

1. $x^2 - x - 6 = 0$

2. $x^2 + 8x + 15 = 0$

3. $x^2 - 144 = 0$

4. $x^2 + 4x = 0$

5. $2x^2 + x - 10 = 0$

6. $3x^2 + 4x - 4 = 0$

7. $5x^2 - 13x + 6 = 0$

8. $2x^2 + 5x - 12 = 0$

9. $15x^2 + 16x = 15$

10. $6x^2 - 25x = -25$

11. $12x^2 + 9 = 24x$

12. $24x^2 + 6 = 24x$

In Exercises 13–20, use the square root property to solve each equation. You may need to factor an expression.

13. $x^2 = 9$ **14.** $x^2 = 20$ **15.** $y^2 - 50 = 0$ **16.** $x^2 - 75 = 0$

17. $(x - 1)^2 = 4$ **18.** $(y + 2)^2 - 49 = 0$

19. $a^2 + 2a + 1 = 9$ **20.** $x^2 - 6x + 9 = 25$

In Exercises 21–32, solve each equation by completing the square.

21. $x^2 - 8x + 15 = 0$ **22.** $x^2 + 10x + 21 = 0$

23. $x^2 + x - 6 = 0$ **24.** $x^2 - 9x + 20 = 0$

25. $x^2 - 25x = 0$ **26.** $x^2 + x = 0$ **27.** $3x^2 - 4 = -4x$ **28.** $5x = 12 - 2x^2$

29. $x^2 + 5 = -5x$ **30.** $x^2 + 1 = -4x$ **31.** $3x^2 = 1 - 4x$ **32.** $2x^2 = 3x + 1$

In Exercises 33–44, solve each equation by using the quadratic formula.

33. $x^2 - 12 = 0$ **34.** $x^2 - 20 = 0$

35. $2x^2 - x - 15 = 0$ **36.** $6x^2 + x - 2 = 0$

37. $5x^2 - 9x - 2 = 0$ **38.** $4x^2 - 4x - 3 = 0$

39. $2x^2 + 2x - 4 = 0$ **40.** $3x^2 + 18x + 15 = 0$

41. $-3x^2 = 5x + 1$ **42.** $2x(x + 3) = -1$

43. $5x\left(x + \dfrac{1}{5}\right) = 3$ **44.** $7x^2 = 2x + 2$

In Exercises 45–54, change each equation to quadratic form and solve by any method.

45. $x + 1 = \dfrac{12}{x}$ **46.** $x - 2 = \dfrac{15}{x}$ **47.** $8x - \dfrac{3}{x} = 10$ **48.** $15x - \dfrac{4}{x} = 4$

49. $\dfrac{5}{x} = \dfrac{4}{x^2} - 6$ **50.** $\dfrac{6}{x^2} + \dfrac{1}{x} = 12$

51. $x\left(30 - \dfrac{13}{x}\right) = \dfrac{10}{x}$ **52.** $x\left(20 - \dfrac{17}{x}\right) - \dfrac{10}{x} = 0$

53. $(a - 2)(a + 4) = 2a(a - 3)$ **54.** $\dfrac{a + 4}{2a} = \dfrac{a - 2}{3}$

2.4 MORE ON QUADRATIC EQUATIONS AND MISCELLANEOUS EQUATIONS

It is possible to discover what types of numbers will be roots of a given quadratic equation without solving that equation. Suppose that the coefficients a, b, and c in the quadratic equation $ax^2 + bx + c = 0$ are all real numbers. The two solutions of this equation are given by the quadratic formula,

$$x = \frac{-b \pm \sqrt{b^2 - 4ac}}{2a}$$

If $b^2 - 4ac$ is a positive number or 0, then the solutions are real numbers. On the other hand, if $b^2 - 4ac$ is a negative number, then the solutions are not real numbers. The expression $b^2 - 4ac$ is called the **discriminant**. Its value determines the nature of the roots of any given quadratic equation. The possibilities are summarized in the following table:

If $b^2 - 4ac$ is ...	the solutions are ...
positive	real numbers and unequal
zero	real numbers and equal
negative	not real numbers
(and if a, b, and c are rational numbers)	
a perfect nonzero square	rational numbers and unequal
positive and not a perfect square	irrational numbers and unequal

Example 1 Determine the nature of the roots of the quadratic equation $3x^2 + 4x + 1 = 0$.

Solution Calculate the discriminant, $b^2 - 4ac$, as follows:

$$b^2 - 4ac = 4^2 - 4(3)(1)$$
$$= 4$$

Because a, b, and c are rational numbers and the discriminant is a perfect square, the two roots of the given quadratic equation are rational and unequal.

∎

Example 2 If k is a constant, many quadratic equations are represented by the equation

$$(k - 2)x^2 + (k + 1)x + 4 = 0$$

What values of k will give an equation with roots that are real and equal?

Solution Compute the discriminant and demand that it be 0.

$$b^2 - 4ac = (k + 1)^2 - 4(k - 2)(4)$$
$$0 = k^2 + 2k + 1 - 16k + 32$$
$$0 = k^2 - 14k + 33$$
$$0 = (k - 3)(k - 11)$$
$$k = 3 \quad \text{or} \quad k = 11$$

If $k = 3$, then the equation $(k - 2)x^2 + (k + 1)x + 4 = 0$ becomes

$$(3 - 2)x^2 + (3 + 1)x + 4 = 0$$

or

$$x^2 + 4x + 4 = 0$$

Solve this equation to show that its roots are real and equal:

$$x^2 + 4x + 4 = 0$$
$$(x + 2)(x + 2) = 0$$
$$x = -2 \quad \text{or} \quad x = -2$$

Similarly, $k = 11$ produces an equation with roots that are real and equal.

■

There are many types of equations that, while not quadratic equations, can be put in quadratic form. These equations can then be solved by using techniques for solving quadratic equations.

Example 3 Solve the equation $x^4 - 5x^2 + 4 = 0$.

Solution This equation is not a quadratic equation because the trinomial on the left of the equals sign is of fourth degree. However, if y is substituted for x^2, the equation takes the form of a quadratic equation and can be solved by factoring.

$$x^4 - 5x^2 + 4 = 0$$
$$(x^2)^2 - 5(x^2) + 4 = 0$$
$$y^2 - 5y + 4 = 0 \qquad \text{Let } y = x^2.$$
$$(y - 4)(y - 1) = 0 \qquad \text{Factor } y^2 - 5y + 4 = 0.$$
$$y - 4 = 0 \quad \text{or} \quad y - 1 = 0$$
$$y = 4 \qquad \qquad y = 1$$

Because $x^2 = y$, it follows that $x^2 = 4$ or $x^2 = 1$. Hence, you have

$$x^2 = 4 \qquad \text{or} \qquad x^2 = 1$$
$$x = 2 \quad \text{or} \quad x = -2 \qquad x = 1 \quad \text{or} \quad x = -1$$

This equation has four solutions. Verify that each one checks.

■

Example 4 Solve the equation $2x^{\frac{2}{5}} - 5x^{\frac{1}{5}} - 3 = 0$.

Solution Let $y = x^{\frac{1}{5}}$. Then the equation can be written as $2(x^{\frac{1}{5}})^2 - 5(x^{\frac{1}{5}}) - 3 = 0$ or $2y^2 - 5y - 3 = 0$. Solve this quadratic equation to find its two roots: $y = -\frac{1}{2}$ or $y = 3$. These roots provide two more equations when y is replaced by its equal, $x^{\frac{1}{5}}$:

$$y = -\frac{1}{2} \quad \text{or} \quad y = 3$$
$$x^{\frac{1}{5}} = -\frac{1}{2} \qquad x^{\frac{1}{5}} = 3$$

Each of these equations can be solved by raising each side of each equation to the fifth power:

$$x^{\frac{1}{5}} = -\frac{1}{2} \quad \text{or} \quad x^{\frac{1}{5}} = 3$$

$$(x^{\frac{1}{5}})^5 = \left(-\frac{1}{2}\right)^5 \qquad (x^{\frac{1}{5}})^5 = 3^5$$

$$x = -\frac{1}{32} \qquad x = 243$$

Verify that each suspected root checks.   ■

Example 5   Solve the equation $\dfrac{1}{x-1} + \dfrac{3}{x+1} = 2$.

Solution   Because the denominator of a fraction cannot be 0, x cannot be 1 or -1. If either 1 or -1 appears as a suspected root, it must be discarded. Solve the equation as follows:

$$\frac{1}{x-1} + \frac{3}{x+1} = 2$$

$$(x+1)(x-1)\left[\frac{1}{x-1} + \frac{3}{x+1}\right] = 2(x+1)(x-1) \qquad \text{Multiply both sides by } (x+1)(x-1).$$

$$(x+1) + 3(x-1) = 2(x^2 - 1) \qquad \text{Remove brackets and simplify.}$$

$$4x - 2 = 2x^2 - 2 \qquad \text{Remove parentheses and simplify.}$$

$$0 = 2x^2 - 4x \qquad \text{Add } 2 - 4x \text{ to both sides.}$$

The resulting equation is a quadratic equation. Factor the binomial $2x^2 - 4x$, set each factor equal to 0, and solve for x:

$$0 = 2x^2 - 4x$$

$$0 = 2x(x - 2)$$

$$2x = 0 \quad \text{or} \quad x - 2 = 0$$

$$x = 0 \quad | \quad x = 2$$

Because the numbers 0 and 2 are in the domain of x, each number is a root of the original equation. Verify this by checking each root.   ■

Radical Equations

Many equations containing radicals can be solved by using the **squaring property of real numbers** to square both sides of the equation.

> **The Squaring Property.** If a and b are real numbers, then
>
> if $a = b$, then $a^2 = b^2$

Example 6   Solve the equation $\sqrt{x + 3} - 4 = 7$.

Solution
$$\sqrt{x + 3} - 4 = 7$$
$$\sqrt{x + 3} = 11 \qquad \text{Add 4 to both sides.}$$
$$x + 3 = 121 \qquad \text{Square both sides.}$$
$$x = 118 \qquad \text{Add } -3 \text{ to both sides.}$$

Squaring both sides of an equation does not guarantee a result that is equivalent to the original equation. Thus, you *must* check the suspected root in the original equation:

$$\sqrt{x + 3} - 4 = 7$$
$$\sqrt{118 + 3} - 4 \overset{?}{=} 7$$
$$\sqrt{121} - 4 \overset{?}{=} 7$$
$$11 - 4 \overset{?}{=} 7$$
$$7 = 7$$

Because it checks, 118 is a solution of $\sqrt{x + 3} - 4 = 7$.   ■

Example 7   Solve the equation $\sqrt{x + 3} = 3x - 1$.

Solution   Rid this equation of the radical by squaring both sides of the equation and then simplify:

$$\sqrt{x + 3} = 3x - 1$$
$$x + 3 = (3x - 1)^2 \qquad \text{Square both sides.}$$
$$x + 3 = 9x^2 - 6x + 1 \qquad \text{Remove parentheses.}$$
$$0 = 9x^2 - 7x - 2 \qquad \text{Add } -x - 3 \text{ to both sides.}$$

The right side of this equation factors as

$$0 = (9x + 2)(x - 1)$$

and the possible solutions can be found by solving two linear equations:

$$9x + 2 = 0 \quad \text{or} \quad x - 1 = 0$$
$$x = -\frac{2}{9} \qquad\qquad x = 1$$

You *must* check each proposed root in the given equation.

$$\sqrt{x+3} \overset{?}{=} 3x - 1 \qquad\qquad \text{or} \quad \sqrt{x+3} \overset{?}{=} 3x - 1$$

$$\sqrt{-\frac{2}{9}+3} \overset{?}{=} 3\left(-\frac{2}{9}\right) - 1 \qquad\qquad \sqrt{1+3} \overset{?}{=} 3(1) - 1$$

$$\sqrt{\frac{25}{9}} \overset{?}{=} -\frac{2}{3} - 1 \qquad\qquad\qquad \sqrt{4} \overset{?}{=} 2$$

$$\frac{5}{3} \neq -\frac{5}{3} \qquad\qquad\qquad\qquad 2 = 2$$

Thus, $-\dfrac{2}{9}$ is not a solution. $\qquad$ Thus, 1 is a solution.

The only solution of the equation $\sqrt{x+3} = 3x - 1$ is the number 1.   ∎

Example 8   Solve the equation $\sqrt{2x+3} + \sqrt{x-2} = 4$.

Solution   To simplify the work, rewrite the equation in the form

$$\sqrt{2x+3} = 4 - \sqrt{x-2}$$

so that its left side contains only one radical. Square both sides to get

$$(\sqrt{2x+3})^2 = (4 - \sqrt{x-2})^2$$
$$2x + 3 = 16 - 8\sqrt{x-2} + (x - 2)$$

Combine terms and add $-x - 14$ to both sides, leaving the single radical by itself on the right side of the equation:

$$2x + 3 = 14 - 8\sqrt{x-2} + x$$
$$x - 11 = -8\sqrt{x-2}$$

Square both sides again, simplify, and solve the resulting equation for x:

$$x^2 - 22x + 121 = 64(x - 2)$$
$$x^2 - 86x + 249 = 0$$
$$(x - 3)(x - 83) = 0$$
$$x = 3 \qquad \text{or} \qquad x = 83$$

Substituting these proposed roots into the given equation shows that 83 does not check. Thus, 83 is an extraneous solution and must be discarded. However, 3 does satisfy the given equation. Hence, 3 is a root.   ∎

Exercise 2.4

In Exercises 1–6, use the discriminant to determine the nature of the roots of each equation. **Do not solve the equations.**

1. $x^2 + 6x + 9 = 0$ $\qquad\qquad\qquad$ **2.** $x^2 - 5x + 2 = 0$

3. $3x^2 - 2x + 5 = 0$ $\qquad\qquad\qquad$ **4.** $9x^2 + 42x + 49 = 0$

5. $10x^2 + 29x = 21$ $\qquad\qquad\qquad$ **6.** $10x^2 + x = 21$

7. Find two values of k such that the equation $x^2 + kx + 3k - 5 = 0$ will have a double root; that is, the roots will be equal.

8. For what value of b will the solutions of the equation $x^2 - 2bx + b^2 = 0$ be equal?

9. Does the equation $1492x^2 + 1984x - 1776 = 0$ have any roots that are real numbers?

10. Does the equation $2004x^2 + 10x + 1985 = 0$ have any roots that are real numbers?

In Exercises 11–46, find all real solutions of each equation by using any method.

11. $x^4 - 13x^2 + 36 = 0$

12. $y^4 - 10y^2 + 9 = 0$

13. $y^4 + 100 = 29y^2$

14. $x^4 = 26x^2 - 25$

15. $2y^4 - 46y^2 + 180 = 0$

16. $2x^4 - 102x^2 + 196 = 0$

17. $4x^4 = 5x^2 + 9$

18. $6x^4 + 384 = 120x^2$

19. $x - 13x^{\frac{1}{2}} + 12 = 0$

20. $x + x^{\frac{1}{2}} - 20 = 0$

21. $2x^{\frac{1}{3}} + 3x^{\frac{1}{6}} - 2 = 0$

22. $6a^{\frac{2}{3}} - a^{\frac{1}{3}} = 2$

23. $(x^2 - 2x)^2 - 11(x^2 - 2x) + 24 = 0$

24. $6(y + 2)^2 - 27(y + 2) + 27 = 0$

25. $\dfrac{1}{x} + \dfrac{3}{x + 2} = 2$

26. $\dfrac{1}{x - 1} + \dfrac{1}{x - 4} = \dfrac{5}{4}$

27. $\dfrac{1}{x + 1} + \dfrac{5}{2x - 4} = 1$

28. $\dfrac{x(2x + 1)}{x - 2} = \dfrac{10}{x - 2}$

29. $x + 1 + \dfrac{x + 2}{x - 1} = \dfrac{3}{x - 1}$

30. $\dfrac{1}{4 - y} = \dfrac{1}{4} + \dfrac{1}{y + 2}$

31. $\sqrt{x - 2} = 5$

32. $3\sqrt{x + 1} = \sqrt{6}$

33. $\sqrt{x^2 + 21} = x + 3$

34. $\sqrt{5 - x^2} = -(x + 1)$

35. $x - \sqrt{7x - 12} = 0$

36. $x - \sqrt{4x - 4} = 0$

37. $\sqrt{\dfrac{x^2 - 1}{x - 2}} = 2\sqrt{2}$

38. $\dfrac{\sqrt{x^2 - 1}}{\sqrt{3x - 5}} = \sqrt{2}$

39. $\sqrt{3x + 1} = x - 1$

40. $\sqrt{x^2 + 1} = \dfrac{\sqrt{-7x + 11}}{\sqrt{6}}$

41. $\sqrt{x + 3} + 2\sqrt{x} = 0$

42. $\sqrt{y + 2} = 4 - y$

43. $3 + \sqrt{x + 7} - \sqrt{x - 4} = 0$

44. $x + 4 = \sqrt{\dfrac{6x + 6}{5}} + 3$

45. $x^{-4} - 13x^{-2} + 36 = 0$

46. $2x^{-4} - 11x^{-2} + 14 = 0$

47. If the general quadratic equation $ax^2 + bx + c = 0$ is divided by x^2, the result is

$$c\left(\frac{1}{x}\right)^2 + b\left(\frac{1}{x}\right) + a = 0$$

This is also a quadratic equation with a new variable, $\frac{1}{x}$. Use the quadratic formula to solve for $\frac{1}{x}$, and find an alternative quadratic formula.

48. Solve $x^2 - 8x + 15 = 0$ by using the formula derived in Exercise 47.

49. Find the mistake in the following "proof" that 1 and 2 are equal.

Let $x = 1$, and proceed as follows:

$$x = 1$$
$$x^2 = x$$
$$x^2 - 1 = x - 1$$
$$(x + 1)(x - 1) = x - 1$$
$$\frac{(x + 1)(x - 1)}{x - 1} = \frac{x - 1}{x - 1}$$
$$x + 1 = 1$$

Because $x = 1$, then

$$1 + 1 = 1$$
$$2 = 1$$

50. If r_1 and r_2 are the two roots of the quadratic equation $ax^2 + bx + c = 0$, find the value of $r_1 + r_2$ and $r_1 r_2$ in terms of a, b, and c. These results can be used to check the roots of a quadratic equation.

51. Find five consecutive integers a, b, c, d, and e such that $a^2 + b^2 + c^2 = d^2 + e^2$.

52. Find three consecutive integers a, b, and c such that $a^2 + b^2 = c^2$.

2.5 APPLICATIONS OF QUADRATIC EQUATIONS

Quadratic equations are used to solve many word problems.

Example 1   The length of a rectangle exceeds its width by 3 feet. The area of the rectangle is 40 square feet. What are its dimensions?

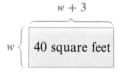

Figure 2-2

Solution   Let w equal the width of the rectangle. Then, $w + 3$ represents its length. See Figure 2-2. Use the formula for the area of a rectangle (*area = length · width*) to express the area as $(w + 3)w$, which is equal to 40. This leads to the equation

The length of the rectangle	·	the width of the rectangle	=	the area of the rectangle.

$$(w + 3) \cdot w = 40$$

Solve this equation for w as follows:

$$w^2 + 3w = 40$$
$$w^2 + 3w - 40 = 0$$
$$(w - 5)(w + 8) = 0$$
$$w - 5 = 0 \quad \text{or} \quad w + 8 = 0$$
$$w = 5 \qquad\qquad w = -8$$

You are asked for the dimensions of the given rectangle. Its width, w, is 5, and its length is $w + 3$, or 8. The solution $w = -8$ must be discarded because a rectangle's width cannot be negative. Verify that this solution is correct by observing that a rectangle with dimensions of 5 feet by 8 feet does have an area of 40 square feet. ∎

Example 2   A man drives 600 miles to a business convention. On the return trip, he increases his speed by 10 miles per hour, and saves 2 hours of driving time. How fast did he go in each direction?

Solution   Let s represent the man's speed (in miles per hour) driving to the convention. On the return trip, his speed was $s + 10$ miles per hour. Recall that the distance traveled by an object moving at a constant rate for a certain time is given by the formula $d = rt$. Thus,

$$t = \frac{d}{r}$$

Organize the information given in this problem in a chart such as Figure 2-3.

	d	$=$	r	$\cdot$	t
outbound trip	600		s		$\dfrac{600}{s}$
return trip	600		$s + 10$		$\dfrac{600}{s + 10}$

Figure 2-3

Although neither the outbound nor the return travel time is given, you know the difference of those two times. This fact can be used to form the equation

The longer time of the outbound trip	$-$	**the shorter time of the return trip**	$=$	**the difference in travel times.**

$$\frac{600}{s} - \frac{600}{s + 10} = 2$$

Multiply both sides of this equation by the common denominator $s(s + 10)$ to clear the equation of fractions, and solve for s.

$$s(s + 10)\left(\frac{600}{s} - \frac{600}{s + 10}\right) = 2s(s + 10)$$
$$600(s + 10) - 600s = 2s(s + 10)$$
$$600s + 6000 - 600s = 2s^2 + 20s$$
$$6000 = 2s^2 + 20s$$
$$0 = 2s^2 + 20s - 6000$$
$$0 = s^2 + 10s - 3000$$
$$0 = (s - 50)(s + 60)$$
$$s - 50 = 0 \quad \text{or} \quad s + 60 = 0$$
$$s = 50 \quad | \quad s = -60$$

The solution $s = -60$ must be discarded because it is negative. The man drove 50 miles per hour to the convention, and $50 + 10$, or 60 miles per hour on the return trip. These answers are correct because a 600-mile trip at 50 miles per hour would take $\frac{600}{50}$ or 12 hours; at 60 miles per hour, that same trip would take only 10 hours—2 hours less. ■

Example 3 If an object is thrown straight up into the air with an initial velocity of 144 feet per second, its height is given by the formula $h = 144t - 16t^2$, where h represents its height (in feet) and t represents the elapsed time (in seconds) since it was thrown. When does the object return to the point from which it was thrown?

Solution When this object returns to its starting point, its height is again 0. Hence, set h equal to 0, and solve for t:

$$h = 144t - 16t^2$$
$$0 = 144t - 16t^2$$
$$0 = 16t(9 - t)$$
$$16t = 0 \quad \text{or} \quad 9 - t = 0$$
$$t = 0 \quad | \quad t = 9$$

At $t = 0$, the object's height is 0 because it is just being released. When $t = 9$, the height is again 0, and the object has returned to its starting point. ■

Exercise 2.5

Solve each word problem.

1. The product of two consecutive even natural numbers is 48. Find the numbers.

2. The product of the first and last of three consecutive odd integers is 45. Find the sum of the three integers.

3. The length of a rectangle is 4 feet longer than its width. Its area is 32 square feet. What are its dimensions?

4. The length of a rectangle is 5 times longer than its width. Its area is 125 square feet. What is its perimeter?

5. The side of a square is 4 centimeters shorter than the side of a second square. The sum of the areas of the squares is 106 square centimeters. Find the length of the side of the larger square.

6. The base of a triangle is one-third as long as its height. How long is the base if the triangle's area is 24 square meters?

7. A boy rides a bicycle a distance of 40 miles. His return trip takes 2 hours longer because his speed decreases by 10 miles per hour. How fast does he ride each way?

8. A farmer drives a tractor from one town to another, a distance of 120 kilometers. He drives 10 kilometers per hour faster on the return trip, cutting 1 hour off the time. How fast does he drive each way?

9. If the speed were increased by 10 miles per hour, a 420-mile trip would take 1 hour less time. How long does the trip take at the slower speed?

10. By increasing her usual speed by 25 kilometers per hour, a bus driver decreases the time on a 25-kilometer trip by 10 minutes. What is the usual speed?

11. The height of a projectile fired upward with an initial velocity of 400 feet per second is given by the formula $h = -16t^2 + 400t$, where h is the height (in feet) and t is the time in seconds. Find the time required for the projectile to return to earth.

12. The height of an object tossed upward with an initial velocity of 104 feet per second is given by the formula $h = 104t - 16t^2$, where h is the height (in feet) and t is the time in seconds. In how many seconds will the object return to its point of departure?

13. An object will fall s feet in t seconds, where $s = 16t^2$. How long will it take a penny to hit the ground if it is dropped from the top of the Sears Tower in Chicago? (*Hint:* The tower is 1454 feet tall.)

14. The height of an object thrown upward with an initial velocity of 32 feet per second is given by the formula $h = 32t - 16t^2$, where t is the time in seconds. How long will it take the object to reach a height of 16 feet?

15. Two pipes are used to fill a water storage tank. The first pipe can fill the tank in 4 hours. The two pipes together can fill the tank in 2 hours less time than the second pipe alone. How long would it take for the second pipe to fill the tank?

16. A hose can fill a swimming pool in 6 hours. Another hose requires 3 more hours to fill the pool than the two hoses together. How long would it take the second hose to fill the pool?

17. Kristy can mow a lawn in 1 hour less time than her brother Steven. Together they can finish in 5 hours. How long would it take Kristy if she worked alone?

18. Working together, Sarah and Heidi can milk the cows in 2 hours. If they work alone, it takes Heidi 3 hours longer than it takes Sarah. How long does it take Heidi to milk the cows alone?

19. Is it possible for a rectangle to have a width that is 3 units shorter than its diagonal and a length that is 4 units longer than its diagonal?

20. If two opposite sides of a square are increased by 10 meters and the other sides are decreased by 8 meters, the area of the rectangle that is formed is 63 square meters. Find the area of the original square.

21. Maude and Matilda have each invested some money for their retirement. Maude invested $1000 more than Matilda, but at an interest rate that was 1% less. Last year Maude received interest of $280 on her investment and Matilda received $240. At what rates were their investments made?

22. Scott and Laura have both invested some money. Scott invested $3000 more than Laura and at a 2% higher interest rate. Scott received $800 interest and Laura received $400. How much did Scott invest?

23. Some mathematics professors at a college would like to purchase a $150 microwave oven for their department workroom. If four of the professors do not contribute, everyone's share will increase by $10. How many mathematics professors teach at the college?

24. A farmer intends to construct a windscreen by planting pine trees in a quarter-mile row. His daughter points out that 44 fewer trees will be needed if they are planted 5 feet farther apart. If the farmer takes her advice, how many trees will he plant? (*Hint*: 1 mile = 5280 feet.)

2.6   LITERAL EQUATIONS

Many equations, called **literal equations**, contain several variables. Often these equations are formulas such as $F = \frac{9}{5}C + 32$, the formula that converts degrees Celsius to degrees Fahrenheit. If we need to change a large number of Fahrenheit readings to degrees Celsius, it is tedious to substitute each value of F into the above formula and then repeatedly solve the equation for C. A more efficient way, especially if we have a calculator, is to solve the formula for C, substitute each value of F (the Fahrenheit readings) into the rearranged formula, and evaluate C directly.

Example 1   Solve the formula $F = \dfrac{9}{5}C + 32$ for C.

Solution   Solve for C by using the techniques for solving linear equations:

$$F = \frac{9}{5}C + 32$$

$$F - 32 = \frac{9}{5}C \qquad \text{Subtract 32 from both sides.}$$

$$\frac{5}{9}(F - 32) = C \qquad \text{Multiply both sides by } \tfrac{5}{9}.$$

This result can be written in the alternate form

$$C = \frac{5F - 160}{9}$$

■

Example 2   The formula $A = p + prt$ is used to determine the amount of money in a savings account at the end of a specific period of time. A represents the amount, p represents the original deposit (the principal), r represents the rate of simple interest per unit of time, and t represents the number of units of time. Solve this formula for p.

Solution   Factor p from both terms on the right side of the equation, and divide both sides by $(1 + rt)$ to solve for p:

$$A = p + prt$$
$$A = p(1 + rt)$$
$$\frac{A}{1 + rt} = p$$

■

Example 3   Solve the equation $y(xy + 3x) + 7 = 0$ for x.

Solution

$$y(xy + 3x) + 7 = 0$$
$$xy^2 + 3xy + 7 = 0 \qquad \text{Remove parentheses.}$$
$$xy^2 + 3xy = -7 \qquad \text{Add } -7 \text{ to both sides.}$$
$$x(y^2 + 3y) = -7 \qquad \text{Factor out } x.$$
$$x = \frac{-7}{y^2 + 3y} \qquad \text{Divide both sides by } y^2 + 3y.$$

■

Example 4   Solve the equation $y(xy + 3x) + 7 = 0$ for y.

Solution   Use the distributive property and remove parentheses first:

$$y(xy + 3x) + 7 = 0$$
$$xy^2 + 3xy + 7 = 0$$

This equation is of the form $ay^2 + by + c = 0$, with $a = x$, $b = 3x$, and $c = 7$. Solve for y by using the quadratic formula:

$$y = \frac{-b \pm \sqrt{b^2 - 4ac}}{2a}$$
$$y = \frac{-3x \pm \sqrt{9x^2 - 28x}}{2x}$$

■

Exercise 2.6

1. The formula $k = 2.2p$ is used to change pounds to kilograms. Solve the formula for p.

2. The formula $p = 2l + 2w$ gives the perimeter of a rectangle with length l and width w. Solve for w.

3. The formula $A = \frac{1}{2} h (b_1 + b_2)$ gives the area of a trapezoid with height h and bases b_1 and b_2. Solve for b_2.

4. The formula $V = \frac{1}{3} \pi r^2 h$ gives the volume of a right circular cone. Solve for h.

5. Solve the formula $V = \dfrac{1}{3} \pi r^2 h$ for r.

6. The formula $z = \dfrac{x - \mu}{\sigma}$ converts a raw score in a normal distribution into a z score, where x is the raw score, μ is the mean, and σ is the standard deviation of the distribution. Solve for μ.

7. The formula $P_n = L + \left(\dfrac{s}{f}\right) i$ is used in statistics to compute percentiles. Solve for s.

8. Solve the formula $P_n = L + \left(\dfrac{s}{f}\right) i$ for f.

9. The formula $\dfrac{1}{r} = \dfrac{1}{r_1} + \dfrac{1}{r_2}$ is used in electronics to calculate the combined resistance of two resistors in parallel. Solve for r.

10. Solve the formula $\dfrac{1}{r} = \dfrac{1}{r_1} + \dfrac{1}{r_2}$ for r_1.

11. The formula $l = a + (n - 1)d$ gives the nth term of an arithmetic sequence. Solve for n.

12. Solve the formula $l = a + (n - 1)d$ for d.

13. In statistics, the formula $\sigma^2 = \dfrac{\sum x^2}{N} - \mu^2$ gives the variance of a normal distribution with N terms. Read $\sum x^2$ as "the sum of the x squares." Solve for N.

14. Solve the equation $S = \pi h(r + h)$ for h.

15. Solve the equation $x + y = \dfrac{7y + 1}{3x}$ for y.   16. Solve the equation $x + y = \dfrac{7y + 1}{3x}$ for x.

17. The formula $S = \dfrac{a - lr}{1 - r}$ gives the sum of the first n terms of a geometric progression. Solve for a.

18. Solve the formula $S = \dfrac{a - lr}{1 - r}$ for r.

19. The total resistance of a parallel circuit with three branches can be computed by using the formula $R = \dfrac{1}{1/r_1 + 1/r_2 + 1/r_3}$, where r_1, r_2, and r_3 are the resistances in each of its three branches. Solve the formula for r_3.

In Exercises 20–32, solve each equation for the indicated variable.

20. $xy = ax - y^2$; for x.   21. $xy = ax - y^2$; for y.   22. $a = (n - 2)\dfrac{180}{n}$, for n.

23. $V = \pi h^2\left(r - \dfrac{h}{3}\right)$; for r.   24. $V = \dfrac{1}{3}(B_1 + B_2 + \sqrt{B_1 B_2})h$; for h.

25. $V = \dfrac{1}{6} h(B + B' + 4M)$; for h.   26. $A = \dfrac{1}{2} r^2(\theta - \phi)$; for θ.

27. $r = \dfrac{x+y}{1-xy}$; for x.

28. $r = \dfrac{x+y}{1-xy}$; for y.

29. $y - y_1 = \dfrac{y_2 - y_1}{x_2 - x_1}(x - x_1)$; for y.

30. $y - y_1 = \dfrac{y_2 - y_1}{x_2 - x_1}(x - x_1)$; for x.

31. $y - y_1 = \dfrac{y_2 - y_1}{x_2 - x_1}(x - x_1)$; for y_1.

32. $y - y_1 = \dfrac{y_2 - y_1}{x_2 - x_1}(x - x_1)$; for x_1.

2.7   LINEAR INEQUALITIES

In this section we will develop techniques for solving inequalities. We begin by defining some terms.

Definition.   If a and b are real numbers, then a is greater than b (written $a > b$) if and only if $a - b$ is positive.
Similarly, a is less than b (written $a < b$) if and only if $b - a$ is positive.

Note that if $a < b$, then $b > a$. If a is greater than or equal to b, then we write $a \geq b$. Similarly, if a is less than or equal to b, then we write $a \leq b$.

Example 1   **a.** $5 > 3$ because $5 - 3$ is a positive real number, which is 2.

b. $2 < 7$ because $7 - 2$ is a positive real number, which is 5.

c. $3 \leq 3$ because 3 is equal to 3.

d. $3 < 3$ is false because $3 - 3$ is 0, and 0 is not positive.

e. $x > x - 1$ because $x - (x - 1) = x - x + 1 = 1$, and 1 is positive.   ∎

We now state without proof several theorems involving $<$. Similar theorems exist for the other relations.

Theorem.   Let a, b, and c represent real numbers.

If $a < b$,   then $a + c < b + c$.

If $a < b$,   then $a - c < b - c$.

The previous theorem states that any real number can be added to (or subtracted from) both sides of an inequality to obtain another inequality with the same order (direction).

Theorem. Let a, b, and c represent real numbers.

If $a < b$ and $c > 0$, then $ca < cb$.

If $a < b$ and $c > 0$, then $\dfrac{a}{c} < \dfrac{b}{c}$.

This theorem states that both sides of an inequality can be multiplied (or divided) by a positive number to obtain another inequality with the same order.

Theorem. Let a, b, and c represent real numbers.

If $a < b$ and $c < 0$, then $ca > cb$.

If $a < b$ and $c < 0$, then $\dfrac{a}{c} > \dfrac{b}{c}$.

This theorem states that both sides of an inequality can be multiplied (or divided) by a negative number to obtain another inequality in the *opposite* order.

These theorems indicate that inequalities are solved like equations. However, *we must always remember to change the order of an inequality when multiplying or dividing both sides of the inequality by a negative number.*

The $<$ relation is neither reflexive (a number cannot be less than itself) nor symmetric (if a is less than b, then b cannot be less than a). However, the relation *is* transitive.

Theorem. If a, b, and c are real numbers with $a < b$ and $b < c$, then $a < c$.

Similarly, the relations $>$, $\leq$, and $\geq$ are transitive.

Inequalities such as $ax + c < 0$ or $ax + c \geq 0$, where $a \neq 0$, are called **linear inequalities**. Numbers that make an inequality true when substituted for the variable are called **solutions** of the inequality. Two inequalities with exactly the same solutions are called **equivalent inequalities**.

Example 2 Solve the inequality $3(x + 2) < 8$.

Solution Proceed as with equations:

$$3(x + 2) < 8$$
$$3x + 6 < 8 \qquad \text{Remove parentheses.}$$
$$3x < 2 \qquad \text{Add } -6 \text{ to both sides.}$$
$$x < \frac{2}{3} \qquad \text{Divide both sides by 3.}$$

Any value of x that is less than $\frac{2}{3}$ is in the solution set of the given inequality. This solution set can be expressed in interval notation as $(-\infty, \frac{2}{3})$. The symbol $-\infty$, read as "minus infinity," indicates the interval is unbounded on the left. The graph of the solution set is shown in Figure 2-4.

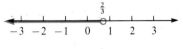

Figure 2-4   ■

Example 3   Solve the inequality $-5(x - 2) < 20 + x$.

Solution   Proceed as with equations, removing the parentheses and solving for x:

$$-5(x - 2) < 20 + x$$
$$-5x + 10 < 20 + x$$
$$-6x + 10 < 20$$
$$-6x < 10$$

Now divide both sides of the inequality by -6, which changes the order (direction) of the inequality:

$$x > -\frac{10}{6}$$

or

$$x > -\frac{5}{3}$$

The graph of the solution set of the inequality $-5(x - 2) < 20 + x$ is shown in Figure 2-5. The interval represented by the graph can be expressed in interval notation as $(-\frac{5}{3}, \infty)$.

Figure 2-5   ■

The statement that x is between two numbers, such as 2 and 5, implies two inequalities: $x > 2$ and $x < 5$. It is customary to write both inequalities as a single expression, $2 < x < 5$, which means that "x is greater than 2, and x is less than 5." The word *and* indicates that the two inequalities hold true simultaneously.

To express that x is *not* between 2 and 5, we must convey the idea that "either x is greater than or equal to 5, or x is less than or equal to 2." Note the either/or in this statement. This either/or statement is written symbolically as "$x \leq 2$ or $x \geq 5$." It is incorrect to string the two together as $2 \geq x \geq 5$ because that would mean that $2 \geq 5$, which is false.

Example 4 Solve the inequality $5 < 3x - 7 \leq 8$.

Solution There are two inequalities involved in this expression. Because a constant can be added to both sides of each inequality, add 7 to each part of the *double inequality* to get

$$5 + 7 < 3x - 7 + 7 \leq 8 + 7$$

or

$$12 < 3x \leq 15$$

Then, divide each part by 3 to obtain

$$4 < x \leq 5$$

This solution is represented graphically on the number line in Figure 2-6. The interval represented by the graph can be expressed in interval notation as $(4, 5]$.

Figure 2-6

Example 5 Solve the inequality $3 + x \leq 3x + 1 < 7x - 2$.

Solution Because it is impossible to isolate the variable x between the inequality signs, you must solve each of the two indicated inequalities separately.

$$
\begin{array}{c|c}
3 + x \leq 3x + 1 & \text{and} \quad 3x + 1 < 7x - 2 \\
3 \leq 2x + 1 & 1 < 4x - 2 \\
2 \leq 2x & 3 < 4x \\
1 \leq x & \dfrac{3}{4} < x
\end{array}
$$

You are interested only in those values of x that are greater than or equal to 1 *and also* greater than $\frac{3}{4}$. Because any values greater than or equal to 1 will also be greater than $\frac{3}{4}$, the solution is $x \geq 1$, which can be represented graphically as shown in Figure 2-7. The interval represented by the graph can be expressed in interval notation as $[1, \infty)$.

Figure 2-7

It is possible for an inequality to be true for all values of its variable. For example, $x < x + 1$ is true for all values of x. It is also possible for an inequality to have no solutions. The inequality $x > x + 1$ is true for no values of x.

Exercise 2.7

In Exercises 1–36, solve each inequality and graph each solution set.

1. $3x + 2 < 5$ **2.** $-2x + 4 < 6$ **3.** $3x + 2 \geq 5$ **4.** $-2x + 4 \geq 6$

5. $-5x + 3 > -2$ **6.** $4x - 3 > -4$ **7.** $-5x + 3 \leq -2$ **8.** $4x - 3 \leq -4$

9. $\dfrac{3}{5}x + 4 > 2$ **10.** $\dfrac{1}{4}x - 3 > 5$

11. $2(x - 3) \leq -2(x - 3)$ **12.** $3(x + 2) \leq 2(x + 5)$

13. $\dfrac{3 + x}{4} \leq \dfrac{2x - 4}{3}$ **14.** $\dfrac{2 + x}{5} > \dfrac{x - 1}{2}$

15. $\dfrac{6(x - 4)}{5} \geq \dfrac{3(x + 2)}{4}$ **16.** $\dfrac{3(x + 3)}{2} < \dfrac{2(x + 7)}{3}$

17. $4 < 2x - 8 \leq 10$ **18.** $3 \leq 2x + 2 < 6$ **19.** $9 \geq \dfrac{x - 4}{2} > 2$ **20.** $5 < \dfrac{x - 2}{6} < 20$

21. $0 \leq \dfrac{4 - x}{3} \leq 5$ **22.** $0 \geq \dfrac{5 - x}{2} \geq -10$

23. $-2 \geq \dfrac{1 - x}{2} \geq -10$ **24.** $-2 \leq \dfrac{1 - x}{2} < 10$

25. $-3x > -2x > -x$ **26.** $-3x < -2x < -x$

27. $x < 2x < 3x$ **28.** $x > 2x > 3x$

29. $2x + 1 < 3x - 2 < 12$ **30.** $2 - x < 3x + 5 < 18$

31. $2 + x < 3x - 2 < 5x + 2$ **32.** $x > 2x + 3 > 4x - 7$

33. $3 + x > 7x - 2 > 5x - 10$ **34.** $2 - x < 3x + 1 < 10x$

35. $x \leq x + 1 \leq 2x + 3$ **36.** $-x \geq -2x + 1 \geq -3x + 1$

In Exercises 37–40, find which values of x make each expression positive.

37. $2(x - 1)$ **38.** $-3(x - 1)$ **39.** $-4(x + 4)$ **40.** $5(x + 4)$

In Exercises 41–44, find which values of x make each expression negative.

41. $-2(x + 4)$ **42.** $3(x - 1)$ **43.** $5(x - 3)$ **44.** $-4(x + 1)$

In Exercises 45–50, solve the word problem.

45. An airline pilot flies at altitudes from 27,720 feet to 33,000 feet. Express this interval in miles. (*Hint:* 1 mile = 5280 feet.)

46. The perimeter of a rectangle is to be between 180 inches and 200 inches. What is the range of values for its length if its width is to be 40 inches?

47. The perimeter of an equilateral triangle is to be between 50 centimeters and 60 centimeters. What is the range of values for the length of one side?

48. The perimeter of a square is to be from 25 meters to 60 meters. What is the range of values for its area?

49. Express the relationship $20 < l < 30$ in terms of P, where $P = 2l + 2w$.

50. Express the relationship $10 < C < 20$ in terms of F, where $F = \frac{9}{5}C + 32$.

2.8 ABSOLUTE VALUE

We now review the definition for the absolute value of a number and examine its consequences in greater detail.

> **Definition.** The **absolute value** of the real number x, denoted by $|x|$, is defined as
>
> If $x \geq 0$, then $|x| = x$.
> If $x < 0$, then $|x| = -x$.

Whether x is positive, zero, or negative, the previous definition enables us to find the absolute value of x. *If x is positive or zero, the absolute value of x is x. If x is negative, the absolute value of x is the positive number $-x$.* Thus, $|x| \geq 0$.

Example 1 Find **a.** $|0|$, **b.** $|7|$, **c.** $|-3|$, and **d.** $|x - 2|$.

Solution **a.** By the first part of the definition $|0| = 0$.

b. Because 7 is positive, by the first part of the definition $|7| = 7$.

c. Because -3 is negative, by the second part of the definition

$$|-3| = -(-3) = 3$$

d. To denote the absolute value of a variable quantity, you must give a conditional answer. Either

$$|x - 2| = x - 2 \quad \text{if } x - 2 \geq 0 \text{ (or if } x \geq 2)$$

or

$$|x - 2| = -(x - 2) = 2 - x \quad \text{if } x - 2 < 0 \text{ (or if } x < 2) \qquad \blacksquare$$

Two theorems are often used when solving inequalities that involve absolute values.

> **Theorem.** If $a > 0$, then the inequality $|x| < a$ is equivalent to the double inequality
>
> $$-a < x < a$$
>
> (The symbol $<$ may be replaced with $\leq$.)

Proof Suppose that $a > 0$ and that $|x| < a$. We consider two cases.

Case 1

Let $x \geq 0$. Then, by the definition of absolute value, $|x| = x$, and the inequality $|x| < a$ is equivalent to $0 \leq x < a$. Because $-a$ is negative, we have

$$-a < x < a$$

Case 2

Let $x < 0$. Then, by the definition of absolute value, $|x| = -x$, and the inequality $|x| < a$ is equivalent to $0 < -x < a$, or $0 > x > -a$. Because a is positive, we have $a > x > -a$, or

$$-a < x < a$$

Amalie Noether (1882–1935) Emmy Noether worked in abstract algebra. Einstein called her the most creative woman mathematical genius since higher education of women began.

Hence, for all x

$$\text{if } \quad |x| < a, \quad \text{then} \quad -a < x < a$$

The theorem is proved. □

> **Theorem.** If $a > 0$, then the inequality $|x| > a$ is equivalent to the statement
>
> $$x > a \qquad \text{or} \qquad x < -a$$
>
> (The symbols $>$ and $<$ may be replaced with $\geq$ and $\leq$, respectively.)

Proof Suppose that $a > 0$ and $|x| > a$. Again, we consider two cases.

Case 1

Let $x \geq 0$. Then, $|x| = x$ and the inequality $|x| > a$ becomes $x > a$.

Case 2

Let $x < 0$. Then, $|x| = -x$ and the inequality $|x| > a$ becomes $-x > a$, or $x < -a$.

Combining the results of Cases 1 and 2, it is apparent that positive values of x must be greater than a, while negative values of x must be less than $-a$.

Hence,

if $|x| > a$,   then either $x > a$ or $x < -a$

The theorem is proved.   □

Theorem.   The equation $|x| = a$ (for $a \geq 0$) is equivalent to the two equations

$$x = a \qquad \text{or} \qquad x = -a$$

The proof of this theorem is similar to those of the two previous theorems. It is left as an exercise.

The graph of the inequality $|x| < a$ (for $a > 0$) includes all points lying *less* than a units from the origin. See Figure 2-8**a**. The graph of the inequality $|x| > a$ (for $a > 0$) includes all points lying *more* than a units from the origin, as in Figure 2-8**b**. The graph of the equation $|x| = a$ (for $a > 0$) includes the two points that lie *exactly* a units from the origin, as in Figure 2-8**c**.

For $a > 0$

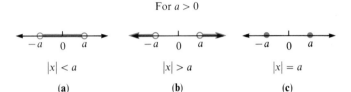

$|x| < a$	$|x| > a$	$|x| = a$
(a)	(b)	(c)

Figure 2-8

By the trichotomy property of real numbers, one of the three statements

$$|x| < a, \qquad |x| > a, \qquad \text{or} \qquad |x| = a$$

must be true. Note in Figure 2-8 that every point on the real number line is included in *exactly one* of these three graphs.

Example 2   Solve the equation $|3x - 5| = 7$.

Solution   The equation $|3x - 5| = 7$ is equivalent to the two equations

$$3x - 5 = 7 \qquad \text{or} \qquad 3x - 5 = -7$$

which can be solved separately as follows:

$$
\begin{array}{lll}
3x - 5 = 7 & \text{or} & 3x - 5 = -7 \\
3x = 12 & & 3x = -2 \\
x = 4 & & x = -\dfrac{2}{3}
\end{array}
$$

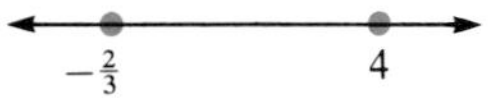

The graph of the solution set consists of the two points shown in Figure 2-9.   ■

Figure 2-9

Example 3 Solve the equation $|2x| = |x - 3|$.

Solution The equation $|2x| = |x - 3|$ will be satisfied if the quantities $2x$ and $x - 3$ are either equal to each other or are the negatives of each other. That idea determines two equations, which can be solved separately:

$$\begin{array}{c|l} 2x = x - 3 \quad \text{or} & 2x = -(x - 3) \\ x = -3 & 2x = -x + 3 \\ & 3x = 3 \\ & x = 1 \end{array}$$

Both $x = -3$ and $x = 1$ satisfy the given equation. ■

Example 4 Find the solution set of the inequality $|x - 2| < 7$, and graph it on a number line.

Solution The inequality $|x - 2| < 7$ is equivalent to the double inequality

$$-7 < x - 2 < 7$$

Add 2 to each part to obtain

$$-5 < x < 9$$

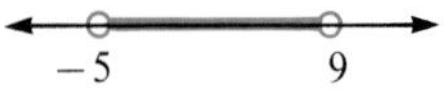

Figure 2-10

The graph of this solution is shown in Figure 2-10. ■

Example 5 Solve the inequality $\left|\frac{2x + 3}{2}\right| \geq 5$, and graph its solution set.

Solution The inequality $\left|\frac{2x + 3}{2}\right| \geq 5$ is equivalent to the following two inequalities, which can be solved separately.

$$\begin{array}{r|r} \dfrac{2x + 3}{2} \geq 5 \quad \text{or} & \dfrac{2x + 3}{2} \leq -5 \\ 2x + 3 \geq 10 & 2x + 3 \leq -10 \\ 2x \geq 7 & 2x \leq -13 \\ x \geq \dfrac{7}{2} & x \leq -\dfrac{13}{2} \end{array}$$

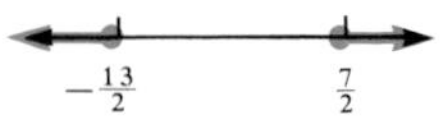

Figure 2-11

The graph of the solution set is shown in Figure 2-11. ■

Example 6 Solve the inequality $0 < |x - 5| \leq 3$, and graph its solution set.

Solution The double inequality $0 < |x - 5| \leq 3$ consists of two inequalities, which may be solved separately. The first inequality, $0 < |x - 5|$, is true for all values of x *except* 5. The second inequality, $|x - 5| \leq 3$, is equivalent to the double

Figure 2-12

inequality $-3 \leq x - 5 \leq 3$, or $2 \leq x \leq 8$. Hence, the solution set consists of all numbers x such that $x \neq 5$ and $2 \leq x \leq 8$. Note that the point $x = 5$ is missing in the graph of the solution set. See Figure 2-12.   ■

Example 7   Express the double inequality $-3 \leq x \leq 7$ as an absolute value inequality.

Solution   Halfway between -3 and 7 is their average,

$$\frac{-3 + 7}{2} = 2$$

Subtracting 2 from each part of the inequality gives

$$-3 \leq x \leq 7$$
$$-3 - 2 \leq x - 2 \leq 7 - 2$$
$$-5 \leq x - 2 \leq 5$$

This statement is equivalent to the absolute value inequality

$$|x - 2| \leq 5$$ ■

Recall that if a is a positive number or zero, then

$$\sqrt{a^2} = a$$

However, if a is unrestricted, we must use absolute value symbols to guarantee that $\sqrt{a^2}$ is nonnegative. In such a case we write

$$\sqrt{a^2} = |a| \quad \text{or} \quad |a| = \sqrt{a^2}$$

This fact can be used to solve certain inequalities.

Example 8   Solve the inequality $|x + 2| > |x + 1|$, and write the solution using interval notation.

Solution
$$|x + 2| > |x + 1|$$
$$\sqrt{(x + 2)^2} > \sqrt{(x + 1)^2} \qquad \text{Use } |a| = \sqrt{a^2}.$$
$$(x + 2)^2 > (x + 1)^2 \qquad \text{Square both sides.}$$
$$x^2 + 4x + 4 > x^2 + 2x + 1$$
$$2x > -3$$
$$x > -\frac{3}{2}$$

In interval notation, the solution is $(-\frac{3}{2}, \infty)$. Check several solutions to verify that this interval does not contain any extraneous solutions.   ■

There are three other properties of absolute values that are sometimes useful.

Properties of Absolute Value.   For all real numbers a and b,

1. $|ab| = |a|\,|b|$

2. $\left|\dfrac{a}{b}\right| = \dfrac{|a|}{|b|}$ $(b \neq 0)$

3. $|a + b| \leq |a| + |b|$ (the triangle inequality)

Property 1 states that the absolute value of a product is the product of the absolute values. Property 2 states that the absolute value of a quotient is the quotient of the absolute values. Property 3 states that the absolute value of a sum is either equal to or less than the sum of the absolute values.

Exercise 2.8

In Exercises 1–8, find the value of each absolute value expression.

1. $|7|$

2. $|-9|$

3. $|0|$

4. $|3 - 5|$

5. $|5| - |-3|$

6. $|-3| + |5|$

7. $|\pi - 2|$

8. $|\pi - 4|$

In Exercises 9–26, solve each equation for x.

9. $|x + 2| = 2$

10. $|2x + 5| = 3$

11. $|3x - 1| = 5$

12. $|7x - 5| = 3$

13. $\left|\dfrac{3x - 4}{2}\right| = 5$

14. $\left|5x + \dfrac{1}{2}\right| = \dfrac{9}{2}$

15. $|x - 3| = -2$

16. $|x - 5| = 0$

17. $|x| = x$

18. $|x| + x = 2$

19. $|x + 3| = |x|$

20. $|x + 5| = |5 - x|$

21. $|x - 3| = |2x + 3|$

22. $|x - 2| = |3x + 8|$

23. $|x + 2| = |x - 2|$

24. $|2x - 3| = |3x - 5|$

25. $\left|\dfrac{x + 3}{2}\right| = |2x - 3|$

26. $\left|\dfrac{x - 2}{3}\right| = |6 - x|$

In Exercises 27–48, solve each inequality and graph each solution set.

27. $|x - 3| < 6$

28. $|x - 2| \geq 4$

29. $|x - 3| \geq 6$

30. $|x - 2| < 4$

31. $|2x + 4| \geq 10$

32. $|5x - 2| \leq 7$

33. $|3x + 5| \leq 10$

34. $|2x - 7| \geq 5$

35. $|x + 3| > 0$

36. $|x - 3| \geq 0$

37. $\left|\dfrac{5x + 2}{3}\right| < 1$

38. $\left|\dfrac{3x + 2}{4}\right| > 2$

39. $\left|\dfrac{x + 3}{2}\right| < \dfrac{2 - x}{4}$

40. $\left|\dfrac{3 - x}{2}\right| > \dfrac{x + 1}{6}$

41. $0 < |2x + 1| < 3$

42. $0 < |2x - 3| < 1$

43. $8 > |3x - 1| > 3$

44. $8 > |4x - 1| > 5$

45. $2 < \left|\dfrac{x - 5}{3}\right| < 4$

46. $3 < \left|\dfrac{x - 3}{2}\right| < 5$

47. $4 \leq \left|\dfrac{x - 2}{2}\right| < 10$

48. $1 < \left|\dfrac{x + 2}{3}\right| \leq 5$

In Exercises 49–56, express each of the inequalities as an absolute value inequality.

49. $3 < x < 7$

50. $-3 < x < 9$

51. $-\dfrac{1}{2} \le x \le \dfrac{9}{2}$

52. $\dfrac{3}{2} \le x \le \dfrac{11}{2}$

53. $x \ne 2$ and $0 < x < 4$

54. $x \ne 3$ and $-1 < x < 7$

55. $x \ne 0$ and $-5 < x < 5$

56. $x \ne -3$ and $-6 \le x \le 0$

In Exercises 57–60, solve each inequality and express the solution using interval notation.

57. $|x + 1| \ge |x|$

58. $|x + 1| < |x + 2|$

59. $|2x + 1| < |2x - 1|$

60. $|3x - 2| \ge |3x + 1|$

2.9 QUADRATIC AND RATIONAL INEQUALITIES

Expressions such as $ax^2 + bx + c < 0$ and $ax^2 + bx + c > 0$, where $a \ne 0$, are called **quadratic inequalities**. We consider two examples to show how to solve these inequalities.

Example 1 Solve the quadratic inequality $x^2 - x - 6 > 0$.

Solution 1 Factor the left side of the inequality:

$$(x - 3)(x + 2) > 0$$

Note that the product $(x - 3)(x + 2)$ will be positive only if both factors are positive, or if both factors are negative.

Case 1

Suppose that both factors are positive. Then,

$$x - 3 > 0 \quad \text{and} \quad x + 2 > 0$$
$$x > 3 \qquad\qquad x > -2$$

Both of the inequalities $x > 3$ and $x > -2$ are satisfied by all values of x that are greater than 3. Hence, Case 1 contributes the solutions $x > 3$.

Case 2

Suppose that both factors are negative. Then,

$$x - 3 < 0 \quad \text{and} \quad x + 2 < 0$$
$$x < 3 \qquad\qquad x < -2$$

Only values of x less than -2 satisfy the inequalities $x < 3$ and $x < -2$ simultaneously. Hence, Case 2 contributes the solutions $x < -2$.

The solutions to the quadratic inequality $x^2 - x - 6 > 0$ are all x such that $x > 3$ (from Case 1) or $x < -2$ (from Case 2). The graph of the solution set is shown in Figure 2-13.

$-2 \qquad 3$

Figure 2-13

Solution 2   This second solution relies exclusively on the number line and a notation that keeps track of the signs of the factors of the quadratic inequality $x^2 - x - 6 > 0$.

At $x = 3$, the factor $x - 3$ is 0. For values of x that are less than 3, the factor $x - 3$ is negative; for values of x that are greater than 3, the factor $x - 3$ is positive. Similarly, the value of the factor $x + 2$ is 0 when $x = -2$. For values of x that are less than -2, the factor $x + 2$ is negative; for values of x that are greater than -2, the factor $x + 2$ is positive.

Because the positiveness of the product $(x - 3)(x + 2)$ depends on the signs of $x - 3$ and $x + 2$, keep track of the positiveness or negativeness of these factors by using plus and minus signs in the **sign graph** shown in Figure 2-14. Only to the left of -2 and to the right of $+3$ do the signs of both factors agree; only there is the product positive.

The graph of the solution set is shown in Figure 2-14.

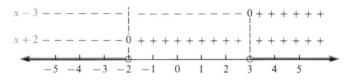

Figure 2-14   ■

Example 2   Solve the quadratic inequality $x(x + 3) < -2$.

Rewrite the inequality so that one side is 0. Remove parentheses and add 2 to both sides to obtain $x^2 + 3x + 2 < 0$.

Solution 1   Factor the left side of the inequality $x^2 + 3x + 2 < 0$:

$$(x + 2)(x + 1) < 0$$

Note that the product $(x + 2)(x + 1)$ will be negative only if one of its factors is positive and one is negative.

Case 1

Suppose that $x + 2$ is positive and $x + 1$ is negative. Then,

$$x + 2 > 0 \quad \text{and} \quad x + 1 < 0$$
$$x > -2 \quad | \quad x < -1$$

The pair of inequalities $x > -2$ and $x < -1$ can be written as $-2 < x < -1$. Hence, all values of x that are between -2 and -1 satisfy the quadratic inequality $x^2 + 3x + 2 < 0$. Case 1 contributes the solutions $-2 < x < -1$.

Case 2

Suppose that $x + 2$ is negative and $x + 1$ is positive. Then,

$$x + 2 < 0 \quad \text{and} \quad x + 1 > 0$$
$$x < -2 \quad | \quad x > -1$$

Figure 2-15

Because no values of x are less than -2 and greater than -1 at the same time, Case 2 contributes no solutions. Hence, the solutions of the quadratic inequality $x^2 + 3x + 2 < 0$ are all values of x such that $-2 < x < -1$ (all from Case 1). The graph of this solution set is shown in Figure 2-15.

Solution 2   Keep track of the positiveness or negativeness of the factors $x + 2$ and $x + 1$ by constructing the sign graph in Figure 2-16. Only between -2 and -1 do the factors have opposite signs. Here, the product is negative. The graph of the solution set is shown in Figure 2-16.

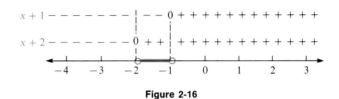

Figure 2-16

An inequality that contains a fraction with a polynomial denominator is called a **rational inequality**. To solve a rational inequality, construct a sign graph.

Example 3   Solve the rational inequality $\dfrac{x^2 - x - 2}{x^2 - 4x + 3} \leq 0$.

Solution   Factor each of the trinomials, and write the given inequality in the form

$$\frac{(x - 2)(x + 1)}{(x - 3)(x - 1)} \leq 0$$

Construct a sign graph to keep track of the signs of each of the four factors. See Figure 2-17. The value of the given rational expression must be negative or 0. The only way this rational expression can be negative is to have an odd number of negative factors; the only way that the expression can be 0 is for its numerator to be 0. From the sign graph, it is apparent that only between -1 and 1 and between 2 and 3 is there an odd number of negative factors. Only there is the rational expression negative. If $x = 2$ or $x = -1$, the numerator is 0, and therefore the fraction is 0 also. The graph of the solution set is shown in Figure 2-17. The circles at $x = -1$ and $x = 2$ are solid to show that these

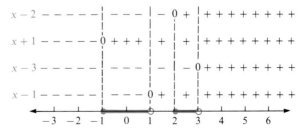

Figure 2-17

two values are included in the solution set. The circles at $x = 1$ and $x = 3$ are left open to indicate that these two values are not included in the solution set. If they were included, the denominator of the fraction would be 0.

Example 4 Solve the inequality $\dfrac{6}{x} > 2$.

Solution Add -2 to both sides of the inequality, and combine terms on the left side as follows:

$$\frac{6}{x} > 2$$

$$\frac{6}{x} - 2 > 0$$

$$\frac{6}{x} - \frac{2x}{x} > 0$$

$$\frac{6 - 2x}{x} > 0$$

The given inequality now has the form of a rational inequality. Construct a sign graph to keep track of the signs of the numerator and denominator. See Figure 2-18. The fraction $\dfrac{6 - 2x}{x}$ will be positive when the signs of the numerator and denominator agree. This is true for all values of x between 0 and 3. Thus, the solutions to the given inequality are all x such that $0 < x < 3$. The graph of the solution set is shown in Figure 2-18.

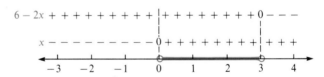

Figure 2-18

Exercise 2.9

In Exercises 1–42, construct a sign graph to find the solution set for each inequality.

1. $x^2 + 7x + 12 < 0$

2. $x^2 - 13x + 12 \le 0$

3. $x^2 - 5x + 6 \ge 0$

4. $6x^2 + 5x - 6 > 0$

5. $x^2 + 5x + 6 < 0$

6. $x^2 + 9x + 20 \ge 0$

7. $6x^2 + 5x + 1 \ge 0$

8. $x^2 + 9x + 20 < 0$

9. $6x^2 - 5x < -1$

10. $9x^2 + 24x > -16$

11. $2x^2 \ge 3 - x$

12. $9x^2 \le 24x - 16$

13. $\dfrac{x + 3}{x - 2} < 0$

14. $\dfrac{x + 3}{x - 2} > 0$

15. $\dfrac{x(x+1)}{x^2-1} > 0$

16. $\dfrac{x^2-4}{x^2-9} < 0$

17. $\dfrac{x^2+5x+6}{x^2+x-6} \geq 0$

18. $\dfrac{x^2+10x+25}{x^2-x-12} \leq 0$

19. $\dfrac{x^2+7x+6}{x^2+8x+15} \leq 0$

20. $\dfrac{x^2-8x+15}{x^2-12x+35} \geq 0$

21. $\dfrac{x^2+13x+36}{x^2+4x+4} > 0$

22. $\dfrac{6x^2-3x-3}{x^2-2x-8} < 0$

23. $\dfrac{3}{x} > 2$

24. $\dfrac{3}{x} < 2$

25. $\dfrac{6}{x} < 4$

26. $\dfrac{6}{x} > 4$

27. $\dfrac{3}{x-2} \leq 5$

28. $\dfrac{3}{x+2} \leq 4$

29. $\dfrac{3}{x+2} \geq 4$

30. $\dfrac{-1}{x-1} \geq 3$

31. $\dfrac{6}{x^2-1} < 1$

32. $\dfrac{6}{x^2-1} > 1$

33. $\dfrac{3}{x} - x > 2$

34. $\dfrac{1}{x} + x < 2$

35. $\dfrac{8}{x} + x < \dfrac{1}{x}$

36. $\dfrac{8}{x} + x > \dfrac{1}{x}$

37. $x(x-1)(x-2) > 0$

38. $(x^2+1)(x^2-x-6) \geq 0$

39. $x^3 \geq x$

40. $(x^2-4)(x^2+5x+6) < 0$

41. $\left| \dfrac{x+2}{x-3} \right| \leq 4$

42. $\left| \dfrac{x}{2x+1} \right| > 5$

REVIEW EXERCISES

In Review Exercises 1–6, find the domain of x.

1. $3x + 7 = 4$

2. $x + \dfrac{1}{x} = 2$

3. $x^2 = x(x-5)$

4. $3(x+2) = 2(x-3)$

5. $\sqrt{x} = 4$

6. $\dfrac{1}{x-2} = \dfrac{2}{x-3}$

In Review Exercises 7–14, solve each equation and classify each one as an identity, a conditional equation, or an equation with no solution.

7. $3(x+4) = 28$

8. $\dfrac{3}{2}x = 7(x+11)$

9. $8(3x-5) - 4(2x+3) = 12$

10. $\dfrac{x+3}{x+4} + \dfrac{x+3}{x+2} = 2$

11. $\dfrac{3}{x-1} = \dfrac{1}{2}$

12. $\dfrac{8x^2+72x}{9+x} = 8x$

13. $\dfrac{3x}{x-1} - \dfrac{5}{x+3} = 3$

14. $x + \dfrac{1}{2x-3} = \dfrac{2x^2}{2x-3}$

In Review Exercises 15–18, solve each equation by factoring.

15. $2x^2 - x - 6 = 0$

16. $12x^2 + 13x = 4$

17. $5x^2 - 8x = 0$

18. $27x^2 = 30x - 8$

In Review Exercises 19–22, solve each equation by completing the square.

19. $x^2 - 8x + 15 = 0$

20. $3x^2 + 18x = -24$

21. $5x^2 - x - 1 = 0$

22. $5x^2 - x = 0$

In Review Exercises 23–26, solve each equation by using the quadratic formula.

23. $x^2 + 5x - 14 = 0$

24. $3x^2 - 25x = 18$

25. $5x^2 = 1 - x$

26. $2x(x + 4) = -5$

In Review Exercises 27–34, solve by any method.

27. $\dfrac{3x}{2} - \dfrac{2x}{x-1} = x - 3$

28. $\dfrac{12}{x} - \dfrac{x}{2} = x - 3$

29. $x^4 - 2x^2 + 1 = 0$

30. $x^4 + 36 = 37x^2$

31. $\sqrt{x-1} + x = 7$

32. $\sqrt{5-x} + \sqrt{5+x} = 4$

33. $\sqrt{a+9} - \sqrt{a} = 3$

34. $\sqrt{y+5} + \sqrt{y} = 1$

35. Find the value of k that will cause the roots of $kx^2 + 4x + 12 = 0$ to be equal.

36. Find the values of k that will cause the roots of $4y^2 + (k + 2)y = 1 - k$ to be equal.

In Review Exercises 37–44, solve each word problem.

37. A liter of fluid is 50% alcohol. How much water must be added to dilute it to a 20% solution?

38. Scott can wash 37 windows in 3 hours, while Bill can wash 27 windows in 2 hours. How long will it take the two of them to wash 100 windows?

39. A tank can be filled in 9 hours by one pipe, and in 12 hours by another. How long will it take both pipes to fill the empty tank?

40. How many ounces of pure zinc must be alloyed with 20 ounces of brass that is 30% zinc and 70% copper, to produce brass that is 40% zinc?

41. A bank loans $10,000, part of it at 11% annual interest, and the rest at 14%. If the annual income is $1265, how much was loaned at each rate?

42. George can paint his summer cottage in 9 hours less time than his son can. Working together, they can paint the house in 20 hours. How long would it take each, if they worked alone?

43. A farmer wishes to enclose a rectangular garden with 300 yards of fencing. A river runs along one side of the garden, so no fencing is needed there.

What will be the dimensions of the rectangle if the area is 10,450 square yards?

44. A jet plane, flying 120 miles per hour faster than a propeller-driven plane, travels 3520 miles in 3 hours less time than the propeller plane requires to fly the same distance. How fast does each plane fly?

In Review Exercises 45–50, solve each equation for the indicated variable.

45. $\dfrac{1}{f} = \dfrac{1}{f_1} + \dfrac{1}{f_2}$; for f_1

46. $\dfrac{1}{C} = \dfrac{1}{C_1} + \dfrac{1}{C_2}$; for C

47. $S = \dfrac{a - lr}{1 - r}$; for l

48. $s(s - a)(s - b)(s - c) = A^2$; for b

49. $\dfrac{y}{x} = \dfrac{x + y}{2}$; for y

50. $\dfrac{y}{x} = \dfrac{x + y}{2}$; for x

In Review Exercises 51–60, find the solution set for each inequality. Graph each solution set on a number line.

51. $2x - 9 < 5$

52. $5x + 3 \geq 2$

53. $\dfrac{3}{7}x + 5 > 10$

54. $\dfrac{2}{3}x - 5 > x$

55. $\dfrac{5(x - 1)}{2} < x$

56. $3(x - 1) > x + 1$

57. $3(x - 5) < x - 1$

58. $0 \leq \dfrac{3 + x}{2} < 4$

59. $-2 \leq \dfrac{2x + 5}{2} < 10$

60. $-x \leq x \leq 2x$

In Review Exercises 61–64, evaluate each expression.

61. $|-8|$

62. $-|3 - 5|$

63. $-|3| \cdot |-3|$

64. $-\dfrac{|3|}{|5 - 2|}$

In Review Exercises 65–70, solve for x.

65. $|x + 3| < 3$

66. $|3x - 7| \geq 1$

67. $|x + 1| = 6$

68. $|2x - 1| = |2x + 1|$

69. $\left|\dfrac{x + 2}{3}\right| < 1$

70. $1 < |2x + 3| < 4$

In Review Exercises 71–76, solve each quadratic inequality.

71. $x^2 - 2x - 3 < 0$

72. $2x^2 + x - 3 > 0$

73. $5x^2 + 32x \geq 21$

74. $7x^2 + 8x + 1 \leq 0$

75. $6x^2 + 19x < -10$

76. $2x^2 - x > 15$

In Review Exercises 77–82, solve each rational inequality.

77. $\dfrac{x^2 + x - 2}{x - 3} \geq 0$

78. $\dfrac{x^2 + 6x - 7}{x + 1} < 0$

79. $\dfrac{2x^2 - 5x - 3}{x + 1} \geq 0$

80. $\dfrac{3x^2 - 5x - 2}{x - 3} \leq 0$

81. $x < \dfrac{2}{x + 1}$

82. $x \leq \dfrac{3}{x - 2}$

3

FUNCTIONS
AND THEIR GRAPHS

Mathematical expressions often indicate relationships between several quantities. For example, the formula $d = rt$ indicates that distance traveled depends on the rate of speed and elapsed time. Because d depends on the values r and t, we say that d is a function of r and t. The concept of function is the topic of this chapter.

3.1 THE CARTESIAN COORDINATE SYSTEM

René Descartes (1596–1650) Descartes merged algebra and geometry into a single subject called analytical geometry.

Any point on a number line is associated with a single real number, called its **coordinate**. René Descartes (1596–1650) is credited with the idea of associating each point in the *plane* with a *pair* of real numbers.

Descartes' idea is based on two perpendicular number lines, usually called the **x-axis** and **y-axis**, which divide the plane into four **quadrants**, numbered as shown in Figure 3-1. These axes intersect at a point called the **origin**, which is the zero point on each number line. The positive direction on the x-axis is to the right, the positive direction on the y-axis is upward, and the same unit distance is used on both axes. These axes determine a **rectangular coordinate system**, or a **Cartesian coordinate system** in honor of its inventor.

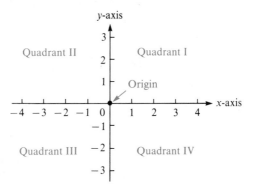

Figure 3-1

To **plot** the point associated with the pair of real numbers $(-3, 2)$, for example, we start at the origin, count 3 units to the left, and then 2 units up. See Figure 3-2. Point P, called the **graph** of the pair $(-3, 2)$, lies in the second quadrant.

The pair $(-3, 2)$ gives the **coordinates** of the point P. Similarly, point Q with coordinates $(2, -3)$ lies in the fourth quadrant. The coordinates of the origin are $(0, 0)$.

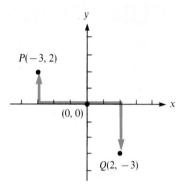

Figure 3-2

Note that the pairs $(-3, 2)$ and $(2, -3)$ represent different points. Because the order of the real numbers of the pair (a, b) is important, such pairs are called **ordered pairs**. The first coordinate, a, of the ordered pair (a, b) is called the **x-coordinate** or the **abscissa**. The second coordinate, b, is called the **y-coordinate**, or the **ordinate**. It is proper to say "the point P with coordinates (a, b)," but it is acceptable to say simply, "the point $P(a, b)$." The set of all points determined by ordered pairs (x, y) is called the **xy-plane** or the **Cartesian plane**.

Example 1   Graph the points $P(2, 3)$, $Q(-4, -2)$, $R(-5, 0)$, $S(0, 4)$, and $T(3, -2)$.

Solution

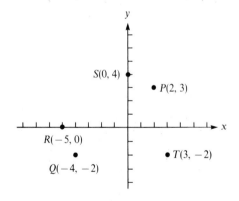

Figure 3-3   ▪

It is possible to draw a "picture" or a graph of an equation in two variables. The **graph of an equation** in the variables x and y is the set of all points on the xy-plane with coordinates (x, y) that satisfy the equation.

Example 2   Graph the equation $x + 2y = 5$.

Solution   Pick some values for either x or y, substitute those values in the equation, and solve for the other variable. For example, if $y = -1$, then you can find x as follows:

$$x + 2y = 5$$
$$x + 2(-1) = 5 \qquad \text{Substitute } -1 \text{ for } y.$$
$$x - 2 = 5 \qquad \text{Simplify.}$$
$$x = 7 \qquad \text{Add 2 to both sides.}$$

Thus, one ordered pair that satisfies the equation is $(7, -1)$.
   If $x = 0$, you have

$$x + 2y = 5$$
$$0 + 2y = 5 \qquad \text{Substitute 0 for } x.$$
$$2y = 5 \qquad \text{Simplify.}$$
$$y = \frac{5}{2} \qquad \text{Divide both sides by 2.}$$

Thus, another ordered pair that satisfies the equation is $(0, \frac{5}{2})$.
   The ordered pairs $(7, -1)$ and $(0, \frac{5}{2})$ and others that satisfy the equation $x + 2y = 5$ are shown in the table of values in Figure 3-4. Plot each of these ordered pairs on a rectangular coordinate system as in the figure. Later, we will show that all points (x, y) that satisfy an equation such as $x + 2y = 5$ lie on a straight line. The line that joins the five points is the graph of the equation.

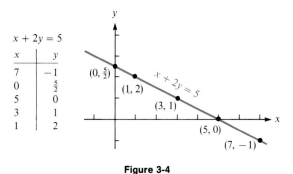

$x + 2y = 5$	
x	y
7	-1
0	$\frac{5}{2}$
5	0
3	1
1	2

Figure 3-4

Because the line in Example 2 intersects the y-axis at the point $(0, \frac{5}{2})$, the number $\frac{5}{2}$ is called the **y-intercept** of the line. Similarly, 5 is the **x-intercept** of the line.

Example 3   Use the x- and y-intercepts to graph the equation $4x - 3y = 12$.

Solution   To find the y-intercept, substitute 0 for x and solve for y:

$$4x - 3y = 12$$
$$4(0) - 3y = 12 \qquad \text{Substitute 0 for } x.$$
$$-3y = 12 \qquad \text{Simplify.}$$
$$y = -4 \qquad \text{Divide both sides by } -3.$$

Because the y-intercept is -4, the line intersects the y-axis at the point $(0, -4)$. To find the x-intercept, substitute 0 for y and solve for x:

$$4x - 3y = 12$$
$$4x - 3(0) = 12 \qquad \text{Substitute 0 for } y.$$
$$4x = 12 \qquad \text{Simplify.}$$
$$x = 3 \qquad \text{Divide both sides by 4.}$$

Because the x-intercept is 3, the line intersects the x-axis at the point $(3, 0)$.

Although these two points are sufficient to draw the line, you should find and plot a third point as a check. If $x = 6$, for example, then $y = 4$. Thus, the line passes through $(6, 4)$. The graph appears in Figure 3-5.

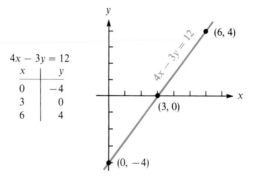

$4x - 3y = 12$

x	y
0	-4
3	0
6	4

Figure 3-5

Example 4   Graph the equation $3(y + 2) = 2x - 3$.

Solution   First solve the equation for y. This will make it easier to find pairs (x, y) that satisfy the equation:

$$3(y + 2) = 2x - 3$$
$$3y + 6 = 2x - 3 \qquad \text{Remove parentheses.}$$
$$3y = 2x - 9 \qquad \text{Add } -6 \text{ to both sides.}$$
$$y = \frac{2}{3}x - 3 \qquad \text{Divide both sides by 3.}$$

Substitute numbers for x, and calculate the corresponding values of y. For example, let $x = 3$:

$$y = \frac{2}{3}x - 3$$

$$y = \frac{2}{3}(3) - 3 \qquad \text{Substitute 3 for } x.$$

$$y = 2 - 3 \qquad \text{Simplify.}$$

$$y = -1$$

Thus, the point $(3, -1)$ lies on the line. To find another point, let $x = 0$ and calculate y:

$$y = \frac{2}{3}x - 3$$

$$y = \frac{2}{3}(0) - 3 \qquad \text{Substitute 0 for } x.$$

$$y = -3 \qquad \text{Simplify.}$$

Thus, the point $(0, -3)$ also lies on the line. These two solutions and several others are plotted in Figure 3-6. The line that passes through all of the points is the graph of the equation.

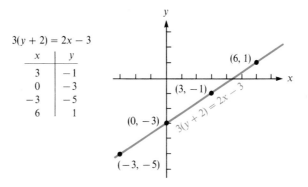

$3(y + 2) = 2x - 3$

x	y
3	−1
0	−3
−3	−5
6	1

Figure 3-6

The Distance Formula

To distinguish between two points on the Cartesian plane, subscript notation is often used. One point might be denoted as $P(x_1, y_1)$, read as "point P with coordinates of x sub 1 and y sub 1," and the second as $Q(x_2, y_2)$. We can derive a formula to calculate the distance between two points $P(x_1, y_1)$ and $Q(x_2, y_2)$.

If $P(x_1, y_1)$ and $Q(x_2, y_2)$ are two points in Figure 3-7, then the right triangle PQR can be formed, with R having coordinates (x_2, y_1). Because the line segment RQ is vertical, the square of its length is $(y_2 - y_1)^2$. Because PR is horizontal, the square of its length is $(x_2 - x_1)^2$. By the Pythagorean theorem, we know that the square of the hypotenuse of right triangle PQR is equal to the

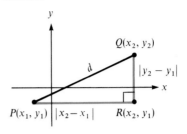

Figure 3-7

sum of the squares of the two legs. Thus, we have

$$d^2 = (x_2 - x_1)^2 + (y_2 - y_1)^2$$

Because equal positive numbers have equal positive square roots, we can take the positive square root of both sides of this equation to obtain a formula called the **distance formula**.

The Distance Formula.   The distance d between points (x_1, y_1) and (x_2, y_2) is given by

$$d = \sqrt{(x_2 - x_1)^2 + (y_2 - y_1)^2}$$

Example 5   Find the distance $d(PQ)$ between the points $P(-1, -2)$ and $Q(-7, 8)$.

Solution   Let (x_1, y_1) be $P(-1, -2)$ and (x_2, y_2) be $Q(-7, 8)$. Substitute these coordinates into the distance formula and simplify:

$$d(PQ) = \sqrt{(x_2 - x_1)^2 + (y_2 - y_1)^2}$$
$$d(PQ) = \sqrt{[-7 - (-1)]^2 + [8 - (-2)]^2}$$
$$= \sqrt{(-6)^2 + (10)^2}$$
$$= \sqrt{36 + 100}$$
$$= \sqrt{136}$$
$$= \sqrt{4 \cdot 34}$$
$$= 2\sqrt{34}$$

The Midpoint Formula

The **midpoint** of a line segment is the point on the segment that is midway between the endpoints. The x-coordinate of the midpoint is found by averaging the x-coordinates of the segment's endpoints. Similarly, the y-coordinate of the midpoint is the average of the y-coordinates of the segment's endpoints. For example, the midpoint of the segment joining $P(3, 4)$ and $Q(5, 7)$ is

$$M\left(\frac{3 + 5}{2}, \frac{4 + 7}{2}\right) \qquad \text{or} \qquad M\left(4, \frac{11}{2}\right)$$

The Midpoint Formula.   The midpoint of the line segment joining points $P(x_1, y_1)$ and $Q(x_2, y_2)$ is the point M with coordinates of

$$\left(\frac{x_1 + x_2}{2}, \frac{y_1 + y_2}{2}\right)$$

Proof   To prove that M is the midpoint of segment PQ, we must show that $d(PM) = d(MQ)$, and that points P, Q, and M all lie on the same line. See Figure 3-8.

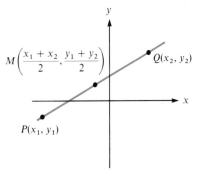

Figure 3-8

To show that $d(PM) = d(MQ)$, we use the distance formula and observe that the distances are equal:

$$d(PM) = \sqrt{\left(\frac{x_1 + x_2}{2} - x_1\right)^2 + \left(\frac{y_1 + y_2}{2} - y_1\right)^2}$$

$$= \sqrt{\left(\frac{x_2 - x_1}{2}\right)^2 + \left(\frac{y_2 - y_1}{2}\right)^2}$$

$$= \sqrt{\frac{(x_2 - x_1)^2}{4} + \frac{(y_2 - y_1)^2}{4}}$$

$$= \frac{1}{2}\sqrt{(x_2 - x_1)^2 + (y_2 - y_1)^2}$$

$$d(MQ) = \sqrt{\left(x_2 - \frac{x_1 + x_2}{2}\right)^2 + \left(y_2 - \frac{y_1 + y_2}{2}\right)^2}$$

$$= \sqrt{\left(\frac{x_2 - x_1}{2}\right)^2 + \left(\frac{y_2 - y_1}{2}\right)^2}$$

$$= \sqrt{\frac{(x_2 - x_1)^2}{4} + \frac{(y_2 - y_1)^2}{4}}$$

$$= \frac{1}{2}\sqrt{(x_2 - x_1)^2 + (y_2 - y_1)^2}$$

Thus, $d(PM) = d(MQ)$. The distance between P and Q is

$$d(PQ) = \sqrt{(x_2 - x_1)^2 + (y_2 - y_1)^2}$$

Because $d(PM) + d(MQ) = d(PQ)$, points P, Q, and M must lie on the same line. The theorem is proved. $\square$

Example 6   The midpoint of the line segment joining $P(-3, 2)$ and point $Q(x_2, y_2)$ is $M(1, 4)$. Find the coordinates of Q.

Solution   Let $P(x_1, y_1)$ be $P(-3, 2)$, and let $M(x_M, y_M)$ be $M(1, 4)$. Find the coordinates x_2 and y_2 of $Q(x_2, y_2)$ as follows:

$$x_M = \frac{x_1 + x_2}{2} \quad \text{and} \quad y_M = \frac{y_1 + y_2}{2}$$

$$1 = \frac{-3 + x_2}{2} \qquad\qquad 4 = \frac{2 + y_2}{2}$$

$$2 = -3 + x_2 \qquad\qquad 8 = 2 + y_2 \qquad \text{Multiply both sides by 2.}$$

$$5 = x_2 \qquad\qquad\qquad 6 = y_2$$

The coordinates of point Q are $(5, 6)$.   ■

Exercise 3.1

In Exercises 1–12, graph each point on the Cartesian plane. Indicate the quadrant in which the point lies or the axis on which it lies.

1. $(2, 5)$	**2.** $(-3, 4)$	**3.** $(-4, -5)$	**4.** $(6, -2)$
5. $(5, 2)$	**6.** $(3, -4)$	**7.** $(4, 0)$	**8.** $(-2, 2)$
9. $(0, 2)$	**10.** $(3, 4)$	**11.** $(-7, 0)$	**12.** $(0, -5)$

In Exercises 13–20, use the x- and y-intercepts to graph each equation.

13. $x + y = 5$	**14.** $x - y = 3$	**15.** $2x - y = 4$	**16.** $3x + y = 9$
17. $3x + 2y = 6$	**18.** $4x - 5y = 20$	**19.** $2x + 7y = 14$	**20.** $3x - 5y = 15$

In Exercises 21–28, first solve each equation for y, and then graph each equation.

21. $y - 2x = 7$	**22.** $y + 3 = -4x$	**23.** $6x - 3y = 10$	**24.** $4x + 8y - 1 = 0$
25. $2(x - y) = 3x + 2$		**26.** $5(x + 2) = 3y - x$	
27. $3x + y = 3(x - 1)$		**28.** $2(y - x) = 3(x + 2)$	

In Exercises 29–34, find the distance between P and the point $(0, 0)$. Assume that a and b are positive numbers.

29. $P(4, -3)$	**30.** $P(-5, 12)$	**31.** $P(-3, 2)$	**32.** $P(5, 0)$
33. $P(a, b)$		**34.** $P(a, -a)$	

In Exercises 35–46, find the distance between P and Q. Assume that all variables represent positive numbers.

35. $P(3, 7)$; $Q(6, 3)$	**36.** $P(4, -6)$; $Q(-1, 6)$
37. $P(0, 5)$; $Q(6, -3)$	**38.** $P(-2, -15)$; $Q(-9, -39)$
39. $P(3, 3)$; $Q(5, 5)$	**40.** $P(6, -3)$; $Q(-3, 2)$

41. $P(3, 7)$; $Q(5, 7)$

42. $P(4, -6)$; $Q(4, -8)$

43. $P(\pi, 2)$; $Q(\pi, 5)$

44. $P(\sqrt{3}, \pi)$; $Q(\pi, \sqrt{3})$

45. $P(x, 0)$; $Q(0, y)$

46. $P(a, b)$; $Q(b, a)$

In Exercises 47–54, find the midpoint of the line segment PQ.

47. $P(2, 4)$; $Q(6, 8)$

48. $P(3, -6)$; $Q(-1, -6)$

49. $P(2, -5)$; $Q(-2, 7)$

50. $P(0, 3)$; $Q(-10, -13)$

51. $P(0, 0)$; $Q(\sqrt{5}, \sqrt{5})$

52. $P(\sqrt{3}, 0)$; $Q(0, \sqrt{3})$

53. $P(a, 2)$; $Q(2, a)$

54. $P(a, b)$; $Q(0, 0)$

In Exercises 55–58, one endpoint P and the midpoint M of line segment PQ are given. Find the coordinates of the other endpoint, Q.

55. $P(1, 4)$; $M(3, 5)$

56. $P(2, -7)$; $M(-5, 6)$

57. $P(5, -5)$; $M(5, 5)$

58. $P(-7, 3)$; $M(0, 0)$

59. Show that a triangle with vertices at $(13, -2)$, $(9, -8)$, and $(5, -2)$ is isosceles (has two equal sides).

60. Show that a triangle with vertices at $(-1, 2)$, $(3, 1)$, and $(4, 5)$ is isosceles.

61. Show that the points $(-3, -1)$, $(3, 1)$, $(1, 7)$, and $(-5, 5)$ are the vertices of a rhombus by showing that its four sides are equal.

62. Use the distance formula to find the radius of a circle with center at $(4, -2)$ and passing through $(6, -9)$.

63. Find the center of a circle if the endpoints of a diameter are $(-17, 6)$ and $(9, -12)$.

64. Find the coordinates of the two points on the y-axis that are $\sqrt{74}$ units from the point $(5, -3)$.

65. The diagonals of a square with area 50 square units lie on the x- and y-axes. Find the coordinates of the vertices.

66. Show that the distance between the points (a, b) and (c, d) is equal to the distance between the point $(a - c, b - d)$ and the origin.

67. In Illustration 1, points M and N are the midpoints of AC and BC, respectively. Find the length of MN.

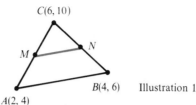

C(6, 10)

M

N

B(4, 6)   Illustration 1

A(2, 4)

68. In Illustration 2, points M and N are the midpoints of AC and BC, respectively. Show that

$$d(MN) = \tfrac{1}{2}[d(AB)]$$

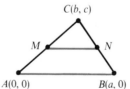

C(b, c)

M

N

A(0, 0)   B(a, 0)   Illustration 2

3.2 THE SLOPE OF A NONVERTICAL LINE

The **slope of a nonvertical line** drawn in the Cartesian plane is a measure of its tilt or inclination. Consider the line l in Figure 3-9**a**, which passes through $P(x_1, y_1)$ and $Q(x_2, y_2)$. If line RQ is perpendicular to the x-axis, and PR is perpendicular to the y-axis, then triangle PRQ is a right triangle, and point R has coordinates of (x_2, y_1). The distance from R to Q, called the **rise**, is the change in the y-coordinates of R and Q: $y_2 - y_1$. The horizontal distance from P to R, called the **run**, is the change in the x-coordinates of points P and R: $x_2 - x_1$. The slope of the line is the *rise* divided by the *run*. The letter m often represents the slope of a line:

$$\text{slope of line } PQ = m = \frac{\text{rise}}{\text{run}} = \frac{y_2 - y_1}{x_2 - x_1}$$

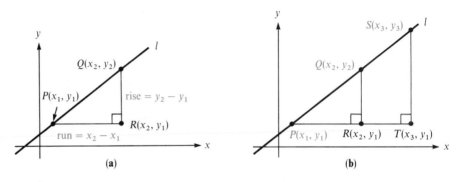

Figure 3-9

In Figure 3-9**b**, point S represents an arbitrary third point on line l. Because triangles PRQ and PTS are similar, their corresponding sides are in proportion. Thus, the ratios of the rise to the run in the two triangles are equal, and we have

$$m = \frac{y_2 - y_1}{x_2 - x_1} = \frac{y_3 - y_1}{x_3 - x_1}$$

This implies that the slope of a nonvertical line is constant and can be calculated using *any* two points on the line. Furthermore, if point P is on a line with slope m, and the ratio of rise to run of a segment PQ is also m, then point Q is on the line.

Definition.   If $P(x_1, y_1)$ and $Q(x_2, y_2)$ are two points on a nonvertical line l in the Cartesian plane, then the slope of l is given by

$$m = \frac{y_2 - y_1}{x_2 - x_1} \quad \text{provided } x_1 \neq x_2$$

If $x_1 = x_2$, then line l is a vertical line and has no defined slope.

Example 1   Find the slope of the line passing through the points $P(-1, -2)$ and $Q(7, 8)$.

Solution   Let $x_1 = -1$, $y_1 = -2$, $x_2 = 7$, and $y_2 = 8$ in the formula that defines slope:

$$m = \frac{y_2 - y_1}{x_2 - x_1} = \frac{8 - (-2)}{7 - (-1)} = \frac{10}{8} = \frac{5}{4}$$

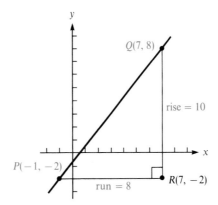

Figure 3-10

Thus, the slope of the line is $\frac{5}{4}$. See Figure 3-10.   ■

Example 2   The graph of the equation $5x + 2y = 10$ is a line. Find the slope of that line.

Solution   Determine the coordinates of two points on the line by finding two ordered pairs (x, y) that satisfy the equation $5x + 2y = 10$. If $x = 0$, for example, then $y = 5$. Thus, the point $(0, 5)$ lies on the line. If $x = -2$, then $y = 10$. Thus, the point $(-2, 10)$ also lies on the line. Let $x_1 = 0$, $y_1 = 5$, $x_2 = -2$, and $y_2 = 10$, substitute into the formula for the slope of a line, and simplify:

$$m = \frac{y_2 - y_1}{x_2 - x_1} = \frac{10 - 5}{-2 - 0} = \frac{5}{-2} = -\frac{5}{2}$$

The slope of the line is $-\frac{5}{2}$.   ■

As a point moves along a nonvertical line l from $P(x_1, y_1)$ to $Q(x_2, y_2)$, its y-coordinate changes by an amount $y_2 - y_1$, often denoted by Δy (read as "the change in y" or "delta y"). Similarly, the difference of the x-coordinates, $x_2 - x_1$, is often denoted by Δx. See Figure 3-11. Thus,

$$m = \frac{y_2 - y_1}{x_2 - x_1} = \frac{\Delta y}{\Delta x}$$

If a line rises as it moves to the right—that is, if increasing x-values result in increasing y-values—then the slope of the line is positive. See Figure 3-12**a**. If a line drops as it moves to the right—that is, if increasing x-values result in decreasing y-values—then the slope of the line is negative. See Figure 3-12**b**.

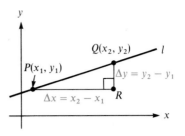

Figure 3-11

If a line is parallel to the x-axis, then the slope of the line is 0 because the difference between the y-coordinates of any two points on the line is 0. See Figure 3-12**c**. If a line is parallel to the y-axis, the slope is undefined because the denominator of the formula that defines slope—the difference of the x-coordinates—is 0. See Figure 3-12**d**.

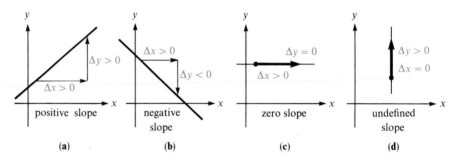

Figure 3-12

A theorem relates parallel lines to their slopes.

Theorem.   Nonvertical parallel lines have the same slope, and lines having the same slope are parallel.

Proof   Suppose that the nonvertical lines l_1 and l_2 of Figure 3-13 have slopes of m_1 and m_2, respectively, and are parallel. Then the right triangles ABC and DEF are similar, and it follows that

$$m_1 = \frac{\text{rise of } l_1}{\text{run of } l_1} = \frac{\text{rise of } l_2}{\text{run of } l_2} = m_2$$

Thus, two nonvertical parallel lines have the same slope.

In the exercises, you will be asked to prove that lines having the same slope are parallel.

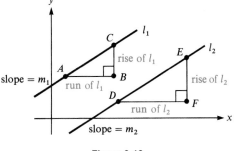

Figure 3-13

It is also true that two lines with no defined slope are vertical lines, and vertical lines are parallel.

Example 3   Find x if the line passing through $P(-2, 3)$ and $Q(3, -1)$ is parallel to the line passing through $R(4, -5)$ and $S(x, 7)$.

Solution   Because the lines PQ and RS are parallel, the slopes of PQ and RS are equal. Find the slope of each line, set them equal to each other, and solve the resulting equation:

$$\text{slope of } PQ = \text{slope of } RS$$

$$\frac{-1-3}{3-(-2)} = \frac{7-(-5)}{x-4}$$

$$\frac{-4}{5} = \frac{12}{x-4} \qquad \text{Simplify.}$$

$$-4(x-4) = 5 \cdot 12 \qquad \text{Multiply both sides by } 5(x-4).$$

$$-4x + 16 = 60 \qquad \text{Remove parentheses and simplify.}$$

$$-4x = 44 \qquad \text{Add } -16 \text{ to both sides.}$$

$$x = -11 \qquad \text{Divide both sides by } -4.$$

Thus, x is -11. The line passing through $P(-2, 3)$ and $Q(3, -1)$ is parallel to the line passing through $R(4, -5)$ and $S(-11, 7)$.   ∎

If the product of two numbers is equal to -1, the numbers are called **negative reciprocals**. The following theorem describes the relation between the slopes of perpendicular lines.

Theorem.   If two nonvertical lines are perpendicular, their slopes are negative reciprocals of each other.

If the slopes of two lines are negative reciprocals, the lines are perpendicular.

Proof   Suppose that l_1 and l_2 are lines with slopes m_1 and m_2 that intersect at the origin. Let $P(a, b)$ be a point on l_1, and let $Q(c, d)$ be a point on l_2. Neither point P nor point Q can be the origin. See Figure 3-14.

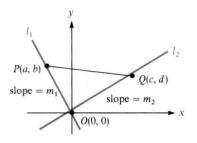

Figure 3-14

First, suppose lines l_1 and l_2 are perpendicular. Then triangle POQ is a right triangle with right angle at O, and by the Pythagorean theorem, $OP^2 + OQ^2 = PQ^2$. Proceed as follows:

$$OP^2 + OQ^2 = PQ^2$$
$$(a - 0)^2 + (b - 0)^2 + (c - 0)^2 + (d - 0)^2 = (a - c)^2 + (b - d)^2$$
$$a^2 + b^2 + c^2 + d^2 = a^2 - 2ac + c^2 + b^2 - 2bd + d^2$$
$$0 = -2ac - 2bd$$

1. $bd = -ac$

The coordinates of P are (a, b) and the coordinates of O are $(0, 0)$. Using the definition of slope, we have

$$m_1 = \frac{b - 0}{a - 0} = \frac{b}{a}$$

or

$$b = am_1$$

Similarly, we have $d = cm_2$. Substitute am_1 for b and cm_2 for d in Equation 1 and simplify:

$$bd = -ac$$
$$(am_1)(cm_2) = -ac \qquad \text{Substitute } am_1 \text{ for } b \text{ and } cm_2 \text{ for } d.$$
$$acm_1m_2 = -ac$$
$$m_1m_2 = -1 \qquad \text{Divide both sides by } ac.$$

or

$$m_1 = -\frac{1}{m_2}$$

Hence, if lines l_1 and l_2 are perpendicular, then they have slopes that are negative reciprocals of each other.

Conversely, suppose that the slopes of lines l_1 and l_2 are negative reciprocals of each other. Because the steps of the proof are reversible, we have that $OP^2 + OQ^2 = PQ^2$. By the Pythagorean theorem, triangle POQ is a right triangle. Thus, l_1 and l_2 are perpendicular. $\square$

It is also true that a line with slope of 0 is horizontal, and thus, perpendicular to a vertical line which has no defined slope.

Example 4 Two lines intersect at the point $P(-5, 3)$. One passes through the point $Q(-1, -3)$ and the other passes through $R(1, 7)$. Are the lines perpendicular?

Solution First find the slopes of lines PQ and PR:

$$\text{slope of } PQ = \frac{\Delta y}{\Delta x} = \frac{-3 - 3}{-1 - (-5)} = -\frac{6}{4} = -\frac{3}{2}$$

$$\text{slope of } PR = \frac{\Delta y}{\Delta x} = \frac{7 - 3}{1 - (-5)} = \frac{4}{6} = \frac{2}{3}$$

Because the slopes of these lines are negative reciprocals of each other, the lines are perpendicular.   ∎

Exercise 3.2

In Exercises 1–10, find the slope of the line passing through each pair of points, if possible.

1. $P(2, 5)$; $Q(3, 10)$

2. $P(3, -1)$; $Q(5, 3)$

3. $P(3, -2)$; $Q(-1, 5)$

4. $P(3, 7)$; $Q(6, 16)$

5. $P(8, -7)$; $Q(4, 1)$

6. $P(5, 17)$; $Q(17, 17)$

7. $P(-4, 3)$; $Q(-4, -3)$

8. $P(2, \sqrt{7})$; $Q(\sqrt{7}, 2)$

9. $P(a + b, c)$; $Q(b + c, a)$ assume $a \neq c$.

10. $P(b, 0)$; $Q(a + b, a)$ assume $a \neq 0$.

In Exercises 11–18, find two points on the line and determine the slope of the line.

11. $y = 3x + 2$

12. $y = 5x - 8$

13. $5x - 10y = 3$

14. $8y + 2x = 5$

15. $3(y + 2) = 2x - 3$

16. $4(x - 2) = 3y + 2$

17. $3(y + x) = 3(x - 2)$

18. $2x + 5 = 2(y + x)$

In Exercises 19–24, determine whether the lines with the given slopes are parallel, perpendicular, or neither.

19. $m_1 = 3$, $m_2 = -\dfrac{1}{3}$

20. $m_1 = \dfrac{2}{3}$, $m_2 = \dfrac{3}{2}$

21. $m_1 = \sqrt{8}$, $m_2 = 2\sqrt{2}$

22. $m_1 = 1$, $m_2 = -1$

23. $m_1 = -\sqrt{2}$, $m_2 = \dfrac{\sqrt{2}}{2}$

24. $m_1 = 2\sqrt{7}$, $m_2 = \sqrt{28}$

In Exercises 25–30, determine if the line through the given points and the line through R(−3, 5) and S(2, 7) are parallel, perpendicular, or neither.

25. *P*(2, 4); *Q*(7, 6)

26. *P*(−3, 8); *Q*(−13, 4)

27. *P*(−4, 6); *Q*(−2, 1)

28. *P*(0, −9); *Q*(4, 1)

29. *P*(*a*, *a*); *Q*(3*a*, 6*a*) and *a* ≠ 0

30. *P*(*b*, *b*); *Q*(−*b*, 6*b*) and *b* ≠ 0

In Exercises 31–32, find the slopes of lines PQ and PR, and determine if points P, Q, and R lie on the same line.

31. *P*(−2, 8); *Q*(−6, 9); *R*(2, 5)

32. *P*(1, −1); *Q*(3, −2); *R*(−3, 0)

In Exercises 33–36, determine which, if any, of the three lines PQ, PR, and QR are perpendicular.

33. *P*(5, 4); *Q*(2, −5); *R*(8, −3)

34. *P*(8, −2); *Q*(4, 6); *R*(6, 7)

35. *P*(0, 0); *Q*(*a*, *b*); *R*(−*b*, *a*)

36. *P*(1, 3); *Q*(1, 9); *R*(7, 3)

37. Show that the three points *A*(−1, −1), *B*(−3, 4), and *C*(4, 1) are the vertices of a right triangle.

38. Show that the three points *D*(0, 1), *E*(−1, 3), and *F*(3, 5) are the vertices of a right triangle.

39. Show that the four points *A*(1, −1), *B*(3, 0), *C*(2, 2), and *D*(0, 1) are the vertices of a square.

40. Show that the four points *E*(−1, −1), *F*(3, 0), *G*(2, 4), and *H*(−2, 3) are the vertices of a square.

41. Show that the four points *A*(−2, −2), *B*(3, 3), *C*(2, 6), and *D*(−3, 1) are the vertices of a parallelogram. (Show that both pairs of opposite sides are parallel.)

42. Show that the four points *E*(1, −2), *F*(5, 1), *G*(3, 4), and *H*(−3, 4) are the vertices of a trapezoid. (Show that only one pair of opposite sides are parallel.)

43. In Illustration 1, points *M* and *N* are midpoints of *CB* and *BA*, respectively. Show that *MN* is parallel to *AC*.

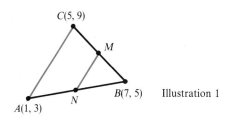

Illustration 1

44. In Illustration 2, *d*(*AB*) = *d*(*AC*). Show that *AD* is perpendicular to *BC*.

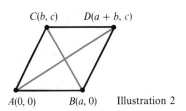

Illustration 2

45. Prove that if two lines have the same slope, they are parallel.

3.3 EQUATIONS OF LINES

Any two points on a nonvertical line can be used to determine the slope of the line. Suppose that the nonvertical line l of Figure 3-15 has a slope of m and passes through the point $P(x_1, y_1)$. If $Q(x, y)$ is another point on that line, then by the definition of *slope*, we have

$$m = \frac{y - y_1}{x - x_1}$$

or

$$y - y_1 = m(x - x_1)$$

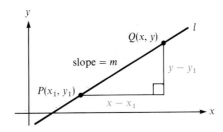

Figure 3-15

This suggests the following theorem.

> **Theorem.** The equation of the line passing through $P(x_1, y_1)$ with slope m is
>
> $$y - y_1 = m(x - x_1)$$

Proof The proof that the equation $y - y_1 = m(x - x_1)$ is the equation of the line l has two parts. We must show

1. that every point on the line has coordinates that satisfy the equation, and
2. that every point whose coordinates satisfy the equation lies on the line.

Part 1. Let $Q(a, b)$ be any point on the line except point $P(x_1, y_1)$. Substitute a for x and b for y in the equation $y - y_1 = m(x - x_1)$ and solve for m:

$$y - y_1 = m(x - x_1)$$
$$b - y_1 = m(a - x_1) \qquad \text{Substitute } a \text{ for } x \text{ and } b \text{ for } y.$$
$$m = \frac{b - y_1}{a - x_1} \qquad \text{Divide both sides by } a - x_1.$$

Because the right side of this equation represents the slope, m, of the line, we have the identity

$$m = m$$

Thus, the coordinates of the point Q satisfy the equation of the line. The coordinates of point $P(x_1, y_1)$ itself satisfy the equation $y - y_1 = m(x - x_1)$, because $y_1 - y_1 = m(x_1 - x_1)$ reduces to $0 = 0$. Thus, *any point on line l has coordinates that satisfy the equation.*

Part 2. Suppose that the coordinates of the point $R(a, b)$ satisfy the equation $y - y_1 = m(x - x_1)$, and that point R is not point P. Then $b - y_1 = m(a - x_1)$, and

$$m = \frac{b - y_1}{a - x_1}$$

Thus, the slope of the line RP is m. Because there is only *one* line with slope m passing through P, and that is line l, point R must lie on line l. Thus, *any point with coordinates that satisfy the equation* $y - y_1 = m(x - x_1)$ *lies on the line l.*

The theorem is proved.   □

Point-Slope Form

Because the equation $y - y_1 = m(x - x_1)$ displays the coordinates of a fixed point on the line and the line's slope, it is called the **point-slope form** of the equation of a line.

The Point-Slope Form of the Equation of a Line.   The equation of the line passing through the point $P(x_1, y_1)$ and having a slope of m is

$$y - y_1 = m(x - x_1)$$

Example 1   Find the equation of the line passing through the point $P(3, -1)$ with a slope of $-\frac{5}{3}$. Then solve that equation for y.

Solution   Substitute 3 for x_1, -1 for y_1, and $-\frac{5}{3}$ for m in the point-slope form of the equation of a line:

$$y - y_1 = m(x - x_1)$$

$$y - (-1) = -\frac{5}{3}(x - 3) \qquad \text{Substitute 3 for } x_1, -1 \text{ for } y_1, \text{ and } -\frac{5}{3} \text{ for } m.$$

$$y + 1 = -\frac{5}{3}x + 5 \qquad \text{Remove parentheses.}$$

$$y = -\frac{5}{3}x + 4 \qquad \text{Add } -1 \text{ to both sides.}$$

■

We can use the concept of slope to help graph the equation obtained in Example 1. We first plot the point $P(3, -1)$ as in Figure 3-16. Because the slope is $-\frac{5}{3}$, $\Delta y = -5$ provided that $\Delta x = 3$. Thus, we can locate another point Q on the line by moving 3 units to the right of P and then 5 units down. The graph of the equation is the line PQ.

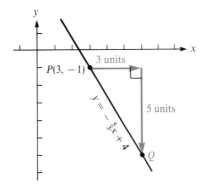

Figure 3-16

Example 2   Find the equation of the line passing through the points $P(3, 7)$ and $Q(-5, 3)$. Solve the equation for y.

Solution   Let $P(x_1, y_1)$ be the point $P(3, 7)$ and $Q(x_2, y_2)$ be the point $Q(-5, 3)$. Then find the slope of the line:

$$m = \frac{y_2 - y_1}{x_2 - x_1} = \frac{3 - 7}{-5 - 3} = \frac{-4}{-8} = \frac{1}{2}$$

Finally, substitute the coordinates of $P(3, 7)$ and $\frac{1}{2}$ for m in the point-slope form of the equation of a line and solve for y:

$$y - y_1 = m(x - x_1)$$
$$y - 7 = \frac{1}{2}(x - 3)$$
$$y = \frac{1}{2}x - \frac{3}{2} + 7$$
$$y = \frac{1}{2}x + \frac{11}{2}$$

■

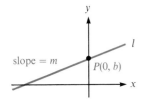

Figure 3-17

Slope-Intercept Form

If the y-intercept of the line with slope m shown in Figure 3-17 is b, the line intersects the y-axis at $P(0, b)$.

We can write the equation of this line by substituting 0 for x_1, and b for y_1

into the point-slope form of the equation of the line and simplifying:

$$y - y_1 = m(x - x_1)$$

$$y - b = m(x - 0) \qquad \text{Substitute 0 for } x_1 \text{ and } b \text{ for } y_1.$$

$$y = mx + b \qquad \text{Simplify and add } b \text{ to both sides.}$$

Because the equation $y = mx + b$ displays both the slope and the y-intercept of a line, it is called the **slope-intercept form** of the equation of a line.

The Slope-Intercept Form of the Equation of a Line.   The equation of the line with slope m and y-intercept b is

$$y = mx + b$$

Example 3   Use the slope-intercept form to find the equation of the line with slope of $\frac{7}{3}$ and y-intercept of -6.

Solution   Substitute $\frac{7}{3}$ for m and -6 for b into the slope-intercept form of the equation of the line and simplify:

$$y = mx + b$$

$$y = \frac{7}{3}x + (-6) \qquad \text{Substitute } \tfrac{7}{3} \text{ for } m \text{ and } -6 \text{ for } b.$$

$$y = \frac{7}{3}x - 6 \qquad \text{Simplify.}$$

■

Example 4   Find the slope and the y-intercept of the line $3(y + 2) = 6x - 1$.

Solution   Write the equation in the form $y = mx + b$ to determine the slope m and the y-intercept b.

$$3(y + 2) = 6x - 1$$

$$3y + 6 = 6x - 1 \qquad \text{Remove parentheses.}$$

$$3y = 6x - 7 \qquad \text{Add } -6 \text{ to both sides.}$$

$$y = 2x - \frac{7}{3} \qquad \text{Divide both sides by 3.}$$

The slope of the line is 2, and the y-intercept is $-\frac{7}{3}$.

■

Example 5   Find the y-intercept of the line having a slope of 2 and passing through the point $P(3, -5)$.

Solution   Because the line passes through the point $P(3, -5)$, the coordinates $x = 3$ and $y = -5$ must satisfy the equation. To find the y-intercept, substitute 3 for x, -5 for y, and 2 for m in the slope-intercept form and solve for b.

$$y = mx + b$$
$$-5 = 2(3) + b$$
$$-5 = 6 + b$$
$$-11 = b$$

The y-intercept is -11.   ∎

Example 6   Find the equation of a line l passing through the point $P(-2, 9)$ and parallel to the line with equation $y = 4x - 73$.

Solution   The slope of the line with equation $y = 4x - 73$ is 4. Because line l is parallel to the given line, it also has a slope of 4. Because line l passes through the point $P(-2, 9)$ and has a slope of 4, substitute -2 for x, 9 for y, and 4 for m into the slope-intercept form of the equation of the line and simplify:

$$y = mx + b$$
$$9 = 4(-2) + b$$
$$9 = -8 + b$$
$$17 = b$$

The equation of line l is $y = 4x + 17$.   ∎

Example 7   Use the point-slope form to find the equation of the line perpendicular to the line $y = \frac{1}{3}x + 7$ and passing through the point $(2, 3)$. Write the result in slope-intercept form.

Solution   The slope of the given line $y = \frac{1}{3}x + 7$ is $\frac{1}{3}$. The slope of the required line must be -3, the negative reciprocal of $\frac{1}{3}$. Use the point-slope form to find the equation of a line with slope of -3 and passing through the point $(2, 3)$ as follows:

$$y - y_1 = m(x - x_1)$$
$$y - 3 = -3(x - 2)$$
$$y - 3 = -3x + 6$$
$$y = -3x + 9$$ ∎

If a line is parallel to the x-axis, its slope m is 0. We can find its equation by substituting 0 for m in the equation $y = mx + b$ and simplifying:

$$y = mx + b$$
$$y = 0x + b$$
$$y = b$$

The equation of a line parallel to the x-axis is $y = b$, the y-intercept is b, and the y-coordinate of *every* point of the line is b. See Figure 3-18.

If a line is parallel to the y-axis, it has no defined slope. Thus, it cannot be the graph of any equation of the form $y = mx + b$. Such a line does have an equation, however. If a vertical line passes through a point with an x-coordinate of a, then the x-coordinate of *every* point on the line is a. The equation of that line is $x = a$. See Figure 3-18.

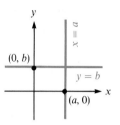

Figure 3-18

Example 8   Find the equation of the line passing through the points $P(-3, 4)$ and $Q(-3, -2)$.

Solution   Because the x-coordinates of points P and Q are equal, the line has no defined slope. It is a vertical line, with equation $x = -3$. See Figure 3-19.

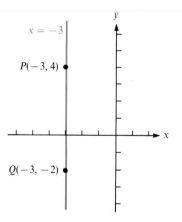

Figure 3-19

General Form

We have shown that the graph of the equation

$$y - y_1 = m(x - x_1)$$

is a line. In this equation, x_1 and y_1 are the coordinates of some fixed point, m is the slope, and x and y are variables. We can rewrite this equation as follows:

$$y - y_1 = m(x - x_1)$$

$$y - y_1 = mx - mx_1 \qquad \text{Remove parentheses.}$$

$$-mx + y = y_1 - mx_1 \qquad \text{Add } y_1 - mx \text{ to both sides.}$$

This final equation is of the form $Ax + By = C$, where A, B, and C are constants: $A = -m$, $B = 1$, and $C = y_1 - mx_1$. In general, for any real numbers A, B, and C (where A and B are not *both* 0), the equation $Ax + By = C$ represents a line and is called the **general form** of the equation of a line. Any equation that can be written in general form is called a **linear equation in x and y.**

The General Form of the Equation of a Line. If A, B, and C are real numbers and B is not 0, then the graph of the equation

$$Ax + By = C$$

is a nonvertical line with slope of $-\dfrac{A}{B}$ and y-intercept of $\dfrac{C}{B}$.

If $B = 0$, then the equation $Ax + By = C$ represents a vertical line with x-intercept of $\frac{C}{A}$. In the exercises, you will be asked to prove this fact and others related to the general form of the equation of a line.

Example 9 Show that the two lines $3x - 2y = 5$ and $-6x + 4y = 7$ are parallel.

Solution To show that two lines are parallel, show that they have the same slope. The first equation, $3x - 2y = 5$, is in general form, with $A = 3$, $B = -2$, and $C = 5$. By the previous theorem, the slope of the line is

$$m_1 = -\frac{A}{B} = -\frac{3}{-2} = \frac{3}{2}$$

Similarly, the second equation, $-6x + 4y = 7$, is in general form, with $A = -6$, $B = 4$, and $C = 7$. The slope of this line is

$$m_2 = -\frac{A}{B} = -\frac{-6}{4} = \frac{3}{2}$$

Because the slopes of the two lines are equal, the lines are parallel. ■

Exercise 3.3

In Exercises 1–6, use the point-slope form to write the equation of the line passing through the given point and having the given slope. Express the answer in slope-intercept form.

1. $P(2, 4)$; $m = 2$

2. $P(3, 5)$; $m = -3$

3. $P\left(-\dfrac{3}{2}, \dfrac{1}{2}\right)$; $m = 2$

4. $P\left(\dfrac{1}{4}, -2\right)$; $m = -6$

5. $P(5, -5)$; $m = -5$

6. $P(0, -8)$; $m = -3$

In Exercises 7–12, use the slope-intercept form to write the equation of the line with the given slope and y-intercept.

7. $m = 3, b = -2$　　　　**8.** $m = -\dfrac{1}{3}, b = \dfrac{2}{3}$　　　　**9.** $m = \sqrt{2}, b = \sqrt{2}$　　　　**10.** $m = \pi, b = \dfrac{1}{\pi}$

11. $m = a, b = \dfrac{1}{a}$　　　　　　　　　　　**12.** $m = a, b = 2a$

In Exercises 13–18, use the slope-intercept form to write the equation of the line passing through the given point and having the given slope. Express the answer in general form.

13. $P(0, 0); m = \dfrac{3}{2}$　　　　　　　　　**14.** $P(-3, -7); m = -\dfrac{2}{3}$

15. $P(-3, 5); m = -3$　　　　　　　　　　**16.** $P(-5, 9); m = 1$
17. $P(0, \sqrt{2}); m = \sqrt{2}$　　　　　　　　　**18.** $P(-\sqrt{3}, 0); m = 2\sqrt{3}$

In Exercises 19–24, write the equation of the line passing through the two given points. Express the answer in general form.

19. $P(3, 2); Q(2, 3)$　　　　　　　　　　**20.** $P(0, 5); Q(-5, 0)$
21. $P(3, -2); Q(4, -8)$　　　　　　　　**22.** $P(-7, 5); Q(-5, -7)$
23. $P(0, 0); Q(-9, 13)$　　　　　　　　　**24.** $P(-9, -5); Q(5, -9)$

In Exercises 25–28, find, if possible, the slope and the y-intercept of each line.

25. $3(14x + 12) = 5(y + 3)$　　　　　　**26.** $-2(3x + 6) = -2y + 15$
27. $2(y - 3) + x = 2y - 7$　　　　　　　**28.** $3(y + 2) + 2x = 2(y + x)$

In Exercises 29–34, find the slope and the y-intercept of the line with the given properties.

29. passes through $P(-3, 5)$ and $Q(7, -5)$
30. passes through $P(15, 3)$ and $Q(-3, -6)$
31. is parallel to line $y = 3(x - 7)$ and passes through $P(-2, 5)$
32. is parallel to line $3y + 1 = 6(x - 2)$ and passes through $P(5, 0)$
33. is perpendicular to line $y + 3x = 8$ and passes through $P(2, 3)$
34. is perpendicular to line $x = 2y - 7$ and passes through $P(0, 8)$

In Exercises 35–51, write the equation of the line with the given properties. Write the answer in general form.

35. passes through $(3, 4)$ and has a slope of $\frac{1}{2}$　　　**36.** passes through $(-2, 6)$ and has a slope of $\frac{3}{5}$
37. passes through $(-2, 4)$ and $(5, -4)$　　　**38.** passes through $(3, 5)$ and $(-5, 3)$
39. passes through the origin and $(-2, 11)$　　　**40.** passes through the origin and $(5, -9)$
41. y-intercept of 7 and passes through $(-3, 4)$　　　**42.** y-intercept of -3 and an x-intercept of 4
43. y-intercept of 3 and an x-intercept of -4　　　**44.** y-intercept of -2 and passes through $(4, -3)$
45. slope of $-\frac{2}{3}$ and a y-intercept of 10　　　**46.** slope of $-\frac{1}{2}$ and an x-intercept of 10
47. passes through $(3, -5)$ and is parallel to the line with equation $y = 5x - 3$
48. has an x-intercept of -4 and is parallel to the line with an equation of $y = 4x$
49. is perpendicular to the line with equation $y = 3x - 17$ and passes through $(0, -5)$

50. is perpendicular to the line with equation $y = -\frac{1}{3}x + 5$ and has the same y-intercept as the line $3x - 2y = 6$

51. is parallel to the line $3y - 5x = 0$ and has the same x-intercept as the line $5x - 2y = 15$

In Exercises 52–55, write the equation of the line with the given properties. Write the answer in slope-intercept form.

52. passing through the origin and is parallel to the line $3x + 2y = 6$

53. passing through $(2, 8)$ and is parallel to the line $2x - 3y = 12$

54. passing through $(-2, 3)$ and is perpendicular to the line $3x + 5y = 25$

55. passing through the origin and is perpendicular to the line $4x - y = 12$

56. Prove that an equation of the line with x-intercept of a and y-intercept of b may be written in the form

$$\frac{x}{a} + \frac{y}{b} = 1$$

57. Prove that, if $B \neq 0$, the graph $Ax + By = C$ has a slope of $-A/B$ and a y-intercept of C/B.

58. Prove that, if $B = 0$, the graph of $Ax + By = C$ is a vertical line with x-intercept of C/A.

59. Show that the lines $Ax + By = C$ and $Bx - Ay = C$ are perpendicular.

60. Show that the graphs of $Ax + By = C$ and $kAx + kBy = D$, where $k \neq 0$, are parallel.

61. Use the distance formula to find the equation of the perpendicular bisector of the line segment joining $A(3, 5)$ and $B(5, -3)$. (*Hint:* Let $P(x, y)$ be a point on the perpendicular bisector. Then $d(PA) = d(PB)$.)

62. Use the point-slope form to find the equation of the perpendicular bisector of the line segment joining $A(3, 5)$ and $B(5, -3)$. (*Hint:* First find the midpoint of AB.)

3.4 FUNCTIONS AND FUNCTION NOTATION

Many equations describe a correspondence between two variables. The equation $y = 3x + 2$, for example, describes a correspondence in which each number x determines one value of y. Such correspondences are called **functions**.

> **Definition.** A **function** is a correspondence that assigns to *each* element x of some set X a *single* value y of a set Y. The set X is called the **domain** of the function. The value y that corresponds to a particular x in the domain is called the **image** of x under the function. The collection of all images of elements in the domain is called the **range** of the function.

Because the value of y depends on the number x that is chosen, y is called the **dependent variable**. The variable x is called the **independent variable**. Unless indicated otherwise, the domain of a function is its **implied domain**—the set of all real numbers x for which the function is defined.

Example 1   Does the equation $y = 5 - 7x$ define y to be a function of x? If so, find its domain and range.

Solution   To determine y in the equation $y = 5 - 7x$, the product of a number x and 7 is subtracted from 5. Because multiplications and subtractions give unique results, a single value of y is produced. Thus, the equation does determine a function.

Because x can represent any real number, the implied domain of the function is the set of real numbers.

Because y can be any real number, the range of the function is the set of real numbers.   ∎

Example 2   Does the equation $y = \dfrac{3}{x - 2}$ define y to be a function of x? If so, find its domain and range.

Solution   Because each number x except 2 does determine exactly one value of y, the equation does define a function.

Because division by zero is not defined, the denominator of the fraction cannot be 0. Thus, x cannot be 2. The domain of the function is the set of all real numbers except 2.

Because the fraction has a nonzero numerator, y cannot be 0. Thus, the range of the function is the set of all real numbers except 0.   ∎

Example 3   Find the domain and the range of the function defined by $y = \sqrt{1 - x^2}$.

Solution   Because y is to be a real value, the radicand $1 - x^2$ must be a nonnegative number. Thus, x^2 cannot be greater than 1. So x itself must be between -1 and 1, inclusive. The domain of the function is the set

$$\{x \mid -1 \le x \le 1\}$$

Because the radicand $1 - x^2$ cannot be greater than 1, the range of the function is the set

$$\{y \mid 0 \le y \le 1\}$$ ∎

To indicate that y depends on x, we can use **function notation** and write $y = f(x)$. Strictly speaking, y is the image of x under the function f, and the correspondence itself is the function. Yet, it is common to read "$y = f(x)$" as "y is a function of x." Functions are often denoted by letters other than f.

Function notation provides a way of indicating the image of a particular number x. If $y = f(x) = 3x + 2$ defines a function f, for example, then the symbol $f(2)$ indicates the image of 2 under f:

$$f(x) = 3x + 2$$
$$f(2) = 3(2) + 2 \qquad \text{Substitute 2 for } x.$$
$$= 8$$

Thus, if $x = 2$, then $y = f(2) = 8$. Similarly, the image of -5 is $f(-5)$, or -13, because

$$f(x) = 3x + 2$$
$$f(-5) = 3(-5) + 2 \qquad \text{Substitute } -5 \text{ for } x.$$
$$= -13$$

Example 4   Let $g(x) = 3x^2 + x - 4$. Find **a.** $g(-3)$, **b.** $g(k)$, and **c.** $g(k + 1)$.

Solution   **a.** $g(x) = 3x^2 + x - 4$

$$g(-3) = 3(-3)^2 + (-3) - 4 \qquad \text{Substitute } -3 \text{ for } x.$$
$$= 3(9) - 3 - 4$$
$$= 20$$

b. $g(x) = 3x^2 + x - 4$

$$g(k) = 3k^2 + k - 4 \qquad \text{Substitute } k \text{ for } x.$$

c. $g(x) = 3x^2 + x - 4$

$$g(k + 1) = 3(k + 1)^2 + (k + 1) - 4 \qquad \text{Substitute } k + 1 \text{ for } x.$$
$$= 3(k^2 + 2k + 1) + k + 1 - 4$$
$$= 3k^2 + 6k + 3 + k + 1 - 4$$
$$= 3k^2 + 7k \qquad \text{Combine terms.} \qquad \blacksquare$$

By definition, a function f is a correspondence from the elements of one set X to the elements of another set Y. We can visualize this correspondence with the diagram in Figure 3-20. A function f that assigns the element $y = f(x)$ to the element x can be represented by an arrow leaving x and pointing to y. The set of all those elements in X from which an arrow originates is the domain of the function. The set of all the images in Y to which arrows point is the range of the function.

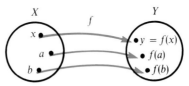

Figure 3-20

A correspondence is still a function if arrows point from several different elements of X to the same element y in the range. See Figure 3-21.

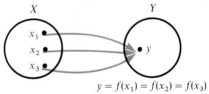

$$y = f(x_1) = f(x_2) = f(x_3)$$

Figure 3-21

However, if arrows leave some element x in the domain and point to several elements in the range as in Figure 3-22, the correspondence is *not* a function because the assignment of y to x is not unique.

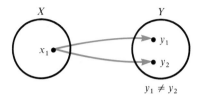

Figure 3-22

Such a correspondence is called a **relation**.

Definition.   A **relation** is a correspondence that assigns to each element x in a set X one or more values of y in a set Y. The set X is called the **domain** of the relation. The **range** of the relation is the set of all y values corresponding to numbers x in the domain.

Note that a function is a relation, but a relation is not necessarily a function.

A *relation* from a subset of the real numbers to a subset of the real numbers determines a set of ordered pairs (x, y), where x is an element in the domain of the relation and y is a corresponding value in the range. The **graph of the relation** is the graph of all these ordered pairs on a rectangular coordinate system. If the relation is a *function* defined by $y = f(x)$, then the **graph of the function** consists of those points and only those points in the xy-plane with coordinates $(x, y) = (x, f(x))$.

There is a test, called the **vertical line test**, that can be used to determine whether a graph of ordered pairs (x, y) represents a function. If each vertical line that intersects the graph does so exactly once, then each number x determines exactly one value of y, and the graph represents a function. See Figure 3-23**a**. If any vertical line intersects the graph more than once, then to some

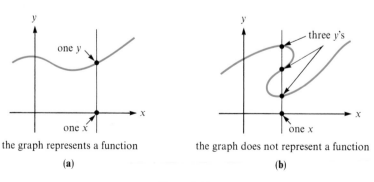

the graph represents a function

(a)

the graph does not represent a function

(b)

Figure 3-23

number x there corresponds more than one value y, and the graph does *not* represent a function. See Figure 3-23**b**.

Example 5 Graph the points (x, y) for which $x = y^2$. Is this correspondence a function?

Solution Several ordered pairs (x, y) that satisfy the equation are listed in the table of values in Figure 3-24. Plot those points and draw the graph as shown in the figure. Because to some numbers x there correspond two y-values, the graph does not pass the vertical line test. Thus, the correspondence is *not* a function. However, it is a relation.

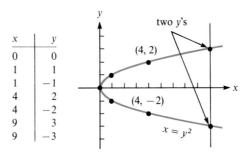

x	y
0	0
1	1
1	-1
4	2
4	-2
9	3
9	-3

Figure 3-24

Example 6 Graph the points (x, y) for which $y = 3x - 5$. Is this correspondence a function?

Solution The equation $y = 3x - 5$ is the equation of a line written in slope-intercept form, $y = mx + b$. The slope of the line is 3, and the y-intercept is -5. The graph of the line is shown in Figure 3-25. Because the graph passes the vertical line test, the correspondence is a function.

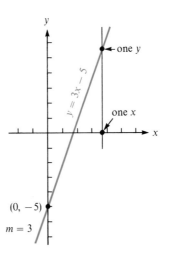

Figure 3-25

Any equation, such as that in Example 6, that can be written in the form $y = mx + b$ determines a function called a **linear function**.

Definition.   A **linear function** is a function determined by an equation of the form

$$f(x) = mx + b$$

Example 7   The cost C of electricity is a linear function of x, the number of kilowatt-hours used. If the cost of 100 kilowatt-hours is \$17 and the cost of 500 kilowatt-hours is \$57, what formula expresses the function?

Solution   Because C is a linear function of x, there are constants m and b such that

$$C = mx + b$$

Because $C = 17$ when $x = 100$, the point $(x_1, C_1) = (100, 17)$ lies on the straight-line graph of $C = mx + b$. Similarly, $C = 57$ when $x = 500$. Thus, the point $(x_2, C_2) = (500, 57)$ also lies on the line. The slope of the line is

$$m = \frac{C_2 - C_1}{x_2 - x_1}$$

$$= \frac{57 - 17}{500 - 100} \qquad \text{Substitute 57 for } C_2, \text{ 17 for } C_1, \text{ 500 for } x_2,$$
$$\text{and 100 for } x_1.$$

$$= \frac{40}{400}$$

$$= 0.10$$

Thus, $m = 0.10$. To find b, substitute **17** for C, **100** for x, and **0.10** for m into the equation $C = mx + b$, and solve for b:

$$C = mx + b$$
$$17 = 0.10(100) + b$$
$$17 = 10 + b$$
$$7 = b$$

Thus, $C = 0.10x + 7$. The electric company charges a flat rate of \$7.00, plus 10¢ per kilowatt-hour used.   ∎

Example 8   Express the area of a square as a function of the perimeter.

Solution   The area A of a square is a function of the length s of a side. This can be expressed as

1. $A = f(s) = s^2$

The perimeter P is a function of the length s of a side:

2. $P = g(s) = 4s$

To express the area A as a function of the perimeter P, you must determine a function h so that $A = h(P)$. To do so, solve Equation 2 for s and substitute into Equation 1.

$$P = 4s$$

$$s = \frac{P}{4} \qquad \text{Divide both sides by 4.}$$

$$A = \left(\frac{P}{4}\right)^2 \qquad \text{Substitute } \tfrac{P}{4} \text{ for } s \text{ in Equation 1.}$$

$$= \frac{P^2}{16}$$

Thus,

$$A = h(P) = \frac{P^2}{16}$$

In the context of this example, the domain of h is not its implied domain of the set of real numbers. Because the perimeter of a square is a positive number, the domain of h is the set of positive real numbers. ∎

Exercise 3.4

In Exercises 1–10, indicate whether the equation determines y as a function of x. Assume that all variables represent real numbers.

1. $y = x$ **2.** $y - 2x = 0$ **3.** $y^2 = x$ **4.** $|y| = x$

5. $y = x^2 + 4$ **6.** $y - 7 = 7$ **7.** $y^2 - 4x = 1$ **8.** $|x - 2| = y$

9. $|x| = |y|$ **10.** $x = 7$

In Exercises 11–18, let function f be defined by the equation $y = f(x)$, where x and y are real numbers. Find the domain and range of each function. Then find $f(3)$.

11. $y = 3x + 5$ **12.** $y = -5x + 2$ **13.** $y = x^2$ **14.** $y = x^2 + 3$

15. $y = \dfrac{3}{x + 1}$ **16.** $y = \dfrac{-7}{x + 3}$ **17.** $y = \sqrt{16 - x^2}$ **18.** $y = \sqrt{x^2 - 9}$

In Exercises 19–26, let function f be defined by the equation $y = f(x)$, where x and y are real numbers. Find the domain of each function, and find $f(-2)$, $f(0)$, and $f(a)$, if possible.

19. $y = \dfrac{2}{x^2 - 2x}$ **20.** $y = \dfrac{3}{x - 2}$ **21.** $y = \sqrt{x^2 - 1}$ **22.** $y = \dfrac{x + 2}{x}$

23. $y = \dfrac{x}{x + 3}$ **24.** $y = \sqrt{25 - x^2}$ **25.** $y = \dfrac{x}{\sqrt{x^2 + 32}}$ **26.** $y = \dfrac{1}{\sqrt{49 - x^2}}$

In Exercises 27–32, some correspondences are illustrated. If the correspondence illustrated could represent y as a function of x, so indicate. If the illustration could not represent such a function, explain why.

27.

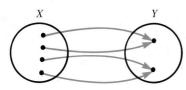

28.

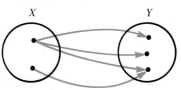

29.

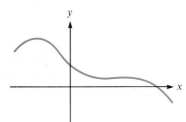

30.

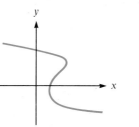

31.

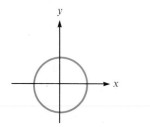

32.

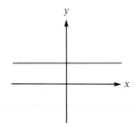

In Exercises 33–54, graph each indicated correspondence. Use the vertical line test to decide if the correspondence defines y as a function of x.

33. $y = 2x + 3$ **34.** $y = 3x + 2$ **35.** $2x = 3y - 3$ **36.** $3x = 2(y + 1)$

37. $y = -x^2$ **38.** $y = 1 - x^2$ **39.** $y = \sqrt{x}$ **40.** $y = -|x|$

41. $y = x^2 + x$ **42.** $|y| = x$

43. The number y is the greatest integer less than or equal to x.

44. The number y is the least integer greater than or equal to x.

45. $x = y^3$ **46.** $x = y^2 + 3$ **47.** $y = |x| + x$ **48.** $|y| = 1 + x$

49. $y = \dfrac{1}{x}$ **50.** $y = \dfrac{1}{x^2}$ **51.** $|x| = 3y + 2$ **52.** $|x| = 7y - 5$

53. $|x| = 3y^2$ **54.** $|x| = 5 + y^2$

55. The Fahrenheit temperature reading F is a linear function of the Celsius reading C. If $C = 0$ when $F = 32$, and $C = 100$ when $F = 212$, express F as a function of C.

56. The velocity v of a falling object is a linear function of the time t it has been falling. If $v = 15$ when $t = 0$, and $v = 79$ when $t = 2$, express v as a function of t.

57. The cost C of water is a linear function of n, the number of gallons used. If 1000 gallons cost $4.70, and 9000 gallons cost $14.30, what equation describes this function?

58. The amount A of money on deposit for t years in an account earning simple interest is a linear function of t. Express that function as an equation if $A = \$272$ when $t = 3$ and $A = \$320$ when $t = 5$.

59. Express the radius of a circle as a function of the circumference.

60. Express the radius of a circle as a function of the area.

61. Express the perimeter of a square as a function of the area.

62. Express the area of a circle as a function of the circumference.

63. Express the volume of a cube as a function of the area of one face.

64. Express the length of the diagonal of a square as a function of the length of an edge.

3.5 POLYNOMIAL FUNCTIONS

A linear function defined by the equation $y = mx + b$ is a polynomial function because its right side is a polynomial in the variable x. In this section, we will discuss functions involving polynomials of higher degree.

> **Definition.** A **polynomial function** in one independent variable, say x, is defined by an equation of the form $y = P(x)$, where $P(x)$ is a polynomial in the variable x.
>   The **degree of the polynomial function** $y = P(x)$ is the degree of $P(x)$.

A polynomial function of second degree is called a **quadratic function**.

> **Definition.** A **quadratic function** is a second-degree polynomial function in one variable. It is defined by an equation of the form $y = ax^2 + bx + c$, where a, b, and c are constants and $a \neq 0$.

Example 1 Graph the quadratic function defined by the equation $y = x^2 - 2x - 3$.

Solution Plot several points whose coordinates satisfy this equation. Connect them with a smooth curve as shown in Figure 3-26.

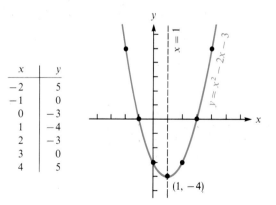

x	y
-2	5
-1	0
0	-3
1	-4
2	-3
3	0
4	5

Figure 3-26

Example 2   Graph the quadratic function defined by $y = -\dfrac{1}{3}x^2 + 3$.

Solution   Plot several points whose coordinates satisfy the equation. Connect them with a smooth curve as shown in Figure 3-27.

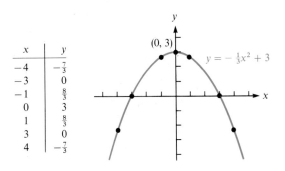

x	y
-4	$-\frac{7}{3}$
-3	0
-1	$\frac{8}{3}$
0	3
1	$\frac{8}{3}$
3	0
4	$-\frac{7}{3}$

$y = -\frac{1}{3}x^2 + 3$

Figure 3-27

The graphs of quadratic functions are called **parabolas**. They open upward when $a > 0$ (as in Example 1) and downward when $a < 0$ (as in Example 2). The "bottom" point of a parabola that opens upward (or the "top" point of a parabola that opens downward) is called the **vertex** of the parabola. The vertex of the parabola in Example 1 is the point $(1, -4)$; the vertex of the parabola in Example 2 is the point $(0, 3)$. The vertical line passing through the vertex is called the **axis of symmetry** because it divides the parabola into two congruent halves. The axis of symmetry in Example 1 is the line $x = 1$, and in Example 2, it is the line $x = 0$ (the y-axis).

If a, h, and k are constants and $a \neq 0$, then the equation

$$y - k = a(x - h)^2$$

also defines a quadratic function. This is because it takes the form of $y = ax^2 + bx + c$ when the right side is expanded and the equation is solved for y. The graph of the equation $y - k = a(x - h)^2$ is a parabola opening upward if $a > 0$ and downward if $a < 0$. This special form of the equation of the parabola displays the coordinates of the vertex of its parabolic graph, as the following discussion will show.

Suppose that $a > 0$ so that the graph of

$$y - k = a(x - h)^2$$

is a parabola opening upward. The vertex of this parabola is that point on the graph that has the least possible y-coordinate. Because a is positive, the smallest value attainable by the right side is 0; this occurs when $x = h$. When $x = h$, the left side of the equation must be 0 also, so the smallest possible value of y is $y = k$. The parabola's vertex, therefore, is the point (h, k). A similar argument holds if $a < 0$.

The previous discussion leads to the following theorem.

> **Theorem.** The graph of the equation
>
> $$y - k = a(x - h)^2 \quad (a \neq 0)$$
>
> is a parabola with its vertex at the point (h, k). The parabola opens upward if $a > 0$ and downward if $a < 0$.

We can use the process of completing the square to transform an equation from the form $y = ax^2 + bx + c$ to the form $y - k = a(x - h)^2$, from which we can read the coordinates of the vertex directly.

Example 3 Find the vertex of the parabola determined by $y = 2x^2 - 5x - 3$.

Solution Complete the square on the right side of the equation:

$$y = 2x^2 - 5x - 3$$

$$y + 3 = 2x^2 - 5x \qquad \text{Add 3 to both sides.}$$

$$y + 3 = 2\left(x^2 - \frac{5}{2}x\right) \qquad \begin{array}{l}\text{Factor a 2 out of the two} \\ \text{terms on the right side.}\end{array}$$

Complete the square within the parentheses by adding the square of one-half of the coefficient of x. This number,

$$\left[\frac{1}{2}\left(-\frac{5}{2}\right)\right]^2 = \frac{25}{16}$$

is added within the parentheses. Because of the factor of 2, however, you are really adding twice $\frac{25}{16}$, or $\frac{25}{8}$, to the right side of the equation. You must also add $\frac{25}{8}$ to the left side.

$$y + 3 = 2\left(x^2 - \frac{5}{2}x \qquad \right)$$

$$y + 3 + \frac{25}{8} = 2\left(x^2 - \frac{5}{2}x + \frac{25}{16}\right) \qquad \text{Add } \tfrac{25}{8} \text{ to both sides.}$$

$$y + \frac{49}{8} = 2\left(x - \frac{5}{4}\right)^2 \qquad \text{Combine terms and factor.}$$

$$y - \left(-\frac{49}{8}\right) = 2\left(x - \frac{5}{4}\right)^2$$

The equation is now in the form

$$y - k = a(x - h)^2$$

with $h = \frac{5}{4}$ and $k = -\frac{49}{8}$. The vertex of the parabola is the point $(h, k) = (\frac{5}{4}, -\frac{49}{8})$.

∎

To find the vertex of a parabola given by an equation of the form $y = ax^2 + bx + c$, we can write the equation in the form $y - k = a(x - h)^2$ by completing the square.

$$y = ax^2 + bx + c$$

$$y = a\left(x^2 + \frac{b}{a}x\right) + c \qquad \text{Factor out } a.$$

$$y + \frac{b^2}{4a} = a\left(x^2 + \frac{b}{a}x + \frac{b^2}{4a^2}\right) + c \qquad \text{Add } \tfrac{b^2}{4a} \text{ to both sides.}$$

$$y - c + \frac{b^2}{4a} = a\left(x^2 + \frac{b}{a}x + \frac{b^2}{4a^2}\right) \qquad \text{Add } -c \text{ to both sides.}$$

$$y - \left(c - \frac{b^2}{4a}\right) = a\left(x - \frac{-b}{2a}\right)^2 \qquad \text{Factor.}$$

Thus, the vertex is the point with coordinates of $\left(-\dfrac{b}{2a}, c - \dfrac{b^2}{4a}\right)$, and we have the following theorem.

Theorem.   The vertex of the parabola determined by the quadratic function

$$y = ax^2 + bx + c$$

is the point with coordinates of $\left(-\dfrac{b}{2a}, c - \dfrac{b^2}{4a}\right)$.

Example 4   A farmer has 400 feet of fencing to enclose a rectangular corral. To save money and fencing, he intends to use one wall of his barn as one boundary of the corral. See Figure 3-28. What dimensions would enclose the largest area?

Figure 3-28

Solution   Let x represent the length of the fenced area, so that $400 - 2x$ represents the width. The area is the product of the length and width. Hence,

$$A = x(400 - 2x)$$

or

$$A = -2x^2 + 400x$$

Because the coefficient of x^2 is negative, the parabola determined by this function opens downward, and its vertex is its "highest" point. The A-coordinate of the vertex is the largest possible area, and the x-coordinate gives the length of the corral of maximum area. The equation

$$A = -2x^2 + 400x$$

defines a quadratic function with $a = -2$, $b = 400$, and $c = 0$. The vertex of the parabola is the point

$$\left(-\frac{b}{2a}, c - \frac{b^2}{4a}\right) = \left(-\frac{400}{2(-2)}, 0 - \frac{400^2}{4(-2)}\right) = (100, 20{,}000)$$

If the farmer's fence runs 100 feet out from the barn, 200 feet parallel to the barn, and 100 feet back to the barn, it will enclose the maximum area of 20,000 square feet. ■

The graph of a parabola is **symmetric** about its axis, making half of the parabola a mirror image of the other half. We now consider three other possible symmetries of graphs.

If the point $(-x, y)$ lies on a graph whenever the point (x, y) does, then the graph is **symmetric about the y-axis**. See Figure 3-29.

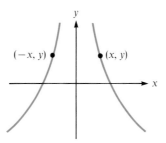

Figure 3-29

It is easy to check for y-axis symmetry of the graph of a function defined by $y = f(x)$. If $y = f(-x)$ is equivalent to $y = f(x)$—that is, if $y = f(x)$ remains unchanged when x is replaced with $-x$—then the graph is symmetric about the y-axis. A function with a graph of this type is called an **even function**.

Definition. A function f with the property that $f(-x) = f(x)$ for all x is called an **even function**.

A graph is **symmetric about the x-axis** if the point $(x, -y)$ lies on the graph whenever the point (x, y) does. See Figure 3-30. Except for the graph of the function $f(x) = 0$, such a graph does not pass the vertical line test and does not represent a function.

Finally, a graph is symmetric about the origin if the point $(-x, -y)$ lies on the graph whenever the point (x, y) does. See Figure 3-31.

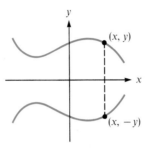

Figure 3-30

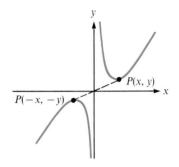

Figure 3-31

To test the graph of a function defined by $y = f(x)$ for symmetry about the origin, replace x with $-x$ and y with $-y$. If the equation $-y = f(-x)$ is equivalent to the equation $y = f(x)$, the graph is symmetric about the origin. This type of function is called an **odd function**.

> **Definition.**   A function f with the property that $f(-x) = -f(x)$ for all x is called an **odd function**.

In summary, we have the following.

> **Tests for Symmetry.**   If the point $(-x, y)$ lies on a graph whenever (x, y) does, the graph is **symmetric about the y-axis**.
>
> If $(x, -y)$ lies on the graph whenever (x, y) does, the graph is **symmetric about the x-axis**.
>
> If $(-x, -y)$ lies on the graph whenever (x, y) does, the graph is **symmetric about the origin**.
>
> If $y = f(x)$ and $y = f(-x)$ are equivalent, then the graph of the function f is symmetric about the y-axis.
>
> If $y = f(x)$ and $-y = f(-x)$ are equivalent, then the graph of the function f is symmetric about the origin.

Example 5   Graph the polynomial function $y = x^3 - x$.

Solution   To check for symmetries, calculate both $f(x)$ and $f(-x)$.

$$y = f(x) = x^3 - x$$
$$y = f(-x) = (-x)^3 - (-x)$$
$$= -x^3 + x$$

Because $y = f(x)$ is not equivalent to $y = f(-x)$, there is no symmetry about the y-axis. Because $y = f(x)$ is a function, there is no symmetry about the x-axis. However, if x and y are replaced with $-x$ and $-y$, respectively, the resulting equation is equivalent to the original equation:

$$y = x^3 - x$$
$$-y = (-x)^3 - (-x) \qquad \text{Replace } x \text{ with } -x, \text{ and } y \text{ with } -y.$$
$$-y = -x^3 + x \qquad \text{Simplify.}$$
$$y = x^3 - x \qquad \text{Multiply both sides by } -1.$$

Because the final result is identical to the original equation, the graph is symmetric with respect to the origin.

It is also useful to know the graph's x-intercepts—the points where the graph intersects the x-axis. The x-intercepts are the numbers x for which $f(x) = 0$. To find them, solve the equation $x^3 - x = 0$.

$$x^3 - x = 0$$
$$x(x^2 - 1) = 0 \qquad \text{Factor out } x.$$
$$x(x + 1)(x - 1) = 0 \qquad \text{Factor } x^2 - 1.$$
$$x = 0 \quad \text{or} \quad x + 1 = 0 \quad \text{or} \quad x - 1 = 0 \qquad \text{Set each factor equal to 0.}$$
$$x = -1 \qquad\qquad x = 1$$

Thus, the x-intercepts are 0, -1, and 1.

Finally, plot a few points for positive x, and use the graph's symmetry to draw the rest of the graph. See Figure 3-32.

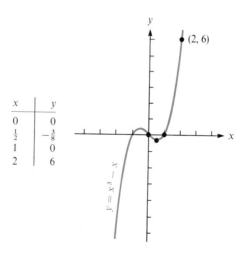

x	y
0	0
$\frac{1}{2}$	$-\frac{3}{8}$
1	0
2	6

Figure 3-32

Example 6   Graph the function defined by $y = x^4 - 5x^2 + 4$.

Solution   Because the variable x appears only with even exponents, $y = f(x)$ is equivalent to $y = f(-x)$, that is, the value of y is the same whether you evaluate $y = f(x)$ or $y = f(-x)$.

$$y = f(x) = x^4 - 5x^2 + 4$$

$$y = f(-x) = (-x)^4 - 5(-x)^2 + 4$$
$$= x^4 - 5x^2 + 4$$

Hence, this graph is symmetric with respect to the y-axis. It is not symmetric to either the x-axis or the origin, however.

The x-intercepts of the graph are those numbers x for which $y = 0$. Set y equal to 0 and solve the equation for x by factoring:

$$y = x^4 - 5x^2 + 4$$
$$0 = x^4 - 5x^2 + 4$$
$$0 = (x^2 - 4)(x^2 - 1)$$
$$0 = (x + 2)(x - 2)(x + 1)(x - 1)$$

$$x + 2 = 0 \quad \text{or} \quad x - 2 = 0 \quad \text{or} \quad x + 1 = 0 \quad \text{or} \quad x - 1 = 0$$
$$x = -2 \quad \mid \quad x = 2 \quad \mid \quad x = -1 \quad \mid \quad x = 1$$

The x-intercepts are $x = -2$, $x = 2$, $x = -1$, and $x = 1$. To graph this function, plot the intercepts and several other points whose coordinates satisfy the equation, and make use of the symmetry with respect to the y-axis. The graph appears in Figure 3-33.

x	y
0	4
1	0
$\frac{3}{2}$	$-\frac{35}{16}$
2	0
3	40

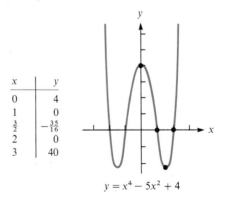

$$y = x^4 - 5x^2 + 4$$

Figure 3-33   ∎

If the values of $f(x)$ increase as x increases on an interval, then we say that the function is **increasing** on that interval. See Figure 3-34**a**. If the values of $f(x)$ decrease as x increases on an interval, then we say the function is **decreasing** on that interval. See Figure 3-34**b**. If the values $f(x)$ remain unchanged as x

increases on an interval, then the function is **constant** on that interval. See Figure 3-34**c**.

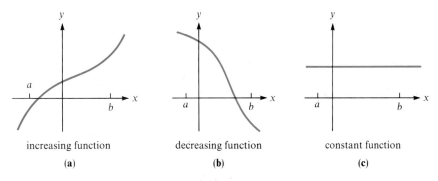

increasing function	decreasing function	constant function
(a)	**(b)**	**(c)**

Figure 3-34

Definition.   A function is **increasing** on the interval (a, b) if $f(x_1) < f(x_2)$ whenever $a < x_1 < x_2 < b$.

A function is **decreasing** on the interval (a, b) if $f(x_1) > f(x_2)$ whenever $a < x_1 < x_2 < b$.

A function f is **constant** on the interval (a, b) if $f(x_1) = f(x_2)$ for all x_1 and x_2 on (a, b).

Example 7   Graph the function $f(x) = x^3 - 3x$ and determine the intervals on which it is increasing and on which it is decreasing.

Solution   The graph of $f(x) = x^3 - 3x$ appears in Figure 3-35. The function is increasing when $x \le -1$ and when $x \ge 1$. The function is decreasing when $-1 \le x \le 1$.

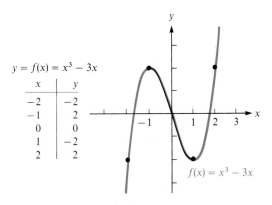

$y = f(x) = x^3 - 3x$

x	y
-2	-2
-1	2
0	0
1	-2
2	2

$f(x) = x^3 - 3x$

Figure 3-35

Exercise 3.5

In Exercises 1–10, graph each quadratic function.

1. $y = x^2 - 4x + 1$

2. $y = -x^2 - 4x + 1$

3. $y = -(x - 1)^2$

4. $y = (x - 1)^2$

5. $y = -3x^2 - 7$

6. $y = 3x^2 - 7$

7. $y = (x - 2)^2$

8. $y = -(x - 2)^2$

9. $y = (2x - 3)^2$

10. $y = 9 - (2x + 3)^2$

In Exercises 11–16, find the vertex of each parabola.

11. $y = x^2 - 4x + 4$

12. $y = x^2 - 10x + 25$

13. $y = x^2 + 6x - 3$

14. $y + 2 = -x^2 + 9x$

15. $-2x^2 + 12x - 17 - y = 0$

16. $2x^2 + 16x - y + 33 = 0$

In Exercises 17–26, find the symmetries, if any, of each curve. If the equation represents a function, indicate whether the function is even, odd, or neither.

17. $y = 3x$

18. $y = x^2 + 2$

19. $y^2 = x$

20. $x^2 + y^2 = 16$

21. $y = 3x - 7$

22. $y = x^3 + 9$

23. $y = x^2 + 2$

24. $y = |x|$

25. $y = (x^3 + 1)(x^3 - 1)$

26. $x = |y|$

27. Draw the graph of $y^2 + 8x - 4y - 28 = 0$. Is the graph a parabola? Is this relation a function?

28. Draw the graph of $y^2 - 2y - 4x + 1 = 0$. Is the graph a parabola? Is this relation a function?

29. A parabolic arch has an equation of $x^2 + 20y - 400 = 0$. Find the maximum height of the arch.

30. An object is thrown from the origin of a coordinate system with x-axis along the ground and y-axis vertical. Its path, or **trajectory**, is given by the equation $y = 400x - 16x^2$. What is the object's maximum height?

In Exercises 31–34, refer to this information. At a time t seconds after an object is tossed vertically upward, it reaches a height s in feet given by the equation $s = 80t - 16t^2$.

31. In how many seconds does the object reach its maximum height?

32. In how many seconds does the object return to the point from which it was thrown?

33. What is the maximum height reached by the object?

34. Show that it takes the same amount of time for the object to reach its maximum height as it does for the object to fall back to the point from which it was thrown.

In Exercises 35–46, graph each polynomial function.

35. $y = x^3$

36. $y = x^4$

37. $y = x^3 + x^2$

38. $y = x^3 - x^2$

39. $y = x^5 - x^3$

40. $y = x^5 + x^3$

41. $y = x^4 - 2x^2 + 1$

42. $y = x^3 + x^2 - 6x$

43. $y = x^3 + 2x^2 - x - 2$

44. $y = x^3 + x^2 - 4x - 4$

45. $y = x^{10}$

46. $y = x^{11}$

47. A farmer wants to partition off a rectangular stall in a corner of his barn. The barn walls form two sides of the stall, and the farmer has 50 feet of fencing for the remaining two sides. What dimensions will maximize the area? See Illustration 1.

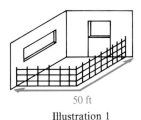

Illustration 1

48. A 24-inch wide sheet of metal is bent into a rectangular trough; a cross section is shown in Illustration 2. What dimensions will maximize the amount of water carried? That is, what dimensions will maximize the cross-sectional area?

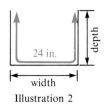

Illustration 2

49. The sum of two numbers is 6, and the sum of the squares of those two numbers is minimum. What are the two numbers?

50. A farmer will use D feet of fencing to enclose a rectangular plot of ground. Show that he will enclose the maximum area if the rectangle is a square.

51. What number most exceeds its square?

52. A rancher wishes to enclose a rectangular partitioned corral with 1800 feet of fencing. See Illustration 3. What dimensions of the corral would enclose the largest possible area? What is the maximum area?

53. Find the dimensions of the largest rectangle that can be inscribed in the right triangle ABC shown in Illustration 4.

In Exercises 54–57, tell where each function is increasing.

54. $y = 2x^3 + 6x^2$

55. $y = 2x^3 - 3x^2$

56. $y = x^3 - 1$

57. $y = x^3 - 1$

Illustration 3

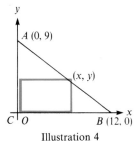

Illustration 4

3.6 RATIONAL FUNCTIONS

We now discuss functions, called **rational functions**, that are defined by equations of the form

$$y = \frac{P(x)}{Q(x)}$$

where $P(x)$ and $Q(x)$ are polynomials. In this section, we assume that the fraction $P(x)/Q(x)$ is in simplified form. Because $Q(x)$ appears in the denominator, the domain of the rational function must exclude all values of x for which $Q(x) = 0$.

Example 1 Find the domain of the rational function defined by $y = f(x) = \dfrac{x^2 - 4}{x^2 - 1}$, and find the symmetries of its graph.

Solution Factor both the numerator and the denominator of the fraction:

$$y = \frac{(x + 2)(x - 2)}{(x - 1)(x + 1)}$$

Observe that the numbers 1 and -1 make the denominator equal to 0. Hence, the domain of the function $y = f(x)$ is the set of all real numbers except 1 and -1.

The graph of the function is symmetric with respect to the y-axis because the variable x appears with even exponents only. Note that the equations $y = f(x)$ and $y = f(-x)$ are equivalent:

$$y = f(x) = \frac{x^2 - 4}{x^2 - 1} \qquad y = f(-x) = \frac{(-x)^2 - 4}{(-x)^2 - 1}$$

$$= \frac{x^2 - 4}{x^2 - 1}$$

∎

Example 2 Graph the rational function defined by $y = \dfrac{x^2 - 4}{x^2 - 1}$.

Solution Because the denominator of a fraction cannot be 0, x cannot be 1 or -1. To discover what happens when x is near 1, calculate values of y for numbers x that are close to 1.

$x < 1$		$x > 1$	
x	y	x	y
0.5	5	1.5	-1.4
0.9	16.8	1.1	-13.3
0.99	151.8	1.01	-148.3
0.999	1501.8	1.001	-1498.3

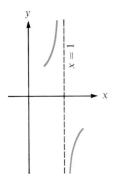

Figure 3-36

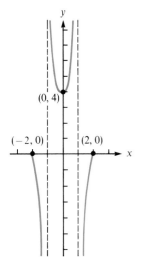

Figure 3-37

Plotting these points suggests a curve that approaches the vertical line $x = 1$ but never touches it. See Figure 3-36. Because there are only even exponents for x in this function, the curve is symmetric to the y-axis. Thus, similar behavior occurs near the line $x = -1$.

The y-intercept of the curve is determined by setting x equal to 0 and computing y; the y-intercept is 4.

The numbers $x = 2$ and $x = -2$ reduce the numerator of the rational expression to 0. The corresponding values of y, therefore, are both 0, and the x-intercepts of the curve are 2 and -2. You can now sketch most of the curve, as shown in Figure 3-37.

To discover the shape of the curve to the right of $x = 2$ and to the left of $x = -2$, examine the behavior of

$$y = \frac{x^2 - 4}{x^2 - 1}$$

as x gets very large. Do this by performing a long division and writing the answer in $quotient + \dfrac{remainder}{divisor}$ form:

$$
\begin{array}{r}
1 \\
x^2 - 1 \overline{\smash{)}\, x^2 - 4} \\
\underline{x^2 - 1} \\
- 3
\end{array}
$$

Hence,

$$y = \frac{x^2 - 4}{x^2 - 1} = 1 + \frac{-3}{x^2 - 1}$$

If x becomes very large, the denominator of the fraction $-3/(x^2 - 1)$ becomes very large also, and the magnitude of the fraction itself becomes very small. Because the term $-3/(x^2 - 1)$ is negative and approaches 0 as x becomes very large, it becomes negligible, and the value of y (which is less than 1) approaches the value 1. As x increases without bound, the value of the function defined by $y = (x^2 - 4)/(x^2 - 1)$ approaches 1 from below. Hence, as x increases, the curve approaches the line $y = 1$ from below. Because of symmetry, the curve also approaches the line $y = 1$ as x decreases without bound. See Figure 3-38.

The broken lines in Figure 3-38 are called **asymptotes**. The function $y = (x^2 - 4)/(x^2 - 1)$ has two vertical asymptotes, the lines $x = 1$ and $x = -1$, and one horizontal asymptote, the line $y = 1$. Although a curve might intersect a horizontal asymptote when $|x|$ is small, it will approach, but never touch, a horizontal asymptote when $|x|$ grows large. The graph of the rational function $y = (x^2 - 4)/(x^2 - 1)$ is shown in Figure 3-38.   ∎

Let's summarize the work done in the previous example. First you found the vertical asymptotes $x = 1$ and $x = -1$ by setting the denominator of the rational expression equal to 0 and solving that equation for x. Then you found

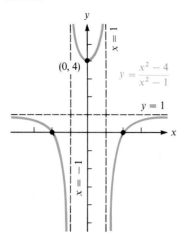

Figure 3-38

the y-intercept 4 by letting x equal 0 in the rational expression and evaluating y. Next you found the x-intercepts by finding the numbers x (2 and -2) that make the numerator of the rational expression equal to 0, and you found the horizontal asymptote $y = 1$ by performing a long division and ignoring the remainder. Finally, you made use of symmetry to graph the part of the curve to the left of the y-axis.

Example 3   Graph the rational function defined by $y = \dfrac{3x}{x - 2}$.

Solution   First look for symmetry about the y-axis. Because x appears to an odd power, the function is not symmetric about the y-axis. The y-intercept is found by setting $x = 0$ and solving for y; the y-intercept is 0. Because $y = 0$ when $x = 0$, the curve passes through the origin.

The x-intercepts are found by setting the numerator of the rational expression equal to 0 and solving for x:

$$3x = 0$$
$$x = 0$$

Thus, the only x-intercept is at 0.

The vertical asymptotes are found by determining which real numbers lead to zeros in the denominator of the rational expression. If x is replaced by 2 in the denominator $x - 2$, that denominator is 0. Hence, the line $x = 2$ is a vertical asymptote.

The horizontal asymptotes, if any, are found by dividing $3x$ by $x - 2$, and expressing the answer in $quotient + \dfrac{remainder}{divisor}$ form.

$$y = \frac{3x}{x - 2} = 3 + \frac{6}{x - 2}$$

As $|x|$ increases without bound, the fraction $6/(x - 2)$ approaches 0. So, the curve

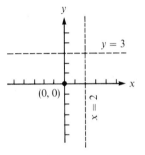

Figure 3-39

approaches the line $y = 3$ as $|x|$ becomes larger. The line $y = 3$ is a horizontal asymptote. The previous results provide a basis for the graph. See Figure 3-39.

To discover what happens to the graph when x is greater than 2, pick a value of x that is greater than 2 and find the corresponding value of y. If $x = 3$, for example, the value of y is 9. After plotting the point $(3, 9)$, you can use the intercepts and asymptotes to sketch the graph. This curve, shown in Figure 3-40, is called a **hyperbola**.

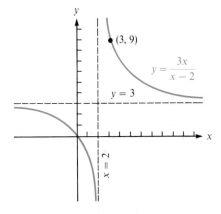

Figure 3-40

Example 4   Graph the rational function defined by $y = \dfrac{x^2 + x - 2}{x - 3}$.

Solution   First factor the numerator of the rational expression:

$$y = \frac{(x - 1)(x + 2)}{x - 3}$$

Most of the information is straightforward:

- The function is not symmetric about the y-axis or origin.
- The y-intercept is $\frac{2}{3}$.
- The x-intercepts are 1 and -2.
- A vertical asymptote is $x = 3$.

Perform the long division and write the rational expression as

$$y = \frac{x^2 + x - 2}{x - 3} = x + 4 + \frac{10}{x - 3}$$

As before, the fraction $10/(x - 3)$ becomes insignificant as x grows large. This time, however, the graph does not approach a constant. Instead, it approaches the line given by $y = x + 4$. Because this line is not horizontal, the graph has no horizontal asymptotes. However, it does have a **slant asymptote** or an **oblique asymptote**: the line $y = x + 4$. Put all of this information together and graph the rational function. See Figure 3-41.

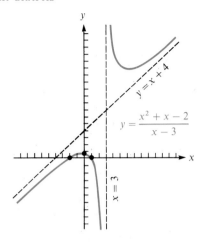

Figure 3-41

Example 5   Discuss the nature of the asymptotes of the following rational functions:

a. $y = \dfrac{x + 2}{x^2 - 1}$ **b.** $y = \dfrac{x^2 + x + 2}{x^2 - 1}$

c. $y = \dfrac{3x^3 + 2x^2 + 2}{x^2 - 1}$ **d.** $y = \dfrac{x^4 + x + 2}{x^2 - 1}$

Solution   All four functions have vertical asymptotes at $x = 1$ and $x = -1$ because at these values the denominators are 0. The nature of the remaining asymptotes must be considered.

a. $y = \dfrac{x + 2}{x^2 - 1}$

Because the degree of the numerator is less than the degree of the denominator, long division is not practical. Try a different approach, dividing both numerator and denominator by x^2, which is the largest power of x in the denominator.

$$y = \frac{x + 2}{x^2 - 1} = \frac{\dfrac{x}{x^2} + \dfrac{2}{x^2}}{\dfrac{x^2}{x^2} - \dfrac{1}{x^2}} = \frac{\dfrac{1}{x} + \dfrac{2}{x^2}}{1 - \dfrac{1}{x^2}}$$

The three fractions in the final result, $\dfrac{1}{x}$, $\dfrac{2}{x^2}$, and $\dfrac{1}{x^2}$, all approach 0 as $|x|$ increases without bound. Hence, y approaches

$$\frac{0 + 0}{1 - 0} = 0$$

The graph of this function has a horizontal asymptote of $y = 0$.

In each of the three remaining functions, a long division can be performed.

b. $y = \dfrac{x^2 + x + 2}{x^2 - 1} = 1 + \dfrac{x + 3}{x^2 - 1}$

As $|x|$ increases without bound, the fraction $(x + 3)/(x^2 - 1)$ approaches 0 (for reasons discussed in part **a**) and y approaches 1. This curve has a horizontal asymptote of $y = 1$.

c. $y = \dfrac{3x^3 + 2x^2 + 2}{x^2 - 1} = 3x + 2 + \dfrac{3x + 4}{x^2 - 1}$

Again, the last fraction approaches 0 as $|x|$ increases without bound, and the curve approaches the slant asymptote $y = 3x + 2$.

d. $y = \dfrac{x^4 + x + 2}{x^2 - 1} = x^2 + 1 + \dfrac{x + 3}{x^2 - 1}$

As $|x|$ increases, the fractional part again approaches 0 and the curve approaches $y = x^2 + 1$. However, $y = x^2 + 1$ does not represent a line. This curve has neither horizontal nor slant asymptotes.   ∎

We generalize the results of Example 5, and summarize the techniques discussed in this section as follows:

Perform the following steps when you graph the rational function $y = \dfrac{P(x)}{Q(x)}$, where $\dfrac{P(x)}{Q(x)}$ is in simplified form.

Check for symmetry.	If the polynomials $P(x)$ and $Q(x)$ involve only even powers of x, the graph is symmetric about the y-axis. Otherwise, y-axis symmetry does not exist.
Look for y-intercepts.	Set x equal to 0. The resulting value of y is the y-intercept of the graph.
Look for x-intercepts.	Set $P(x)$ equal to 0. The solutions of the equation $P(x) = 0$ (if any) are the x-intercepts of the graph.
Look for vertical asymptotes.	Set $Q(x)$ equal to 0. The solutions of the equation $Q(x) = 0$ (if any) determine the vertical asymptotes of the graph.
Look for horizontal asymptotes.	If the degree of $P(x)$ is less than the degree of $Q(x)$, then the line $y = 0$ is a horizontal asymptote.

(continued)

(continued)

If the degrees of $P(x)$ and $Q(x)$ are equal, then the line $y = p/q$, where p and q are the lead coefficients of $P(x)$ and $Q(x)$, is a horizontal asymptote. (Be sure that $P(x)$ and $Q(x)$ are written in descending powers of x before applying this rule.)

If the degree of $P(x)$ is greater than the degree of $Q(x)$, then there is no horizontal asymptote.

Look for slant asymptotes.

If the degree of $P(x)$ is exactly one greater than the degree of $Q(x)$, there is a slant asymptote. To find it, perform the long division $Q(x) \overline{)\, P(x)}$ and ignore the remainder.

Exercise 3.6

In Exercises 1–24, find all asymptotes (vertical, horizontal, and slant asymptotes) and then graph each rational function.

1. $y = \dfrac{1}{x - 2}$

2. $y = \dfrac{3}{x + 3}$

3. $y = \dfrac{x}{x - 1}$

4. $y = \dfrac{x}{x + 2}$

5. $y = \dfrac{x + 1}{x + 2}$

6. $y = \dfrac{x - 1}{x - 2}$

7. $y = \dfrac{2x - 1}{x - 1}$

8. $y = \dfrac{3x + 2}{x - 1}$

9. $y = \dfrac{x^2 - 9}{x^2 - 4}$

10. $y = \dfrac{x^2 - 4}{x^2 - 9}$

11. $y = \dfrac{x^2 - 5x + 6}{x^2 - 2x + 1}$

12. $y = \dfrac{x^2 + 7x + 12}{x^2 - 4x + 4}$

13. $y = \dfrac{3x^2}{x^2 + 1}$

14. $y = \dfrac{x^2 - 9}{2x^2 + 1}$

15. $y = \dfrac{2x^2 - 2}{x^2 - 25}$

16. $y = \dfrac{x^2 - 4}{3x^2 - 27}$

17. $y = \dfrac{2x^2 - 3x - 2}{x^2 + x - 2}$

18. $y = \dfrac{3x^2 - 4x + 1}{2x^2 + 3x + 1}$

19. $y = \dfrac{x^2 - 9}{2x^2 - 8}$

20. $y = \dfrac{3x^2 - 12}{x^2}$

21. $y = \dfrac{x^2 - 2x - 8}{x - 1}$

22. $y = \dfrac{x^2 + x - 6}{x + 2}$

23. $y = \dfrac{x^3 + x^2 + 6x}{x^2 - 1}$

24. $y = \dfrac{x^3 - 2x^2 + x}{x^2 - 4}$

25. Can a rational function have more than one horizontal asymptote? Explain.

26. The graph of a rational function cannot cross a vertical asymptote. Can it cross a horizontal asymptote? Explain.

3.7 ALGEBRA AND COMPOSITION OF FUNCTIONS

It is possible to perform some arithmetic with functions.

Definition. If the ranges of functions f and g are subsets of the real numbers, then four new functions can be defined.

The sum of f and g, denoted as **$f + g$**, is defined by

$$(f + g)(x) = f(x) + g(x)$$

The difference of f and g, denoted as **$f - g$**, is defined by

$$(f - g)(x) = f(x) - g(x)$$

The product of f and g, denoted as **$f \cdot g$**, is defined by

$$(f \cdot g)(x) = f(x) \cdot g(x)$$

The quotient of f and g, denoted as **f/g**, is defined by

$$(f/g)(x) = \frac{f(x)}{g(x)} \quad \text{provided } g(x) \neq 0$$

The domain of each of these functions (unless otherwise restricted) is the set of all real numbers x that are in the domain of *both* f and g, and, in the case of the quotient f/g, there is the further restriction that $g(x) \neq 0$.

Example 1   Let $f(x) = 3x + 1$ and $g(x) = 2x - 3$. Find **a.** $f + g$ and **b.** $f \cdot g$, and give the domain of each.

Solution   **a.** $(f + g)(x) = f(x) + g(x)$

$\qquad\qquad\qquad = (3x + 1) + (2x - 3)$ Substitute $3x + 1$ for $f(x)$ and $2x - 3$ for $g(x)$.

$\qquad\qquad\qquad = 5x - 2$ Combine terms.

Thus, the function $f + g$ is defined by $(f + g)(x) = 5x - 2$.

The domain of $f + g$ is the set of all real numbers that are in the domain of both f and g. Because the domain of both f and g is the set of real numbers, the domain of $f + g$ is also the set of real numbers.

b. $(f \cdot g)(x) = f(x) \cdot g(x)$

$\qquad\qquad\qquad = (3x + 1)(2x - 3)$ Substitute $3x + 1$ for $f(x)$ and $2x - 3$ for $g(x)$.

$\qquad\qquad\qquad = 6x^2 - 7x - 3$

Thus, the function $f \cdot g$ is defined by $(f \cdot g)(x) = 6x^2 - 7x - 3$.

The domain of $f \cdot g$ is the set of real numbers.   ∎

Example 2   Let $f(x) = x^2 - 4$ and $g(x) = \sqrt{x}$. If all calculations involve real numbers only, find the functions **a.** $f + g$, **b.** $f \cdot g$, **c.** f/g, and **d.** g/f, and give the domain of each.

Solution   Because any real number can be squared, the domain of f is the set of real numbers. Because all calculations are to involve real numbers only, the domain of g is the set of nonnegative real numbers: $\{x \mid x \geq 0\}$.

a. $(f + g)(x)$ is defined as $f(x) + g(x)$. Hence,

$$(f + g)(x) = x^2 - 4 + \sqrt{x}$$

The domain consists of the numbers x that are common to the domains of f and g. The domain of $f + g$ is $\{x \mid x \geq 0\}$.

b. $(f \cdot g)(x)$ is defined as $f(x) \cdot g(x)$. Hence,

$$(f \cdot g)(x) = (x^2 - 4) \cdot \sqrt{x}$$
$$= x^2\sqrt{x} - 4\sqrt{x}$$

The domain consists of those numbers that are common to the domains of f and g. The domain of $f \cdot g$ is $\{x \mid x \geq 0\}$.

c. $(f/g)(x)$ is defined as $\dfrac{f(x)}{g(x)}$. Hence,

$$(f/g)(x) = \frac{x^2 - 4}{\sqrt{x}}$$

Its domain consists of those numbers common to the domains of f and g except 0, because division by 0 is not defined. The domain of f/g is $\{x \mid x > 0\}$.

d. $(g/f)(x)$ is defined as $\dfrac{g(x)}{f(x)}$. Hence,

$$(g/f)(x) = \frac{\sqrt{x}}{x^2 - 4}$$

The domain consists of those numbers common to the domains of f and g except the number 2 because division by 0 is not defined. Thus, the domain of g/f is $\{x \mid x \geq 0 \quad \text{and} \quad x \neq 2\}$.   ∎

Example 3   Find $(f + g)(3)$ if $f(x) = x^2 + 1$ and $g(x) = 2x + 1$.

Solution   First find $(f + g)(x)$.

$$(f + g)(x) = f(x) + g(x)$$
$$= x^2 + 1 + 2x + 1$$
$$= x^2 + 2x + 2$$

Then find $(f + g)(3)$.

$$(f + g)(3) = 3^2 + 2(3) + 2$$
$$= 9 + 6 + 2$$
$$= 17$$ ∎

Example 4   Let $h(x) = x^2 + 3x + 2$. Find functions f and g such that **a.** $h = f + g$ and **b.** $h = f \cdot g$.

Solution   **a.**   Several answers are possible. One answer is $f(x) = x^2$ and $g(x) = 3x + 2$. Then

$$h(x) = x^2 + 3x + 2$$
$$= (x^2) + (3x + 2)$$
$$= f(x) + g(x)$$
$$= (f + g)(x)$$

Another answer is $f(x) = x^2 + 2x$ and $g(x) = x + 2$.

b.   Several answers are possible. One answer is suggested by factoring the polynomial $x^2 + 3x + 2$:

$$h(x) = x^2 + 3x + 2 = (x + 1)(x + 2)$$

Let $f(x) = x + 1$ and $g(x) = x + 2$. Then

$$h(x) = x^2 + 3x + 2$$
$$= (x + 1)(x + 2)$$
$$= f(x) \cdot g(x)$$
$$= (f \cdot g)(x)$$

Another answer is $f(x) = 3$ and $g(x) = \dfrac{x^2}{3} + x + \dfrac{2}{3}$. ∎

Composition of Functions

Often one quantity is a function of a second quantity that depends, in turn, on a third quantity. A farmer's income, for example, is a function of the number of bushels of grain harvested. But the harvest, in turn, is a function of the number of inches of rainfall received (among other things). Such chains of dependence can be analyzed mathematically by considering the **composition of functions**.

We suppose that $y = f(x)$ and $y = g(x)$ define two functions. Any number x in the domain of g will produce the corresponding value $g(x)$ in the range of g. If this value $g(x)$ is in the domain of the function f, then $g(x)$ is an acceptable input into the function f, and a corresponding value $f(g(x))$ is determined. This two-step process defines a new function, called a **composite function**, denoted by $f \circ g$.

To visualize the domain of the composite function $f \circ g$, refer to Figure 3-42. Note that the value of the independent variable x must be in the domain of the function g because x is the number used as input into the function g. Also note that $g(x)$ must be in the domain of function f because $g(x)$ is the number used as input to the function f. Hence, the domain of $f \circ g$ consists of all those elements x that are permissible inputs into g, and for which each $g(x)$ is a permissible input into f.

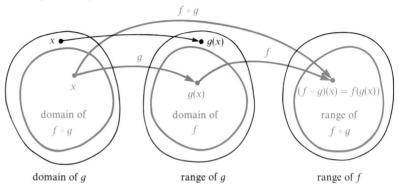

Figure 3-42

Definition.   The **composite function $f \circ g$** is defined by

$$(f \circ g)(x) = f(g(x))$$

The **domain of the composite function $f \circ g$** consists of all those elements in the domain of g for which $g(x)$ is an element in the domain of f.

Example 5   If $f(x) = x + 1$ and $g(x) = x^2$, find $(f \circ g)(2)$.

Solution   To compute $(f \circ g)(2)$, you must calculate $f(g(2))$. Because $g(x) = x^2$, $g(2) = 2^2 = 4$. Because 4 is in the domain of f, calculate $f(4)$, which is $4 + 1 = 5$. Hence,

$$(f \circ g)(2) = f(g(2)) = f(4) = 5$$ ∎

Example 6   If $f(x) = 2x + 7$ and $g(x) = 4x + 1$, find **a.** $(f \circ g)(x)$ and **b.** $(g \circ f)(x)$.

Solution   **a.** $(f \circ g)(x) = f(g(x)) = f(4x + 1)$
$$= 2(4x + 1) + 7$$
$$= 8x + 9$$

b. $(g \circ f)(x) = g(f(x)) = g(2x + 7)$
$$= 4(2x + 7) + 1$$
$$= 8x + 29$$ ∎

Example 6 illustrates that composition of functions is *not* commutative. In general, $(f \circ g)(x)$ is not equal to $(g \circ f)(x)$.

Example 7 Find two functions f and g such that $(f \circ g)(x) = \sqrt{x^2 + 2}$.

Solution Let $g(x) = x^2 + 2$ and $f(x) = \sqrt{x}$. Then

$$(f \circ g)(x) = f(g(x)) = f(x^2 + 2)$$
$$= \sqrt{x^2 + 2}$$

Thus, $g(x) = x^2 + 2$ and $f(x) = \sqrt{x}$ are two functions such that $(f \circ g)(x) = \sqrt{x^2 + 2}$.
∎

Example 8 Let $f(x) = \sqrt{x}$ and $g(x) = x - 3$. Find the domain of functions **a.** f, **b.** g, **c.** $f \circ g$, and **d.** $g \circ f$.

Solution **a.** Because the range is to be a subset of the real numbers, the domain of f is the set of nonnegative real numbers: $\{x \mid x \geq 0\}$.

b. The domain of g is the set of real numbers.

c. Because $g(x)$ must be in the domain of f, the value $g(x)$ must be nonnegative. Thus,

$$g(x) \geq 0$$
$$x - 3 \geq 0 \qquad \text{Because } g(x) = x - 3.$$
$$x \geq 3$$

Hence, the domain of $f \circ g$ is $\{x \mid x \geq 3\}$.

d. The domain of $g \circ f$ contains those numbers x for which $f(x)$ is also in the domain of g. Because the domain of g includes all real numbers, the domain of $g \circ f$ is just the domain of f itself: $\{x \mid x \geq 0\}$.
∎

Example 9 Let $f(x) = \sqrt{1 - x}$ and $g(x) = x^2$. If all expressions represent real numbers, find **a.** $(f \circ g)(x)$ and **b.** $(g \circ f)(x)$, and give the domain of each function.

Solution Begin by finding the domain of f. Because all expressions represent real numbers, the radicand $1 - x$ cannot be negative. So, x must be less than or equal to 1. The domain of f is the set $\{x \mid x \leq 1\}$.

Now find the domain of g. Because the square of any real number is still a real number, the domain of g is the set of real numbers.

a. $(f \circ g)(x) = f(g(x))$
$$= f(x^2)$$
$$= \sqrt{1 - x^2}$$

The domain of $f \circ g$ consists of those numbers x for which $g(x)$ is in the domain of f. Thus, the value $g(x)$ must be less than or equal to 1.

$$g(x) \leq 1$$
$$x^2 \leq 1 \qquad \text{Because } g(x) = x^2.$$

Hence, the domain of $f \circ g$ is $\{x \mid -1 \leq x \leq 1\}$.

b. $(g \circ f)(x) = g(f(x))$
$$= g(\sqrt{1-x})$$
$$= (\sqrt{1-x})^2$$
$$= 1 - x$$

The domain of $g \circ f$ consists of those numbers x for which $f(x)$ is in the domain of g. Because any real number is acceptable to g, the domain of $g \circ f$ is just the domain of f: $\{x \mid x \leq 1\}$.   ∎

Exercise 3.7

In Exercises 1–4, let $f(x) = 2x + 1$ and $g(x) = 3x - 2$. Find each function, and determine its domain.

1. $f + g$ **2.** $f - g$ **3.** $f \cdot g$ **4.** f/g

In Exercises 5–8, let $f(x) = x^2 + x$ and $g(x) = x^2 - 1$. Find each function, and determine its domain.

5. $f - g$ **6.** $f + g$ **7.** f/g **8.** $f \cdot g$

In Exercises 9–16, let $f(x) = x^2 - 1$ and $g(x) = 3x - 2$. Find each value, if possible.

9. $(f + g)(2)$ **10.** $(f + g)(-3)$ **11.** $(f - g)(0)$ **12.** $(f - g)(-5)$

13. $(f \cdot g)(2)$ **14.** $(f \cdot g)(-1)$ **15.** $(f/g)(\frac{2}{3})$ **16.** $(f/g)(1)$

In Exercises 17–24, find two functions f and g such that the given correspondence can be expressed as the function indicated. Several answers are possible.

17. $y = 3x^2 + 2x; f + g$ **18.** $y = 3x^2; f \cdot g$

19. $y = \dfrac{3x^2}{x^2 - 1}; f/g$ **20.** $y = 5x + x^4; f - g$

21. $y = x(3x^2 + 1); f - g$ **22.** $y = (3x - 2)(3x + 2); f + g$

23. $y = x^2 + 7x - 18; f \cdot g$ **24.** $y = 5x^5; f/g$

In Exercises 25–28, let $f(x) = 2x - 5$ and $g(x) = 5x - 2$. Find each value.

25. $(f \circ g)(2)$ **26.** $(g \circ f)(-3)$ **27.** $(f \circ f)(-\frac{1}{2})$ **28.** $(g \circ g)(\frac{3}{5})$

In Exercises 29–32, let $f(x) = 3x^2 - 2$ and $g(x) = 4(x + 1)$. Find each value.

29. $(f \circ g)(-3)$ **30.** $(g \circ f)(3)$ **31.** $(f \circ f)(\sqrt{3})$ **32.** $(g \circ g)(-4)$

In Exercises 33–36, let $f(x) = 3x$ and $g(x) = x + 1$. Find each composite function, and determine its domain.

33. $f \circ g$ **34.** $g \circ f$ **35.** $f \circ f$ **36.** $g \circ g$

In Exercises 37–40, let $f(x) = x^2$ and $g(x) = 2x$. Find each composite function, and determine its domain.

37. $g \circ f$ **38.** $f \circ g$ **39.** $f \circ f$ **40.** $g \circ g$

In Exercises 41–44, let $f(x) = \sqrt{x}$ *and* $g(x) = x^2 + 1$. *Find each composite function, and determine its domain.*

41. $f \circ g$ **42.** $g \circ f$ **43.** $f \circ f$ **44.** $g \circ g$

In Exercises 45–48, let $f(x) = \sqrt{x + 1}$ *and* $g(x) = x^2 - 1$. *Find each composite function, and determine its domain.*

45. $g \circ f$ **46.** $f \circ g$ **47.** $g \circ g$ **48.** $f \circ f$

In Exercises 49–60, find two functions f *and* g *such that the composition* $f \circ g$ *expresses the given correspondence. Several answers are possible.*

49. $y = 3x - 2$ **50.** $y = 7x - 5$ **51.** $y = x^2 - 2$ **52.** $y = x^3 - 3$

53. $y = (x - 2)^2$ **54.** $y = (x - 3)^3$ **55.** $y = \sqrt{x + 2}$ **56.** $y = \dfrac{1}{x - 5}$

57. $y = \sqrt{x} + 2$ **58.** $y = \dfrac{1}{x} - 5$ **59.** $y = x$ **60.** $y = 3$

61. Let $f(x) = 3x$. Show that $(f + f)(x) = f(x + x)$.

62. Let $g(x) = x^2$. Show that $(g + g)(x) \neq g(x + x)$.

63. Let $f(x) = \dfrac{x - 1}{x + 1}$. Find $(f \circ f)(x)$.   **64.** Let $g(x) = \dfrac{x}{x - 1}$. Find $(g \circ g)(x)$.

3.8 INVERSE FUNCTIONS

By definition, each element x in the domain of a function has a *single* image y. For some functions, distinct numbers x in the domain have the *same* image. See Figure 3-43. For other functions, called **one-to-one functions**, distinct numbers x have distinct images. See Figure 3-44.

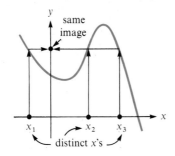

not a one-to-one function

Figure 3-43

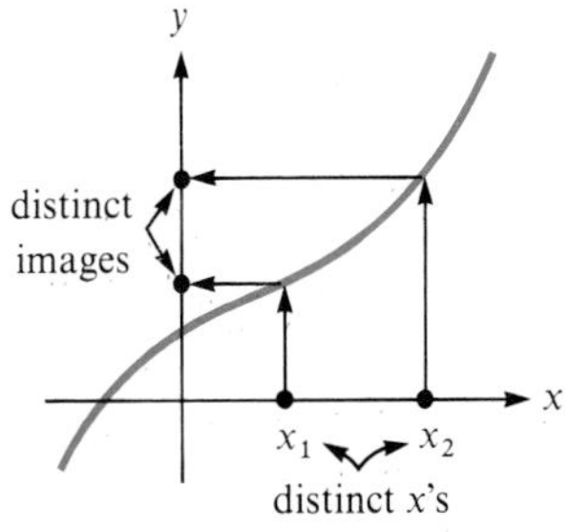

a one-to-one function

Figure 3-44

Definition. A function f from a set X to a set Y is called **one-to-one** if and only if different elements in the domain of f have different images in the range of f: if x_1 and x_2 are two elements in the domain of f and $x_1 \neq x_2$, then $f(x_1) \neq f(x_2)$.

Example 1   Determine if the functions **a.** $f(x) = x^2$ and **b.** $f(x) = x^3$ are one-to-one.

Solution   **a.**   The function $f(x) = x^2$ is *not* one-to-one because different elements in the domain (2 and -2, for example) have the same image ($f(2) = f(-2) = 4$).

b.   The function $f(x) = x^3$ *is* one-to-one because two distinct numbers x produce different images $f(x)$—two different numbers have different cubes.   ∎

There is a test, called the **horizontal line test**, that can be used to determine whether a graph represents a one-to-one function. If any horizontal line intersects the graph of a function more than once, the function is *not* one-to-one. See Figure 3-45. If every horizontal line that intersects the graph does so only once, the function *is* one-to-one. See Figure 3-46.

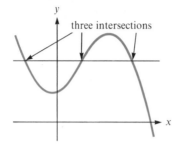

not a one-to-one function

Figure 3-45

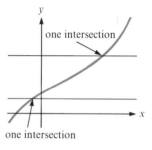

a one-to-one function

Figure 3-46

Figure 3-47 illustrates a function f from a set X to a set Y. Because several arrows point to a single y, the function f is not one-to-one. If the arrows of Figure 3-47 were reversed, the diagram would not represent a function, because to some y of set Y would correspond several values x in set X. If the arrows of the one-to-one function f in Figure 3-48 were reversed, the diagram would still represent a function.

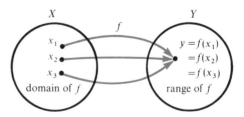

Figure 3-47

This "backwards function" is called the **inverse** of the original function f and is denoted by the symbol f^{-1}. The -1 indicates the inverse of function f and is *not* an exponent.

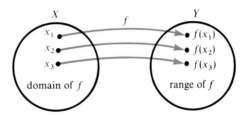

Figure 3-48

Refer to the one-to-one function f and its inverse f^{-1} shown in Figure 3-49. To the element x in the domain of f there corresponds its image $f(x)$ in the range of f. This element $f(x)$, however, is also in the domain of f^{-1}. The image of $f(x)$ under the function f^{-1} is $f^{-1}(f(x))$, which is the original number x. Thus, $(f^{-1} \circ f)(x) = f^{-1}(f(x)) = x$. Similarly, if y is an element of the domain of f^{-1}, then $(f \circ f^{-1})(y) = f(f^{-1}(y)) = y$.

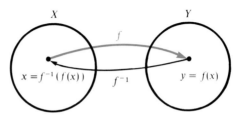

Figure 3-49

Theorem.   If f is a one-to-one function with domain X and range Y, then there is a one-to-one function f^{-1} with domain Y and range X such that

$$(f^{-1} \circ f)(x) = x$$

and

$$(f \circ f^{-1})(y) = y$$

To show that one function is the inverse of another, we must show that their composition is the **identity function**—the function that assigns x itself as the image of each real number x.

Example 2   Show that the function $f(x) = x^3$ is the inverse of $g(x) = \sqrt[3]{x}$.

Solution   Show that the composition of f and g (in both directions) is the identity function.

$$(f \circ g)(x) = f(g(x)) = f(\sqrt[3]{x}) = (\sqrt[3]{x})^3 = x$$
$$(g \circ f)(x) = g(f(x)) = g(x^3) = \sqrt[3]{x^3} = x$$

∎

If f is a one-to-one function defined by $y = f(x)$, then f^{-1} reverses the correspondence of f. That is, if $f(a) = b$, then $f^{-1}(b) = a$. To determine f^{-1}, we interchange the variables x and y in the equation $y = f(x)$. The resulting equation, $x = f(y)$, defines the inverse function, f^{-1}. If we can solve $x = f(y)$ for y, we will have expressed the inverse as the equation $y = f^{-1}(x)$.

Example 3   Find the inverse of the function defined by $y = \dfrac{3}{2}x + 2$, and verify the result.

Solution   To find the inverse of $y = \dfrac{3}{2}x + 2$, interchange the x and y to obtain

$$x = \frac{3}{2}y + 2$$

To express this equation in the form $y = f^{-1}(x)$, solve it for y:

$$x = \frac{3}{2}y + 2$$
$$2x = 3y + 4$$
$$2x - 4 = 3y$$
$$y = \frac{2x - 4}{3}$$

Thus, the inverse of the function defined by $y = \dfrac{3}{2}x + 2$ is defined by $y = \dfrac{2x - 4}{3}$.

To verify that

$$f(x) = \frac{3}{2}x + 2 \qquad \text{and} \qquad f^{-1}(x) = \frac{2x - 4}{3}$$

are inverses of each other, show that their composition is the identity function:

$$(f \circ f^{-1})(x) = f(f^{-1}(x)) = f\left(\frac{2x - 4}{3}\right) = \frac{3}{2}\left(\frac{2x - 4}{3}\right) + 2 = x - 2 + 2 = x$$

Similarly,

$$(f^{-1} \circ f)(x) = f^{-1}(f(x)) = f^{-1}\left(\frac{3}{2}x + 2\right) = \frac{2\left(\dfrac{3}{2}x + 2\right) - 4}{3} = \frac{3x + 4 - 4}{3} = x \qquad ∎$$

Because x and y are interchanged when finding the inverse, the point (b, a) lies on the graph of $y = f^{-1}(x)$ whenever the point (a, b) lies on the graph of $y = f(x)$. Thus, the graph of a function and its inverse are reflections of each other in the line $y = x$.

Example 4   Find the inverse of $y = f(x) = x^3 + 3$, and graph both the function and its inverse on a single set of coordinate axes.

Solution The inverse of the function $y = x^3 + 3$ is found by interchanging x and y. Hence, the inverse is defined by

$$x = y^3 + 3$$

Solve this equation for y to obtain

$$x - 3 = y^3$$
$$y = \sqrt[3]{x - 3}$$

Thus, $f^{-1}(x) = \sqrt[3]{x - 3}$.

The graphs appear in Figure 3-50. Note that $y = x$ is the axis of symmetry.

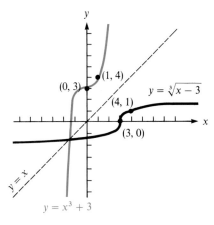

Figure 3-50

Example 5 The function defined by the equation $f(x) = x^2 + 3$ is not one-to-one. However, the function becomes one-to-one if the domain is restricted to a carefully chosen subset of the real numbers, such as the set $\{x \mid x \le 0\}$. Find **a.** the range of f, **b.** the inverse of f along with its domain and range, and **c.** graph each function.

Solution **a.** The function f is defined by $f(x) = y = x^2 + 3$, with domain $\{x \mid x \le 0\}$. If x is replaced with numbers from this domain, y ranges over the values 3 and above. Thus, the range of f is the set $\{y \mid y \ge 3\}$.

b. To find the inverse of f, interchange x and y in the equation that defines f, and solve for y:

$$y = x^2 + 3 \quad \text{where } x \le 0$$
$$x = y^2 + 3 \quad \text{where } y \le 0 \qquad \text{Interchange } x \text{ and } y.$$
$$x - 3 = y^2 \qquad \text{where } y \le 0 \qquad \text{Add } -3 \text{ to both sides.}$$

To solve this equation for y, take the square root of both sides. Because $y \le 0$, you have

$$-\sqrt{x - 3} = y \qquad \text{where } y \le 0$$

Thus, the inverse of f is defined by the equation

$$y = f^{-1}(x) = -\sqrt{x - 3}$$

It has domain $\{x \mid x \geq 3\}$ and range $\{y \mid y \leq 0\}$.

c. The graphs of these two functions appear in Figure 3-51. Note that the line of symmetry is $y = x$.

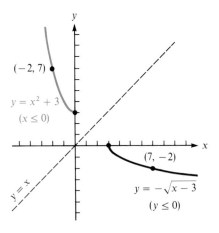

$(-2, 7)$

$y = x^2 + 3$
$(x \leq 0)$

$(7, -2)$

$y = -\sqrt{x - 3}$
$(y \leq 0)$

$y = x$

Figure 3-51

If a function is defined by the equation $y = f(x)$, we can often find the domain of f by inspection. Finding the range can be more difficult. One way to find the range of f is to find the domain of f^{-1}.

Example 6   Find the domain and the range of the function defined by

$$y = f(x) = \frac{2}{x} + 3$$

Solution   Because x cannot be 0, the domain of f is $\{x \mid x \text{ is a real number and } x \neq 0\}$. To find the range of f, find the domain of f^{-1}. Find f^{-1} as follows:

$$y = \frac{2}{x} + 3$$

$$x = \frac{2}{y} + 3 \qquad \text{Interchange } x \text{ and } y.$$

$$yx = 2 + 3y \qquad \text{Multiply both sides by } y.$$

$$yx - 3y = 2 \qquad \text{Add } -3y \text{ to both sides.}$$

$$y(x - 3) = 2 \qquad \text{Factor out } y.$$

$$y = \frac{2}{x - 3} \qquad \text{Divide both sides by } x - 3.$$

This final equation defines f^{-1} whose domain is $\{x \mid x$ is a real number and $x \neq 3\}$. Because the range of f is equal to the domain of f^{-1}, the range of f is

$$\{y \mid y \text{ is a real number and } y \neq 3\}$$ ∎

Exercise 3.8

In Exercises 1–10, determine if each function is one-to-one.

1. $y = 3x$

2. $y = \dfrac{1}{2}x$

3. $y = x^2 + 3$

4. $y = x^4 - x^2$

5. $y = x^3 - x$

6. $y = x^2 - x$

7. $y = |x|$

8. $y = |x - 3|$

9. $y = 5$

10. $x = \sqrt{y - 5}$ and $y \geq 5$

In Exercises 11–14, use the horizontal line test to determine if the graph represents a one-to-one function.

11.

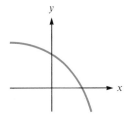

12.

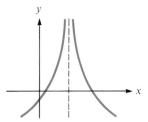

13.

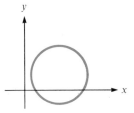

14.

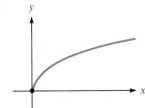

In Exercises 15–18, verify that the functions are inverses by determining that $f \circ g$ and $g \circ f$ are the identity function.

15. $f(x) = 5x;\ g(x) = \dfrac{1}{5}x$

16. $f(x) = 3x + 2;\ g(x) = \dfrac{x - 2}{3}$

17. $f(x) = \dfrac{x + 1}{x};\ g(x) = \dfrac{1}{x - 1}$

18. $f(x) = \dfrac{x + 1}{x - 1};\ g(x) = \dfrac{x + 1}{x - 1}$

In Exercises 19–26, each equation defines a one-to-one function f. Determine f^{-1} and verify that $f \circ f^{-1}$ and $f^{-1} \circ f$ are the identity function.

19. $y = 3x$

20. $y = \dfrac{1}{3}x$

21. $y = 3x + 2$

22. $y = 2x - 5$

23. $y = \dfrac{1}{x + 3}$

24. $y = \dfrac{1}{x - 2}$

25. $y = \dfrac{1}{2x}$

26. $y = \dfrac{1}{x^3}$

In Exercises 27–36, find the inverse of each one-to-one function, and graph both the given function and its inverse on one set of coordinate axes.

27. $y = 5x$

28. $y = \dfrac{3}{2}x$

29. $y = 2x - 4$

30. $y = \dfrac{3}{2}x - 2$

31. $y = \dfrac{1}{2x}$

32. $y = \dfrac{1}{x - 3}$

33. $2x + y = 4$

34. $3x + 2y = 6$

35. $x - y = 2$

36. $x + y = 0$

In Exercises 37–42, the function f defined by the given equation is one-to-one on the given domain. Find $f^{-1}(x)$.

37. $f(x) = x^2 - 3$ $\{x \mid x \le 0\}$

38. $f(x) = \dfrac{1}{x^2}$ $\{x \mid x > 0\}$

39. $f(x) = x^4 - 8$ $\{x \mid x \ge 0\}$

40. $f(x) = \dfrac{-1}{x^4}$ $\{x \mid x < 0\}$

41. $f(x) = \sqrt{4 - x^2}$ $\{x \mid 0 \le x \le 2\}$

42. $f(x) = \sqrt{x^2 - 1}$ $\{x \mid x \le -1\}$

In Exercises 43–46, find the domain and the range of f. Find the range by finding the domain of f^{-1}.

43. $f(x) = \dfrac{x}{x - 1}$

44. $f(x) = \dfrac{x - 2}{x + 3}$

45. $f(x) = \dfrac{1}{x} - 2$

46. $f(x) = \dfrac{3}{x} - \dfrac{1}{2}$

3.9 PROPORTION AND VARIATION

Quantities may be compared by using the concept of **ratio**. We might say, for example, that the *ratio* of girls to boys is 3 to 2, or that gasoline and oil are to be mixed in a *ratio* of 50 to 1.

> **Definition.** A **ratio** is the comparison of two numbers by their indicated quotient.

This definition implies that a ratio is a fraction. Thus, the denominator of a ratio cannot be 0. Some examples of ratios are

$$\frac{3}{5} \qquad \frac{x + 1}{9} \qquad \frac{a}{b} \qquad \frac{x^2 - 4}{x + 5}$$

> **Definition.** A **proportion** is a statement indicating that two ratios are equal.

Some examples of proportions are

$$\frac{2}{3} = \frac{4}{6} \qquad \frac{x}{y} = \frac{3}{5} \qquad \frac{x^2 + 8}{2(x + 3)} = \frac{17(x + 3)}{2}$$

In the proportion $\frac{a}{b} = \frac{c}{d}$, the numbers b and c are called the **means** of the proportion, and the numbers a and d are called the **extremes**. Because the equality $\frac{a}{b} = \frac{c}{d}$ is equivalent to $ad = bc$, we have the following theorem.

> **Theorem.**   In any proportion, the product of the means is equal to the product of the extremes.

Example 1   Solve the proportion $\dfrac{x}{5} = \dfrac{2}{x+3}$ for the variable x.

Solution

$$\frac{x}{5} = \frac{2}{x+3}$$

$$x(x+3) = 5 \cdot 2 \qquad \text{The product of the means equals the product of the extremes.}$$

$$x^2 + 3x = 10 \qquad \text{Remove parentheses and simplify.}$$

$$x^2 + 3x - 10 = 0 \qquad \text{Add } -10 \text{ to both sides.}$$

$$(x-2)(x+5) = 0 \qquad \text{Factor.}$$

$$x - 2 = 0 \quad \text{or} \quad x + 5 = 0 \qquad \text{Set each factor equal to 0.}$$

$$x = 2 \qquad\qquad x = -5$$

Thus, $x = 2$ or $x = -5$.   ■

Example 2   Gasoline and oil for a lawn mower are to be mixed in a 50 to 1 ratio. How many ounces of oil should be mixed with 6 gallons of gasoline?

Solution   First, express 6 gallons as $6 \cdot 128$, or 768 ounces. Let x represent the number of ounces of oil needed. Set up and solve the proportion

$$\frac{50}{1} = \frac{768}{x}$$

$$50x = 768 \qquad \text{The product of the means equals the product of the extremes.}$$

$$x = \frac{768}{50} \qquad \text{Divide both sides by 50.}$$

$$x \approx 15 \qquad \text{Read } \approx \text{ as ``is approximately equal to.''}$$

Thus, approximately 15 ounces of oil should be added to 6 gallons of gasoline.   ■

Two variables are said to be **directly proportional** if their ratio is a constant. The variables x and y are directly proportional if

$$\frac{y}{x} = k \qquad \text{or} \qquad y = kx$$

where k is a constant.

> **Definition.** The words "y **varies directly** as x," or "y is **directly propor-tional** to x," mean that $y = kx$ for some real constant k. The number k is called the **constant of proportionality**.

Example 3 Distance traveled in a given time is directly proportional to the speed. If a car travels 70 miles at 30 miles per hour, how far will it travel in the same time at 45 miles per hour?

Solution Let d represent distance traveled, and let s represent the speed. "Distance is di-rectly proportional to the speed" translates into the formula $d = ks$. The constant of proportionality k can be evaluated by using the fact that $d = 70$ when $s = 30$:

$$d = ks$$
$$70 = k \cdot 30$$
$$\frac{7}{3} = k$$

Substitute $\frac{7}{3}$ for k in the formula $d = ks$. To evaluate the distance traveled at 45 miles per hour, let $s = 45$ in the formula $d = \frac{7}{3}s$ and simplify:

$$d = \frac{7}{3}s$$
$$= \frac{7}{3}(45)$$
$$= 105$$

In the time it took to go 70 miles at 30 miles per hour, the car could travel 105 miles at 45 miles per hour. ∎

Example 4 The area of a circle varies directly as the square of the circle's radius. What is the constant of proportionality?

Solution Let A represent the area and r represent the radius. "The area varies directly as the square of the radius" translates into the formula $A = kr^2$. You know that $A = \pi r^2$ is the formula for the area of any circle. The constant of proportionality k is the number π. ∎

> **Definition.** The words "y **varies inversely** with x" or "y is **inversely pro-portional** to x" means that $y = \dfrac{k}{x}$ for some real constant k.

Example 5 The intensity of illumination from a light source varies inversely with the square of the distance from the source. If the intensity is 100 lumens at a distance of 20 feet, what will be the intensity at 30 feet?

Solution If I is the intensity and d is the distance from the light source, then the phrase "intensity varies inversely as the square of the distance" translates into the formula

$$I = \frac{k}{d^2}$$

Evaluate k by using the fact that $I = 100$ when $d = 20$. Substitute these values into the formula $I = \frac{k}{d^2}$ and solve for k:

$$I = \frac{k}{d^2}$$

$$100 = \frac{k}{20^2}$$

$$40{,}000 = k$$

Use $k = 40{,}000$ in the formula to find the intensity at 30 feet:

$$I = \frac{40{,}000}{d^2}$$

$$= \frac{40{,}000}{30^2}$$

$$= \frac{40{,}000}{900}$$

$$= \frac{400}{9}$$

At 30 feet, the intensity would be $\frac{400}{9}$ lumens per square centimeter. ∎

Definition. The words "y **varies jointly** as w and x" mean that $y = kwx$ for some real constant k.

Example 6 The energy an object has because of its motion is called **kinetic energy**. The kinetic energy of an object varies jointly as its mass and the square of its velocity. A 25-gram mass moving at a rate of 30 centimeters per second has a kinetic energy of 11,250 dyne-centimeters. What would be the kinetic energy of a 10-gram mass that is moving at 40 centimeters per second?

Solution Let E, m, and v represent the kinetic energy, mass, and velocity, respectively. The phrase "energy varies jointly as its mass and the square of its velocity" translates into the formula

$$E = kmv^2$$

The constant k can be evaluated by using the given information: $E = 11{,}250$, $m = 25$, and $v = 30$. Substitute these values into the formula $E = kmv^2$, and solve for k:

$$E = kmv^2$$
$$11{,}250 = k \cdot 25 \cdot 30^2$$
$$11{,}250 = k \cdot 22{,}500$$
$$\frac{1}{2} = k$$

Substitute $\frac{1}{2}$ for k in the formula to evaluate E when $m = 10$ and $v = 40$:

$$E = kmv^2$$
$$= \frac{1}{2} mv^2$$
$$= \frac{1}{2} \cdot 10 \cdot 40^2$$
$$= 8000$$

A 10-gram mass that is moving at 40 centimeters per second has a kinetic energy of 8000 dyne-centimeters.   ∎

The preceding terminology can be used in various combinations. In each of the following statements, the wording on the left translates into the formula on the right:

y varies directly as x and inversely as z:	$y = k\dfrac{x}{z}$
y varies jointly as the square of x and the cube root of z:	$y = kx^2 \sqrt[3]{z}$
y varies jointly as x and the square root of z and inversely as the cube root of t:	$y = \dfrac{kx\sqrt{z}}{\sqrt[3]{t}}$
y varies inversely as the product of x and z:	$y = \dfrac{k}{xz}$

Exercise 3.9

In Exercises 1–4, solve each proportion.

1. $\dfrac{4}{x} = \dfrac{2}{7}$

2. $\dfrac{5}{2} = \dfrac{x}{6}$

3. $\dfrac{x}{2} = \dfrac{3}{x+1}$

4. $\dfrac{x+5}{6} = \dfrac{7}{8-x}$

In Exercises 5–6, set up and solve a proportion to answer each question.

5. The ratio of girls to boys in class is 3 to 5. If there are 30 boys in the class, how many girls are there?

6. The ratio of lime to sand in mortar is 3 to 7. How much lime must be mixed with 21 bags of sand?

In Exercises 7–12, determine the constant of proportionality.

7. y is directly proportional to x. If $x = 30$, then $y = 15$.

8. z is directly proportional to t. If $t = 7$, then $z = 21$.

9. I is inversely proportional to R. If $R = 20$, then $I = 50$.

10. R is inversely proportional to the square of I. If $I = 25$, then $R = 100$.

11. E varies jointly as I and R. If $R = 25$ and $I = 5$, then $E = 125$.

12. z is directly proportional to the sum of x and y. If $x = 2$ and $y = 5$, then $z = 28$.

13. y is directly proportional to x. If $y = 15$ when $x = 4$, find y when $x = \frac{7}{5}$.

14. w is directly proportional to z. If $w = -6$ when $z = 2$, find w when $z = -3$.

15. P varies jointly with r and s. If $P = 16$ when $r = 5$ and $s = -8$, find P when $r = 2$ and $s = 10$.

16. m varies jointly as the square of n and the square root of q. If $m = 24$ when $n = 2$ and $q = 4$, find m when $n = 5$ and $q = 9$.

17. The volume of a gas varies directly as the temperature and inversely as the pressure. When the temperature of a certain gas is $330°$, the pressure is 40 pounds per square inch and the volume is 20 cubic feet. Find the volume if the pressure increases 10 pounds per square inch and the temperature decreases to $300°$.

18. The force required to stretch a spring a distance d is directly proportional to d. A force of 5 newtons stretches a spring 0.2 meters. What force will stretch the spring 0.35 meters?

19. The distance that an object falls in t seconds varies directly as the square of t. An object falls 16 feet in 1 second. How long will it take to fall 144 feet?

20. The power, in watts, dissipated as heat in a resistor varies jointly as the resistance, in ohms, and the square of the current, in amperes. A 10-ohm resistor carrying a current of 1 ampere dissipates 10 watts. How much power is dissipated in a 5-ohm resistor carrying a current of 3 amperes?

21. The power, in watts, dissipated as heat in a resistor varies directly as the square of the voltage and inversely as the resistance. If 20 volts are placed across a 20-ohm resistor, they will dissipate 20 watts. What voltage across a 10-ohm resistor will dissipate 40 watts?

22. The time required for one complete swing of a pendulum is called the **period** of the pendulum. The period varies directly as the square root of its length. A 1-meter pendulum has a period of 1 second. What will be the length of a pendulum with a period of 2 seconds?

23. The pitch, or frequency, of a vibrating string varies directly as the square root of the tension. If a string vibrates at a frequency of 144 hertz due to a tension of 2 pounds, what would be its frequency if the tension was 18 pounds?

24. The gravitational attraction between two massive objects varies jointly as their masses and inversely as the square of the distance between them. What happens to this force if each mass is tripled, and the distance between them is doubled?

25. The area of an equilateral triangle varies directly as the square of a side. What is the constant of proportionality?

26. The diagonal of a cube varies directly as a side. What is the constant of proportionality?

3.10 INEQUALITIES IN TWO VARIABLES

It is possible to use graphing techniques to find solution sets of **inequalities in two variables**.

Example 1 Graph the inequality $3x + 2y < 6$.

Solution Because of the trichotomy property of real numbers, one and only one of the following expressions is true for each ordered pair (x, y):

$$3x + 2y < 6 \quad \text{or} \quad 3x + 2y = 6 \quad \text{or} \quad 3x + 2y > 6$$

The graph of the function defined by the equation $3x + 2y = 6$ is a straight line. The graph of each of the two inequalities is a half-plane, one on each side of that line. For this reason, think of the graph of the equation as a boundary separating the two half-planes. The graph associated with the equation $3x + 2y = 6$ is shown in Figure 3-52.

To determine which half-plane is the graph of the inequality $3x + 2y < 6$, substitute into the inequality the coordinates of some point that does not lie on the boundary. In this case, the origin is a convenient choice:

$$3x + 2y < 6$$
$$3(0) + 2(0) < 6$$
$$0 < 6$$

Because the coordinates of the origin satisfy the inequality, the origin is a member of the half-plane represented by $3x + 2y < 6$. The graph of the inequality $3x + 2y < 6$ is shown in Figure 3-53. Note that the boundary, $3x + 2y = 6$, is drawn with a *broken* line because it is *not* included in the graph of $3x + 2y < 6$.

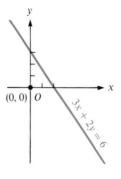

Figure 3-52

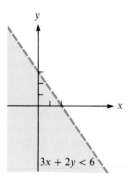

Figure 3-53

Example 2   Graph the inequality $3x + 2y \geq 6$.

Solution   The boundary of this inequality is the same line as in the previous example. If you substitute the coordinates of the origin into the given inequality $3x + 2y \geq 6$, you find that they do *not* satisfy that inequality: $0 \geq 6$ is a false statement. Therefore, the origin is *not* in the half-plane represented by the inequality $3x + 2y \geq 6$. The graph of $3x + 2y \geq 6$ is shown in Figure 3-54. Because the boundary *is* included in the graph of this solution set, use a solid line, rather than a broken line, to indicate this boundary.

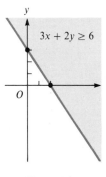

Figure 3-54   ■

In a similar fashion, we can graph inequalities in two unknowns with terms of higher degree than 1.

Example 3   Graph the inequality $y < x^2 - 2x - 3$.

Solution   First consider the function $y = x^2 - 2x - 3$, graphing the parabola with a broken curve because it is not part of the solution set of the inequality $y < x^2 - 2x - 3$. See Figure 3-55. Substitute the coordinates of the origin $(0, 0)$ into the inequality:

$$y < x^2 - 2x - 3$$
$$0 < (0)^2 - 2(0) - 3$$
$$0 < -3$$

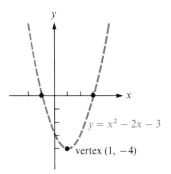

Figure 3-55

Because the statement $0 < -3$ is false, the origin is *not* in the graph of $y < x^2 - 2x - 3$, and the graph of the inequality is *below* the broken curve, as shown in Figure 3-56. ■

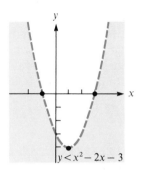

Figure 3-56

Example 4 Graph the inequality $y \geq \dfrac{1}{x^2}$.

Solution The graph of the rational function $y = \frac{1}{x^2}$ is shown in Figure 3-57. It is drawn as a solid curve because it *is* included in the solution set of $y \geq \frac{1}{x^2}$. To determine which regions are part of the solution set, substitute into the inequality the coordinates of some points, such as points P and Q in Figure 3-57, that do not lie on the curve. Point $P(2, 2)$ is a part of the solution set because its coordinates satisfy the inequality: $2 \geq \frac{1}{4}$ is a true statement. Similarly, point $Q(-2, 2)$ is also in the solution set. The graph of the inequality $y \geq \frac{1}{x^2}$ is shown in Figure 3-58.

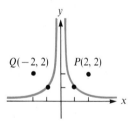

Figure 3-57

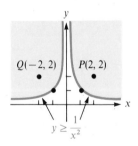

Figure 3-58 ■

Exercise 3.10

In Exercises 1–40, graph each inequality.

1. $y < 3x + 7$

2. $y > x + 2$

3. $y > 2x - y$

4. $y < 3 - x$

5. $y \geq \dfrac{1}{2}x + 1$

6. $y \leq \dfrac{1}{3}x + 2$

7. $2x - 3y \leq 12$

8. $3x + 4y \geq 12$

9. $2x + 5y > 10$

10. $3x - 5y < 15$

11. $y > 2x^2$

12. $y < 5x^2$

13. $x(x + 5) \leq x^2 + y$

14. $3x(x + 1) \geq 3x^2 + y$

15. $y > 0$ **16.** $x < 1$ **17.** $x \leq 2$ **18.** $y \leq 0$

19. $y \leq x^2 - 2$ **20.** $y \geq x^2 + 2$ **21.** $y > x^2 - 1$ **22.** $y < 2 - x^2$

23. $y > x^3$ **24.** $y \leq x^3$ **25.** $x^2 - x + y > 0$ **26.** $x^2 - 3x - y < 0$

27. $y < x^3 - x$ **28.** $y > x^3 - x^2$ **29.** $y < |x|$ **30.** $y > |x|$

31. $y \geq |x - 1|$ **32.** $y < |x - 1|$ **33.** $|y| < |x|$ **34.** $|y| > |x|$

35. $y < \dfrac{1}{x^2}$ **36.** $y < \dfrac{1}{x}$ **37.** $y \leq \dfrac{1}{x}$ **38.** $y \geq \dfrac{1}{x + 1}$

39. $xy < 0$ **40.** $xy > 0$

REVIEW EXERCISES

*In Review Exercises 1–4, find **a.** the length and **b.** the midpoint of the line segment PQ.*

1. $P(-3, 7); Q(3, -1)$ **2.** $P(0, 5); Q(-12, 10)$

3. $P(-\sqrt{3}, 9); Q(-\sqrt{3}, -7)$ **4.** $P(a, -a); Q(-a, a)$ $(a > 0)$

In Review Exercises 5–8, find the slope of the line PQ, if possible.

5. $P(3, -5); Q(1, 7)$ **6.** $P(2, 7); Q(-5, -7)$

7. $P(b, a); Q(a, b)$ **8.** $P(a + b, b); Q(b, b - a)$

In Review Exercises 9–18, write the equation of the line with the given properties. Express the answer in general form.

9. The line passes through the origin and the point $(-5, 7)$.

10. The line passes through $(-2, 1)$ and has a slope of -4.

11. The line passes through $(7, -5)$ and $(4, 1)$.

12. The line has a slope of $\frac{2}{3}$ and a y-intercept of 3.

13. The line has a slope of 0 and passes through $(-5, 17)$.

14. The line has no defined slope and passes through $(-5, 17)$.

15. The line passes through $(8, -2)$ and is parallel to the line segment joining $(2, 4)$ and $(4, -10)$.

16. The line passes through $(8, -2)$ and is perpendicular to the line segment joining $(2, 4)$ and $(4, -10)$.

17. The line is parallel to $3x - 4y = 7$ and passes through $(2, 0)$.

18. The line passes through $(7, 0)$ and is perpendicular to the line $3y + x - 4 = 0$.

In Review Exercises 19–24, determine if the given equation determines y as a function of x. For those that do, give the domain and range, and graph the function.

19. $y = 7x - 2$ **20.** $y = |x - 2|$

21. $y = \sqrt{x - 1}$ **22.** $y^2 = 2x + 7$

23. $y = x^2 - x + 1$ **24.** $y = \dfrac{x}{|x|}$

In Review Exercises 25–32, let $f(x) = 2x^2 - x$ and $g(x) = x + 3$.

25. find $f(2)$ **26.** find $g(-3)$

27. find $(f + g)(2)$ **28.** find $(f - g)(-1)$

29. find $(f \cdot g)(0)$ **30.** find $(f/g)(2)$

31. find $(f \circ g)(-2)$ **32.** find $(g \circ f)(4)$

In Review Exercises 33–38, let $f(x) = \sqrt{x + 1}$ and $g(x) = x^2 - 1$.

33. find $(f + g)(x)$ **34.** find $(f \cdot g)(x)$

35. find $(f \circ g)(x)$ **36.** find $(g \circ f)(x)$

37. determine the domain of f/g.   **38.** determine the domain of $f \circ g$.

In Review Exercises 39–42, graph each polynomial function. If the graph is a parabola, find the coordinates of its vertex.

39. $y = x^2 - 3x - 2$ **40.** $x^2 + y - 2x + 4 = 0$

41. $y = x^4 + 2x^2 + 1$ **42.** $y = x^3 - 1$

In Review Exercises 43–48, graph each rational function.

43. $y = \dfrac{2}{x - 4}$ **44.** $y = \dfrac{x^2 - 9}{x^2 - 4}$

45. $y = \dfrac{4x}{x - 3}$ **46.** $y = \dfrac{x^2 + 5x + 6}{x - 1}$

47. $y = \dfrac{x - 2}{x^2 - 3x - 4}$ **48.** $y = \dfrac{x^2 + 4}{x^2 - 3x - 4}$

In Review Exercises 49–54, determine if the given function is one-to-one. If so, determine its inverse function.

49. $y = 7x$ **50.** $y = 7x^2$ **51.** $y = \dfrac{1}{x - 1}$ **52.** $y = \dfrac{3}{x^3}$

53. $y = \dfrac{x + 2}{x - 3}$ **54.** $y = |x| + 3$

In Review Exercises 55–58, solve each word problem.

55. Hooke's law states that the force required to stretch a spring is proportional to the amount of stretch. If a 3-pound force stretches a spring 5 inches, what force would stretch the spring 3 inches?

56. The volume of gas in a balloon varies directly as its temperature, and inversely as the pressure. If the volume is 400 cc when the temperature is 300°K and the pressure is 25 dynes per sq cm, find the volume when the temperature is 200°K, and the pressure is 20 dynes per sq cm.

57. A moving body has a kinetic energy proportional to the square of its velocity. By what factor does the kinetic energy of an automobile increase if its speed increases from 30 mph to 50 mph?

58. The electrical resistance of a wire varies directly as the length of the wire, and inversely as the square of its diameter. A 1000-ft length of wire, 0.05 inches in diameter, has a resistance of 200 ohms. What would be the resistance of a 1500-ft length of wire, 0.08 inches in diameter?

In Review Exercises 59–60, find the domain and the range of each function. Find the range by finding the domain of the function's inverse.

59. $y = f(x) = \dfrac{x+1}{x-1}$

60. $y = f(x) = \dfrac{2}{x} + \dfrac{1}{2}$

In Review Exercises 61–66, graph each inequality.

61. $y \le 3x + 7$

62. $2x - y \ge 3y - 5$

63. $y > 0.5x^2$

64. $y = \dfrac{|x|}{x}$

65. $x^2 + 3x + y > 0$

66. $|y| < x$

4

EXPONENTIAL AND LOGARITHMIC FUNCTIONS

In this chapter, we discuss two functions that are important in certain applications of mathematics. The *exponential function* can be used, for example, to compute compound interest and to provide a model for population growth and radioactivity. The *logarithmic function* can be used to simplify calculations, measure the acidity of a solution or the intensity of an earthquake, and determine safe noise levels for factory workers.

4.1 EXPONENTIAL FUNCTIONS

In the discussion of exponential functions, we will consider expressions such as 3^x where x is a *real* number. Because we have only defined 3^x where x is a *rational* number, we must now give meaning to 3^x where x is an *irrational* number.

To this end, we consider the expression $3^{\sqrt{2}}$, where $\sqrt{2}$ is the irrational number $1.414213562\ldots$. Because $1 < \sqrt{2} < 2$, it can be shown that $3^1 < 3^{\sqrt{2}} < 3^2$. Similarly, $1.4 < \sqrt{2} < 1.5$, so $3^{1.4} < 3^{\sqrt{2}} < 3^{1.5}$.

The value of $3^{\sqrt{2}}$ is bounded by two numbers involving only rational powers of 3, as shown in the following list. As the list continues, $3^{\sqrt{2}}$ gets squeezed into a smaller and smaller interval:

$$3^1 = 3 \qquad < 3^{\sqrt{2}} < 9 \qquad = 3^2$$
$$3^{1.4} \approx 4.656 \qquad < 3^{\sqrt{2}} < 5.196 \qquad \approx 3^{1.5}$$
$$3^{1.41} \approx 4.7070 \quad < 3^{\sqrt{2}} < 4.7590 \quad \approx 3^{1.42}$$
$$3^{1.414} \approx 4.727695 < 3^{\sqrt{2}} < 4.732892 \approx 3^{1.415}$$

There is exactly one real number that is larger than any of the increasing numbers on the left of the previous list and less than all of the decreasing numbers on the right. By definition, that number is $3^{\sqrt{2}}$.

To find an approximation for $3^{\sqrt{2}}$, we can press the following keys on a calculator:

$$3 \quad \boxed{y^x} \quad 2 \quad \boxed{\sqrt{}} \quad \boxed{=}$$

The display will show 4.7288044. Thus,

$$3^{\sqrt{2}} \approx 4.7288044$$

In general, if b is a *positive* real number and x is *any* real number, then the exponential expression b^x represents a unique positive real number. If $b > 0$ and $b \neq 1$, the function defined by the equation $y = f(x) = b^x$ is called an **exponential function**.

> **Definition.**   The **exponential function with base b** is defined by the equation
>
> $$y = f(x) = b^x$$
>
> where $b > 0$ and $b \neq 1$.
>   The domain of the exponential function with base b is the set of real numbers. Its range is the set of positive real numbers.

Example 1   Graph the exponential functions $y = 2^x$ and $y = 7^x$.

Solution   Calculate several pairs (x, y) that satisfy each equation. Plot the points and join them with a smooth curve. The graph of $y = 2^x$ appears in Figure 4-1a, and the graph of $y = 7^x$ appears in Figure 4-1b.

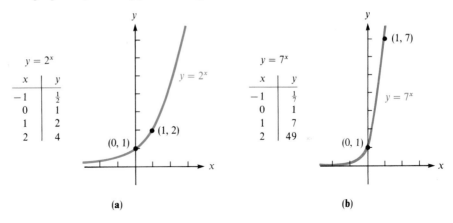

Figure 4-1

Note that $y = 2^x$ and $y = 7^x$ are *increasing* functions, that each graph passes through the point $(0, 1)$, and that the x-axis is an asymptote of each graph. Note also that the graph of $y = 2^x$ passes through the point $(1, 2)$, and that of $y = 7^x$ passes through the point $(1, 7)$.   ∎

Example 2   Graph the exponential functions $y = \left(\dfrac{1}{2}\right)^x$ and $y = \left(\dfrac{1}{7}\right)^x$.

Solution   Calculate and plot several pairs (x, y) that satisfy each equation. The graph of $y = (\frac{1}{2})^x$ appears in Figure 4-2a, and the graph of $y = (\frac{1}{7})^x$ in Figure 4-2b.

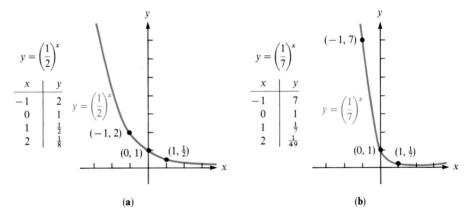

Figure 4-2

Note that $y = (\frac{1}{2})^x$ and $y = (\frac{1}{7})^x$ are *decreasing* functions, that each graph passes through the point $(0, 1)$, and that the x-axis is an asymptote of each graph. Note also that the graph of $y = (\frac{1}{2})^x$ passes through the point $(1, \frac{1}{2})$, and that of $y = (\frac{1}{7})^x$ passes through $(1, \frac{1}{7})$.   ∎

Examples 1 and 2 suggest that an exponential function with base b is either increasing (for $b > 1$) or decreasing (for $0 < b < 1$). Thus, distinct real numbers x will determine distinct values b^x. The exponential function, therefore, is one-to-one. This fact is the basis of an important fact involving exponential expressions.

If $b > 0$, $b \neq 1$ and $b^r = b^s$, then $r = s$.

Example 3   On the same set of coordinate axes, graph $y = \left(\dfrac{3}{2}\right)^x$ and $y = \left(\dfrac{2}{3}\right)^x$.

Solution   Plot several pairs (x, y) that satisfy each equation and draw each graph as in Figure 4-3.

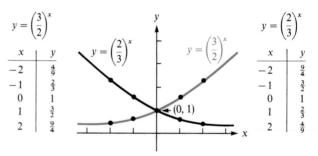

Figure 4-3

Note that $\frac{3}{2}$ and $\frac{2}{3}$ are reciprocals of each other and that the graphs are reflections of each other in the y-axis. This follows from the properties of exponents. If the number x in the first equation $y = (\frac{3}{2})^x$ is replaced with $-x$, the result is the second equation, $y = (\frac{2}{3})^x$.

$$y = \left(\frac{3}{2}\right)^x$$

$$y = \left(\frac{3}{2}\right)^{-x} \qquad \text{Replace } x \text{ with } -x.$$

$$= \left(\frac{2}{3}\right)^x$$

The symmetry discussed in this example is not the same as the y-axis symmetry discussed in Chapter 3. There, you considered curves that are reflections of themselves in the y-axis. Here, *two* curves are reflections of each other. ■

We summarize the properties of the exponential function with base b as follows:

If $b > 1$, then $y = b^x$ defines an *increasing* function.
If $0 < b < 1$, then $y = b^x$ defines a *decreasing* function.
The graph of $y = b^x$ passes through the points $(0, 1)$ and $(1, b)$.
The x-axis is an asymptote of the graph of $y = b^x$.
The graphs of $y = b^x$ and $y = b^{-x}$ are reflections of each other in the y-axis.
The exponential function defined by $y = b^x$ is one-to-one.

Example 4 Graph the function $y = 2(3^{\frac{x}{2}})$, and determine its domain and range.

Solution Plot several pairs (x, y) that satisfy the equation $y = 2(3^{\frac{x}{2}})$, and join them with a smooth curve. The graph appears in Figure 4-4. From the graph, you can

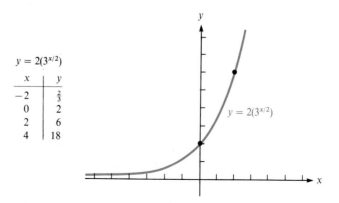

$y = 2(3^{x/2})$

x	y
-2	$\frac{2}{3}$
0	2
2	6
4	18

$y = 2(3^{x/2})$

Figure 4-4

see that the domain of the function is the set of real numbers, and the range is the set of positive real numbers. Note that the y-intercept is 2.   ■

A mathematical description of an observed event is called a **model** of that event. Many observed events can be modeled by functions defined by equations of the form

$$y = f(x) = ab^{kx}$$ Remember that ab^{kx} means $a(b^{kx})$.

where a, b, and k are constants. If f is an increasing function such as the one in Example 4, then y is said to **grow exponentially**. If f is a decreasing function, then y **decays exponentially**.

A decreasing function determined by an equation of the form $y = ab^{kx}$ provides a model for a process called **radioactive decay**. The atomic structure of radioactive material changes as the material emits radiation. Uranium, for example, decays into thorium, then into radium, and eventually into lead.

Experiments have determined the time it takes for *half* of a given amount of radioactive material to decompose. This time, called the **half-life**, is constant for any given substance. The amount A of radioactive material present decays exponentially according to the model

Radioactive Decay
Formula

$$A = A_0 2^{-\frac{t}{h}}$$

where A_0 is the amount present at $t = 0$ and h is the material's half-life.

Example 5   If the half-life of radium is 1600 years, how much of a 1-gram sample will remain after 660 years?

Solution   In this example, $A_0 = 1$, $h = 1600$, and $t = 660$. Substitute these values into the equation $A = A_0 2^{-\frac{t}{h}}$ and simplify:

$$A = A_0 2^{-\frac{t}{h}}$$

$$A = 1 \cdot 2^{-\frac{660}{1600}}$$

$$= 1 \cdot 2^{-0.4125}$$ Use a calculator.

$$\approx 0.75$$

After 660 years, approximately 0.75 gram of radium will remain.   ■

One application of exponential growth in banking is **compound interest**. If the interest earned on money in a savings account is allowed to accumulate, then that interest also earns interest. The amount in the account grows exponentially according to the equation

Compound Interest
Formula

$$A = A_0 \left(1 + \frac{r}{k}\right)^{kt}$$

where A represents the amount in the account after t years, with interest paid k times a year at an annual rate of r percent on an initial deposit A_0.

Example 6   If $1000 is deposited in an account that earns 12% interest compounded quarterly, how much will be in the account after 20 years?

Solution   Calculate A using the formula

$$A = A_0 \left(1 + \frac{r}{k}\right)^{kt}$$

with $A_0 = 1000$, $r = 0.12$, and $t = 20$. Because quarterly interest payments occur four times a year, $k = 4$.

$$A = A_0 \left(1 + \frac{r}{k}\right)^{kt}$$

$$A = 1000 \left(1 + \frac{0.12}{4}\right)^{4 \cdot 20}$$

$$= 1000(1.03)^{80}$$

$$= 10{,}640.89 \qquad \text{Use a calculator.}$$

In 20 years, the account will contain $10,640.89.   ∎

Exercise 4.1

In Exercises 1–8, graph each exponential function.

1. $y = 2^x$

2. $y = \left(\dfrac{1}{2}\right)^x$

3. $y = \left(\dfrac{1}{3}\right)^x$

4. $y = 3^x$

5. $y = 2.5^x$

6. $y = \left(\dfrac{2}{5}\right)^x$

7. $y = 10^x$

8. $y = (0.1)^x$

In Exercises 9–18, graph the function defined by each equation.

9. $y = 5(2^x)$

10. $y = 2(5^x)$

11. $y = 3(2^x)$

12. $y = 4(5^x)$

13. $y = 2^{x+1}$

14. $y = 2^{x-3}$

15. $y = 2 + 3^x$

16. $y = 3^x - 3$

17. $y = 2^{|x|}$

18. $y = 2^{-x}$

In Exercises 19–24, find the value of b, if any, that would cause the graph of $y = b^x$ to look like the graph indicated.

19.

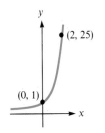

20.

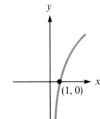

21.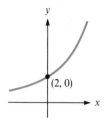

22.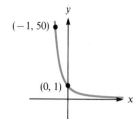

23.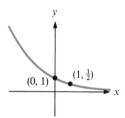

24.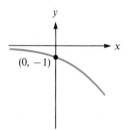

25. A radioactive material decays according to the formula $A = A_0(\frac{2}{3})^t$, where A_0 is the amount present initially and t is measured in years. What amount will be present in 5 years?

26. Tritium, a radioactive isotope of hydrogen, has a half-life of 12.4 years. Of an initial sample of 0.05 gram, how much will remain after 100 years?

27. The half-life of radioactive carbon-14 is 5700 years. How much of an initial sample will remain after 3000 years?

28. The **biological half-life** of the asthma medication theophylline is 4.5 hours for smokers and 8 hours for nonsmokers. Twelve hours after administering equal doses, what is the ratio of drug retained in a smoker's system to that in a nonsmoker's?

In Exercises 29–32, assume that there are no deposits or withdrawals.

29. An initial deposit of $500 earns 10% interest compounded quarterly. How much will be in the account in 10 years?

30. An initial deposit of $1000 earns 12% interest compounded monthly. How much will be in the account in $4\frac{1}{2}$ years?

31. If $1 had been invested in 1776 at 5% interest compounded annually, what would it be worth in 2076?

32. Some financial institutions pay daily interest, compounded by the **360/365 method**, by using the formula

$$A = A_0\left(1 + \frac{r}{360}\right)^{365t} \quad (t \text{ is in years})$$

Using this method, what will an initial investment of $1000 be worth in 5 years assuming a 12% annual interest rate?

33. A colony of 6 million bacteria is growing in a culture medium. The population P after t hours is modeled by the formula $P = (6 \times 10^6)(2.3)^t$. What is the population after 4 hours?

34. The population of North Rivers is growing exponentially according to the model $P = 375(1.3)^t$, where t is measured in years from the present date. What will be the population in 3 years?

35. A bacteria culture grows exponentially according to the model $P = P_0\, 2^{\frac{t}{24}}$, where P_0 is the initial population and t is measured in hours. By what factor will it have increased in 36 hours?

36. The charge remaining in a battery is decreasing exponentially according to the formula $C = C_0(0.7)^t$, where C is the charge remaining after t days, and C_0 is the initial charge. If a charge of 2.471×10^{-5} coulombs remains after 7 days, what was the battery's initial charge?

4.2 BASE-e EXPONENTIAL FUNCTIONS

In mathematical models of natural events, one number appears often as the base of an exponential function. This number is **e**. The symbol *e* was first used by Leonhard Euler (1707–1783). We introduce this important number by allowing *k* in the compound interest formula

$$A = A_0 \left(1 + \frac{r}{k}\right)^{kt}$$

to become very large. To see what happens, we let $k = rp$, where p is a new variable, and proceed as follows:

$$A = A_0 \left(1 + \frac{r}{k}\right)^{kt}$$

$$= A_0 \left(1 + \frac{r}{rp}\right)^{rpt} \qquad \text{Substitute } rp \text{ for } k.$$

$$= A_0 \left(1 + \frac{1}{p}\right)^{rpt} \qquad \text{Simplify } \frac{r}{rp}.$$

$$= A_0 \left[\left(1 + \frac{1}{p}\right)^{p}\right]^{rt} \qquad \text{Remember that } (x^m)^n = x^{mn}.$$

Because r is a positive constant and $k = rp$, it follows that as k becomes very large, then so does p. The question of what happens to the value of A becomes tied to the question: What happens to the value of $(1 + \frac{1}{p})^p$ as p becomes very large?

Some results calculated for increasing values of p appear in Table 4-1.

Table 4-1

p	$(1 + \frac{1}{p})^p$
1	2
10	2.5937
1000	2.7169
1,000,000	2.7182805
$\vdots$	$\vdots$

The results in the table suggest that as p increases, the value of $(1 + \frac{1}{p})^p$ approaches a fixed number. This number is e, an irrational number with a decimal representation of 2.71828182845904. . . .

If interest on an amount A_0 is compounded more and more often, the number p grows large without bound and the formula

$$A = A_0 \left[\left(1 + \frac{1}{p} \right)^p \right]^{rt}$$

becomes

Continuous Compound Interest Formula

$$A = A_0 e^{rt}$$

When the amount invested grows exponentially according to the formula $A = A_0 e^{rt}$, interest is said to be **compounded continuously**.

Example 1   If \$1000 accumulates interest at an annual rate of 12% compounded continuously, how much money will be in the account in 20 years?

Solution $A = A_0 e^{rt}$

$A = 1000 e^{(0.12)(20)}$ Substitute 1000 for A_0, 0.12 for r, and 20 for t.

$\approx 1000(11.02318)$ Use a calculator.

$\approx 11{,}023.18$

In 20 years, the account will contain \$11,023.18.   ■

The exponential function $y = e^x$ is so important that it is often called *the* exponential function.

Example 2   Graph the exponential function.

Solution   Use a calculator to find several pairs (x, y) that satisfy the equation $y = e^x$. Plot them and join them with a smooth curve. The graph appears in Figure 4-5.

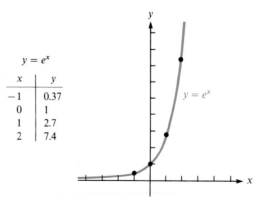

$y = e^x$	
x	y
-1	0.37
0	1
1	2.7
2	7.4

Figure 4-5   ■

Example 3   Graph $y = 3e^{-\frac{x}{2}}$.

Solution   Plot several pairs (x, y) that satisfy the equation $y = 3e^{-\frac{x}{2}}$ and join them with a smooth curve. The graph appears in Figure 4-6.

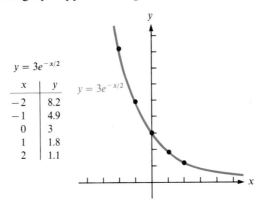

$y = 3e^{-x/2}$

x	y
-2	8.2
-1	4.9
0	3
1	1.8
2	1.1

$y = 3e^{-x/2}$

Figure 4-6

A function similar to the exponential function provides a model for **population growth**. A population (of people, fish, bacteria, or other living organisms) changes with time. Many factors—such as birth and death rates, immigration, pollution, diet, wars, plagues, and famines—affect the population. Models of population growth that account for several factors are very complex. A simpler model, called the **Malthusian model of population growth**, assumes a constant birth rate B and constant death rate D and incorporates no other factors. In that model, the population P grows exponentially according to the formula

Population Growth
Formula

$$P = P_0 e^{kt}$$

where P_0 is the population at $t = 0$, and $k = B - D$. If t is measured in years, then k is called the **annual growth rate**.

Example 4   The annual birth rate in a certain country is 19 per 1000 population, and the death rate is 7 per 1000. What would the Malthusian model predict the population of the country to be in 50 years if the current population is 2.3 million?

Solution   Use the Malthusian model of population growth

$$P = P_0 e^{kt}$$

The number k is the difference between the birth and death rates. The birth rate, B, is $\frac{19}{1000}$, or 0.019. The death rate, D, is $\frac{7}{1000}$, or 0.007. Thus,

$$k = B - D$$
$$k = 0.019 - 0.007$$
$$= 0.012$$

Substitute 2.3×10^6 for P_0, 50 for t, and 0.012 for k in the equation $P = P_0e^{kt}$ and simplify:

$$P = P_0e^{kt}$$
$$P = (2.3 \times 10^6)e^{(0.012)(50)}$$
$$= (2.3 \times 10^6)(1.82)$$
$$= 4.2 \times 10^6$$

After 50 years, the population will exceed 4 million.   ∎

Example 5   A population of 1000 bacteria doubles in 8 hours. Assuming the Malthusian model, what will be the population in 12 hours?

Solution   The population P grows according to the formula $P = P_0e^{kt}$. Let $P_0 = 1000$, $P = 2000$, and $t = 8$. Then proceed as follows:

$$P = P_0e^{kt}$$
$$2000 = 1000e^{k8} \qquad \text{Substitute 2000 for } P, \text{ 1000 for } P_0, \text{ and 8 for } t.$$
$$2 = e^{k \cdot 8} \qquad \text{Divide both sides by 1000.}$$
$$2^{\frac{1}{8}} = (e^{k \cdot 8})^{\frac{1}{8}} \qquad \text{Raise both sides to the } \tfrac{1}{8} \text{ power.}$$
$$2^{\frac{1}{8}} = e^{k}$$

We know that the population grows according to the formula

$$P = 1000e^{kt}$$
$$= 1000(2^{\frac{1}{8}})^t \qquad \text{Substitute } 2^{\frac{1}{8}} \text{ for } e^k.$$
$$= 1000(2^{\frac{t}{8}})$$

To find the population after 12 hours, substitute 12 for t and simplify:

$$P = 1000(2^{\frac{t}{8}})$$
$$= 1000(2^{\frac{12}{8}})$$
$$= 1000(2^{\frac{3}{2}})$$
$$\approx 1000(2.8284) \qquad \text{Use a calculator.}$$
$$\approx 2800$$

After 12 hours, there are approximately 2800 bacteria.   ∎

Exercise 4.2

In Exercises 1–8, graph the function defined by each equation. **Use a calculator.**

1. $y = -e^x$ **2.** $y = e^{-x}$ **3.** $y = e^{-0.5x}$ **4.** $y = -e^{2x}$

5. $y = 2e^{-x}$ **6.** $y = -3e^x$ **7.** $y = e^x + 1$ **8.** $y = 2 - e^x$

In Exercises 9–16, tell whether the graph of $y = e^x$ could look like the graph indicated.

9.

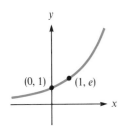

10.

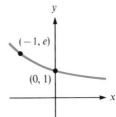

11.

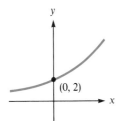

12.

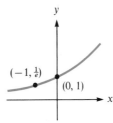

13.

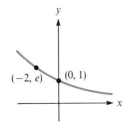

14.

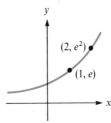

15.

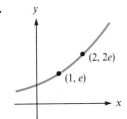

16.

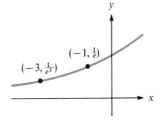

In Exercises 17–21, assume no deposits or withdrawals.

17.   An initial investment of $5000 earns 11.2% interest compounded continuously. What will the investment be worth in 12 years?

18.   An initial deposit of $2000 earns 8% interest compounded continuously. How much will be in the account in 15 years?

19.   An account now contains $11,180. It has been accumulating interest at 13%, compounded continuously, for 7 years. What was the initial deposit?

20.   An account now contains $3610. It has been accumulating interest at $10\frac{1}{2}$% compounded continuously. How much was in the account 1 year ago?

21. An initial deposit grows at a continuously compounded annual rate of 14%. If $5000 is in the account after 2 years, how much will be in the account after 6 years?

22. The growth of a population is modeled by

$$P = 173e^{0.03t}$$

How large will the population be when $t = 20$?

23. The decline of a population is modeled by

$$P = 1.2 \times 10^6 e^{-0.008t}$$

How large will the population be when $t = 30$?

24. The world population is approximately 4.2 billion and is growing at an annual rate of 1.9%. Assuming a Malthusian growth model, what will be the world's population in 30 years?

25. Assuming a Malthusian model and an annual growth rate of 1.9%, by what factor will the world's current population increase in 50 years?

26. A country's population is now 2×10^5 people and is expected to double every 20 years. Assuming a Malthusian model, what will be the population in 35 years?

27. The population of a small town is presently 140 and is expected to grow exponentially, tripling every 15 years. Assuming a Malthusian model, what can the city planners expect the population to be in 5 years?

28. The amount A of a drug remaining in a person's system after t hours is given by the formula

$$A = A_0 e^{kt}$$

where A_0 is the initial dose. After 2.3 hours, one-half of an initial dose of triazolam, a drug for treating insomnia, will remain. What percent will remain after 24 hours?

29. On a sheet of graph paper, graph the function $y = \frac{1}{2}(e^x + e^{-x})$ for values of x between -2 and 2. You'll need to calculate and plot about five or six points before joining them with a smooth curve. The graph looks like a parabola, but it is not. It is called a **catenary** and is important in the design of power distribution networks because it represents the shape of a cable drooping between its supporting poles.

30. The value of e can be calculated to any degree of accuracy by adding the first several terms of the following list.

$$1, 1, \frac{1}{2}, \frac{1}{2 \cdot 3}, \frac{1}{2 \cdot 3 \cdot 4}, \cdots, \frac{1}{2 \cdot 3 \cdots \cdot n}, \cdots$$

The more terms that are added, the closer the sum is to the actual value of e. Calculate an approximation of the value of e by adding the first eight values in the list above. To how many decimal places is your sum accurate?

4.3 LOGARITHMIC FUNCTIONS

Because an exponential function defined by $y = b^x$ is one-to-one, it has an inverse that is defined by the equation $x = b^y$. To express this inverse function in the form $y = f^{-1}(x)$, we must solve the equation $x = b^y$ for y. To do so, we need the following definition.

> **Definition.**   The **logarithmic function with base b** is defined by the equation
>
> $$y = \log_b x$$
>
> where $b > 0$ and $b \neq 1$. This equation is equivalent to the exponential equation
>
> $$x = b^y$$
>
> The domain of the logarithmic function is the set of positive real numbers. Its range is the set of real numbers.

Because the function $y = \log_b x$ is the inverse of the one-to-one exponential function $y = b^x$, the function $y = \log_b x$ is one-to-one also.

The previous definition implies that any pair (x, y) that satisfies the equation $x = b^y$ also satisfies the equation $y = \log_b x$ (or $\log_b x = y$). Thus,

$$\log_5 25 = 2 \quad \text{because} \quad 25 = 5^2$$
$$\log_7 1 = 0 \quad \text{because} \quad 1 = 7^0$$
$$\log_{16} 4 = \frac{1}{2} \quad \text{because} \quad 4 = 16^{\frac{1}{2}}$$

and

$$\log_2 \frac{1}{8} = -3 \quad \text{because} \quad \frac{1}{8} = 2^{-3}$$

Note that in each case, the logarithm of a number is an exponent.

Because the domain of the logarithmic function is the set of positive real numbers, it is impossible to find the logarithm of 0 or of a negative number.

Example 1   Find the value of y in each of the following equations: **a.** $\log_5 1 = y$, **b.** $\log_2 8 = y$, and **c.** $\log_7 \frac{1}{7} = y$.

Solution   **a.**   Change the equation $\log_5 1 = y$ into the equivalent exponential form $1 = 5^y$. Because $1 = 5^0$, it follows that $y = 0$. Hence, $\log_5 1 = 0$.

b. $\log_2 8 = y$ is equivalent to $8 = 2^y$. Because $8 = 2^3$, it follows that $y = 3$. Hence, $\log_2 8 = 3$.

c. $\log_7 \frac{1}{7} = y$ is equivalent to $\frac{1}{7} = 7^y$. Because $\frac{1}{7} = 7^{-1}$, it follows that $y = -1$. Hence, $\log_7 \frac{1}{7} = -1$.   ∎

Example 2   Find the value of a in each equation: **a.** $\log_3 \frac{1}{9} = a$, **b.** $\log_a 32 = 5$, and **c.** $\log_9 a = -\frac{1}{2}$.

Solution   **a.** $\log_3 \frac{1}{9} = a$ is equivalent to $\frac{1}{9} = 3^a$. Because $\frac{1}{9} = 3^{-2}$, it follows that $a = -2$.

b. $\log_a 32 = 5$ is equivalent to $a^5 = 32$. Because $2^5 = 32$, it follows that $a = 2$.

c. $\log_9 a = -\frac{1}{2}$ is equivalent to $9^{-\frac{1}{2}} = a$. Because $9^{-\frac{1}{2}} = \frac{1}{3}$, it follows that $a = \frac{1}{3}$.   ∎

Example 3   Graph the logarithmic function defined by $y = \log_2 x$.

Solution   The equation $y = \log_2 x$ is equivalent to the equation $x = 2^y$. Calculate and plot pairs (x, y) that satisfy the equation $x = 2^y$ and connect them with a smooth curve. The graph appears in Figure 4-7.

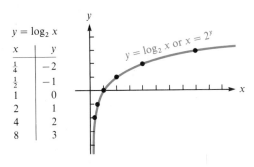

$y = \log_2 x$

x	y
$\frac{1}{4}$	-2
$\frac{1}{2}$	-1
1	0
2	1
4	2
8	3

$y = \log_2 x$ or $x = 2^y$

Figure 4-7

Example 4   Graph the logarithmic function defined by $y = \log_{\frac{1}{2}} x$.

Solution   Rewrite $y = \log_{\frac{1}{2}} x$ as $x = (\frac{1}{2})^y$ and proceed as in Example 3. The graph appears in Figure 4-8.

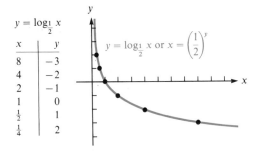

$y = \log_{\frac{1}{2}} x$

x	y
8	-3
4	-2
2	-1
1	0
$\frac{1}{2}$	1
$\frac{1}{4}$	2

$y = \log_{\frac{1}{2}} x$ or $x = \left(\frac{1}{2}\right)^y$

Figure 4-8

Examples 3 and 4 suggest that the graphs of logarithmic functions are similar (to those in Figure 4-9. If $b > 1$, the logarithmic function is *increasing* as in Figure 4-9**a**, and if $0 < b < 1$, the logarithmic function is *decreasing* as in Figure 4-9**b**. Note that each graph of $y = \log_b x$ passes through the points $(1, 0)$ and $(b, 1)$, and that the y-axis is an asymptote to the curve.

Because the logarithmic function is one-to-one, we have the following property of logarithms:

If $\log_b r = \log_b s$, then $r = s$.

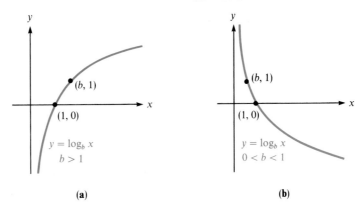

Figure 4-9

Several other properties of logarithms can be found by expressing the properties of exponents in logarithmic form.

Properties of Logarithms.   If M, N, p, and b are positive numbers, and $b \neq 1$, then

1. $\log_b 1 = 0$ **5.** $\log_b MN = \log_b M + \log_b N$

2. $\log_b b = 1$

6. $\log_b \dfrac{M}{N} = \log_b M - \log_b N$

3. $\log_b b^x = x$

4. $b^{\log_b x} = x$ **7.** $\log_b M^p = p \log_b M$

Proof   Properties 1 through 4 follow directly from the definition. To prove Property 5, let $m = \log_b M$ and $n = \log_b N$ Using the definition of logarithm gives these equations in the form

$$M = b^m \qquad \text{and} \qquad N = b^n$$

We multiply equal quantities by equal quantities to get

$$MN = b^m b^n$$

or

$$MN = b^{m+n}$$

By the definition of a logarithm, this equation is equivalent to

$$\log_b MN = m + n$$

We substitute the values of m and n to complete the proof of Property 5.

$$\log_b MN = \log_b M + \log_b N \qquad \square$$

The proofs of Properties 6 and 7 are similar and are left as exercises.

Property 5 of logarithms asserts that the logarithm of the *product* of two numbers is equal to the *sum* of their logarithms. The logarithm of a *sum* usually does not simplify. In general,

$$\log_b(M + N) \neq \log_b M + \log_b N$$

Similarly,

$$\log_b(M - N) \neq \log_b M - \log_b N$$

Example 5   If A, B, C, and b are positive numbers, then

a. $\log_b \dfrac{AB}{C} = \log_b(AB) - \log_b C$ Use Property 6.

$\qquad\qquad = \log_b A + \log_b B - \log_b C$ Use Property 5.

b. $\log_b(A^3B^2) = \log_b A^3 + \log_b B^2$ Use Property 5.

$\qquad\qquad = 3\log_b A + 2\log_b B$ Use Property 7.

c. $\log_b b^2 = 2$ Use Property 3.

d. $b^{\log_b 3} = 3$ Use Property 4. ∎

Example 6   Given that $\log_{10} 2 \approx 0.3010$ and $\log_{10} 3 \approx 0.4771$, find approximate values for **a.** $\log_{10} 18$ and **b.** $\log_{10} 2.5$.

Solution   **a.** $\log_{10} 18 = \log_{10}(2 \cdot 3^2)$ Factor 18.

$\qquad\qquad = \log_{10} 2 + \log_{10} 3^2$ Use Property 5.

$\qquad\qquad = \log_{10} 2 + 2\log_{10} 3$ Use Property 7.

$\qquad\qquad \approx 0.3010 + 2(0.4771)$ Substitute the value of each logarithm.

$\qquad\qquad \approx 1.2552$ Simplify.

b. $\log_{10} 2.5 = \log_{10}\left(\dfrac{5}{2}\right)$ Write 2.5 as $\frac{5}{2}$.

$\qquad\qquad = \log_{10} 5 - \log_{10} 2$ Use Property 6.

$\qquad\qquad = \log_{10} \dfrac{10}{2} - \log_{10} 2$ Write 5 as $\frac{10}{2}$.

$\qquad\qquad = \log_{10} 10 - \log_{10} 2 - \log_{10} 2$ Use Property 6.

$\qquad\qquad = 1 - 2\log_{10} 2$ Use Property 2 and combine terms.

$\qquad\qquad \approx 1 - 2(0.3010)$ Substitute 0.3010 for $\log_{10} 2$.

$\qquad\qquad \approx 0.3980$ Simplify. ∎

Logarithms are easiest to find by using a calculator. Most scientific calculators provide both base 10 and base e logarithms.

Logarithms with the base of 10 are called **common logarithms**. In this book, if the base b is not indicated in the notation $\log_b x$, assume that b is 10:

$$\log A \quad \text{means} \quad \log_{10} A$$

Because the number e appears often in mathematical models of events in nature, base-e logarithms are called **natural logarithms**. They are also called **Naperian logarithms** after John Napier (1550–1617). Natural logarithms are usually denoted by the symbol $\ln x$, rather than $\log_e x$.

$$\ln x \quad \text{means} \quad \log_e x$$

Example 7 Use a calculator to find **a.** log 2.34 and **b.** ln 2.34.

Solution **a.** To find log 2.34, enter the number 2.34 and press the $\boxed{\log}$ key. (You may have to press a $\boxed{2^{nd}}$ function key first.) The display should read .3692158574. Hence,

$$\log 2.34 \approx 0.3692$$

b. To find ln 2.34, enter the number 2.34 and press the $\boxed{\ln x}$ key. The display should read .850150929. Hence,

$$\ln 2.34 \approx 0.8502 \qquad \blacksquare$$

Example 8 Find the value of x in each equation: **a.** log $x = 0.7482$ and **b.** ln $x = 1.335$.

Solution **a.** $\log x = 0.7482$ is equivalent to $10^{0.7482} = x$. To find x, enter the number 10, press the $\boxed{y^x}$ key, enter the number .7482, and press $\boxed{=}$. The display reads 5.6001544. Hence,

$$x \approx 5.6$$

If your calculator has a $\boxed{10^x}$ key, simply enter .7482 and press it to get the same result. (You might have to press a $\boxed{2^{nd}}$ function key.)

b. $\ln x = 1.335$ is equivalent to $e^{1.335} = x$. To find x, enter the number 1.335 and press the $\boxed{e^x}$ key. The display reads 3.79999595. Hence,

$$x \approx 3.8 \qquad \blacksquare$$

Just as $y = e^x$ represented a special exponential function, so does $y = \ln x$ represent a special logarithmic function.

Example 9 Graph the function determined by $y = \ln x$.

Solution Find and plot several pairs (x, y) that satisfy the equation $y = \ln x$, and join them with a smooth curve. The graph appears in Figure 4-10.

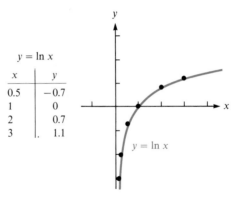

$y = \ln x$

x	y
0.5	-0.7
1	0
2	0.7
3	1.1

$y = \ln x$

Figure 4-10

Applications of Base-10 Logarithms

Example 10   In chemistry, common logarithms are used to express the acidity of solutions. The more acidic a solution, the greater the concentration of hydrogen ions. This concentration is indicated indirectly by the **pH scale**, or **hydrogen-ion index**. The pH of a solution is defined by the equation

$$pH = -\log[H^+]$$

where $[H^+]$ is the hydrogen-ion concentration in gram-ions per liter. Pure water has a few free hydrogen ions—$[H^+]$ is approximately 10^{-7} gram-ions per liter. The pH of pure water is

$$pH = -\log 10^{-7}$$
$$= -(-7)\log 10 \qquad \text{Use Property 7.}$$
$$= -(-7) \qquad \text{Use Property 2.}$$
$$= 7$$

Seawater has a pH of approximately 8.5, and its hydrogen-ion concentration is found by solving the equation $8.5 = -\log_{10}[H^+]$ for $[H^+]$.

$$8.5 = -\log[H^+]$$
$$-8.5 = \log[H^+]$$
$$[H^+] = 10^{-8.5} \qquad \text{Change the equation from logarithmic to exponential form.}$$

Use a calculator to find that

$$[H^+] \approx 3.2 \times 10^{-9} \text{ gram-ions per liter}$$

Example 11   In electrical engineering, common logarithms are used to express the voltage gain (or loss) of an electronic device such as an amplifier or a length of transmission line. The unit of gain (or loss), called the **decibel**, is defined by a logarithmic relation. If E_O is the output voltage of a device, and E_I is the input voltage, the decibel voltage gain is defined as

$$\text{decibel voltage gain} = 20 \log \frac{E_O}{E_I}$$

If, for example, the input to an amplifier is 0.5 volt and the output is 40 volts, the decibel voltage gain is calculated by substituting these values into the formula:

$$\text{decibel voltage gain} = 20 \log \frac{E_O}{E_I}$$

$$\text{decibel voltage gain} = 20 \log \frac{40}{0.5}$$

$$= 20 \log 80$$

$$\approx 20(1.9031)$$

$$\approx 38$$

The amplifier provides a 38 decibel voltage gain.   ■

Example 12   In seismology, common logarithms are used to measure the intensity of earthquakes on the **Richter scale**. The intensity R is given by

$$R = \log \frac{A}{P}$$

where A is the amplitude of the tremor (measured in micrometers) and P is the period of the tremor (the time of one oscillation of the earth's surface, measured in seconds). To calculate the intensity of an earthquake with an amplitude of 10,000 micrometers (1 centimeter) and a period of 0.1 second, substitute 10,000 for A and 0.1 for P in the formula and simplify:

$$R = \log \frac{A}{P}$$

$$R = \log \frac{10,000}{0.1}$$

$$= \log 100,000$$

$$= \log 10^5$$

$$= 5 \log 10 \qquad \text{Use Property 7.}$$

$$= 5 \qquad \text{Use Property 2.}$$

The earthquake measures 5 on the Richter scale.   ■

Applications of Base-e Logarithms

Example 13   In electronics, a mathematical model of the time required for charging a battery uses the natural logarithm function. A battery charges at a rate that depends on how close it is to being fully charged; it charges fastest when it is most

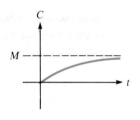

Figure 4-11

discharged. The charge C at any instant t is modeled by the formula

$$C = M(1 - e^{-kt})$$

where M is the theoretical maximum charge that the battery can hold, and k is a positive constant that depends on the battery and the charger. Plotting the variable C against t gives a curve like that in Figure 4-11. Notice that the full charge M is never attained; the actual charge can come very close to M, however, if the battery is charged long enough. To determine how long it will take a battery to reach a given charge C, solve the equation $C = M(1 - e^{-kt})$ for t as follows:

$$C = M(1 - e^{-kt})$$

$$\frac{C}{M} = 1 - e^{-kt} \qquad \text{Divide both sides by } M.$$

$$\frac{C}{M} - 1 = -e^{-kt} \qquad \text{Add } -1 \text{ to both sides.}$$

$$1 - \frac{C}{M} = e^{-kt} \qquad \text{Multiply both sides by } -1.$$

$$\ln\left(1 - \frac{C}{M}\right) = -kt \qquad \begin{array}{l}\text{Change the exponential equation}\\\text{to logarithmic form.}\end{array}$$

$$-\frac{1}{k}\ln\left(1 - \frac{C}{M}\right) = t \qquad \text{Multiply both sides by } -\tfrac{1}{k}.$$

The formula that determines the time t required to charge a battery to a given level C is

$$t = -\frac{1}{k}\ln\left(1 - \frac{C}{M}\right)$$

■

Example 14   In physiology, experiments suggest that the relationship of loudness and intensity of sound is a logarithmic one known as the **Weber-Fechner law**: the apparent loudness L of a sound is proportional to the natural logarithm of its actual intensity I. In symbols,

$$L = k \ln I$$

For example, what actual increase in intensity will cause a doubling of the apparent loudness? If the original loudness is L_0, caused by an actual intensity I_0, then $L_0 = k \ln I_0$. To double the apparent loudness, multiply both sides of the equation by 2 and use Property 7 of logarithms:

$$2L_0 = 2k \ln I_0$$
$$= k \ln(I_0)^2$$

Thus, to double the apparent volume of a sound, the actual intensity must be squared.

■

Exercise 4.3

In Exercises 1–24, find the value of x. A calculator is of no value.

1. $\log_2 8 = x$

2. $\log_{\frac{1}{2}} \frac{1}{8} = x$

3. $\log_{\frac{1}{2}} 8 = x$

4. $\log_{25} 5 = x$

5. $\log_5 25 = x$

6. $\log_8 x = 2$

7. $\log_x 8 = 3$

8. $\log_7 x = 0$

9. $\log_7 x = 1$

10. $\log_4 x = \frac{1}{2}$

11. $\log_x \frac{1}{16} = -2$

12. $\log_{125} x = \frac{2}{3}$

13. $\log_{100} \frac{1}{1000} = x$

14. $\log_{\frac{5}{2}} \frac{4}{25} = x$

15. $\log_{27} 9 = x$

16. $\log_{12} x = 0$

17. $\log_x 5^3 = 3$

18. $\log_x 5 = 1$

19. $\log_x \frac{9}{4} = 2$

20. $\log_x \frac{\sqrt{3}}{3} = \frac{1}{2}$

21. $\log_{\sqrt{3}} x = -4$

22. $\log_{\pi} x = 3$

23. $\log_{2\sqrt{2}} x = 2$

24. $\log_4 8 = x$

In Exercises 25–32, graph the function determined by each equation.

25. $y = \log_5 x$

26. $y = \log_{\frac{1}{5}} x$

27. $y = \log_{\frac{1}{3}} x$

28. $y = \log_3 x$

29. $y = \ln x^2$

30. $y = \ln 2x$

31. $y = (\ln x) - 1$

32. $y = \ln(x - 1)$

In Exercises 33–38, graph each pair of equations on one set of coordinate axes.

33. $y = \log_3 x$ and $y = \log_3(3x)$

34. $y = \log_3 x$ and $y = \log_3\left(\dfrac{x}{3}\right)$

35. $y = \log_2 x$ and $y = \log_2(x + 1)$

36. $y = \log_2 x$ and $y = \log_2(-x)$

37. $y = \log_5 x$ and $y = 5^x$

38. $y = \ln x$ and $y = e^x$

In Exercises 39–44, find the value of b, if any, that would cause the graph of $y = \log_b x$ to look like the graph indicated.

39.

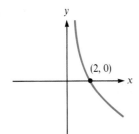

(2, 0)

40.

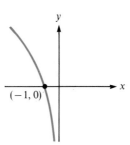

(−1, 0)

41.

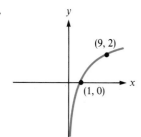

(9, 2)

(1, 0)

42.

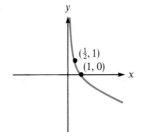

$(\frac{1}{2}, 1)$

(1, 0)

43.

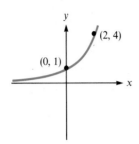

44.

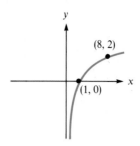

In Exercises 45–48, tell whether the graph of $y = \ln x$ could look like the graph indicated.

45.

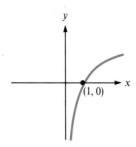

46.

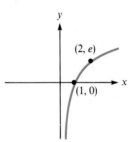

47.

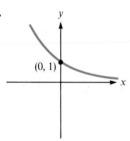

48.

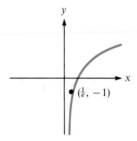

In Exercises 49–70, tell if the given statement is true. If it is not true, so indicate.

49. $\log_b ab = \log_b a + 1$

50. $\log_b \dfrac{1}{a} = -\log_b a$

51. $\log_b 0 = 1$

52. $\log_b 2 = \log_2 b$

53. $\log_b(x + y) \ne \log_b x + \log_b y$

54. $\log_b xy = (\log_b x)(\log_b y)$

55. If $\log_a b = c$, then $\log_b a = c$.

56. If $\log_a b = c$, then $\log_b a = \dfrac{1}{c}$.

57. $\log_7 7^7 = 7$

58. $7^{\log_7 7} = 7$

59. $\log_b(-x) = -\log_b x$

60. If $\log_b a = c$, then $\log_b a^p = pc$.

61. $\dfrac{\log A}{\log B} = \log A - \log B$

62. $\log(A - B) = \dfrac{\log A}{\log B}$

63. $\log \dfrac{1}{5} = -\log 5$

64. $3 \log_b \sqrt[3]{a} = \log_b a$

65. $\dfrac{1}{3} \log_b a^3 = \log_b a$

66. A logarithm cannot be negative.

67. $\log 10^3 = 3(10^{\log 3})$

68. If x lies between 0 and 1, $\log_b x$ is negative.

69. $\log_{\frac{4}{3}} y = -\log_{\frac{3}{4}} y$

70. $\log_b y + \log_{\frac{1}{b}} y = 0$

In Exercises 71–80, assume that $\log_{10} 4 \approx 0.6021$, $\log_{10} 7 \approx 0.8451$, and $\log_{10} 9 \approx 0.9542$. Use these values and the properties of logarithms to find the approximate value of each quantity.

71. $\log_{10} 28$

72. $\log_{10} \dfrac{7}{4}$

73. $\log_{10} 2.25$

74. $\log_{10} 36$

75. $\log_{10} \dfrac{63}{4}$

76. $\log_{10} \dfrac{4}{63}$

77. $\log_{10} 252$

78. $\log_{10} 49$

79. $\log_{10} 112$

80. $\log_{10} 324$

In Exercises 81–86, use a calculator to find the value of the variable. Express all answers to four decimal places.

81. $\log 3.25 = x$

82. $\ln 7.39 = x$

83. $\ln y = 4.24$

84. $\log y = 0.926$

85. $\log M = \ln 8$

86. $\ln M = \log 7$

87. Find the pH of a solution with a hydrogen-ion concentration of 1.7×10^{-5} gram-ions per liter.

88. What is the hydrogen-ion concentration of a saturated solution of calcium hydroxide whose pH is 13.2?

89. The pH of apples can range from 2.9 to 3.3. What is the range in the hydrogen-ion concentration?

90. The hydrogen-ion concentration of sour pickles is 6.31×10^{-4}. What is the pH?

91. The decibel voltage gain of an amplifier is 29. If the output is 20 volts, what is the input voltage?

92. The decibel voltage gain of an amplifier is 35. If the input signal is 0.05 volt, what is the output voltage?

93. The power output (or input) of an amplifier is directly proportional to the square of the voltage output (or input). Show that the formula for decibel voltage gain is

$$\text{decibel voltage gain} = 10 \log \frac{P_O}{P_I}$$

where P_O is the power output and P_I is the power input.

94. An amplifier produces an output of 30 watts when driven by an input signal of 0.1 watt. What is the amplifier's voltage gain?

95. An earthquake has an amplitude of 5000 micrometers and a period of 0.2 second. What does it measure on the Richter scale?

96. An earthquake with amplitude of 8000 micrometers measures 6 on the Richter scale. What is its period?

97. An earthquake with a period of $\frac{1}{4}$ second measures 4 on the Richter scale. What is its amplitude?

98. By what factor must the period of an earthquake change to increase its severity by 1 point on the Richter scale? Assume that the amplitude remains constant.

99. If a battery can reach half of its full charge in 6 hours, how long will it take the battery to reach a 90% charge? Assume that the battery was fully discharged when it began charging.

100. A battery reaches 80% of a full charge in 8 hours. If it started charging when it was fully discharged, how long did it take to reach a 40% charge?

101. If the intensity of a sound is doubled, what is the apparent change in loudness?

102. What increase in intensity of sound will cause an apparent tripling of the loudness?

103. If the intensity of a sound is tripled, what is the apparent change in loudness?

104. Prove that $e^{x \ln a} = a^x$.

105. Prove that $e^{\ln x} = x$.

106. Prove that $\ln(e^x) = x$.

107. Show that the equation $t = -\dfrac{1}{k} \ln\left(1 - \dfrac{C}{M}\right)$ can be written in the form $t = \ln\left(\dfrac{M}{M-C}\right)^{\frac{1}{k}}$.

108. Prove Property 6 of logarithms.

109. Prove Property 7 of logarithms.

4.4 EXPONENTIAL AND LOGARITHMIC EQUATIONS

An **exponential equation** is one that contains the variable in an exponent. A **logarithmic equation** is one that involves logarithms of expressions that contain the variable.

Example 1 Solve the exponential equation $3^x = 5$.

Solution Because the logarithms of equal numbers are equal, you can take the common logarithm of both sides of the equation. Property 7 of logarithms then provides a means for moving the variable x from its position as an exponent to a position as a factor.

$$3^x = 5$$
$$\log 3^x = \log 5 \qquad \text{Take the common logarithm of both sides.}$$
$$x \log 3 = \log 5 \qquad \text{Use Property 7 of logarithms.}$$
1. $\qquad x = \dfrac{\log 5}{\log 3} \qquad \text{Divide both sides by log 3.}$
$$\approx \dfrac{0.6990}{0.4771} \qquad \text{Substitute values for log 5 and log 3.}$$
$$\approx 1.465$$

Thus, $x \approx 1.465$.

A careless reading of Equation 1 can lead to a common error. Because $\log \dfrac{A}{B} = \log A - \log B$, you may think that the expression $\dfrac{\log 5}{\log 3}$ also involves subtraction. It does not. The expression $\dfrac{\log 5}{\log 3}$ calls for division. ∎

Example 2 Solve the exponential equation $6^{x-3} = 2^x$.

Solution

$$6^{x-3} = 2^x$$

$\log 6^{x-3} = \log 2^x$	Take the common logarithm of both sides.
$(x - 3) \log 6 = x \log 2$	Use Property 7 of logarithms.
$x \log 6 - 3 \log 6 = x \log 2$	Remove parentheses.
$x \log 6 - x \log 2 = 3 \log 6$	Add $3 \log 6 - x \log 2$ to both sides.
$x(\log 6 - \log 2) = 3 \log 6$	Factor out x from the left side.
$x = \dfrac{3 \log 6}{\log 6 - \log 2}$	Divide both sides by $\log 6 - \log 2$.
$x \approx 4.893$	Substitute values for $\log 6$ and $\log 2$ and simplify. ∎

Example 3 Solve the logarithmic equation $\log x + \log(x - 3) = 1$.

Solution

$\log x + \log(x - 3) = 1$	
$\log x(x - 3) = 1$	Use Property 5 of logarithms.
$x(x - 3) = 10^1$	Use the definition of logarithm to change the equation to exponential form.
$x^2 - 3x - 10 = 0$	Remove parentheses and add -10 to both sides.
$(x + 2)(x - 5) = 0$	Factor $x^2 - 3x - 10$.
$x + 2 = 0 \quad$ or $\quad x - 5 = 0$	
$x = -2 \quad | \quad x = 5$	

Check: The number -2 is not a solution because it does not satisfy the equation: A negative number does not have a logarithm. Check the remaining number 5:

$\log x + \log(x - 3) = 1$	
$\log 5 + \log(5 - 3) \overset{?}{=} 1$	Substitute 5 for x.
$\log 5 + \log 2 \overset{?}{=} 1$	
$\log(5 \cdot 2) \overset{?}{=} 1$	Use Property 5 of logarithms.
$\log 10 \overset{?}{=} 1$	
$1 = 1$	Use Property 2 of logarithms.

The solution 5 does check. ∎

Example 4   Solve the logarithmic equation $\log_b(3x + 2) - \log_b(2x - 3) = 0$.

Solution $\log_b(3x + 2) - \log_b(2x - 3) = 0$

$\log_b(3x + 2) = \log_b(2x - 3)$	Add $\log_b(2x - 3)$ to both sides.
$3x + 2 = 2x - 3$	If $\log_b r = \log_b s$, then $r = s$.
$x = -5$	Add $-2x - 2$ to both sides.

Check: $\log_b(3x + 2) - \log_b(2x - 3) = 0$

$$\log_b[3(-5) + 2] - \log_b[2(-5) - 3] \stackrel{?}{=} 0$$
$$\log_b(-13) - \log_b(-13) \stackrel{?}{=} 0$$

Because the logarithm of a negative number does not exist, the number -5 is not a solution. The given equation has no solutions.   ∎

If we know the base-a logarithm of a number, we can find the logarithm of that number to some other base b by using a formula called the **change-of-base formula**.

The Change-of-Base Formula.

$$\log_b x = \frac{\log_a x}{\log_a b}$$

Proof   We begin with the equation $\log_b x = y$ and proceed as follows:

1. $\log_b x = y$

$b^y = x$	Change the equation from logarithmic to exponential form.
$\log_a b^y = \log_a x$	Take the base-a logarithm of both sides.
$y \log_a b = \log_a x$	Use Property 7 of logarithms.
$y = \dfrac{\log_a x}{\log_a b}$	Divide both sides by $\log_a b$.
$\log_b x = \dfrac{\log_a x}{\log_a b}$	Refer to Equation 1 and substitute $\log_b x$ for y.

□

If we know logarithms to base a (for example, $a = 10$), we can find the logarithm of x to a new base b. To do so, we divide the base-a logarithm of x by the base-a logarithm of b.

Example 5   Use the change-of-base formula to find $\log_3 5$.

Solution   Use the change-of-base formula with $b = 3$, $a = 10$, and $x = 5$:

$$\log_b x = \frac{\log_a x}{\log_a b}$$

$$\log_3 5 = \frac{\log_{10} 5}{\log_{10} 3} \qquad \text{Substitute 3 for } b, \text{ 10 for } a, \text{ and 5 for } x.$$

$$\approx \frac{0.6990}{0.4771} \qquad \text{Substitute values for log 5 and log 3.}$$

$$\approx 1.465 \qquad\qquad\qquad\qquad\qquad\qquad\qquad ∎$$

Applications of Exponential and Logarithmic Equations

Example 6   When a living organism dies, the oxygen/carbon dioxide cycle common to all living things ceases and carbon-14, a radioactive isotope with a half-life of 5700 years, is no longer absorbed. By measuring the amount of carbon-14 present in an ancient object, archeologists can estimate the object's age and answer questions such as the following: How old is a wooden statue that contains only $\frac{1}{3}$ of its original carbon-14 content?

Solution   The amount A of radioactive material present at time t is given by the model

$$A = A_0 2^{-\frac{t}{h}}$$

where A_0 is the amount present initially and h is the material's half-life.

To determine the time t when A is $\frac{1}{3}$ of A_0, substitute $\dfrac{A_0}{3}$ for A, **5700** for h, and solve for t:

$$A = A_0 2^{-\frac{t}{h}}$$

$$\frac{A_0}{3} = A_0 2^{-\frac{t}{5700}} \qquad \text{Substitute } \tfrac{A_0}{3} \text{ for } A \text{ and 5700 for } h.$$

$$1 = 3 \cdot 2^{-\frac{t}{5700}} \qquad \text{Multiply both sides by } \tfrac{3}{A_0}.$$

$$\log 1 = \log(3 \cdot 2^{-\frac{t}{5700}}) \qquad \text{Take the common logarithm of both sides.}$$

$$0 = \log 3 + \log 2^{-\frac{t}{5700}} \qquad \text{Use Properties 1 and 5 of logarithms.}$$

$$-\log 3 = \frac{t}{5700}\log 2 \qquad \begin{array}{l}\text{Add} -\log 3 \text{ to both sides and use Property} \\ \text{7 of logarithms.}\end{array}$$

$$t = 5700\left(\frac{\log 3}{\log 2}\right) \qquad \text{Multiply both sides by } -\tfrac{5700}{\log 2}.$$

$$\approx 9034.29$$

The wooden statue is approximately 9000 years old.   ∎

Example 7   When there is sufficient food supply and space, populations of living organisms tend to increase exponentially according to the Malthusian population growth model

$$P = P_0 e^{kt}$$

where P_0 is the initial population (at $t = 0$), and k depends on the rate of growth.

The bacteria population in a laboratory culture increased from an initial population of 500 to 1500 in 3 hours. Determine the time it will take the population to reach 10,000.

Solution $P = P_0 e^{kt}$

$1500 = 500(e^{k \cdot 3})$	Substitute 1500 for P, 500 for P_0, and 3 for t.
$3 = e^{3k}$	Divide both sides by 500.
$3k = \ln 3$	Change the equation from exponential to logarithmic form.
$k = \dfrac{\ln 3}{3}$	Divide both sides by 3.

To find when the population will reach 10,000, substitute 10,000 for P, 500 for P_0, and $\dfrac{\ln 3}{3}$ for k in the equation $P = P_0 e^{kt}$ and solve for t:

$$P = P_0 e^{kt}$$

$10,000 = 500 e^{\left(\frac{\ln 3}{3}\right)t}$	
$20 = e^{\left(\frac{\ln 3}{3}\right)t}$	Divide both sides by 500.
$\left(\dfrac{\ln 3}{3}\right)t = \ln 20$	Change the equation from exponential to logarithmic form.
$t = \dfrac{3 \ln 20}{\ln 3}$	Multiply both sides by $\frac{3}{\ln 3}$.
≈ 8.18	

The culture will reach the 10,000 mark in approximately 8 hours.   ∎

Exercise 4.4

In Exercises 1–12, solve each exponential equation.

1. $4^x = 5$ **2.** $7^x = 12$ **3.** $13^{x-1} = 2$ **4.** $5^{x+1} = 3$

5. $2^{x+1} = 3^x$ **6.** $5^{x-3} = 3^{2x}$ **7.** $2^x = 3^x$ **8.** $3^{2x} = 4^x$

9. $7^{x^2} = 10$ **10.** $8^{x^2} = 11$ **11.** $8^{x^2} = 9^x$ **12.** $5^{x^2} = 2^{5x}$

In Exercises 13–30, solve each logarithmic equation.

13. $\log(2x - 3) = \log(x + 4)$ **14.** $\log(3x + 5) - \log(2x + 6) = 0$

15. $\log \dfrac{4x + 1}{2x + 9} = 0$ **16.** $\log \dfrac{5x + 2}{2(x + 7)} = 0$

17. $\log x^2 = 2$

18. $\log x^3 = 3$

19. $\log x + \log(x - 48) = 2$

20. $\log x + \log(x + 9) = 1$

21. $\log x + \log(x - 15) = 2$

22. $\log x + \log(x + 21) = 2$

23. $\log(x + 90) = 3 - \log x$

24. $\log(x - 6) - \log(x - 2) = \log \dfrac{5}{x}$

25. $\log(x - 1) - \log 6 = \log(x - 2) - \log x$

26. $\log(2x - 3) - \log(x - 1) = 0$

27. $\log_{10} x^2 = (\log_{10} x)^2$

28. $\log_{10}(\log_{10} x) = 1$

29. $\log_3 x = \log_3 \left(\dfrac{1}{x}\right) + 4$

30. $\log_5(7 + x) + \log_5(8 - x) - \log_5 2 = 2$

In Exercises 31–34, find the logarithm with the indicated base.

31. $\log_3 7$

32. $\log_7 3$

33. $\log_{\sqrt{2}} \sqrt{5}$

34. $\log_\pi e$

35. The half-life of tritium is 12.4 years. How long will it take for 25% of a sample of tritium to decompose?

36. In 2 years 20% of a newly discovered radioactive element decays. What is its half-life?

37. An isotope of thorium, ^{227}Th, has a half-life of 18.4 days. How long will it take 80% of a sample to decompose?

38. An isotope of lead, ^{201}Pb, has a half-life of 8.4 hours. How many hours ago was there 30% more of the substance?

39. A parchment fragment is found in a newly discovered ancient tomb. It contains 60% of the carbon-14 that it is assumed to have had initially. Approximately how old is the fragment?

40. Only 10% of the carbon-14 in a small wooden bowl remains. How old is the bowl?

41. If $500 is deposited into an account paying 12% interest compounded semiannually, how long will it take for the account to increase to $800? How long will it take if the interest is compounded continuously?

42. If $1300 is deposited into an account paying 14% interest compounded quarterly, how long will it take to increase the amount to $2100?

43. A sum of $5000 deposited in an account grows to $7000 in 5 years. Assuming annual compounding, what interest rate is paid?

44. A quick rule of thumb for determining how long it takes an investment to double is known as the "rule of seventy": Divide 70 by the rate (as a percent). At 5%, for example, it requires $\frac{70}{5} = 14$ years to double the capital. At 7%, it takes $\frac{70}{7} = 10$ years. Why does this formula work?

45. A bacteria culture grows according to the formula

$$P = P_0 a^t$$

If it takes 5 days for the culture to triple in size, how long does it take to double in size?

46. The intensity I of light a distance x meters beneath the surface of a lake decreases exponentially. If the light intensity at 6 meters is 70% of the intensity at the surface, at what depth will the intensity be 20%?

In Review Exercises 1–4, graph the function defined by each equation.

1. $y = \left(\dfrac{6}{5}\right)^x$

2. $y = \left(\dfrac{3}{4}\right)^x$

3. $y = \log x$

4. $y = \ln x$

In Review Exercises 5–8, graph each pair of equations on one set of coordinate axes.

5. $y = \left(\dfrac{1}{3}\right)^x$ and $y = \log_{\frac{1}{3}} x$

6. $y = \left(\dfrac{2}{5}\right)^x$ and $y = \log_{\frac{2}{5}} x$

7. $y = 4^x$ and $y = \log_4 x$

8. $y = 3^x$ and $y = \log_3 x$

In Review Exercises 9–30, solve each equation for x.

9. $\log_2 x = 3$

10. $\log_3 x = -2$

11. $\log_x 9 = 2$

12. $\log_x 0.125 = -3$

13. $\log_7 7 = x$

14. $\log_3 \sqrt{3} = x$

15. $\log_8 \sqrt{2} = x$

16. $\log_6 36 = x$

17. $\log_{\frac{1}{3}} 9 = x$

18. $\log_{\frac{1}{2}} 1 = x$

19. $\log_x 3 = \dfrac{1}{3}$

20. $\log_x 25 = -2$

21. $\log_2 x = 5$

22. $\log_{\sqrt{3}} x = 4$

23. $\log_{\sqrt{3}} x = 6$

24. $\log_{0.1} 10 = x$

25. $\log_x 2 = -\dfrac{1}{3}$

26. $\log_x 32 = 5$

27. $\log_{0.25} x = -1$

28. $\log_{0.125} x = -\dfrac{1}{3}$

29. $\log_{\sqrt{2}} 32 = x$

30. $\log_{\sqrt{5}} x = -4$

In Review Exercises 31–34, use a calculator to find the value of the variable, if possible. Express each answer to four decimal places.

31. $\log 735.4 = x$

32. $\log(-0.002345) = y$

33. $\ln \dfrac{2}{15} = z$

34. $\ln M = 5.345$

In Review Exercises 35–44, solve for x, if possible.

35. $3^x = 7$

36. $1.2 = (3.4)^{5.6x}$

37. $2^x = 3^{x-1}$

38. $\log x + \log(29 - x) = 2$

39. $\log_2 x + \log_2(x - 2) = 3$

40. $\log_2(x + 2) + \log_2(x - 1) = 2$

41. $e^{x \ln 2} = 9$

42. $\ln x = \ln(x - 1)$

43. $\ln x = \ln(x - 1) + 1$

44. $\ln x = \log_{10} x$ (*Hint:* Use the change-of-base formula.)

45. A wooden statue excavated from the sands of Egypt has a carbon-14 content that is $\frac{1}{3}$ of that found in living wood. The half-life of carbon-14 is 5700 years. How old is the statue?

46. The pH of grapefruit juice is approximately 3.1. What is its hydrogen-ion concentration?

47. Some chemistry texts define the pH of a solution as the common logarithm of the reciprocal of the hydrogen-ion concentration:

$$pH = \log_{10} \frac{1}{[H^+]}$$

Show that this definition is equivalent to the one given in the text.

48. What is the half-life of a radioactive material if $\frac{1}{3}$ of it decays in 20 years?

5

TRIGONOMETRIC FUNCTIONS

Trigonometry is used in both theoretical and practical applications. It is used theoretically in the natural sciences, such as physics, biology, and chemistry, and practically in surveying, navigation, engineering, electronics, and carpentry.

5.1 FUNCTIONS OF TRIGONOMETRIC ANGLES

In plane geometry, an angle is defined as a figure formed by two rays originating from a common point, called the **vertex**. A trigonometric angle is defined differently.

> **Definition.** Consider two rays with common initial point O. A **trigonometric angle** is the rotation required to move one ray, called the **initial side** of the angle, into coincidence with the other, called the **terminal side**.
> Angles that are rotations in a counterclockwise direction are considered to be positive; those in a clockwise direction are considered to be negative.

Figure 5-1

This definition of a trigonometric angle is an extension of the definition of a geometric angle because one geometric angle can represent many, indeed infinitely many, different trigonometric angles. For example, the two rays forming the geometric angle in Figure 5-1 could represent any of the trigonometric angles represented in Figure 5-2.

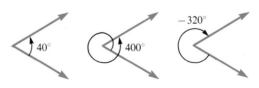

Figure 5-2

To draw a trigonometric angle, we must include an arrow curved from the angle's initial side to its terminal side to indicate the direction of rotation.

> **Definition.** A trigonometric angle is in **standard position** if and only if it is drawn in a Cartesian coordinate system with its vertex at the origin and its initial side along the positive *x*-axis.
>   The angle is called a **first-**, **second-**, **third-**, or **fourth-quadrant angle** according to whether its terminal side lies in the first, second, third, or fourth quadrant. If the terminal side lies on the *x*- or *y*-axis, then the angle is called a **quadrantal angle**.

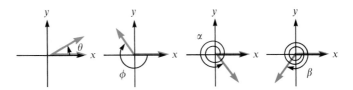

Figure 5-3

All of the angles in Figure 5-3 are in standard position because each one has its initial side along the positive *x*-axis and its vertex at the origin.

Trigonometric angles are often denoted by a Greek letter next to the curved arrow. Figure 5-3 illustrates angle θ (theta), ϕ (phi), α (alpha), and β (beta). Note that angles θ and α are positive angles and that angles ϕ and β are negative angles.

The angles in Figure 5-4 are not in standard position. In Figure 5-4**a**, the initial side is not on the positive *x*-axis. In Figure 5-4**b** the vertex is not at the origin.

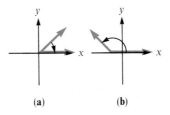

(a)   (b)

Figure 5-4

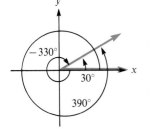

Figure 5-5

> **Definition.** If the terminal sides of two trigonometric angles in standard position coincide, the angles are called **coterminal**.

Because the three angles of 30°, 390°, and $-330°$ shown in Figure 5-5 are in standard position and their terminal sides coincide, they are coterminal.

Example 1   Find three positive and three negative angles that are coterminal with an angle in standard position that measures 100°.

Solution   A trigonometric angle in standard position is shown in Figure 5-6. Three positive angles that are coterminal with 100° are shown in Figure 5-7. They are

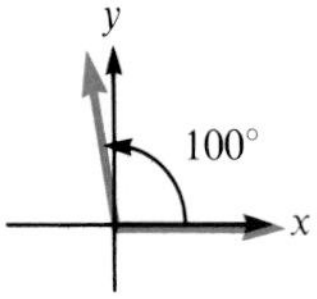

$$100° + 360° \quad \text{or} \quad 460° \quad (100° \text{ plus one revolution})$$
$$100° + 2(360°) \quad \text{or} \quad 820° \quad (100° \text{ plus two revolutions})$$
$$100° + 3(360°) \quad \text{or} \quad 1180° \quad (100° \text{ plus three revolutions})$$

Three negative angles that are coterminal with 100° are also shown in Figure 5-7. They are

Figure 5-6

$$100° - 360° \quad \text{or} \quad -260° \quad (100° \text{ minus one revolution})$$
$$100° - 2(360°) \quad \text{or} \quad -620° \quad (100° \text{ minus two revolutions})$$
$$100° - 3(360°) \quad \text{or} \quad -980° \quad (100° \text{ minus three revolutions})$$

In fact, if n is a whole number, any angle of $100° \pm n360°$ is coterminal with an angle of 100°.

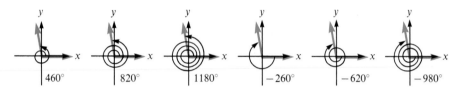

Figure 5-7

   Most of the work in trigonometry is based on the following six trigonometric functions.

Definition.   Let θ be a trigonometric angle in standard position as shown in Figure 5-8. Let $P(x, y)$ be any point (except the origin) on the terminal side of angle θ. Then $r = \sqrt{x^2 + y^2}$ is the length of line segment OP. The **six trigonometric functions of the angle θ** are:

$$\sin \theta = \frac{y}{r} \quad \text{read as "sine theta"}$$

$$\cos \theta = \frac{x}{r} \quad \text{read as "cosine theta"}$$

$$\tan \theta = \frac{y}{x} \quad \text{read as "tangent theta"}$$

$$\csc \theta = \frac{r}{y} \quad \text{read as "cosecant theta"}$$

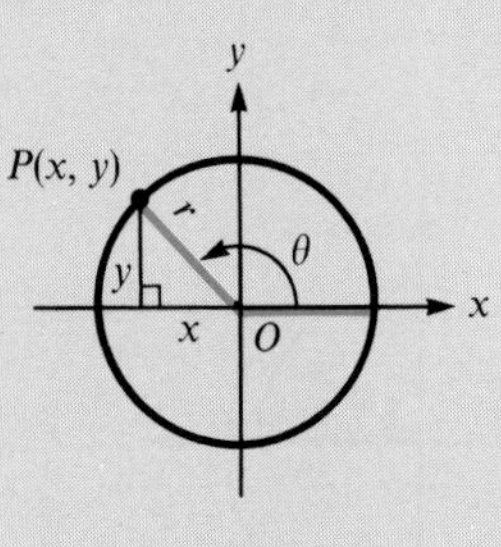

Figure 5-8

(continued)

(continued)

$$\sec \theta = \frac{r}{x} \quad \text{read as "secant theta"}$$

$$\cot \theta = \frac{x}{y} \quad \text{read as "cotangent theta"}$$

Remember that division by 0 is not permitted. If $x = 0$, then $\tan \theta$ and $\sec \theta$ are undefined. If $y = 0$, then $\csc \theta$ and $\cot \theta$ are undefined.

This definition implies that the values of trigonometric functions of coterminal angles are equal.

Example 2 The point $(-3, 4)$ is on the terminal side of angle ϕ. Draw angle ϕ in standard position, and calculate the six trigonometric functions of ϕ.

Solution Refer to Figure 5-9. The point $P(x, y) = P(-3, 4)$ lies at a distance $r = \sqrt{(-3)^2 + 4^2} = \sqrt{25} = 5$ units from point O, and ϕ is a second-quadrant angle. From the definitions,

$$\sin \phi = \frac{y}{r} = \frac{4}{5}$$

$$\cos \phi = \frac{x}{r} = \frac{-3}{5} = -\frac{3}{5}$$

$$\tan \phi = \frac{y}{x} = \frac{4}{-3} = -\frac{4}{3}$$

$$\csc \phi = \frac{r}{y} = \frac{5}{4}$$

$$\sec \phi = \frac{r}{x} = \frac{5}{-3} = -\frac{5}{3}$$

$$\cot \phi = \frac{x}{y} = \frac{-3}{4} = -\frac{3}{4}$$

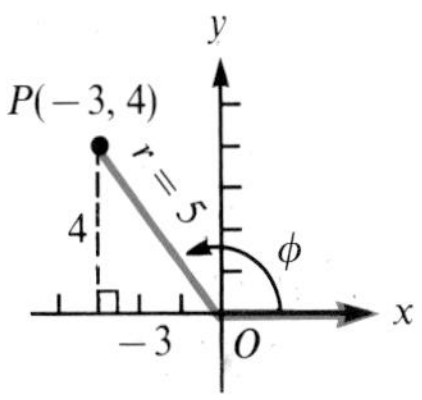

Figure 5-9

Any angle coterminal with angle ϕ will produce the same values for each of the six trigonometric functions. ∎

The following theorem guarantees that *any* point on the terminal side of an angle θ in standard position can be used to compute the values for six trigonometric functions of θ.

Theorem. If angle θ is in standard position and if P is some point on the terminal side of angle θ, then the values of the trigonometric functions of θ are independent of the choice of point P.

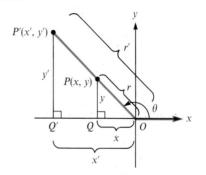

Figure 5-10

Proof Let $P(x, y)$ and $P'(x', y')$ be two points on the terminal side of angle θ. Let $OP = r$ and $OP' = r'$. Points P, O, and Q and points P', O, and Q' determine two triangles: triangle OPQ and triangle $OP'Q'$. See Figure 5-10. Since these triangles are similar, all ratios of corresponding sides are equal. In particular,

$$\sin \theta = \frac{y}{r} = \frac{y'}{r'}$$

and

$$\cos \theta = \frac{x}{r} = \frac{x'}{r'}$$

This shows that the values of $\sin \theta$ and $\cos \theta$ are independent of the choice of point P. It can be shown in a similar manner that the values of the remaining four trigonometric functions are also independent of the choice of point P. ☐

Example 3 Assume that $\sin \alpha = \frac{12}{13}$ and α is in quadrant II (QII). Find $\cos \alpha$.

Solution Because any point P on the terminal side of α can be used, choose a point with y-coordinate of 12 and $r = 13$, for then $\sin \alpha = \ = \frac{12}{13}$. To find the x-coordinate of P, first solve the equation $r^2 = x^2 + y^2$ for x.

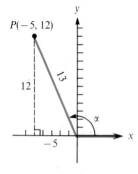

$$x^2 = r^2 - y^2$$
$$x = \pm\sqrt{r^2 - y^2}$$

Then substitute **12** for y and **13** for r.

$$x = \pm\sqrt{13^2 - 12^2}$$
$$= \pm\sqrt{25}$$
$$= \pm 5$$

Because angle α is in QII, the x-coordinate of P must be negative, so $x = -5$. See Figure 5-11. Thus,

Figure 5-11

$$\cos \alpha = \frac{x}{r} = \frac{-5}{13} = -\frac{5}{13}$$

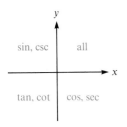

Figure 5-12

It is worthwhile to note in which quadrants each of the trigonometric functions is positive. Because r is always positive, the sine and cosecant functions are positive whenever y is positive. This occurs in QI and QII. Because x is positive in QI and QIV, the cosine and secant functions are positive in these quadrants also. The tangent and cotangent functions are positive when x and y agree in sign, and this happens in QI and QIII. Figure 5-12 will help you remember this information.

Example 4 If angle β is *not* in QII and $\cos \beta = -\frac{3}{4}$, find $\sin \beta$ and $\tan \beta$.

Solution Because $\cos \beta$ is negative, β must be in QII or QIII. You are told that β is not in QII, so β is a third-quadrant angle. See Figure 5-13. Choose point $P(x, y)$ so that $x = -3$ and $r = 4$ (remember that r is always positive). Then,

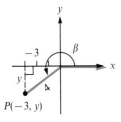

Figure 5-13

$$y = -\sqrt{r^2 - x^2}$$
$$= -\sqrt{4^2 - (-3)^2}$$
$$= -\sqrt{7}$$

The minus sign preceding the above radicals is required because y is negative in QIII. Use these values of x, y, and r to compute $\sin \beta$ and $\tan \beta$:

$$\sin \beta = \frac{y}{r} = \frac{-\sqrt{7}}{4} = -\frac{\sqrt{7}}{4}$$

$$\tan \beta = \frac{y}{x} = \frac{-\sqrt{7}}{-3} = \frac{\sqrt{7}}{3}$$

Note that the tangent is positive in QIII, as expected. ∎

Exercise 5.1

In Exercises 1–8, tell whether the angle is in standard position. Indicate whether the angle is positive or negative.

1.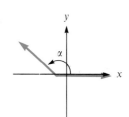

2.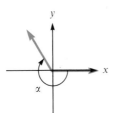

3.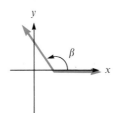

4.

5.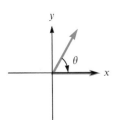

6.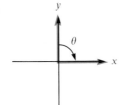

7.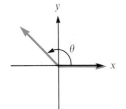

8.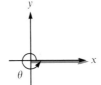

In Exercises 9–20, tell whether θ is a QI, QII, QIII, or QIV angle. Assume that θ is in standard position.

9.

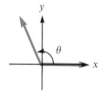

10.

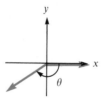

11.

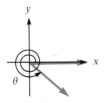

12.

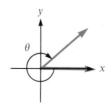

13. $\theta = 170°$ **14.** $\theta = 350°$ **15.** $\theta = 612°$ **16.** $\theta = 460°$

17. $\theta = -70°$ **18.** $\theta = -212°$ **19.** $\theta = -1190°$ **20.** $\theta = -1470°$

In Exercises 21–28, tell whether the given angles are coterminal. Assume that all angles are in standard position.

21. $40°, 400°$ **22.** $90°, -270°$ **23.** $135°, 270°$ **24.** $135°, -135°$

25. $740°, 380°$ **26.** $-340°, -700°$ **27.** $1035°, -405°$ **28.** $10°, -10°$

In Exercises 29–40, point P is on the terminal side of angle θ, which is in standard position. Draw angle θ, with θ positive, and calculate the six trigonometric functions of angle θ.

29. $P(3, 4)$ **30.** $P(-5, -12)$ **31.** $P(-9, 40)$ **32.** $P(9, -40)$

33. $P(1, 1)$ **34.** $P(-3, 3)$ **35.** $P(-3, 4)$ **36.** $P(-1, -1)$

37. $P(3, 5)$ **38.** $P(24, 10)$ **39.** $P(24, -10)$ **40.** $P(-3, -4)$

In Exercises 41–50, find the remaining trigonometric functions of angle θ. Assume that θ is positive and in standard position.

41. $\sin \theta = \dfrac{3}{5}$; θ in QI

42. $\tan \theta = 1$; θ not in QI

43. $\cot \theta = \dfrac{5}{12}$; $\cos \theta = -\dfrac{5}{13}$

44. $\cos \theta = \dfrac{\sqrt{5}}{5}$; $\csc \theta = -\dfrac{\sqrt{5}}{2}$

45. $\sin \theta = -\dfrac{9}{41}$; θ in QIV

46. $\tan \theta = -1$; θ not in QII

47. $\sec \theta = -\dfrac{5}{3}$; $\csc \theta = \dfrac{5}{4}$

48. $\tan \theta = -\dfrac{3}{5}$; $\sin \theta = \dfrac{3\sqrt{34}}{34}$

49. $\tan \theta = -\dfrac{40}{9}$; $\cos \theta = \dfrac{9}{41}$

50. $\sin \theta = -\dfrac{\sqrt{7}}{4}$; $\cos \theta = -\dfrac{3}{4}$

5.2 THE EIGHT FUNDAMENTAL TRIGONOMETRIC IDENTITIES

The six trigonometric functions can be grouped in reciprocal pairs. For example, $\cos \theta$ and $\sec \theta$ are reciprocals because

$$\frac{x}{r} \quad \text{and} \quad \frac{r}{x}$$

are reciprocals. Similarly, $\sin \theta$ and $\csc \theta$ are reciprocals, as are $\tan \theta$ and $\cot \theta$.

Reciprocal Relationships.   For any angle θ for which the functions are defined,

$$\sin \theta \csc \theta = 1 \qquad \sin \theta = \frac{1}{\csc \theta} \qquad \csc \theta = \frac{1}{\sin \theta}$$

$$\cos \theta \sec \theta = 1 \qquad \cos \theta = \frac{1}{\sec \theta} \qquad \sec \theta = \frac{1}{\cos \theta}$$

$$\tan \theta \cot \theta = 1 \qquad \tan \theta = \frac{1}{\cot \theta} \qquad \cot \theta = \frac{1}{\tan \theta}$$

Each of these nine equations is true for all values of the variable for which the expressions are defined. Recall that such equations are called **identities**.

There are many other identities involving the trigonometric functions.

Example 1   Verify the identity $\dfrac{\sin \theta}{\cos \theta} = \tan \theta$.

Solution   Use the definitions of the sine and cosine functions, as follows:

$$\frac{\sin \theta}{\cos \theta} = \frac{\dfrac{y}{r}}{\dfrac{x}{r}} = \frac{y}{r} \cdot \frac{r}{x} = \frac{y}{x} = \tan \theta$$

∎

Similarly, we can establish the identity

$$\frac{\cos \theta}{\sin \theta} = \cot \theta$$

By convention, $\sin^2 \theta$ is standard notation for $(\sin \theta)^2$. Therefore

$$\sin^2 \theta = \sin \theta \cdot \sin \theta$$

Powers of the other trigonometric functions are denoted in a similar way.

The six trigonometric functions involve the three variables x, y, and r, which are related by the formula $x^2 + y^2 = r^2$. This formula can be used to establish three more important relationships.

Example 2   Use the result $y^2 + x^2 = r^2$ to show that $\sin^2 \theta + \cos^2 \theta = 1$.

Solution   Divide $y^2 + x^2 = r^2$ by r^2 to obtain

$$\frac{y^2}{r^2} + \frac{x^2}{r^2} = \frac{r^2}{r^2}$$

or

$$\left(\frac{y}{r}\right)^2 + \left(\frac{x}{r}\right)^2 = 1$$

Then, use the definitions of the trigonometric functions to obtain

$$\sin^2 \theta + \cos^2 \theta = 1 \qquad\qquad ■$$

Example 3   Use the result $y^2 + x^2 = r^2$ to show that $\tan^2 \theta + 1 = \sec^2 \theta$.

Solution   Divide $y^2 + x^2 = r^2$ by x^2 and use the definitions of the trigonometric functions to obtain

$$\frac{y^2}{x^2} + \frac{x^2}{x^2} = \frac{r^2}{x^2}$$

$$\left(\frac{y}{x}\right)^2 + 1 = \left(\frac{r}{x}\right)^2$$

or

$$\tan^2 \theta + 1 = \sec^2 \theta \qquad\qquad ■$$

By dividing both sides of $x^2 + y^2 = r^2$ by y^2, we can establish yet another identity:

$$\cot^2 \theta + 1 = \csc^2 \theta$$

The identities discussed so far are summarized as follows:

The Eight Fundamental Identities.   For any trigonometric angle θ for which the functions are defined,

$$\sec \theta = \frac{1}{\cos \theta} \qquad\qquad \csc \theta = \frac{1}{\sin \theta}$$

$$\cot \theta = \frac{1}{\tan \theta} \qquad\qquad \tan \theta = \frac{\sin \theta}{\cos \theta}$$

$$\cot \theta = \frac{\cos \theta}{\sin \theta} \qquad\qquad \sin^2 \theta + \cos^2 \theta = 1$$

$$\tan^2 \theta + 1 = \sec^2 \theta \qquad\qquad \cot^2 \theta + 1 = \csc^2 \theta$$

Other identities can also be established.

Example 4   Show that $\tan \theta = \dfrac{\sec \theta}{\csc \theta}$.

Solution   Make use of the fundamental identities and proceed as follows:

$$\tan \theta = \frac{\sin \theta}{\cos \theta} = \frac{\dfrac{1}{\csc \theta}}{\dfrac{1}{\sec \theta}} = \frac{1}{\csc \theta} \cdot \frac{\sec \theta}{1} = \frac{\sec \theta}{\csc \theta}$$

∎

Example 5   Assume that θ is in QII and that $\sin \theta = \frac{4}{5}$. Use the eight fundamental identities to find the values of the other five trigonometric functions.

Solution   Because the sine and cosecant of an angle are reciprocals,

$$\csc \theta = \frac{1}{\sin \theta} = \frac{5}{4}$$

Because $\sin^2 \theta + \cos^2 \theta = 1$,

$$\cos^2 \theta = 1 - \sin^2 \theta$$
$$\cos \theta = \pm\sqrt{1 - \sin^2 \theta}$$

Because θ is in QII, $\cos \theta$ must be negative. Therefore,

$$\cos \theta = -\sqrt{1 - \sin^2 \theta}$$
$$= -\sqrt{1 - \left(\frac{4}{5}\right)^2} \qquad \text{Substitute } \tfrac{4}{5} \text{ for } \sin \theta.$$
$$= -\sqrt{\frac{9}{25}}$$
$$= -\frac{3}{5}$$

Because $\sec \theta$ is the reciprocal of $\cos \theta$,

$$\sec \theta = -\frac{5}{3}$$

Use the relationship $\tan \theta = \dfrac{\sin \theta}{\cos \theta}$ to find $\tan \theta$:

$$\tan \theta = \frac{\sin \theta}{\cos \theta}$$
$$= \frac{\dfrac{4}{5}}{-\dfrac{3}{5}} \qquad \text{Substitute } \tfrac{4}{5} \text{ for } \sin \theta \text{ and } -\tfrac{3}{5} \text{ for } \cos \theta.$$
$$= -\frac{4}{3}$$

Finally, because $\cot \theta$ is the reciprocal of $\tan \theta$,

$$\cot \theta = -\frac{3}{4}$$

■

Example 6   Given that $\cos \theta = \frac{1}{3}$ and that $\tan \theta$ is negative, use the fundamental identities to find $\tan \theta$.

Solution   First decide in which quadrant the terminal side of angle θ lies. Because $\cos \theta$ is positive, θ must be a QI or QIV angle. Because $\tan \theta$ is negative, θ must be a QII or QIV angle. Putting these facts together, you know that θ is a QIV angle. See Figure 5-14. From the eight basic trigonometric relationships, it follows that

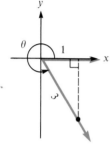

$$\sec \theta = \frac{1}{\cos \theta}$$

$$= \frac{1}{\frac{1}{3}} \qquad \text{Substitute } \tfrac{1}{3} \text{ for } \cos \theta.$$

$$= 3$$

Figure 5-14

Thus, $\sec \theta = 3$. Because $\tan^2 \theta = \sec^2 \theta - 1$, you have $\tan \theta = \pm\sqrt{\sec^2 \theta - 1}$. Because $\tan \theta$ is negative,

$$\tan \theta = -\sqrt{\sec^2 \theta - 1}$$

$$= -\sqrt{3^2 - 1} \qquad \text{Substitute 3 for } \sec \theta.$$

$$= -\sqrt{8}$$

$$= -2\sqrt{2} \qquad \text{Remember that } \sqrt{8} = \sqrt{4}\sqrt{2} = 2\sqrt{2}.$$

■

We now consider six more identities. Consider angles θ and $-\theta$ drawn in standard position as shown in Figure 5-15. Points P and Q are on the terminal sides of the respective angles. Let (x, y) be coordinates of point P, and let $(x, -y)$ be the coordinates of point Q. The trigonometric functions of these angles are

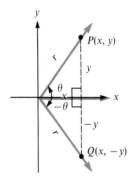

$$\left\{ \begin{array}{l} \sin(-\theta) = \dfrac{-y}{r} \\[2mm] \sin \theta = \dfrac{y}{r} \end{array} \right.
\left\{ \begin{array}{l} \cos(-\theta) = \dfrac{x}{r} \\[2mm] \cos \theta = \dfrac{x}{r} \end{array} \right.
\left\{ \begin{array}{l} \tan(-\theta) = \dfrac{-y}{x} \\[2mm] \tan \theta = \dfrac{y}{x} \end{array} \right.$$

$$\left\{ \begin{array}{l} \csc(-\theta) = \dfrac{r}{-y} \\[2mm] \csc \theta = \dfrac{r}{y} \end{array} \right.
\left\{ \begin{array}{l} \sec(-\theta) = \dfrac{r}{x} \\[2mm] \sec \theta = \dfrac{r}{x} \end{array} \right.
\left\{ \begin{array}{l} \cot(-\theta) = \dfrac{x}{-y} \\[2mm] \cot \theta = \dfrac{x}{y} \end{array} \right.$$

Figure 5-15

From these results, it follows that

Trigonometric Functions of $(-\theta)$.

$$\sin(-\theta) = -\sin\theta \qquad \csc(-\theta) = -\csc\theta$$
$$\cos(-\theta) = \cos\theta \qquad \sec(-\theta) = \sec\theta$$
$$\tan(-\theta) = -\tan\theta \qquad \cot(-\theta) = -\cot\theta$$

Recall that f is an odd function if $f(-x) = -f(x)$ for all x, and that f is an even function if $f(-x) = f(x)$ for all x. Thus, the sine, tangent, cosecant, and cotangent functions are odd functions; the cosine and secant functions are even functions.

Example 7   Show that the function $f(\theta) = \cos\theta - \sin\theta$ is neither an even function nor an odd function.

Solution   The function is an even function if and only if $f(-\theta) = f(\theta)$ for all θ. The function is an odd function if and only if $f(-\theta) = -f(\theta)$ for all θ. Calculate $f(-\theta)$ and $-f(\theta)$, and compare the results with $f(\theta)$. Note that

$$f(\theta) = \cos\theta - \sin\theta$$
$$f(-\theta) = \cos(-\theta) - \sin(-\theta)$$
$$= \cos\theta + \sin\theta$$

and

$$-f(\theta) = -(\cos\theta - \sin\theta)$$
$$= -\cos\theta + \sin\theta$$

Because $f(-\theta) \neq f(\theta)$, the function $f(\theta) = \cos\theta - \sin\theta$ is not an even function. Because $f(-\theta) \neq -f(\theta)$, the function $f(\theta)$ is not an odd function. Hence, it is neither even nor odd.   ∎

Exercise 5.2

In Exercises 1–8, use the letters x, y, and r and the definitions of the trigonometric functions to verify the following equations.

1. $\dfrac{\cos\theta}{\sin\theta} = \cot\theta$

2. $\cot^2\theta + 1 = \csc^2\theta$

3. $\tan\theta = \dfrac{1}{\cos\theta\,\csc\theta}$

4. $\cot\theta = \cos\theta\,\csc\theta$

5. $\sin^2\theta + \sin^2\theta\,\cot^2\theta = 1$

6. $\cot\theta = \dfrac{\csc\theta}{\sec\theta}$

7. $\cot^2\theta + \sin^2\theta = \csc^2\theta - \cos^2\theta$

8. $\tan^2\theta + \cos^2\theta = \sec^2\theta - \sin^2\theta$

In Exercises 9–16, use one or more of the eight fundamental trigonometric relationships to verify the following equations.

9. $\dfrac{\cos \theta}{\sin \theta} = \dfrac{1}{\tan \theta}$

10. $\csc^2 \theta - 1 = \cot^2 \theta$

11. $\tan \theta = \dfrac{1}{\cos \theta \csc \theta}$

12. $\cot \theta = \cos \theta \csc \theta$

13. $\sin^2 \theta + \sin^2 \theta \cot^2 \theta = 1$

14. $\cot \theta = \dfrac{\csc \theta}{\sec \theta}$

15. $\cot^2 \theta + \sin^2 \theta = \csc^2 \theta - \cos^2 \theta$

16. $\tan^2 \theta + \cos^2 \theta = \sec^2 \theta - \sin^2 \theta$

In Exercises 17–26, use one or more of the eight fundamental trigonometric relationships to find the values of the remaining trigonometric functions.

17. $\sin \theta = \dfrac{4}{5}$; θ in QI

18. $\sin \theta = -\dfrac{4}{5}$; θ in QIII

19. $\cos \theta = -\dfrac{5}{13}$; θ in QII

20. $\cos \theta = -\dfrac{5}{13}$; θ in QIII

21. $\tan \theta = -\dfrac{4}{3}$; θ in QII

22. $\tan \theta = -\dfrac{3}{4}$; θ in QIV

23. $\cot \theta = \dfrac{9}{40}$; θ in QI

24. $\csc \theta = \dfrac{13}{5}$; θ in QII

25. $\sec \theta = -\dfrac{13}{5}$; θ in QIII

26. $\sec \theta = -\dfrac{13}{5}$; θ in QII

27. Let θ be a second-quadrant angle. Draw a figure similar to Figure 5-15 and show that $\sin(-\theta) = -\sin \theta$, $\cos(-\theta) = \cos \theta$, and $\tan(-\theta) = -\tan \theta$.

28. Let θ be a third-quadrant angle. Draw a figure similar to Figure 5-15 and show that $\sin(-\theta) = -\sin \theta$, $\cos(-\theta) = \cos \theta$, and $\tan(-\theta) = -\tan \theta$.

29. Let θ be a fourth-quadrant angle. Draw a figure similar to Figure 5-15 and show that $\sin(-\theta) = -\sin \theta$, $\cos(-\theta) = \cos \theta$, and $\tan(-\theta) = -\tan \theta$.

30. If θ is a second-quadrant angle, is $\sin(-\theta)$ positive or negative?

31. If θ is a third-quadrant angle, is $\cos(-\theta)$ positive or negative?

32. Identify the product $\sin(-\theta) \cos(-\theta) \tan(-\theta)$ as positive or negative if θ is a second-quadrant angle.

33. Identify the product $\sin(-\theta) \cos(-\theta) \tan(-\theta)$ as positive or negative if θ is a third-quadrant angle.

34. Identify the product $\sin(-\theta) \cos(-\theta) \tan(-\theta)$ as positive or negative if θ is a fourth-quadrant angle.

In Exercises 35–40, tell whether the given function is odd, even, or neither.

35. $f(\theta) = \sin \theta + \tan \theta$

36. $f(\theta) = \cot \theta + \sin \theta$

37. $f(\theta) = \sec \theta + \cos \theta$

38. $f(\theta) = \csc \theta + \sec \theta$

39. $f(\theta) = \sin \theta - \tan \theta$

40. $f(\theta) = \cos \theta - \sec \theta$

5.3   TRIGONOMETRIC FUNCTIONS OF ANGLES

Exact values for the trigonometric functions of many angles can be found by using elementary geometry. We begin by considering the quadrantal angles.

An angle of $0°$ placed in standard position has both its initial side and its terminal side along the positive x-axis. See Figure 5-16. The point $P(1, 0)$ lies on the terminal side, and r, the distance of P from the origin, is 1 unit. Four of the six trigonometric functions of $0°$ are

$$\sin 0° = \frac{y}{r} = \frac{0}{1} = 0$$

$$\cos 0° = \frac{x}{r} = \frac{1}{1} = 1$$

$$\tan 0° = \frac{y}{x} = \frac{0}{1} = 0$$

$$\sec 0° = \frac{r}{x} = \frac{1}{1} = 1$$

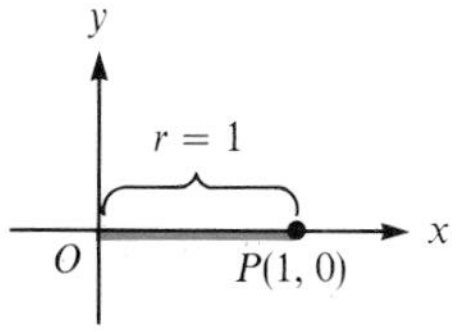

Figure 5-16

Values for cot $0°$ and csc $0°$ are undefined because division by 0 is not permitted.

As another illustration, we consider an angle of $270°$ in standard position and the point $P(0, -1)$ on its terminal side. Again, four functions have values and two functions do not. See Figure 5-17.

$$\sin 270° = \frac{y}{r} = \frac{-1}{1} = -1$$

$$\cos 270° = \frac{x}{r} = \frac{0}{1} = 0$$

$$\cot 270° = \frac{x}{y} = \frac{0}{-1} = 0$$

$$\csc 270° = \frac{r}{y} = \frac{1}{-1} = -1$$

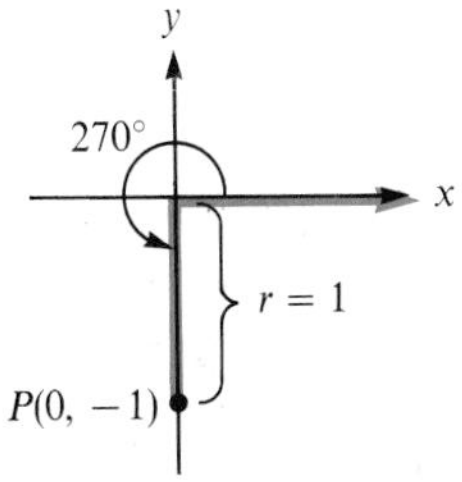

Figure 5-17

Values for tan $270°$ and sec $270°$ are undefined.

You will be asked in the exercises to determine the values of the trigonometric functions of $90°$ and $180°$.

Example 1   Find $\sin 990°$.

Solution   Because $990°$ is greater than $360°$, find an angle that is less than $360°$ but coterminal with $990°$. This can be done by repeatedly subtracting $360°$ from $990°$ until an angle less than $360°$ is found. Then, find the sine of that angle.

$$\sin 990° = \sin[990° - 2(360°)]$$
$$= \sin 270°$$
$$= -1$$

∎

If $P(x, y)$ is a point on the terminal side of a 30° angle placed in standard position, and r is the distance from P to the origin, then triangle OPA is a 30°-60° right triangle. See Figure 5-18. Recall that the leg opposite the 30° angle measures one-half of the length of the hypotenuse, and the leg opposite the 60° angle is $\sqrt{3}$ times the leg opposite the 30° angle. Therefore, $r = 2y$ and $x = \sqrt{3}y$.

$$\sin 30° = \frac{y}{r} = \frac{y}{2y} = \frac{1}{2}$$

$$\cos 30° = \frac{x}{r} = \frac{\sqrt{3}y}{2y} = \frac{\sqrt{3}}{2}$$

$$\tan 30° = \frac{y}{x} = \frac{y}{\sqrt{3}y} = \frac{1}{\sqrt{3}} = \frac{\sqrt{3}}{3}$$

$$\csc 30° = 2$$

$$\sec 30° = \frac{2\sqrt{3}}{3}$$

$$\cot 30° = \sqrt{3}$$

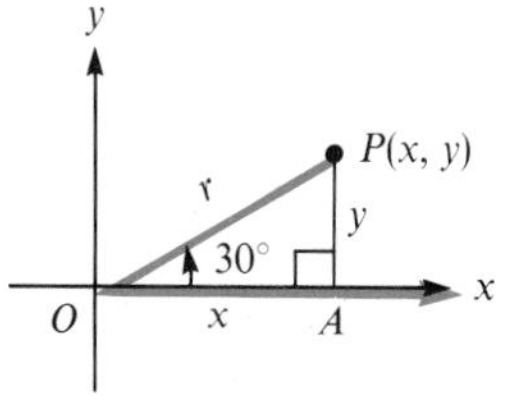

Figure 5-18

The values of the cosecant, secant, and cotangent functions were found by taking the reciprocals of the values of the sine, cosine, and tangent functions, respectively.

A similar argument determines the trigonometric functions of a 60° angle. See Figure 5-19. Triangle OPA is again a 30°-60° right triangle with $r = 2x$ and $y = \sqrt{3}x$. Thus,

$$\sin 60° = \frac{y}{r} = \frac{\sqrt{3}x}{2x} = \frac{\sqrt{3}}{2}$$

$$\cos 60° = \frac{x}{r} = \frac{x}{2x} = \frac{1}{2}$$

$$\tan 60° = \frac{y}{x} = \frac{\sqrt{3}x}{x} = \sqrt{3}$$

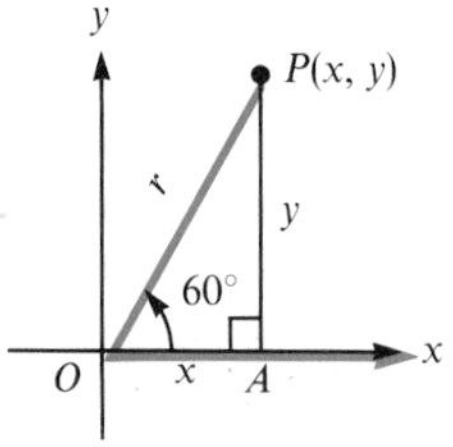

Figure 5-19

Use the reciprocal relationships to determine values for the three remaining trigonometric functions.

Because an angle of 45° is a base angle of the isosceles right triangle in Figure 5-20, it follows that $x = y$ and that $r = \sqrt{2}x = \sqrt{2}y$. Thus,

$$\sin 45° = \frac{y}{r} = \frac{y}{\sqrt{2}y} = \frac{1}{\sqrt{2}} = \frac{\sqrt{2}}{2}$$

$$\cos 45° = \frac{x}{r} = \frac{x}{\sqrt{2}x} = \frac{1}{\sqrt{2}} = \frac{\sqrt{2}}{2}$$

$$\tan 45° = \frac{y}{x} = 1$$

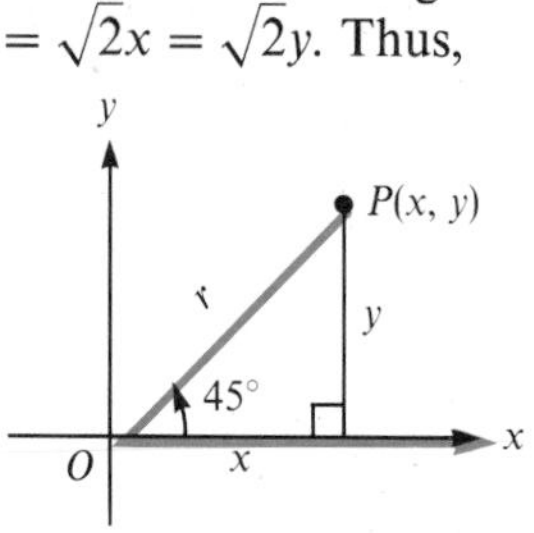

Figure 5-20

Use the reciprocal relationships to determine values for the remaining three trigonometric functions.

We summarize these results in the following chart:

θ	$\sin \theta$	$\cos \theta$	$\tan \theta$	$\csc \theta$	$\sec \theta$	$\cot \theta$
0°	0	1	0	undefined	1	undefined
30°	$\dfrac{1}{2}$	$\dfrac{\sqrt{3}}{2}$	$\dfrac{\sqrt{3}}{3}$	2	$\dfrac{2\sqrt{3}}{3}$	$\sqrt{3}$
45°	$\dfrac{\sqrt{2}}{2}$	$\dfrac{\sqrt{2}}{2}$	1	$\sqrt{2}$	$\sqrt{2}$	1
60°	$\dfrac{\sqrt{3}}{2}$	$\dfrac{1}{2}$	$\sqrt{3}$	$\dfrac{2\sqrt{3}}{3}$	2	$\dfrac{\sqrt{3}}{3}$
90°	1	0	undefined	1	undefined	0
180°	0	-1	0	undefined	-1	undefined
270°	-1	0	undefined	-1	undefined	0

It is now possible to determine the values of trigonometric functions of certain angles in any quadrant.

Example 2   Find $\sin 120°$, $\cos 120°$, and $\tan 120°$.

Solution   Draw an angle of 120° in standard position and mark a point $P(x, y)$ on the terminal side at a distance r from the origin. See Figure 5-21. Draw segment PA so that it is perpendicular to the x-axis. This forms triangle OPA, which is called the **reference triangle**. Because P is in the second quadrant, x is negative and y is positive; r is always positive. Because the reference triangle OPA is a 30°-60° right triangle, $r = -2x$ and $y = -\sqrt{3}x$. (Remember that if x is negative, $-2x$ and $-\sqrt{3}x$ are positive.) Thus,

$$\sin 120° = \frac{y}{r} = \frac{-\sqrt{3}x}{-2x} = \frac{\sqrt{3}}{2}$$

$$\cos 120° = \frac{x}{r} = \frac{x}{-2x} = -\frac{1}{2}$$

$$\tan 120° = \frac{y}{x} = \frac{-\sqrt{3}x}{x} = -\sqrt{3}$$

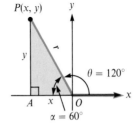

Figure 5-21

Note that angle α (not in standard position) formed by the terminal side of θ and the x-axis is 60°. Also note that the values of the functions of 120° equal the values of the functions of a 60° angle, except for an occasional minus sign.

Example 3   Find csc 225°, sec 225°, and cot 225°.

Solution   Draw an angle of 225° in standard position. Place point $P(x, y)$ on the terminal side at a distance r from the origin and draw segment PA perpendicular to the x-axis to form the reference triangle OPA. See Figure 5-22. Because point P is in the third quadrant, x and y are both negative (and r is, of course, positive). Triangle OPA is an isosceles right triangle, so $x = y$ and $r = -\sqrt{2}x = -\sqrt{2}y$. Thus,

$$\csc 225° = \frac{r}{y}$$

$$= \frac{-\sqrt{2}y}{y}$$

$$= -\sqrt{2}$$

$$\sec 225° = \frac{r}{x}$$

$$= \frac{-\sqrt{2}x}{x}$$

$$= -\sqrt{2}$$

$$\cot 225° = \frac{x}{y}$$

$$= \frac{y}{y}$$

$$= 1$$

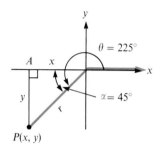

Figure 5-22

Note that the acute angle α between the terminal side of angle θ and the x-axis is 45°. Also note that the values of the functions of 225° equal the values of the functions of 45° except for sign.   ■

A pattern emerges from Examples 2 and 3. Associated with any angle θ is an acute angle in the reference triangle called the **reference angle**. The reference angle is that acute angle formed by the terminal side of angle θ and the x-axis. The reference angle is not always in standard position. However, if it were, it would be a first-quadrant angle with six positive values for the six trigonometric functions. The useful fact is this:

> **The six trigonometric functions of any angle θ are equal to those of the reference angle of θ, except possibly for sign.**

The appropriate sign can be determined independently by considering the quadrant in which the terminal side of θ lies.

Example 4   If $\theta = -405°$, find $\sin \theta$ and $\cos \theta$.

Solution   Sketch a $-405°$ angle in standard position, as in Figure 5-23. Angle θ is a QIV angle with a $45°$ reference angle. Only the cosine and secant of a QIV angle are positive. Thus,

$$\sin(-405°) = -\sin 45°$$

$$= -\frac{\sqrt{2}}{2}$$

$$\cos(-405°) = \cos 45°$$

$$= \frac{\sqrt{2}}{2}$$

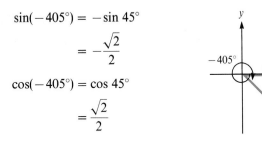

Figure 5-23

Example 5   If $\sin \theta = -\dfrac{1}{2}$ and $\cos \theta = \dfrac{\sqrt{3}}{2}$, find θ.

Solution   Because $\sin \theta$ is negative and $\cos \theta$ is positive, θ must be a QIV angle. See Figure 5-24. Because

$$\sin 30° = \frac{1}{2} \qquad \text{and} \qquad \cos 30° = \frac{\sqrt{3}}{2}$$

the reference angle must be $30°$. There are infinitely many fourth-quadrant angles that have reference angles of $30°$. Two examples are $\theta = -30°$ and $\theta = 330°$. All such angles θ are of the form

$$330° \pm n \cdot 360°$$

where n is a whole number.

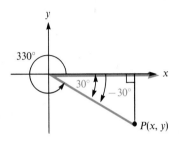

Figure 5-24

To find approximate values for the trigonometric functions of general angles, we use either tables or a calculator.

Using Tables

Example 6   Use Table A in Appendix III to find the sine, cosine, tangent, and cotangent of 43.5°.

Solution   A portion of Table A is presented below. In the **Degrees** column on the left side of the table, locate the number 43.5. Move to the right in that row and read the entries in the columns headed by **Sin**, **Cos**, **Tan**, and **Cot**.

Radians	**Degrees**	**Sin**	**Cos**	**Tan**	**Cot**		
.7575	43.4°	.6871	.7266	.9457	1.057	46.6°	.8133
.7592	43.5°	.6884	.7254	.9490	1.054	46.5°	.8116
.7610	43.6°	.6896	.7242	.9523	1.050	46.4°	.8098
		Cos	**Sin**	**Cot**	**Tan**	**Degrees**	**Radians**

$$\sin 43.5° \approx 0.6884 \qquad \cos 43.5° \approx 0.7254$$
$$\tan 43.5° \approx 0.9490 \qquad \cot 43.5° \approx 1.054$$

■

Example 7   Use Table A in Appendix III to find the sine, cosine, tangent, and cotangent of an angle of 46.5°.

Solution   The **Degrees** column on the left side of the table ends at 45°. However, 46.5° can be found on the right side in a column that is footed by **Degrees**. See the table in Example 6. Move to the left in the row containing 46.5° and read the entries in the columns footed by **Cos**, **Sin**, **Cot**, and **Tan**.

$$\sin 46.5° \approx 0.7254 \qquad \cos 46.5° \approx 0.6884$$
$$\tan 46.5° \approx 1.054 \qquad \cot 46.5° \approx 0.9490$$

■

Note that the same row of the table was used in both Examples 6 and 7; it did double duty for the complementary angles 43.5° and 46.5°. (Remember that if the sum of two acute angles is 90°, they are called **complementary angles**.) Note also that

$$\sin 43.5° = \cos 46.5° \qquad \text{and} \qquad \tan 43.5° = \cot 46.5°$$

are true statements. Because $\sin 43.5° = \cos 46.5°$, their reciprocals must be equal also. Therefore,

$$\csc 43.5° = \sec 46.5°$$

> **Definition.** The trigonometric functions of sine and cosine are called **cofunctions.** The tangent and cotangent functions are cofunctions, as are the secant and cosecant functions.

It is always true that the sine of an acute angle θ is equal to the cosine of the complement of θ. In like manner, the tangent of an acute θ is equal to the cotangent of the complement of θ, and the secant of an acute angle θ is equal to the cosecant of the complement of θ.

> **Theorem.** If θ is any acute angle, any trigonometric function of θ is equal to the cofunction of the complement of θ:
>
> $$\sin \theta = \cos(90° - \theta) \quad \text{and} \quad \cos \theta = \sin(90° - \theta)$$
> $$\tan \theta = \cot(90° - \theta) \quad \text{and} \quad \cot \theta = \tan(90° - \theta)$$
> $$\csc \theta = \sec(90° - \theta) \quad \text{and} \quad \sec \theta = \csc(90° - \theta)$$

Later, you will see that these relationships can be extended to all values of θ.

Example 8   Use Table A in Appendix III to find the values of the sine, cosine, tangent, and cotangent of an angle of 107°.

Solution   Although an angle of 107° does not appear in the table, its reference angle, 73°, does. Remember that any trigonometric function of an angle in standard position can differ only in sign from that same trigonometric function of its reference angle. Because a 107° angle is a second-quadrant angle, only the sine and cosecant functions are positive; the rest are negative:

$$\sin 107° = +\sin 73° \approx 0.9563$$
$$\cos 107° = -\cos 73° \approx -0.2924$$
$$\tan 107° = -\tan 73° \approx -3.271$$
$$\cot 107° = -\cot 73° \approx -0.3057$$

∎

Using Calculators

Example 9   Use a calculator to find the values of **a.** $\sin 313.27°$, **b.** $\cos(-28.2°)$, and **c.** $\tan 90°$.

Solution   Set your calculator for degree measure of angles.

a.   To evaluate $\sin 313.27°$, enter the number 313.27 and press the $\boxed{\text{SIN}}$ key. The display should read -0.7281317515. To the nearest ten-thousandth, $\sin 313.27° = -0.7281$.

b.   To evaluate $\cos(-28.2°)$, enter the number 28.2. Press the $\boxed{+/-}$ and $\boxed{\text{COS}}$ keys. The display should read 0.8813034521. To the nearest ten-thousandth, $\cos(-28.2°) = 0.8813$.

c. Finally, to attempt to evaluate tan 90°, enter 90 and press the $\boxed{\text{TAN}}$ key. The display will either blink 9's at you or read $\boxed{\text{ERROR}}$. Either way, it is telling you that tan 90° is undefined. ∎

Example 10   Find the value of sec 43°.

Solution   Neither tables nor calculators will allow you to evaluate the secant function directly. You must use the property that sec 43° is the reciprocal of cos 43°. To find sec 43° on a calculator, set your calculator for degrees, enter the number 43, and then press in order the $\boxed{\text{COS}}$ and $\boxed{1/x}$ keys. This gives the reciprocal of cos 43°, which is sec 43° ≈ 1.367327461. To the nearest ten-thousandth, sec 43° = 1.3673. ∎

Example 11   If θ is an acute angle and cos θ = 0.7660, find angle θ.

Solution   Angle θ can be found by using either a calculator or Table A.

If you use a calculator, be sure it is set for degrees. Enter the number .7660 and press the $\boxed{\text{INV}}$ and $\boxed{\text{COS}}$ keys. If your calculator does not have an $\boxed{\text{INV}}$ key, consult your owner's manual. To the nearest tenth, θ = 40.0°.

If you use Table A, find the column headed by **Cos** at the top of the page. Run your finger down the column, moving to successive pages if necessary, until you find the number .7660. Move to the left in that row to find the value of 40.0° in the degree column. ∎

Example 12   If θ is between 180° and 270° and sin θ = −0.9397, find angle θ.

Solution   See Figure 5-25. Any trigonometric function of θ has the same value as that same trigonometric function of θ's reference angle α, except possibly for sign. Enter the number .9397 in your calculator and press the $\boxed{\text{INV}}$ and $\boxed{\text{SIN}}$ keys. The display will give the value of the acute reference angle α. To the nearest tenth, α = 70.0°. Because θ is a third-quadrant angle, add 180° to α to find angle θ. Thus, θ = 180° + 70.0° = 250.0°. Table A can be used to find angle α if you do not have a calculator.

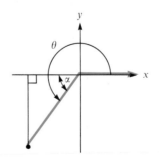

Figure 5-25 ∎

Exercise 5.3

In Exercises 1–8, find the exact value of the sine, cosine, and tangent of each angle, if possible. **Do not use a calculator.**

1. $135°$ **2.** $630°$ **3.** $450°$ **4.** $-30°$

5. $-240°$ **6.** $300°$ **7.** $540°$ **8.** $-315°$

In Exercises 9–16, find the exact value of the cosecant, secant, and cotangent of each angle, if possible. **Do not use a calculator.**

9. $225°$ **10.** $-1260°$ **11.** $1080°$ **12.** $-225°$

13. $-210°$ **14.** $-480°$ **15.** $585°$ **16.** $150°$

In Exercises 17–28, evaluate each expression. **Do not use a calculator.**

17. $\sin 0° + \cos 0° \tan 45°$

18. $\sin^2 90° + \cos 180° \tan 0°$

19. $\cos^2 90° + \cos 90° \sin^2 180°$

20. $\cos^2 0° + \sin^2 90° + \cot^2 90°$

21. $\sin^2 270° + \csc^2 270° + \cot^2 270°$

22. $\cos 180° \sin 180° - \tan^2 180°$

23. $\sin 30° \cos 60° - 2 \tan^2 60°$

24. $\sin^2 120° \cos 45° + \tan 45° \sin 90°$

25. $\sin 45° \cos 330° - \tan 150° \tan 60°$

26. $\cos 30° \tan 60° + \cos^3 45° \tan 45°$

27. $\csc^2 210° \sec 30° - \sec 315° \cot 60°$

28. $\csc 90° \csc 210° + \csc 45° \sin 135°$

In Exercises 29–42, use the given information to find values of θ, where $0° \leq \theta < 360°$. **Do not use a calculator.**

29. $\tan \theta = \dfrac{\sqrt{3}}{3}; \sin \theta = \dfrac{1}{2}$

30. $\tan \theta = -1; \cos \theta = \dfrac{-\sqrt{2}}{2}$

31. $\tan \theta = -\sqrt{3}; \cos \theta = \dfrac{1}{2}$

32. $\cot \theta = \sqrt{3}; \cos \theta = \dfrac{-\sqrt{3}}{2}$

33. $\sin \theta = -\dfrac{1}{2}; \sec \theta = \dfrac{-2\sqrt{3}}{3}$

34. $\cos \theta = \dfrac{\sqrt{3}}{2}; \csc \theta = 2$

35. $\tan \theta = -1; \sec \theta = \sqrt{2}$

36. $\tan \theta = -\sqrt{3}; \cos \theta = -\dfrac{1}{2}$

37. $\sec \theta = -\sqrt{2}; \cot \theta = -1$

38. $\sin \theta = -1$

39. $\tan \theta$ is undefined.

40. $\csc \theta = -\sqrt{2}; \cot \theta = -1$

41. $\cos \theta = \dfrac{\sqrt{3}}{2}; \sin \theta = -\dfrac{1}{2}$

42. $\sin \theta = -\dfrac{\sqrt{3}}{2}; \cos \theta = \dfrac{1}{2}$

In Exercises 43–52, use the given information to find all values of α, where possible, if $0° \leq \alpha < 360°$.

43. $\sin \alpha = \dfrac{1}{2}$

44. $\cos \alpha = -\dfrac{\sqrt{3}}{2}$

45. $\tan \alpha = \dfrac{\sqrt{3}}{3}$

46. $\cot \alpha = -\sqrt{3}$

47. $\sec \alpha = -2$

48. $\csc \alpha = -2$

49. $\sin \alpha = -\dfrac{1}{2}$; α in QII   **50.** $\tan \alpha = -1$; α not in QII

51. $\cot \alpha = \sqrt{3}$; α not in QI   **52.** $\cos \alpha = \dfrac{1}{2}$; α in QIII

In Exercises 53–58, use Table A in Appendix III to find the values of the sine, cosine, and tangent of the given angle.

53. $17°$ **54.** $55.2°$ **55.** $73°$ **56.** $34.8°$ **57.** $89°$ **58.** $0.6°$

59. Determine the values of the trigonometric functions of $90°$.

60. Determine the values of the trigonometric functions of $180°$.

In Exercises 61–72, use a calculator to find each value to four decimal places.

61. $\sin 23.1°$ **62.** $\sin 57.8°$ **63.** $\cos 133.7°$ **64.** $\cos 211.7°$

65. $\tan 223.5°$ **66.** $\tan(-223.5°)$ **67.** $\csc 312.4°$ **68.** $\csc 129.2°$

69. $\sec(-47.4°)$ **70.** $\sec 11.3°$ **71.** $\cot 640.6°$ **72.** $\cot 302.2°$

In Exercises 73–84, angle θ ($0° \leq \theta \leq 360°$) is in a given quadrant and the value of a trigonometric function is known. Use a calculator to find angle θ to the nearest tenth of a degree.

73. QI; $\tan \theta = 0.2493$ **74.** QI; $\sin \theta = 0.9986$

75. QII; $\cos \theta = -0.3420$ **76.** QII; $\cos \theta = -0.9063$

77. QIII; $\sin \theta = -0.4540$ **78.** QIII; $\cos \theta = -0.7193$

79. QIV; $\tan \theta = -5.6713$ **80.** QIV; $\sin \theta = -0.1908$

81. QI; $\csc \theta = 1.3250$ **82.** QII; $\sec \theta = -57.2987$

83. QIII; $\cot \theta = 1.1918$ **84.** QIV; $\csc \theta = -11.4737$

5.4   RADIAN MEASURE

Degree measure of angles is very common, but a different system, called *radian measure*, is more convenient in many situations.

> **Definition.**   Consider the circle in Figure 5-26. The **radian measure** of the central angle θ is given by
>
> $$\theta = \frac{s}{r}$$
>
>
>
> **Figure 5-26**
>
> where r is the radius of the circle and s is the length of the intercepted arc.

Figure 5-27

Figure 5-28

Suppose that the central angle θ of Figure 5-27 intercepts an arc of length equal to the circle's radius. By the previous definition, the radian measure of θ is

$$\theta = \frac{s}{r} = \frac{r}{r} = 1$$

Thus, a central angle of 1 radian intercepts an arc with length equal to the radius of the circle.

To determine how the radian measure of an angle is related to its degree measure, consider the circle in Figure 5-28 with radius r and a central angle of 180°. Because the length of the arc intercepted by a 180° angle is one-half of the circumference of the circle, or πr, the radian measure of a 180° angle is

$$180° = \frac{\pi r}{r} \text{ radians}$$

or

$$180° = \pi \text{ radians}$$

This result determines two equations that we can use to convert angular measure from one system to the other:

$$1° = \frac{\pi}{180} \text{ radian} \quad \text{and} \quad 1 \text{ radian} = \frac{180°}{\pi}$$

Example 1   Change to degrees: **a.** $\frac{5\pi}{6}$ radians and **b.** 7 radians.

Solution   **a.**
$$1 \text{ radian} = \frac{180°}{\pi}$$

$$\frac{5\pi}{6} \cdot 1 \text{ radian} = \frac{5\pi}{6} \cdot \frac{180°}{\pi}$$

$$\frac{5\pi}{6} \text{ radians} = 150°$$

b.
$$1 \text{ radian} = \frac{180°}{\pi}$$

$$7 \cdot 1 \text{ radian} = 7 \cdot \frac{180°}{\pi}$$

$$7 \text{ radians} = \frac{1260°}{\pi}$$ ∎

Example 2   Change to radians: **a.** 120° and **b.** −315°.

Solution   **a.**
$$1° = \frac{\pi}{180} \text{ radian}$$

$$120 \cdot 1° = 120 \cdot \frac{\pi}{180} \text{ radians}$$

$$120° = \frac{2\pi}{3} \text{ radians}$$

b.
$$1° = \frac{\pi}{180} \text{ radian}$$

$$-315 \cdot 1° = -315 \cdot \frac{\pi}{180} \text{ radians}$$

$$-315° = -\frac{7\pi}{4} \text{ radians}$$ ∎

The following table gives the degree measure and the corresponding radian measure of five common angles:

Degree measure	0°	30°	45°	60°	90°
Radian measure	0	$\dfrac{\pi}{6}$	$\dfrac{\pi}{4}$	$\dfrac{\pi}{3}$	$\dfrac{\pi}{2}$

The information in this table enables us to convert angles such as 120° and 225° from degree measure to radian measure in the following way:

$$120° = 2(60°) = 2\left(\frac{\pi}{3}\right) = \frac{2\pi}{3} \qquad 225° = 5(45°) = 5\left(\frac{\pi}{4}\right) = \frac{5\pi}{4}$$

We can also use this information to change angles such as $\frac{7\pi}{4}$ and $\frac{11\pi}{6}$ from radian measure to degree measure:

$$\frac{7\pi}{4} = 7\left(\frac{\pi}{4}\right) = 7(45°) = 315° \qquad \frac{11\pi}{6} = 11\left(\frac{\pi}{6}\right) = 11(30°) = 330°$$

By using radians, the arc length in a circle can be calculated easily. If we multiply both sides of the equation $\theta = \frac{s}{r}$ by r, the following formula results:

Figure 5-29

Formula for Arc Length. If a central angle θ in a circle with radius r is in radians, the length s of the intercepted arc is given by the formula

$$s = r\theta$$

See Figure 5-29.

Example 3 What is the length of the arc intercepted by a central angle of $\frac{5\pi}{6}$ radians in a circle with a radius of 18 meters?

Solution By the formula $s = r\theta$ and the known values $r = 18$ and $\theta = \frac{5\pi}{6}$, it follows that

$$s = 18\left(\frac{5\pi}{6}\right) = 15\pi \approx 47$$

The arc length is approximately 47 meters. ■

Example 4 A pulley with a diameter of exactly 8 inches drives another pulley with diameter of exactly 6 inches. If the larger pulley turns through 1 revolution, through what angle does the smaller pulley turn?

Solution The radii of the two pulleys are exactly 4 inches and 3 inches. When the larger pulley turns through 1 revolution (2π radians), a point on the edge of the pulley

moves

$$s = r\theta = 4(2\pi) = 8\pi \text{ inches}$$

If the belt does not slip, that motion is transferred to the smaller pulley, which turns through an angle

$$\theta = \frac{s}{r} = \frac{8\pi}{3} \approx 8.38 \text{ radians}$$
■

Example 5   Scott knows that the moon is about 237,000 miles from the earth, but he has forgotten its diameter. If the angle between his lines of sight to either side of the moon is 0.52°, how can Scott estimate its diameter? Give the answer to the nearest hundred.

Solution   Because the moon is so far from the earth, the length of the arc AB (with the earth as its center, and intercepted by a diameter of the moon) is a good estimate of the length of the diameter. See Figure 5-30. Because the moon is approximately 237,000 miles from the earth, side EB of triangle EAB is approximately 237,000 miles long as well. First, change 0.52° to radians:

$$1° = \frac{\pi}{180} \text{ radian}$$

$$0.52(1°) = 0.52\left(\frac{\pi}{180}\right) \text{ radian}$$

$$0.52° \approx 0.0091 \text{ radian}$$

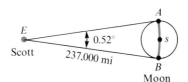

Figure 5-30

Then, substitute **237,000** for r and **0.0091** for θ in the formula $s = r\theta$ to obtain

$$s = r\theta$$
$$\approx 237,000(0.0091)$$
$$\approx 2160$$

Scott's estimate of the diameter of the moon is 2200 miles.   ■

The shaded area in Figure 5-31 is called a **sector** of the circle. To find the area of the sector, we first suppose that θ is measured in radians and then set up a proportion indicating that the area A of the sector is to the area of the circle as the length of arc s is to the circumference of the circle:

Figure 5-31

$$\frac{A}{\pi r^2} = \frac{s}{2\pi r}$$

We solve this proportion for A to obtain

$$A = \frac{1}{2}rs$$

Because $s = r\theta$, we can substitute $r\theta$ for s in the formula to obtain

$$A = \frac{1}{2}r^2\theta$$

Thus, we have

Formula for the Area of a Sector.   If a sector of a circle with radius r has a central angle θ in radians, then the area of the sector is given by the formula

$$A = \frac{1}{2}r^2\theta$$

Example 6   A sector of a circle has a central angle of $50.0°$ and an area of 605 square centimeters. Find the radius of the circle.

Solution   First, change the angle of $50.0°$ to radians:

$$1° = \frac{\pi}{180} \text{ radian}$$

$$50.0° = \frac{5\pi}{18} \text{ radian} \qquad \text{Multiply both sides by 50.0 and simplify.}$$

Now substitute **605** for A and $\frac{5\pi}{18}$ for θ in the formula $A = \frac{1}{2}r^2\theta$ and solve for r:

$$A = \frac{1}{2}r^2\theta$$

$$605 = \frac{1}{2}r^2\frac{5\pi}{18}$$

$$\frac{605(2)(18)}{5\pi} = r^2$$

$$1386.56 \approx r^2$$

$$\sqrt{1386.56} \approx r$$

$$37.2 \approx r$$

The radius of the circle is approximately 37.2 centimeters.   ∎

Linear and Angular Velocity

The question "How fast is that train moving?" might be answered "70 miles per hour" or "90 feet per second." These answers indicate the train's **linear**

velocity, a measure of how far the train will travel in 1 unit of time. The question "How fast is that phonograph record turning?" might be answered "$33\frac{1}{3}$ revolutions per minute" or "45 revolutions per minute." These answers indicate the record's **angular velocity**, a measure of the angle through which the record rotates in 1 unit of time.

Linear velocity and angular velocity are often related. Because a wheel of a car, for example, moves with the car, it has linear velocity. Because the wheel is spinning, it has angular velocity as well. To see how linear velocity, angular velocity, and the radius of the wheel are all related, we consider a wheel of radius r as it rolls a distance s without slipping. Any point A on the circumference of the wheel also moves a distance s, measured along the arc of the wheel. See Figure 5-32. The wheel rotates through an angle θ, and s and θ are related by the formula

$$s = r\theta \quad (\theta \text{ in radians})$$

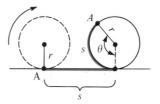

Figure 5-32

If this movement is accomplished in a length of time t, then $\frac{s}{t}$ is the linear velocity of the wheel and $\frac{\theta}{t}$ is the angular velocity. Dividing both sides of the equation $s = r\theta$ by t gives a formula that relates linear and angular velocity:

$$\frac{s}{t} = r\left(\frac{\theta}{t}\right)$$

If we denote the linear velocity $\frac{s}{t}$ by v, and the angular velocity $\frac{\theta}{t}$ by ω (the Greek letter omega), the previous equation becomes

$$v = r\omega$$

where ω is in units of radians per unit time.

Example 7   What is the linear velocity of a point on the Equator in miles per hour?

Solution   Assume that the radius of the earth is approximately 4000 miles. Because the earth rotates once in 24 hours, its angular velocity is

$$\omega = \frac{2\pi \text{ rad}}{24 \text{ h}}$$

$$= \frac{\pi \text{ rad}}{12 \text{ h}}$$

To find the linear velocity, substitute 4000 for r and $\frac{\pi}{12}$ for ω in the formula $v = r\omega$ and simplify:

$$v = r\omega$$

$$= 4000 \cdot \frac{\pi}{12}$$

$$\approx 1000$$

The linear velocity of a point on the Equator is approximately 1000 miles per hour. ■

Example 8   An 8-inch-diameter pulley drives a 6-inch-diameter pulley. The larger pulley makes 15 revolutions per second. What is the angular velocity of the smaller pulley in revolutions per second?

Solution   The angular velocity of the drive pulley is

$$15 \, \frac{\text{rev}}{\text{sec}} \cdot 2\pi \, \frac{\text{rad}}{\text{rev}} = 30\pi \, \frac{\text{rad}}{\text{sec}}$$

Because the belt that connects the two pulleys does not slip, the linear velocities of points on either circumference are the same—the product $r_1\omega_1$ for one pulley is equal to the product $r_2\omega_2$ for the second pulley. Thus,

$$r_1\omega_1 = v = r_2\omega_2$$
$$4(30\pi) = 3(\omega_2) \qquad \text{Substitute 4 for } r_1, \, 30\pi \text{ for } \omega_1, \text{ and 3 for } r_2.$$
$$\frac{4(30\pi)}{3} = \omega_2$$
$$40\pi = \omega_2$$

The angular velocity of the smaller pulley is $40\pi \, \frac{\text{rad}}{\text{sec}}$. To convert to revolutions per second, multiply by $\frac{1}{2\pi} \frac{\text{rev}}{\text{rad}}$ and simplify:

$$40\pi \, \frac{\text{rad}}{\text{sec}} \cdot \frac{1}{2\pi} \frac{\text{rev}}{\text{rad}} = 20 \, \frac{\text{rev}}{\text{sec}}$$

The angular velocity of the smaller pulley is $20 \, \frac{\text{rev}}{\text{sec}}$. ■

Exercise 5.4

In Exercises 1–12, change each angle to radians.

1. $15°$ **2.** $75°$ **3.** $120°$ **4.** $150°$ **5.** $210°$ **6.** $240°$

7. $300°$ **8.** $330°$ **9.** $780°$ **10.** $660°$ **11.** $-520°$ **12.** $-880°$

In Exercises 13–24, each angle is expressed in radians. Change each angle to degrees.

13. $\frac{3}{4}\pi$ **14.** 3π **15.** $\frac{5}{2}\pi$ **16.** $\frac{7}{3}\pi$ **17.** $\frac{4}{3}\pi$ **18.** $\frac{11}{6}\pi$

19. 6 **20.** 8 **21.** -10 **22.** -5 **23.** $12\frac{1}{2}$ **24.** -15.3

25. Find the radius of a circle if a central angle of 25° intercepts an arc of 17 centimeters.

26. Find the central angle in radians that intercepts an arc of 10 centimeters in a circle with diameter of 10 centimeters.

In Exercises 27–30, use 3960 miles for the radius of the earth.

27. The latitude of Manchester, New Hampshire, is 43.0° N. How far is Manchester from the Equator? See Illustration 1. Give the answer to the nearest 10 miles.

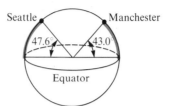

Illustration 1

28. The latitude of Seattle, Washington, is 47.6° N. How far is Seattle from the Equator? See Illustration 1. Give the answer to the nearest 10 miles.

29. St. Louis, Missouri, is 2670 miles north of the Equator. Find its latitude to the nearest tenth of a degree.

30. Pittsburgh, Pennsylvania, is 2800 miles north of the Equator. Find its latitude to the nearest tenth of a degree.

31. Find the area, to the nearest hundredth, of a sector of a circle if the sector has a central angle of exactly 30° and the circle has a radius of exactly 20 units.

32. If a circle contains a sector with central angle of $\frac{2\pi}{3}$ and area of 30 square meters, find the diameter of the circle.

33. A regular octagon is inscribed in a circle 30 centimeters in diameter. How long is an arc intercepted by one of the sides of the octagon?

34. A railroad track curves along a 17° arc of a circle. If the radius of the circle is 250 meters, how long is the track?

35. The earth is approximately 93 million miles from the sun. In one day, the earth moves through an arc of 0.986°. How many miles does the earth travel through space in one week?

36. A **nautical mile** is defined as $\frac{1}{60}$° of arc on the Equator. If the radius of the earth is 3960 statute miles, find the number of statute miles contained in one nautical mile. Give your answer to the nearest hundredth.

In Exercises 37–42, find the angular velocity of the object in the unit specified.

37. The minute hand of a clock in radians per hour.

38. The second hand of a clock in radians per second.

39. The minute hand of a clock in radians per second.

40. The earth in its orbit in radians per month.

41. The moon in its orbit in radians per day. Assume that the moon circles the earth in 29.5 days.

42. The earth about its axis in radians per minute.

43. A car is traveling 88 $\frac{\text{ft}}{\text{sec}}$. How fast are its 3-foot-diameter tires spinning in radians per second?

44. A car is traveling 40 $\frac{\text{ft}}{\text{sec}}$. Find the angular velocity of its 2-foot-diameter tires in radians per second.

45. The 30-inch-diameter tires of a truck are turning at 400 rpm. What is the linear velocity of the truck in feet per minute?

46. The 27-inch tires of a bicycle are turning at the rate of 125 rpm. What is the linear velocity of the bicycle in inches per minute?

47. A wheel that is driven by a belt is making 1 $\frac{\text{rev}}{\text{sec}}$. If the wheel is 6 inches in radius, what is the linear velocity of the belt?

48. A wheel that is 8.0 centimeters in diameter is driven by a belt with a linear velocity of 40 $\frac{\text{cm}}{\text{sec}}$. How many revolutions per minute is the wheel making?

49. An idler pulley 3 inches in diameter is making 95 rpm. What is the linear velocity of the belt driving the pulley in inches per second?

50. A belt drives two pulleys. One pulley is 10 inches in diameter and the other is 12 inches in diameter. How many revolutions per minute is the large pulley making if the small pulley is turning at 20 rpm?

51. A belt drives two wheels that are 15 and 20 inches in diameter. If the 20-inch wheel is turning at 10 rpm, how fast is the other wheel turning in revolutions per minute?

52. What is the linear velocity due to the rotation of the earth of Green Bay, Wisconsin (latitude of 44.5° N)? Assume the radius of the earth to be 3960 miles. See Illustration 2.

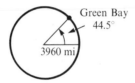

Illustration 2

53. What is the linear velocity due to the rotation of the earth of Miami, Florida (latitude of 25.8° N)? Assume the radius of the earth to be 3960 miles. See Illustration 3.

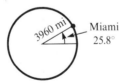

Illustration 3

54. What is the latitude of Pocatello, Idaho, if its linear velocity due to the rotation of the earth is 759 mph? Assume the radius of the earth to be 3960 miles.

5.5 THE CIRCULAR FUNCTIONS

An alternative definition of the trigonometric functions makes use of the **unit circle**, the circle that is centered at the origin and has a radius of 1 unit.

Definition. Let θ be a central angle of the unit circle shown in Figure 5-33. Let $P(x, y)$ be the point where the terminal side of angle θ intersects the circle. Then, $r = OP = 1$. The **six trigonometric functions of angle θ** are

$$\sin \theta = y$$

$$\cos \theta = x$$

$$\tan \theta = \frac{y}{x} \quad (x \neq 0)$$

$$\csc \theta = \frac{1}{y} \quad (y \neq 0)$$

$$\sec \theta = \frac{1}{x} \quad (x \neq 0)$$

$$\cot \theta = \frac{x}{y} \quad (y \neq 0)$$

Figure 5-33

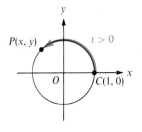

Figure 5-34

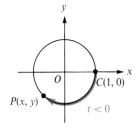

Figure 5-35

Note that the unit-circle definition of the trigonometric functions is a special case of the angle-in-standard-position definition. In the unit-circle definition, the radius r is required to be 1 unit in length. In the angle-in-standard-position definition, r can be any positive number.

Recall that the domain of the function defined by the equation $y = f(x)$ is the set of all admissible values of x, and the range is the set of all values of y. For virtually all functions previously studied, both the domain and the range of the function have been subsets of the real number system. Thus far, however, the domains of the trigonometric functions have been sets of angles rather than sets of real numbers. The unit-circle approach will enable us to show that the trigonometric functions can also be thought of as functions with real number domains.

The argument begins by noting that any real number can represent the length of exactly one arc on the unit circle. If t is a positive number, we can find the arc of length t by measuring a distance t in a counterclockwise direction along an arc of the unit circle beginning at the point $C(1, 0)$. This determines arc CP of length t. See Figure 5-34. If t is a negative number, we can find the arc of length t by measuring a distance $|t|$ in a clockwise direction along an arc of the unit circle beginning at the point $C(1, 0)$. This determines arc CP of length t. See Figure 5-35.

In each case, marking off the arc determines the unique point P with coordinates (x, y) that corresponds to the real number t. Now suppose that we let t be any real number and place an arc of length $|t|$ on the unit circle, as in Figure 5-36. From the previous section we know that if s is an arc intercepted by a central angle θ and θ is measured in radians, we have

$$s = r\theta$$

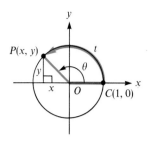

Figure 5-36

We can substitute t for s and 1 for r in the formula $s = r\theta$ to obtain

$$t = 1\theta$$

or

$$\theta = t$$

Thus, when the measure of an arc on the unit circle is the real number t, then t is also the radian measure of the central angle determined by that arc. This gives

$$\sin \theta = \sin t \qquad \csc \theta = \csc t$$
$$\cos \theta = \cos t \qquad \sec \theta = \sec t$$
$$\tan \theta = \tan t \qquad \cot \theta = \cot t$$

where θ is an angle measured in radians and t is a real number. Hence, we can think of each trigonometric expression as being either a trigonometric function of an angle measured in radians or as a trigonometric function of a real number. The important point is this: *The trigonometric functions can now be thought of as functions that have domains and ranges that are subsets of the real numbers.* Trigonometric functions of a real variable are called the **circular functions**.

Example 1 Find **a.** $\sin\left(\dfrac{\pi}{3}\right)$ and **b.** $\cos\left(\dfrac{5\pi}{2}\right)$.

Solution   **a.** $\sin\left(\dfrac{\pi}{3}\right) = \sin\left(\dfrac{\pi}{3} \text{ rad}\right)$

$$= \dfrac{\sqrt{3}}{2}$$

b. $\cos\left(\dfrac{5\pi}{2}\right) = \cos\left(\dfrac{5\pi}{2} \text{ rad}\right)$

$$= \cos\left(\dfrac{\pi}{2} \text{ rad}\right) \qquad \tfrac{\pi}{2} \text{ is coterminal with } \tfrac{5\pi}{2}.$$

$$= 0 \qquad\qquad\qquad\qquad\qquad\qquad \blacksquare$$

Example 2 Find **a.** $\tan\left(\dfrac{3\pi}{4}\right)$ and **b.** $\csc\left(\dfrac{7\pi}{6}\right)$.

Solution   **a.** $\tan\left(\dfrac{3\pi}{4}\right) = \tan\left(\dfrac{3\pi}{4}\ \text{rad}\right)$

$\qquad\qquad = -\tan\left(\dfrac{\pi}{4}\ \text{rad}\right)$ $\quad\frac{3\pi}{4}$ is a QII angle with reference angle $\frac{\pi}{4}$.

$\qquad\qquad = -1$

b. $\csc\left(\dfrac{7\pi}{6}\right) = \csc\left(\dfrac{7\pi}{6}\ \text{rad}\right)$

$\qquad\qquad = \dfrac{1}{\sin\left(\dfrac{7\pi}{6}\ \text{rad}\right)}$

$\qquad\qquad = \dfrac{1}{-\sin\left(\dfrac{\pi}{6}\ \text{rad}\right)}$ $\quad\frac{7\pi}{6}$ is a QIII angle with reference angle $\frac{\pi}{6}$.

$\qquad\qquad = -2$ ■

Example 3   Find the coordinates (x, y) of the point P on the unit circle that correspond to the real numbers **a.** π, **b.** $\dfrac{\pi}{6}$, and **c.** $\dfrac{-5\pi}{4}$.

Solution   **a.**   From the definition, you have

$\qquad\qquad x = \cos\theta \qquad\text{and}\qquad y = \sin\theta$

But if θ is measured in radians, you also have

$\qquad\qquad x = \cos t \qquad\text{and}\qquad y = \sin t$

where t is the real number that is the radian measure of central angle θ. Hence, the coordinates (x, y) of the point P that correspond to the real number π are

$\qquad\qquad x = \cos\pi = \cos(\pi\ \text{rad}) = -1$

$\qquad\qquad y = \sin\pi = \sin(\pi\ \text{rad}) = 0$

Thus, point P has coordinates $(-1, 0)$.

Note that because the circumference of the unit circle is 2π, the number π is the measure of half that circumference. From Figure 5-37, you can see that the coordinates of the point P must be $(-1, 0)$.

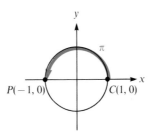

Figure 5-37

b. The point P on the unit circle that corresponds to $\dfrac{\pi}{6}$ has coordinates of

$$x = \cos\left(\frac{\pi}{6}\right) = \cos\left(\frac{\pi}{6}\text{ rad}\right) = \frac{\sqrt{3}}{2}$$

$$y = \sin\left(\frac{\pi}{6}\right) = \sin\left(\frac{\pi}{6}\text{ rad}\right) = \frac{1}{2}$$

Thus, the point P has coordinates $\left(\dfrac{\sqrt{3}}{2}, \dfrac{1}{2}\right)$.

c. The point P on the unit circle that corresponds to $\dfrac{-5\pi}{4}$ has coordinates of

$$x = \cos\left(\frac{-5\pi}{4}\right) = \cos\left(\frac{-5\pi}{4}\text{ rad}\right) = -\frac{\sqrt{2}}{2}$$

$$y = \sin\left(\frac{-5\pi}{4}\right) = \sin\left(\frac{-5\pi}{4}\text{ rad}\right) = \frac{\sqrt{2}}{2}$$

Thus, the point P has coordinates $\left(-\dfrac{\sqrt{2}}{2}, \dfrac{\sqrt{2}}{2}\right)$. ■

Example 4 Find the coordinates of the point P on the unit circle that correspond to the real number -1.37.

Solution The coordinates of the point P are

$$x = \cos(-1.37) = \cos(-1.37\text{ rad})$$
$$y = \sin(-1.37) = \sin(-1.37\text{ rad})$$

Use your calculator, set in radian mode, to determine that

$$x \approx \quad 0.1994$$
$$y \approx -0.9799$$

Thus, point P has approximate coordinates of $(0.1994, -0.9799)$. ■

Simple Harmonic Motion

When the domains of the trigonometric functions are considered as real numbers, many applications are possible. One such application is in the study of *simple harmonic motion*.

If an object is suspended from a ceiling by a spring, and then pushed up and released, it will start to bounce. See Figure 5-38. Its position above (positive) or below (negative) the equilibrium position is given by a formula involving a circular function:

$$y = A \cos\left(\sqrt{\frac{k}{m}}\, t\right)$$

The coefficient A, called the **amplitude**, represents the distance the object is pushed above the equilibrium position. The constant k, called the **spring constant**, depends on the stiffness of the spring, and m is the mass of the object. The variable t represents the time in seconds since the object was released.

The first cycle is completed when

$$\sqrt{\frac{k}{m}}\, t = 2\pi$$

or when

$$t = 2\pi \sqrt{\frac{m}{k}}$$

The value

$$2\pi \sqrt{\frac{m}{k}}$$

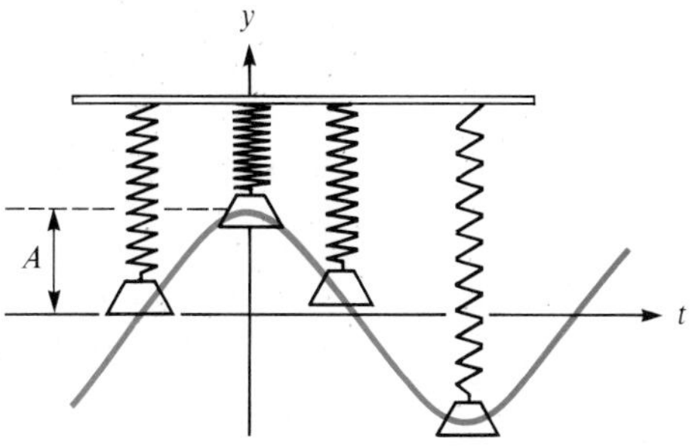

Figure 5-38

is called the **period of oscillation** and indicates the number of seconds per complete cycle. Its reciprocal

$$\frac{1}{2\pi} \sqrt{\frac{k}{m}}$$

is the **frequency of the oscillation**, indicating the number of cycles per second.

Example 5 A mass of 4 grams, attached to a spring with a spring constant of 9 g/cm, is raised above the equilibrium position and released. Find the period and the frequency of the oscillation.

Solution The period of the oscillation is given by the value

$$2\pi \sqrt{\frac{m}{k}}$$

Substituting 4 for m and 9 for k gives

$$\text{period} = 2\pi \sqrt{\frac{m}{k}} = 2\pi \sqrt{\frac{4}{9}} = 2\pi \cdot \frac{2}{3} = \frac{4\pi}{3}$$

It takes the mass $\frac{4\pi}{3}$ seconds, or about 4.2 seconds, to return to its starting point. The frequency of the oscillation is the reciprocal of the period. Thus,

$$\text{frequency} = \frac{1}{\text{period}} = \frac{3}{4\pi}$$

The frequency is $\frac{3}{4\pi}$, or approximately 0.24 hertz (cycles per second). ■

Any motion described by an equation of the form

$$y = A \sin kt \qquad \text{or} \qquad y = A \cos kt$$

is called **simple harmonic motion**. We give two more examples of such motion.

A pendulum swinging through a small arc makes an angle θ with the vertical where θ is given by the formula

$$\theta = A \sin \sqrt{\frac{g}{l}} \, t$$

The coefficient A is the maximum amplitude of the swing, l is the length of the pendulum, and g is a constant related to the force of gravity. The motion described by the pendulum is simple harmonic motion.

If a floating object is pushed under water and then released, it will oscillate as it bobs up and down. For certain shapes, the position y relative to the equilibrium position is given by the equation

$$y = A \cos \sqrt{\frac{k}{w}} \, t$$

where A is the maximum displacement from equilibrium, w is the object's weight, and k depends on the density of the water and the force of gravity. The bobbing action of such a floating object also describes simple harmonic motion.

Exercise 5.5

In Exercises 1–12, evaluate each given expression. **Do not use a calculator.**

1. $\sin \dfrac{\pi}{6}$

2. $\cos \dfrac{\pi}{6}$

3. $\cos \left(-\dfrac{5\pi}{6} \right)$

4. $\sin \left(-\dfrac{5\pi}{6} \right)$

5. $\tan \dfrac{5\pi}{4}$

6. $\cot \dfrac{5\pi}{4}$

7. $\csc \dfrac{5\pi}{6}$

8. $\sec \dfrac{5\pi}{6}$

9. $\sin \left(-\dfrac{8\pi}{3} \right)$

10. $\cos \dfrac{8\pi}{3}$

11. $\tan \dfrac{15\pi}{4}$

12. $\csc \dfrac{7\pi}{3}$

In Exercises 13–24, use a calculator to evaluate each given expression to four decimal places. Remember to set your calculator in radian mode.

13. $\sin 2$

14. $\sin 3$

15. $\cos 8$

16. $\tan 5$

17. $\sin \dfrac{3}{\pi}$

18. $\cos \left(-\dfrac{3}{\pi} \right)$

19. $\sec 5$

20. $\csc 4$

21. $\cot 1$

22. $\cot(-1)$

23. $\sin(3 + \pi)$

24. $\cos(3 - \pi)$

In Exercises 25–48, find the coordinates of the point P on the unit circle that correspond to each given real number. **Do not use a calculator.**

25. $\dfrac{3\pi}{2}$

26. $\dfrac{-\pi}{2}$

27. $-\pi$

28. 2π

29. 3π

30. 4π

31. $\dfrac{-7\pi}{2}$

32. $\dfrac{9\pi}{2}$

33. $\dfrac{\pi}{4}$

34. $\dfrac{-\pi}{4}$

35. $\dfrac{-3\pi}{4}$

36. $\dfrac{5\pi}{4}$

37. $\dfrac{\pi}{3}$

38. $\dfrac{-\pi}{6}$

39. $\dfrac{-2\pi}{3}$

40. $\dfrac{2\pi}{3}$

41. $\dfrac{4\pi}{3}$

42. $\dfrac{-4\pi}{3}$

43. $\dfrac{-5\pi}{3}$ **44.** $\dfrac{5\pi}{6}$ **45.** $\dfrac{11\pi}{6}$ **46.** $\dfrac{17\pi}{6}$ **47.** $\dfrac{23\pi}{6}$ **48.** $\dfrac{-17\pi}{6}$

49. Show that the area of triangle OAR in Illustration 1 is given by the formula

$$A = \frac{1}{2}\cos\theta\sin\theta$$

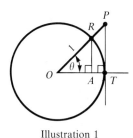

Illustration 1

50. Show that the area of triangle OTP in Illustration 1 is given by the formula

$$A = \frac{1}{2}\tan\theta$$

51. A spring with a spring constant of 6 newtons per meter hangs from a ceiling and supports a mass of 24 kilograms. The mass is pulled to a starting point a few centimeters below the equilibrium position and released. How long will it take to return to the starting point?

52. What is the frequency of the oscillation in Exercise 51?

53. A spring supports a mass of 12 kilograms. The frequency of its oscillation is $\frac{1}{\pi}$ hertz (cycles per second). What is the spring constant? The units of your answer will be newtons per meter.

54. A pendulum 1 meter long is set to swinging through a small amplitude. Let $g = 9.8$ m/sec^2. Find the period and the frequency of the oscillation.

55. A cubical box floats on its side and oscillates with a period of $\frac{1}{3}$ second. What is its weight? Let $k = 200{,}000$ lb/sec^2.

56. For a given spring, what is the effect on the period of its oscillation if a mass suspended by the spring is doubled?

5.6   GRAPHS OF FUNCTIONS INVOLVING SIN X AND COS X

In this section we shall discuss the graphs of the sine and cosine functions. We begin by noting that the sine of an angle θ is equal to the sine of any angle that is coterminal with angle θ. Similarly, the cosine of an angle θ is equal to the cosine of any angle that is coterminal with angle θ. Therefore, for any real number x,

$$\sin x = \sin(x \pm 2\pi) = \sin(x \pm 4\pi) = \ldots$$
$$\cos x = \cos(x \pm 2\pi) = \cos(x \pm 4\pi) = \ldots$$

The sine and cosine functions are called **periodic functions** because as x increases, the values of $\sin x$ and $\cos x$ repeat in a predictable way.

> **Definition.** A function f is said to be **periodic with period p** if p is the smallest positive number for which
>
> $$f(x) = f(x + p)$$
>
> for all x in the domain of f.

The sine function has a period of 2π because $\sin x = \sin(x + 2\pi)$ for all x, and $\sin x = \sin(x + p)$ is true for no positive number p that is less than 2π. The cosine function also has a period of 2π because $\cos x = \cos(x + 2\pi)$ for all x, and $\cos x = \cos(x + p)$ is true for no positive number p that is less than 2π.

To graph the sine function, use a calculator or Table A in Appendix III to find pairs (x, y) that satisfy the equation $y = \sin x$. Remember that x, often called the **argument** of the function, can be thought of either as an angle in radians or as a real number.

The table of values in Figure 5-39 shows that as x increases from 0 to $\frac{\pi}{2}$, the values of $\sin x$ increase from 0 to 1. As x continues to increase from $\frac{\pi}{2}$ to π, the values of $\sin x$ decrease, heading back to 0. As x goes from π to $\frac{3\pi}{2}$, the values of $\sin x$ continue to decrease until they reach a value of -1 at $x = \frac{3\pi}{2}$. As x continues to increase from $\frac{3\pi}{2}$ to 2π, the values of $\sin x$ increase, reaching the value of 0 when $x = 2\pi$. If we plot the pairs of values $(x, \sin x)$ shown in the table of values, we can draw the graph of $y = \sin x$.

The graph of $y = \sin x$ doesn't "start" at 0, nor does it "end" at 2π. It continues to oscillate in both directions forever. Because the function is periodic, however, if we know its behavior through one period, we know its behavior everywhere. Note that the values of $\sin x$ are always between -1 and 1 inclusive.

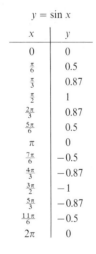

$y = \sin x$

x	y
0	0
$\frac{\pi}{6}$	0.5
$\frac{\pi}{3}$	0.87
$\frac{\pi}{2}$	1
$\frac{2\pi}{3}$	0.87
$\frac{5\pi}{6}$	0.5
π	0
$\frac{7\pi}{6}$	-0.5
$\frac{4\pi}{3}$	-0.87
$\frac{3\pi}{2}$	-1
$\frac{5\pi}{3}$	-0.87
$\frac{11\pi}{6}$	-0.5
2π	0

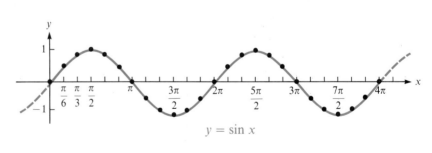

Figure 5-39

The graph of the cosine function is similar to that of the sine function. Because $\cos 0 = 1$, we shall draw its graph beginning at the point $(0, 1)$. As x increases from 0 to π, the values of $\cos x$ decrease, dropping to a value of -1 when $x = \pi$.

As x increases from π to 2π, the values of cos x increase back to 1 when $x = 2\pi$. The graph of $y = \cos x$ appears in Figure 5-40. As with the values of sin x, the values of cos x are always between -1 and 1 inclusive.

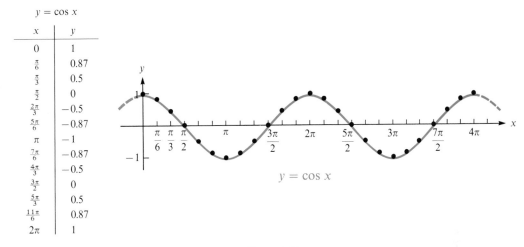

x	y
0	1
$\frac{\pi}{6}$	0.87
$\frac{\pi}{3}$	0.5
$\frac{\pi}{2}$	0
$\frac{2\pi}{3}$	-0.5
$\frac{5\pi}{6}$	-0.87
π	-1
$\frac{7\pi}{6}$	-0.87
$\frac{4\pi}{3}$	-0.5
$\frac{3\pi}{2}$	0
$\frac{5\pi}{3}$	0.5
$\frac{11\pi}{6}$	0.87
2π	1

Figure 5-40

Example 1   Graph the function defined by $y = 3 \sin x$.

Solution   Because sin x has values between -1 and 1 as x increases from 0 to 2π, the values of 3 sin x must be between -3 and 3 in that same interval. As x increases from 0 to $\frac{\pi}{2}$, the values of 3 sin x increase from 0 to 3. As x increases from $\frac{\pi}{2}$ to $\frac{3\pi}{2}$, the values of 3 sin x decrease from 3 to -3. As x increases from $\frac{3\pi}{2}$ to 2π, the values of sin x increase back to their starting value of 0. Thus, the graph of $y = 3 \sin x$ finishes one complete cycle as x grows from 0 to 2π. The graph of $y = 3 \sin x$ is shown in Figure 5-41. (The graph of $y = \sin x$ is included for reference.)

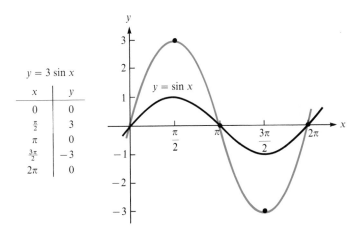

x	y
0	0
$\frac{\pi}{2}$	3
π	0
$\frac{3\pi}{2}$	-3
2π	0

Figure 5-41

Example 2   Graph $y = 2 \cos x$ and $y = -2 \cos x$ on the same set of coordinate axes.

Solution   As x increases from 0 to 2π, the values of $2 \cos x$ begin at 2, decrease to -2 at $x = \pi$, and then increase back to 2 at $x = 2\pi$. The graph of $y = 2 \cos x$ appears in Figure 5-42.

As x increases from 0 to 2π, the values of $-2 \cos x$ begin at -2, increase to 2 at $x = \pi$, and then decrease back to -2 at $x = 2\pi$. The graph of $y = -2 \cos x$ also appears in Figure 5-42. Note that the two graphs are reflections of each other, with the x-axis as the axis of symmetry.

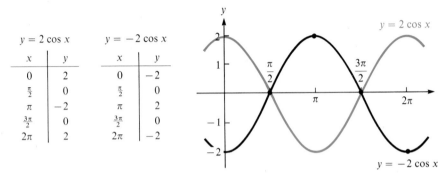

$y = 2 \cos x$

x	y
0	2
$\frac{\pi}{2}$	0
π	-2
$\frac{3\pi}{2}$	0
2π	2

$y = -2 \cos x$

x	y
0	-2
$\frac{\pi}{2}$	0
π	2
$\frac{3\pi}{2}$	0
2π	-2

Figure 5-42

Example 3   Graph $y = \sin 2x$.

Solution   Because the graph of $y = \sin x$ completes one cycle as x increases from 0 to 2π, the graph of $y = \sin 2x$ must complete one cycle as $2x$ increases from 0 to 2π, or as x increases from 0 to π. Thus, the period of $y = \sin 2x$ is π. Note that the graph of $y = \sin 2x$ oscillates twice as fast as the graph of $y = \sin x$; that is, it completes two cycles into the same space required for one cycle of the graph of $y = \sin x$. The graph of $y = \sin 2x$, along with a table of values, is shown in Figure 5-43. Note that the coefficient of x is a number greater than 1 and that the curve is compressed.

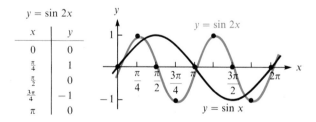

$y = \sin 2x$

x	y
0	0
$\frac{\pi}{4}$	1
$\frac{\pi}{2}$	0
$\frac{3\pi}{4}$	-1
π	0

Figure 5-43

Example 4   Graph $y = \cos \dfrac{1}{3} x$.

Solution   As x increases from 0 to 2π, the graph of $y = \cos x$ completes one cycle. As $\frac{1}{3}x$ increases from 0 to 2π (that is, as x increases from 0 to 6π), the graph of $y = \cos \frac{1}{3}x$ completes one cycle. Thus, the period of the graph of $y = \cos \frac{1}{3}x$ is 6π. The graph, along with a table of values, is shown in Figure 5-44. Note that the coefficient of x is a fraction less than 1 and that the curve is stretched.

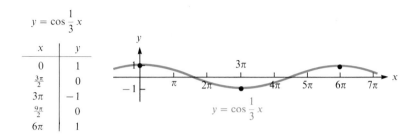

$y = \cos \dfrac{1}{3}x$

x	y
0	1
$\frac{3\pi}{2}$	0
3π	-1
$\frac{9\pi}{2}$	0
6π	1

Figure 5-44

Example 5   Graph $y = \cos \pi x$.

Solution   As x increases from 0 to 2π, the graph of $y = \cos x$ completes one cycle. As πx increases from 0 to 2π (that is, as x increases from 0 to 2), the graph of $y = \cos \pi x$ completes one cycle. Thus, the period of the graph of $y = \cos \pi x$ is 2. The graph of $y = \cos \pi x$ appears in Figure 5-45.

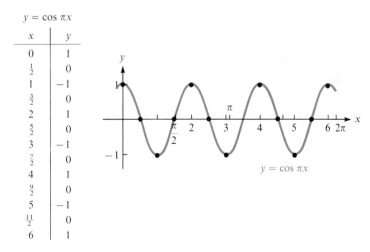

$y = \cos \pi x$

x	y
0	1
$\frac{1}{2}$	0
1	-1
$\frac{3}{2}$	0
2	1
$\frac{5}{2}$	0
3	-1
$\frac{7}{2}$	0
4	1
$\frac{9}{2}$	0
5	-1
$\frac{11}{2}$	0
6	1

Figure 5-45

One cycle of the graph of $y = a \sin bx$ is completed as bx increases from 0 to 2π. When $bx = 0$, then x must be 0. When $bx = 2\pi$, then $x = \frac{2\pi}{b}$. It follows that one cycle of the curve $y = a \sin bx$ is completed as x itself increases from 0 to $\left|\frac{2\pi}{b}\right|$. Thus, the period of the graph of $y = a \sin bx$ is $\left|\frac{2\pi}{b}\right|$.

The largest value attained by sin bx is 1. Therefore, the largest value that can be attained by a sin bx is $|a| \cdot 1 = |a|$. This value is called the **amplitude** of the graph of $y = a$ sin bx.

A similar argument applies to determine the period and the amplitude of the graph of $y = a \cos bx$.

Period and Amplitude.   The **period** of the graph of $\begin{Bmatrix} y = a \sin bx \\ y = a \cos bx \end{Bmatrix}$ is $\left| \dfrac{2\pi}{b} \right|$, and the **amplitude** is $|a|$.

Example 6   Graph $y = 5 \sin 7x$.

Solution   The amplitude is 5 and the period is $\frac{2\pi}{7}$. One cycle of $y = 5 \sin 7x$ appears in Figure 5-46.

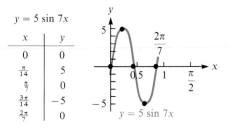

x	y
0	0
$\frac{\pi}{14}$	5
$\frac{\pi}{7}$	0
$\frac{3\pi}{14}$	-5
$\frac{2\pi}{7}$	0

Figure 5-46

Example 7   Graph $y = 2 \cos \dfrac{1}{2} x$.

Solution   The amplitude is 2 and the period is

$$\frac{2\pi}{\frac{1}{2}} = 4\pi$$

$y = 2 \cos \dfrac{1}{2} x$

x	y
0	2
π	0
2π	-2
3π	0
4π	2

$y = 2 \cos \dfrac{1}{2} x$

Figure 5-47

Because $2 \cos \frac{1}{2}x$ is zero when $x = \ldots, -\pi, \pi, 3\pi, \ldots$, the curve intersects the x-axis at these points. One cycle of $y = 2 \cos \frac{1}{2}x$ appears in Figure 5-47. Note that the coefficient of x (which is a fraction less than 1) stretches the curve.

Exercise 5.6

In Exercises 1–16, find the amplitude and period of each function. **Do not construct a graph.**

1. $y = 2 \sin x$

2. $y = 3 \cos x$

3. $y = \cos 9x$

4. $y = -\sin 11x$

5. $y = \sin \dfrac{1}{3} x$

6. $y = \cos \dfrac{1}{4} x$

7. $y = -\cos 0.2x$

8. $y = \sin 0.25x$

9. $y = 3 \sin \dfrac{1}{2} x$

10. $y = \dfrac{1}{2} \cos \dfrac{1}{3} x$

11. $y = -\dfrac{1}{2} \cos \pi x$

12. $y = 17 \sin 2\pi x$

13. $y = 3 \sin 2\pi x$

14. $y = 8 \cos \pi x$

15. $y = -\dfrac{1}{3} \sin \dfrac{3x}{\pi}$

16. $y = -\dfrac{5}{3} \cos \dfrac{x}{\pi}$

In Exercises 17–26, graph each pair of functions over the indicated interval.

17. $y = \sin x$ and $y = 2 \sin x$, $0 \le x \le 2\pi$

18. $y = \cos x$ and $y = -3 \cos x$, $0 \le x \le 2\pi$

19. $y = \cos x$ and $y = -\dfrac{1}{3} \cos x$, $0 \le x \le 2\pi$

20. $y = \sin x$ and $y = \dfrac{1}{2} \sin x$, $0 \le x \le 2\pi$

21. $y = \sin x$ and $y = \sin 2x$, $0 \le x \le 2\pi$

22. $y = \cos x$ and $y = \cos 3x$, $0 \le x \le 2\pi$

23. $y = \cos x$ and $y = \cos \dfrac{1}{3} x$, $0 \le x \le 6\pi$

24. $y = \sin x$ and $y = \sin \dfrac{1}{2} x$, $0 \le x \le 4\pi$

25. $y = \sin x$ and $y = \sin \pi x$, $0 \le x \le 2\pi$

26. $y = \cos x$ and $y = \cos \pi x$, $0 \le x \le 2\pi$

In Exercises 27–40, graph each given function over an interval that is at least one period long.

27. $y = 3 \cos x$

28. $y = 4 \sin x$

29. $y = -\sin x$

30. $y = -\cos x$

31. $y = \cos 2x$

32. $y = \sin 3x$

33. $y = -\sin \dfrac{x}{4}$

34. $y = \cos \dfrac{x}{4}$

35. $y = 3 \sin \pi x$

36. $y = -2 \cos \pi x$

37. $y = \dfrac{1}{2} \cos 4x$

38. $y = -4 \sin 3x$

39. $y = -4 \sin \dfrac{x}{2}$

40. $y = \dfrac{1}{3} \cos \dfrac{x}{2}$

5.7 GRAPHS OF FUNCTIONS INVOLVING TAN X, COT X, CSC X, AND SEC X

Because the reference angles of x and $x + \pi$ are equal (see Figure 5-48) and because the tangents of angles in nonadjacent quadrants agree in sign, we have

$$\tan x = \tan(x + \pi)$$

for all x. Furthermore, the number π is the smallest positive number p for which $\tan x = \tan(x + p)$. Therefore, the period of the tangent function is π.

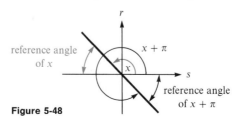

Figure 5-48

Because $\tan 0 = 0$, the graph of $y = \tan x$ passes through the origin. See Figure 5-49. As x gets very close to $\frac{\pi}{2}$, the value of $|\tan x|$ becomes very large. Since a value for $\tan \frac{\pi}{2}$ does not exist, the graph of $y = \tan x$ cannot intersect the line $x = \frac{\pi}{2}$. However, the graph of $y = \tan x$ does approach the vertical line $x = \frac{\pi}{2}$ as x approaches $\frac{\pi}{2}$. The vertical line $x = \frac{\pi}{2}$ is an **asymptote.** Other vertical asymptotes of the tangent function are the vertical lines $x = \frac{3\pi}{2}$, $x = \frac{5\pi}{2}$, and so on. The graph of $y = \tan x$ approaches these lines, but it never touches them.

For all x between $\frac{\pi}{2}$ and π, the values of $\tan x$ are negative, returning to 0 when $x = \pi$. The graph of $y = \tan x$ is shown in Figure 5-49. In the exercises, you will be asked to explain why the tangent function has no defined amplitude.

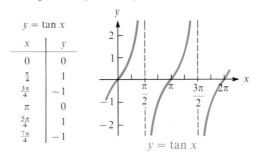

$y = \tan x$	
x	y
0	0
$\frac{\pi}{4}$	1
$\frac{3\pi}{4}$	-1
π	0
$\frac{5\pi}{4}$	1
$\frac{7\pi}{4}$	-1

Figure 5-49

Because π is the smallest positive number p for which

$$\cot x = \cot(x + p)$$

the period of the cotangent function is also π.

The asymptotes of the graph of $y = \cot x$ are the vertical lines determined by the equation

$$x = \pm k\pi \qquad \text{where } k \text{ is a nonnegative integer}$$

because $\cot(\pm k\pi)$ is undefined. The x-intercepts are numbers of the form

$$\frac{\pi}{2} \pm k\pi$$

because $\cot(\frac{\pi}{2} \pm k\pi) = 0$. The graph of $y = \cot x$ is shown in Figure 5-50. In the exercises, you will be asked to explain why the cotangent function has no defined amplitude.

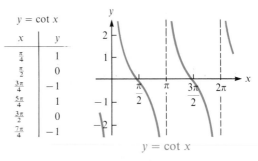

$y = \cot x$	
x	y
$\frac{\pi}{4}$	1
$\frac{\pi}{2}$	0
$\frac{3\pi}{4}$	-1
$\frac{5\pi}{4}$	1
$\frac{3\pi}{2}$	0
$\frac{7\pi}{4}$	-1

Figure 5-50

Because csc x is the reciprocal of sin x, it has the same period as sin x and is undefined whenever sin x is 0—at $x = 0$, at $x = \pi$, at $x = 2\pi$, and so on. These values determine the vertical asymptotes for the graph of $y = \csc x$. The graph of $y = \csc x$ appears in Figure 5-51. In the exercises, you will be asked to explain why the cosecant function has no amplitude.

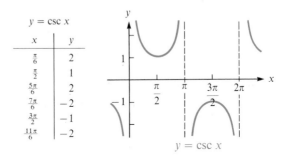

Figure 5-51

The graph of $y = \sec x$ has a period of 2π because sec x is the reciprocal of cos x and cos x has a period of 2π. There is no value for the secant function at $x = \frac{\pi}{2}$, at $x = \frac{3\pi}{2}$, and so on because the cosine of these values is 0. Thus, the lines $x = \frac{\pi}{2}$, $x = \frac{3\pi}{2}$, ... are vertical asymptotes for the graph of $y = \sec x$. The graph is shown in Figure 5-52. In the exercises, you will be asked to explain why the secant function has no amplitude.

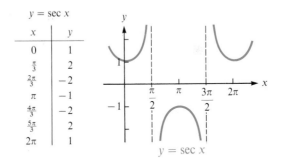

Figure 5-52

Example 1   Graph $y = \tan 3x$.

Solution   The graph of $y = \tan 3x$ intersects the x-axis when $\tan 3x = 0$. This is true when $3x = 0$, when $3x = \pi$, when $3x = 2\pi$, and so on. Hence, the graph intersects the x-axis at $x = 0$, at $x = \frac{\pi}{3}$, at $x = \frac{2\pi}{3}$, and so on. The graph of $y = \tan 3x$ completes one cycle as $3x$ increases from 0 to π—that is, as x itself increases from 0 to $\frac{\pi}{3}$. No value for tan $3x$ is defined at $3x = \frac{\pi}{2}$, or at $x = \frac{\pi}{6}$. Thus, the period is $\frac{\pi}{3}$ and the vertical asymptotes are $x = \frac{\pi}{6}$, $x = \frac{\pi}{2}$, $x = \frac{5\pi}{6}$, and so on. The graph is shown in Figure 5-53.

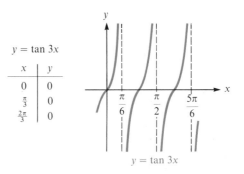

$y = \tan 3x$

x	y
0	0
$\frac{\pi}{3}$	0
$\frac{2\pi}{3}$	0

$y = \tan 3x$

Figure 5-53   ∎

Example 2   Graph $y = 2 \csc x$ on the interval from 0 to 2π.

Solution   Because csc x is the reciprocal of sin x, it has the same period as sin x and is undefined wherever sin x is 0. Thus, in the interval from 0 to 2π, no value for $2 \csc x$ is defined at $x = 0$, at $x = \pi$, and at $x = 2\pi$. These values determine vertical asymptotes for the graph of $y = 2 \csc x$. A table of values and the graph of $y = 2 \csc x$ appear in Figure 5-54.

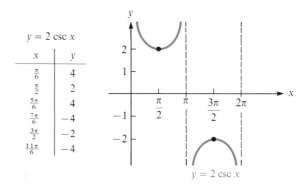

$y = 2 \csc x$

x	y
$\frac{\pi}{6}$	4
$\frac{\pi}{2}$	2
$\frac{5\pi}{6}$	4
$\frac{7\pi}{6}$	-4
$\frac{3\pi}{2}$	-2
$\frac{11\pi}{6}$	-4

$y = 2 \csc x$

Figure 5-54   ∎

Example 3   Graph $y = -2 \sec 2x$ on the interval from 0 to 2π.

Solution   Because sec $2x$ is the reciprocal of cos $2x$, it has the same period as cos $2x$. Thus, the period of $y = -2 \sec 2x$ is $\frac{2\pi}{2}$, or π. Furthermore, no value for sec $2x$ is defined wherever cos $2x$ is 0. In the interval from 0 to 2π, this occurs at $x = \frac{\pi}{4}$, at $x = \frac{3\pi}{4}$, at $x = \frac{5\pi}{4}$, and at $x = \frac{7\pi}{4}$. These values determine the vertical asymptotes for the graph of $y = -2 \sec 2x$. The graph of $y = -2 \sec 2x$ appears in Figure 5-55.

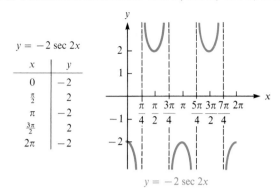

$y = -2 \sec 2x$

x	y
0	-2
$\frac{\pi}{2}$	2
π	-2
$\frac{3\pi}{2}$	2
2π	-2

$y = -2 \sec 2x$

Figure 5-55

We summarize the results of this section.

> The **period** of $\begin{cases} y = a \tan bx \\ y = a \cot bx \end{cases}$ is $\left|\dfrac{\pi}{b}\right|$.
>
> The **period** of $\begin{cases} y = a \csc bx \\ y = a \sec bx \end{cases}$ is $\left|\dfrac{2\pi}{b}\right|$.
>
> The tangent, cotangent, cosecant, and secant functions have no defined amplitude.

Exercise 5.7

In Exercises 1–18, give the period of each of the given functions. **Do not construct a graph.**

1. $y = 3 \tan x$

2. $y = 2 \csc x$

3. $y = \dfrac{1}{2} \sec x$

4. $y = \dfrac{1}{3} \cot x$

5. $y = \dfrac{1}{3} \tan 3x$

6. $y = \dfrac{1}{2} \sec 2x$

7. $y = -2 \csc \pi x$

8. $y = -3 \tan \pi x$

9. $y = 3 \sec \dfrac{x}{3}$

10. $y = -2 \tan \dfrac{\pi x}{3}$

11. $y = \dfrac{7}{2} \cot \dfrac{2\pi x}{3}$

12. $y = -\dfrac{2}{3} \csc \dfrac{\pi x}{3}$

13. $y = 3 \csc \dfrac{\pi x}{2}$

14. $y = -4 \sec \dfrac{2\pi x}{5}$

15. $y = -\cot \dfrac{x}{2\pi}$

16. $y = 7 \csc \dfrac{x}{4\pi}$

17. $y = -\dfrac{2}{5} \sec \dfrac{3x}{\pi}$

18. $y = \dfrac{7}{9} \cot \dfrac{2x}{\pi}$

In Exercises 19–30, graph each given function over the indicated interval.

19. $y = 2 \tan x$, $\dfrac{-\pi}{2} < x < \dfrac{3\pi}{2}$

20. $y = 2 \csc x$, $0 < x < 2\pi$

21. $y = -3 \sec x$, $0 \le x \le 2\pi$

22. $y = -\csc 2x$, $0 < x < \dfrac{3\pi}{2}$

23. $y = \cot 2x, \quad 0 < x < \pi$

24. $y = \sec 3x, \quad 0 \le x \le \dfrac{2\pi}{3}$

25. $y = -2 \tan \dfrac{x}{2}, \quad 0 < x < 2\pi$

26. $y = -3 \csc \dfrac{x}{2}, \quad 0 < x < 4\pi$

27. $y = 2 \sec 2x, \quad 0 \le x \le 2\pi$

28. $y = 2 \csc 2x, \quad 0 < x < 2\pi$

29. $y = -2 \cot \dfrac{\pi}{4} x, \quad 0 < x < 4$

30. $y = -2 \sec \dfrac{\pi}{4} x, \quad 0 \le x \le 8$

31. Explain why the tangent, cotangent, secant, and cosecant functions have no amplitude.

32. For what values of x, if any, are the values of $\sin x$ and $\csc x$ equal?

33. For what values of x, if any, are the values of $\cos x$ and $\sec x$ equal?

34. For what values of x, if any, are the values of $\tan x$ and $\cot x$ equal?

5.8 VERTICAL AND HORIZONTAL TRANSLATIONS; ADDITION OF ORDINATES

In applied work we often encounter graphs of trigonometric functions that have been shifted from standard position in either a vertical or a horizontal direction. Such shifts are called **translations**. The graph of a function can be shifted vertically by adding a constant value to the function, and horizontally by adding a constant value to the argument of the function.

Example 1   Graph $y = 2 + \cos x$.

Solution   You already know that $y = \cos x$ has a period of 2π, has values that lie between -1 and 1, and has x-intercepts at $x = \frac{\pi}{2}$, at $x = \frac{3\pi}{2}$, at $x = \frac{5\pi}{2}$, and so on. The values of $2 + \cos x$ will be similar to the values of $\cos x$ except that each value will be increased by 2. Thus, the graph of $y = 2 + \cos x$ has a period of 2π and has values that lie between $2 + (-1)$ and $2 + 1$, or between 1 and 3. The curve will intersect the line $y = 2$ at $x = \frac{\pi}{2}$, at $x = \frac{3\pi}{2}$, at $x = \frac{5\pi}{2}$, and so on. A table of values and the graph of $y = 2 + \cos x$ are shown in Figure 5-56.

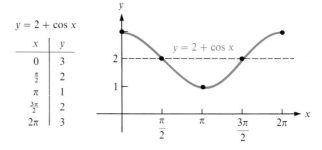

$y = 2 + \cos x$

x	y
0	3
$\frac{\pi}{2}$	2
π	1
$\frac{3\pi}{2}$	2
2π	3

Figure 5-56

Example 2   Graph $y = -3 + \tan \dfrac{x}{2}$.

Solution   The function $y = \tan \frac{x}{2}$ has a period of $\frac{\pi}{\frac{1}{2}}$, or 2π, and has vertical asymptotes at $x = \pi$, at $x = 3\pi$, at $x = 5\pi$, and so on. The values of $-3 + \tan \frac{x}{2}$ will be similar to the values of $\tan \frac{x}{2}$ except that each value of $\tan \frac{x}{2}$ will be decreased by 3. The graph of $y = -3 + \tan \frac{x}{2}$ appears in Figure 5-57.

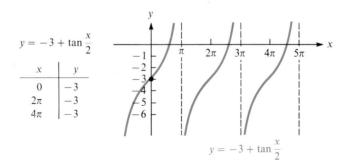

$y = -3 + \tan \dfrac{x}{2}$

x	y
0	-3
2π	-3
4π	-3

$y = -3 + \tan \dfrac{x}{2}$

Figure 5-57

The preceding two examples suggest the following facts:

Vertical Shift.   The graph of $y = k + a \sin bx$ is identical to the graph of $y = a \sin bx$ except that it is shifted $|k|$ units

$$\left.\begin{matrix} \text{up} \\ \text{down} \end{matrix}\right\} \text{ if } k \text{ is } \left\{\begin{matrix} \text{positive} \\ \text{negative} \end{matrix}\right\}$$

A similar statement is true for each of the other trigonometric functions.

Example 3   Graph $y = \sin\left(x + \dfrac{\pi}{6}\right)$ and $y = \sin x$ on the same set of coordinate axes.

Solution   One complete cycle of $y = \sin x$ is described as x increases from 0 to 2π. Similarly, one cycle of $y = \sin(x + \frac{\pi}{6})$ is completed when $x + \frac{\pi}{6}$ increases from 0 to 2π—that is, when x itself increases from $-\frac{\pi}{6}$ to $(2\pi - \frac{\pi}{6})$. At $x = -\frac{\pi}{6}$, $\sin(x + \frac{\pi}{6})$ is zero. The graph of $y = \sin(x + \frac{\pi}{6})$ looks just like the graph of $y = \sin x$ except that it is shifted to the left by a distance of $\frac{\pi}{6}$. The graph appears in Figure 5-58.

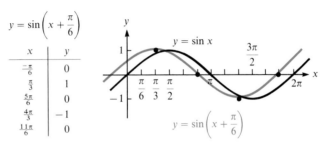

$y = \sin\left(x + \dfrac{\pi}{6}\right)$

x	y
$\frac{-\pi}{6}$	0
$\frac{\pi}{3}$	1
$\frac{5\pi}{6}$	0
$\frac{4\pi}{3}$	-1
$\frac{11\pi}{6}$	0

$y = \sin x$

$y = \sin\left(x + \dfrac{\pi}{6}\right)$

Figure 5-58

The distance that a graph is shifted to the left or right is called the **phase shift** of the graph.

Phase Shift. The graph of $y = a \sin b(x + c)$ is identical to the graph of $y = a \sin bx$ except that it is shifted $|c|$ units to the

$$\begin{Bmatrix} \text{left} \\ \text{right} \end{Bmatrix} \text{ if } c \text{ is } \begin{Bmatrix} \text{positive} \\ \text{negative} \end{Bmatrix}$$

The number $|c|$ is called the **phase shift** of the graph. A similar statement is true for $y = a \cos b(x + c)$.

Example 4   Graph $y = 3 \cos\left(2x - \dfrac{\pi}{3}\right)$.

Solution   Rewrite

$$y = 3 \cos\left(2x - \frac{\pi}{3}\right)$$

in the form $y = a \cos b(x + c)$ by factoring out a 2 from the binomial $2x - \frac{\pi}{3}$. Thus,

$$y = 3 \cos\left(2x - \frac{\pi}{3}\right)$$
$$= 3 \cos 2\left(x - \frac{\pi}{6}\right)$$
$$= 3 \cos 2\left(x + \frac{-\pi}{6}\right)$$

$y = 3 \cos\left(2x - \dfrac{\pi}{3}\right)$

x	y
$\frac{\pi}{6}$	3
$\frac{5\pi}{12}$	0
$\frac{2\pi}{3}$	−3
$\frac{11\pi}{12}$	0
$\frac{7\pi}{6}$	3

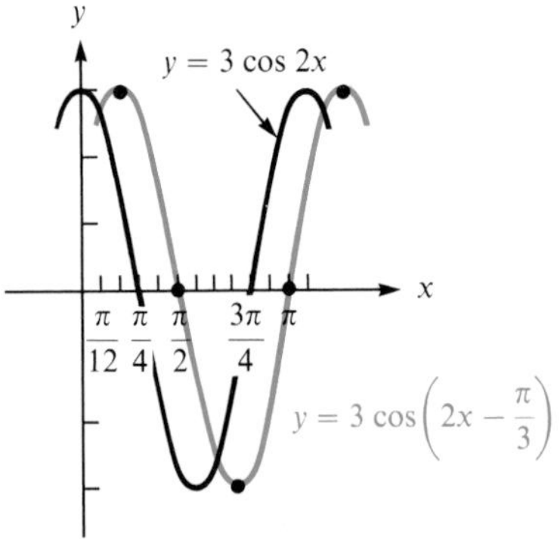

Figure 5-59

From the equation $y = 3 \cos 2(x + \frac{-\pi}{6})$, you can see that the amplitude is 3, the period is $\frac{2\pi}{2} = \pi$, and the graph looks like that of $y = 3 \cos 2x$. However, it is shifted $\frac{\pi}{6}$ units to the right. With $y = 3 \cos 2x$ included for reference, the graph appears in Figure 5-59.

Example 5   Graph $y = \tan\left(x - \dfrac{\pi}{4}\right)$.

Solution   One cycle of $y = \tan x$ is completed as x increases from 0 to π. Thus, one cycle of $y = \tan(x - \frac{\pi}{4})$ is completed when $x - \frac{\pi}{4}$ increases from 0 to π—that is, when x itself increases from $\frac{\pi}{4}$ to $\frac{5\pi}{4}$. At $x = \frac{\pi}{4}$, the value of $\tan(x - \frac{\pi}{4})$ is zero. The graph of $y = \tan(x - \frac{\pi}{4})$ looks just like the graph of $y = \tan x$ except that it is shifted to the right by a distance of $\frac{\pi}{4}$. The graph appears in Figure 5-60.

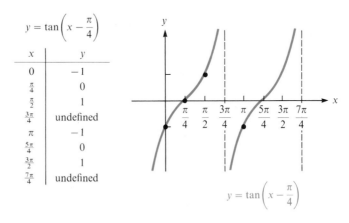

$y = \tan\left(x - \dfrac{\pi}{4}\right)$

x	y
0	-1
$\frac{\pi}{4}$	0
$\frac{\pi}{2}$	1
$\frac{3\pi}{4}$	undefined
π	-1
$\frac{5\pi}{4}$	0
$\frac{3\pi}{2}$	1
$\frac{7\pi}{4}$	undefined

$y = \tan\left(x - \dfrac{\pi}{4}\right)$

Figure 5-60   ■

The following statement summarizes the facts involving phase shifts of the tangent and the cotangent functions.

Phase Shift.   The graph of $\begin{cases} y = a \tan b(x + c) \\ y = a \cot b(x + c) \end{cases}$ is identical to the graph

of $\begin{cases} y = a \tan bx \\ y = a \cot bx \end{cases}$ except that it is shifted $|c|$ units to the

$\begin{cases} \text{left} \\ \text{right} \end{cases}$ if c is $\begin{cases} \text{positive} \\ \text{negative} \end{cases}$

The number $|c|$ is called the **phase shift** of the graph.

Example 6   Graph $y = \csc\left(3x - \dfrac{\pi}{2}\right)$.

Solution   Factor 3 out of $3x - \frac{\pi}{2}$ to write the equation in the form $y = \csc b(x + c)$.

$$y = \csc\left(3x - \frac{\pi}{2}\right)$$

$$= \csc 3\left[x + \left(-\frac{\pi}{6}\right)\right]$$

As with the sine function, the period of the cosecant function is $\left|\frac{2\pi}{b}\right|$ or in this case $\frac{2\pi}{3}$. The graph looks like that of $y = \csc 3x$, but it is shifted $\frac{\pi}{6}$ units to the right. The graph appears in Figure 5-61.

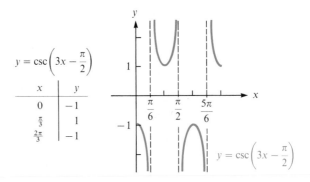

x	y
0	-1
$\frac{\pi}{3}$	1
$\frac{2\pi}{3}$	-1

Figure 5-61

The following statement summarizes the facts involving phase shifts of the cosecant and secant functions:

Phase Shift.   The graph of $\begin{cases} y = a \csc b(x + c) \\ y = a \sec b(x + c) \end{cases}$ is identical to the graph

of $\begin{cases} y = a \csc bx \\ y = a \sec bx \end{cases}$ except that it is shifted $|c|$ units to the

$\begin{cases} \text{left} \\ \text{right} \end{cases}$ if c is $\begin{cases} \text{positive} \\ \text{negative} \end{cases}$

The number $|c|$ is called the **phase shift** of the graph.

Addition of Ordinates

We can often graph the sum of two functions by graphing each function separately and then adding their corresponding y-values. This technique is called **addition of ordinates**.

Example 7 Graph $y = x + \cos x$.

Solution It is possible to make an extensive table of values, plot the points, and draw the curve. However, there is an easier way. Use your knowledge of the functions $y = x$ and $y = \cos x$, and sketch each one separately, as in Figure 5-62. Then pick several numbers x, such as x_1, and add the values of x_1 and $\cos x_1$ together to obtain the y-value of a point on the graph of $y = x + \cos x$. For example, if $x_1 = 0$, the value of $\cos x_1 = 1$. Thus, the point $(0, 0 + 1)$ or $(0, 1)$ is on the graph of $y = x + \cos x$. As another example, if $x_1 = 1$, the value of $\cos x_1 \approx 0.54$. Thus, the point $(1, 1 + 0.54)$ or $(1, 1.54)$ is on the graph of $y = x + \cos x$.

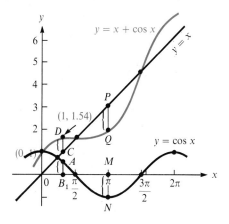

Figure 5-62

It is easy to use a compass to add these values of y. Set your compass so that it spans segment BA and transfer that length to form segment CD, locating point D on the desired graph. However, if $x_1 = \pi$, the value of $\cos x_1$ is negative. In this case, you must subtract the magnitudes of the y-values. Do this by using a compass to transfer segment MN to position PQ to locate point Q on the desired graph. Repeat this process of adding y-values to determine several points. Then join them with a smooth curve to obtain the graph of $y = x + \cos x$.

■

Example 8 Graph $y = \cos x + 2 \sin 2x$.

Solution Proceed as in Example 7. For values of x between 0 and $\frac{\pi}{2}$, such as $\frac{\pi}{4}$, the values of $\cos x$ and $2 \sin 2x$ are both positive. In this case, you must add the y-values. Do this by transferring segments such as MN to new positions such as PQ. This locates point Q, which is on the graph of $y = \cos x + 2 \sin 2x$. For values of x that are between π and $\frac{3\pi}{2}$, such as $\frac{5\pi}{4}$, the y-value of $2 \sin 2x$ is positive while the y-value of $\cos x$ is negative. In this case, subtract the magnitudes of the y-values of the two functions. Do this by transferring segments such as AB to new positions such as CD to locate point D on the graph of $y = \cos x + 2 \sin 2x$. See Figure 5-63.

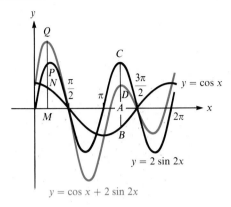

Figure 5-63

Exercise 5.8

In Exercises 1–12, give the number of units and the direction (either up or down) that each trigonometric function has been shifted. Also give the period. **Do not draw the graph.**

1. $y = 2 + \sin x$ **2.** $y = -4 + \cos x$

3. $y = \tan x - 1$ **4.** $y = \csc x + 3$

5. $y = 7 + 9 \sec 5x$ **6.** $y = 7 + 9 \cot 5x$

7. $y = 3 - \sin x$ **8.** $y = -2 - \cos x$

9. $y + 5 = \csc 2x$ **10.** $y + 5 = \cot 2x$

11. $y = 2(3 + \tan \pi x)$ **12.** $y = -4(1 - \sec \pi x)$

In Exercises 13–30, give the period and the phase shift (including direction), if any, of each given function. **Do not draw the graph.**

13. $y = \sin\left(x - \dfrac{\pi}{3}\right)$ **14.** $y = \cos\left(x + \dfrac{\pi}{4}\right)$

15. $y = \cos\left(x + \dfrac{\pi}{6}\right)$ **16.** $y = -\sin\left(x - \dfrac{\pi}{2}\right)$

17. $y + 2 = 3 \cos 2\pi x$ **18.** $y = 3 \sin \dfrac{2x}{\pi}$

19. $y = \tan(x - \pi)$ **20.** $y = \csc\left(x + \dfrac{\pi}{6}\right)$

21. $y = -\sec\left(x + \dfrac{\pi}{4}\right)$ **22.** $y = -2 \sec\left(x - \dfrac{\pi}{3}\right)$

23. $y = \sin 2\left(x + \dfrac{\pi}{2}\right)$ **24.** $y = \cos 3\left(x - \dfrac{\pi}{6}\right)$

25. $y = \tan\left(\dfrac{\pi x}{2} + \dfrac{\pi}{4}\right)$

26. $y = \csc\left(\dfrac{2\pi x}{3} + \dfrac{\pi}{9}\right)$

27. $y = 2\sec\left(\dfrac{1}{3}x - 6\pi\right)$

28. $y = 2\cot\left(\dfrac{\pi}{10} + \dfrac{x}{5}\right)$

29. $2y = 3\cot\left(7x - \dfrac{21}{2}\pi\right)$

30. $17y = \sec\left(\dfrac{x}{5} + \dfrac{\pi}{4}\right)$

In Exercises 31–46, graph each function through at least one period.

31. $y = -4 + \sin x$

32. $y + 2 = \tan\dfrac{x}{2}$

33. $y = 3 - \sec x$

34. $y = 1 + \csc x$

35. $y = \cot\dfrac{x}{2} - 2$

36. $y = 1 - 2\cos x$

37. $y = \sin\left(x + \dfrac{\pi}{2}\right)$

38. $y = -\cos\left(x - \dfrac{\pi}{2}\right)$

39. $y = \tan\left(x - \dfrac{\pi}{2}\right)$

40. $y = \csc\left(x + \dfrac{\pi}{4}\right)$

41. $y = \cos(2x + \pi)$

42. $y = \sin(3x - \pi)$

43. $y = \sec\left(3x + \dfrac{\pi}{2}\right)$

44. $y = \tan\left(\dfrac{x}{2} - \dfrac{\pi}{2}\right)$

45. $y - 1 = \csc\left(2x - \dfrac{\pi}{6}\right)$

46. $y + 2 = \cot\left(\dfrac{x}{3} - \dfrac{\pi}{2}\right)$

In Exercises 47–52, use addition of ordinates to graph the given function from 0 to 4π.

47. $y = \sin x + \cos x$

48. $y = 2\sin x + \cos x$

49. $y = \sin\dfrac{x}{2} + \sin x$

50. $y = x + \sin x$

51. $y = -x + \cos x$

52. $y = x - \cos x$

**REVIEW
EXERCISES**

In Review Exercises 1–4, tell whether the given angle is in standard position.

1.

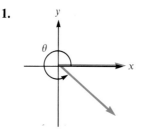

2.

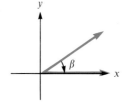

3.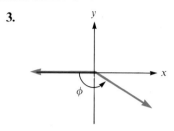

4.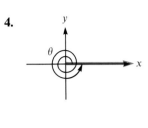

5. Are angles of 360° and 410° coterminal?

6. Are angles of 190° and 820° coterminal?

In Review Exercises 7–10, find the values of the remaining trigonometric functions of angle θ, which is in standard position.

7. $\sin \theta = \dfrac{-7}{10}$; θ in QIII

8. $\tan \theta = \dfrac{7}{9}$; θ not in QI

9. $\cos \theta = \dfrac{-7}{10}$; θ in QII

10. $\cot \theta = \dfrac{-9}{8}$; θ in QIV

11. Use the eight fundamental relationships to show that

$$\frac{1}{\sec \theta} = \sin \theta \cot \theta$$

12. Use the eight fundamental relationships to show that $\cos \theta \csc \theta = \cot \theta$.

In Review Exercises 13–16, evaluate each trigonometric expression. **Do not use a calculator.**

13. $\sin 45° \cos 30°$

14. $\cos 120° \tan 135°$

15. $\tan^2 225° \cos^2 30° \sin^2 300°$

16. $\sec 30° \csc 30° + \sec 330° \csc 330°$

In Review Exercises 17–20, find the value of the sine, cosine, and tangent of the given angle. **Do not use a calculator.**

17. 930°   **18.** 1380°   **19.** −300°   **20.** −585°

In Review Exercises 21–26, use a calculator or Table A in Appendix III to evaluate the sine, cosine, and tangent of the given angle. Give your answer to four decimal places.

21. 15°   **22.** 160°   **23.** 265°   **24.** 340°

25. −160°   **26.** −340°

In Review Exercises 27–32, use a calculator or Table A in Appendix III to find angle α (0° ≤ α < 360°) in the given quadrant. Check your work with a calculator.

27. $\sin \alpha = 0.8746$; α in QII

28. $\tan \alpha = 0.6009$; α in QIII

29. $\cos \alpha = 0.7314$; α in QIV

30. $\sec \alpha = 1.871$; α in QI

31. $\cot \alpha = -0.1763$; α in QII

32. $\csc \alpha = -1.046$; α in QIV

In Review Exercises 33–36, change each angle to radians.

33. $105°$ **34.** $325°$ **35.** $318°$ **36.** $-105°$

In Review Exercises 37–40, each angle is expressed in radians. Change each angle to degrees.

37. $\dfrac{19}{6}\pi$ **38.** $-\dfrac{5}{6}\pi$ **39.** 7π **40.** 8

In Review Exercises 41–44, find the values of the given function without using a calculator. Then check your work with a calculator.

41. $\sin\dfrac{5\pi}{6}$ **42.** $\cos\left(-\dfrac{13}{6}\pi\right)$

43. $\tan\left(-\dfrac{\pi}{3}\right)$ **44.** $\csc\dfrac{\pi}{6}$

45. The latitude of Springfield, Illinois, is $39.8°$ N. How far is Springfield from the Equator? Use 3960 miles as the radius of the earth.

46. Des Moines, Iowa, is approximately 2870 miles north of the Equator. If the radius of the earth is about 3960 miles, find the latitude of Des Moines.

47. Find the area of a sector of a circle if the sector has a central angle of $15°$ and the circle has a radius of 12 centimeters.

48. Find the angular velocity of the earth in radians per second.

49. A truck is traveling $50\,\frac{\text{ft}}{\text{sec}}$. How fast are its 32-inch-diameter tires spinning in revolutions per minute?

50. A vehicle has 40-inch-diameter tires that are making 100 rpm. How fast is the vehicle going in feet per minute?

*In Review Exercises 51–52, find the coordinates of the point P on the unit circle that corresponds to each given real number. **Do not use a calculator.***

51. $\dfrac{7\pi}{6}$ **52.** $\dfrac{13\pi}{4}$

In Review Exercises 53–56, use a calculator to find the values of the given functions, if possible. Give answers to the nearest ten-thousandth.

53. $\cos 7$ **54.** $\tan(2+\pi)$ **55.** $\csc 3\pi$ **56.** $\sin(\pi^3)$

In Review Exercises 57–60, find the amplitude and the period of the given function.

57. $y = 4\sin 3x$ **58.** $y = \dfrac{\cos 4x}{8}$

59. $y = -\dfrac{1}{3}\cos\dfrac{x}{3}$ **60.** $y = 0.875\sin\dfrac{1}{4}x$

In Review Exercises 61–64, find the vertical shift and phase shift, if any, of each of the given functions.

61. $y = 2 + \tan x$

62. $y = \csc\left(2x + \dfrac{\pi}{3}\right)$

63. $y - 4 = 3\sin\left(\dfrac{x}{7} + \dfrac{3}{2}\right)$

64. $y = -\cos\left(\dfrac{1}{5}x - \dfrac{1}{2}\pi\right) - 1$

In Review Exercises 65–74, graph the given function.

65. $y = 4\sin x$

66. $y = 0.5\cos x$

67. $y = \cos\dfrac{x}{4}$

68. $y = \tan\dfrac{\pi x}{3}$

69. $y = 3 + \sin x$

70. $y = -2 + \tan x$

71. $y = 2\sin\left(x - \dfrac{5\pi}{6}\right)$

72. $y = \tan\left(x - \dfrac{2\pi}{3}\right)$

73. $y = 2\cos x + \sin\dfrac{x}{2}$

74. $y = \dfrac{x}{2} + \cos\dfrac{x}{2}$

6

APPLICATIONS OF TRIGONOMETRY

In this chapter, we shall discuss many applications of trigonometry.

6.1 RIGHT TRIANGLE TRIGONOMETRY

The word *trigonometry* means "measuring triangles (trigons)." The possibility of solving triangles—determining all sides and all angles when only some are known—has made trigonometry indispensable in astronomy, navigation, and surveying.

Example 1   A right triangle ABC has an acute angle A of $27°$ and a hypotenuse c of length 14 feet. Solve the triangle by finding the other acute angle and the lengths of the two unknown sides.

Solution   Let the lengths of the two unknown sides of right triangle ABC be represented by a and b. Then, triangle ABC may be drawn in QI of a coordinate system as in Figure 6-1. In triangle ABC, $27°$ is the reference angle, $r = c = 14$, and point $B(x, y) = B(b, a)$. Because the triangle is a right triangle, angle A and angle B are complementary. Thus, angle $B = 63°$. The values of a and b can be found by using trigonometric functions:

$$\sin 27° = \frac{a}{c} = \frac{a}{14}$$

$$\cos 27° = \frac{b}{c} = \frac{b}{14}$$

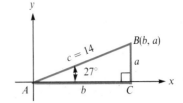

Figure 6-1

Solve these equations for a and b:

$a = 14 \sin 27°$

$\quad \approx 14(0.4539904997)$ Use a calculator to find $\sin 27°$.

$\quad \approx 6.355866996$

$\quad \approx 6.4$

Similarly,

$$b = 14 \cos 27°$$

$$\approx 14(0.8910065242) \qquad \text{Use a calculator to find } \cos 27°.$$

$$\approx 12.47409134$$

$$\approx 12$$

The remaining two sides of the triangle are 6.4 feet and 12 feet.   ■

You may wonder why the answers in Example 1 were rounded off to two digits (called **significant digits**). If the hypotenuse were *exactly* 14 feet and the acute angle *exactly* 27°, rounding off to two digits would be unnecessary. More likely, however, 14 feet and 27° are not exact, but are only approximate measurements. In that case, the impressive strings of decimal digits are unwarranted.

Calculators routinely provide answers to 8, 10, or 12 digits. You must decide how many of these are significant. A good rule of thumb for determining acceptable accuracy is provided in the following table:

Accuracy in measurements of sides	Accuracy in angles
Two significant digits	Nearest degree
Three significant digits	Nearest tenth of a degree
Four significant digits	Nearest hundredth of a degree

When solving triangles, remember that answers can be only as accurate as the least accurate of the given data. However, if a calculation requires several intermediate steps, do not round off until you have the final answer.

It is not always easy to decide how many significant digits a number has. For example, is the number 140 accurate to two or three significant digits? If the number is rounded to the nearest ten, the zero is merely a placeholder and 140 has two significant digits. On the other hand, if the number has been rounded to the nearest unit, the zero is significant and 140 has three significant digits. In this book, we shall assume the greatest possible number of significant digits unless stated otherwise—therefore 140 has three significant digits.

3234 has four significant digits.

104 has three significant digits.

140.00 has five significant digits.

0.00012 has two significant digits.

0.000120 has three significant digits.

0.0003 has one significant digit.

1.0003 has five significant digits.

To solve right triangles, it is helpful to view the definitions of the trigonometric functions in a different light. We place right triangle *ABC* on a coordinate system so it is the reference triangle for the acute angle *A*. See Figure 6-2. Let *a* be the length of *BC*, the side opposite angle *A*. Let *b* be the length of *AC*, the side adjacent to angle *A*. Finally, let the hypotenuse have length *c*. The six trigonometric functions of the acute angle *A* can be defined as ratios involving sides of the right triangle *ABC*.

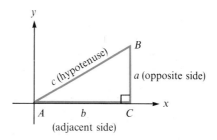

Figure 6-2

Definition.   If angle *A* is an acute angle in right triangle *ABC*, then

$$\sin A = \frac{\text{opposite side}}{\text{hypotenuse}} \qquad \csc A = \frac{\text{hypotenuse}}{\text{opposite side}}$$

$$\cos A = \frac{\text{adjacent side}}{\text{hypotenuse}} \qquad \sec A = \frac{\text{hypotenuse}}{\text{adjacent side}}$$

$$\tan A = \frac{\text{opposite side}}{\text{adjacent side}} \qquad \cot A = \frac{\text{adjacent side}}{\text{opposite side}}$$

Remember that the terminology of adjacent and opposite sides pertains to acute angles of right triangles only.

In many trigonometry problems, angles are given by schemes that may be unfamiliar to you. Some terminology may require explanation.

If an observer looks up at an object such as an airplane or a cloud, the angle that the observer's line of sight makes with the horizontal is called the **angle of elevation**. If the observer looks down to see the object, the angle made with the horizontal is called the **angle of depression**. See Figure 6-3.

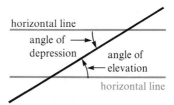

Figure 6-3

In nautical navigation and surveying, the concept of *bearing* is used. See Figure 6-4. The **bearing** of point A from point O (the observer) is the acute angle measured from the north-south line to the line segment OA. This bearing is denoted as N 30° E and is read as "north 30° east" (or "30° east of north"). The bearing of point B from O is N 75° W, the bearing of point C from O is S 20° W, and the bearing of point D from O is S 80° E.

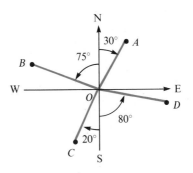

Figure 6-4

Example 2   From a location 23.0 meters from a flagpole's base, the angle of elevation to its top is 37.5°. How tall is the flagpole?

Solution   Consider the right triangle in Figure 6-5. The flagpole is the side opposite the 37.5° angle, and 23.0 meters is the length of the adjacent side. Because the tangent of the angle involves the opposite and the adjacent sides,

$$\tan 37.5° = \frac{\text{opposite side}}{\text{adjacent side}}$$

$$= \frac{h}{23.0}$$

Solve for h:

$$h = 23.0 \tan 37.5°$$

$$\approx (23.0)(0.7673)$$

$$\approx 17.6479$$

$$\approx 17.6$$

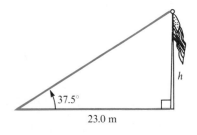

Figure 6-5

The flagpole is 17.6 meters tall. Note that the answer has been properly rounded to three significant digits, which is consistent with the accuracy of the data.

■

Example 3   A circus tightrope walker ascends from the ground to a platform 75 feet above the arena by walking a taut cable that is 92 feet long. At what angle is the cable from the horizontal?

Solution   See Figure 6-6. You are given the lengths of the hypotenuse and a side and must find angle θ. First, use the sine ratio to find the value of $\sin \theta$:

$$\sin \theta = \frac{\text{opposite side}}{\text{hypotenuse}}$$

$$\sin \theta = \frac{75}{92}$$

$$\approx 0.8152$$

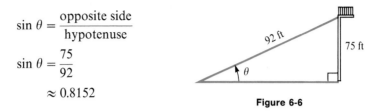

Figure 6-6

Now if you look up .8152 in the body of Table A in Appendix III or enter .8152 and press ⎡INV⎤ ⎡SIN⎤ on a calculator, you will find that

$$\theta \approx 55°$$

This result is accurate to the nearest degree and is consistent with the accuracy of the given sides.   ∎

Example 4   Perryville is 25.0 miles due south of Rock City, and Prairie Town is 90.0 miles due east of Rock City. What is the bearing of Prairie Town from Perryville?

Solution   See Figure 6-7. To find the bearing, find angle θ by using the tangent ratio:

$$\tan \theta = \frac{\text{opposite side}}{\text{adjacent side}}$$

$$\tan \theta = \frac{90}{25}$$

$$= 3.6$$

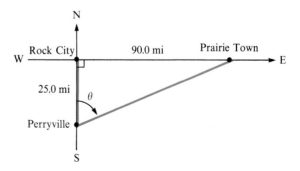

Figure 6-7

The angle with tangent of 3.6 is 74.5° (from either a calculator or the tables). Because the distances are accurate to three significant digits, the angle is given to the nearest tenth of a degree. The bearing of Prairie Town from Perryville is N 74.5° E. Incidentally, the bearing of Perryville from Prairie Town is S 74.5° W. ∎

Example 5   A television tower stands on top of a building. From a point 75.3 feet from the base of the building, the angles of elevation to the top and to the base of the tower are 60.1° and 47.4°, respectively. How tall is the tower?

Solution   See Figure 6-8 and find the total height H of the structure by using the tangent ratio:

$$\tan 60.1° = \frac{H}{75.3}$$
$$75.3(\tan 60.1°) = H$$
$$75.3(1.7391) \approx H$$
$$130.95 \approx H$$

The total height of the structure is approximately 130.95 feet.

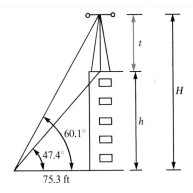

60.1°

47.4°

75.3 ft

t

h

H

Figure 6-8

Then, find the height h of the building by using the tangent ratio:

$$\tan 47.4° = \frac{h}{75.3}$$
$$75.3(\tan 47.4°) = h$$
$$75.3(1.0875) \approx h$$
$$81.89 \approx h$$

The height of the building is approximately 81.89 feet.

The height t of the tower is the difference between the total height of the structure and the height of the building. Thus,

$$t = H - h$$
$$t \approx 130.95 - 81.89$$
$$\approx 49.06$$

The height of the tower is approximately 49.1 feet. Note that the final answer is rounded off to the proper degree of accuracy.   ∎

Example 6 An observer A notices that the Space Needle of Seattle is due north and that the angle of elevation to its top is 44.4°. A second observer B, 706 feet due east of A, notices that the bearing of the Space Needle is N 48.8° W. How tall is the Space Needle?

Solution In this problem, two triangles must be solved. Triangle PAB lies on the ground, while triangle APQ sits on its edge with the Space Needle as one of its sides. See Figure 6-9. Not enough information is given about triangle APQ to determine the height, h, from that triangle alone. If the distance x were known, however, h could be computed. Note that

$$\tan 44.4° = \frac{h}{x}$$

or

$$h = x \tan 44.4°$$

The distance x can be found by working with the triangle on the ground. Triangle PAB is a right triangle with a right angle at A. Angle PBA is $90° - 48.8° = 41.2°$. Form the equation

$$\tan 41.2° = \frac{x}{706}$$

or

$$x = 706 \tan 41.2°$$

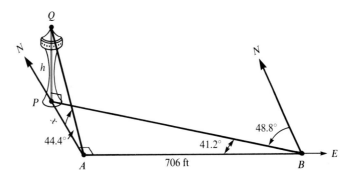

Figure 6-9

Putting these facts together gives

$$h = 706 \tan 41.2° \tan 44.4°$$
$$\approx 706(0.8754)(0.9793)$$
$$\approx 605.2391293$$
$$\approx 605$$

Seattle's Space Needle is approximately 605 feet tall.   ■

Exercise 6.1

In Exercises 1–4, solve each triangle by finding the unknown sides and angles.

1.

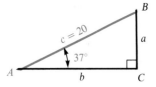

2.

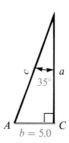

3.

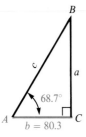

4.

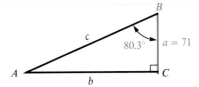

5. If the angle of elevation to the top of a flagpole from a point 40 feet from its base is 20°, find the height of the flagpole.

6. A person on the edge of a cliff looks down at a boat on a lake. The angle of depression of the person's line of sight is 10.0°, and the line-of-sight distance from the person to the boat is 555 feet. How high is the cliff?

7. A car drives up a long hill on a road that makes an angle of 7.5° with the horizontal. How far will the car travel to reach the top if the horizontal distance traveled is 715 feet?

8. On an approach to an airport a plane is descending at an angle of 3.5°. How much altitude is lost as the plane travels a horizontal distance of 21.2 miles?

9. A plane loses 2750 feet in altitude as it travels a horizontal distance of 39,300 feet. What is its angle of descent to the nearest tenth of a degree?

10. A train rises 230 feet as it travels 1 mile up a steep grade. What is its angle of ascent? (*Hint:* 5280 feet = 1 mile.)

11. The angle of depression from a point in the top of a tree to a point on the ground is 57°. Find the height of the tree if the line-of-sight distance from the top of the tree to the point on the ground is 34 feet.

12. An observer noted that the angle of elevation to a plane passing over a landmark was 32°. If the landmark was 1500 meters from the observer, what was the altitude of the plane?

13. A ship leaves from a port of call on a bearing of S 12.7° E. How far south has the ship traveled after a trip of 327 miles?

14. The bearing of Madison, Wisconsin, from Stevens Point, Wisconsin, is S 4.1° E. The distance between the cities is 108 miles. How much farther east is Madison than Stevens Point?

15. A ship leaves port and sails 8800 kilometers due west. It then sails 4500 kilometers due south. To the nearest tenth of a degree, what is the ship's bearing from its port?

16. A ship is 3.3 miles from a lighthouse. It is also due north of a buoy that is 2.5 miles due east of the lighthouse. Find the bearing of the ship from the lighthouse.

17. Two lighthouses are on an east-west line. The bearing of a ship from one lighthouse is N 59° E, and the bearing from the other lighthouse is N 31° W. How far apart are the lighthouses if the ship is 5.0 miles from the first lighthouse?

18. Two lookout stations are on a north-south line. The bearing of a forest fire from one lookout is S 67° E, and the bearing of the fire from the second lookout is N 23° E. If the fire is 3.5 kilometers from the second lookout station, how far is the fire from the other lookout station?

19. A plane flying at 18,100 feet passes directly over an observer. Thirty seconds later, the observer notes that the plane's angle of elevation is 31.0°. How fast is the plane going in miles per hour? (*Hint*: 5280 feet = 1 mile.)

20. A plane flying horizontally at 650 mph passes directly over a small city. One minute later the pilot notes that the angle of depression to that city is 13°. What is the plane's altitude in feet?

21. Use the information in Illustration 1 to compute the height of George Washington's face on Mount Rushmore.

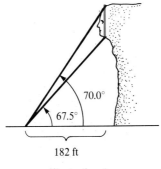

70.0°
67.5°
182 ft

Illustration 1

22. A boat is 537 meters from a lighthouse and has a bearing from the lighthouse of N 33.7° W. A second boat is 212 meters from the same lighthouse and has a bearing from the lighthouse of S 20.1° W. How many meters north of the second boat is the first?

23. A plane is flying at an altitude of 5120 feet. As it approaches an island, the navigator determines the angles of depression as in Illustration 2. What is the length of the island in feet and in miles?

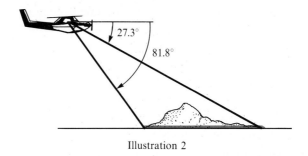

Illustration 2

24. From an observation point 6500 meters from a launch site, an observer watches the vertical flight of a rocket. At one instant, the angle of elevation of the rocket is 15°. How far will the rocket ascend in the time it takes the angle of elevation to increase by 57°?

25. Compute the height, h, of the Sears Tower using the information given in Illustration 3.

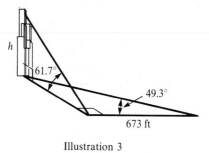

Illustration 3

26. Compute the height, h, of the Empire State Building using the information given in Illustration 4.

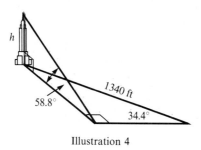

Illustration 4

27. Refer to Illustration 5 and find θ.

28. Refer to Illustration 5 and find ϕ.

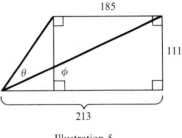

Illustration 5

6.2 INTRODUCTION TO VECTORS

Quantities that have only magnitude are called **scalar quantities**. However, there is often a need to represent quantities that have both magnitude and direction. The enrollment in an art class, for example, is a scalar quantity that possesses magnitude only, but the flight of an airplane must be described by both a speed and a direction. "Thirty-seven students" adequately describes the enrollment, whereas "350 miles per hour northeast" describes the flight. Quantities with both magnitude and direction are called **vector quantities** and are represented by mathematical entities called **vectors**.

> **Definition.** A **vector** is a directed line segment. The direction of the vector is indicated by the angle it makes with some convenient reference line.
>   The **norm**, or **magnitude**, of a vector is the length of the line segment. If a vector is denoted by **v**, then the norm of the vector is denoted by $|\mathbf{v}|$.

A vector quantity does not have location; any two directed line segments with the same length and the same direction are regarded as **equal vectors**. When drawing diagrams, it is usual to position a vector in the most convenient location. A force of 30 pounds, for example, exerted in a northwesterly direction may be represented by the directed line segment in Figure 6-10. The length of the arrow is 30 units, and because it is the terminal side of a 135° angle, it "points" northwest.

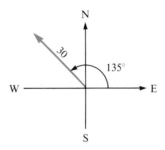

Figure 6-10

A 20-mph wind blowing from the east may be represented by the vector in Figure 6-11.

An airplane flying 350 mph on a **heading** (an intended direction of travel measured clockwise from the north line) of 240° can be represented by the vector in Figure 6-12.

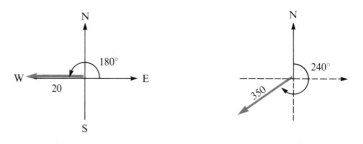

Figure 6-11   Figure 6-12

Vector quantities can be added, but the process must take into account both their norms and directions. For example, two forces of 40 pounds exerted on the same object might not combine to be an 80-pound force. If they acted in opposite directions, the net force would be zero.

The sum of two vectors is another vector, called the **resultant**. Vector quantities may be added by using the **parallelogram law**. If two vectors originating at a common point are adjacent sides of a parallelogram, their resultant vector (or vector sum) is that parallelogram's diagonal drawn from the common point. In Figure 6-13, the sum of vectors **AB** and **AD** is the vector **AC**, given by the diagonal of the parallelogram *ABCD*. Note that vector **DC** has the same magnitude and direction as vector **AB**.

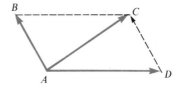

Figure 6-13

Example 1   A boat capable of 8.0 mph in still water attempts to go directly across a river with a current of 3.0 mph. By what angle is the boat pushed off its intended path? What is the effective speed of the boat?

Solution   The two given velocities can be represented by vectors as shown in Figure 6-14. The direction the boat travels, called its **course**, is represented by a vector sum, which is the diagonal of the rectangle. Angle θ specifies the direction in which the boat is forced to travel, and the length of OP specifies the effective speed of the boat. Because AP is also 3.0 units, $\tan \theta = \frac{3}{8}$. From this relationship, you can calculate θ and find that the river pushes the boat approximately 21° off its intended path. Then use the Pythagorean Theorem to find the length of OP and, hence, the speed of the boat:

$$OP = \sqrt{8^2 + 3^2}$$
$$= \sqrt{64 + 9}$$
$$= \sqrt{73}$$
$$\approx 8.5$$

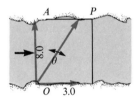

Figure 6-14

The effective speed of the boat is approximately 8.5 mph.   ∎

Example 2   An airplane capable of a speed of 270 mph in still air sets a heading of 75°. A very strong wind is blowing in the direction of 165° and forces the plane onto a course that is due east. What is the velocity of the wind, and what is the **groundspeed** (the speed relative to the ground) of the plane?

Solution   The velocities involved are represented in Figure 6-15. Vector **w** represents the wind velocity, and vector **v** represents the resultant velocity, or groundspeed, of the plane. Because $165° - 75° = 90°$, the vector parallelogram is a rectangle with angle α equal to 90°. Because each triangle formed by the diagonal is a right triangle,

$$\tan 15° = \frac{|\mathbf{w}|}{270}$$

Multiplying both sides by 270 and simplifying gives

$$|\mathbf{w}| = 270 \tan 15°$$
$$\approx 72$$

Also,

$$\cos 15° = \frac{270}{|\mathbf{v}|}$$
$$|\mathbf{v}| \cos 15° = 270 \qquad \text{Multiply both sides by } |\mathbf{v}|.$$
$$|\mathbf{v}| = \frac{270}{\cos 15°} \qquad \text{Divide both sides by } \cos 15°.$$
$$\approx 280 \qquad \text{Simplify.}$$

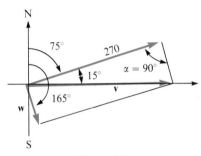

Figure 6-15

The wind speed is approximately 72 mph, and the groundspeed of the plane (the resultant of the plane's airspeed and the wind speed) is approximately 280 mph.   ■

Instead of two vectors adding to form a single resultant, it is possible to separate a single vector into several components. Suppose that a car weighing 3000 pounds is parked on a hill. It is pulled directly downward by gravity with a force of 3000 pounds. Part of this force appears as a tendency to roll the car down the hill, and another part presses the car against the road. Just how the 3000 pounds is apportioned depends on the angle of the hill. If there were no hill, there would be no tendency to roll; if the hill were very steep, only a small force would hold the car to the road. The weight of the car is said to be **resolved** into two components—one directed down the hill, and the other directed into the hill. These vectors obey the parallelogram law. See Figure 6-16. Note that the angle of the hill, α, is also the angle between two of the vectors because both of these angles are complementary to angle β.

Figure 6-16

Example 3   A 3000-pound car sits on a 23.0° incline. What force is required to prevent the car from rolling down the hill? With what force is it held to the roadway?

Solution   You must find the norms of vectors **t** and **n** in Figure 6-17. Because the figure $OACB$ is a rectangle, the opposite sides are equal and angle $B = 90°$. Hence,

$$\sin 23.0° = \frac{|\mathbf{t}|}{3000}$$

$$|\mathbf{t}| = 3000 \sin 23.0°$$

$$\approx 1170$$

Also,

$$\cos 23.0° = \frac{|\mathbf{n}|}{3000}$$

$$|\mathbf{n}| = 3000 \cos 23.0°$$

$$\approx 2760$$

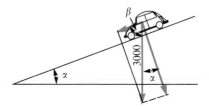

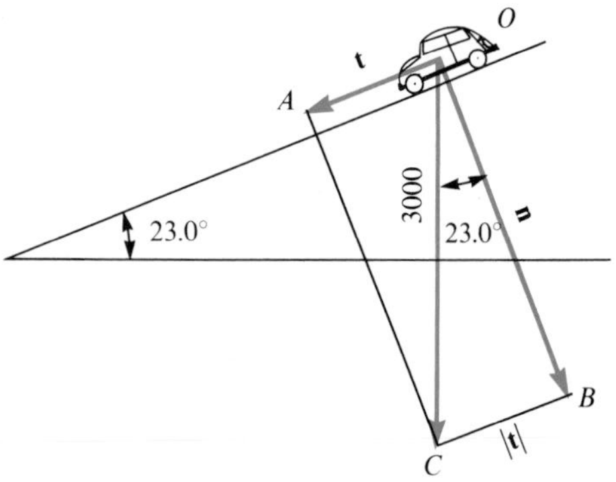

Figure 6-17

A force of approximately 1170 pounds is required to prevent rolling, and a force of approximately 2760 pounds keeps the car on the hill. Note that because the angles are to the nearest tenth of a degree, only three-place accuracy is permitted for the sides. In this example, the lengths of the vectors have only three significant digits. ∎

Exercise 6.2

1. A boat capable of a speed of 6 mph in still water attempts to go directly across a river. As the boat crosses the river, it drifts 30° from its intended path. How strong is the current? What is the effective speed of the boat?

2. A boat capable of a speed of 11 mph in still water attempts to go directly across a river with a current of 5.6 mph. By what angle is the boat pushed off its intended path? What is the effective speed of the boat?

3. Laura can row a boat $\frac{1}{2}$ mph in still water. She attempts to row straight across a river that has a current of 1 mph. If she must row for 2 hours to cross the river, by what angle is the current pushing her off her intended path? Give your answer to the nearest degree.

4. A boat attempts to go directly across a river with a current of 3.7 mph. The current causes the boat to drift 23° from its intended path. How far will the boat travel if the trip takes 10 minutes?

5. A plane has a heading of 260.0° and is flying at 357 mph. If a southerly wind causes the plane's course (the direction it is actually going) to be due west, find the groundspeed of the plane (its speed relative to the ground).

6. A plane has an airspeed of 411 mph and a heading of 90.0°. A wind from the north is blowing at 31.0 mph. By how many degrees is the plane blown off its heading? Find the groundspeed of the plane (its speed relative to the ground).

7. A plane leaves an airport with a heading of 45.0° and an airspeed of 201 mph. At the same time, another plane leaves the same airport with a heading of 135.0° and an airspeed of 305 mph. At the end of 2 hours, what is the bearing of the first plane from the second?

8. A rifle with a muzzle velocity of 4100 $\frac{ft}{sec}$ is fixed at an angle of elevation of 32°. What is the horizontal component of the bullet's velocity?

9. A 317-pound weight is hanging from the ceiling on a long rope. A man pushes horizontally against the weight to rotate the rope through an angle of 11.2°. What resultant force is being counteracted by the rope?

10. A plane leaves an airport with a heading of 170°. At the same time, a second plane leaves the same airport with a heading of 260°. One hour later, the first plane is 1300 miles directly southeast of the second plane. How fast is the first plane going?

11. What force is required to keep a 2210-pound car from rolling down a ramp that makes a 10.0° angle with the horizontal?

12. A force of 25 pounds is necessary to hold a barrel in place on a ramp that makes an angle of 7° with the horizontal. How much does the barrel weigh?

13. A vehicle presses against a roadway with a force of 1100 pounds. How much does the vehicle weigh if the roadway has an 18.0° grade?

14. A board will break if it is subjected to a force greater than 350 pounds. Will the board hold a 450-pound piano supported by a single dolly as it slides up the board and into a truck? Assume that the board makes an angle of 35.0° with the horizontal.

15. A garden tractor weighing 351 pounds is being driven up a ramp onto a trailer. If the tractor presses against the ramp with a 341-pound force, what angle does the ramp make with the horizontal?

16. If a force of 21.3 pounds is necessary to keep a 50.1-pound barrel from rolling down an inclined plane, what angle does the inclined plane make with the horizontal?

17. A 201-pound force is directed due east. What force, directed due north, is needed to produce a resultant force of 301 pounds? What angle is formed by the vectors representing the 201-pound and the 301-pound force?

18. A 312-pound force is directed due west. What force, directed exactly southeast, would cause the resultant force to be directed due south?

19. It requires 35 pounds of force to keep two children from sliding down the chute shown in Illustration 1. If one child weighs 110 pounds and the other 51 pounds, what angle does the chute make with the horizontal?

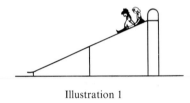

Illustration 1

20. A 50-pound girl is playing in a tire swing hanging from the limb of a large tree. See Illustration 2. To get her started, a friend pulls the swing backward until it makes an angle of 40° with the vertical. What horizontal force **F** is required to hold the swing in this position?

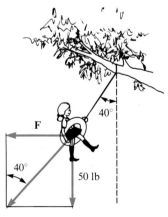

Illustration 2

21. A weight of 160 pounds is supported by a string as in Illustration 3. Find the vertical and horizontal components of force F_1. (*Hint:* The vertical component of F_1 supports half the weight.)

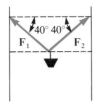

Illustration 3

22. A weight of 220 pounds is supported by a cable as shown in Illustration 4. Find the vertical and horizontal components of force F_1.

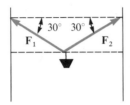

Illustration 4

In Exercises 23–24, you must draw some lines to create right triangles.

23. A plane leaves an airport at noon with a heading of 60.0° and an airspeed of 451 mph. One hour later, another plane leaves the same airport with a heading of 290.0° and an airspeed of 611 mph. What is the bearing of the first plane from the second at 2:00 P.M.? Assume no wind.

24. A plane leaves an airport at 3:00 P.M. with a heading of 70.0° and an airspeed of 512 mph. Two hours later, another plane leaves the same airport with a heading of 100.0° and an airspeed of 621 mph. How far apart are the two planes at 6:00 P.M.? Assume no wind.

In Exercises 25–26, you must recall some theorems about rhombuses.

25. Two forces of 30 pounds make an angle of 30° with each other. What is the norm of their resultant?

26. Two equal forces of **F** pounds make an angle of θ with each other. What is the norm of their resultant?

6.3   THE LAW OF COSINES

Given two sides or one acute angle and one side of a *right* triangle, it is always possible to compute the remaining parts. If the given triangle is *not* a right triangle, it is called an **oblique triangle**, and the triangle-solving techniques previously discussed no longer apply.

In the discussion of oblique triangles, we shall use the following convention: capital letters will be used to name the vertices of a triangle, and the corresponding lowercase letters will name the sides opposite those vertices. Thus, side a is opposite angle A, side b is opposite angle B, and side c is opposite angle C.

In elementary geometry, you learned that if two sides and the included angle of one triangle are equal to two sides and the included angle of a second triangle, then the two triangles are congruent. This fact (often abbreviated as SAS) indicates that if two sides and the angle between them are given, the triangle is uniquely determined. Because the triangle is "fixed," it is possible to compute the other side and the remaining two angles.

Similarly, if the three sides of a triangle are given (SSS), the triangle is determined and the three angles of the triangle can be computed. If two angles and any side of a triangle are known (ASA or AAS), the triangle is "fixed" and all remaining parts can be calculated.

The law of cosines is useful in solving oblique triangles when the given information is in SSS or SAS form. Place triangle ABC in a coordinate system with vertex A at the origin, as indicated in Figure 6-18. Because angle A is in standard position and point B is on its terminal side, the cosine of A is the ratio of the x-coordinate of B to the distance that B is from the origin. Thus,

$$\cos A = \frac{x\text{-coordinate of } B}{c}$$

or

$$x\text{-coordinate of } B = c \cos A$$

Similarly,

$$\sin A = \frac{y\text{-coordinate of } B}{c}$$

or

$$y\text{-coordinate of } B = c \sin A$$

Thus, the coordinates of point B are ($c \cos A$, $c \sin A$), the coordinates of A are (0, 0), and the coordinates of C are (b, 0).

We can now use the distance formula to compute a^2:

$$\begin{aligned}
a^2 &= (c \cos A - b)^2 + (c \sin A - 0)^2 \\
&= c^2 \cos^2 A - 2bc \cos A + b^2 + c^2 \sin^2 A \\
&= c^2(\cos^2 A + \sin^2 A) + b^2 - 2bc \cos A \\
&= c^2 + b^2 - 2bc \cos A
\end{aligned}$$

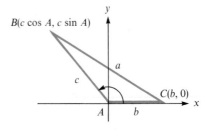

Figure 6-18

Thus,

1. $a^2 = b^2 + c^2 - 2bc \cos A$

Although the triangle in Figure 6-18 is obtuse, this derivation is valid for any triangle. Also, it need not be vertex A that is placed at the origin. If point B had been placed at the origin instead of point A, the following formula would have been derived.

2. $b^2 = c^2 + a^2 - 2ca \cos B$

If point C had been placed at the origin, a third formula would have been derived:

3. $c^2 = a^2 + b^2 - 2ab \cos C$

These three formulas are called the **law of cosines**.

The Law of Cosines.   The square of any side of any triangle is equal to the sum of the squares of the remaining two sides, minus twice the product of these two sides and the cosine of the angle between them.

$$a^2 = b^2 + c^2 - 2bc \cos A$$
$$b^2 = c^2 + a^2 - 2ca \cos B$$
$$c^2 = a^2 + b^2 - 2ab \cos C$$

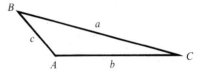

Figure 6-19

Example 1   Oblique triangle ABC has $b = 27$, $c = 14$, and $A = 43°$. Find side a.

Solution   Sides b and c and the included angle A are given (SAS). See Figure 6-20. Use the law of cosines.

$$a^2 = b^2 + c^2 - 2bc \cos A$$
$$a^2 = 27^2 + 14^2 - 2(27)(14) \cos 43°$$
$$\approx 729 + 196 - 552.90$$
$$\approx 372.10$$
$$a \approx 19.29$$
$$\approx 19$$

The value of a is approximately 19 units.

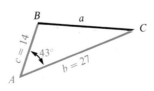

Figure 6-20

Example 2   In the oblique triangle ABC in Figure 6-21, $a = 5.2$, $b = 3.7$, and $c = 7.1$ units. Find angle B.

Solution   Use the form of the law of cosines that involves angle B:

$$b^2 = a^2 + c^2 - 2ac \cos B$$

Solve this formula for $\cos B$, substitute the values for a, b, and c, and calculate angle B:

$$\cos B = \frac{a^2 + c^2 - b^2}{2ac}$$

$$\cos B = \frac{5.2^2 + 7.1^2 - 3.7^2}{2(5.2)(7.1)}$$

$$\approx 0.8635$$

$$B \approx 30.29°$$

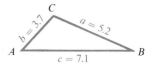

Figure 6-21

To the nearest degree, angle $B = 30°$.   ∎

Example 3   A farmer uses two horses and two ropes to pull a tractor out of the mud. When pulled tight, the ropes form an angle of $27°$, and each horse exerts a pull of 950 pounds. What force is applied to the tractor?

Solution   By the parallelogram law for adding vectors, the combined force is length b in the diagram of Figure 6-22. Because opposite sides of a parallelogram are equal, all four sides are 950 units in length. Because consecutive angles of a parallelogram are supplementary, the obtuse angle at B is $180° - 27° = 153°$. Apply the law of cosines to triangle ABC:

$$b^2 = a^2 + c^2 - 2ac \cos B$$
$$b^2 = 950^2 + 950^2 - 2(950)(950) \cos 153°$$
$$\approx 3{,}413{,}266.8$$
$$b \approx 1847.5$$
$$\approx 1800$$

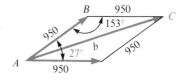

Figure 6-22

The combined pull of the horses is approximately 1800 pounds.   ∎

Example 4   An airplane flies N 86.2° W for a distance of 143 kilometers. The pilot, experiencing some engine problems, alters course and flies 79.5 kilometers in the direction S 32.7° E in an attempt to find a place to land. He crash lands safely in a cornfield. How far and in what direction will the rescue team need to travel?

Solution   The course diagram appears in Figure 6-23. To find distance d, use the law of cosines:

$$d^2 = 79.5^2 + 143^2 - 2(79.5)(143) \cos 53.5°$$
$$\approx 13{,}244.8$$
$$d \approx 115.1$$
$$\approx 115$$

The pilot is about 115 kilometers from the airport.

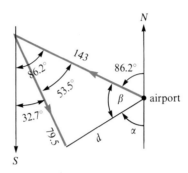

Figure 6-23

As a first step toward finding the direction in which the rescue team must travel, use the law of cosines again to find angle β:

$$79.5^2 = 143^2 + 115^2 - 2(143)(115) \cos \beta$$
$$\cos \beta = \frac{143^2 + 115^2 - 79.5^2}{2(143)(115)}$$
$$\approx 0.8317$$
$$\beta \approx 33.7°$$

Angle α is $180° - 86.2° - 33.7°$, or $60.1°$. The rescue team must bear S $60.1°$ W. ∎

Example 5   A 100-pound weight is suspended by two cables as in Figure 6-24. The tension on the left cable is 55 pounds and on the right cable, 75 pounds. What angle does each cable make with the horizontal?

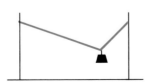

Figure 6-24

Solution   The forces in the two ropes are such that their resultant force is 100 pounds, directed upward to exactly counter the 100-pound downward pull. The force diagram appears in Figure 6-25. The angles that the cables make with the horizontal are the complements of angles α and β. Angles α and β can be found by using the law of cosines:

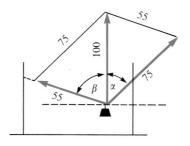

Figure 6-25

$$55^2 = 75^2 + 100^2 - 2(75)(100) \cos \alpha$$
$$\cos \alpha = \frac{75^2 + 100^2 - 55^2}{2(75)(100)}$$
$$= 0.8400$$
$$\alpha \approx 32.9°$$

Similarly,

$$75^2 = 55^2 + 100^2 - 2(55)(100) \cos \beta$$
$$\cos \beta = \frac{55^2 + 100^2 - 75^2}{2(55)(100)}$$
$$\approx 0.6727$$
$$\beta \approx 47.7°$$

The left cable makes an angle of $90° - 47.7° \approx 42°$ with the horizontal. The right cable is $90° - 32.9° \approx 57°$ from the horizontal.   ■

Exercise 6.3

In Exercises 1–12, refer to Illustration 1 and find the value requested. Use a calculator.

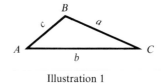

Illustration 1

1. $a = 42.9$ cm, $c = 37.2$ cm, $B = 99.0°$; find b.
2. $b = 192$ m, $c = 86.9$ m, $A = 21.2°$; find a.
3. $a = 2730$ km, $b = 3520$ km, $C = 21.7°$; find c.
4. $b = 2.1$ km, $c = 1.3$ km, $A = 14°$; find a.
5. $a = 91.1$ cm, $c = 87.6$ cm, $B = 43.2°$; find b.
6. $a = 107$ cm, $b = 205$ cm, $C = 86.5°$; find c.
7. $a = 19$ km, $b = 23$ km, $c = 18$ km; find A.
8. $a = 14.3$ km, $b = 29.7$ km, $c = 21.3$ km; find B.
9. $a = 30$ ft, $b = 40$ ft, $c = 50$ ft; find C.
10. $a = 130$ mi, $b = 50$ mi, $c = 120$ mi; find A.
11. $a = 1580$, $b = 2137$, $c = 3152$; find B.
12. $a = 0.0031$, $b = 0.0047$, $c = 0.0093$; find C.
13. Two men are pulling on ropes attached to the bumper of a car that is stuck in a snowdrift. If one man pulls with a force of 114 pounds and the other with a force of 97 pounds and the angle between the ropes is 13°, what is the force exerted on the car?
14. Two forces, one of 75 pounds and the other of 90 pounds, are exerted at an angle of 102° from each other. What is the magnitude of the resultant force?
15. In Exercise 13, what is the angle between the resultant force and the rope pulled by the stronger man?
16. In Exercise 14, what is the angle between the resultant force and the direction of the 90-pound force?
17. A donkey and a horse are tied to a large stone. The horse pulls with a force of 950 pounds; the donkey lazily tugs with a force of 150 pounds. The angle between their tethers is 19.5°. With what force do they pull on the stone?
18. A ship sails 21.2 nautical miles in a direction of N 42.0° W and then turns onto a course of S 15.0° E and sails 19.0 nautical miles. How far is the ship from its starting point?
19. A ship sails 14.3 nautical miles in a direction of S 28.0° W and then turns onto a course of S 52.0° W and sails 23.2 nautical miles. How far is the ship from its starting point?
20. To measure the length of a lake, a surveyor determines the measurements shown in Illustration 2. How long is the lake?

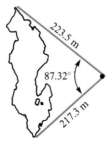

223.5 m

87.32°

217.3 m

Illustration 2

21. To estimate the cost of building a tunnel, a surveyor must find the distance through a hill. The surveyor determines the measurements shown in Illustration 3. How long must the tunnel be to pass through the hill?

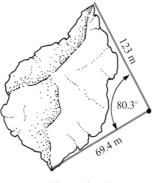

123 m

80.3°

69.4 m

Illustration 3

22. The three circles in Illustration 4 have radii of 4.0 centimeters, 7.0 centimeters, and 9.0 centimeters. If the circles are externally tangent to each other, what are the angles of the triangle that joins their centers?

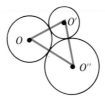

O O' O''

Illustration 4

23. The three circles in Illustration 4 have radii of 21.2 centimeters, 19.3 centimeters, and 31.2 centimeters. If the circles are externally tangent to each other, what are the angles of the triangle that joins their centers?

24. Show that the Pythagorean Theorem is a special case of the law of cosines.

25. Consider triangle ABC in Figure 6-18. Rotate triangle ABC so that point B is at the origin. Prove that $b^2 = c^2 + a^2 - 2ac \cos B$.

26. Consider triangle ABC in Figure 6-18. Rotate triangle ABC so that point C is at the origin. Prove that $c^2 = a^2 + b^2 - 2ab \cos C$.

27. To determine whether two interior walls meet at a right angle, carpenters often mark a point 3 feet from the corner on one wall and a point (at the same height) on the other wall 4 feet from the corner. If the straight-line distance between those points is 5 feet, the walls are square. At what angle do the walls meet if the distance measures 4 feet 10 inches?

28. To build a counter top for a kitchen, a cabinetmaker must determine the angle at which two walls meet. The method of Exercise 27 is used. What is the angle between the walls if the measured distance is 5 feet 3 inches?

29. Triangle *ABC* is formed by points *A*(3, 4), *B*(1, 5), and *C*(5, 9). Find angle *A* to the nearest tenth of a degree.

30. In Exercise 29, find angle *B* to the nearest tenth of a degree.

31. In Illustration 5, *D* is the midpoint of *BC*. Find angle 2 to the nearest tenth of a degree.

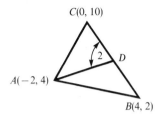

Illustration 5

32. In Exercise 31, find angle *DAB* to the nearest tenth of a degree.

33. A lighthouse is 15.0 nautical miles N 23.0° W of a dock. A ship leaves the dock heading due east at 26.3 knots. How long will it take for the ship to reach a distance of 35.0 nautical miles from the lighthouse?

6.4 THE LAW OF SINES

The law of cosines is not useful in the ASA or AAS case. Another set of formulas, called the **law of sines**, is required.

In the triangles of Figure 6-26 and Figure 6-27, we draw a line segment *CD* perpendicular to side *AB* or to an extension of *AB*. Let *h* be the length of segment *CD*. In the two right triangles of Figure 6-26, the following formulas are true:

$$h = b \sin A$$

and

$$h = a \sin B$$

Because both $b \sin A$ and $a \sin B$ are equal to h, they are equal to each other:

$$b \sin A = a \sin B$$

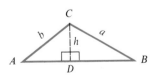

Figure 6-26

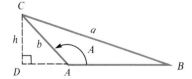

Figure 6-27

or

$$\frac{a}{\sin A} = \frac{b}{\sin B}$$

In the two right triangles of Figure 6-27 these equations hold:

$$h = a \sin B$$

and

$$h = b \sin(180° - A)$$
$$\quad = b \sin A$$

Because $a \sin B$ and $b \sin A$ both equal h, they are equal to each other:

$$b \sin A = a \sin B$$

or

$$\frac{a}{\sin A} = \frac{b}{\sin B}$$

In either of these triangles, the perpendicular need not be drawn from C to AB. If drawn from another vertex, similar reasoning yields

$$\frac{a}{\sin A} = \frac{c}{\sin C}$$

or

$$\frac{b}{\sin B} = \frac{c}{\sin C}$$

Because of the transitive law of equality, it follows that

$$\frac{a}{\sin A} = \frac{c}{\sin C} = \frac{b}{\sin B}$$

These results are summed up in a formula called the **law of sines**.

The Law of Sines.   The sides in any triangle are proportional to the sines of the angles opposite those sides.

$$\frac{a}{\sin A} = \frac{b}{\sin B} = \frac{c}{\sin C}$$

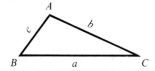

Figure 6-28

Example 1   In Figure 6-29, $a = 14$ km, $A = 21°$, and $B = 35°$. Find side b.

Solution   Note that the given information, in the pattern AAS, fills three of the four spots in the law of sines.

$$\frac{a}{\sin A} = \frac{b}{\sin B}$$

$$\frac{14}{\sin 21°} = \frac{b}{\sin 35°}$$

$$b = \frac{14 \sin 35°}{\sin 21°}$$

$$b \approx 22.4$$

$$\approx 22$$

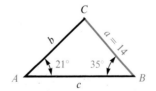

Figure 6-29

Thus, b is approximately 22 kilometers.   ■

Example 2   In Figure 6-30, $a = 29$ m, $B = 42°$, and $C = 31°$. Find side c.

Solution   Note that the given information is of the pattern ASA. However, the law of sines is not useful until you find angle A:

$$A = 180° - B - C$$

$$= 180° - 42° - 31°$$

$$= 107°$$

Now use the law of sines:

$$\frac{a}{\sin A} = \frac{c}{\sin C}$$

$$\frac{29}{\sin 107°} = \frac{c}{\sin 31°}$$

$$c = \frac{29 \sin 31°}{\sin 107°}$$

$$c \approx 15.62$$

$$\approx 16$$

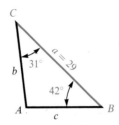

Figure 6-30

Thus, c is approximately 16 meters.   ■

Example 3   A ship is sailing due east. The skipper observes a lighthouse with a bearing of N 37.5° E. After the ship has sailed 4.70 nautical miles, the bearing to the lighthouse is N 9.0° W. How close to the lighthouse did the ship pass?

Solution   The information of the problem is illustrated in Figure 6-31. Solve for b first. Then the required distance d can be obtained by solving right triangle ACD. The law of sines provides a way to compute b:

$$\frac{b}{\sin B} = \frac{c}{\sin C}$$

$$\frac{b}{\sin 81.0°} = \frac{4.70}{\sin 46.5°}$$

$$b = \sin 81° \left(\frac{4.70}{\sin 46.5°} \right)$$

$$\approx 6.3996$$

$$\approx 6.40$$

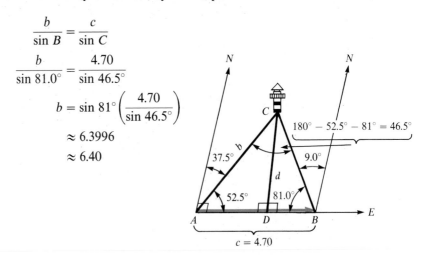

Figure 6-31

In right triangle ACD, it follows that

$$\sin A = \frac{d}{b}$$

$$d = b \sin A$$

$$d \approx 6.40 \sin 52.5°$$

$$\approx 5.0775$$

$$\approx 5.08$$

The ship's closest approach to the lighthouse is 5.08 nautical miles   ■

Example 4   On a bright moonlit night, the moon is directly overhead in city A. In city B, 2500 miles to the north, the moonlight strikes a pole at an angle of 36.57°. If the radius of the earth is 4000 miles (to two significant digits), find the distance from the earth to the moon.

Solution   Refer to Figure 6-32. Because 2500 miles is $\frac{1}{10}$ of the earth's circumference, angle BOA is $\frac{1}{10}$ of a complete revolution, or 36°. Angle OBM is supplementary to 36.57°. Hence, angle OBM is 143.43°. Because the sum of the angles in any triangle must be 180°, angle M is 0.57°. Use the law of sines to set up the following proportion and solve for x:

$$\frac{x}{\sin 143.43°} = \frac{4000}{\sin 0.57°}$$

$$x \approx \frac{0.5958(4000)}{0.00995}$$

$$\approx 239,518$$

$$\approx 240,000$$

The estimate of the distance to the moon is 240,000 − 4000 = 236,000 miles. To two significant digits, the distance is 240,000 miles.

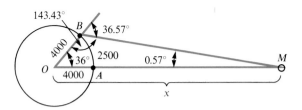

Figure 6-32

Example 5   A pilot wishes to fly in the direction of 20.0° east of north against a 45.0-mph wind blowing from the east. The airspeed of the plane is to be 185 mph. What should be the pilot's heading, and what will be the plane's groundspeed?

Solution   Refer to Figure 6-33. The pilot's intended direction of travel is represented by the vector **OA**. The direction of vector **OA** is 20.0° east of north. The length of vector **OH** represents the plane's airspeed of 185 mph. To find the pilot's heading, you must find angle θ, which is the direction of vector **OH**.

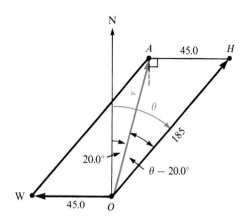

Figure 6-33

In triangle OAH, side AH has a length of 45.0 miles. Angle OAH is $90.0° + 20.0°$, or $110.0°$. Use the law of sines on triangle OAH to find $\theta - 20.0°$. You can then find θ:

$$\frac{185}{\sin 110.0°} = \frac{45.0}{\sin(\theta - 20.0°)}$$

$$\sin(\theta - 20.0°) = \frac{45.0 \sin 110.0°}{185}$$

$$\approx 0.2286$$

$$\theta - 20.0° \approx 13.2°$$

$$\theta \approx 33.2°$$

To go in his intended direction of travel, the pilot must set a heading of $33.2°$.

In the figure, the groundspeed of the plane is represented by the length, v, of vector **OA**. To determine this length, calculate angle H and use the law of cosines. Note that

$$\text{angle } H \approx 180.0° - 13.2° - 110.0°$$

$$\approx 56.8°$$

By the law of cosines,

$$v^2 \approx 45^2 + 185^2 - 2(45)(185) \cos 56.8°$$

$$\approx 27{,}133$$

$$v \approx 164.7$$

$$\approx 165$$

The plane's groundspeed is approximately 165 mph.   ■

The Ambiguous Case: SSA

A triangle might or might not be determined if two sides and a nonincluded angle are given. Thus, the SSA case is called the **ambiguous case**.

When given information in the SSA form, we can carefully sketch a scale drawing of the triangle and use common sense to determine whether two, one, or no triangles exist. If a scale drawing indicates a situation that is too close to call, we can let the law of sines decide the outcome.

Example 6   Three parts of a triangle are exactly $a = 1$, $b = 2$, and $A = 20°$. Find angles B and C to the nearest tenth of a degree and side c to the nearest hundredth.

Solution   A sketch indicates that there are two possible triangles. See Figure 6-34. By the law of sines, it follows that

$$\frac{a}{\sin A} = \frac{b}{\sin B}$$

$$\sin B = \frac{b \sin A}{a}$$

$$= \frac{2 \sin 20^\circ}{1}$$

$$\approx 0.6840$$

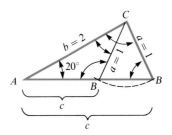

Figure 6-34

There are two possible values of B, one acute (a first-quadrant angle) and the other obtuse (a second-quadrant angle):

$$B \approx 43.2^\circ$$

or

$$B \approx 180^\circ - 43.2^\circ = 136.8^\circ$$

The third angle C has two possibilities also:

$$C = 180^\circ - A - \mathbf{B}$$
$$\approx 180^\circ - 20^\circ - \mathbf{43.2^\circ}$$
$$\approx 116.8^\circ$$

or

$$C = 180^\circ - A - \mathbf{B}$$
$$C \approx 180^\circ - 20^\circ - \mathbf{136.8^\circ}$$
$$\approx 23.2^\circ$$

The third side c can also be found by using the law of sines. Side c has two possibilities:

$$\frac{c}{\sin C} = \frac{a}{\sin A}$$

$$c \approx \frac{1 \cdot \sin 116.8^\circ}{\sin 20^\circ} \approx 2.61$$

or

$$c \approx \frac{1 \cdot \sin 23.2^\circ}{\sin 20^\circ} \approx 1.15$$

The two triangles are shown in Figure 6-35.

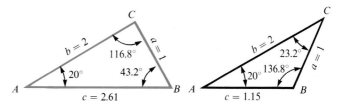

Figure 6-35

Example 7   In triangle ABC, $a = 1$, $b = 2$, and $A = 45°$. If possible, solve the triangle.

Solution   By the law of sines, it follows that

$$\frac{a}{\sin A} = \frac{b}{\sin B}$$

$$\frac{1}{\sin 45°} = \frac{2}{\sin B}$$

$$\sin B = 2 \sin 45°$$

$$= 2\left(\frac{\sqrt{2}}{2}\right)$$

$$= \sqrt{2}$$

Because $\sqrt{2}$ is greater than 1, and $\sin B$ cannot be greater than 1, there is no triangle that satisfies the given conditions.   ∎

Example 8   A vertical tower 255 feet tall stands on a hill. From a point 700 feet down the hill, the angle between the hill and an observer's line of sight to the top of the tower is 12.0°. What is the angle of inclination of the hill (the angle the ground makes with the horizontal)?

Solution   The given information is used to draw Figure 6-36. By the law of sines, it follows that

$$\frac{255}{\sin 12.0°} = \frac{700}{\sin B}$$

$$\sin B = \frac{700 \cdot \sin 12.0°}{255}$$

$$\approx 0.5707$$

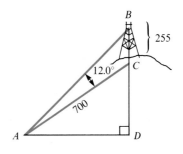

Figure 6-36

There are two possibilities for angle B—one acute and one obtuse. However, from the diagram, you want only the acute angle. Because $\sin B \approx 0.5707$,

$$B \approx 34.8°$$

Once you know the measure of angle B, it is easy to compute angle C (angle ACB):

$$C = 180° - A - B$$
$$\approx 180° - 12.0° - 34.8°$$
$$\approx 133.2°$$

The angle the hill makes with the horizontal is $C - 90°$, which is approximately 43.2°.   ∎

Exercise 6.4

In Exercises 1–16, refer to Illustration 1. Use the law of sines and your calculator to find the required value.

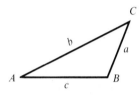

Illustration 1

1. $A = 12°$, $B = 97°$, $a = 14$ km; find b.
2. $A = 19°$, $C = 102°$, $c = 37$ ft; find a.
3. $A = 21.3°$, $B = 19.2°$, $a = 143$ m; find c.
4. $A = 28.8°$, $C = 9.3°$, $c = 135$ m; find b.
5. $B = 8.6°$, $C = 9.2°$, $c = 2.73$ m; find b.
6. $A = 86.3°$, $C = 7.6°$, $a = 43.0$ km; find c.
7. $A = 99.8°$, $C = 43.2°$, $b = 186$ m; find a.
8. $A = 14.61°$, $B = 87.10°$, $c = 1437$ ft; find b.
9. $A = \dfrac{\pi}{7}$, $C = \dfrac{\pi}{2}$, $b = 44.3$ cm; find c.
10. $A = \dfrac{2\pi}{9}$, $B = \dfrac{\pi}{11}$, $c = 56.7$ cm; find a.
11. $B = 107°$, $C = 11°$, $a = 0.96$ mi; find b.
12. $A = 141°$, $B = 5°$, $c = 0.037$; find a.
13. $B = 32.0°$, $b = 120$, $a = c$; find a.
14. $C = 12°$, $c = 5.4$, $b = a$; find b.
15. $A = x°$, $C = 2x°$, $B = 3x°$, $b = 7.93$; find a.
16. $A = x°$, $C = 3x°$, $B = 5x°$, $c = 12.5$; find a.
17. To measure the distance up a steep hill, Mary determines the measurements shown in Illustration 2. What is the distance d?

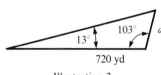

Illustration 2

18. A ship sails 3.2 nautical miles on a bearing of N 33° E. After reaching a lighthouse, the ship turns and sails 6.7 nautical miles to a position that is due east of the starting point. What is the bearing of the lighthouse from the ship's final position?

19. Points A and B are on opposite sides of a river. See Illustration 3. A tree at point C is 310 feet from point A. Angle A measures 125°, and angle C measures 32°. How wide is the river?

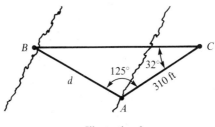

Illustration 3

20. Observers at points A and B are directly in line with a hot-air balloon and are themselves 215 feet apart. See Illustration 4. The angles of elevation of the balloon from A and B are as shown in the illustration. How high is the balloon? (*Hint*: First use the law of sines to find b.)

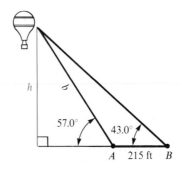

Illustration 4

21. A radio tower 175 feet high is located on top of a hill. At a point 800 feet down the hill, the angle of elevation to the top of the tower is 19.0°. What angle does the hill make with the horizontal? See Illustration 5.

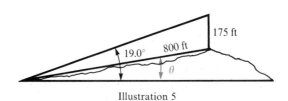

Illustration 5

22. Two children are on one riverbank, 120 feet apart. They each sight the same tree on the opposite bank. See Illustration 6. Angle A measures 79°, and angle B measures 63°. How wide is the river?

23. A ship sails due north at 3.2 knots (nautical miles per hour). At 2:00 P.M. the skipper sights a lighthouse in the direction of N 43° W. One hour later, the lighthouse bears S 78° W. How close to the lighthouse did the ship sail?

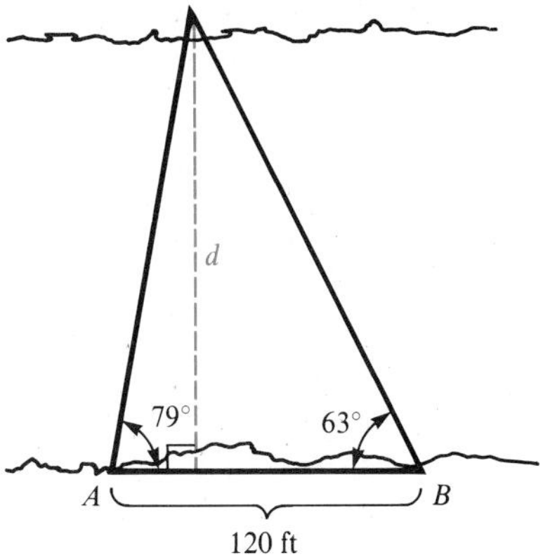

Illustration 6

24. A ship sails on a course bearing N 21° E at a speed of 14 knots (nautical miles per hour). At noon, the first mate sights an island in the direction N 35° E and one hour later sights the same island due east. If the ship continues on its course, how close will it approach the island?

In Exercises 25–32, calculate all possibilities for the indicated value. If no triangle is possible, so indicate.

25. $A = 42.0°$, $a = 123$ ft, $b = 96.0$ ft; find B.

26. $B = 56.2°$, $b = 13.5$ yd, $c = 15.3$ yd; find C.

27. $C = 98.6°$, $a = 42.1$ cm, $c = 47.3$ cm; find A.

28. $B = 17.5°$, $a = 0.063$ m, $b = 0.152$ m; find A.

29. $A = 57°$, $b = 13$ m, $a = 12$ m; find c.

30. $C = 48°$, $b = 29$ km, $c = 26$ km; find a.

31. $B = 87°$, $a = 35$ cm, $b = 32$ cm; find c.

32. $B = 38°$, $a = 12$ cm, $b = 40$ cm; find c.

33. The triangular piece of land owned by Farmer Brown is bounded by three straight highways. The angle between two of them—U.S. 45 and County M—is 43°. Brown's property runs for 2500 feet along County M, and for 2000 feet along the third highway—scenic Silo Drive. How much land might Brown own fronting on U.S. 45?

34. From the roof of Farmer Brown's barn, the angle of elevation to the top of a ranger lookout tower is 17°. From the barn's ground level 43 feet below, the angle of elevation of the tower is 21°. How far above ground level is the top of the tower?

35. A 210-foot television tower stands on the top of an office building. From a point on level ground, the angles of elevation to the top and base of the tower are 25.2° and 21.1°. How tall is the office building?

36. A pilot leaves point A and flies 800 kilometers with a heading of 320° to point B. From B she flies due south to a point C, which is 700 kilometers from point A. How long is the BC leg of the trip?

6.5   AREAS OF TRIANGLES

The area of any triangle is given by the formula

$$A = \frac{1}{2} bh$$

where b is the base and h is the height of the triangle. In any of the triangles of Figure 6-37, the base is 6 and the height is 3. The area of each triangle is

$$A = \frac{1}{2} bh$$

$$A = \frac{1}{2}(6)(3)$$

$$= 9 \text{ square units}$$

If we do not know the height of a triangle, we can still compute its area provided that we know three parts that uniquely determine the triangle.

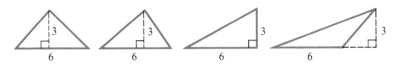

Figure 6-37

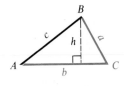

Figure 6-38

If two sides and the included angle are given (SAS), the height can be calculated. In triangle ABC of Figure 6-38, assume that b, a, and angle C are given. The height h is $a \sin C$ and the area K is

$$K = \frac{1}{2} bh$$

Hence,

$$K = \frac{1}{2} ba \sin C \qquad \text{Substitute } a \sin C \text{ for } h.$$

If c, b, and angle A are given, a similar argument gives

$$K = \frac{1}{2} cb \sin A$$

Finally, if a, c, and angle B are given, we have

$$K = \frac{1}{2} ac \sin B$$

Example 1   Find the area of the triangle in Figure 6-39.

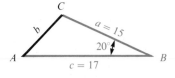

Figure 6-39

Solution Use the formula $K = \dfrac{1}{2} ab \sin C$.

$$K = \frac{1}{2} ab \sin C$$

$$K = \frac{1}{2} (15)(17) \sin 20°$$

$$\approx 43.6076$$

$$\approx 44 \text{ square units}$$

The area of the triangle is approximately 44 square units.   ■

Example 2 What is the area of an equilateral triangle of side s?

Solution Each angle of an equilateral triangle is 60°. Hence,

$$K = \frac{1}{2} ab \sin C$$

$$K = \frac{1}{2} ss \sin 60°$$

$$= \frac{1}{2} s^2 \left(\frac{\sqrt{3}}{2} \right)$$

$$= \frac{\sqrt{3}}{4} s^2$$

The area of the equilateral triangle is $\dfrac{\sqrt{3}}{4} s^2$ square units.   ■

If two angles and a side are given (AAS or ASA), the previous formulas can be adjusted to provide the area. In the triangle of Figure 6-40,

$$K = \frac{1}{2} cb \sin A$$

By the law of sines,

$$b = \frac{c \sin B}{\sin C}$$

Substituting $\dfrac{c \sin B}{\sin C}$ for b, we have

$$K = \frac{1}{2} c \frac{c \sin B}{\sin C} \sin A$$

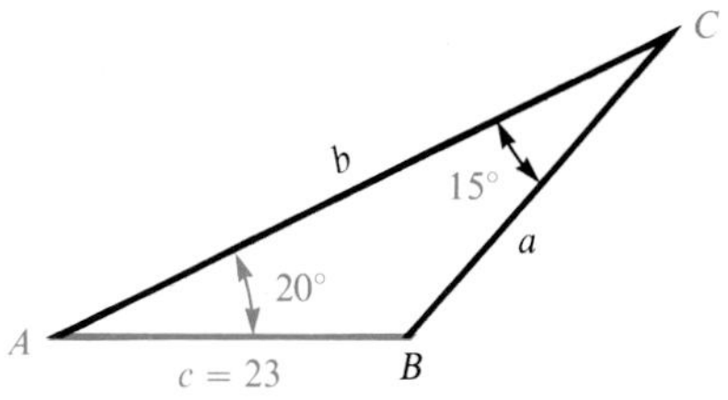

Figure 6-40

or

$$K = \frac{c^2 \sin A \sin B}{2 \sin C}$$

A similar argument produces two more formulas:

$$K = \frac{a^2 \sin B \sin C}{2 \sin A}$$

and

$$K = \frac{b^2 \sin C \sin A}{2 \sin B}$$

Note the cyclic change of the letters in the above three formulas.

Example 3   Find the area of the triangle in Figure 6-40.

Solution   Angle $B = 180° - 20° - 15° = 145°$ and

$$K = \frac{c^2 \sin A \sin B}{2 \sin C}$$

$$K = \frac{23^2 \sin 20° \sin 145°}{2 \sin 15°}$$

$$\approx 200.4806$$

$$\approx 200 \text{ square units}$$ ∎

Example 4   Find the area of the isosceles triangle in Figure 6-41.

Solution   The vertex angle B is $180° - 2\theta$ and

$$K = \frac{b^2 \sin A \sin C}{2 \sin B}$$

$$K = \frac{b^2 \sin \theta \sin \theta}{2 \sin(180° - 2\theta)}$$

$$= \frac{b^2 \sin^2 \theta}{2 \sin 2\theta}$$

$$= \frac{b^2 \sin^2 \theta}{2(2 \sin \theta \cos \theta)}$$

$$= \frac{b^2 \sin \theta}{4 \cos \theta}$$

$$= \frac{b^2}{4} \tan \theta$$ ∎

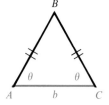

Figure 6-41

Finally, if three sides of a triangle are given (SSS), the triangle is determined and the area may be calculated by a formula attributed to Heron (Hero) of Alexandria (circa 240 A.D.). We state the formula without proof.

Heron's Formula. If a, b, and c are the three sides of a triangle and

$$s = \frac{a + b + c}{2} \quad \text{(s is the \textbf{semiperimeter})}$$

then the area of the triangle is given by

$$K = \sqrt{s(s - a)(s - b)(s - c)}$$

Example 5 Find the area of a triangle with sides of exactly 5, 7, and 10 centimeters.

Solution Let $a = 5$, $b = 7$, and $c = 10$. Then

$$s = \frac{5 + 7 + 10}{2} = 11$$

Use Heron's formula to find the area:

$$K = \sqrt{s(s - a)(s - b)(s - c)}$$
$$K = \sqrt{11(11 - 5)(11 - 7)(11 - 10)}$$
$$= \sqrt{11 \cdot 6 \cdot 4 \cdot 1}$$
$$= \sqrt{264}$$
$$\approx 16.248$$

The area is approximately 16.248 square centimeters. ∎

Exercise 6.5

In Exercises 1–16, find the area of the triangle whose parts are given, if possible.

1. $b = 23$ ft, $a = 17$ ft, $C = 80°$
2. $c = 1.7$ yd, $b = 3.5$ yd, $A = 60°$
3. $a = 32.3$ cm, $c = 21.5$ cm, $B = 120.0°$
4. $B = 33.2°$, $a = 101$ km, $c = 97.3$ km
5. $a = 3.0$, $b = 5.0$, $c = 7.0$
6. $a = 2.1$, $b = 3.2$, $c = 5.7$
7. $a = 3$, $b = 4$, $c = 5$
8. $a = 1.2$, $b = 2.3$, $c = 3.4$
9. $A = 55°$, $B = 45°$, $c = 12$
10. $A = 102°$, $C = 47°$, $b = 82$
11. $B = 15°$, $A = 70°$, $b = 23$
12. $C = 41°$, $B = 62°$, $c = 17$
13. $a = 0.06$ mm, $b = 0.05$ mm, $c = 0.07$ mm
14. $a = 0.017$ cm, $b = 0.032$ cm, $c = 0.055$ cm
15. $a = 976$ km, $b = 728$ km, $c = 543$ km
16. $a = 1860$ m, $b = 2150$ m, $c = 1590$ m
17. To find the area of a triangular lot, the owner starts at one corner and walks due west 205 feet to a second corner. After turning through an angle of 87.3°, he walks 307 feet to the third corner. What is the area of the lot in square feet?

18. A painter wishes to estimate the area of the gable end of a house. What is its area in square feet if the triangle has dimensions as shown in Illustration 1?

Illustration 1

19. A printer wishes to make a sign in the form of an isosceles triangle with base angles of 70° and a side of 15 meters. Find the area of the triangle.

20. Point C has a bearing of N 20° E from point A and a bearing of N 10° E from point B. What is the area of triangle ABC if B is due east of A and 17 kilometers from C?

21. Three circles with radii of exactly 3, 5, and 9 centimeters are externally tangent. What is the area of the triangle joining their centers?

22. Three circles have diameters of 7.8, 5, and 11.4 centimeters. If they are tangent externally, what is the area of the triangle joining their centers?

23. A Boy Scout walks 520 feet, turns and walks 490 feet, turns again and walks 670 feet, returning to his starting point. What area did his walk encompass?

24. A Girl Scout hikes 523 meters, turns and jogs 412 meters, turns again and runs 375 meters, returning to her starting point. What area did her trip encompass?

25. Prove that in any triangle,

$$\cos^2 \frac{A}{2} = \frac{s(s-a)}{bc}$$

where s is one-half of the perimeter.

26. Prove that in any triangle,

$$\sin^2 \frac{A}{2} = \frac{(s-b)(s-c)}{bc}$$

where s is one-half of the perimeter.

27. Prove that the area of a parallelogram is one-half the product of the diagonals and the sine of the angle between the diagonals.

28. Three externally tangent circles have radii of 5, 7, and 8, as shown in Illustration 2. What is the area of the curve-sided "triangle" they enclose?

Illustration 2

29. Find the area of an isosceles triangle with vertex angle α and base b.

30. Find the area of an isosceles triangle with vertex angle α and one of the equal sides of length a.

31. Find the area of an isosceles triangle if the base b is one-half the length of one of the equal sides.

32. Derive a formula for the area of the segment of the circle (the shaded area) in Illustration 3.

Illustration 3

6.6 MORE ON VECTORS

Recall that a *vector* is a directed line segment that can be denoted by a boldface letter such as **V**. (In handwritten work, we often express a vector as a letter with an arrow above it, such as $\vec{V}$. If a vector starts at point A and ends at point B, that vector can be denoted either as **AB** or as $\overrightarrow{AB}$.)

> **Definition.** Two **vectors** are equal if and only if they have the same length and the same direction.

Recall that the length of a vector **V** is called its **norm** and is denoted by $|\mathbf{V}|$. Because the norm of a vector is a numerical value, it is a scalar quantity.

In Figure 6-42 vectors **AB** and **CD** have the same length and direction. Thus $|\mathbf{AB}| = |\mathbf{CD}|$ and **AB** = **CD**. In Figure 6-43, $|\mathbf{OA}| = |\mathbf{OB}|$, but because **OA** and **OB** have different directions, **OA** ≠ **OB**.

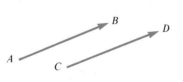

Figure 6-42

Figure 6-43

Vectors are easier to handle if they are placed on a coordinate system. The vector **V** in Figure 6-44, for example, is placed on a coordinate system so that it starts at the origin and ends at the point (3, 2). If we assume that *all* vectors start at the origin, then each is completely determined by its endpoint. Thus, we can denote the vector shown in Figure 6-44 by the ordered pair of numbers $\langle 3, 2 \rangle$. We use corner brackets $\langle \; \rangle$ to distinguish the vector from the point (3, 2). Note that the distance formula can be used to find the norm of **V**:

$$|\mathbf{V}| = \sqrt{3^2 + 2^2}$$
$$= \sqrt{13}$$

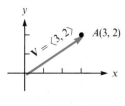

Figure 6-44

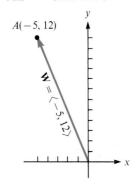

$A(-5, 12)$

$\mathbf{W} = \langle -5, 12 \rangle$

Figure 6-45

Similarly, in Figure 6-45, $\mathbf{W} = \langle -5, 12 \rangle$, and

$$|\mathbf{W}| = \sqrt{(-5)^2 + 12^2}$$
$$= \sqrt{169}$$
$$= 13$$

In general,

Theorem.   If $\mathbf{V}$ is placed on a coordinate system and is represented by $\langle a, b \rangle$, then the norm of $\mathbf{V}$ is given by

$$|\mathbf{V}| = \sqrt{a^2 + b^2}$$

Example 1   If vector $\mathbf{V} = \langle 3, -5 \rangle$, find $|\mathbf{V}|$.

Solution $|\mathbf{V}| = \sqrt{3^2 + (-5)^2}$
$$= \sqrt{9 + 25}$$
$$= \sqrt{34}$$ ∎

The addition of vectors is easy using this ordered-pair notation. The coordinates of point C in the parallelogram of Figure 6-46 are found by adding the corresponding coordinates of points A and B. Thus, if $\mathbf{OA} = \langle -3, 1 \rangle$ and $\mathbf{OB} = \langle 6, 1 \rangle$, then

$$\mathbf{OA} + \mathbf{OB} = \langle -3, 1 \rangle + \langle 6, 1 \rangle$$
$$= \langle -3 + 6, 1 + 1 \rangle$$
$$= \langle 3, 2 \rangle$$
$$= \mathbf{OC}$$

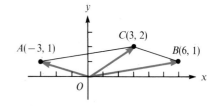

Figure 6-46

The previous example suggests the following definition.

Definition.   If vector $\mathbf{V} = \langle a, b \rangle$ and vector $\mathbf{W} = \langle c, d \rangle$, then

$$\mathbf{V} + \mathbf{W} = \langle a + c, b + d \rangle$$

Example 2 If $V = \langle 3, -5 \rangle$ and $W = \langle 1, 2 \rangle$, find $V + W$.

Solution $\begin{aligned} V + W &= \langle 3, -5 \rangle + \langle 1, 2 \rangle \\ &= \langle 3 + 1, -5 + 2 \rangle \\ &= \langle 4, -3 \rangle \end{aligned}$ ∎

If k is a real number and V is a vector, we can define a type of multiplication called **scalar multiplication**.

Definition. If k is a scalar and V is the vector $\langle a, b \rangle$, then

$$kV = k\langle a, b \rangle = \langle ka, kb \rangle$$

Example 3 If $V = \langle 3, -5 \rangle$, find **a.** $2V$ and **b.** $-8V$.

Solution   **a.** $\begin{aligned} 2V &= 2\langle 3, -5 \rangle \\ &= \langle 6, -10 \rangle \end{aligned}$

b. $\begin{aligned} -8V &= -8\langle 3, -5 \rangle \\ &= \langle -24, 40 \rangle \end{aligned}$ ∎

Note that if k is a real number and V is a vector, then the product kV is also a vector. Its norm is $|k|$ times the norm of V itself. If k is a positive real number, then kV has the same direction as V. If k is a negative real number, then kV has the opposite direction of V. See Figure 6-47. The vector kV is called a **scalar multiple** of V.

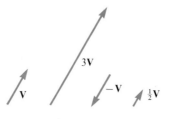

Figure 6-47

Let i be the vector $\langle 1, 0 \rangle$. Note that i is a vector of length 1 unit pointing in the positive x-direction. Similarly, let j be the vector $\langle 0, 1 \rangle$. The vector j is of length 1 unit pointing in the positive y-direction. Any vector can be written as the sum of scalar multiples of these vectors i and j. For example, the vector $\langle 5, 2 \rangle$ can be written in this form by proceeding as follows:

$$\begin{aligned} \langle 5, 2 \rangle &= \langle 5, 0 \rangle + \langle 0, 2 \rangle \\ &= 5\langle 1, 0 \rangle + 2\langle 0, 1 \rangle \\ &= 5i + 2j \end{aligned}$$

The two vectors $5\mathbf{i}$ and $2\mathbf{j}$ are called the **x-** and the **y-components** of the vector $\langle 5, 2 \rangle$. See Figure 6-48.

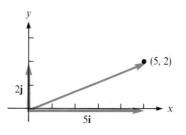

Figure 6-48

Definition.   The vectors $\mathbf{i} = \langle 1, 0 \rangle$ and $\mathbf{j} = \langle 0, 1 \rangle$ are called **unit coordinate vectors**.

Any vector $\langle x, y \rangle$ is **resolved** into its **x-** or **horizontal component** $x\mathbf{i}$ and its **y-** or **vertical component** $y\mathbf{j}$ when it is written in the form $x\mathbf{i} + y\mathbf{j}$.

Example 4   Let $\mathbf{V} = 2\mathbf{i} + 3\mathbf{j}$ and $\mathbf{W} = 4\mathbf{i} - \mathbf{j}$. Calculate **a.** $5\mathbf{V} + 3\mathbf{W}$ and **b.** $\mathbf{V} - \mathbf{W}$.

Solution   **a.** $5\mathbf{V} + 3\mathbf{W} = 5(2\mathbf{i} + 3\mathbf{j}) + 3(4\mathbf{i} - 1\mathbf{j})$
$= (10\mathbf{i} + 15\mathbf{j}) + (12\mathbf{i} - 3\mathbf{j})$
$= 22\mathbf{i} + 12\mathbf{j}$

b. $\mathbf{V} - \mathbf{W} = \mathbf{V} + (-1)\mathbf{W}$
$= (2\mathbf{i} + 3\mathbf{j}) + (-1)(4\mathbf{i} - 1\mathbf{j})$
$= (2\mathbf{i} + 3\mathbf{j}) + (-4\mathbf{i} + 1\mathbf{j})$
$= -2\mathbf{i} + 4\mathbf{j}$ ∎

The definition of scalar multiplication provides the way to multiply a vector by a real number. We now define a way, called the **dot product**, to multiply one vector by another.

Definition.   The **dot product** of vectors $\mathbf{V}$ and $\mathbf{W}$ is the *scalar*

$$\mathbf{V} \cdot \mathbf{W} = |\mathbf{V}||\mathbf{W}| \cos \theta$$

where θ is the angle between $\mathbf{V}$ and $\mathbf{W}$.

It is not convenient to calculate the dot product of two vectors by using the previous definition. However, we can calculate a dot product by using a theorem, which is stated without proof.

> **Theorem.**   Let $\mathbf{V} = a\mathbf{i} + b\mathbf{j}$ and $\mathbf{W} = c\mathbf{i} + d\mathbf{j}$. Then
>
> $$\mathbf{V} \cdot \mathbf{W} = ac + bd$$

Example 5   If $\mathbf{A} = 2\mathbf{i} + 3\mathbf{j}$ and $\mathbf{B} = 5\mathbf{i} - 4\mathbf{j}$, calculate $\mathbf{A} \cdot \mathbf{B}$.

Solution
$$\begin{aligned}
\mathbf{A} \cdot \mathbf{B} &= (2\mathbf{i} + 3\mathbf{j}) \cdot (5\mathbf{i} - 4\mathbf{j}) \\
&= 2 \cdot 5 + 3 \cdot (-4) \\
&= -2
\end{aligned}$$ ∎

Example 6   Find the angle between $\mathbf{A} = 3\mathbf{i} + 4\mathbf{j}$ and $\mathbf{B} = 5\mathbf{i} - 12\mathbf{j}$.

Solution   See Figure 6-49. By definition, $\mathbf{A} \cdot \mathbf{B} = |\mathbf{A}|\,|\mathbf{B}| \cos \theta$, where θ is the angle between the vectors. Solve for $\cos \theta$ and proceed as follows:

$$\mathbf{A} \cdot \mathbf{B} = |\mathbf{A}|\,|\mathbf{B}| \cos \theta$$

$$\cos \theta = \frac{\mathbf{A} \cdot \mathbf{B}}{|\mathbf{A}|\,|\mathbf{B}|}$$

$$= \frac{(3\mathbf{i} + 4\mathbf{j}) \cdot (5\mathbf{i} - 12\mathbf{j})}{\sqrt{3^2 + 4^2}\,\sqrt{5^2 + (-12)^2}}$$

$$= \frac{15 - 48}{5 \cdot 13}$$

$$\cos \theta = \frac{-33}{65}$$

$$\cos \theta \approx -0.5077$$

$$\theta \approx 120.5°$$ ∎

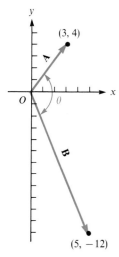

Figure 6-49

If the dot product, $|\mathbf{A}|\,|\mathbf{B}| \cos \theta$, of two nonzero vectors $\mathbf{A}$ and $\mathbf{B}$ is 0, then $\cos \theta$ must be 0. If $\cos \theta = 0$, then $\theta = 90°$ and the two vectors must be perpendicular. Thus, the dot product provides a test for the perpendicularity of two vectors.

> **Theorem.**   Two nonzero vectors are perpendicular if and only if their dot product is zero.

Example 7   Are the vectors $\mathbf{A} = 6\mathbf{i} - 2\mathbf{j}$ and $\mathbf{B} = \mathbf{i} + 3\mathbf{j}$ perpendicular?

Solution   Calculate $\mathbf{A} \cdot \mathbf{B}$. If $\mathbf{A} \cdot \mathbf{B} = 0$, the vectors are perpendicular. If $\mathbf{A} \cdot \mathbf{B} \neq 0$, they are not.

$$\mathbf{A} \cdot \mathbf{B} = (6\mathbf{i} - 2\mathbf{j}) \cdot (\mathbf{i} + 3\mathbf{j})$$
$$= 6 \cdot 1 + (-2)(+3)$$
$$= 6 - 6$$
$$= 0$$

Because the dot product is 0, the vectors **A** and **B** are perpendicular. ∎

Example 8   What are the horizontal and vertical components of a 2.0-pound force that makes an angle of 30° with the x-axis? Express the 2.0-pound force in $a\mathbf{i} + b\mathbf{j}$ form.

Solution   The horizontal component of the given force is vector **OA**, which is one leg of the right triangle OAC. See Figure 6-50.

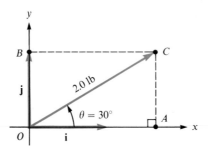

Figure 6-50

The norm of **OA** is found as follows:

$$|\mathbf{OA}| = |\mathbf{OC}| \cos \theta$$
$$= 2.0(\cos 30°)$$
$$= 2.0\left(\frac{\sqrt{3}}{2}\right)$$
$$= \sqrt{3}$$

The horizontal component is $\sqrt{3}$ pounds.
    Similarly, the vertical component **OB** is found as follows:

$$|\mathbf{OB}| = |\mathbf{OC}| \sin \theta$$
$$= 2.0(\sin 30°)$$
$$= 1.0$$

The vertical component is 1.0 pound.
    Thus, you have $\mathbf{OC} = \sqrt{3}\mathbf{i} + \mathbf{j}$. ∎

Example 9   A force of 2.0 pounds makes an angle of 30° with the horizontal. What is the component of this force in the direction $\mathbf{OB} = 12\mathbf{i} + 5\mathbf{j}$?

Solution You must find the component of the given force in a direction other than that of an axis. See Figure 6-51.

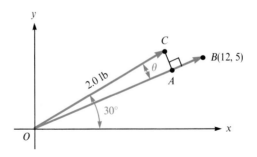

Figure 6-51

You must find the norm of vector **OA**. In right triangle *OAC*, the following relation holds:

$$|\mathbf{OA}| = |\mathbf{OC}| \cos \theta$$

Use the dot product to make this calculation by proceeding as follows:

$$
\begin{aligned}
|\mathbf{OA}| &= |\mathbf{OC}| \cos \theta \\
&= \frac{|\mathbf{OC}| |\mathbf{OB}| \cos \theta}{|\mathbf{OB}|} \qquad &&\text{Multiply and divide by } |\mathbf{OB}|. \\
&= \frac{\mathbf{OC} \cdot \mathbf{OB}}{|\mathbf{OB}|} \qquad &&\text{Use the definition of dot product.} \\
&= \frac{(\sqrt{3}\mathbf{i} + \mathbf{j}) \cdot (12\mathbf{i} + 5\mathbf{j})}{\sqrt{12^2 + 5^2}} \qquad &&\text{Use the result from Example 8, and } \mathbf{OB} = 12\mathbf{i} + 5\mathbf{j}. \\
&= \frac{12\sqrt{3} + 5}{13}
\end{aligned}
$$

The component of the 2.0-pound force in the direction of **OB** is $\dfrac{12\sqrt{3} + 5}{13}$.

■

Exercise 6.6

In Exercises 1–12, let $\mathbf{U} = \langle 2, -3 \rangle$, $\mathbf{V} = \langle 5, -2 \rangle$, *and* $\mathbf{W} = \langle -1, 1 \rangle$. *Calculate each quantity.*

1. $\mathbf{U} + \mathbf{V}$	**2.** $\mathbf{V} + \mathbf{W}$	**3.** $3\mathbf{U}$	**4.** $5\mathbf{V}$
5. $2\mathbf{U} + \mathbf{V}$	**6.** $3\mathbf{U} - \mathbf{W}$	**7.** $|\mathbf{U}|$	**8.** $|3\mathbf{U}|$
9. $|\mathbf{U} + \mathbf{W}|$	**10.** $|\mathbf{V} - \mathbf{W}|$	**11.** $|\mathbf{U}| + |\mathbf{W}|$	**12.** $|3\mathbf{V} - 5\mathbf{W}|$

*In Exercises 13–18, resolve each vector into its horizontal and vertical components by writing each vector in a**i** + b**j** form.*

13. $\langle 3, 5 \rangle + \langle 5, 3 \rangle$ **14.** $\langle -2, 7 \rangle + \langle 2, 3 \rangle$

15. A vector of length 10, making an angle of 30° with the x-axis

16. A vector of length 10, making an angle of 45° with the x-axis

17. A vector of length 23.3, making an angle of 37.2° with the x-axis

18. A vector of length 19.1, making an angle of 183.7° with the x-axis

In Exercises 19–24, find the dot product of the two given vectors.

19. $\langle 2, -3 \rangle$ and $\langle 3, -1 \rangle$ 20. $\langle 1, -5 \rangle$ and $\langle 5, 1 \rangle$

21. $2\mathbf{i} + 5\mathbf{j}$ and $\mathbf{i} + \mathbf{j}$ 22. $3\mathbf{i} - 3\mathbf{j}$ and $2\mathbf{i} + \mathbf{j}$

23. $\mathbf{i}$ and $\mathbf{j}$ 24. $2\mathbf{i}$ and $3\mathbf{i}$

In Exercises 25–30, find the angle between the given vectors.

25. $\langle 2, 2 \rangle$ and $\langle 5, 0 \rangle$ 26. $\langle \sqrt{3}, 1 \rangle$ and $\langle 3, 3 \rangle$

27. $\langle \sqrt{3}, -1 \rangle$ and $\langle -1, \sqrt{3} \rangle$ 28. $2\mathbf{i} + 3\mathbf{j}$ and $-3\mathbf{i} + 2\mathbf{j}$

29. $3\mathbf{i} - \mathbf{j}$ and $3\mathbf{i} + \mathbf{j}$ 30. $3\mathbf{i} - 4\mathbf{j}$ and $5\mathbf{i} + 12\mathbf{j}$

In Exercises 31–36, indicate whether the given vectors are perpendicular.

31. $\langle 2, 3 \rangle$ and $\langle -3, 2 \rangle$ 32. $\langle 2, 3 \rangle$ and $\langle 3, 2 \rangle$

33. $5\mathbf{i} + \mathbf{j}$ and $\mathbf{i} + \mathbf{j}$ 34. $6\mathbf{i} - 2\mathbf{j}$ and $\mathbf{i} + 3\mathbf{j}$

35. $\mathbf{i}$ and $\mathbf{j}$ 36. $-\mathbf{i}$ and $3\mathbf{i}$

In Exercises 37–40, find the component of the first vector in the direction of the second vector.

37. $\langle 3, 4 \rangle, \langle 5, 12 \rangle$ 38. $\langle 1, 1 \rangle, \langle 3, 2 \rangle$

39. $6\mathbf{i} + 8\mathbf{j}, 4\mathbf{i} - 3\mathbf{j}$ 40. $\mathbf{i}, \mathbf{i} + \mathbf{j}$

41. Find an example to illustrate that $|\mathbf{U} + \mathbf{V}| \neq |\mathbf{U}| + |\mathbf{V}|$.

42. Find an example to support the distributive law, $(a + b)\mathbf{V} = a\mathbf{V} + b\mathbf{V}$.

43. Find an example to support the distributive law, $a(\mathbf{V} + \mathbf{W}) = a\mathbf{V} + a\mathbf{W}$.

44. Find an example to support the associative law, $a(\mathbf{V} \cdot \mathbf{W}) = (a\mathbf{V}) \cdot \mathbf{W}$.

45. Let $\mathbf{V} = a\mathbf{i} + b\mathbf{j}$. Prove that $\mathbf{V} \cdot \mathbf{V} = |\mathbf{V}|^2$.

REVIEW EXERCISES

1. From a location 32.1 meters from the base of a flagpole, the angle of elevation to its top is α. Find α if the flagpole is 10.0 meters tall.

2. The angle of depression from a window in a building to a point on the ground is 17.7°. If the point on the ground is 187 feet from the base of the building, how high is the observer?

3. Owatonna, Minnesota, is approximately 55 miles due south of Minneapolis, and the bearing of Winona, Minnesota, from Minneapolis is about S 45° E. How far is Winona from Owatonna if Owatonna is due west of Winona?

4. Assume that the bearing of South Bend, Indiana, from Fort Wayne, Indiana, is N 48° W and that the distance between the cities is about 71 miles. Further assume that South Bend is due north of Indianapolis

and that the bearing of Fort Wayne from Indianapolis is N 21° E. How far is Fort Wayne from Indianapolis?

5. A barrel weighing 60 pounds rests on a ramp that makes an angle of 10° with the horizontal. How much force is necessary to keep the barrel from rolling down the ramp?

6. A 2500-pound car rests on a hill. A force of 500 pounds is required to keep the car from rolling down the hill. How steep is the grade?

In Review Exercises 7–18, consider the given parts of triangle ABC. Use a calculator to solve for the required value, if possible. If more than one value is possible, give both.

7. $a = 12$, $c = 15$, $B = 30°$; find b.

8. $b = 23$, $a = 13$, $C = 125°$; find c.

9. $c = 0.5$, $b = 0.8$, $A = 50°$; find a.

10. $a = 28.7$, $b = 37.8$, $C = 11.2°$; find c.

11. $a = 12$, $c = 18$, $C = 40°$; find A.

12. $b = 17$, $a = 12$, $A = 25°$; find B.

13. $c = 31.5$, $b = 27.5$, $B = 16.2°$; find C.

14. $a = 315.2$, $b = 457.8$, $A = 32.51°$; find B.

15. $A = 24.3°$, $B = 56.8°$, $a = 32.3$; find b.

16. $B = 10.3°$, $C = 59.4°$, $c = 341$; find b.

17. $b = 17$, $c = 21$, $B = 42°$; find A.

18. $c = 189$, $a = 150$, $C = 85.3°$; find B.

In Review Exercises 19–26, find the area of the triangle with the given parts, if possible.

17.9°   94.5° l
542 ft

Illustration 1

19. $a = 32$, $C = 47°$, $b = 55$

20. $b = 29$, $A = 96°$, $c = 85$

21. $B = 33°$, $C = 25°$, $a = 17$

22. $A = 85°$, $C = 80°$, $b = 7.5$

23. $A = 130°$, $B = 20°$, $a = 3.5$

24. $C = 15°$, $A = 110°$, $c = 91$

25. $a = 57$, $b = 85$, $c = 110$

26. $a = 8.50$, $b = 17.4$, $c = 22.3$

27. Two airplanes leave an airport at 2:00 P.M., one with a heading of 30.0° and groundspeed of 425 mph and the other with a heading of 85.0° and groundspeed of 375 mph. How far apart are they at 3:30 P.M.?

28. Find the angles of the triangle with vertices of (0, 0), (5, 0), and (7, 8).

29. From a point 542 feet away from the base of the Leaning Tower of Pisa, the angle of elevation to its top is 17.9°. If the tower makes an angle with the ground of 94.5° (see Illustration 1), how tall is the tower?

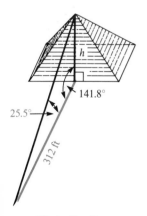

25.5°   141.8° h
312 ft

Illustration 2

30. From a point 312 feet from the base of the Great Pyramid of Khufu (Cheops) at Gizeh, the angle of elevation to its top is 25.5°. If the pyramid makes an angle with the ground of 141.8° (see Illustration 2), find its slant height.

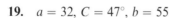

31. Find the area of a triangular lot with sides of 21, 32, and 47 meters.

32. Find the area of the triangle joining points with coordinates of (0, 0), (0, 8), and (6, 14).

In Review Exercises 33–36, assume that $\mathbf{V} = \langle 3, 7 \rangle$ *and* $\mathbf{W} = \langle -2, 5 \rangle$. *Calculate each quantity.*

33. $2\mathbf{V} + 3\mathbf{W}$ **34.** $|3\mathbf{V}| - |\mathbf{V}|$ **35.** $5(\mathbf{V} - \mathbf{W})$ **36.** $|3\mathbf{V} - \mathbf{W}|$

In Review Exercises 37–40, find the angle between the two given vectors.

37. $\langle 0, 5 \rangle$ and $\langle 2, 0 \rangle$ **38.** $\langle 8, 2 \rangle$ and $\langle 4, 1 \rangle$

39. $\sqrt{3}\mathbf{i} + \mathbf{j}$ and $\mathbf{i} - \mathbf{j}$ **40.** $2\mathbf{i} - 2\sqrt{3}\mathbf{j}$ and $\mathbf{i} + \sqrt{3}\mathbf{j}$

7 TRIGONOMETRIC IDENTITIES AND EQUATIONS

We have used trigonometry to solve applied problems such as finding heights of flagpoles and distances across rivers. Trigonometry can also be used in a more theoretical way. Because the trigonometric functions are related to each other and to certain algebraic expressions, they often enable us to simplify very complicated mathematical expressions. In this chapter, we shall explore some of these relationships by verifying trigonometric identities and solving trigonometric equations.

7.1 VERIFYING IDENTITIES

Recall the following fundamental identities:

$$\csc \theta = \frac{1}{\sin \theta} \qquad \tan \theta = \frac{\sin \theta}{\cos \theta} \qquad \tan^2 \theta + 1 = \sec^2 \theta$$

$$\qquad\qquad\qquad\qquad\qquad\qquad\qquad\qquad \cot^2 \theta + 1 = \csc^2 \theta$$

$$\sec \theta = \frac{1}{\cos \theta} \qquad \cot \theta = \frac{\cos \theta}{\sin \theta} \qquad \sin(-\theta) = -\sin \theta$$

$$\qquad\qquad\qquad\qquad\qquad\qquad\qquad\qquad \cos(-\theta) = \cos \theta$$

$$\cot \theta = \frac{1}{\tan \theta} \qquad \sin^2 \theta + \cos^2 \theta = 1 \qquad \tan(-\theta) = -\tan \theta$$

These identities and the rules of algebra enable us to show that many more equations are identities. In doing so, we usually work with one side of the equation only, manipulating it until it is transformed into the other side. Here are some suggestions for verifying identities:

1. Memorize the fundamental identities. Whenever you see one side of one of these identities, the other side should come to mind immediately.
2. Start with the more complicated side of the equation and try to transform it into the other side.
3. Trigonometric fractions may be added, multiplied, and simplified just as the fractions of algebra. Trigonometric expressions may also be factored and similar terms combined.
4. As you work on one side of the equation, keep an eye on the other side. It is easier to hit a target you can see.
5. If one side of the equation contains only a single trigonometric function, eliminate all other functions, if possible, from the other side.

Johannes Müller (Regiomontanus) (1436–1476) Müller first developed trigonometry into a separate subject.

6. It is sometimes helpful to change all functions into sine and cosine functions before proceeding.

7. If the numerator or denominator of a fraction contains a factor of $1 + \sin x$ (or $1 - \sin x$), consider multiplying both the numerator and the denominator by the conjugate $1 - \sin x$ (or $1 + \sin x$). This operation creates a factor of $1 - \sin^2 x$, which may be replaced with $\cos^2 x$. The same idea applies to factors of $1 \pm \cos x$, $\sec x \pm 1$, and $\sec x \pm \tan x$.

Example 1 Verify that $\dfrac{\tan x}{\sin x} = \sec x$ is an identity.

Solution Work on the left side because it is the more complicated.

$\dfrac{\tan x}{\sin x}$	$\sec x$
$= \dfrac{\dfrac{\sin x}{\cos x}}{\dfrac{\sin x}{1}}$	Replace $\tan x$ with $\frac{\sin x}{\cos x}$.
$= \dfrac{\sin x}{\cos x} \cdot \dfrac{1}{\sin x}$	Invert and multiply.
$= \dfrac{1}{\cos x}$	Divide out the $\sin x$.
$= \sec x$	Replace $\frac{1}{\cos x}$ with $\sec x$.

Because the left side has been transformed into the right side, the identity is verified. ∎

Example 2 Verify the identity $\tan x + \cot x = \csc x \sec x$.

Solution Work with the left side of this equality and change $\tan x$ and $\cot x$ into expressions containing $\sin x$ and $\cos x$. Then proceed as follows:

$\tan x + \cot x$	$\csc x \sec x$
$= \dfrac{\sin x}{\cos x} + \dfrac{\cos x}{\sin x}$	
$= \dfrac{\sin x \sin x}{\sin x \cos x} + \dfrac{\cos x \cos x}{\sin x \cos x}$	Change each fraction to a fraction with a common denominator.
$= \dfrac{\sin^2 x + \cos^2 x}{\sin x \cos x}$	Add the fractions.
$= \dfrac{1}{\sin x \cos x}$	Replace $\sin^2 x + \cos^2 x$ with 1.
$= \dfrac{1}{\sin x} \cdot \dfrac{1}{\cos x}$	Rewrite as two fractions.
$= \csc x \sec x$	

Because the left side has been transformed into the right side, the identity is verified. ■

Example 3 Verify that $1 + \tan x = \sec x(\cos x + \sin x)$.

Solution Use the distributive property on the right side and remember that $\sec x \cos x = 1$:

$$
\begin{array}{l|ll}
1 + \tan x & \sec x(\cos x + \sin x) & \\
& = \sec x \cos x + \sec x \sin x & \text{Remove parentheses.} \\
& = 1 + \sec x \sin x & \text{Recall that } \sec x \cos x = 1. \\
& = 1 + \dfrac{\sin x}{\cos x} & \text{Recall that } \sec x = \frac{1}{\cos x}. \\
& = 1 + \tan x & \text{Recall that } \frac{\sin x}{\cos x} = \tan x.
\end{array}
$$

The identity is verified. ■

Example 4 Verify that $(1 + \tan^2 x)\cos^2 x = 1$.

Solution Change the terms of the left side into sines and cosines and remove parentheses:

$$
\begin{array}{l|ll}
(1 + \tan^2 x)\cos^2 x & 1 & \\
= \left(1 + \dfrac{\sin^2 x}{\cos^2 x}\right)\cos^2 x & & \text{Recall that } \tan^2 x = \frac{\sin^2 x}{\cos^2 x}. \\
= \cos^2 x + \sin^2 x & & \text{Remove parentheses.} \\
= 1 & & \text{Recall that } \cos^2 x + \sin^2 x = 1.
\end{array}
$$

The identity is verified. ■

Example 5 Verify that $\cos^4 x - \sin^4 x = 1 - 2\sin^2 x$.

Solution The left side is the difference of two squares and can be factored. Because the expression on the right side involves only the sine function, eliminate the cosine function from the left side:

$$
\begin{array}{l|ll}
\cos^4 x - \sin^4 x & 1 - 2\sin^2 x & \\
= (\cos^2 x + \sin^2 x)(\cos^2 x - \sin^2 x) & & \\
= 1 \cdot (\cos^2 x - \sin^2 x) & & \text{Remember that} \\
= \cos^2 x - \sin^2 x & & \cos^2 x + \sin^2 x = 1. \\
= 1 - \sin^2 x - \sin^2 x & & \text{Remember that} \\
= 1 - 2\sin^2 x & & \cos^2 x = 1 - \sin^2 x.
\end{array}
$$

The identity is verified. ■

Example 6   Verify that $\dfrac{1}{\sec x - \tan x} - \dfrac{1}{\sec x + \tan x} = 2 \tan x$.

Solution   Find the common denominator for the fractions on the left side and add them:

$$\dfrac{1}{\sec x - \tan x} - \dfrac{1}{\sec x + \tan x} \qquad\qquad \Big| \quad 2 \tan x$$

$$= \dfrac{(\sec x + \tan x)}{(\sec x - \tan x)(\sec x + \tan x)}$$

$$- \dfrac{(\sec x - \tan x)}{(\sec x + \tan x)(\sec x - \tan x)}$$

$$= \dfrac{\sec x + \tan x - \sec x + \tan x}{\sec^2 x - \tan^2 x}$$

$$= \dfrac{2 \tan x}{1}$$

$$= 2 \tan x$$

The identity is verified.   ∎

Sometimes mathematicians work on both sides of an identity independently until each side is transformed into a common third expression. When following this strategy, it is important that each step in the process be reversible so that each side of the identity can be derived from the common third expression. It is also important to note that each side must be worked on independently. It is incorrect to multiply or divide both sides of an equation whose truth you are trying to establish by an expression containing a variable, because the resulting equation might not be equivalent to the given equation.

Example 7   Verify that $\dfrac{1 - \cos x}{1 + \cos x} = (\csc x - \cot x)^2$.

Solution   In this example, change the left side and the right side of the equation to a common third expression. The left side can be changed as follows:

$$\dfrac{1 - \cos x}{1 + \cos x} = \dfrac{(1 - \cos x)(1 - \cos x)}{(1 + \cos x)(1 - \cos x)} \qquad \text{Multiply both numerator and denominator by } 1 - \cos x.$$

$$= \dfrac{(1 - \cos x)^2}{1 - \cos^2 x}$$

$$= \dfrac{(1 - \cos x)^2}{\sin^2 x}$$

$$= \left(\dfrac{1 - \cos x}{\sin x} \right)^2$$

The right side can be changed as follows:

$$(\csc x - \cot x)^2 = \left(\frac{1}{\sin x} - \frac{\cos x}{\sin x}\right)^2$$

$$= \left(\frac{1 - \cos x}{\sin x}\right)^2$$

Each side has been transformed independently into the expression

$$\left(\frac{1 - \cos x}{\sin x}\right)^2$$

Because each step is reversible, it follows that the given equation is an identity. ∎

Example 8 For what values of x is $\sqrt{1 - \cos^2 x} = \sin x$?

Solution Take the square root of both sides of the identity $1 - \cos^2 x = \sin^2 x$:

$$\sqrt{1 - \cos^2 x} = \sqrt{\sin^2 x}$$

$$= |\sin x|$$

The equation $\sqrt{1 - \cos^2 x} = |\sin x|$ is an identity, but because $\sin x$ is sometimes negative, $\sqrt{1 - \cos^2 x} = \sin x$ is *not* an identity. If you think of x as a real number from 0 to 2π, $\sin x$ is nonnegative for $0 \le x \le \pi$. Hence, $\sqrt{1 - \cos^2 x} = \sin x$ only if x is a number from 0 to π, from 2π to 3π, from 4π to 5π, and so on. ∎

When verifying identities, remember to write the variable associated with a trigonometric function. The notation "cos x" represents a numerical value, but the notation "cos" is meaningless.

Exercise 7.1

In Exercises 1–10, indicate whether the statement is an identity. If so, verify it. If not, explain why. Remember that if an equation is false for one value of its variable, it cannot be an identity.

1. $\sin x + \sin x = 2 \sin x$

2. $\sec^2 x - \tan^2 x = 1$

3. $\sin x + \cos x = 1$

4. $\dfrac{1}{\tan x} = \dfrac{\cos x}{\sin x}$

5. $\tan \alpha \cos \alpha \csc \alpha = 1$

6. $\cot \beta \sec \beta \cos \beta = 1$

7. $\sqrt{1 - \sin^2 x} = \cos x$

8. $\sqrt{1 - \cos^2 x} = |\sin x|$

9. $(\sin x + 1)^2 = \sin^2 x + 2 \sin x + 1$

10. $\sin^2 x - 1 = \cos^2 x$

In Exercises 11–60, verify each identity.

11. $\dfrac{1 - \cos^2 x}{\sin x} = \sin x$

12. $\dfrac{\cot x}{\csc x} = \cos x$

13. $\dfrac{1 + \tan^2 x}{\sec^2 x} = 1$

14. $\tan x \csc x = \sec x$

15. $(\sin x + \cos x)^2 = 1 + 2 \sin x \cos x$

16. $(\sin x + 1)^2 = \sin^2 x + 2 \sin x + 1$

17. $(1 + \cos x)(1 - \cos x) = \sin^2 x$

18. $(\sin x + 1)(\sin x - 1) = -\cos^2 x$

19. $(\sec x + 1)(\sec x - 1) = \tan^2 x$

20. $\sin^4 x - \cos^4 x = \sin^2 x - \cos^2 x$

21. $\sin x(\sin x + \cot x \cos x) = 1$

22. $(\csc x - 1)(\csc x + 1) = \cot^2 x$

23. $\dfrac{1}{1 - \cos^2 x} = 1 + \cot^2 x$

24. $\cos x(\cos x + \sin x \tan x) = 1$

25. $\dfrac{\sin x}{\cos^2 x - 1} = -\csc x$

26. $\dfrac{1}{1 - \sin^2 x} = 1 + \tan^2 x$

27. $\dfrac{1 - \cos^2 x}{1 - \sin^2 x} = \tan^2 x$

28. $\sin^2 x + \cos^2 x = \cos^2 x \sec^2 x$

29. $\dfrac{1 - \sin^2 x}{1 + \tan^2 x} = \cos^4 x$

30. $\tan x \sin x = \dfrac{\csc x}{\cot x + \cot^3 x}$

31. $\dfrac{\cos x(\cos x + 1)}{\sin x} = \cos x \cot x + \cot x$

32. $\cos^2 x \csc x - \csc x = -\sin x$

33. $\sin^2 x \sec x - \sec x = -\cos x$

34. $(\sin x - \cos x)(1 + \sin x \cos x) = \sin^3 x - \cos^3 x$

35. $\sin^2 x - \tan^2 x = -\sin^2 x \tan^2 x$

36. $\dfrac{1 + \cot x}{\csc x} = \sin x + \cos x$

37. $\dfrac{1 - \csc x}{\cot x} = \tan x - \sec x$

38. $\dfrac{\cos^2 x - \tan^2 x}{\sin^2 x} = \cot^2 x - \sec^2 x$

39. $\dfrac{1}{\sec x - \tan x} = \tan x + \sec x$

40. $\csc^2 x + \sec^2 x = \csc^2 x \sec^2 x$

41. $\dfrac{\cos x}{\cot x} + \dfrac{\sin x}{\tan x} = \sin x + \cos x$

42. $\dfrac{\cos x}{1 + \sin x} = \sec x - \tan x$

43. $\dfrac{\cos x}{1 - \sin x} = \dfrac{1 + \sin x}{\cos x}$

44. $\cos^2 x + \sin x \cos x = \dfrac{\cos x(\cot x + 1)}{\csc x}$

45. $(\sin x + \cos x)^2 + (\sin x - \cos x)^2 = 2$

46. $\dfrac{\sec x + 1}{\tan x} = \dfrac{\tan x}{\sec x - 1}$

47. $\dfrac{\cos x - \cot x}{\cos x \cot x} = \dfrac{\sin x - 1}{\cos x}$

48. $\cos^4 x - \sin^4 x = 2 \cos^2 x - 1$

49. $\dfrac{1}{1 + \sin x} + \dfrac{1}{1 - \sin x} = 2 \sec^2 x$

50. $\sqrt{\dfrac{1 - \sin x}{1 + \sin x}} = \sec x - \tan x \quad \left(0 < x < \dfrac{\pi}{2}\right)$

51. $\sqrt{\dfrac{1 - \sin x}{1 + \sin x}} = \dfrac{1 - \sin x}{\cos x} \quad$ with $0 < x < \dfrac{\pi}{2}$

52. $-\sqrt{\dfrac{\sec x - 1}{\sec x + 1}} = \dfrac{1 - \sec x}{\tan x} \quad$ with $0 < x < \dfrac{\pi}{2}$

53. $\sqrt{\dfrac{\csc x - \cot x}{\csc x + \cot x}} = \dfrac{\sin x}{1 + \cos x} \quad$ with $0 < x < \dfrac{\pi}{2}$

54. $\dfrac{\cos x}{1 - \sin x} - \dfrac{1}{\cos x} = \tan x$

55. $\dfrac{1}{\sec x(1 + \sin x)} = \sec x(1 - \sin x)$

56. $\dfrac{\cot x - \cos x}{\cot x \cos x} = \dfrac{1 - \sin x}{\cos x}$

57. $\dfrac{1}{\tan x(\csc x + 1)} = \tan x(\csc x - 1)$

58. $\dfrac{(\cos x + 1)^2}{\sin^2 x} = 2 \csc^2 x + 2 \csc x \cot x - 1$

59. $\dfrac{\csc x}{\sec x - \csc x} = \dfrac{\cot^2 x + \csc x \sec x + 1}{\tan^2 x - \cot^2 x}$

60. $(\sin x + \cos x)^2 - (\sin x - \cos x)^2 = 4 \sin x \cos x$

In Exercises 61–76, verify each identity.

61. $\dfrac{\sin^2 x \tan x + \sin^2 x}{\sec x - 2 \sin x \tan x} = \dfrac{\sin^2 x}{\cos x - \sin x}$

62. $\dfrac{\sin x + \cos x + 1}{\sin x + \cos x - 1} = \csc x \sec x + \csc x + \sec x + 1$

63. $\dfrac{\sec x + \tan x}{\csc x + \cot x} = \dfrac{\cot x - \csc x}{\tan x - \sec x}$

64. $\dfrac{\cos x + \sin x}{\cos x - \sin x} = \dfrac{\csc x + 2 \cos x}{\csc x - 2 \sin x}$

65. $\dfrac{3 \cos^2 x + 11 \sin x - 11}{\cos^2 x} = \dfrac{3 \sin x - 8}{1 + \sin x}$

66. $\dfrac{3 \sin^2 x + 5 \cos x - 5}{\sin^2 x} = \dfrac{3 \cos x - 2}{1 + \cos x}$

67. $\dfrac{\cos x + \sin x + 1}{\cos x - \sin x - 1} = \dfrac{1 + \cos x}{-\sin x}$

68. $\dfrac{1 + \sin x + \cos x}{1 + \sin x - \cos x} = \dfrac{\sin x}{1 - \cos x}$

69. $\dfrac{1 + \sin x}{\cos x} = \dfrac{\sin x - \cos x + 1}{\sin x + \cos x - 1}$

70. $\dfrac{\tan x + \sec x + 1}{\tan x + \sec x - 1} = \dfrac{1 + \cos x}{\sin x}$

71. $\dfrac{\csc x + 1 + \cot x}{\csc x + 1 - \cot x} = \dfrac{1 + \cos x}{\sin x}$

72. $\dfrac{3 \cot x \csc x - 2 \csc^2 x}{\csc^2 x + \cot x \csc x} - 3 = -5(\csc^2 x - \csc x \cot x)$

73. $\dfrac{\sin^3 x \cos x + \cos x - \sin^2 x \cos x - \cos x \sin x}{\cos^4 x} = \sec x - \tan x$

74. $\dfrac{\sin x + \sin x \cos x - \sin x \cos^2 x - \sin x \cos^3 x}{\cos^4 x - 2 \cos^2 x + 1} = \csc x + \cot x$

75. $\dfrac{1}{\sec x + \csc x - \sec x \csc x} = \dfrac{\sin x + \cos x + 1}{2}$

76. $\dfrac{2(\sin x + 1)}{1 + \cot x + \csc x} = \sin x + 1 - \cos x$

7.2   IDENTITIES INVOLVING SUMS AND DIFFERENCES OF TWO ANGLES

We shall often encounter expressions that contain a trigonometric function of either a sum or a difference of two angles. It is tempting to believe that an expression such as $\cos(A + B)$, for example, is equal to $\cos A + \cos B$. However, this is *not* true, as the following work shows:

$$\cos(30° + 60°) = \cos 90° = 0$$

$$\cos 30° + \cos 60° = \frac{\sqrt{3}}{2} + \frac{1}{2} = \frac{\sqrt{3} + 1}{2}$$

Hence,

$$\cos(30° + 60°) \neq \cos 30° + \cos 60°$$

In general, a trigonometric function of the sum (or difference) of two angles is not equal to the sum (or difference) of the trigonometric functions of each angle. It is possible, however, to develop formulas for finding a trigonometric function of the sum (or difference) of two angles. We shall use the distance formula to derive a formula to evaluate the expression $\cos(A + B)$.

We draw angles A and $-B$ in standard position on the unit circle, as in Figure 7-1. We locate point R on the circle so that angle POR is equal to angle B. We form triangles POQ and ROS. Because these two isosceles triangles have equal vertex angles, they are congruent. Hence, $RS = PQ$. The coordinates of points P, Q, R, and S are:

P: $(\cos A, \sin A)$

Q: $(\cos[-B], \sin[-B]) = (\cos B, -\sin B)$

R: $(\cos[A + B], \sin[A + B])$

S: $(1, 0)$

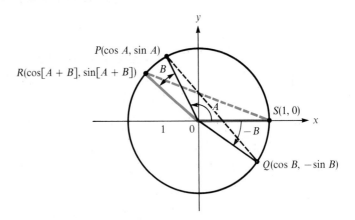

Figure 7-1

Because $RS = PQ$, their squares are also equal. Hence, $RS^2 = PQ^2$. By the distance formula,

$$[\cos(A + B) - 1]^2 + [\sin(A + B) - 0]^2$$
$$= (\cos A - \cos B)^2 + (\sin A + \sin B)^2$$

or

$$\cos^2(A + B) - 2\cos(A + B) + 1 + \sin^2(A + B)$$
$$= \cos^2 A - 2\cos A \cos B + \cos^2 B + \sin^2 A + 2\sin A \sin B + \sin^2 B$$

Because $\cos^2 \theta + \sin^2 \theta = 1$ for any angle θ, the previous equation can be written as

$$1 - 2\cos(A + B) + 1 = 1 - 2\cos A \cos B + 1 + 2\sin A \sin B$$
$$2 - 2\cos(A + B) = 2 - 2\cos A \cos B + 2\sin A \sin B$$
$$-2\cos(A + B) = -2\cos A \cos B + 2\sin A \sin B$$

Dividing both sides of the equation above by -2 gives the identity*

$$\cos(A + B) = \cos A \cos B - \sin A \sin B$$

Example 1   Find $\cos 75°$ without using tables or a calculator.

Solution   Note that $\cos 75° = \cos(45° + 30°)$ and use the formula for the cosine of the sum of two angles:

$$\cos(A + B) = \cos A \cos B - \sin A \sin B$$
$$\cos(45° + 30°) = \cos 45° \cos 30° - \sin 45° \sin 30°$$
$$= \frac{\sqrt{2}}{2} \cdot \frac{\sqrt{3}}{2} - \frac{\sqrt{2}}{2} \cdot \frac{1}{2}$$
$$\cos 75° = \frac{\sqrt{6} - \sqrt{2}}{4}$$

■

The identity for the cosine of the difference of two angles follows from the fact that $A - B = A + (-B)$.

$$\cos(A - B) = \cos[A + (-B)]$$
$$= \cos A \cos(-B) - \sin A \sin(-B)$$

Because $\cos(-B) = \cos B$ and $\sin(-B) = -\sin B$, this result can be rewritten as

$$\cos(A - B) = \cos A \cos B + \sin A \sin B$$

Example 2   Find the value of $\cos \dfrac{\pi}{12}$ without using tables or a calculator.

---

* All formulas developed in this chapter are also true if A and B represent real numbers.

Solution   Because $\frac{\pi}{12} = \frac{\pi}{4} - \frac{\pi}{6}$, it follows that

$$\cos\frac{\pi}{12} = \cos\left(\frac{\pi}{4} - \frac{\pi}{6}\right)$$

$$= \cos\frac{\pi}{4}\cos\frac{\pi}{6} + \sin\frac{\pi}{4}\sin\frac{\pi}{6}$$

$$= \frac{\sqrt{2}}{2}\cdot\frac{\sqrt{3}}{2} + \frac{\sqrt{2}}{2}\cdot\frac{1}{2}$$

$$= \frac{\sqrt{6} + \sqrt{2}}{4}$$

■

We have seen that any trigonometric function of an acute angle θ is equal to the cofunction of the complement of θ. This property can be extended to all angles. For example,

$$\cos(90° - \theta) = \cos 90° \cos\theta + \sin 90° \sin\theta$$

$$= 0\cdot\cos\theta + 1\cdot\sin\theta$$

$$= \sin\theta$$

Thus, for any angle θ,

$$\sin\theta = \cos(90° - \theta)$$

Now note that $\theta = 90° - (90° - \theta)$ and find a value for $\cos\theta$:

$$\cos\theta = \cos[90° - (90° - \theta)]$$

$$= \cos 90° \cos(90° - \theta) + \sin 90° \sin(90° - \theta)$$

$$= \sin(90° - \theta)$$

Thus, for any angle θ,

$$\cos\theta = \sin(90° - \theta)$$

In the exercises, you will be asked to show that

$$\tan\theta = \cot(90° - \theta)$$

To develop a formula for $\sin(A + B)$, we substitute $A + B$ for θ in the equation $\sin\theta = \cos(90° - \theta)$ and proceed as follows:

$$\sin(A + B) = \cos[90° - (A + B)]$$

$$= \cos[(90° - A) - B]$$

$$= \cos(90° - A)\cos B + \sin(90° - A)\sin B$$

Because $\cos(90° - A) = \sin A$ and $\sin(90° - A) = \cos A$, it follows that

$$\sin(A + B) = \sin A \cos B + \cos A \sin B$$

To find a formula for $\sin(A - B)$, we proceed as follows:

$$\sin(A - B) = \sin[A + (-B)]$$
$$= \sin A \cos(-B) + \cos A \sin(-B)$$
$$= \sin A \cos B - \cos A \sin B$$

This gives the identity

$$\sin(A - B) = \sin A \cos B - \cos A \sin B$$

Example 3   Derive a formula for $\sin(A + B + C)$.

Solution
$$\sin(A + B + C) = \sin[(A + B) + C]$$
$$= \sin(A + B) \cos C + \cos(A + B) \sin C$$
$$= (\sin A \cos B + \cos A \sin B) \cos C$$
$$+ (\cos A \cos B - \sin A \sin B) \sin C$$
$$= \sin A \cos B \cos C + \cos A \sin B \cos C$$
$$+ \cos A \cos B \sin C - \sin A \sin B \sin C \qquad \blacksquare$$

To find identities for the tangent of the sum and difference of two angles, we make use of the identity

$$\tan x = \frac{\sin x}{\cos x}$$

and substitute $A + B$ for x:

$$\tan(A + B) = \frac{\sin(A + B)}{\cos(A + B)}$$
$$= \frac{\sin A \cos B + \cos A \sin B}{\cos A \cos B - \sin A \sin B}$$

To simplify this result, we divide both the numerator and denominator of the fraction by $\cos A \cos B$:

$$\tan(A + B) = \frac{\dfrac{\sin A \cos B}{\cos A \cos B} + \dfrac{\cos A \sin B}{\cos A \cos B}}{\dfrac{\cos A \cos B}{\cos A \cos B} - \dfrac{\sin A \sin B}{\cos A \cos B}}$$
$$= \frac{\tan A + \tan B}{1 - \tan A \tan B}$$

This gives the identity

$$\tan(A + B) = \frac{\tan A + \tan B}{1 - \tan A \tan B}$$

If we substitute $-B$ for B in the previous equation and simplify, we obtain an identity for the tangent of the difference of two angles.

$$\tan(A - B) = \frac{\tan A - \tan B}{1 + \tan A \tan B}$$

Example 4   Given that $\sin \alpha = \dfrac{12}{13}$, α in QI, and $\cos \beta = -\dfrac{4}{5}$, β in QII, find $\sin(\beta - \alpha)$.

Solution   Because $\sin \alpha = \frac{12}{13}$ and α is in QI, you can draw Figure 7-2**a** and determine that $\cos \alpha = \frac{5}{13}$. Because $\cos \beta = -\frac{4}{5}$ and β is in QII, you can draw Figure 7-2**b** and determine that $\sin \beta = \frac{3}{5}$. You can substitute $\frac{3}{5}$ for $\sin \beta$, $\frac{5}{13}$ for $\cos \alpha$, $-\frac{4}{5}$ for $\cos \beta$, and $\frac{12}{13}$ for $\sin \alpha$ in the following identity and simplify:

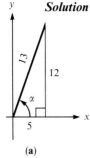

(a)

$$\sin(\beta - \alpha) = \sin \beta \cos \alpha - \cos \beta \sin \alpha$$

$$= \frac{3}{5}\left(\frac{5}{13}\right) - \left(-\frac{4}{5}\right)\left(\frac{12}{13}\right)$$

$$= \frac{15}{65} + \frac{48}{65}$$

$$= \frac{63}{65}$$

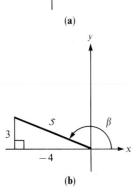

(b)

Figure 7-2

Thus,

$$\sin(\beta - \alpha) = \frac{63}{65}$$

Example 5   Verify the identity $\tan(\theta + 45°) = \dfrac{1 + \tan \theta}{1 - \tan \theta}$.

Solution   Use the formula for the tangent of the sum of two angles:

$$\tan(\theta + 45°)\qquad\qquad \left|\ \frac{1 + \tan \theta}{1 - \tan \theta}\right.$$

$$= \frac{\tan \theta + \tan 45°}{1 - \tan \theta \tan 45°}$$

$$= \frac{\tan \theta + 1}{1 - \tan \theta}$$

$$= \frac{1 + \tan \theta}{1 - \tan \theta}$$

The identity is verified.

Exercise 7.2

1. Find $\sin 195°$ from the trigonometric functions of $45°$ and $150°$.

2. Find $\cos 195°$ from the trigonometric functions of $45°$ and $150°$.

3. Find $\tan 195°$ from the trigonometric functions of $225°$ and $30°$.

4. Find $\tan 165°$ from the trigonometric functions of $210°$ and $45°$.

5. Find $\cos \dfrac{11\pi}{12}$ from the trigonometric functions of $\dfrac{\pi}{6}$ and $\dfrac{3\pi}{4}$.

6. Find $\sin \dfrac{11\pi}{12}$ from the trigonometric functions of $\dfrac{\pi}{6}$ and $\dfrac{3\pi}{4}$.

7. Find $\cos \dfrac{19\pi}{12}$ from the trigonometric functions of $\dfrac{11\pi}{6}$ and $\dfrac{\pi}{4}$.

8. Find $\tan \dfrac{19\pi}{12}$ from the trigonometric functions of $\dfrac{11\pi}{6}$ and $\dfrac{\pi}{4}$.

In Exercises 9–16, choose two appropriate angles and use an identity of the section to evaluate each expression. **Do not use a calculator.**

9. $\sin 255°$

10. $\cos 285°$

11. $\tan 105°$

12. $\cot 255°$

13. $\cos \dfrac{\pi}{12}$

14. $\sin \dfrac{7\pi}{12}$

15. $\sin \dfrac{5\pi}{12}$

16. $\cos \dfrac{13\pi}{12}$

17. Show that $\sin(60° + \theta) = \dfrac{\sqrt{3}}{2} \cos \theta + \dfrac{1}{2} \sin \theta$.

18. Show that $\cos\left(\dfrac{\pi}{2} + x\right) = -\sin x$.

19. Show that $\tan(\pi + x) = \tan x$.

20. Show that $\sin\left(\dfrac{3\pi}{2} - x\right) = -\cos x$.

21. Show that $\cos(\pi - x) = -\cos x$.

22. Show that $\tan\left(\dfrac{\pi}{4} - x\right) = \dfrac{1 - \tan x}{1 + \tan x}$.

In Exercises 23–30, express each quantity as a single function of one angle.

23. $\sin 10° \cos 30° + \cos 10° \sin 30°$

24. $\cos 20° \cos 30° - \sin 20° \sin 30°$

25. $\dfrac{\tan 75° + \tan 40°}{1 - \tan 75° \tan 40°}$

26. $\sin 100° \cos 80° - \cos 100° \sin 80°$

27. $\cos 120° \cos 40° + \sin 120° \sin 40°$

28. $\dfrac{\tan 37° - \tan 125°}{1 + \tan 37° \tan 125°}$

29. $\sin x \cos 2x + \sin 2x \cos x$

30. $\cos 2y \cos 3y - \sin 2y \sin 3y$

31. Given that $\sin \alpha = \frac{3}{5}$, α in QII, and $\cos \beta = -\frac{12}{13}$, β in QII, find $\sin(\alpha + \beta)$ and $\cos(\alpha - \beta)$.

32. Given that $\sin \alpha = \frac{7}{25}$, α in QI, and $\sin \beta = \frac{15}{17}$, β in QI, find $\sin(\alpha - \beta)$ and $\cos(\alpha + \beta)$.

33. Given that $\tan \alpha = -\frac{5}{12}$, α in QII, and $\tan \beta = \frac{15}{8}$, β in QI, find $\tan(\alpha + \beta)$ and $\tan(\alpha - \beta)$.

34. Given that $\sin \alpha = \frac{12}{13}$, α in QI, and $\sin(\alpha + \beta) = \frac{24}{25}$, $\alpha + \beta$ in QII, find $\sin \beta$ and $\cos \beta$.

35. Given that $\cos \beta = -\frac{15}{17}$, β in QII, and $\sin(\alpha - \beta) = -\frac{24}{25}$, $\alpha - \beta$ in QIV, find $\sin \alpha$ and $\cos \alpha$.

36. Given that $\sin(\alpha + \beta) = \frac{3}{5}$, $\alpha + \beta$ in QI, and $\cos(\alpha - \beta) = \frac{12}{13}$, $\alpha - \beta$ in QIV, find $\sin 2\alpha$ and $\cos 2\beta$.

In Exercises 37–51, verify each identity.

37. $\sin(30° + \theta) - \cos(60° + \theta) = \sqrt{3}\sin\theta$

38. $\sin(60° + \theta) - \cos(30° + \theta) = \sin\theta$

39. $\sin(30° + \theta) + \cos(60° + \theta) = \cos\theta$

40. $\sin(A + B) - \sin(A - B) = 2\cos A\sin B$

41. $\cos(A + B) + \cos(A - B) = 2\cos A\cos B$

42. $\sin(A + B)\sin(A - B) = \sin^2 A - \sin^2 B$

43. $\cos(A + B)\cos(A - B) = \cos^2 A + \cos^2 B - 1$

44. $\cot(A + B) = \dfrac{\cot A\cot B - 1}{\cot A + \cot B}$

45. $\cot(A - B) = \dfrac{\cot A\cot B + 1}{\cot B - \cot A}$

46. $\cos(A + B) - \cos(A - B) = -2\sin A\cos B$

47. $\sin(A + B) + \sin(A - B) = 2\sin A\cos B$

48. $-\tan A = \cot\left(A + \dfrac{\pi}{2}\right)$

49. $\dfrac{\tan A + \tan B}{1 + \tan A\tan B} = \dfrac{\sin(A + B)}{\cos(A - B)}$

50. $\dfrac{\cos(A + B)}{\sin(A - B)} = \dfrac{\cot A - \tan B}{1 - \cot A\tan B}$

51. $\dfrac{\cos(A - B)}{\sin(A + B)} = \dfrac{1 + \tan A\tan B}{\tan A + \tan B}$

52. Derive a formula for $\cos(A + B + C)$.

53. Derive a formula for $\sin(A - B - C)$.

54. Derive a formula for $\cos(A - B - C)$.

55. Derive a formula for $\tan(A + B + C)$.

56. Show that $\tan\theta = \cot(90° - \theta)$.

7.3 THE DOUBLE-ANGLE IDENTITIES

If angles A and B are equal, the formulas for the sine of the sum of two angles and the cosine of the sum of two angles may be transformed into formulas called the **double-angle identities**.

Let $A = B$ in the identity $\sin(A + B) = \sin A\cos B + \cos A\sin B$ and simplify:

$$\sin(A + A) = \sin A\cos A + \cos A\sin A$$

$$\sin 2A = 2\sin A\cos A$$

Similarly, let $A = B$ in the identity $\cos(A + B) = \cos A\cos B - \sin A\sin B$ and simplify:

$$\cos(A + A) = \cos A\cos A - \sin A\sin A$$

$$\cos 2A = \cos^2 A - \sin^2 A$$

There are two variations of this identity for $\cos 2A$. To obtain one of them, we substitute $1 - \cos^2 A$ for $\sin^2 A$ and simplify:

$$\cos 2A = \cos^2 A - \sin^2 A$$

$$= \cos^2 A - (1 - \cos^2 A)$$

Thus,

$$\cos 2A = 2\cos^2 A - 1$$

To obtain the other, we substitute $1 - \sin^2 A$ for $\cos^2 A$ and simplify:

$$\cos 2A = \cos^2 A - \sin^2 A$$
$$= 1 - \sin^2 A - \sin^2 A$$

Thus,

$$\cos 2A = 1 - 2\sin^2 A$$

To find the double-angle identity for the tangent function, we let $A = B$ in the identity

$$\tan(A + B) = \frac{\tan A + \tan B}{1 - \tan A \tan B}$$

and simplify:

$$\tan(A + A) = \frac{\tan A + \tan A}{1 - \tan A \tan A}$$

$$\tan 2A = \frac{2\tan A}{1 - \tan^2 A}$$

Example 1   Simplify the expression $\sin 5A \cos 5A$.

Solution   Because the given expression is one-half of $2\sin 5A \cos 5A$, it is one-half of $\sin 2(5A)$:

$$\sin 5A \cos 5A = \frac{1}{2}(2\sin 5A \cos 5A)$$

$$= \frac{1}{2}\sin 2(5A)$$

$$= \frac{1}{2}\sin 10A \qquad \blacksquare$$

Example 2   Use a double-angle identity to evaluate $\tan 2(60°)$.

Solution   Use the identity $\tan 2A = \dfrac{2\tan A}{1 - \tan^2 A}$ and substitute $60°$ for A:

$$\tan 2A = \frac{2\tan A}{1 - \tan^2 A}$$

$$\tan 2(60°) = \frac{2\tan(60°)}{1 - \tan^2(60°)}$$

$$= \frac{2\sqrt{3}}{1 - 3}$$

$$= \frac{2\sqrt{3}}{-2}$$

$$= -\sqrt{3} \qquad \blacksquare$$

Example 3   Find the **a.** sine, **b.** cosine, and **c.** tangent of 2θ if $\cos \theta = \dfrac{12}{13}$ and θ is in QIV.

Solution   Use the information that $\cos \theta = \frac{12}{13}$ and that θ is in QIV to sketch the angle in standard position, as in Figure 7-3. From the figure, it follows that

$$\sin \theta = \frac{-5}{13} = -\frac{5}{13}$$

$$\tan \theta = \frac{-5}{12} = -\frac{5}{12}$$

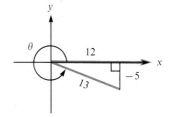

Figure 7-3

Use the double-angle identities to find the functions of 2θ.

a. $\sin 2\theta = 2 \sin \theta \cos \theta$

$$= 2\left(-\frac{5}{13}\right)\left(\frac{12}{13}\right)$$

$$= -\frac{120}{169}$$

b. $\cos 2\theta = \cos^2 \theta - \sin^2 \theta$

$$= \left(\frac{12}{13}\right)^2 - \left(-\frac{5}{13}\right)^2$$

$$= \frac{119}{169}$$

c. $\tan 2\theta = \dfrac{2\left(-\dfrac{5}{12}\right)}{1 - \left(-\dfrac{5}{12}\right)^2} = \dfrac{-\dfrac{10}{12}}{1 - \dfrac{25}{144}} = \dfrac{-\dfrac{5}{6}}{\dfrac{119}{144}}$

$$= -\frac{5}{\underset{1}{6}} \cdot \frac{\overset{24}{144}}{119} = -\frac{120}{119}$$

Example 4   Verify the identity $\cos 2A = \cos^4 A - \sin^4 A$.

Solution   Because the right side factors, work on that side:

$$
\begin{array}{c|l}
\cos 2A & \cos^4 A - \sin^4 A \\
 & = (\cos^2 A + \sin^2 A)(\cos^2 A - \sin^2 A) \\
 & = 1 \cdot (\cos^2 A - \sin^2 A) \\
 & = \cos 2A
\end{array}
$$

Example 5   Verify the identity $\cos A \sin 2A = 2 \sin A - 2 \sin^3 A$.

Solution   Although the right side looks more complicated, work with the left side because it involves a double angle.

$$
\begin{array}{l|l}
\cos A \sin 2A & 2 \sin A - 2 \sin^3 A \\
= \cos A(2 \sin A \cos A) & \\
= 2 \cos^2 A \sin A & \\
= 2(1 - \sin^2 A) \sin A & \\
= 2 \sin A - 2 \sin^3 A &
\end{array}
$$

∎

Example 6   Verify the identity $\cos 2A \sec^2 A = 2 - \sec^2 A$.

Solution   Work on the left side to eliminate the double angle:

$$
\begin{array}{l|l}
\cos 2A \sec^2 A & 2 - \sec^2 A \\
= (2 \cos^2 A - 1) \sec^2 A & \\
= 2 \cos^2 A \sec^2 A - \sec^2 A & \\
= 2 - \sec^2 A &
\end{array}
$$

∎

Example 7   Verify that $\sin(A + B) + \sin(A - B) = 2 \sin A \cos B$.

Solution

$$
\begin{array}{l|l}
\sin(A + B) + \sin(A - B) & 2 \sin A \cos B \\
= \sin A \cos B + \cos A \sin B & \\
\quad + \sin A \cos B - \cos A \sin B & \\
= \sin A \cos B + \sin A \cos B & \\
= 2 \sin A \cos B &
\end{array}
$$

∎

Example 8   Verify the identity $2 \cot 2A = \cot A - \tan A$.

Solution   Work with the left side of the equation:

$$
\begin{array}{l|l}
2 \cot 2A & \cos A - \tan A \\
= 2\left(\dfrac{1}{\tan 2A}\right) & \\
= 2\left(\dfrac{1 - \tan^2 A}{2 \tan A}\right) & \\
= \dfrac{1 - \tan^2 A}{\tan A} & \\
= \dfrac{1}{\tan A} - \dfrac{\tan^2 A}{\tan A} & \\
= \cot A - \tan A &
\end{array}
$$

∎

Exercise 7.3

In Exercises 1–26, write the given expression in terms of a single trigonometric function of twice the given angle.

1. $2 \sin \alpha \cos \alpha$

2. $2 \cos^2 \alpha - 1$

3. $2 \sin 3\theta \cos 3\theta$

4. $2 \cos^2 2A - 1$

5. $\cos^2 \beta - \sin^2 \beta$

6. $1 - 2 \sin^2 \beta$

7. $2 \cos^2 \dfrac{B}{2} - 1$

8. $\sin 5\theta \cos 5\theta$

9. $4 \sin \theta \cos \theta$

10. $\cos^2 \dfrac{\theta}{2} - \sin^2 \dfrac{\theta}{2}$

11. $4 \sin^2 2\theta \cos^2 2\theta$

12. $2 - 4 \sin^2 6B$

13. $\cos^2 \alpha - \dfrac{1}{2}$

14. $2 - 4 \sin^2 \dfrac{\alpha}{4}$

15. $\cos^2 9\theta - \sin^2 9\theta$

16. $4 \sin 4B \cos 4B$

17. $4 \sin^2 5\theta \cos^2 5\theta$

18. $\dfrac{2 \tan A}{1 - \tan^2 A}$

19. $\dfrac{2 \tan 4C}{1 - \tan^2 4C}$

20. $3 - 6 \cos^2 6x$

21. $\dfrac{\tan \dfrac{A}{2}}{\dfrac{1}{2} - \dfrac{1}{2} \tan^2 \dfrac{A}{2}}$

22. $\dfrac{2 \tan \alpha}{2 - \sec^2 \alpha}$

23. $\cos^4 4x - \sin^4 4x$

24. $\dfrac{1}{2} \sec \theta \csc \theta$

25. $1 - 2 \cos^2 5x$

26. $\dfrac{1 - \tan^2 3x}{2 \tan 3x}$

In Exercises 27–38, use a double-angle identity to find the value of the given expression. Then, check your work by evaluating the expression directly.

27. $\sin 2(30°)$

28. $\cos 2(30°)$

29. $\tan 2(45°)$

30. $\cot 2(135°)$

31. $\cos 2(240°)$

32. $\sin 2(150°)$

33. $\cot 2(225°)$

34. $\tan 2(315°)$

35. $\sin 2\left(\dfrac{\pi}{3}\right)$

36. $\cos 2\left(\dfrac{7\pi}{6}\right)$

37. $\cos 2\left(\dfrac{11\pi}{6}\right)$

38. $\sin 2\left(\dfrac{5\pi}{3}\right)$

In Exercises 39–50, find the exact value of the sine, cosine, and tangent of 2θ using the given information.

39. $\sin \theta = \dfrac{12}{13}; \theta$ in QI

40. $\cos \theta = \dfrac{5}{13}; \theta$ in QIV

41. $\tan \theta = \dfrac{12}{5}; \theta$ in QIII

42. $\sin \theta = \dfrac{3}{5}; \theta$ in QII

43. $\cos \theta = -\dfrac{4}{5}; \theta$ in QIII

44. $\cot \theta = -\dfrac{15}{8}; \theta$ in QIV

45. $\sin \theta = -\dfrac{24}{25}; \theta$ in QIV

46. $\cos \theta = -\dfrac{7}{25}; \theta$ in QIII

47. $\sin \theta = \dfrac{40}{41}; \cos \theta$ is positive

48. $\sin \theta = -\dfrac{40}{41}; \tan \theta$ is negative

49. $\cos \theta = \dfrac{40}{41}; \tan \theta$ is negative

50. $\cos \theta = 0$

In Exercises 51–77, verify the given identity.

51. $\dfrac{\tan 2x}{\sin 2x} = \sec 2x$

52. $\dfrac{\cot 2x}{\cos 2x} = \csc 2x$

53. $2 \csc 2A = \sec A \csc A$

54. $\sin 2A + 2 \sin A = \dfrac{-2 \sin^3 A}{\cos A - 1}$

55. $\cos 2x - \dfrac{\sin 2x}{\cos x} + 2 \sin^2 x = 1 - 2 \sin x$

56. $2 \cos^4 \theta + 2 \sin^2 \theta \cos^2 \theta - 1 = \cos 2\theta$

57. $\sin 2A - 2 \sin A = -\dfrac{2 \sin^3 A}{\cos A + 1}$

58. $2 \cos A - 1 = \dfrac{\cos 2A + \cos A}{\cos A + 1}$

59. $2 \sin^4 \theta + 2 \sin^2 \theta \cos^2 \theta - 1 = -\cos 2\theta$

60. $\sin 2\theta - \tan \theta = \cos 2\theta \tan \theta$

61. $2 \cos A + 1 = \dfrac{\cos 2A - \cos A}{\cos A - 1}$

62. $(\sin A + \cos A)^2 = 1 + \sin 2A$

63. $\sec 2\theta = \dfrac{\tan \theta + \cot \theta}{\cot \theta - \tan \theta}$

64. $\sin 4x = 4 \cos x \sin x - 8 \cos x \sin^3 x$

65. $4 \csc^2 2A = \sec^2 A + \csc^2 A$

66. $1 - \dfrac{1}{2} \sin 2x = \dfrac{\sin^3 x + \cos^3 x}{\sin x + \cos x}$

67. $\cos 4x = 8 \cos^4 x - 8 \cos^2 x + 1$

68. $-\tan 2x = \dfrac{2}{\tan x - \cot x}$

69. $2 \tan 2x = \dfrac{\sin x - \cos x}{\sin x + \cos x} - \dfrac{\sin x + \cos x}{\sin x - \cos x}$

70. $\sin^4 \theta = \dfrac{\cos 4\theta}{8} - \dfrac{\cos 2\theta}{2} + \dfrac{3}{8}$

71. $\dfrac{\sin x + \sin 2x}{\cos x - \cos 2x} = \dfrac{\sec x + 2}{\csc x - \cot x + \tan x}$

72. $\dfrac{\cos x - \sin x}{\cos 2x} = \dfrac{\cos x + \sin x}{1 + \sin 2x}$

73. $\dfrac{1 - \tan^2 x}{1 + \tan^2 x} = \cos 2x$

74. $\dfrac{\sec^2 x}{\sec 2x} = 2 - \sec^2 x$

75. $\dfrac{1 + \tan x}{1 - \tan x} = \dfrac{\cos 2x}{1 - \sin 2x}$

76. $\cos 2x = \dfrac{1}{\tan 2x \tan x + 1}$

77. $\tan A \sin 2A + \cos 2A = 1$

7.4 THE HALF-ANGLE IDENTITIES

In this section we develop formulas that are called the **half-angle identities**. We begin by solving the identity $\cos 2\theta = 2 \cos^2 \theta - 1$ for $\cos \theta$.

$$\cos 2\theta = 2 \cos^2 \theta - 1$$

$$2 \cos^2 \theta = 1 + \cos 2\theta$$

$$\cos^2 \theta = \frac{1 + \cos 2\theta}{2}$$

$$\cos \theta = \pm \sqrt{\frac{1 + \cos 2\theta}{2}}$$

With an appropriate choice of the sign preceding the radical, this identity is true for all values of θ. To derive the first half-angle identity, we let $\theta = \frac{A}{2}$:

$$\cos \frac{A}{2} = \pm \sqrt{\frac{1 + \cos A}{2}}$$

The sign preceding the radical in the previous equation is determined by the quadrant in which $\frac{A}{2}$ lies.

In a similar fashion, we solve $\cos 2\theta = 1 - 2 \sin^2 \theta$ for $\sin \theta$ and let $\theta = \frac{A}{2}$ to obtain the half-angle identity for the sine function:

$$\cos 2\theta = 1 - 2 \sin^2 \theta$$

$$2 \sin^2 \theta = 1 - \cos 2\theta$$

$$\sin^2 \theta = \frac{1 - \cos 2\theta}{2}$$

$$\sin \theta = \pm \sqrt{\frac{1 - \cos 2\theta}{2}}$$

$$\sin \frac{A}{2} = \pm \sqrt{\frac{1 - \cos A}{2}}$$

Again, the $+$ or $-$ sign is chosen by the quadrant in which $\frac{A}{2}$ lies.

Example 1   Use a half-angle identity to find $\sin 15°$.

Solution   Because $15°$ is $\frac{1}{2}(30°)$, it follows that

$$\sin 15° = \sin \frac{30°}{2} = +\sqrt{\frac{1 - \cos 30°}{2}}$$

$$= +\sqrt{\frac{1 - \frac{\sqrt{3}}{2}}{2}}$$

$$= +\sqrt{\frac{2 - \sqrt{3}}{4}}$$

$$= +\frac{\sqrt{2 - \sqrt{3}}}{2}$$

The $+$ sign is chosen because $15°$ is a first-quadrant angle and the sine of a first-quadrant angle is positive.   ∎

Example 2   Use a half-angle identity to find $\cos \frac{7\pi}{12}$.

Solution   Because $\frac{7\pi}{12}$ is $\frac{1}{2}(\frac{7\pi}{6})$, it follows that

$$\cos\frac{7\pi}{12} = \cos\left(\frac{1}{2}\cdot\frac{7\pi}{6}\right) = -\sqrt{\frac{1+\cos\dfrac{7\pi}{6}}{2}}$$

$$= -\sqrt{\frac{1-\dfrac{\sqrt{3}}{2}}{2}}$$

$$= -\frac{\sqrt{2-\sqrt{3}}}{2}$$

Here, the $-$ sign is chosen because $\frac{7\pi}{12}$ is a second-quadrant angle and the cosine of a second-quadrant angle is negative.   ■

To develop the identity involving $\tan\frac{A}{2}$, we substitute $\frac{A}{2}$ for θ in the relationship

$$\tan\theta = \frac{\sin\theta}{\cos\theta}$$

and proceed as follows:

$$\tan\frac{A}{2} = \frac{\sin\dfrac{A}{2}}{\cos\dfrac{A}{2}} = \frac{\pm\sqrt{\dfrac{1-\cos A}{2}}}{\pm\sqrt{\dfrac{1+\cos A}{2}}}$$

$$= \pm\sqrt{\frac{1-\cos A}{1+\cos A}}$$

If the fraction within the radical is multiplied by 1 written in the form

$$\frac{1-\cos A}{1-\cos A}$$

further simplification results:

$$\tan\frac{A}{2} = \pm\sqrt{\frac{(1-\cos A)(1-\cos A)}{(1+\cos A)(1-\cos A)}}$$

$$= \pm\sqrt{\frac{(1-\cos A)^2}{1-\cos^2 A}}$$

$$= \pm\sqrt{\frac{(1-\cos A)^2}{\sin^2 A}}$$

$$\tan\frac{A}{2} = \frac{1-\cos A}{\sin A}$$

If we use the previous identity, it is not necessary to choose the appropriate + or − sign. The selection is automatic because $1 - \cos A$ is never negative and $\sin A$ and $\tan \frac{A}{2}$ always agree in sign. You will be asked to prove this in an exercise.

Another form is also possible. Because

$$\tan \frac{A}{2} = \frac{1 - \cos A}{\sin A}$$

it follows that

$$\tan \frac{A}{2} = \frac{(1 - \cos A)(1 + \cos A)}{\sin A(1 + \cos A)}$$

$$= \frac{1 - \cos^2 A}{\sin A(1 + \cos A)}$$

$$= \frac{\sin^2 A}{\sin A(1 + \cos A)}$$

$$\tan \frac{A}{2} = \frac{\sin A}{1 + \cos A}$$

Example 3   Use the identity $\tan \dfrac{A}{2} = \dfrac{1 - \cos A}{\sin A}$ to find $\tan \dfrac{\pi}{8}$.

Solution   Because $\frac{\pi}{8}$ is $\frac{1}{2}(\frac{\pi}{4})$, it follows that

$$\tan \frac{\pi}{8} = \tan \frac{\dfrac{\pi}{4}}{2}$$

$$= \frac{1 - \cos \dfrac{\pi}{4}}{\sin \dfrac{\pi}{4}}$$

$$= \frac{1 - \dfrac{\sqrt{2}}{2}}{\dfrac{\sqrt{2}}{2}}$$

$$= \frac{2 - \sqrt{2}}{\sqrt{2}} \qquad \text{Multiply numerator and denominator by 2.}$$

$$= \sqrt{2} - 1 \qquad \text{Multiply numerator and denominator by } \sqrt{2}. \qquad ∎$$

Example 4   Use the identity $\tan \dfrac{A}{2} = \dfrac{\sin A}{1 + \cos A}$ to find $\tan 157.5°$.

Solution   Because $157.5°$ is $\frac{315°}{2}$, then

$$\tan 157.5° = \tan \frac{315°}{2} = \frac{\sin 315°}{1 + \cos 315°}$$

$$= \frac{-\dfrac{\sqrt{2}}{2}}{1 + \dfrac{\sqrt{2}}{2}}$$

$$= \frac{-\dfrac{\sqrt{2}}{2}}{\dfrac{2 + \sqrt{2}}{2}}$$

$$= \frac{-\sqrt{2}}{2 + \sqrt{2}}$$

$$= \frac{-\sqrt{2}}{(2 + \sqrt{2})} \cdot \frac{(2 - \sqrt{2})}{(2 - \sqrt{2})}$$

$$= \frac{-\sqrt{2}(2 - \sqrt{2})}{2}$$

$$= 1 - \sqrt{2}$$

This number is negative, as it must be for the tangent of a second-quadrant angle.   ■

Example 5   Find the **a.** sine, **b.** cosine, and **c.** tangent of $\dfrac{\theta}{2}$, if $\sin \theta = \dfrac{3}{5}$ and $\dfrac{\pi}{2} < \theta < \pi$.

Solution   Use the information that $\sin \theta = \frac{3}{5}$ and that θ is in QII to sketch the angle in standard position, as in Figure 7-4. The value of $\cos \theta$ can be read from the figure:

$$\cos \theta = -\frac{4}{5}$$

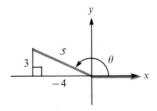

Figure 7-4

Use the half-angle identities to find the functions of $\frac{\theta}{2}$:

a. $\sin\dfrac{\theta}{2} = \sqrt{\dfrac{1 - \left(\dfrac{-4}{5}\right)}{2}} = \sqrt{\dfrac{9}{10}} = \dfrac{3\sqrt{10}}{10}$

b. $\cos\dfrac{\theta}{2} = \sqrt{\dfrac{1 + \left(\dfrac{-4}{5}\right)}{2}} = \sqrt{\dfrac{1}{10}} = \dfrac{\sqrt{10}}{10}$

c. $\tan\dfrac{\theta}{2} = \dfrac{\sin\dfrac{\theta}{2}}{\cos\dfrac{\theta}{2}} = \dfrac{\dfrac{3\sqrt{10}}{10}}{\dfrac{\sqrt{10}}{10}} = 3$

Choose the radicals to be positive, because if $\frac{\pi}{2} < \theta < \pi$, then $\frac{\theta}{2}$ is in QI.   ∎

Example 6   Write $\dfrac{\sin(-20A)}{-\cos 20A - 1}$ as a trigonometric function of $10A$.

Solution   Because $\sin(-20A) = -\sin 20A$, it follows that

$$\frac{\sin(-20A)}{-\cos 20A - 1} = \frac{-\sin 20A}{-(1 + \cos 20A)} = \frac{\sin 20A}{1 + \cos 20A}$$

One of the identities for $\tan\frac{\theta}{2}$ is

$$\tan\frac{\theta}{2} = \frac{\sin\theta}{1 + \cos\theta}$$

Hence,

$$\frac{\sin 20A}{1 + \cos 20A} = \tan\frac{20A}{2} = \tan 10A$$ ∎

Example 7   Verify the identity $\tan\dfrac{\theta}{2} = \csc\theta - \cot\theta$.

Solution   Work on the left side to get rid of the half-angle:

$$
\begin{array}{c|c}
\tan\dfrac{\theta}{2} & \csc\theta - \cot\theta \\[2ex]
= \dfrac{1 - \cos\theta}{\sin\theta} & \\[2ex]
= \dfrac{1}{\sin\theta} - \dfrac{\cos\theta}{\sin\theta} & \\[2ex]
= \csc\theta - \cot\theta &
\end{array}
$$ ∎

Example 8   Verify the identity $2\sin^2\dfrac{x}{2}\tan x = \tan x - \sin x$.

Solution   Work on the left side:

$$2 \sin^2 \frac{x}{2} \tan x \qquad \Big| \qquad \tan x - \sin x$$

$$= 2 \cdot \frac{1 - \cos x}{2} \cdot \frac{\sin x}{\cos x}$$

$$= \frac{\sin x - \sin x \cos x}{\cos x}$$

$$= \frac{\sin x}{\cos x} - \sin x$$

$$= \tan x - \sin x \qquad \qquad \qquad \qquad \blacksquare$$

Example 9   Verify the identity $\dfrac{2 \tan \dfrac{x}{2}}{\sin x} = \sec^2 \dfrac{x}{2}$.

Solution   Again, work on the left side:

$$\frac{2 \tan \dfrac{x}{2}}{\sin x} \qquad \Big| \qquad \sec^2 \frac{x}{2}$$

$$= \frac{2 \cdot \dfrac{\sin x}{1 + \cos x}}{\sin x}$$

$$= \frac{2}{1 + \cos x}$$

$$= \frac{1}{\dfrac{1 + \cos x}{2}}$$

$$= \frac{1}{\cos^2 \dfrac{x}{2}}$$

$$= \sec^2 \frac{x}{2} \qquad \qquad \qquad \qquad \blacksquare$$

Exercise 7.4

In Exercises 1–12, use half-angle identities to find the required values. ***Do not use a calculator.***

1. $\cos 15°$

2. $\tan 15°$

3. $\tan 105°$

4. $\sin 105°$

5. $\sin \dfrac{\pi}{8}$

6. $\cos \dfrac{\pi}{8}$

7. $\cos \dfrac{\pi}{12}$

8. $\sin \dfrac{\pi}{12}$

9. $\tan 165°$

10. $\cos 165°$

11. $\cot \dfrac{5\pi}{4}$

12. $\tan \dfrac{7\pi}{4}$

In Exercises 13–24, use the given information to find the exact value of the sine, cosine, and tangent of $\frac{\theta}{2}$. Assume that $0° \leq \theta < 360°$.

13. $\sin \theta = \frac{3}{5}$; θ in QI

14. $\cos \theta = \frac{12}{13}$; θ in QI

15. $\tan \theta = \frac{4}{3}$; θ in QIII

16. $\tan \theta = \frac{3}{4}$; θ in QIII

17. $\cos \theta = \frac{8}{17}$; θ in QIV

18. $\sin \theta = -\frac{7}{25}$; θ in QIII

19. $\cot \theta = \frac{40}{9}$; θ in QI

20. $\sec \theta = -\frac{41}{40}$; θ in QII

21. $\csc \theta = \frac{17}{8}$; θ in QII

22. $\csc \theta = -\frac{5}{3}$; θ in QIV

23. $\sec \theta = \frac{3}{2}$; θ in QIV

24. $\cos \theta = -0.1$; θ in QIII

In Exercises 25–36, write the given expression as a single trigonometric function of half the given angle.

25. $\sqrt{\dfrac{1 + \cos 30°}{2}}$

26. $\sqrt{\dfrac{1 - \cos 30°}{2}}$

27. $\dfrac{1 - \cos 200°}{\sin 200°}$

28. $\dfrac{\sin 50°}{1 + \cos 50°}$

29. $\csc 80° - \cot 80°$

30. $\dfrac{2 \tan 140°}{\sin 280°}$ (280° is the given angle)

31. $\sqrt{\dfrac{1 - \cos 2\pi}{1 + \cos 2\pi}}$

32. $\dfrac{1 - \cos 4\theta}{\sin 4\theta}$

33. $\dfrac{1 - \cos \dfrac{x}{2}}{\sin \dfrac{x}{2}}$

34. $\dfrac{1 - \cos 2x}{1 + \cos 2x}$

35. $\dfrac{\sin 10A}{1 + \cos 10A}$

36. $\dfrac{1 + \cos 4x}{1 - \cos 4x}$

In Exercises 37–50, verify the given identity.

37. $\sin^2 \dfrac{\theta}{2} = \dfrac{1}{2}(1 - \cos \theta)$

38. $\cos^2 \dfrac{\theta}{2} = \dfrac{1}{2}(1 + \cos \theta)$

39. $\sec^2 \dfrac{\theta}{2} = \dfrac{2}{1 + \cos \theta}$

40. $\csc^2 \dfrac{\theta}{2} = \dfrac{2}{1 - \cos \theta}$

41. $\cot \dfrac{\theta}{2} = \dfrac{1 + \cos \theta}{\sin \theta}$

42. $\cot \dfrac{\theta}{2} = \dfrac{\sin \theta}{1 - \cos \theta}$

43. $\csc^2 \dfrac{\theta}{2} = 2 \csc^2 \theta + 2 \cot \theta \csc \theta$

44. $\sin^2 \dfrac{\theta}{2} = \dfrac{\sec \theta - 1}{2 \sec \theta}$

45. $-\sec B = \dfrac{\sec^2 \dfrac{B}{2}}{\sec^2 \dfrac{B}{2} - 2}$

46. $\tan\left(\dfrac{\pi}{4} + \dfrac{\theta}{2}\right) = \dfrac{1 + \cos\theta + \sin\theta}{1 + \cos\theta - \sin\theta}$

47. $\csc\theta = \dfrac{1}{2}\csc\dfrac{\theta}{2}\sec\dfrac{\theta}{2}$

48. $\dfrac{1}{2}\sin x \tan\dfrac{x}{2}\csc^2\dfrac{x}{2} = 1$

49. $\tan\dfrac{B}{2}\cos B + \tan\dfrac{B}{2} = \sin B$

50. $\left(\cos\dfrac{\alpha}{2} - \sin\dfrac{\alpha}{2}\right)^2 = 1 - \sin\alpha$

51. Show that $\sin A$ and $\tan\dfrac{A}{2}$ always agree in sign.

7.5 MORE IDENTITIES

If the identities for $\sin(x + y)$ and $\sin(x - y)$ are added, we obtain more identities.

1. $\sin(x + y) = \sin x \cos y + \cos x \sin y$

2. $\sin(x - y) = \sin x \cos y - \cos x \sin y$

Adding Equations 1 and 2 causes the $\cos x \sin y$ term to drop out. The result is shown in Equation 3:

3. $\sin(x + y) + \sin(x - y) = 2\sin x \cos y$

We divide both sides of Equation 3 by 2 to get

$$\sin x \cos y = \frac{1}{2}[\sin(x + y) + \sin(x - y)]$$

This identity is used to convert a product of the sine and cosine functions into a sum.

Example 1 Calculate the value of $\sin 67.5° \cos 22.5°$.

Solution Let $x = 67.5°$ and $y = 22.5°$. Substitute these values into the product-to-sum formula and simplify:

$$\sin x \cos y = \frac{1}{2}[\sin(x + y) + \sin(x - y)]$$

$$\sin 67.5° \cos 22.5° = \frac{1}{2}[\sin(67.5° + 22.5°) + \sin(67.5° - 22.5°)]$$

$$= \frac{1}{2}(\sin 90° + \sin 45°)$$

$$= \frac{1}{2}\left(1 + \frac{\sqrt{2}}{2}\right)$$

$$= \frac{1}{2} + \frac{\sqrt{2}}{4}$$

∎

To develop a formula to convert the sum of two sines into a product, we let $A = x + y$ and $B = x - y$ and solve for x and y. We first solve

$$\begin{cases} A = x + y \\ B = x - y \end{cases}$$

for x by adding the equations:

$$A + B = 2x$$

$$x = \frac{1}{2}(A + B)$$

To find y, we subtract one equation from the other:

$$A - B = 2y$$

$$y = \frac{1}{2}(A - B)$$

We substitute

$$A \text{ for } x + y, \quad B \text{ for } x - y, \quad \frac{A + B}{2} \text{ for } x, \quad \text{and} \quad \frac{A - B}{2} \text{ for } y$$

in Equation 3 to get

$$\sin A + \sin B = 2 \sin \frac{A + B}{2} \cos \frac{A - B}{2}$$

Example 2 Verify the identity $\sin 3\theta + \sin \theta = 2 \sin 2\theta \cos \theta$.

Solution Use the sum-to-product identity with $A = 3\theta$ and $B = \theta$. Work on the left side:

$$\sin 3\theta + \sin \theta \qquad \bigg| \qquad 2 \sin 2\theta \cos \theta$$

$$= 2 \sin \frac{3\theta + \theta}{2} \cos \frac{3\theta - \theta}{2}$$

$$= 2 \sin \frac{4\theta}{2} \cos \frac{2\theta}{2}$$

$$= 2 \sin 2\theta \cos \theta$$

$\blacksquare$

Example 3 Find the value of $\sin 75° + \sin 15°$.

Solution Use the sum-to-product identity with $A = 75°$ and $B = 15°$:

$$\sin A + \sin B = 2 \sin \frac{A + B}{2} \cos \frac{A - B}{2}$$

$$\sin 75° + \sin 15° = 2 \sin \frac{75° + 15°}{2} \cos \frac{75° - 15°}{2}$$

$$= 2 \sin \frac{90°}{2} \cos \frac{60°}{2}$$

$$= 2 \sin 45° \cos 30°$$

$$= 2 \cdot \frac{\sqrt{2}}{2} \cdot \frac{\sqrt{3}}{2}$$

$$= \frac{\sqrt{6}}{2}$$

∎

If Equations 1 and 2 are subtracted, more identities result:

1. $\qquad\qquad \sin(x + y) = \sin x \cos y + \cos x \sin y$

2. $\qquad\qquad \sin(x - y) = \sin x \cos y - \cos x \sin y$

3. $\quad \sin(x + y) - \sin(x - y) = 2 \cos x \sin y$

Dividing both sides of Equation 3 by 2 gives a new identity:

$$\cos x \sin y = \frac{1}{2}\left[\sin(x + y) - \sin(x - y)\right]$$

This identity is used to convert a product into a difference. If we let $x + y = A$ and $x - y = B$, then

$$x = \frac{A + B}{2} \qquad \text{and} \qquad y = \frac{A - B}{2}$$

We can substitute the values for $x + y$, $x - y$, x, and y into Equation 3 to obtain the next identity:

$$\sin A - \sin B = 2 \cos \frac{A + B}{2} \sin \frac{A - B}{2}$$

This formula is used to convert the difference of the sines of two angles into a product.

If the formulas for the cosines of the sum and the difference of two angles are added, still more identities result:

4. $\quad \begin{cases} \cos(x + y) = \cos x \cos y - \sin x \sin y \\ \cos(x - y) = \cos x \cos y + \sin x \sin y \end{cases}$
5.

After adding, the $\sin x \sin y$ terms drop out, as shown in Equation 6:

6. $\quad \cos(x + y) + \cos(x - y) = 2 \cos x \cos y$

We divide both sides of Equation 6 by 2 to obtain a new identity:

$$\cos x \cos y = \frac{1}{2}\left[\cos(x + y) + \cos(x - y)\right]$$

This identity is used when a product of cosines needs to be changed to a sum. To derive an identity to convert from sums to products, we let $x + y = A$ and $x - y = B$. Then

$$x = \frac{A + B}{2} \quad \text{and} \quad y = \frac{A - B}{2}$$

We use Equation 6 and substitute to get

$$\cos A + \cos B = 2 \cos \frac{A + B}{2} \cos \frac{A - B}{2}$$

We obtain the final identities when we subtract Equation 4 from Equation 5 and make the usual substitutions:

$$\sin x \sin y = \frac{1}{2} \left[\cos(x - y) - \cos(x + y) \right]$$

$$\cos A - \cos B = -2 \sin \frac{A + B}{2} \sin \frac{A - B}{2}$$

We list all of the previous formulas for easy reference.

The Product-to-Sum Formulas.

$$\sin A \cos B = \frac{1}{2} \left[\sin(A + B) + \sin(A - B) \right]$$

$$\cos A \sin B = \frac{1}{2} \left[\sin(A + B) - \sin(A - B) \right]$$

$$\sin A \sin B = \frac{1}{2} \left[\cos(A - B) - \cos(A + B) \right]$$

$$\cos A \cos B = \frac{1}{2} \left[\cos(A + B) + \cos(A - B) \right]$$

The Sum-to-Product Formulas.

$$\sin A + \sin B = 2 \sin \frac{A + B}{2} \cos \frac{A - B}{2}$$

$$\sin A - \sin B = 2 \cos \frac{A + B}{2} \sin \frac{A - B}{2}$$

$$\cos A + \cos B = 2 \cos \frac{A + B}{2} \cos \frac{A - B}{2}$$

$$\cos A - \cos B = -2 \sin \frac{A + B}{2} \sin \frac{A - B}{2}$$

Example 4   Write $\cos 2\theta + \cos 6\theta$ as a product of two functions.

Solution   Substitute 2θ for A and 6θ for B in the identity for $\cos A + \cos B$, and simplify:

$$\cos 2\theta + \cos 6\theta = 2 \cos \frac{2\theta + 6\theta}{2} \cos \frac{2\theta - 6\theta}{2}$$

$$= 2 \cos 4\theta \cos(-2\theta)$$

$$= 2 \cos 4\theta \cos 2\theta \qquad\qquad \text{Remember that} \\ \cos(-2\theta) = \cos 2\theta. \qquad ■$$

Example 5   Verify the identity $\tan 3A = \dfrac{\sin 4A + \sin 2A}{\cos 4A + \cos 2A}$.

Solution   Work on the right side of the equation:

$$\tan 3A \ \left| \ \frac{\sin 4A + \sin 2A}{\cos 4A + \cos 2A} \right.$$

$$= \frac{2 \sin \dfrac{4A + 2A}{2} \cos \dfrac{4A - 2A}{2}}{2 \cos \dfrac{4A + 2A}{2} \cos \dfrac{4A - 2A}{2}}$$

$$= \frac{2 \sin 3A \cos A}{2 \cos 3A \cos A}$$

$$= \frac{\sin 3A}{\cos 3A}$$

$$= \tan 3A \qquad\qquad ■$$

Sums of the Form $A \sin x + B \cos x$

The graph of $y = A \sin x + B \cos x$ is a sine curve of the form $y = k \sin(x + \phi)$. The amplitude k and the phase shift angle ϕ are determined by the values of A and B. Some algebraic manipulations will establish this result and provide the proper values for k and ϕ.

From the terms of the expression $A \sin x + B \cos x$, factor out a common factor of $\sqrt{A^2 + B^2}$:

$$A \sin x + B \cos x = \sqrt{A^2 + B^2}\left(\frac{A}{\sqrt{A^2 + B^2}} \sin x + \frac{B}{\sqrt{A^2 + B^2}} \cos x\right)$$

Because the sum of the squares of the coefficients

$$\frac{A}{\sqrt{A^2 + B^2}} \quad \text{and} \quad \frac{B}{\sqrt{A^2 + B^2}}$$

is 1, one of these coefficients is sin ϕ and one is cos ϕ for some angle ϕ. We let ϕ be an angle such that

$$\sin \phi = \frac{B}{\sqrt{A^2 + B^2}} \qquad \text{and} \qquad \cos \phi = \frac{A}{\sqrt{A^2 + B^2}}$$

Then the previous equation can be written as

$$A \sin x + B \cos x = \sqrt{A^2 + B^2}(\cos \phi \sin x + \sin \phi \cos x)$$
$$= \sqrt{A^2 + B^2} \sin(x + \phi)$$

and we have the following theorem.

Theorem. $A \sin x + B \cos x = k \sin(x + \phi)$ where $k = \sqrt{A^2 + B^2}$ and ϕ is any angle for which

$$\sin \phi = \frac{B}{\sqrt{A^2 + B^2}} \qquad \text{and} \qquad \cos \phi = \frac{A}{\sqrt{A^2 + B^2}}$$

Example 6   Express $3 \sin x + 4 \cos x$ in the form $k \sin(x + \phi)$:

Solution   Begin by evaluating k:

$$k = \sqrt{A^2 + B^2} = \sqrt{3^2 + 4^2} = \sqrt{25} = 5$$

ϕ is an angle such that $\sin \phi = \frac{4}{5} = 0.8$ and $\cos \phi = \frac{3}{5} = 0.6$. Use a calculator to determine that $\phi \approx 53.1°$. Thus,

$$3 \sin x + 4 \cos x \approx 5 \sin(x + 53.1°) \qquad \blacksquare$$

Example 7   Express $5 \sin 3x - 12 \cos 3x$ as a single expression involving the sine function only.

Solution   The expression $5 \sin 3x - 12 \cos 3x$ can be written as $k \sin(3x + \phi)$, where $k = \sqrt{5^2 + (-12)^2} = \sqrt{25 + 144} = \sqrt{169} = 13$. The angle ϕ is such that $\sin \phi = -\frac{12}{13} \approx -0.9231$ and $\cos \phi = \frac{5}{13} \approx 0.3846$. Because $\sin \phi$ is negative and $\cos \phi$ is positive, ϕ must lie in the fourth quadrant; $\phi \approx -67.4°$. Hence,

$$5 \sin 3x - 12 \cos 3x \approx 13 \sin(3x - 67.4°) \qquad \blacksquare$$

Example 8   Graph the function $y = \cos x - \sin x$.

Solution   Note that $\cos x - \sin x$ or $-\sin x + \cos x$ can be changed to $k \sin(x + \phi)$ with

$$k = \sqrt{(-1)^2 + 1^2} = \sqrt{2}, \quad \sin \phi = \frac{1}{\sqrt{2}} = \frac{\sqrt{2}}{2}, \quad \text{and} \quad \cos \phi = \frac{-1}{\sqrt{2}} = -\frac{\sqrt{2}}{2}$$

Because $\sin \phi$ is positive and $\cos \phi$ is negative, angle ϕ is a second-quadrant angle. The second-quadrant angle with a sine of $\frac{\sqrt{2}}{2}$ is $\frac{3\pi}{4}$ radians. Hence, $y = \cos x - \sin x$ is equivalent to

$$y = \sqrt{2} \sin\left(x + \frac{3\pi}{4}\right)$$

The graph of this function is a simple sine curve with amplitude of $\sqrt{2}$ and a phase shift of $\frac{3\pi}{4}$ to the left. The graph appears in Figure 7-5. Note that the function $y = \cos x - \sin x$ could be graphed using the method of addition of ordinates.

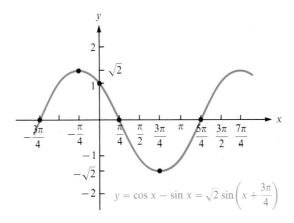

Figure 7-5

Exercise 7.5

In Exercises 1–12, express each product as a sum or difference and find its value. **Do not use a calculator or tables.**

1. $\cos 75° \cos 15°$

2. $\sin 15° \cos 75°$

3. $\sin 165° \sin 105°$

4. $\sin 15° \cos 15°$

5. $\cos 22.5° \cos 67.5°$

6. $\cos 105° \sin 15°$

7. $\sin \dfrac{\pi}{12} \sin \dfrac{5\pi}{12}$

8. $\sin \dfrac{5\pi}{12} \cos \dfrac{13\pi}{12}$

9. $\cos \dfrac{7\pi}{12} \cos \dfrac{5\pi}{12}$

10. $\cos \dfrac{7\pi}{12} \sin \dfrac{13\pi}{12}$

11. $\cos \dfrac{\pi}{12} \sin \dfrac{5\pi}{12}$

12. $\sin \dfrac{\pi}{12} \cos \dfrac{5\pi}{12}$

In Exercises 13–24, express each quantity as a product and find its value. **Do not use a calculator or tables.**

13. $\cos 75° + \cos 15°$

14. $\sin 15° + \sin 75°$

15. $\sin 165° - \sin 105°$

16. $\sin 15° - \sin 75°$

17. $\cos 165° - \cos 105°$

18. $\cos 105° - \cos 15°$

19. $\sin \dfrac{\pi}{12} + \sin \dfrac{5\pi}{12}$

20. $\sin \dfrac{5\pi}{12} - \sin \dfrac{13\pi}{12}$

21. $\cos \dfrac{7\pi}{12} + \cos \dfrac{5\pi}{12}$

22. $\sin \dfrac{7\pi}{12} + \sin \dfrac{13\pi}{12}$

23. $\sin \dfrac{5\pi}{12} - \sin \dfrac{\pi}{12}$

24. $\cos \dfrac{11\pi}{12} - \cos \dfrac{7\pi}{12}$

In Exercises 25–36, write the given expression in the form $k \sin(x + \phi)$.

25. $6 \sin x + 8 \cos x$

26. $12 \sin x + 5 \cos x$

27. $6 \sin x - 8 \cos x$

28. $12 \sin x - 5 \cos x$

29. $2 \sin x + \cos x$

30. $\sin x - \cos x$

31. $\sin x + \cos x$

32. $2 \sin x - \cos x$

33. $-\sin x + 5 \cos x$

34. $\sqrt{3} \sin x + 3 \cos x$

35. $\sqrt{3} \sin x - 3 \cos x$

36. $-\sin x - 5 \cos x$

In Exercises 37–42, use the method of Example 8 to graph each function.

37. $y = \sin x + \cos x$

38. $y = \sin x - \cos x$

39. $y = \sin x - \sqrt{3} \cos x$

40. $y = \sqrt{3} \sin x + \cos x$

41. $y = \sin 2x + \sqrt{3} \cos 2x$

42. $y = \sin 2x - \sqrt{3} \cos 2x$

In Exercises 43–52, verify each identity.

43. $\dfrac{\sin A + \sin B}{\sin A - \sin B} = \tan \dfrac{1}{2}(A + B) \cot \dfrac{1}{2}(A - B)$

44. $\dfrac{\sin A + \sin B}{\cos A + \cos B} = \tan \dfrac{1}{2}(A + B)$

45. $\dfrac{\sin A + \sin B}{\cos A - \cos B} = -\cot \dfrac{1}{2}(A - B)$

46. $\dfrac{\cos A + \cos B}{\sin A - \sin B} = \cot \dfrac{1}{2}(A - B)$

47. $\dfrac{\cos A + \cos 5A}{\cos A - \cos 5A} = \cot 3A \cot 2A$

48. $\sin^2 A - \sin^2 B = \sin(A + B) \sin(A - B)$

49. $\cos^2 A - \cos^2 B = \sin(B + A) \sin(B - A)$

50. $\cos 2A(1 + 2 \cos A) = \cos A + \cos 2A + \cos 3A$

51. $2 \cos 5A \sin 2A = \sin 7A - \sin 3A$

52. $\cot 7A \cot 5A = \dfrac{\cos 12A + \cos 2A}{\cos 2A - \cos 12A}$

7.6   TRIGONOMETRIC EQUATIONS

The equation $\sin 2x = 2 \sin x$ is not an identity because it is not true for all values of x. It is true, however, for *some* values of x. For example, the equation is true if $x = 0$, as the following check shows:

$$\sin 2x = 2 \sin x$$
$$\sin 2(0) \overset{?}{=} 2 \sin 0$$
$$\sin 0 \overset{?}{=} 2 \cdot 0$$
$$0 = 0$$

Other values of x, such as π, 2π, 3π, and so on also satisfy this equation. The process of finding *all* such values is called **solving** the equation. To solve a trigonometric equation, we must apply techniques such as combining terms, adding terms to both sides of an equation, factoring, and so on. We must also use the trigonometric identities previously discussed. Example 1 provides a formal solution of the equation $\sin 2x = 2 \sin x$.

Example 1   Solve $\sin 2x = 2 \sin x$ for all values of x, where x is a real number.

Solution   Use the identity for $\sin 2x$ to rewrite the left side of the equation:

$$\sin 2x = 2 \sin x$$
$$2 \sin x \cos x = 2 \sin x$$

Divide both sides of the previous equation by 2, subtract $\sin x$ from both sides, and proceed as follows. Do not begin by dividing both sides of the equation by $\sin x$. Dividing by an expression that might be zero could cause you to lose a solution of the equation.

$\sin x \cos x = \sin x$	Divide both sides by 2.
$\sin x \cos x - \sin x = 0$	Subtract $\sin x$ from both sides.
$\sin x(\cos x - 1) = 0$	Factor out $\sin x$.
$\sin x = 0$ or $\cos x - 1 = 0$	Set each factor equal to 0.
$\cos x = 1$	

The solutions to the equation $\sin 2x = 2 \sin x$ are all real numbers whose sine is 0 or whose cosine is 1. Numbers with a sine of 0 are

$$\ldots, -4\pi, -3\pi, -2\pi, -\pi, 0, \pi, 2\pi, 3\pi, 4\pi, \ldots$$

Numbers with a cosine of 1 are

$$\ldots, -4\pi, -2\pi, 0, 2\pi, 4\pi, \ldots$$

Thus, the solutions of the given equation are the real numbers

$$\ldots, -4\pi, -3\pi, -2\pi, -\pi, 0, \pi, 2\pi, 3\pi, 4\pi, \ldots$$

Verify that several of these solutions satisfy the given equation.   ∎

Example 2   Solve $2 \sin^2 \theta + \sin \theta = 1$ for all values of θ, where $0° \le \theta < 360°$.

Solution   After subtracting 1 from both sides of the equation, factor the left-hand side. Then, set each factor equal to zero and solve each equation for θ:

$$2 \sin^2 \theta + \sin \theta = 1$$
$$2 \sin^2 \theta + \sin \theta - 1 = 0$$
$$(2 \sin \theta - 1)(\sin \theta + 1) = 0$$

$2 \sin \theta - 1 = 0$	or $\sin \theta + 1 = 0$
$2 \sin \theta = 1$	$\sin \theta = -1$
$\sin \theta = \dfrac{1}{2}$	
$\theta = 30°, 150°$	$\theta = 270°$

These are the three solutions that lie in the designated interval from 0° to 360°. Verify that all three values satisfy the given equation.   ∎

Example 3   Solve $\sin 3x = \dfrac{1}{2}$ for all x, where $0 \le x < 2\pi$.

Solution   There are infinitely many values of $3x$ that will satisfy the equation $\sin 3x = \frac{1}{2}$. Because you are given that $0 \le x < 2\pi$, you must find all such values of $3x$ where

$$3(0) \le 3x < 3(2\pi)$$

or

$$0 \le 3x < 6\pi$$

There are six such values:

$$3x = \frac{\pi}{6}, \frac{5\pi}{6}, \frac{13\pi}{6}, \frac{17\pi}{6}, \frac{25\pi}{6}, \frac{29\pi}{6}$$

The values of x are found by dividing each of the possible values of $3x$ by 3. Thus,

$$x = \frac{\pi}{18}, \frac{5\pi}{18}, \frac{13\pi}{18}, \frac{17\pi}{18}, \frac{25\pi}{18}, \frac{29\pi}{18}$$

Note that the largest of these, $\frac{29\pi}{18}$, is still less than 2π. All six roots do satisfy the given equation.   ∎

Example 4   Solve $\sin \theta = \cos \theta$ for θ, where $0° \le \theta < 360°$.

Solution   If $\cos \theta \ne 0$, you may divide both sides of the equation by $\cos \theta$:

$$\sin \theta = \cos \theta$$
$$\frac{\sin \theta}{\cos \theta} = 1$$
$$\tan \theta = 1$$
$$\theta = 45°, 225°$$

Both values satisfy the given equation. If $\cos \theta = 0$, then $\theta = 90°$ or $270°$. These values do not satisfy the given equation. Hence, the solutions are $45°$ and $225°$.

∎

Example 5   Solve the equation $2 \cos^3 \theta = \cos \theta$ for θ, where $0° \le \theta < 360°$.

Solution   Subtract $\cos \theta$ from both sides of the equation and factor out $\cos \theta$. Then, proceed as follows:

$$2 \cos^3 \theta = \cos \theta$$

$$2 \cos^3 \theta - \cos \theta = 0$$

$$\cos \theta (2 \cos^2 \theta - 1) = 0$$

$$\cos \theta = 0 \qquad \text{or} \qquad 2 \cos^2 \theta - 1 = 0$$

$$\cos^2 \theta = \frac{1}{2}$$

$$\cos \theta = \pm \frac{1}{\sqrt{2}} = \pm \frac{\sqrt{2}}{2}$$

$$\theta = 90°, 270° \qquad\qquad \theta = 45°, 135°, 225°, 315°$$

All six values satisfy the given equation. Note that if both sides of the equation had been divided by $\cos \theta$, the solutions $90°$ and $270°$ would have been lost.

∎

Example 6   Solve $4 \sin^2 \dfrac{x}{2} = 1$ for x, where $0 \le x < 2\pi$.

Solution   Rearrange the terms, factor, and set each factor equal to 0:

$$4 \sin^2 \frac{x}{2} = 1$$

$$4 \sin^2 \frac{x}{2} - 1 = 0$$

$$\left(2 \sin \frac{x}{2} - 1 \right) \left(2 \sin \frac{x}{2} + 1 \right) = 0$$

$$2 \sin \frac{x}{2} - 1 = 0 \qquad \text{or} \qquad 2 \sin \frac{x}{2} + 1 = 0$$

$$\sin \frac{x}{2} = \frac{1}{2} \qquad\qquad \sin \frac{x}{2} = -\frac{1}{2}$$

$$\frac{x}{2} = \frac{\pi}{6}, \frac{5\pi}{6} \qquad\qquad \frac{x}{2} = \frac{7\pi}{6}, \frac{11\pi}{6}$$

$$x = \frac{\pi}{3}, \frac{5\pi}{3} \qquad\qquad x = \frac{7\pi}{3}, \frac{11\pi}{3}$$

Because $\frac{7\pi}{3}$ and $\frac{11\pi}{3}$ are greater than 2π, they must be excluded. The only solutions are $x = \frac{\pi}{3}$ and $x = \frac{5\pi}{3}$. Verify that each of these solutions satisfies the given equation.

∎

Example 7   Solve $4 \sin^2 \dfrac{x}{2} = 1$ for x, where $0 \le x < 2\pi$, by using a half-angle identity.

Solution   From the half-angle identity for $\sin \frac{x}{2}$, it follows that

$$\sin^2 \frac{x}{2} = \frac{1 - \cos x}{2}$$

Substitute $\frac{1 - \cos x}{2}$ for $\sin^2 \frac{x}{2}$ in the original equation to get

$$4 \sin^2 \frac{x}{2} = 1$$

$$4\left(\frac{1 - \cos x}{2}\right) = 1$$

$$1 - \cos x = \frac{1}{2}$$

$$\cos x = \frac{1}{2}$$

$$x = \frac{\pi}{3}, \frac{5\pi}{3}$$

Verify that each value satisfies the given equation.   ■

Example 8   Solve $2 \sin x \cos x + \cos x - 2 \sin x - 1 = 0$, where $0 \le x < 2\pi$.

Solution   Use the technique of *factoring by grouping*. The four terms on the left side share no common factors. However, the first two terms share a common factor of $\cos x$, and the last two terms share a common factor of -1. Proceed as follows:

$$2 \sin x \cos x + \cos x - 2 \sin x - 1 = 0$$

$$\cos x(2 \sin x + 1) - 1(2 \sin x + 1) = 0$$

$$(2 \sin x + 1)(\cos x - 1) = 0 \qquad \text{Factor out the common factor of } 2 \sin x + 1.$$

$$2 \sin x + 1 = 0 \qquad \text{or} \qquad \cos x - 1 = 0 \qquad \text{Set each factor}$$

$$2 \sin x = -1 \qquad\qquad\qquad \cos x = 1 \qquad \text{equal to 0.}$$

$$\sin x = \frac{-1}{2}$$

$$x = \frac{7\pi}{6}, \frac{11\pi}{6} \qquad\qquad\qquad x = 0$$

Verify that each value satisfies the given equation.   ■

Example 9   Solve $\sin \theta + \cos \theta = 1$ for θ, where $0° \le \theta < 360°$.

Solution   Use the identity

$$A \sin \theta + B \cos \theta = \sqrt{A^2 + B^2} \, \sin(\theta + \alpha)$$

to write $\sin \theta + \cos \theta$ in the form $k \sin(\theta + \alpha)$:

$$\sin \theta + \cos \theta = 1$$
$$\sqrt{1^2 + 1^2} \sin(\theta + \alpha) = 1$$
$$\sqrt{2} \sin(\theta + \alpha) = 1$$

Remember that α is an angle for which $\sin \alpha = \cos \alpha = \dfrac{1}{\sqrt{2}} = \dfrac{\sqrt{2}}{2}$. Thus, $\alpha = 45°$ and you have

$$\sqrt{2} \sin(\theta + 45°) = 1$$
$$\sin(\theta + 45°) = \frac{1}{\sqrt{2}} = \frac{\sqrt{2}}{2}$$

$$\theta + 45° = 45° \quad \text{or} \quad \theta + 45° = 135°$$
$$\theta = 0° \qquad\qquad\quad \theta = 90°$$

Verify that each value satisfies the given equation.   ■

Exercise 7.6

In Exercises 1–4, solve for all values of the variable. Assume that x is a real number and that θ is an angle in degrees. **Do not use a calculator or tables.**

1. $\sin \theta = \dfrac{\sqrt{3}}{2}$ **2.** $\cos x = \dfrac{-\sqrt{2}}{2}$ **3.** $\sin 2x = 0$ **4.** $\cos 2\theta = 1$

In Exercises 5–46, solve for all values of the variable between $0°$ and $360°$, including $0°$. **Do not use a calculator or tables.**

5. $\cos \dfrac{\theta}{2} = \dfrac{1}{2}$ **6.** $\sin \dfrac{\theta}{2} = \dfrac{\sqrt{3}}{2}$

7. $\sin \theta \cos \theta = 0$ **8.** $\sin \theta \cos \theta - \sin \theta = 0$

9. $\cos \theta \sin \theta - \cos \theta = 0$ **10.** $\cos \theta \sin \theta + \cos \theta = 0$

11. $\sin \theta \cos \theta + \sin \theta = 0$ **12.** $\cos^2 \theta = 1$

13. $\sin^2 \theta = \dfrac{1}{2}$ **14.** $\sin^2 \theta - \dfrac{1}{2} \sin \theta - \dfrac{1}{2} = 0$

15. $\cos^2 \theta - \dfrac{1}{2} \cos \theta - \dfrac{1}{2} = 0$ **16.** $\sin^2 \theta - 3 \sin \theta + 2 = 0$

17. $4 \sin^2 \theta + 4 \sin \theta + 1 = 0$ **18.** $\sin 2\theta - \cos \theta = 0$

19. $\sin 2A + \cos A = 0$ **20.** $\cos^2 A + 4 \cos A + 3 = 0$

21. $\cos B = \sin B$ **22.** $\cos B = \sqrt{3} \sin B$

23. $\cos \theta = \cos \dfrac{\theta}{2}$ **24.** $\cos \theta = \sin \dfrac{\theta}{2}$

25. $\cos 2\theta = \cos \theta$ **26.** $\cos 2\theta = \sin \theta$

27. $\tan A = -\sin A$ **28.** $\cot A = -\cos A$

29. $\cos^2 C - \sin^2 C = \dfrac{1}{2}$

30. $\cos^2 C + \sin^2 C = \dfrac{1}{2}$

31. $\sin \theta + \cos \theta = \sqrt{2}$
(*Hint*: Square both sides.)

32. $\sin \theta - \cos \theta = \sqrt{2}$

33. $\cos 2A = 1 - \sin A$

34. $\cos 2A = \cos A - 1$

35. $\cos 2B + \cos B + 1 = 0$

36. $\cos^2 B + \sin B - 1 = 0$

37. $\sin 4\theta = \sin 2\theta$

38. $\cos 4\theta = \cos 2\theta$

39. $\cos B = 1 + \sqrt{3} \sin B$

40. $\tan 2B + \sec 2B = 1$

41. $9 \cos^4 A = \sin^4 A$

42. $6 \cos^2 A = -9 \sin A$

43. $\sec C = \tan C + \cos C$

44. $1 - \tan C = \sqrt{2} \sec C$

45. $2 \cos x \sin x - \sqrt{2} \cos x = \sqrt{2} \sin x - 1$

46. $4 \sin x \cos x - 2 \sin x = 2 \cos x - 1$

In Exercises 47–52, solve for θ, where $0° \le \theta < 360°$. Recall that the expression $a \sin x + b \cos x$ can be written as $k \sin(x + \phi)$.

47. $\dfrac{\sqrt{3}}{2} \sin \theta + \dfrac{1}{2} \cos \theta = \dfrac{1}{2}$

48. $\sqrt{2} \sin \theta + \sqrt{2} \cos \theta = \sqrt{3}$

49. $\dfrac{1}{2} \sin \theta + \dfrac{\sqrt{3}}{2} \cos \theta = 1$

50. $\cos \theta - \sin \theta = \sqrt{2}$

51. $\cos \theta - \sqrt{3} \sin \theta = 1$

52. $\sin \theta + \cos \theta = -\sqrt{2}$

In Exercises 53–62, solve for x, where $0 \le x < 2\pi$. Consider using a sum-to-product or a product-to-sum identity.

53. $\sin x \cos x = \dfrac{1}{2}$

54. $\cos x \sin x = \dfrac{\sqrt{3}}{4}$

55. $2 \sin \dfrac{3x}{2} \cos \dfrac{x}{2} = \sin x$

56. $2 \cos \dfrac{3x}{2} \cos \dfrac{x}{2} = \cos 2x$

57. $\sin 3x \cos x = \dfrac{1}{2} \sin 2x$

58. $\cos 3x \cos x = \cos^2 2x$

59. $\sin 4x = -\sin 2x$

60. $\cos 4x = -\cos 2x$

61. $\cos 9x = -\cos 3x$

62. $\sin 12x = -\sin 4x$

In Exercises 63–76, solve for x, where $0 \le x < 2\pi$.

63. $2 \sin^4 x - 9 \sin^2 x + 4 = 0$

64. $2 \sin^3 x + \sin^2 x = \sin x$

65. $4 \cos x \sin^2 x - \cos x = 0$

66. $\tan^2 x = 1 + \sec x$

67. $\csc^4 x = 2 \csc^2 x - 1$

68. $\tan^2 x - 5 \tan x + 6 = 0$

69. $4 \sin^2 x + 4 \sin x + 1 = 0$

70. $2 \sin 5x = 1$

71. $2 \sin^2 5x = 1$

72. $2 \sin^2 2x + \sin 2x - 5 = 0$

73. $\tan^2 x - \tan x = 0$

74. $2 \sin x - \sqrt{2} \tan x = \sqrt{2} \sec x - 2$

75. $\cot^2 x \cos x + \cot^2 x - 3 \cos x - 3 = 0$

76. $\tan x + \cot x = -2$

77. Suppose that $\sin \sqrt{2}\,\pi t = 0$. Show that the values of $\cos \sqrt{2}\,\pi t$ are 1 or -1.

78. Suppose that $\cos \dfrac{t}{2} = 0$. Show that the values of $\sin \dfrac{t}{2}$ are 1 or -1.

79. Suppose that $\cos t - \sin t = 0$. Show that the values of $\cos t + \sin t$ are $\sqrt{2}$ and $-\sqrt{2}$.

80. Suppose that $\cos 2t = 1$. Show that the value of $2 \sin t \cos t$ is 0.

7.7 INVERSES OF THE TRIGONOMETRIC FUNCTIONS

The set of ordered pairs of real numbers (x, y) that satisfy the equation $y = \sin x$ determines a function because every real number x gives a single value y. The graph of $y = \sin x$ appears in Figure 7-6.

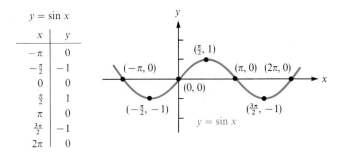

Figure 7-6

Interchanging x and y in the function $y = \sin x$ forms the inverse relation $x = \sin y$, whose graph appears in Figure 7-7.

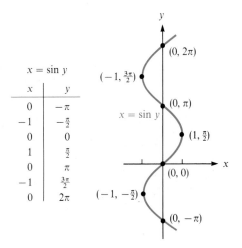

Figure 7-7

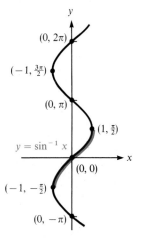

Figure 7-8

Because the graph of $x = \sin y$ does not pass the vertical line test, it cannot represent a function. However, the portion of the graph lying between $y = -\frac{\pi}{2}$ and $y = \frac{\pi}{2}$ does pass the vertical line test and does determine a function. The colored portion of the graph in Figure 7-8 is the graph of a function called the **inverse sine function**, sometimes called the **arcsine function**.

Definition. The **inverse sine** (or arcsine) **function**, denoted by $y = \sin^{-1} x$ (or $y = \arcsin x$), has a domain of $\{x \mid -1 \le x \le 1\}$ and a range of $\{y \mid -\frac{\pi}{2} \le y \le \frac{\pi}{2}\}$.

$$y = \sin^{-1} x \text{ if and only if } x = \sin y \text{ and } -\frac{\pi}{2} \le y \le \frac{\pi}{2}$$

The solid black curve in Figure 7-9 is that portion of $y = \sin x$ for which $-\frac{\pi}{2} \le x \le \frac{\pi}{2}$. That portion represents an increasing function, which has an inverse function and, therefore, is one-to-one. Its inverse is the colored curve in the figure, which is the graph of the function $y = \sin^{-1} x$. Note that the black and the red curves are reflections of each other, with the line $y = x$ as the axis of symmetry.

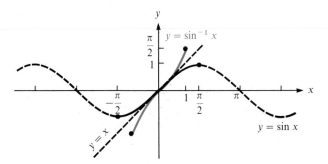

Figure 7-9

Although the domain and range of the inverse sine function are defined to be sets of real numbers, it is often convenient to think of $\sin^{-1} x$ as the "angle whose sine is x." In this situation, the range of the inverse sine function may be thought of as a set of angles measured in radians.

Example 1 Find **a.** $\sin^{-1} \frac{1}{2}$, **b.** $\sin^{-1}\left(-\frac{\sqrt{2}}{2}\right)$, **c.** $\sin^{-1} \pi$, **d.** $\sin^{-1} \frac{\pi}{4}$, and **e.** $\arcsin 0.8330$.

Solution **a.** The expression $\sin^{-1} \frac{1}{2}$ represents the number between $-\frac{\pi}{2}$ and $\frac{\pi}{2}$, inclusive, whose sine is $\frac{1}{2}$. Because $\frac{\pi}{6}$ is the only number in this interval whose sine is $\frac{1}{2}$,

$$\sin^{-1} \frac{1}{2} = \frac{\pi}{6}$$

b. The expression $\sin^{-1}(-\frac{\sqrt{2}}{2})$ represents the number between $-\frac{\pi}{2}$ and $\frac{\pi}{2}$, inclusive, that has a sine of $-\frac{\sqrt{2}}{2}$. Because $\sin(-\frac{\pi}{4}) = -\frac{\sqrt{2}}{2}$ and $-\frac{\pi}{2} < -\frac{\pi}{4} < \frac{\pi}{2}$,

$$\sin^{-1}\left(-\frac{\sqrt{2}}{2}\right) = -\frac{\pi}{4}$$

c. The expression $\sin^{-1}\pi$ represents the number between $-\frac{\pi}{2}$ and $\frac{\pi}{2}$, inclusive, with a sine of π. Because no number has a sine of π, $\sin^{-1}\pi$ is undefined.

d. To find $\sin^{-1}\frac{\pi}{4}$, use a calculator set for radian measure. Divide π by 4 and then press the $\boxed{\text{INV}}$ and $\boxed{\text{SIN}}$ keys in that order. The angle whose sine is $\frac{\pi}{4}$ is approximately 0.9033 radian, as displayed on the calculator.

e. To find arcsin 0.8330, use a calculator set for radian measure. Enter .833 and press the $\boxed{\text{INV}}$ and $\boxed{\text{SIN}}$ keys in succession. The angle whose sine is 0.8330 is approximately 0.9845 radian, as displayed on the calculator.

■

Because $y = \sin x$ (restricted to $-\frac{\pi}{2} \le x \le \frac{\pi}{2}$) and $y = \sin^{-1}x$ are inverse functions of each other, the effect of performing the functions in succession is worth considering. Because $\sin^{-1}x$ means "the number whose sine is x," $\sin(\sin^{-1}x)$ means "the sine of the number whose sine is x." The obvious answer to this question is x.

If $-1 \le x \le 1$, then $\sin(\sin^{-1}x) = x$

Similarly,

If $-\dfrac{\pi}{2} \le x \le \dfrac{\pi}{2}$, then $\sin^{-1}(\sin x) = x$

However, if x is not restricted to the proper interval, then $\sin^{-1}(\sin x)$ might *not* be x. For example,

$$\sin^{-1}\left(\sin\frac{5\pi}{6}\right) = \sin^{-1}\left(\frac{1}{2}\right) = \frac{\pi}{6}$$

Similar considerations produce inverse functions of the remaining five trigonometric functions.

Definition. The **inverse cosine** (or **arccosine**) **function**, denoted by $y = \cos^{-1}x$ (or $y = $ arccos x), has a domain of $\{x \mid -1 \le x \le 1\}$ and a range of $\{y \mid 0 \le y \le \pi\}$.

$y = \cos^{-1}x$ if and only if $x = \cos y$ and $0 \le y \le \pi$

The graph of $y = \cos^{-1}x$, shown in Figure 7-10, is the reflection of a portion of the cosine curve with the line $y = x$ as an axis of symmetry.

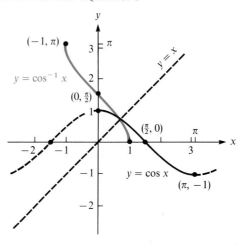

Figure 7-10

> **Definition.**   The **inverse tangent** (or arctangent) **function**, denoted by $y = \tan^{-1} x$ (or $y = \arctan x$), has a domain of the real numbers and a range of $\{y \mid -\frac{\pi}{2} < y < \frac{\pi}{2}\}$.
>
> $y = \tan^{-1} x$ if and only if $x = \tan y$ and $-\frac{\pi}{2} < y < \frac{\pi}{2}$

The graph of $y = \tan^{-1} x$, shown in Figure 7-11, is the reflection of a portion of the tangent curve with the line $y = x$ as an axis of symmetry.

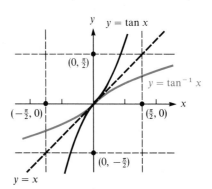

Figure 7-11

Example 2   Find **a.** $\tan^{-1}(-1)$, **b.** $\cos^{-1}\left(\dfrac{\sqrt{3}}{2}\right)$, and **c.** $\tan^{-1}\left(\dfrac{\pi}{4}\right)$.

Solution   **a.**   Because $\tan(-\frac{\pi}{4}) = -1$ and because $-\frac{\pi}{4}$ is in the interval $-\frac{\pi}{2} \leq x \leq \frac{\pi}{2}$,

$$\tan^{-1}(-1) = -\frac{\pi}{4}$$

b. Because $\cos(\frac{\pi}{6}) = \frac{\sqrt{3}}{2}$ and because $0 \leq \frac{\pi}{6} \leq \pi$,

$$\cos^{-1}\left(\frac{\sqrt{3}}{2}\right) = \frac{\pi}{6}$$

c. To evaluate $\tan^{-1}\frac{\pi}{4}$, use a calculator set in radian mode. Divide π by 4. Then, press the $\boxed{\text{INV}}$ and $\boxed{\text{TAN}}$ keys to obtain

$$\tan^{-1}\left(\frac{\pi}{4}\right) \approx 0.6658$$ ∎

Example 3 Find $\cos(\sin^{-1} 1)$.

Solution Because $\frac{\pi}{2}$ is the angle whose sine is 1, it follows that

$$\cos(\sin^{-1} 1) = \cos\left(\frac{\pi}{2}\right) = 0$$ ∎

Example 4 Find $\cos(\sin^{-1} x)$ given that $0 < x \leq 1$.

Solution Because the value of x is unknown, you don't know what angle has a sine of x. However, you do know that $\sin^{-1} x$ is a first- or fourth-quadrant angle, because $-\frac{\pi}{2} \leq \sin^{-1} x \leq \frac{\pi}{2}$. In Figure 7-12, the angle denoted by $\sin^{-1} x$ has a sine of $x(x > 0)$, because the side opposite $\sin^{-1} x$ is labeled x and the hypotenuse is 1. Use the Pythagorean Theorem to find that the remaining side of the triangle has length of $\sqrt{1 - x^2}$. You can then read the value of the cosine of $\sin^{-1} x$ from the figure.

$$\cos(\sin^{-1} x) = \frac{\text{adjacent side}}{\text{hypotenuse}}$$
$$= \frac{\sqrt{1 - x^2}}{1}$$
$$= \sqrt{1 - x^2}$$

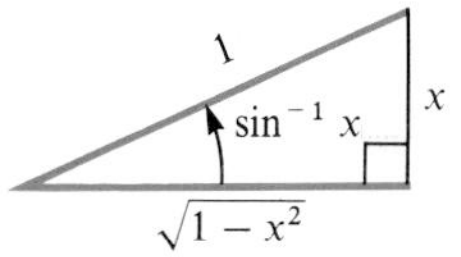

Figure 7-12

Because $-\frac{\pi}{2} \leq \sin^{-1} x \leq \frac{\pi}{2}$, the expression $\cos(\sin^{-1} x)$ is never negative and the positive value of the radical is correct. ∎

Example 5 Find $\tan(\cos^{-1} x)$, where $-1 \leq x \leq 1$.

Solution In this example, there are two cases: when $0 \leq x \leq 1$ and when $-1 \leq x < 0$.

If x is positive or zero, then $\cos^{-1} x$ is an acute angle that may be drawn in standard position, as in Figure 7-13. Because $\cos^{-1} x$ represents the angle whose cosine is x, the side adjacent to the angle may be labeled x and the hypotenuse labeled 1. The remaining side of the right triangle is determined by the Pythagorean Theorem. Then, you have

$$\tan(\cos^{-1} x) = \frac{\sqrt{1 - x^2}}{x}$$

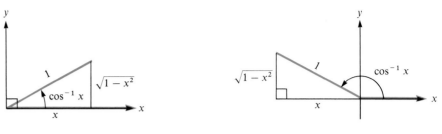

Figure 7-13   Figure 7-14

The tangent of an acute angle is positive. Because x is also positive, the radical was chosen to be positive.

Now consider the case when x is negative. If x is negative, then $\cos^{-1} x$ is a second-quadrant angle that may be drawn in standard position, as in Figure 7-14. Because $\cos^{-1} x$ is the angle whose cosine is x, the adjacent side may be labeled x and the hypotenuse labeled 1. Use the Pythagorean Theorem to determine that the remaining side has length of $\sqrt{1 - x^2}$.

Thus,

$$\tan(\cos^{-1} x) = \frac{\sqrt{1 - x^2}}{x}$$

The tangent of a second-quadrant angle is negative. Because x is negative, the radical must again be chosen positive. Taken together, the results of these two cases determine that, for all x in the interval $-1 \le x \le 1$,

$$\tan(\cos^{-1} x) = \frac{\sqrt{1 - x^2}}{x}$$ ■

Example 6   Write $\sin(\sin^{-1} x + \sin^{-1} y)$ as an algebraic expression in the variables x and y.

Solution   Note that both $\sin^{-1} x$ and $\sin^{-1} y$ are angles, and use the identity $\sin(A + B) = \sin A \cos B + \cos A \sin B$ to remove parentheses.

$$\sin(\sin^{-1} x + \sin^{-1} y) = \sin(\sin^{-1} x) \cos(\sin^{-1} y)$$
$$+ \cos(\sin^{-1} x) \sin(\sin^{-1} y)$$

Because $\sin(\sin^{-1} z) = z$ and $\cos(\sin^{-1} z) = \sqrt{1 - z^2}$ (see Example 4), it follows that

$$\sin(\sin^{-1} x + \sin^{-1} y) = x\sqrt{1 - y^2} + \sqrt{1 - x^2} \cdot y$$
$$= x\sqrt{1 - y^2} + y\sqrt{1 - x^2}$$ ■

The domain, range, and graph of the remaining three inverse trigonometric functions are shown in Table 7-1. There is no agreed-upon definition of the ranges of the inverse secant and inverse cosecant functions. We have given definitions that correspond to the ranges of the inverse cosine and inverse sine functions. Because the inverse cotangent, inverse secant, and inverse cosecant functions are seldom used, we shall not discuss them in detail.

Table 7-1

Function	Domain	Range	Graph
$y = \cot^{-1} x$	all real numbers x	$0 < y < \pi$	
$y = \sec^{-1} x$	$x \le -1$ or $x \ge 1$	$0 \le y \le \pi$* and $y \ne \dfrac{\pi}{2}$	
$y = \csc^{-1} x$	$x \le -1$ or $x \ge 1$	$-\dfrac{\pi}{2} \le y \le \dfrac{\pi}{2}$† and $y \ne 0$	

* Some books restrict y to the intervals $-\pi \le y < -\frac{\pi}{2}$ and $0 \le y < \frac{\pi}{2}$.

† Some books restrict y to the intervals $-\pi < y \le -\frac{\pi}{2}$ and $0 < y \le \frac{\pi}{2}$.

Exercise 7.7

In Exercises 1–12, find the value of x, if any. **Do not use a calculator or tables.** Note that each answer should be a real number.

1. $\sin^{-1} \dfrac{1}{2} = x$

2. $\cos^{-1} \dfrac{\sqrt{3}}{2} = x$

3. $\cos^{-1} 0 = x$

4. $\sin^{-1} \sqrt{3} = x$

5. $\tan^{-1} 1 = x$ **6.** $\tan^{-1} 0 = x$

7. $\sin^{-1} 3 = x$ **8.** $\tan^{-1}(-\sqrt{3}) = x$

9. $\cos^{-1}\left(-\dfrac{\sqrt{2}}{2}\right) = x$ **10.** $\sin^{-1}\dfrac{\sqrt{2}}{2} = x$

11. $\arcsin\dfrac{\sqrt{3}}{2} = x$ **12.** $\arccos\dfrac{1}{2} = x$

In Exercises 13–24, find the sine, cosine, and tangent of θ. **Do not use a calculator or tables.**

13. $\sin^{-1}\dfrac{1}{2} = \theta$ **14.** $\cos^{-1}\dfrac{\sqrt{3}}{2} = \theta$

15. $\tan^{-1} 0 = \theta$ **16.** $\tan^{-1} 1 = \theta$

17. $\cos^{-1}\left(-\dfrac{\sqrt{3}}{2}\right) = \theta$ **18.** $\sin^{-1}\left(-\dfrac{\sqrt{3}}{2}\right) = \theta$

19. $\arcsin 1 = \theta$ **20.** $\arctan 1 = \theta$

21. $\cos^{-1}(-1) = \theta$ **22.** $\sin^{-1}\dfrac{\sqrt{2}}{2} = \theta$

23. $\arccos\dfrac{\sqrt{2}}{2} = \theta$ **24.** $\arcsin 0 = \theta$

In Exercises 25–36, find the value of x. **Do not use a calculator or tables.**

25. $\sin\left(\sin^{-1}\dfrac{1}{2}\right) = x$ **26.** $\cos\left(\cos^{-1}\dfrac{1}{2}\right) = x$

27. $\tan(\tan^{-1} 1) = x$ **28.** $\sin(\sin^{-1} 0) = x$

29. $\cos(\cos^{-1} 1) = x$ **30.** $\tan(\tan^{-1} 0) = x$

31. $\sin\left(\cos^{-1}\dfrac{\sqrt{3}}{2}\right) = x$ **32.** $\cos\left(\sin^{-1}\dfrac{1}{2}\right) = x$

33. $\tan\left[\sin^{-1}\left(-\dfrac{\sqrt{3}}{2}\right)\right] = x$ **34.** $\cot\left(\cos^{-1}\dfrac{\sqrt{2}}{2}\right) = x$

35. $\cos(\arctan 1) = x$ **36.** $\sin(\arctan 0) = x$

In Exercises 37–48, evaluate each expression without using a calculator or tables.

37. $\cos\left(\sin^{-1}\dfrac{4}{5}\right)$ **38.** $\sin\left(\cos^{-1}\dfrac{3}{5}\right)$

39. $\sin\left(\cos^{-1}\dfrac{5}{13}\right)$ **40.** $\cos\left[\sin^{-1}\left(-\dfrac{5}{13}\right)\right]$

41. $\tan\left[\sin^{-1}\left(-\dfrac{4}{5}\right)\right]$ **42.** $\tan\left(\cos^{-1}\dfrac{3}{5}\right)$

43. $\tan\left(\cos^{-1}\dfrac{5}{13}\right)$ **44.** $\tan\left[\sin^{-1}\left(-\dfrac{12}{13}\right)\right]$

45. $\cos\left[\tan^{-1}\left(-\dfrac{5}{12}\right)\right]$

46. $\sin\left(\tan^{-1}\dfrac{12}{5}\right)$

47. $\sin\left(\arccos\dfrac{9}{41}\right)$

48. $\cos\left(\arcsin\dfrac{40}{41}\right)$

In Exercises 49–60, find the required value without using a calculator or tables.

49. $\sin\left(\sin^{-1}\dfrac{1}{2}+\cos^{-1}\dfrac{1}{2}\right)$

50. $\sin\left(\sin^{-1}\dfrac{1}{2}-\cos^{-1}\dfrac{1}{2}\right)$

51. $\cos\left(\sin^{-1}\dfrac{4}{5}-\cos^{-1}\dfrac{4}{5}\right)$

52. $\cos\left(\sin^{-1}\dfrac{5}{13}+\cos^{-1}\dfrac{5}{13}\right)$

53. $\sin 2\left(\sin^{-1}\dfrac{\sqrt{2}}{2}\right)$

54. $\cos 2\left(\sin^{-1}\dfrac{\sqrt{2}}{2}\right)$

55. $\tan 2\left(\sin^{-1}\dfrac{\sqrt{2}}{2}\right)$

56. $\cot 2\left(\sin^{-1}\dfrac{\sqrt{2}}{2}\right)$

57. $\sin\dfrac{1}{2}\left(\cos^{-1}\dfrac{1}{2}\right)$

58. $\cos\dfrac{1}{2}\left(\cos^{-1}\dfrac{1}{2}\right)$

59. $\tan\dfrac{1}{2}\left(\arccos\dfrac{15}{17}\right)$

60. $\cot\dfrac{1}{2}\left(\arcsin\dfrac{8}{17}\right)$

In Exercises 61–72, rewrite each value as an algebraic expression in the variable x.

61. $\sin(\tan^{-1}x)$ **62.** $\cos(\tan^{-1}x)$ **63.** $\tan(\sin^{-1}x)$ **64.** $\tan(\cos^{-1}x)$

65. $\sin(\cos^{-1}x)$ **66.** $\cos(\sin^{-1}x)$ **67.** $\sin(2\sin^{-1}x)$ **68.** $\cos(2\cos^{-1}x)$

69. $\tan(2\arctan x)$ **70.** $\sin(2\arccos x)$ **71.** $\cos(2\sin^{-1}x)$ **72.** $\sin\left(\dfrac{1}{2}\sin^{-1}x\right)$

In Exercises 73–82, evaluate each of the expressions.

73. $\sin^{-1}\left(\sin\dfrac{11\pi}{6}\right)$

74. $\cos^{-1}\left(\cos\dfrac{7\pi}{6}\right)$

75. $\cos^{-1}\left(\sin\dfrac{3\pi}{4}\right)$

76. $\sin^{-1}\left(\cos\dfrac{5\pi}{4}\right)$

77. $\cot^{-1}\sqrt{3}$

78. $\cot^{-1}\left(-\dfrac{\sqrt{3}}{3}\right)$

79. $\sec^{-1}(-1)$

80. $\sec^{-1}(-2)$

81. $\csc^{-1}(-1)$

82. $\csc^{-1}2$

REVIEW EXERCISES

In Review Exercises 1–4, verify each identity.

1. $\dfrac{\sin\theta+\cos\theta\tan\theta}{\tan\theta}=2\cos\theta$

2. $\dfrac{\sin\theta}{\sec\theta}=\dfrac{1}{\tan\theta+\cot\theta}$

3. $2\tan\theta=\dfrac{\sin\theta}{\cos\theta}+\dfrac{\sec\theta}{\csc\theta}$

4. $\sec\theta-\tan\theta=\dfrac{1}{\sec\theta+\tan\theta}$

In Review Exercises 5–8, express each quantity as a single function of one angle.

5. $\sin 20° \cos 51° + \cos 20° \sin 51°$

6. $\dfrac{\tan 20° - \tan 51°}{1 + \tan 20° \tan 51°}$

7. $\cos \dfrac{3\pi}{11} \cos \dfrac{\pi}{11} - \sin \dfrac{3\pi}{11} \sin \dfrac{\pi}{11}$

8. $1 - 2 \sin^2 \dfrac{\pi}{3}$

In Review Exercises 9–12, verify each identity.

9. $\cos(60° + \theta) = \dfrac{1}{2} (\cos \theta - \sqrt{3} \sin \theta)$

10. $\tan(180° - \theta) = -\tan \theta$

11. $\sin x = \dfrac{\sin 2x}{2 \cos x}$

12. $\cos \theta = \pm \sqrt{\sin^2 \theta + \cos 2\theta}$

In Review Exercises 13–18, use the given information to find the sine, cosine, and tangent of 2θ.

13. $\cos \theta = -\dfrac{12}{13}$; θ in QII

14. $\sin \theta = \dfrac{-5}{13}$; θ in QIII

15. $\tan \theta = \dfrac{-20}{21}$; θ in QIV

16. $\cos \theta = \dfrac{21}{29}$; θ in QI

17. $\sin \theta = -\dfrac{4}{5}$; θ in QIII

18. $\cot \theta = -\dfrac{3}{4}$; θ in QII

In Review Exercises 19–24, use the given information to find the sine, cosine, and tangent of $\frac{\theta}{2}$. Assume that $0° \le \theta < 360°$.

19. $\cos \theta = \dfrac{12}{13}$; θ in QI

20. $\sin \theta = -\dfrac{5}{13}$; θ in QIV

21. $\tan \theta = -\dfrac{5}{12}$; θ in QIV

22. $\cos \theta = -\dfrac{3}{5}$; θ in QII

23. $\sin \theta = \dfrac{4}{5}$; θ in QII

24. $\cot \theta = \dfrac{3}{4}$; θ in QIII

*In Review Exercises 25–28, evaluate each of the products. **Do not use a calculator or tables.***

25. $\sin 285° \cos 15°$

26. $\cos 285° \sin 15°$

27. $\sin \dfrac{5\pi}{4} \sin \dfrac{\pi}{12}$

28. $\cos \dfrac{\pi}{12} \cos \dfrac{5\pi}{4}$

In Review Exercises 29–32, express each quantity as a product.

29. $\sin 5° + \sin 7°$

30. $\sin 312° - \sin 140°$

31. $\cos \dfrac{3\pi}{5} - \cos \dfrac{\pi}{5}$

32. $\cos \dfrac{2\pi}{7} + \cos \dfrac{3\pi}{7}$

In Review Exercises 33–36, express each quantity as a product and evaluate without using a calculator or tables.

33. $\sin 285° + \sin 15°$

34. $\sin 15° - \sin 285°$

35. $\cos \dfrac{5\pi}{12} - \cos \dfrac{\pi}{12}$

36. $\cos \dfrac{\pi}{12} + \cos \dfrac{5\pi}{12}$

In Review Exercises 37–38, write each expression in the form $k \sin(x + \phi)$.

37. $y = \sin x + 2 \cos x$

38. $y = -\sin x + \cos x$

In Review Exercises 39–42, solve each equation for all values of the variable between 0 and 2π, including 0. **Do not use a calculator or tables.**

39. $\sin 2x = 1$

40. $\sin x + 1 = \tan x + \cos x$

41. $\cos^2 x - \sin^2 x = 0$

42. $\csc x \sec x = \sec x + \cot x$

In Review Exercises 43–48, find the number x, if it exists. **Do not use a calculator or tables.**

43. $\sin^{-1} \dfrac{\sqrt{3}}{2} = x$

44. $\cos^{-1} \left(-\dfrac{1}{2} \right) = x$

45. $\tan^{-1}(-\sqrt{3}) = x$

46. $\cos^{-1} \sqrt{3} = x$

47. $\cot^{-1} 0 = x$

48. $\csc^{-1}(0.5) = x$

In Review Exercises 49–54, find each value of θ, if any. Express all answers in radians. **Do not use a calculator or tables.**

49. $\sin^{-1} \left(-\dfrac{1}{2} \right) = \theta$

50. $\tan^{-1} \dfrac{\sqrt{3}}{3} = \theta$

51. $\cos^{-1} 0 = \theta$

52. $\operatorname{arcsec}(-1) = \theta$

53. $\arcsin 0 = \theta$

54. $\arccos 1 = \theta$

In Review Exercises 55–58, find the required value without using a calculator or tables.

55. $\sin \left[\sin^{-1} \left(-\dfrac{1}{2} \right) + \cos^{-1} \left(-\dfrac{1}{2} \right) \right]$

56. $\sin 2 \left(\sin^{-1} \dfrac{1}{2} \right)$

57. $\sin \dfrac{1}{2} \left[\arccos \left(-\dfrac{1}{2} \right) \right]$

58. $\tan 2(\arcsin 1)$

In Review Exercises 59–60, rewrite each value as an algebraic expression in the variable u.

59. $\sin(\cos^{-1} u)$

60. $\tan(2 \tan^{-1} u)$

8

LINEAR SYSTEMS

Any collection of m equations in n variables is called a **system of equations**. For example, the equations

$$\begin{cases} 3x + y = 1 \\ -x + 2y = 9 \end{cases}$$

form a system of two linear equations in two variables. In this chapter, we discuss several methods for solving such systems.

8.1 SYSTEMS OF LINEAR EQUATIONS

Consider the system of linear equations

$$\begin{cases} 3x + y = 1 \\ -x + 2y = 9 \end{cases}$$

There are infinitely many ordered pairs of real numbers (x, y) that satisfy the first equation, and there are infinitely many ordered pairs of real numbers (x, y) that satisfy the second equation. However, there is only one ordered pair of real numbers (x, y) that satisfies both equations simultaneously. This ordered pair is called the **simultaneous solution** or just the **solution** of the system of equations. The process of finding a solution to a system of equations is called **solving the system**. We begin by discussing three methods for solving the previous system of equations.

The Graphing Method

Use the following steps to solve a system of two equations in two variables by graphing.

---

1. On a single coordinate grid, graph each equation.
2. Read the coordinates of the point or points where *all* of the graphs intersect. These coordinates give (often approximately) the solutions of the system.
3. If no point is common to all of the graphs, the system has no solution.

The results of Example 1 illustrate the various possibilities that can occur when graphing two linear equations in two variables.

Example 1   Use the graphing method to solve each system:

a. $\begin{cases} 3x + y = 1 \\ -x + 2y = 9 \end{cases}$ **b.** $\begin{cases} 2x - 3y = 4 \\ 4x = -4 + 6y \end{cases}$ **c.** $\begin{cases} y = 4 - x \\ 2x + 2y = 8 \end{cases}$

Solution   **a.**   The graph of each equation is a straight line. See Figure 8-1**a**. The solution of this system is given by the coordinates of the point where the lines intersect, $(-1, 4)$. Thus, the solution of the system is $x = -1$ and $y = 4$. Verify that these values satisfy each equation.

b.   The graph of each equation is a straight line, and the lines are parallel. See Figure 8-1**b**. Thus, this system has no solutions.

c.   The graph of each equation is a straight line, and they coincide. See Figure 8-1**c**. Thus, this system has infinitely many solutions. All ordered pairs whose coordinates satisfy one of the equations satisfy the other also.

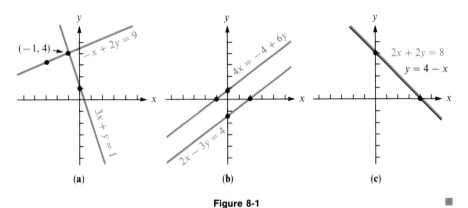

(a)   **(b)**   **(c)**

Figure 8-1   ■

If a system of equations has at least one solution, as in part **a** of Example 1, the system is called **consistent**. If it has no solutions, as in part **b**, it is called **inconsistent**.

If a system of n linear equations in n variables has exactly one solution, as in part **a**, or no solutions, as in part **b**, the equations in the system are called **independent**. If a system of linear equations has infinitely many solutions, as in part **c**, the equations of the system are called **dependent**.

The Substitution Method

Use the following steps to solve a system of two equations in two variables by substitution.

1. Solve one equation for a variable, say y.
2. Substitute the expression obtained for y for all occurrences of y in the other equation.
3. Solve the resulting equation.
4. Substitute the solution found in Step 3 in the equation found in Step 1 and solve for y.

Example 2 Use the substitution method to solve the system

$$\begin{cases} 3x + y = 1 \\ -x + 2y = 9 \end{cases}$$

Solution Solve the first equation for y to obtain $y = 1 - 3x$. Then, substitute $1 - 3x$ for y in the second equation.

$$\begin{cases} 3x + y = 1 \longrightarrow y = \boxed{1 - 3x} \\ -x + 2y = 9 \end{cases}$$

This substitution gives one linear equation in one variable, which can be solved for x:

$$\begin{aligned} -x + 2(1 - 3x) &= 9 \\ -x + 2 - 6x &= 9 \\ -7x &= 7 \\ x &= -1 \end{aligned}$$

To find y, substitute -1 for x in the equation $y = 1 - 3x$ and simplify:

$$\begin{aligned} y &= 1 - 3x \\ &= 1 - 3(-1) \\ &= 1 + 3 \\ &= 4 \end{aligned}$$

The solution to the system is $(-1, 4)$. ■

The Addition Method

The addition method is based on three algebraic manipulations that will transform a system of equations into an **equivalent system**, which has the same solutions.

1. The positions of any two equations of the system can be interchanged.
2. Both sides of any equation of the system can be multiplied or divided by a nonzero constant.
3. Any equation of the system can be altered by adding to its sides a constant multiple of the corresponding sides of another equation of the system.

Example 3 Use the addition method to solve the system

$$\begin{cases} 3x + y = 1 \\ -x + 2y = 9 \end{cases}$$

Solution Multiply both sides of the second equation by 3, and add the result to the first equation to eliminate the variable x:

$$\begin{aligned} 3x + y &= 1 \\ -3x + 6y &= 27 \\ \hline 7y &= 28 \\ y &= 4 \end{aligned} \quad \text{Divide both sides by 7.}$$

Substitute 4 for y into either of the original equations and solve for x. For example, use the first equation:

$$\begin{aligned} 3x + y &= 1 \\ 3x + 4 &= 1 \\ 3x &= -3 \\ x &= -1 \end{aligned}$$

Thus, the solution of the given system is $(-1, 4)$. ■

Example 4 Use the addition method to solve the system

$$\begin{cases} x + 2y = 3 \\ 2x + 4y = 6 \end{cases}$$

Solution Multiply both sides of the first equation by -2, and add the results to the second equation:

$$\begin{cases} -2x - 4y = -6 \\ \underline{2x + 4y = 6} \\ 0 = 0 \end{cases}$$

Although the result $0 = 0$ is true, it does not lead to a solution for y. Because the second equation is twice the first, the two equations in this system are essentially the same; they are equivalent. If each equation were graphed, the resulting lines would coincide. The pairs of coordinates of all points on the *one* line described by this system make up the infinite set of solutions to the given system. The system is consistent, and the equations are dependent. ■

Example 5 Solve the system

$$\begin{cases} x + y = 3 \\ x + y = 2 \end{cases}$$

Solution Multiply both sides of the second equation by -1, and add the results to the first equation:

$$\begin{cases} x + y = 3 \\ \underline{-x - y = -2} \\ 0 = 1 \end{cases}$$

Because this result is impossible, the system has no solutions. If each equation in this system were graphed, the graphs would be parallel lines. This system is inconsistent, and the equations are independent.   ■

Example 6 shows how to use the substitution and addition methods to solve a system of three equations in three variables.

Example 6   Solve the system

$$\begin{cases} \textbf{1.} & x + 2y + z = 8 \\ \textbf{2.} & 2x + y - z = 1 \\ \textbf{3.} & x + y - 2z = -3 \end{cases}$$

Solution   Use Equations 1 and 2 to eliminate the variable z. Then use Equations 1 and 3 to eliminate the variable z. You can then solve the resulting system of two equations in the two variables x and y.

Add Equations 1 and 2 to eliminate the variable z:

$$\begin{array}{ll} \textbf{1.} & x + 2y + z = 8 \\ \textbf{2.} & \underline{2x + y - z = 1} \\ & 3x + 3y = 9 \end{array}$$

Divide both sides of the resulting equation by 3 to obtain

$$\textbf{4.} \qquad x + y = 3$$

The variable z can be eliminated again by using Equations 1 and 3. Multiply both sides of Equation 1 by the constant 2, and add the result to Equation 3:

$$\begin{array}{ll} & 2x + 4y + 2z = 16 \\ \textbf{3.} & \underline{x + y - 2z = -3} \\ \textbf{5.} & 3x + 5y = 13 \end{array}$$

Equations 4 and 5 form a system of two equations in two variables. Solve this system by substitution as follows:

$$\begin{array}{ll} \textbf{4.} & x + y = 3 \longrightarrow y = \boxed{3 - x} \\ \textbf{5.} & 3x + 5y = 13 \end{array}$$

$$\begin{aligned} 3x + 5(3 - x) &= 13 \\ 3x + 15 - 5x &= 13 \\ -2x &= -2 \\ x &= 1 \end{aligned}$$

Substitute 1 for x in the equation $y = 3 - x$ to find y:

$$y = 3 - 1$$
$$y = 2$$

To find z, substitute 1 for x, and 2 for y in any of the original equations. You will find that $z = 3$. The solution to the given system is the triple

(1, 2, 3)

Because this solution is unique, the given system is consistent and the equations in the system are independent. Verify that $x = 1$, $y = 2$, and $z = 3$ satisfy each equation in the original system. ∎

Example 7   An airplane flies 600 miles with the wind for 2 hours. The return trip against the wind takes 3 hours. Find the speed of the wind and the airspeed of the plane.

Solution   Let a represent the airspeed of the plane, and w represent the speed of the wind. The groundspeed of the plane on the outbound trip is the combined speed $a + w$. On the return trip, against a head wind, the groundspeed is $a - w$. The information of this problem, organized in the chart in Figure 8-2, gives a system of two equations in the two variables a and w.

	d	$=$	r	$\cdot$	t
outbound trip	600		$a + w$		2
return trip	600		$a - w$		3

Figure 8-2

Because $d = rt$, you have

$$\begin{cases} 600 = 2(a + w) \\ 600 = 3(a - w) \end{cases}$$

or

$$\begin{cases} 300 = a + w \\ 200 = a - w \end{cases}$$

Add the equations together to get

$$500 = 2a$$

or

$$a = 250$$

To calculate w, substitute 250 for a in a previous equation such as $300 = a + w$ and solve for w:

$$300 = a + w$$
$$300 = 250 + w$$
$$w = 50$$

The plane could do 250 miles per hour in still air. With a 50-mile-per-hour tail wind, the groundspeed would be 300 miles per hour, and the 600-mile trip would take 2 hours. With a 50-mile-per-hour head wind, the groundspeed would be 200 miles per hour, and the 600-mile trip would take 3 hours. The answers check.   ∎

Exercise 8.1

In Exercises 1–4, solve each system of equations by the graphing method.

1. $\begin{cases} 3x + y = 5 \\ x - 2y = -3 \end{cases}$
2. $\begin{cases} x - 2y = -3 \\ 3x + y = -9 \end{cases}$
3. $\begin{cases} \dfrac{3x}{2} + y = 1 \\ -\dfrac{x}{8} + \dfrac{3y}{16} = 1 \end{cases}$
4. $\begin{cases} x + y = 0 \\ 7x + 8y = 1 \end{cases}$

In Exercises 5–12, solve each system of equations by the substitution method, if possible.

5. $\begin{cases} y = x \\ y = 2x \end{cases}$
6. $\begin{cases} 2y = x \\ 3y = 2x \end{cases}$
7. $\begin{cases} 2x + 3y = 0 \\ y = 3x - 11 \end{cases}$
8. $\begin{cases} 2x + y = 3 \\ y = 5x - 11 \end{cases}$

9. $\begin{cases} \dfrac{5}{2}x + y = 8 \\ 2x - \dfrac{2}{3}y = 2 \end{cases}$
10. $\begin{cases} \dfrac{x + y}{3} - \dfrac{x - y}{2} = 1 \\ x = 2y \end{cases}$

11. $\begin{cases} x + 3y = 1 \\ 2x + 6y = 3 \end{cases}$
12. $\begin{cases} x - 3y = 14 \\ 3(x - 12) = 9y \end{cases}$

In Exercises 13–24, solve each system of equations by the addition method, if possible.

13. $\begin{cases} 5x - 3y = 12 \\ 2x - 3y = 3 \end{cases}$
14. $\begin{cases} 2x + 3y = 8 \\ -5x + y = -3 \end{cases}$

15. $\begin{cases} x - 7y = -11 \\ 8x + 2y = 28 \end{cases}$
16. $\begin{cases} 3x + 9y = 9 \\ -x + 5y = -3 \end{cases}$

17. $\begin{cases} 3(x - y) = y - 9 \\ 5(x + y) = -15 \end{cases}$
18. $\begin{cases} 2(x + y) = y + 1 \\ 3(x + 1) = y - 3 \end{cases}$

19. $\begin{cases} 2 = \dfrac{1}{x + y} \\ 2 = \dfrac{3}{x - y} \end{cases}$
20. $\begin{cases} \dfrac{1}{x + y} = 12 \\ \dfrac{3x}{y} = -4 \end{cases}$

21. $\begin{cases} 5x = 10y \\ -x + 2y = 0 \end{cases}$
22. $\begin{cases} -0.3x + 0.1y = -0.1 \\ 6x - 2y = 2 \end{cases}$

23. $\begin{cases} \dfrac{x + y}{2} + \dfrac{x - y}{5} = 2 \\ 2x = y + 2 \end{cases}$
24. $\begin{cases} \dfrac{3}{2}x + \dfrac{1}{3}y = 2 \\ \dfrac{2}{3}x + \dfrac{1}{9}y = 1 \end{cases}$

In Exercises 25–40, solve each system of equations, if possible.

25.
$$\begin{cases} x + y + z = 3 \\ 2x + y + z = 4 \\ 3x + y - z = 5 \end{cases}$$

26.
$$\begin{cases} x - y - z = 0 \\ x + y - z = 0 \\ x - y + z = 2 \end{cases}$$

27.
$$\begin{cases} x - y + z = 0 \\ x + y + 2z = -1 \\ -x - y + z = 0 \end{cases}$$

28.
$$\begin{cases} 2x + y - z = 7 \\ x - y + z = 2 \\ x + y - 3z = 2 \end{cases}$$

29.
$$\begin{cases} 2x + y = 4 \\ x - z = 2 \\ y + z = 1 \end{cases}$$

30.
$$\begin{cases} 3x + y + z = 0 \\ 2x - y + z = 0 \\ 2x + y + z = 0 \end{cases}$$

31.
$$\begin{cases} x + y + z = 6 \\ 2x + y + 3z = 17 \\ x + y + 2z = 11 \end{cases}$$

32.
$$\begin{cases} x + y + z = 3 \\ 2x + y + z = 6 \\ x + 2y + 3z = 2 \end{cases}$$

33.
$$\begin{cases} x + y + z = 3 \\ x + z = 2 \\ 2x + 2y + 2z = 3 \end{cases}$$

34.
$$\begin{cases} x + y + z = 3 \\ x + z = 2 \\ 2x + y + 2z = 5 \end{cases}$$

35.
$$\begin{cases} x + y = 2 \\ y + z = 2 \\ x - z = 0 \end{cases}$$

36.
$$\begin{cases} x + y = 2 \\ y + z = 2 \\ 3x + 3y = 2 \end{cases}$$

37.
$$\begin{cases} x + y + z = 4 \\ 2x + y + z = 5 \\ 3x + 2y + z = 7 \end{cases}$$

38.
$$\begin{cases} x + 2y - z = 2 \\ 2x - y = -1 \\ 3x + y + z = 1 \end{cases}$$

39.
$$\begin{cases} (x + y) + (y + z) + (z + x) = 6 \\ (x - y) + (y - z) + (z - x) = 0 \\ x + y + 2z = 4 \end{cases}$$

40.
$$\begin{cases} (x + y) + (y + z) = 1 \\ (x + z) + (x + y) = 3 \\ (x - y) - (x - z) = -1 \end{cases}$$

41. Suppose that a retailer obtains a product from three sources, *A*, *B*, and *C*. The retailer buys as many units from source *A* as from the other two combined. She must pay source *A* at the rate of $4 per unit and sources *B* and *C* at the rate of $5 per unit. If she requires 100 units per month to satisfy customer demand, and if her cost for one month's supply is $450, how many units does she buy from each source?

42. Flutter Hi-Fi, Wow Stereo, and Rumble Electronics buy a total of 175 cassette tape decks from High Hiss Distributors each month. Because Rumble Electronics buys 25 more tape decks than the other two stores combined, Rumble's cost is only $160 per unit. The decks cost Wow Stereo $165 each and Flutter Hi-Fi $170 each. How many decks does each retailer buy each month if High Hiss receives $28,500 each month from the sale of tape decks to these companies?

43. A college student earns $99.25 per week working three part-time jobs. Half of the student's 30-hour work week is spent cooking hamburgers at a fast food chain, earning $2.85 per hour. In addition, the student earns $3.15 per hour working at a gas station and $5 per hour doing janitorial work. How many hours per week does the student work at each job?

44. A factory manufactures widgets, gidgets, and gadgets at a monthly cost of $6850 for 2150 units. It costs $2 to make a widget, $3 to make a gidget, and $4 to make a gadget. A widget sells for $3, a gidget for $4.50, and a gadget for $5.50. The monthly profit is $2975. How many of each item are manufactured?

45. A collection of nickels, dimes, and quarters has a value of $3.40. There are twice as many dimes as quarters, and there are 32 coins in all. How many of each kind are there?

46. The sum of the angles of a triangle is 180°. The largest angle is 20° greater than the sum of the other two, and is 10° greater than 3 times the smallest. How large is each angle?

8.2 GAUSSIAN ELIMINATION AND MATRIX METHODS

There is a method, called **Gaussian elimination**, that can be used to solve systems of linear equations. In this method, a system of equations is transformed into an equivalent system that can be solved by a process called **back substitution**.

In Gaussian elimination, it is convenient to solve a system of equations by working with only the coefficients of the variables. All of the information needed to find a solution of

$$\begin{cases} x + 2y + z = 8 \\ 2x + y - z = 1 \\ x + y - 2z = -3 \end{cases}$$

for example, is contained in the following rectangular array of numbers, called a **matrix**:

$$\begin{bmatrix} 1 & 2 & 1 & | & 8 \\ 2 & 1 & -1 & | & 1 \\ 1 & 1 & -2 & | & -3 \end{bmatrix}$$

Each row in this matrix represents one of the equations of the system. The first row, for example, represents the equation $x + 2y + z = 8$.

Because the matrix above has three rows and four columns, it is called a 3×4 (read as "3 by 4") matrix. The 3×3 matrix to the left of the broken line is called the **coefficient matrix**. The entire matrix is called the **augmented matrix**.

Example 1 Use Gaussian elimination to solve the system

1. $\begin{cases} x + 2y + z = 8 \\ \textbf{2. } 2x + y - z = 1 \\ \textbf{3. } x + y - 2z = -3 \end{cases}$

Solution First multiply each term in Equation 1 by -2, and add the result to Equation 2 to obtain Equation 4 below. Then multiply each term of Equation 1 by -1, and add the result to Equation 3 to obtain Equation 5. This gives an equivalent system with the same solution as the original system. The system is shown both in equation form and in matrix form. Note that the variable x does not appear in Equation 4 or 5:

1. $\begin{cases} x + 2y + z = 8 \\ -3y - 3z = -15 \\ -y - 3z = -11 \end{cases}$ $\begin{bmatrix} 1 & 2 & 1 & | & 8 \\ 0 & -3 & -3 & | & -15 \\ 0 & -1 & -3 & | & -11 \end{bmatrix}$

4.

5.

Divide both sides of Equation 4 by -3 to obtain Equation 6:

1. $\begin{cases} x + 2y + z = 8 \\ y + z = 5 \\ -y - 3z = -11 \end{cases}$ $\begin{bmatrix} 1 & 2 & 1 & | & 8 \\ 0 & 1 & 1 & | & 5 \\ 0 & -1 & -3 & | & -11 \end{bmatrix}$

6.

5.

Add Equation 6 to Equation 5 to obtain Equation 7:

1. $\begin{cases} x + 2y + z = 8 \\ y + z = 5 \\ -2z = -6 \end{cases}$ $\begin{bmatrix} 1 & 2 & 1 & | & 8 \\ 0 & 1 & 1 & | & 5 \\ 0 & 0 & -2 & | & -6 \end{bmatrix}$

6.

7.

Divide both sides of Equation 7 by -2 to obtain the system

1. $\begin{cases} x + 2y + z = 8 \\ y + z = 5 \\ z = 3 \end{cases}$ $\begin{bmatrix} 1 & 2 & 1 & | & 8 \\ 0 & 1 & 1 & | & 5 \\ 0 & 0 & 1 & | & 3 \end{bmatrix}$

6.

The system can now be solved by back substitution. Because $z = 3$, you can substitute 3 for z in Equation 6, and solve for y:

6. $y + z = 5$

$y + 3 = 5$

$y = 2$

Substitute 2 for y and 3 for z in Equation 1, and solve for x:

1. $x + 2y + z = 8$

$x + 2(2) + 3 = 8$

$x = 1$

The solution of the given system of equations is the ordered triple

$(1, 2, 3)$

Verify that $x = 1$, $y = 2$, and $z = 3$ satisfy each of the original three equations.

■

Note the triangular formation of 0s in the final matrix in Example 1. A matrix such as

$$\begin{bmatrix} 1 & 2 & 1 & | & 8 \\ 0 & 1 & 1 & | & 5 \\ 0 & 0 & 1 & | & 3 \end{bmatrix}$$

is said to be in **triangular form** if all entries below the diagonal running from the upper left to the lower right are 0.

Example 2   Use matrices to solve the system

$$\begin{cases} x + 2y + 3z = 4 \\ 2x - y - 2z = 0 \\ x - 3y - 3z = -2 \end{cases}$$

Solution   This system is represented by the augmented matrix

$$\begin{bmatrix} 1 & 2 & 3 & \vdots & 4 \\ 2 & -1 & -2 & \vdots & 0 \\ 1 & -3 & -3 & \vdots & -2 \end{bmatrix}$$

Change the individual rows of this matrix by adding multiples of one row to another. When combining rows in this way, you are adding multiples of the coefficients of one equation to the corresponding coefficients of another, always obtaining coefficients that form an equivalent set of equations. The goal is to produce a matrix in triangular form that represents an equivalent system of equations.

To begin, use the 1 in the upper left-hand corner of the augmented matrix to "zero out" the rest of the first column. The notation "$(-2)R1 + R2 \rightarrow R2$" means "multiply row one by -2 and add the result to row two to get a new row two."

$$(-2)R1 + R2 \rightarrow R2$$

$$\begin{bmatrix} 1 & 2 & 3 & \vdots & 4 \\ 2 & -1 & -2 & \vdots & 0 \\ 1 & -3 & -3 & \vdots & -2 \end{bmatrix} \Leftrightarrow \begin{bmatrix} 1 & 2 & 3 & \vdots & 4 \\ 0 & -5 & -8 & \vdots & -8 \\ 1 & -3 & -3 & \vdots & -2 \end{bmatrix}$$
Read $\Leftrightarrow$ as "is equivalent to."

Next, multiply row one by -1, and add the result to row three to get a new row three:

$$(-1)R1 + R3 \rightarrow R3$$

$$\begin{bmatrix} 1 & 2 & 3 & \vdots & 4 \\ 0 & -5 & -8 & \vdots & -8 \\ 1 & -3 & -3 & \vdots & -2 \end{bmatrix} \Leftrightarrow \begin{bmatrix} 1 & 2 & 3 & \vdots & 4 \\ 0 & -5 & -8 & \vdots & -8 \\ 0 & -5 & -6 & \vdots & -6 \end{bmatrix}$$

Because each row of a matrix represents an equation, multiplying an entire row by a constant is equivalent to multiplying both sides of an equation by a constant. Hence, multiply row two by -1 to eliminate some minus signs and add the result to row three to get

$$(-1)R2 \rightarrow R2 \qquad\qquad R2 + R3 \rightarrow R3$$

$$\begin{bmatrix} 1 & 2 & 3 & \vdots & 4 \\ 0 & -5 & -8 & \vdots & -8 \\ 0 & -5 & -6 & \vdots & -6 \end{bmatrix} \Leftrightarrow \begin{bmatrix} 1 & 2 & 3 & \vdots & 4 \\ 0 & 5 & 8 & \vdots & 8 \\ 0 & -5 & -6 & \vdots & -6 \end{bmatrix} \Leftrightarrow \begin{bmatrix} 1 & 2 & 3 & \vdots & 4 \\ 0 & 5 & 8 & \vdots & 8 \\ 0 & 0 & 2 & \vdots & 2 \end{bmatrix}$$

Finally, multiply the third row by $\frac{1}{2}$:

$$(\tfrac{1}{2})R3 \rightarrow R3$$

$$\begin{bmatrix} 1 & 2 & 3 & \vdots & 4 \\ 0 & 5 & 8 & \vdots & 8 \\ 0 & 0 & 2 & \vdots & 2 \end{bmatrix} \Leftrightarrow \begin{bmatrix} 1 & 2 & 3 & \vdots & 4 \\ 0 & 5 & 8 & \vdots & 8 \\ 0 & 0 & 1 & \vdots & 1 \end{bmatrix}$$

This final matrix is in triangular form and represents the system

1. $\quad x + 2y + 3z = 4$
2. $\quad\quad\quad 5y + 8z = 8$
3. $\quad\quad\quad\quad\quad z = 1$

To solve this system by back substitution, substitute **1** for z in Equation 2, and solve for y:

2. $\quad 5y + 8z = 8$
$\quad\quad 5y + 8(1) = 8$
$\quad\quad\quad\quad\quad y = 0$

Then, substitute **0** for y and **1** for z in Equation 1, and solve for x:

1. $\quad\quad x + 2y + 3z = 4$
$\quad\quad x + 2(0) + 3(1) = 4$
$\quad\quad\quad\quad\quad\quad x = 1$

The solution of the original system is the ordered triple

$$(1, 0, 1)$$

Verify that $x = 1$, $y = 0$, and $z = 1$ satisfy each of the original equations.   ■

In the preceding example, one matrix was transformed into another by certain manipulations called **elementary row operations**.

Elementary Row Operations.

Type 1 row operation: two rows of a matrix can be interchanged.

Type 2 row operation: the elements of a row of a matrix can be multiplied by a nonzero constant.

Type 3 row operation: a row of a matrix can be altered by adding to it a multiple of any other row.

Each type of elementary row operation yields a new matrix representing a system of equations with the same solution as the original system. A type 1 row operation is equivalent to writing the equations of a system in a different order. A type 2 row operation is equivalent to multiplying both sides of an equation by a nonzero constant, and a type 3 row operation is equivalent to adding a multiple of one equation to another.

In a type 3 row operation, it is important to remember which row is being changed and which row is causing that change. For example, if twice row three is added to row one, it is row one that is being altered. Row three, which was used to accomplish the change, stays the same.

Any matrix that can be obtained from another matrix by a sequence of elementary row operations is called **row equivalent** to the original matrix. Thus, the symbol $\Leftrightarrow$ can be read as "is row equivalent to."

If a matrix has the following properties, it is said to be in **echelon form**:

1. The lead entry (the first nonzero entry) of each row is 1.
2. Lead entries appear farther to the right as you move down the rows of the matrix.
3. Rows containing only 0s are at the bottom of the matrix.

Example 3   Use matrix methods to solve the following system of four equations in three variables:

$$\begin{cases} x + 2y - z = 6 \\ x - y + z = -2 \\ 2x \quad + z = 1 \\ x + y + z = 2 \end{cases}$$

Solution   Note that this system has more equations than it has variables. To solve it, form the augmented matrix and use elementary row operations to write the augmented matrix in echelon form.

$$(-1)R1 + R2 \rightarrow R2$$

$$\begin{bmatrix} 1 & 2 & -1 & | & 6 \\ 1 & -1 & 1 & | & -2 \\ 2 & 0 & 1 & | & 1 \\ 1 & 1 & 1 & | & 2 \end{bmatrix} \Leftrightarrow \begin{bmatrix} 1 & 2 & -1 & | & 6 \\ 0 & -3 & 2 & | & -8 \\ 2 & 0 & 1 & | & 1 \\ 1 & 1 & 1 & | & 2 \end{bmatrix}$$

$$(-2)R1 + R3 \rightarrow R3$$
$$(-1)R1 + R4 \rightarrow R4$$

$$\Leftrightarrow \begin{bmatrix} 1 & 2 & -1 & | & 6 \\ 0 & -3 & 2 & | & -8 \\ 0 & -4 & 3 & | & -11 \\ 0 & -1 & 2 & | & -4 \end{bmatrix}$$

$$(-1)R4 \rightarrow R4$$

$$\Leftrightarrow \begin{bmatrix} 1 & 2 & -1 & | & 6 \\ 0 & -3 & 2 & | & -8 \\ 0 & -4 & 3 & | & -11 \\ 0 & 1 & -2 & | & 4 \end{bmatrix}$$

$(4)R4 + R3 \rightarrow R3$
$(3)R4 + R2 \rightarrow R2$

$$\Leftrightarrow \begin{bmatrix} 1 & 2 & -1 & | & 6 \\ 0 & 0 & -4 & | & 4 \\ 0 & 0 & -5 & | & 5 \\ 0 & 1 & -2 & | & 4 \end{bmatrix}$$

$(-\frac{1}{4})R2 \rightarrow R2$
$(-\frac{1}{5})R3 \rightarrow R3$

$$\Leftrightarrow \begin{bmatrix} 1 & 2 & -1 & | & 6 \\ 0 & 0 & 1 & | & -1 \\ 0 & 0 & 1 & | & -1 \\ 0 & 1 & -2 & | & 4 \end{bmatrix}$$

$R2 \leftrightarrow R4$ (exchange $R2$ and $R4$)

$$\Leftrightarrow \begin{bmatrix} 1 & 2 & -1 & | & 6 \\ 0 & 1 & -2 & | & 4 \\ 0 & 0 & 1 & | & -1 \\ 0 & 0 & 1 & | & -1 \end{bmatrix}$$

$(-1)R3 + R4 \rightarrow R4$

$$\Leftrightarrow \begin{bmatrix} 1 & 2 & -1 & | & 6 \\ 0 & 1 & -2 & | & 4 \\ 0 & 0 & 1 & | & -1 \\ 0 & 0 & 0 & | & 0 \end{bmatrix}$$

This final matrix is in echelon form. All lead entries are 1, the lead entries appear farther to the right as you move down the rows of the matrix, and the row of 0s is last. This matrix represents the following system of equations, which can be solved by back substitution:

1. $\quad x + 2y - z = 6$
2. $\quad\quad\quad y - 2z = 4$
3. $\quad\quad\quad\quad\quad z = -1$
4. $\quad 0x + 0y - 0z = 0$

Because Equation 4 is satisfied by *all* numbers x, y, and z, it is unnecessary and can be ignored. From Equation 3, you know that $z = -1$. So, substitute -1 for z in Equation 2 and solve for y:

2. $\quad\quad y - 2z = 4$
$$y - 2(-1) = 4$$
$$y + 2 = 4$$
$$y = 2$$

Then, substitute -1 for z and 2 for y in Equation 1 and solve for x:

1. $x + 2y - z = 6$

$x + 2(2) - (-1) = 6$

$x + 4 + 1 = 6$

$x = 1$

Thus, the solution of the original system of equations is the ordered triple

$$(x, y, z) = (1, 2, -1)$$

The system is consistent because it has a solution. Check this solution to verify that it satisfies each of the four original equations.   ∎

Example 4   Use matrices to solve the system

$$\begin{cases} x + 2y + z = 8 \\ 2x + y - z = 1 \\ x - y - 2z = -7 \end{cases}$$

Solution   Set up the augmented matrix for the system, and use row operations to reduce it to echelon form:

$$(-2)R1 + R2 \rightarrow R2$$
$$(-1)R1 + R3 \rightarrow R3$$

$$\begin{bmatrix} 1 & 2 & 1 & | & 8 \\ 2 & 1 & -1 & | & 1 \\ 1 & -1 & -2 & | & -7 \end{bmatrix} \Leftrightarrow \begin{bmatrix} 1 & 2 & 1 & | & 8 \\ 0 & -3 & -3 & | & -15 \\ 0 & -3 & -3 & | & -15 \end{bmatrix}$$

$$(-\tfrac{1}{3})R2 \rightarrow R2$$
$$(-\tfrac{1}{3})R3 \rightarrow R3$$

$$\Leftrightarrow \begin{bmatrix} 1 & 2 & 1 & | & 8 \\ 0 & 1 & 1 & | & 5 \\ 0 & 1 & 1 & | & 5 \end{bmatrix}$$

$$(-1)R2 + R3 \rightarrow R3$$

$$\Leftrightarrow \begin{bmatrix} 1 & 2 & 1 & | & 8 \\ 0 & 1 & 1 & | & 5 \\ 0 & 0 & 0 & | & 0 \end{bmatrix}$$

This matrix is in echelon form and represents the system

1. $\begin{cases} x + 2y + z = 8 \\ y + z = 5 \\ 0x + 0y + 0z = 0 \end{cases}$
2.
3.

As before, the bottom equation can be ignored. Solve this system by back substitution. First, solve Equation 2 for the variable y:

$$y = 5 - z$$

Then, substitute $5 - z$ for y in Equation 1 and solve for x:

1.
$$x + 2y + z = 8$$
$$x + 2(5 - z) + z = 8$$
$$x + 10 - z = 8$$
$$x = -2 + z$$

The solution of this system is

$$(x, y, z) = (-2 + z, 5 - z, z)$$

There is no unique solution for this system. The variable z can be any real number, but once you pick a number z, the numbers x and y are determined. For example, if z was equal to 3, then x would equal $-2 + 3$, or 1, and y would equal $5 - 3$, or 2. Thus, one possible solution to this system is $x = 1$, $y = 2$, and $z = 3$. Picking $z = 2$ gives another solution: $(0, 3, 2)$. Because this system has infinitely many solutions, it is consistent, but the equations are dependent. ∎

Example 5   Use matrices to solve the system

$$\begin{cases} w + x + y + z = 3 \\ x + 2y - z = -2 \\ w - y - z = 2 \\ 2w + x = 2 \end{cases}$$

Solution   Row reduce the augmented matrix as follows:

$$(-1)R1 + R3 \rightarrow R3$$

$$\left[\begin{array}{cccc|c} 1 & 1 & 1 & 1 & 3 \\ 0 & 1 & 2 & -1 & -2 \\ 1 & 0 & -1 & -1 & 2 \\ 2 & 1 & 0 & 0 & 2 \end{array}\right] \Leftrightarrow \left[\begin{array}{cccc|c} 1 & 1 & 1 & 1 & 3 \\ 0 & 1 & 2 & -1 & -2 \\ 0 & -1 & -2 & -2 & -1 \\ 2 & 1 & 0 & 0 & 2 \end{array}\right]$$

$$(-1)R3 \rightarrow R3$$
$$(-2)R1 + R4 \rightarrow R4$$

$$\Leftrightarrow \left[\begin{array}{cccc|c} 1 & 1 & 1 & 1 & 3 \\ 0 & 1 & 2 & -1 & -2 \\ 0 & 1 & 2 & 2 & 1 \\ 0 & -1 & -2 & -2 & -4 \end{array}\right]$$

$$R3 + R4 \to R4$$

$$\Leftrightarrow \begin{bmatrix} 1 & 1 & 1 & 1 & | & 3 \\ 0 & 1 & 2 & -1 & | & -2 \\ 0 & 1 & 2 & 2 & | & 1 \\ 0 & 0 & 0 & 0 & | & -3 \end{bmatrix}$$

The last row of the final matrix represents the equation

$$0w + 0x + 0y + 0z = -3$$

Obviously, *no* values of w, x, y, and z could make $0 = -3$. Because a solution must satisfy *each* equation, the given system has no solution. Hence, it is inconsistent.   ∎

Exercise 8.2

In Exercises 1–28, solve each system of equations using matrix methods.

1. $\begin{cases} 2x + y = 3 \\ x - 3y = 5 \end{cases}$

2. $\begin{cases} x + 2y = -1 \\ 3x - 5y = 19 \end{cases}$

3. $\begin{cases} x - 7y = -2 \\ 5x - 2y = -10 \end{cases}$

4. $\begin{cases} 3x - y = 3 \\ 2x + y = -3 \end{cases}$

5. $\begin{cases} 2x - y = 5 \\ x + 3y = 6 \end{cases}$

6. $\begin{cases} 3x - 5y = -25 \\ 2x + y = 5 \end{cases}$

7. $\begin{cases} x - 2y = 3 \\ -2x + 4y = 6 \end{cases}$

8. $\begin{cases} 3x - y = 7 \\ -x + \dfrac{1}{3}y = -\dfrac{7}{3} \end{cases}$

9. $\begin{cases} x - y + z = 3 \\ 2x - y + z = 4 \\ x + 2y - z = -1 \end{cases}$

10. $\begin{cases} 2x + y - z = 1 \\ x + y - z = 0 \\ 3x + y + 2z = 2 \end{cases}$

11. $\begin{cases} x + y - z = -1 \\ 3x + y = 4 \\ y - 2z = -4 \end{cases}$

12. $\begin{cases} 3x + y = 7 \\ x - z = 0 \\ y - 2z = -8 \end{cases}$

13. $\begin{cases} x - y + z = 2 \\ 2x + y + z = 5 \\ 3x - 4z = -5 \end{cases}$

14. $\begin{cases} x + z = -1 \\ 3x + y = 2 \\ 2x + y + 5z = 3 \end{cases}$

15. $\begin{cases} x + y + 2z = 4 \\ -x - y - 3z = -5 \\ 2x + y + z = 2 \end{cases}$

16. $\begin{cases} 2x - y + z = 6 \\ 3x + y - z = 2 \\ -x + 3y - 3z = 8 \end{cases}$

17. $\begin{cases} x + y = -2 \\ 3x - y = 6 \\ 2x + 2y = -4 \\ x - y = 4 \end{cases}$

18. $\begin{cases} x - y = -3 \\ 2x + y = -3 \\ 3x - y = -7 \\ 4x + y = -7 \end{cases}$

19. $\begin{cases} x + 2y + z = 4 \\ 3x - y - z = 2 \end{cases}$

20. $\begin{cases} x + 2y - 3z = -5 \\ 5x + y - z = -11 \end{cases}$

21. $\begin{cases} w + x - y + z = 2 \\ 2w - x - 2y + z = 0 \\ w - 2x - y + z = -1 \end{cases}$

22. $\begin{cases} w + x = 1 \\ w + y = 0 \\ x + z = 0 \end{cases}$

23. $\begin{cases} x + 2y + z = 4 \\ x - y + z = 1 \\ 2x + y + 2z = 2 \end{cases}$

24. $\begin{cases} x + y = 3 \\ 2x + y = 1 \\ 3x + 2y = 2 \end{cases}$

25. $\begin{cases} 2x - 2y + 3z + t = 2 \\ x + y + z + t = 5 \\ -x + 2y - 3z + 2t = 2 \\ x + y + 2z - t = 4 \end{cases}$

26. $\begin{cases} x + y + 2z + t = 1 \\ x + 2y + z + t = 2 \\ 2x + y + z + t = 4 \\ x + y + z + 2t = 3 \end{cases}$

27. $\begin{cases} x + y + t = 4 \\ x + z + t = 2 \\ 2x + 2y + z + 2t = 8 \\ x - y + z - t = -2 \end{cases}$

28. $\begin{cases} x - y + 2z + t = 3 \\ 3x - 2y - z - t = 4 \\ 2x + y + 2z - t = 10 \\ x + 2y + z - 3t = 8 \end{cases}$

In Exercises 29–32, solve each system of equations using matrix methods.

29. $\begin{cases} x^2 + y^2 + z^2 = 14 \\ 2x^2 + 3y^2 - 2z^2 = -7 \\ x^2 - 5y^2 + z^2 = 8 \end{cases}$

(*Hint:* Solve first as a linear system in x^2, y^2, and z^2.)

30. $\begin{cases} \dfrac{3}{x} + \dfrac{1}{y} + \dfrac{1}{z} = 4 \\[2mm] \dfrac{1}{x} - \dfrac{3}{y} - \dfrac{2}{z} = -3 \\[2mm] \dfrac{7}{x} - \dfrac{9}{y} + \dfrac{3}{z} = -14 \end{cases}$

31. $\begin{cases} 5\sqrt{x} + 2\sqrt{y} + \sqrt{z} = 22 \\ \sqrt{x} + \sqrt{y} - \sqrt{z} = 5 \\ 3\sqrt{x} - 2\sqrt{y} - 3\sqrt{z} = 10 \end{cases}$

32. $\begin{cases} \sqrt{x} + \dfrac{2}{y} - z^2 = 0 \\[2mm] 2\sqrt{x} - \dfrac{5}{y} + z^2 = 3 \\[2mm] -\sqrt{x} + \dfrac{1}{y} + 2z^2 = 7 \end{cases}$

8.3 MATRIX ALGEBRA

In this section we discuss how to add, subtract, and multiply matrices.

> **Definition.** An **$m \times n$ matrix** is a rectangular array of $m \cdot n$ numbers arranged in m rows and n columns.

We will use any of the following notations to denote the matrix A:

$$A_{m \times n}, \quad A, \quad [a_{ij}], \quad \begin{bmatrix} a_{11} & a_{12} & a_{13} & \cdots & a_{1n} \\ a_{21} & a_{22} & a_{23} & \cdots & a_{2n} \\ \vdots & & & & \vdots \\ a_{m1} & a_{m2} & a_{m3} & \cdots & a_{mn} \end{bmatrix}$$

The symbol a_{23} represents the entry in row two, column three, of the matrix A. Similarly, the symbol a_{ij} represents the entry in the ith row and the jth column.

Arthur Cayley (1821–1895) Cayley was a major founder of the theory of matrices.

Definition of Equal Matrices. If $A = [a_{ij}]$ and $B = [b_{ij}]$ are both $m \times n$ matrices, then

$$A = B \text{ if and only if } a_{ij} = b_{ij}$$

for all i and j, where $i = 1, 2, 3, \ldots, m$ and $j = 1, 2, 3, \ldots, n$.

The previous definition points out that two matrices must be identical to be equal. They must be the same size and have the same corresponding entries.

The Sum of Two Matrices. Let $A = [a_{ij}]$ and $B = [b_{ij}]$ be two $m \times n$ matrices. The sum, $A + B$, is the $m \times n$ matrix C, found by adding the corresponding entries of matrices A and B:

$$A + B = C = [c_{ij}]$$

where $c_{ij} = a_{ij} + b_{ij}$, for $i = 1, 2, 3, \ldots, m$ and $j = 1, 2, 3, \ldots, n$.

Example 1   Add the matrices

$$\begin{bmatrix} 2 & 1 & 3 \\ 1 & -1 & 0 \end{bmatrix} \quad \text{and} \quad \begin{bmatrix} 1 & -1 & 2 \\ -1 & 1 & 5 \end{bmatrix}$$

Solution   Because each matrix is 2×3, their sum is defined and can be calculated by adding their corresponding elements:

$$\begin{bmatrix} 2 & 1 & 3 \\ 1 & -1 & 0 \end{bmatrix} + \begin{bmatrix} 1 & -1 & 2 \\ -1 & 1 & 5 \end{bmatrix} = \begin{bmatrix} 2+1 & 1-1 & 3+2 \\ 1-1 & -1+1 & 0+5 \end{bmatrix}$$

$$= \begin{bmatrix} 3 & 0 & 5 \\ 0 & 0 & 5 \end{bmatrix} \quad \blacksquare$$

Example 2   If possible, add the matrices

$$\begin{bmatrix} 2 & 4 & 3 \\ 1 & 1 & 1 \end{bmatrix} \quad \text{and} \quad \begin{bmatrix} 1 & 2 \\ 2 & 3 \end{bmatrix}$$

Solution   The first matrix is 2×3 and the second is 2×2. Because these matrices are of different sizes, they cannot be added.   ∎

Several of the field properties discussed in Chapter 1 apply to matrices also.

Theorem.   The addition of two $m \times n$ matrices is commutative.

Proof   Let $A = [a_{ij}]$ and $B = [b_{ij}]$ be $m \times n$ matrices. Then $A + B = C = [c_{ij}]$, where $c_{ij} = a_{ij} + b_{ij}$ for each $i = 1, 2, 3, \ldots, m$ and $j = 1, 2, 3, \ldots, n$. On the other hand, $B + A = D = [d_{ij}]$, where $d_{ij} = b_{ij} + a_{ij}$ for each i and j.

Because each entry in each matrix is a real number and the addition of real numbers is commutative, it follows that

$$c_{ij} = a_{ij} + b_{ij} = b_{ij} + a_{ij} = d_{ij}$$

for all i and j, $i = 1, 2, 3, \ldots, m$ and $j = 1, 2, 3, \ldots, n$.

By the definition of equality of matrices, $C = D$, and therefore $A + B = B + A$.   □

Theorem.   The addition of three $m \times n$ matrices is associative.

The proof of the previous theorem is left as an exercise.

In the collection of all $m \times n$ matrices, there is a matrix called the **zero matrix**.

Definition.   Let A be any $m \times n$ matrix. There is an $m \times n$ matrix **0**, called the **zero matrix**, or the **additive identity matrix**, for which

$$A + 0 = 0 + A = A$$

The matrix **0** consists of m rows and n columns of 0s.

To illustrate the above definition, we note that the matrix

$$\begin{bmatrix} 0 & 0 & 0 \\ 0 & 0 & 0 \\ 0 & 0 & 0 \end{bmatrix}$$

is the 3×3 zero matrix, and that

$$\begin{bmatrix} 0 & 0 & 0 \\ 0 & 0 & 0 \\ 0 & 0 & 0 \end{bmatrix} + \begin{bmatrix} 1 & 2 & 3 \\ 4 & 5 & 6 \\ 7 & 8 & 9 \end{bmatrix} = \begin{bmatrix} 1 & 2 & 3 \\ 4 & 5 & 6 \\ 7 & 8 & 9 \end{bmatrix}$$

Matrices are similar to real numbers in another respect: every matrix has an additive inverse.

> **Definition.**   Any $m \times n$ matrix A has an **additive inverse**, an $m \times n$ matrix $-A$ with the property that the sum of A and $-A$ is the zero matrix:
>
> $$A + (-A) = (-A) + A = \mathbf{0}$$
>
> The entries of $-A$ are the negatives of the corresponding entries of A.

The additive inverse of the 2×3 matrix $A = \begin{bmatrix} 1 & -3 & 2 \\ 0 & 1 & -5 \end{bmatrix}$ is the matrix

$$-A = \begin{bmatrix} -1 & 3 & -2 \\ 0 & -1 & 5 \end{bmatrix}$$

because their sum is the zero matrix:

$$A + (-A) = \begin{bmatrix} 1 & -3 & 2 \\ 0 & 1 & -5 \end{bmatrix} + \begin{bmatrix} -1 & 3 & -2 \\ 0 & -1 & 5 \end{bmatrix}$$

$$= \begin{bmatrix} 1-1 & -3+3 & 2-2 \\ 0+0 & 1-1 & -5+5 \end{bmatrix}$$

$$= \begin{bmatrix} 0 & 0 & 0 \\ 0 & 0 & 0 \end{bmatrix}$$

> **The Difference of Two Matrices.**   If A and B are $m \times n$ matrices, their difference, $A - B$, is the sum of A and the additive inverse of B:
>
> $$A - B = A + (-B)$$

Example 3

$$\begin{bmatrix} 2 & -5 \\ 3 & 1 \end{bmatrix} - \begin{bmatrix} 4 & -5 \\ -3 & 9 \end{bmatrix} = \begin{bmatrix} 2 & -5 \\ 3 & 1 \end{bmatrix} + \begin{bmatrix} -4 & 5 \\ 3 & -9 \end{bmatrix}$$

$$= \begin{bmatrix} -2 & 0 \\ 6 & -8 \end{bmatrix} \qquad \blacksquare$$

We illustrate how to find the product of two matrices by computing the product of a 2×3 matrix A and a 3×3 matrix B. The result is the 2×3 matrix C.

$$A \cdot B = \begin{bmatrix} 1 & 2 & 3 \\ 4 & 5 & 6 \end{bmatrix} \cdot \begin{bmatrix} a & b & c \\ d & e & f \\ g & h & i \end{bmatrix} = C$$

Each entry of matrix C is the result of a calculation that involves a row of A and a column of B. For example, the first-row, third-column entry of matrix

C is found by keeping a running total of the products of corresponding entries of the first row of A and the third column of B:

$$\begin{bmatrix} 1 & 2 & 3 \\ 4 & 5 & 6 \end{bmatrix} \begin{bmatrix} a & b & c \\ d & e & f \\ g & h & i \end{bmatrix} = \begin{bmatrix} ? & ? & 1c + 2f + 3i \\ ? & ? & ? \end{bmatrix}$$

Similarly, the second-row, second-column entry of matrix C is formed by a calculation involving the second row of A and the second column of B.

$$\begin{bmatrix} 1 & 2 & 3 \\ 4 & 5 & 6 \end{bmatrix} \begin{bmatrix} a & b & c \\ d & e & f \\ g & h & i \end{bmatrix} = \begin{bmatrix} ? & ? & 1c + 2f + 3i \\ ? & 4b + 5e + 6h & ? \end{bmatrix}$$

To calculate the first-row, first-column entry of matrix C, we use the first row of A and the first column of B.

$$\begin{bmatrix} 1 & 2 & 3 \\ 4 & 5 & 6 \end{bmatrix} \cdot \begin{bmatrix} a & b & c \\ d & e & f \\ g & h & i \end{bmatrix} = \begin{bmatrix} 1a + 2d + 3g & ? & 1c + 2f + 3i \\ ? & 4b + 5e + 6h & ? \end{bmatrix}$$

The complete product C is

$$\begin{bmatrix} 1 & 2 & 3 \\ 4 & 5 & 6 \end{bmatrix} \begin{bmatrix} a & b & c \\ d & e & f \\ g & h & i \end{bmatrix} = \begin{bmatrix} 1a + 2d + 3g & 1b + 2e + 3h & 1c + 2f + 3i \\ 4a + 5d + 6g & 4b + 5e + 6h & 4c + 5f + 6i \end{bmatrix}$$

For the product $A \cdot B$ to exist, the number of columns of A must equal the number of rows of B. If the product exists, it will have as many rows as A and as many columns as B:

$$\begin{array}{ccc} A & \cdot & B & = & C \\ m \times n & & n \times p & & m \times p \end{array}$$

These must agree.

The product is $m \times p$.

More formally, we have the following definition.

The Product of Two Matrices. Let $A = [a_{ij}]$ be an $m \times n$ matrix and $B = [b_{ij}]$ be an $n \times p$ matrix. The product, AB, is the $m \times p$ matrix C found as follows:

$AB = C = [c_{ij}]$, where c_{ij} is the sum of the products of the corresponding entries in ith row of A and the jth column of B, where $i = 1, 2, 3, \ldots, m$ and $j = 1, 2, 3, \ldots, p$

Example 4 Find C if $A \cdot B = \begin{bmatrix} 1 & 2 \\ 3 & 4 \\ 5 & 6 \end{bmatrix} \begin{bmatrix} a & b \\ c & d \end{bmatrix} = C$.

Solution Because the first matrix is 3×2 and the second matrix is 2×2, the product C exists, and it is a 3×2 matrix. The first-row, first-column entry of C is the total of the products of corresponding entries in the first row of A and the first column of B: $c_{11} = 1a + 2c$. Similarly, c_{12} is computed by using the first row of A and the second column of B: $c_{12} = 1b + 2d$. The entire product is

$$\begin{bmatrix} 1 & 2 \\ 3 & 4 \\ 5 & 6 \end{bmatrix} \begin{bmatrix} a & b \\ c & d \end{bmatrix} = \begin{bmatrix} 1a + 2c & 1b + 2d \\ 3a + 4c & 3b + 4d \\ 5a + 6c & 5b + 6d \end{bmatrix}$$ ■

Example 5 Find the product $\begin{bmatrix} 1 & -1 & 2 \\ 1 & 3 & 0 \\ 0 & 1 & 1 \end{bmatrix} \begin{bmatrix} 2 & 1 \\ 1 & 3 \\ 0 & 1 \end{bmatrix}$.

Solution Because the matrices are 3×3 and 3×2, the product is a 3×2 matrix.

$$\begin{bmatrix} 1 & -1 & 2 \\ 1 & 3 & 0 \\ 0 & 1 & 1 \end{bmatrix} \begin{bmatrix} 2 & 1 \\ 1 & 3 \\ 0 & 1 \end{bmatrix}$$

$$= \begin{bmatrix} 1 \cdot 2 + (-1) \cdot 1 + 2 \cdot 0 & 1 \cdot 1 + (-1) \cdot 3 + 2 \cdot 1 \\ 1 \cdot 2 + 3 \cdot 1 + 0 \cdot 0 & 1 \cdot 1 + 3 \cdot 3 + 0 \cdot 1 \\ 0 \cdot 2 + 1 \cdot 1 + 1 \cdot 0 & 0 \cdot 1 + 1 \cdot 3 + 1 \cdot 1 \end{bmatrix}$$

$$= \begin{bmatrix} 1 & 0 \\ 5 & 10 \\ 1 & 4 \end{bmatrix}$$ ■

Example 6 Find the product $\begin{bmatrix} 1 & 2 & 3 \end{bmatrix} \begin{bmatrix} 4 \\ 5 \\ 6 \end{bmatrix}$.

Solution Because the first matrix is 1×3 and the second matrix is 3×1, the product is a 1×1 matrix:

$$\begin{bmatrix} 1 & 2 & 3 \end{bmatrix} \begin{bmatrix} 4 \\ 5 \\ 6 \end{bmatrix} = \begin{bmatrix} 1 \cdot 4 + 2 \cdot 5 + 3 \cdot 6 \end{bmatrix} = \begin{bmatrix} 32 \end{bmatrix}$$ ■

Example 7 If $A = \begin{bmatrix} 1 & 1 \\ 0 & 0 \end{bmatrix}$ and $B = \begin{bmatrix} 0 & 1 \\ 0 & 1 \end{bmatrix}$, calculate AB and BA and thereby show that multiplication of matrices is not commutative.

Solution
$$AB = \begin{bmatrix} 1 & 1 \\ 0 & 0 \end{bmatrix}\begin{bmatrix} 0 & 1 \\ 0 & 1 \end{bmatrix} = \begin{bmatrix} \mathbf{0} & \mathbf{2} \\ \mathbf{0} & \mathbf{0} \end{bmatrix}$$

$$BA = \begin{bmatrix} 0 & 1 \\ 0 & 1 \end{bmatrix}\begin{bmatrix} 1 & 1 \\ 0 & 0 \end{bmatrix} = \begin{bmatrix} \mathbf{0} & \mathbf{0} \\ \mathbf{0} & \mathbf{0} \end{bmatrix}$$

Because the products are not equal, matrix multiplication is not commutative. ∎

Example 8 Find values for x, y, and z such that

$$\begin{bmatrix} 1 & 2 & 3 \\ 2 & -1 & -2 \\ 1 & -3 & -3 \end{bmatrix}\begin{bmatrix} x \\ y \\ z \end{bmatrix} = \begin{bmatrix} 4 \\ 0 \\ -2 \end{bmatrix}$$

Solution Find the product of the first two matrices and set it equal to the third matrix.

$$\begin{bmatrix} 1 & 2 & 3 \\ 2 & -1 & -2 \\ 1 & -3 & -3 \end{bmatrix}\begin{bmatrix} x \\ y \\ z \end{bmatrix} = \begin{bmatrix} 1x + 2y + 3z \\ 2x - 1y - 2z \\ 1x - 3y - 3z \end{bmatrix} = \begin{bmatrix} 4 \\ 0 \\ -2 \end{bmatrix}$$

The product will be equal to the third matrix if and only if their corresponding components are equal. Set the corresponding components equal to get

$$\begin{cases} x + 2y + 3z = 4 \\ 2x - y - 2z = 0 \\ x - 3y - 3z = -2 \end{cases}$$

This system was solved in Example 2 of Section 8.2. Its solution is $x = 1$, $y = 0$, and $z = 1$. ∎

The number 1 is called the identity for multiplication because multiplying a number by 1 does not change that number. There is a **multiplicative identity matrix** with a similar property.

Definition. Let A be an $n \times n$ matrix. There is an $n \times n$ **identity matrix** I for which

$$AI = IA = A$$

The matrix I consists of 1s on its diagonal and 0s elsewhere.

$$I = \begin{bmatrix} 1 & 0 & 0 & \cdots & 0 \\ 0 & 1 & 0 & \cdots & 0 \\ 0 & 0 & 1 & \cdots & 0 \\ \vdots & \vdots & \vdots & & \vdots \\ 0 & 0 & 0 & \cdots & 1 \end{bmatrix}$$

Note that an identity matrix is a square matrix—it has the same number of rows and columns.

Example 9 illustrates the previous definition for the 3×3 identity matrix.

Example 9 Find **a.** $\begin{bmatrix} 1 & 0 & 0 \\ 0 & 1 & 0 \\ 0 & 0 & 1 \end{bmatrix} \begin{bmatrix} 1 & 2 & 3 \\ 4 & 5 & 6 \\ 7 & 8 & 9 \end{bmatrix}$ and **b.** $\begin{bmatrix} 1 & 2 & 3 \\ 4 & 5 & 6 \\ 7 & 8 & 9 \end{bmatrix} \begin{bmatrix} 1 & 0 & 0 \\ 0 & 1 & 0 \\ 0 & 0 & 1 \end{bmatrix}$.

Solution **a.** $\begin{bmatrix} 1 & 0 & 0 \\ 0 & 1 & 0 \\ 0 & 0 & 1 \end{bmatrix} \begin{bmatrix} 1 & 2 & 3 \\ 4 & 5 & 6 \\ 7 & 8 & 9 \end{bmatrix}$

$$= \begin{bmatrix} 1 \cdot 1 + 0 \cdot 4 + 0 \cdot 7 & 1 \cdot 2 + 0 \cdot 5 + 0 \cdot 8 & 1 \cdot 3 + 0 \cdot 6 + 0 \cdot 9 \\ 0 \cdot 1 + 1 \cdot 4 + 0 \cdot 7 & 0 \cdot 2 + 1 \cdot 5 + 0 \cdot 8 & 0 \cdot 3 + 1 \cdot 6 + 0 \cdot 9 \\ 0 \cdot 1 + 0 \cdot 4 + 1 \cdot 7 & 0 \cdot 2 + 0 \cdot 5 + 1 \cdot 8 & 0 \cdot 3 + 0 \cdot 6 + 1 \cdot 9 \end{bmatrix}$$

$$= \begin{bmatrix} 1 & 2 & 3 \\ 4 & 5 & 6 \\ 7 & 8 & 9 \end{bmatrix}$$

b. $\begin{bmatrix} 1 & 2 & 3 \\ 4 & 5 & 6 \\ 7 & 8 & 9 \end{bmatrix} \begin{bmatrix} 1 & 0 & 0 \\ 0 & 1 & 0 \\ 0 & 0 & 1 \end{bmatrix} = \begin{bmatrix} 1 & 2 & 3 \\ 4 & 5 & 6 \\ 7 & 8 & 9 \end{bmatrix}$ ∎

Exercise 8.3

In Exercises 1–8, find values of x and y, if any, that will make the two matrices equal.

1. $\begin{bmatrix} x & y \\ 1 & 3 \end{bmatrix} = \begin{bmatrix} 2 & 5 \\ 1 & 3 \end{bmatrix}$

2. $\begin{bmatrix} x & 5 \\ 3 & y \end{bmatrix} = \begin{bmatrix} 0 & 5 \\ 3 & 2 \end{bmatrix}$

3. $\begin{bmatrix} x & y \\ 1 & 3 \end{bmatrix} = \begin{bmatrix} 2 & 5 \\ 1 & 4 \end{bmatrix}$

4. $\begin{bmatrix} x & y \\ 1 & x + y \end{bmatrix} = \begin{bmatrix} 2 & 1 \\ 1 & 2 \end{bmatrix}$

5. $\begin{bmatrix} x + y & 3 + x \\ -2 & 5y \end{bmatrix} = \begin{bmatrix} 3 & 4 \\ -2 & 10 \end{bmatrix}$

6. $\begin{bmatrix} x + y & x - y \\ 2x & 3y \end{bmatrix} = \begin{bmatrix} -x & x - 2 \\ -y & 8 - y \end{bmatrix}$

7. $\begin{bmatrix} x & 3x \\ y & x + 1 \end{bmatrix} = \begin{bmatrix} y & 6 \\ 2 & 3 \end{bmatrix}$

8. $\begin{bmatrix} x \\ y \end{bmatrix} = \begin{bmatrix} 1 & 2 \\ 3 & 4 \end{bmatrix}$

In Exercises 9–18, perform the indicated operation, if possible.

9. $\begin{bmatrix} 2 & 1 & -1 \\ -3 & 2 & 5 \end{bmatrix} + \begin{bmatrix} -3 & 1 & 2 \\ -3 & -2 & -5 \end{bmatrix}$

10. $\begin{bmatrix} 3 & 1 \\ 2 & 2 \end{bmatrix} + \begin{bmatrix} 2 & 1 \\ -1 & 0 \end{bmatrix} + \begin{bmatrix} -5 & -2 \\ -1 & -2 \end{bmatrix}$

11. $\begin{bmatrix} 3 & 2 & 1 \\ -2 & 3 & -3 \\ -4 & -2 & -1 \end{bmatrix} - \begin{bmatrix} -2 & 6 & -2 \\ 5 & 7 & -1 \\ -4 & -6 & 7 \end{bmatrix}$

12. $\begin{bmatrix} -2 & 7 & -3 \\ 3 & 6 & -7 \\ -9 & -2 & -5 \end{bmatrix} + \begin{bmatrix} -5 & -4 & -3 \\ -1 & 2 & 10 \\ -1 & -3 & -4 \end{bmatrix}$

13. $\begin{bmatrix} 1 & 3 & -1 \\ 2 & 1 & 5 \\ 1 & 3 & 0 \end{bmatrix} + \begin{bmatrix} 2 \\ 0 \\ -3 \end{bmatrix}$

14. $\begin{bmatrix} 3 \\ 2 \\ 3 \end{bmatrix} - \begin{bmatrix} -3 & 5 & -6 \\ -3 & -5 & -6 \\ 4 & 6 & -6 \end{bmatrix}$

15. $\begin{bmatrix} 1 & 2 & 3 \end{bmatrix} + \begin{bmatrix} 4 & 5 & 6 \end{bmatrix}$

16. $\begin{bmatrix} 1 \\ 2 \\ 3 \end{bmatrix} + \begin{bmatrix} 4 & -5 & -6 \end{bmatrix}$

17. $\begin{bmatrix} 1 & 3 & -4 \\ 2 & -1 & 3 \\ 1 & 5 & 7 \end{bmatrix} + \begin{bmatrix} 3 & 2 & -8 \\ 9 & 11 & 17 \\ 2 & 1 & 3 \end{bmatrix} - \begin{bmatrix} 1 & 3 & -5 \\ 2 & -9 & 5 \\ 3 & 10 & 11 \end{bmatrix}$

18. $\begin{bmatrix} -3 & -2 & 15 \\ 2 & -5 & 9 \end{bmatrix} - \begin{bmatrix} 3 & 2 & -15 \\ -2 & 5 & -9 \end{bmatrix} + \begin{bmatrix} 6 & 4 & -30 \\ -3 & 12 & -15 \end{bmatrix}$

In Exercises 19–30, find each product, if possible.

19. $\begin{bmatrix} 2 & 3 \\ 3 & -2 \end{bmatrix}\begin{bmatrix} 1 & 2 \\ 0 & -2 \end{bmatrix}$

20. $\begin{bmatrix} -2 & 3 \\ 3 & -2 \end{bmatrix}\begin{bmatrix} 2 & 4 \\ -5 & 7 \end{bmatrix}$

21. $\begin{bmatrix} -4 & -2 \\ 21 & 0 \end{bmatrix}\begin{bmatrix} -5 & 6 \\ 21 & -1 \end{bmatrix}$

22. $\begin{bmatrix} -5 & 4 \\ 4 & -5 \end{bmatrix}\begin{bmatrix} 6 & -2 \\ 1 & 3 \end{bmatrix}$

23. $\begin{bmatrix} 2 & 1 & 3 \\ 1 & 2 & -1 \\ 0 & 1 & 0 \end{bmatrix}\begin{bmatrix} 1 & 2 & 3 \\ 2 & -2 & 1 \\ 0 & 0 & 1 \end{bmatrix}$

24. $\begin{bmatrix} 2 & 1 & 1 \\ 1 & 1 & 2 \\ 1 & -2 & -1 \end{bmatrix}\begin{bmatrix} 1 & 2 & 3 \\ 1 & 2 & -3 \\ -1 & -1 & 3 \end{bmatrix}$

25. $\begin{bmatrix} 1 & -2 & -3 \end{bmatrix}\begin{bmatrix} 4 \\ -5 \\ -6 \end{bmatrix}$

26. $\begin{bmatrix} 1 \\ -2 \\ -3 \end{bmatrix}\begin{bmatrix} 4 & -5 & -6 \end{bmatrix}$

27. $\begin{bmatrix} 1 & 2 & 3 \end{bmatrix}\begin{bmatrix} 4 & 5 & 6 \end{bmatrix}$

28. $\begin{bmatrix} 2 & 3 & 4 \\ 1 & 2 & 3 \\ -2 & 2 & 2 \end{bmatrix}\begin{bmatrix} -1 \\ 2 \\ 3 \end{bmatrix}$

29. $\begin{bmatrix} 1 & 2 & 3 \end{bmatrix}\begin{bmatrix} 1 & 2 & 3 \\ 4 & 5 & 6 \\ 7 & 8 & 9 \end{bmatrix}$

30. $\begin{bmatrix} 1 & 2 & 3 \\ 1 & 2 & 1 \\ 1 & -1 & -1 \end{bmatrix}\begin{bmatrix} 1 & 2 \\ 2 & 1 \\ 1 & 1 \end{bmatrix}$

In Exercises 31–36, perform the indicated operations.

31. $\begin{bmatrix} 1 & 2 \\ 2 & 3 \end{bmatrix}\left(\begin{bmatrix} 2 & 1 & -5 \\ 1 & 1 & 2 \end{bmatrix} + \begin{bmatrix} -2 & -1 & 6 \\ 0 & -1 & -1 \end{bmatrix}\right)$

32. $\begin{bmatrix} 1 & 2 \\ 2 & 3 \end{bmatrix}\begin{bmatrix} 2 & 1 & -5 \\ 1 & 1 & 2 \end{bmatrix} + \begin{bmatrix} 1 & 2 \\ 2 & 3 \end{bmatrix}\begin{bmatrix} -2 & -1 & 6 \\ 0 & -1 & -1 \end{bmatrix}$

33. $\begin{bmatrix} 1 & 2 & 3 \\ 2 & 3 & 1 \\ 1 & 2 & 1 \end{bmatrix}\begin{bmatrix} 2 & 1 & 1 \\ 3 & -1 & -1 \\ 2 & -2 & 2 \end{bmatrix} + \begin{bmatrix} -2 & 3 & 4 \\ 1 & 1 & 1 \\ 0 & 1 & 0 \end{bmatrix}$

34. $\begin{bmatrix} 2 & 1 & 0 \\ 1 & -2 & -1 \\ 1 & 1 & -1 \end{bmatrix} \left(\begin{bmatrix} 1 & 0 & 1 \\ 1 & 1 & 2 \\ 1 & 2 & -1 \end{bmatrix} + \begin{bmatrix} -1 & -1 & 2 \\ 0 & 0 & 1 \\ 1 & 0 & -1 \end{bmatrix} \right)$

35. $\left(\begin{bmatrix} 1 & 2 \\ 2 & 3 \end{bmatrix} \begin{bmatrix} 1 \\ -3 \end{bmatrix} + \begin{bmatrix} -2 \\ 1 \end{bmatrix} \right) \left(\begin{bmatrix} 1 & 2 \end{bmatrix} \begin{bmatrix} 1 \\ -3 \end{bmatrix} + [4] \right)$

36. $\begin{bmatrix} 1 \\ 2 \end{bmatrix} \begin{bmatrix} -3 & -4 \end{bmatrix} - \begin{bmatrix} 0 & 3 \\ 2 & 1 \end{bmatrix} \begin{bmatrix} 2 & 0 \\ 1 & -1 \end{bmatrix}$

In Exercises 37–42, let $A = \begin{bmatrix} 1 & 3 \\ 2 & 5 \end{bmatrix}$, $B = \begin{bmatrix} -1 \\ 3 \end{bmatrix}$, and $C = \begin{bmatrix} 3 & 2 \end{bmatrix}$. Perform, if possible, the indicated operations.

37. $A - BC$ **38.** $AB + B$ **39.** $CB - AB$ **40.** CAB

41. ABC **42.** $CA + C$

43. Let $A = \begin{bmatrix} 1 & 1 \\ 1 & 1 \end{bmatrix}$. Find A^7. (*Hint:* Calculate A^2 and A^3. Do you see a pattern?)

44. Let a, b, and c be real numbers. If $ab = ac$ and $a \neq 0$, then $b = c$. Find 2×2 matrices A, B, and C, where $A \neq 0$, to show that such a law does not hold for all matrices.

45. In the real number system the numbers 0 and 1 are the only numbers that equal their own squares: if $a^2 = a$, then $a = 0$ or $a = 1$. Find a 2×2 matrix A that is neither the zero matrix nor the identity matrix, such that $A^2 = A \cdot A = A$.

46. Another property of the real numbers is that, if $ab = 0$, then either $a = 0$ or $b = 0$. To show that this property is not true for matrices, find two nonzero 2×2 matrices, A and B, such that $AB = 0$.

47. Multiplication of three $n \times n$ matrices is associative: $(AB)C = A(BC)$. Verify this by an example chosen from the set of 2×2 matrices.

48. If A is an $m \times n$ matrix and B and C are each $n \times p$ matrices, then $A(B + C) = AB + AC$. Illustrate this with an example showing that matrix multiplication distributes over matrix addition.

49. Prove that the addition of matrices is associative.

8.4 MATRIX INVERSION

Two real numbers are called **multiplicative inverses** if their product is the multiplicative identity 1. Some matrices have multiplicative inverses also.

> **Definition.** If A and B are $n \times n$ matrices, I is the $n \times n$ identity matrix, and
>
> $$A \cdot B = B \cdot A = I$$
>
> then A and B are called **multiplicative inverses**. Matrix A is the **inverse** of B, and B is the **inverse** of A.

It can be shown that the inverse of a matrix A, if it exists, is unique. The inverse of A is denoted by A^{-1}.

Example 1 If $A = \begin{bmatrix} 1 & 1 & 0 \\ 4 & 3 & 0 \\ 2 & 1 & -1 \end{bmatrix}$ and $B = \begin{bmatrix} -3 & 1 & 0 \\ 4 & -1 & 0 \\ -2 & 1 & -1 \end{bmatrix}$, show that A and B are inverses.

Solution Multiply the matrices in each order to show that the product is the identity matrix.

$$AB = \begin{bmatrix} 1 & 1 & 0 \\ 4 & 3 & 0 \\ 2 & 1 & -1 \end{bmatrix} \begin{bmatrix} -3 & 1 & 0 \\ 4 & -1 & 0 \\ -2 & 1 & -1 \end{bmatrix}$$

$$= \begin{bmatrix} -3+4 & 1-1 & 0 \\ -12+12 & 4-3 & 0 \\ -6+4+2 & 2-1-1 & 1 \end{bmatrix} = \begin{bmatrix} 1 & 0 & 0 \\ 0 & 1 & 0 \\ 0 & 0 & 1 \end{bmatrix}$$

$$BA = \begin{bmatrix} -3 & 1 & 0 \\ 4 & -1 & 0 \\ -2 & 1 & -1 \end{bmatrix} \begin{bmatrix} 1 & 1 & 0 \\ 4 & 3 & 0 \\ 2 & 1 & -1 \end{bmatrix} = \begin{bmatrix} 1 & 0 & 0 \\ 0 & 1 & 0 \\ 0 & 0 & 1 \end{bmatrix} \quad \blacksquare$$

If a matrix has an inverse, it is called a **nonsingular matrix**. Otherwise, it is called a **singular matrix**. The following theorem, stated without proof, provides a way of calculating the inverse of a nonsingular matrix.

Theorem. If a sequence of elementary row operations performed on the $n \times n$ matrix A reduces A to the $n \times n$ identity matrix I, then those same row operations, performed in the same order on the identity matrix I, will transform I into A^{-1}. Furthermore, if *no* sequence of row operations will reduce A to I, then A is singular.

To use the previous theorem, we perform elementary row operations on matrix A to change it to the identity matrix I. At the same time, we perform these elementary row operations on the identity matrix I. This changes I into A^{-1}.

A notation for this process uses an n-row by $2n$-column matrix, with matrix A as the left half and matrix I as the right half. If A is nonsingular, the proper row operations performed on $[A \mid I]$ will transform it into $[I \mid A^{-1}]$.

Example 2 Find the inverse of matrix A if $A = \begin{bmatrix} 2 & -4 \\ 4 & -7 \end{bmatrix}$.

Solution   Set up a 2×4 matrix with A on the left and I on the right of the broken line:

$$[A \mid I] = \begin{bmatrix} 2 & -4 & | & 1 & 0 \\ 4 & -7 & | & 0 & 1 \end{bmatrix}$$

Perform row operations on the entire matrix to transform the left half into I:

$$(\tfrac{1}{2})R1 \to R1$$
$$(-2)R1 + R2 \to R2$$

$$\begin{bmatrix} 2 & -4 & | & 1 & 0 \\ 4 & -7 & | & 0 & 1 \end{bmatrix} \Leftrightarrow \begin{bmatrix} 1 & -2 & | & \tfrac{1}{2} & 0 \\ 0 & 1 & | & -2 & 1 \end{bmatrix}$$

$$(2)R2 + R1 \to R1$$

$$\Leftrightarrow \begin{bmatrix} 1 & 0 & | & -\tfrac{7}{2} & 2 \\ 0 & 1 & | & -2 & 1 \end{bmatrix}$$

Matrix A has been transformed into I. Thus, the right side of the previous matrix is A^{-1}. Verify this by finding AA^{-1} and $A^{-1}A$ and showing that each product is I:

$$AA^{-1} = \begin{bmatrix} 2 & -4 \\ 4 & -7 \end{bmatrix} \begin{bmatrix} -\tfrac{7}{2} & 2 \\ -2 & 1 \end{bmatrix} = \begin{bmatrix} 1 & 0 \\ 0 & 1 \end{bmatrix}$$

$$A^{-1}A = \begin{bmatrix} -\tfrac{7}{2} & 2 \\ -2 & 1 \end{bmatrix} \begin{bmatrix} 2 & -4 \\ 4 & -7 \end{bmatrix} = \begin{bmatrix} 1 & 0 \\ 0 & 1 \end{bmatrix}$$ ∎

Example 3   Find the inverse of matrix A if $A = \begin{bmatrix} 1 & 1 & 0 \\ 1 & 2 & 1 \\ 2 & 3 & 2 \end{bmatrix}$.

Solution   Set up a 3×6 matrix with A on the left and I on the right of the broken line:

$$[A \mid I] = \begin{bmatrix} 1 & 1 & 0 & | & 1 & 0 & 0 \\ 1 & 2 & 1 & | & 0 & 1 & 0 \\ 2 & 3 & 2 & | & 0 & 0 & 1 \end{bmatrix}$$

Perform row operations on the entire matrix to transform the left half into I.

$$(-1)R1 + R2 \to R2$$
$$(-2)R1 + R3 \to R3$$

$$\begin{bmatrix} 1 & 1 & 0 & | & 1 & 0 & 0 \\ 1 & 2 & 1 & | & 0 & 1 & 0 \\ 2 & 3 & 2 & | & 0 & 0 & 1 \end{bmatrix} \Leftrightarrow \begin{bmatrix} 1 & 1 & 0 & | & 1 & 0 & 0 \\ 0 & 1 & 1 & | & -1 & 1 & 0 \\ 0 & 1 & 2 & | & -2 & 0 & 1 \end{bmatrix}$$

$$(-1)R2 + R1 \rightarrow R1$$
$$(-1)R2 + R3 \rightarrow R3$$

$$\Leftrightarrow \begin{bmatrix} 1 & 0 & -1 & | & 2 & -1 & 0 \\ 0 & 1 & 1 & | & -1 & 1 & 0 \\ 0 & 0 & 1 & | & -1 & -1 & 1 \end{bmatrix}$$

$$R3 + R1 \rightarrow R1$$
$$(-1)R3 + R2 \rightarrow R2$$

$$\Leftrightarrow \begin{bmatrix} 1 & 0 & 0 & | & 1 & -2 & 1 \\ 0 & 1 & 0 & | & 0 & 2 & -1 \\ 0 & 0 & 1 & | & -1 & -1 & 1 \end{bmatrix}$$

The left half has been transformed into the identity matrix, and the right half has become A^{-1}. Thus,

$$A^{-1} = \begin{bmatrix} 1 & -2 & 1 \\ 0 & 2 & -1 \\ -1 & -1 & 1 \end{bmatrix}$$

■

Example 4 Find the inverse of $A = \begin{bmatrix} 1 & 2 \\ 2 & 4 \end{bmatrix}$, if possible.

Solution Form the 2×4 matrix

$$[A \mid I] = \begin{bmatrix} 1 & 2 & | & 1 & 0 \\ 2 & 4 & | & 0 & 1 \end{bmatrix}$$

and begin to transform the left side of the matrix into the identity matrix I:

$$(-2)R1 + R2 \rightarrow R2$$

$$\begin{bmatrix} 1 & 2 & | & 1 & 0 \\ 2 & 4 & | & 0 & 1 \end{bmatrix} \Leftrightarrow \begin{bmatrix} 1 & 2 & | & 1 & 0 \\ 0 & 0 & | & -2 & 1 \end{bmatrix}$$

In obtaining the second-row, first-column position of A, the entire second row of A is "zeroed out." Because it is impossible to transform A to the identity, matrix A is singular and has no inverse. ■

The next example shows how the inverse of a nonsingular matrix can be used to solve a system of equations.

Example 5 Solve the system

$$\begin{cases} x + y & = 3 \\ x + 2y + z = -2 \\ 2x + 3y + 2z = 1 \end{cases}$$

Solution   This system can be written as a single equation involving three matrices.

1. $$\begin{bmatrix} 1 & 1 & 0 \\ 1 & 2 & 1 \\ 2 & 3 & 2 \end{bmatrix} \begin{bmatrix} x \\ y \\ z \end{bmatrix} = \begin{bmatrix} 3 \\ -2 \\ 1 \end{bmatrix}$$

The 3×3 matrix on the left is the matrix whose inverse was found in Example 3. Multiply each side of Equation 1 on the left by this inverse to obtain an equivalent system of equations. The solution of this system can be read directly from the matrix to the right of the equals sign:

$$\begin{bmatrix} 1 & -2 & 1 \\ 0 & 2 & -1 \\ -1 & -1 & 1 \end{bmatrix} \begin{bmatrix} 1 & 1 & 0 \\ 1 & 2 & 1 \\ 2 & 3 & 2 \end{bmatrix} \begin{bmatrix} x \\ y \\ z \end{bmatrix} = \begin{bmatrix} 1 & -2 & 1 \\ 0 & 2 & -1 \\ -1 & -1 & 1 \end{bmatrix} \begin{bmatrix} 3 \\ -2 \\ 1 \end{bmatrix}$$

$$\begin{bmatrix} 1 & 0 & 0 \\ 0 & 1 & 0 \\ 0 & 0 & 1 \end{bmatrix} \begin{bmatrix} x \\ y \\ z \end{bmatrix} = \begin{bmatrix} 8 \\ -5 \\ 0 \end{bmatrix}$$

$$\begin{bmatrix} x \\ y \\ z \end{bmatrix} = \begin{bmatrix} 8 \\ -5 \\ 0 \end{bmatrix}$$

The solution of this system of equations is $x = 8$, $y = -5$, $z = 0$. Verify that these results satisfy all three of the original equations.   ∎

The equations of Example 5 can be thought of as the matrix equation $AX = B$, where A is the coefficient matrix,

$$A = \begin{bmatrix} 1 & 1 & 0 \\ 1 & 2 & 1 \\ 2 & 3 & 2 \end{bmatrix}$$

X is a column matrix of the variables,

$$X = \begin{bmatrix} x \\ y \\ z \end{bmatrix}$$

and B is a column matrix of the constants from the right sides of the equations,

$$B = \begin{bmatrix} 3 \\ -2 \\ 1 \end{bmatrix}$$

When each side of $AX = B$ is multiplied on the left by A^{-1}, the solution of the system appears as the column of numbers in the matrix that is the product of A^{-1} and B:

$$A^{-1}AX = A^{-1}B$$
$$IX = A^{-1}B$$
$$X = A^{-1}B$$

This method is especially useful for finding solutions of several systems of equations that differ from each other *only* in the column matrix B. If the coefficient matrix A remains unchanged from one system of equations to the next, then A^{-1} needs to be found only once. The solution of each system is found by a single matrix multiplication, $A^{-1}B$.

Exercise 8.4

In Exercises 1–16, find the inverse of each given matrix, if possible.

1. $\begin{bmatrix} 3 & -4 \\ -2 & 3 \end{bmatrix}$
2. $\begin{bmatrix} 2 & 3 \\ 3 & 5 \end{bmatrix}$
3. $\begin{bmatrix} 3 & 7 \\ 2 & 5 \end{bmatrix}$
4. $\begin{bmatrix} 1 & -2 \\ 2 & -5 \end{bmatrix}$

5. $\begin{bmatrix} 1 & 2 & 3 \\ 2 & 5 & 3 \\ 1 & 0 & 8 \end{bmatrix}$
6. $\begin{bmatrix} 2 & 1 & -1 \\ 2 & 2 & -1 \\ -1 & -1 & 1 \end{bmatrix}$
7. $\begin{bmatrix} 3 & 2 & 1 \\ 1 & 1 & -1 \\ 4 & 3 & 1 \end{bmatrix}$
8. $\begin{bmatrix} -2 & 1 & -3 \\ 2 & 3 & 0 \\ 1 & 0 & 1 \end{bmatrix}$

9. $\begin{bmatrix} 1 & 3 & 5 \\ 0 & 1 & 6 \\ 1 & 4 & 11 \end{bmatrix}$
10. $\begin{bmatrix} 1 & 1 & 1 \\ 2 & 2 & 2 \\ 3 & 3 & 3 \end{bmatrix}$
11. $\begin{bmatrix} 1 & 2 & 3 \\ 0 & 1 & 2 \\ 0 & 0 & 1 \end{bmatrix}$
12. $\begin{bmatrix} 1 & 2 & 3 \\ 0 & 1 & 1 \\ 0 & -1 & 0 \end{bmatrix}$

13. $\begin{bmatrix} 1 & 6 & 4 \\ 1 & -2 & -5 \\ 2 & 4 & -1 \end{bmatrix}$
14. $\begin{bmatrix} 1 & 1 & 1 \\ 1 & 0 & -1 \\ 1 & 2 & 3 \end{bmatrix}$

15. $\begin{bmatrix} 1 & 2 & 3 & 4 \\ 0 & 1 & 2 & 3 \\ 0 & 0 & 1 & 2 \\ 0 & 0 & 0 & 1 \end{bmatrix}$
16. $\begin{bmatrix} 1 & 0 & 0 & 0 \\ 1 & 1 & 0 & 0 \\ 1 & 1 & 1 & 0 \\ 1 & 2 & 2 & 1 \end{bmatrix}$

In Exercises 17–26, use the method of Example 5 to solve each system of equations.

17. $\begin{cases} 3x - 4y = 1 \\ -2x + 3y = 5 \end{cases}$
18. $\begin{cases} 2x + 3y = 7 \\ 3x + 5y = -5 \end{cases}$
19. $\begin{cases} 3x + 7y = 0 \\ 2x + 5y = -10 \end{cases}$
20. $\begin{cases} x - 2y = 12 \\ 2x - 5y = 13 \end{cases}$

21. $\begin{cases} x + 2y + 3z = 1 \\ 2x + 5y + 3z = 3 \\ x \phantom{{}+2y} + 8z = -2 \end{cases}$
22. $\begin{cases} 2x + y - z = 3 \\ 2x + 2y - z = -1 \\ -x - y + z = 4 \end{cases}$

23. $\begin{cases} 3x + 2y + z = 2 \\ x + y - z = -1 \\ 4x + 3y + z = 0 \end{cases}$
24. $\begin{cases} -2x + y - 3z = 5 \\ 2x + 3y \phantom{{}-3z} = 1 \\ x \phantom{{}+3y} + z = -2 \end{cases}$

25. $\begin{cases} x + 2y + 3z = 1 \\ 2x + 2y + 2z = 2x + y \\ z = 3 \end{cases}$
26. $\begin{cases} x + 2y + 3z = 2 \\ x + y + z = x \\ x - y = x \end{cases}$

27. If the $n \times n$ matrix A is nonsingular, and if B and C are $n \times n$ matrices such that $AB = AC$, prove that $B = C$.

28. If B is an $n \times n$ matrix that behaves as an identity ($AB = BA = A$, for any $n \times n$ matrix A), prove that $B = I$.

29. If $A = \begin{bmatrix} 0 & 1 \\ 1 & 0 \end{bmatrix}$, compute $A^2, A^3, A^4, \ldots$. Give a general rule for A^n, where n is a natural number.

30. If A and B are 2×2 matrices, is $(AB)^2 = A^2B^2$? Support your answer.

31. Prove that $\begin{bmatrix} a & b \\ c & d \end{bmatrix}$ has an inverse if and only if $ad - bc \neq 0$.

32. If $A = \begin{bmatrix} 1 & 1 \\ 0 & 1 \end{bmatrix}$, compute A^n for various values of n ($n = 2, 3, 4, \ldots$). What do you notice?

33. If $A = \begin{bmatrix} 1 & 0 \\ 1 & 1 \end{bmatrix}$, compute A^n for various values of n ($n = 2, 3, 4, \ldots$). What do you notice?

34. For what value of x will $\begin{bmatrix} 3 & 8 \\ 6 & x \end{bmatrix}$ not have a multiplicative inverse? (*Hint*: See Exercise 31.)

35. For what values of x will $\begin{bmatrix} x & 8 \\ 2 & x \end{bmatrix}$ not have a multiplicative inverse?

36. Does $(AB)^{-1} = A^{-1}B^{-1}$? Support your answer with an example chosen from 2×2 matrices.

37. Use an example chosen from 2×2 matrices to illustrate that $(AB)^{-1} = B^{-1}A^{-1}$.

38. Let A be any 3×3 matrix. Find a 3×3 matrix E such that the product EA is the result of performing the row operation $R1 \leftrightarrow R2$ on matrix A (the row operation that exchanges rows one and two of matrix A). What is E^{-1}?

39. Let A be any 3×3 matrix. Find another 3×3 matrix E such that the product EA is the result of performing the row operation $(3)R1 + R3 \rightarrow R3$ on matrix A. What is E^{-1}?

8.5 SOLUTION OF SYSTEMS OF EQUATIONS BY DETERMINANTS

Gabriel Cramer (1704–1752) Although other mathematicians had worked with determinants, it was the work of Cramer that popularized them.

There is a function, called the **determinant function**, that associates a numerical value with every square matrix. For any square matrix A, the symbol $\det(A)$ or the symbol $|A|$ represents the determinant of matrix A. We begin by defining the determinant of a 2×2 matrix.

Definition. If a, b, c, and d are numbers, then the determinant of

$$A = \begin{bmatrix} a & b \\ c & d \end{bmatrix} \text{ is}$$

$$\det(A) = \begin{vmatrix} a & b \\ c & d \end{vmatrix} = ad - bc$$

The determinant of a 2×2 matrix A is the number that is equal to the product of the entries on the major diagonal

$$\begin{vmatrix} a & b \\ c & d \end{vmatrix}$$

minus the product of the entries on the other diagonal

$$\begin{vmatrix} a & b \\ c & d \end{vmatrix}$$

Example 1   **a.** $\begin{vmatrix} 1 & 2 \\ 3 & 4 \end{vmatrix} = 1 \cdot 4 - 2 \cdot 3 = 4 - 6 = -2$

b. $\begin{vmatrix} -1 & -3 \\ 2 & 0 \end{vmatrix} = (-1) \cdot 0 - (-3) \cdot (2) = 0 - (-6) = 6$

c. $\begin{vmatrix} -2 & 3 \\ -\pi & \frac{1}{2} \end{vmatrix} = (-2) \cdot \left(\frac{1}{2}\right) - (3) \cdot (-\pi) = -1 + 3\pi$ ∎

To see how determinants can be used to solve a system of equations, we consider the system

$$\begin{cases} ax + by = e \\ cx + dy = f \end{cases}$$

By multiplying the first equation by d, the second equation by $-b$, and adding, the terms that involve y drop out:

$$\begin{aligned} a\,dx + b\,dy &= e\,d \\ -b cx - b dy &= -bf \\ \hline adx - bcx &= ed - bf \end{aligned}$$

We can solve the resulting equation for x:

$$adx - bcx = ed - bf$$
$$(ad - bc)x = ed - bf$$

1. $x = \dfrac{ed - bf}{ad - bc}$ provided $ad - bc \neq 0$

Solving this system for y gives

2. $y = \dfrac{af - ec}{ad - bc}$ provided $ad - bc \neq 0$

The numerators and denominators of Equations 1 and 2 can be expressed as determinants:

$$x = \frac{\begin{vmatrix} e & b \\ f & d \end{vmatrix}}{\begin{vmatrix} a & b \\ c & d \end{vmatrix}} = \frac{ed - bf}{ad - bc} \qquad y = \frac{\begin{vmatrix} a & e \\ c & f \end{vmatrix}}{\begin{vmatrix} a & b \\ c & d \end{vmatrix}} = \frac{af - ec}{ad - bc}$$

If these formulas are compared with the original system of equations,

$$\begin{cases} ax + by = e \\ cx + dy = f \end{cases}$$

it is apparent that the denominator determinant consists of the coefficients of the variables of each equation:

$$\text{denominator determinant} = \begin{vmatrix} a & b \\ c & d \end{vmatrix}$$

Each numerator determinant is a modified copy of the denominator determinant. The column of coefficients of the variable for which we are solving is replaced with the column of constants that appears to the right of the equal signs. Thus, when solving for x, the coefficients of x (a and c) are replaced in the numerator determinant by the constants e and f:

$$x = \frac{\begin{vmatrix} e & b \\ f & d \end{vmatrix}}{\begin{vmatrix} a & b \\ c & d \end{vmatrix}} \qquad \begin{cases} ax + by = e \\ cx + dy = f \end{cases}$$

Similarly, when solving for y, the coefficients of y (b and d) are replaced in the numerator determinant by the constants e and f.

$$y = \frac{\begin{vmatrix} a & e \\ c & f \end{vmatrix}}{\begin{vmatrix} a & b \\ c & d \end{vmatrix}} \qquad \begin{cases} ax + by = e \\ cx + dy = f \end{cases}$$

The method of using determinants to solve systems of equations is called **Cramer's rule.**

Cramer's Rule for Two Equations in Two Variables.   The solution of the system of equations

$$\begin{cases} ax + by = e \\ cx + dy = f \end{cases}$$

is the pair

$$x = \frac{\begin{vmatrix} e & b \\ f & d \end{vmatrix}}{\begin{vmatrix} a & b \\ c & d \end{vmatrix}} \qquad \text{and} \qquad y = \frac{\begin{vmatrix} a & e \\ c & f \end{vmatrix}}{\begin{vmatrix} a & b \\ c & d \end{vmatrix}} \qquad \text{provided } \begin{vmatrix} a & b \\ c & d \end{vmatrix} \neq 0$$

If the denominators and the numerators of these fractions are all 0, the system is consistent but the equations are dependent.

If the denominators are 0 and one numerator is not 0, the system is inconsistent.

Example 2   Use determinants to solve the system

$$\begin{cases} 3x + 2y = 7 \\ -x + 5y = 9 \end{cases}$$

Solution   Use Cramer's rule:

$$x = \frac{\begin{vmatrix} 7 & 2 \\ 9 & 5 \end{vmatrix}}{\begin{vmatrix} 3 & 2 \\ -1 & 5 \end{vmatrix}} = \frac{7 \cdot 5 - 2 \cdot 9}{3 \cdot 5 - 2(-1)} = \frac{35 - 18}{15 + 2} = \frac{17}{17} = 1$$

$$y = \frac{\begin{vmatrix} 3 & 7 \\ -1 & 9 \end{vmatrix}}{\begin{vmatrix} 3 & 2 \\ -1 & 5 \end{vmatrix}} = \frac{3 \cdot 9 - 7(-1)}{3 \cdot 5 - 2(-1)} = \frac{27 + 7}{15 + 2} = \frac{34}{17} = 2$$

The solution of the given system is $x = 1$ and $y = 2$. The pair $(1, 2)$ satisfies both of the equations in the given system.   ∎

Cramer's rule can be used to solve many systems of n linear equations in n variables where $n > 2$. But to do so, we must develop a method of evaluating determinants larger than 2×2 that will give the correct solutions. Such a method is called **expansion by minors**.

Definition.   Let $|A|$ be a determinant of an $n \times n$ matrix A, and let a_{ij} be the element in the ith row and jth column of A. The **minor** of a_{ij} is the determinant of the $n - 1 \times n - 1$ matrix formed by those elements of A that do not lie in row i or in column j.

To find the minor of any chosen element of a determinant, we cross out the row and the column of that element. The minor is the determinant of the square array of numbers that remains.

Definition.   The **cofactor** of a_{ij} is the minor of a_{ij} if $i + j$ is even, or the negative of the minor of a_{ij} if $i + j$ is odd.

Example 3   Find the cofactor of **a.** 7 and **b.** 2 in the matrix $\begin{bmatrix} 1 & 2 & 3 \\ 4 & 5 & 6 \\ 7 & 8 & 9 \end{bmatrix}$.

Solution   **a.**   In the determinant of the 3×3 matrix A

$$\begin{vmatrix} 1 & 2 & 3 \\ 4 & 5 & 6 \\ 7 & 8 & 9 \end{vmatrix}$$

the minor of the 7 that appears in the third row, first column, is the determinant of the 2×2 matrix

$$\begin{vmatrix} 2 & 3 \\ 5 & 6 \end{vmatrix}$$

Because 7 is in the third row, first column, its row number plus the column number is even $(3 + 1 = 4)$. Thus, the cofactor of 7 is its minor:

$$\begin{vmatrix} 2 & 3 \\ 5 & 6 \end{vmatrix} = 2 \cdot 6 - 3 \cdot 5 = 12 - 15 = -3$$

b.   The minor of 2, which appears in the first row, second column of A is

$$\begin{vmatrix} 4 & 6 \\ 7 & 9 \end{vmatrix} = 4 \cdot 9 - 6 \cdot 7 = 36 - 42 = -6$$

This is found by ignoring the row and column in which the 2 is located. Because the sum of the row number and column number is odd $(1 + 2 = 3)$, the cofactor of the element 2 is the *negative* of its minor. Thus, the cofactor of 2 is $-(-6) = +6$.   ∎

Definition.   The value of the determinant of any square matrix is the sum of the products of the elements in the first row of the matrix and the cofactors of those elements.

It can be proved that the value of the determinant of any square matrix is the sum of the products of the elements of *any* row (or column) of the matrix and the cofactors of those elements.

Example 4   Evaluate the determinant $\begin{vmatrix} 1 & 2 & -3 \\ -1 & 0 & 1 \\ -2 & 2 & 1 \end{vmatrix}$ along **a.** the first row, **b.** the third row, and **c.** the second column.

Solution   **a.**   Multiply each element in the first row by its cofactor and form the sum of these three products:

$$\begin{vmatrix} 1 & 2 & -3 \\ -1 & 0 & 1 \\ -2 & 2 & 1 \end{vmatrix} = 1 \begin{vmatrix} 0 & 1 \\ 2 & 1 \end{vmatrix} - 2 \begin{vmatrix} -1 & 1 \\ -2 & 1 \end{vmatrix} + (-3) \begin{vmatrix} -1 & 0 \\ -2 & 2 \end{vmatrix}$$

Evaluate the three 2×2 determinants and simplify:

$$\begin{vmatrix} 1 & 2 & -3 \\ -1 & 0 & 1 \\ -2 & 2 & 1 \end{vmatrix} = 1(-2) - 2(1) - 3(-2)$$

$$= -2 - 2 + 6$$

$$= 2$$

b. Multiply each element in the third row by its cofactor and form the sum of these three products:

$$\begin{vmatrix} 1 & 2 & -3 \\ -1 & 0 & 1 \\ -2 & 2 & 1 \end{vmatrix} = -2 \begin{vmatrix} 2 & -3 \\ 0 & 1 \end{vmatrix} - 2 \begin{vmatrix} 1 & -3 \\ -1 & 1 \end{vmatrix} + 1 \begin{vmatrix} 1 & 2 \\ -1 & 0 \end{vmatrix}$$

$$= -2(2) - 2(-2) + 1(2)$$

$$= -4 + 4 + 2$$

$$= 2$$

c. Multiply each element of the second column by its cofactor and form the sum of these three products:

$$\begin{vmatrix} 1 & 2 & -3 \\ -1 & 0 & 1 \\ -2 & 2 & 1 \end{vmatrix} = -2 \begin{vmatrix} -1 & 1 \\ -2 & 1 \end{vmatrix} + 0 \begin{vmatrix} 1 & -3 \\ -2 & 1 \end{vmatrix} - 2 \begin{vmatrix} 1 & -3 \\ -1 & 1 \end{vmatrix}$$

$$= -2(1) + 0(-5) - 2(-2)$$

$$= -2 + 4$$

$$= 2$$

Note that the same result is obtained in each part of this example. In fact, the same result will be obtained if you expand the determinant along *any* row or column. ∎

Example 5 Evaluate the determinant $\begin{vmatrix} 0 & 0 & 2 & 0 \\ 1 & 2 & 17 & -3 \\ -1 & 0 & 28 & 1 \\ -2 & 2 & -37 & 1 \end{vmatrix}$.

Solution Because the determinant can be evaluated along *any* row or column, choose the row or column with the most 0s. So, expand this determinant along its first row. Note that three of the four minors need not be evaluated because each will be multiplied by 0.

$$\begin{vmatrix} 0 & 0 & 2 & 0 \\ 1 & 2 & 17 & -3 \\ -1 & 0 & 28 & 1 \\ -2 & 2 & -37 & 1 \end{vmatrix} = 0 \begin{vmatrix} \text{Who} \\ \text{cares?} \end{vmatrix} - 0 \begin{vmatrix} \text{Who} \\ \text{cares?} \end{vmatrix} + 2 \begin{vmatrix} 1 & 2 & -3 \\ -1 & 0 & 1 \\ -2 & 2 & 1 \end{vmatrix} - 0 \begin{vmatrix} \text{Who} \\ \text{cares?} \end{vmatrix}$$

$$= 2(2) \qquad \text{See Example 4.}$$

$$= 4 \qquad \blacksquare$$

Example 5 suggests the following theorem.

> **Theorem.**   If any row or column of a square matrix consists entirely of 0s, the value of the determinant of the matrix is 0.

Proof   A determinant can be evaluated along any of its rows or columns. Evaluate the determinant along its row or column of 0s. Each entry in that row or column is 0, and the product of each entry with its cofactor is also 0. The value of the determinant is the sum of these products. Hence, the value of the determinant is 0.   □

The next example illustrates how Cramer's rule can be used to solve a system of three equations in three variables.

Example 6   Use Cramer's rule to solve the following system of equations:

$$\begin{cases} 2x - y + 2z = 3 \\ x - y + z = 2 \\ x + y + 2z = 3 \end{cases}$$

Solution   Each of the values x, y, and z is the ratio of two 3×3 determinants. The denominator of each quotient is the determinant consisting of the nine coefficients of the variables. The numerators for x, y, and z are modified copies of this denominator determinant. The column of constants is substituted for the coefficients of the variable for which you are solving.

$$\begin{cases} 2x - y + 2z = 3 \\ x - y + z = 2 \\ x + y + 2z = 3 \end{cases}$$

$$x = \frac{\begin{vmatrix} 3 & -1 & 2 \\ 2 & -1 & 1 \\ 3 & 1 & 2 \end{vmatrix}}{\begin{vmatrix} 2 & -1 & 2 \\ 1 & -1 & 1 \\ 1 & 1 & 2 \end{vmatrix}} = \frac{3\begin{vmatrix} -1 & 1 \\ 1 & 2 \end{vmatrix} - (-1)\begin{vmatrix} 2 & 1 \\ 3 & 2 \end{vmatrix} + 2\begin{vmatrix} 2 & -1 \\ 3 & 1 \end{vmatrix}}{2\begin{vmatrix} -1 & 1 \\ 1 & 2 \end{vmatrix} - (-1)\begin{vmatrix} 1 & 1 \\ 1 & 2 \end{vmatrix} + 2\begin{vmatrix} 1 & -1 \\ 1 & 1 \end{vmatrix}} = \frac{2}{-1} = -2$$

$$y = \frac{\begin{vmatrix} 2 & 3 & 2 \\ 1 & 2 & 1 \\ 1 & 3 & 2 \end{vmatrix}}{\begin{vmatrix} 2 & -1 & 2 \\ 1 & -1 & 2 \\ 1 & 1 & 2 \end{vmatrix}} = \frac{2\begin{vmatrix} 2 & 1 \\ 3 & 2 \end{vmatrix} - 3\begin{vmatrix} 1 & 1 \\ 1 & 2 \end{vmatrix} + 2\begin{vmatrix} 1 & 2 \\ 1 & 3 \end{vmatrix}}{-1} = \frac{1}{-1} = -1$$

$$z = \frac{\begin{vmatrix} 2 & -1 & 3 \\ 1 & -1 & 2 \\ 1 & 1 & 3 \end{vmatrix}}{\begin{vmatrix} 2 & -1 & 2 \\ 1 & -1 & 1 \\ 1 & 1 & 2 \end{vmatrix}} = \frac{2\begin{vmatrix} -1 & 2 \\ 1 & 3 \end{vmatrix} - (-1)\begin{vmatrix} 1 & 2 \\ 1 & 3 \end{vmatrix} + 3\begin{vmatrix} 1 & -1 \\ 1 & 1 \end{vmatrix}}{-1} = \frac{-3}{-1} = 3$$

The triple $(-2, -1, 3)$ satisfies each of the equations in the given system. ■

Type 3 row operations can be used to simplify the calculations involved in determinant expansion.

Theorem. A type 3 elementary row operation performed on a square matrix does not alter the value of its determinant.

Justification.   The following discussion will illustrate the previous theorem for one particular row operation on a 3×3 matrix.

Evaluating $|D|$ by expanding along the first row, we get

$$|D| = \begin{vmatrix} a & b & c \\ d & e & f \\ g & h & i \end{vmatrix} = a\begin{vmatrix} e & f \\ h & i \end{vmatrix} - b\begin{vmatrix} d & f \\ g & i \end{vmatrix} + c\begin{vmatrix} d & e \\ g & h \end{vmatrix}$$

If we perform the type 3 row operation $kR3 + R1 \to R1$ on matrix D, the resulting modified determinant, $|D'|$, has a new first row. We expand that determinant on its first row:

$$|D'| = \begin{vmatrix} a+kg & b+kh & c+ki \\ d & e & f \\ g & h & i \end{vmatrix} = (a+kg)\begin{vmatrix} e & f \\ h & i \end{vmatrix} - (b+kh)\begin{vmatrix} d & f \\ g & i \end{vmatrix} + (c+ki)\begin{vmatrix} d & e \\ g & h \end{vmatrix}$$

Using the distributive law to remove the parentheses and rearranging the terms gives

$$|D'| = a\begin{vmatrix} e & f \\ h & i \end{vmatrix} - b\begin{vmatrix} d & f \\ g & i \end{vmatrix} + c\begin{vmatrix} d & e \\ g & h \end{vmatrix} + kg\begin{vmatrix} e & f \\ h & i \end{vmatrix} - kh\begin{vmatrix} d & f \\ g & i \end{vmatrix} + ki\begin{vmatrix} d & e \\ g & h \end{vmatrix}$$

The first three terms of the above expansion are identical to the original determinant $|D|$. The original determinant and the modified determinant differ by the amount that is represented by the last three terms of the expansion of $|D'|$. Hence,

$$|D'| = |D| + k\left(g\begin{vmatrix} e & f \\ h & i \end{vmatrix} - h\begin{vmatrix} d & f \\ g & i \end{vmatrix} + i\begin{vmatrix} d & e \\ g & h \end{vmatrix} \right)$$

$$|D'| = |D| + k(gei - ghf - hdi + hgf + idh - ige)$$

All terms within the parentheses subtract out, and we have

$$|D'| = |D| + k \cdot 0$$
$$|D'| = |D|$$

□

A similar result holds for type 3 column operations.

> **Theorem.** If any column of a square matrix is altered by adding to it any multiple of another column, the value of the determinant of the matrix is unchanged.

These two theorems provide a way to reduce the work involved in evaluating a large determinant.

Example 7   Evaluate $|A|$ if $|A| = \begin{vmatrix} 1 & 2 & -1 & 2 \\ 2 & 1 & 1 & 1 \\ 1 & 2 & -3 & 2 \\ 2 & -1 & -1 & 1 \end{vmatrix}$

Solution   Expanding the given determinant along any row or column leads to four determinants involving 3×3 matrices. The row operation $(-1)R3 + R1 \to R1$ gives three 0s in the first row. The introduction of these 0s simplifies the work because only one 3×3 determinant needs to be evaluated. Expand the new determinant along the first row.

$$|A| = \begin{vmatrix} 0 & 0 & 2 & 0 \\ 2 & 1 & 1 & 1 \\ 1 & 2 & -3 & 2 \\ 2 & -1 & -1 & 1 \end{vmatrix} = 2\begin{vmatrix} 2 & 1 & 1 \\ 1 & 2 & 2 \\ 2 & -1 & 1 \end{vmatrix}$$

To introduce more 0s, perform the column operation $(-2)C3 + C1 \to C1$ on the 3×3 determinant $\begin{vmatrix} 2 & 1 & 1 \\ 1 & 2 & 2 \\ 2 & -1 & 1 \end{vmatrix}$ and expand that result on its first column:

$$|A| = 2\begin{vmatrix} 0 & 1 & 1 \\ -3 & 2 & 2 \\ 0 & -1 & 1 \end{vmatrix} = 2\left[-(-3)\begin{vmatrix} 1 & 1 \\ -1 & 1 \end{vmatrix} \right]$$
$$= 2 \cdot 3[1 - (-1)]$$
$$= 2 \cdot 3 \cdot 2$$
$$= 12$$

■

Two more theorems describe the effect of type 2 and type 1 row operations on a determinant.

Theorem. If any row or column of a square matrix is multiplied by a constant k, the value of the determinant of the matrix is multiplied by k.

Theorem. If two rows or columns of a square matrix are interchanged, the value of the determinant of the matrix is multiplied by -1.

Exercise 8.5

In Exercises 1–16, evaluate each determinant.

1. $\begin{vmatrix} 2 & 1 \\ -2 & 3 \end{vmatrix}$

2. $\begin{vmatrix} -3 & -6 \\ 2 & -5 \end{vmatrix}$

3. $\begin{vmatrix} 2 & -3 \\ -3 & 5 \end{vmatrix}$

4. $\begin{vmatrix} 5 & 8 \\ -6 & -2 \end{vmatrix}$

5. $\begin{vmatrix} 2 & -3 & 5 \\ -2 & 1 & 3 \\ 1 & 3 & -2 \end{vmatrix}$

6. $\begin{vmatrix} 1 & 3 & 1 \\ -2 & 5 & 3 \\ 3 & -2 & -2 \end{vmatrix}$

7. $\begin{vmatrix} 1 & -1 & 2 \\ 2 & 1 & 3 \\ 1 & 1 & -1 \end{vmatrix}$

8. $\begin{vmatrix} 1 & 3 & 1 \\ 2 & 1 & -1 \\ 2 & -1 & 1 \end{vmatrix}$

9. $\begin{vmatrix} 2 & 1 & -1 \\ 1 & 3 & 5 \\ 2 & -5 & 3 \end{vmatrix}$

10. $\begin{vmatrix} 3 & 1 & -2 \\ -3 & 2 & 1 \\ 1 & 3 & 0 \end{vmatrix}$

11. $\begin{vmatrix} 0 & 1 & -3 \\ -3 & 5 & 2 \\ 2 & -5 & 3 \end{vmatrix}$

12. $\begin{vmatrix} 1 & -7 & -2 \\ -2 & 0 & 3 \\ -1 & 7 & 1 \end{vmatrix}$

13. $\begin{vmatrix} 1 & 2 & 1 & 3 \\ -2 & 1 & -3 & 1 \\ -1 & 0 & 1 & -2 \\ 2 & -1 & -1 & 3 \end{vmatrix}$

14. $\begin{vmatrix} -1 & 3 & -2 & 5 \\ 2 & 1 & 0 & 1 \\ 1 & 3 & -2 & 5 \\ 2 & -1 & 0 & -1 \end{vmatrix}$

15. $\begin{vmatrix} 1 & 2 & 3 & 4 & 5 \\ 0 & 1 & 2 & 3 & 4 \\ 0 & 0 & 1 & 2 & 3 \\ 0 & 0 & 0 & 1 & 2 \\ 0 & 0 & 0 & 0 & 1 \end{vmatrix}$

16. $\begin{vmatrix} 1 & 1 & 1 & 1 & 1 \\ 1 & 1 & 1 & 1 & 2 \\ 1 & 1 & 1 & 2 & 2 \\ 1 & 1 & 2 & 2 & 2 \\ 1 & 2 & 2 & 2 & 2 \end{vmatrix}$

In Exercises 17–28, use Cramer's rule to find the solution to each system of equations, if possible.

17. $\begin{cases} 3x + 2y = 7 \\ 2x - 3y = -4 \end{cases}$

18. $\begin{cases} x - 5y = -6 \\ 3x + 2y = -1 \end{cases}$

19. $\begin{cases} x - y = 3 \\ 3x - 7y = 9 \end{cases}$

20. $\begin{cases} 2x - y = -6 \\ x + y = 0 \end{cases}$

21. $\begin{cases} x + 2y + z = 2 \\ x - y + z = 2 \\ x + y + 3z = 4 \end{cases}$

22. $\begin{cases} x + 2y - z = -1 \\ 2x + y - z = 1 \\ x - 3y - 5z = 17 \end{cases}$

23. $\begin{cases} 2x - y + z = 5 \\ 3x - 3y + 2z = 10 \\ x + 3y + z = 0 \end{cases}$

24. $\begin{cases} x - y - z = 2 \\ x + y + z = 2 \\ -x - y + z = -4 \end{cases}$

25. $\begin{cases} \dfrac{x}{2} + \dfrac{y}{3} + \dfrac{z}{2} = 11 \\[2mm] \dfrac{x}{3} + y - \dfrac{z}{6} = 6 \\[2mm] \dfrac{x}{2} + \dfrac{y}{6} + z = 16 \end{cases}$

26. $\begin{cases} \dfrac{x}{2} + \dfrac{y}{5} + \dfrac{z}{3} = 17 \\[2mm] \dfrac{x}{5} + \dfrac{y}{2} + \dfrac{z}{5} = 32 \\[2mm] x + \dfrac{y}{3} + \dfrac{z}{2} = 30 \end{cases}$

27. $\begin{cases} 2p - q + 3r - s = 3 \\ p + q \quad\;\; - 2s = 0 \\ 3p \quad\;\; - r \quad\;\; = 2 \\ p - q \quad\;\; + 3s = 3 \end{cases}$

28. $\begin{cases} a + b + c + d = -1 \\ a + b + c + 2d = 0 \\ a + b + 2c + 3d = 1 \\ a + 2b + 3c + 4d = 0 \end{cases}$

29. Use the method of addition to solve the system $\begin{cases} ax + by = e \\ cx + dy = f \end{cases}$ for y, and thereby show that

$$y = \frac{af - ec}{ad - bc}$$

30. Use an example chosen from 2×2 matrices to show that the determinant of the product of two matrices is the product of the determinants of those two matrices.

31. Find an example among 2×2 determinants to show that the determinant of a sum of two matrices is not equal to the sum of the determinants of those matrices.

32. Find a 2×2 matrix A for which $|A| \neq 0$. Then find A^{-1}, and verify that $|A^{-1}| = \dfrac{1}{|A|}$.

33. Show that $\begin{vmatrix} a & b & c \\ 0 & d & e \\ 0 & 0 & f \end{vmatrix} = adf$.

34. Show that $\begin{vmatrix} a & b & c & d \\ 0 & e & f & g \\ 0 & 0 & h & i \\ 0 & 0 & 0 & j \end{vmatrix} = aehj$.

35. Show that multiplying the first row of $\begin{vmatrix} 2 & -1 & 3 \\ 1 & 2 & -1 \\ 3 & 2 & -1 \end{vmatrix}$ by 3 multiplies the value of the determinant by 3.

36. Interchange two rows of the determinant in Exercise 35 and show that this has the effect of multiplying the value of the determinant by -1.

37. Interchange two columns of the determinant in Exercise 35 and show that this has the effect of multiplying the value of the determinant by -1.

38. Find the value of k if $\begin{vmatrix} 2a & 2b & 2c \\ 3d & 3e & 3f \\ 5g & 5h & 5i \end{vmatrix} = k \begin{vmatrix} a & b & c \\ d & e & f \\ g & h & i \end{vmatrix}$.

In Exercises 39–42, expand the determinants and solve for x.

39. $\begin{vmatrix} 3 & x \\ 1 & 2 \end{vmatrix} = \begin{vmatrix} 2 & -1 \\ x & -5 \end{vmatrix}$

40. $\begin{vmatrix} 4 & x^2 \\ 1 & -1 \end{vmatrix} = \begin{vmatrix} x & 4 \\ 2 & 3 \end{vmatrix}$

41. $\begin{vmatrix} 3 & x & 1 \\ x & 0 & -2 \\ 4 & 0 & 1 \end{vmatrix} = \begin{vmatrix} 2 & x \\ x & 4 \end{vmatrix}$

42. $\begin{vmatrix} x & -1 & 2 \\ -2 & x & 3 \\ 4 & -3 & -1 \end{vmatrix} = \begin{vmatrix} 2 & 2 \\ 5 & x \end{vmatrix}$

43. A determinant is a function that associates a number with every square matrix. Give the domain and the range of that function.

44. Select 2×2 matrices A and B and show that $|AB| = |A||B|$.

45. If A and B are matrices and if $|AB| = 0$, must $|A| = 0$ or $|B| = 0$? Support your answer.

46. If A and B are matrices and if $|AB| = 0$, must $A = \mathbf{0}$ or $B = \mathbf{0}$? Support your answer.

8.6 PARTIAL FRACTIONS

In this section, we discuss how to express a complicated fraction as the sum of several simpler fractions. This process of decomposing a fraction into **partial fractions** is used in calculus. We begin by reviewing the process of adding fractions.

Example 1 Find the sum: $\dfrac{2}{x} + \dfrac{6}{x+1} + \dfrac{-1}{(x+1)^2}$.

Solution Write each fraction in a form with the least common denominator, $x(x+1)^2$, and add:

$$\frac{2}{x} + \frac{6}{x+1} + \frac{-1}{(x+1)^2} = \frac{2(x+1)^2}{x(x+1)^2} + \frac{6x(x+1)}{(x+1)x(x+1)} + \frac{-1x}{(x+1)^2x}$$

$$= \frac{2x^2 + 4x + 2 + 6x^2 + 6x - x}{x(x+1)^2}$$

$$= \frac{8x^2 + 9x + 2}{x(x+1)^2} \qquad \blacksquare$$

Example 2 Express the fraction $\dfrac{3x^2 - x + 1}{x(x-1)^2}$ as the sum of several fractions with denominators of the smallest degree possible.

Solution Example 1 leads you to suspect that constants A, B, and C can be found such that

$$\frac{3x^2 - x + 1}{x(x-1)^2} = \frac{A}{x} + \frac{B}{x-1} + \frac{C}{(x-1)^2}$$

After you write the terms on the right side as fractions with the common denominator $x(x - 1)^2$, combine them:

$$\frac{3x^2 - x + 1}{x(x - 1)^2} = \frac{A(x - 1)^2}{x(x - 1)^2} + \frac{Bx(x - 1)}{x(x - 1)(x - 1)} + \frac{Cx}{(x - 1)^2 x}$$

$$= \frac{Ax^2 - 2Ax + A + Bx^2 - Bx + Cx}{x(x - 1)^2}$$

$$= \frac{(A + B)x^2 + (-2A - B + C)x + A}{x(x - 1)^2}$$

Because the fractions are equal, the numerator $3x^2 - x + 1$ must equal the numerator $(A + B)x^2 + (-2A - B + C)x + A$. These quantities are equal provided their coefficients are equal. Thus,

$$\begin{cases} A + B & = & 3 & \text{The coefficients of } x^2. \\ -2A - B + C & = & -1 & \text{The coefficients of } x. \\ A & = & 1 & \text{The constants.} \end{cases}$$

This system of three equations in three variables can be solved by substitution. The solutions are $A = 1$, $B = 2$, and $C = 3$. Hence,

$$\frac{3x^2 - x + 1}{x(x - 1)^2} = \frac{A}{x} + \frac{B}{x - 1} + \frac{C}{(x - 1)^2} = \frac{1}{x} + \frac{2}{x - 1} + \frac{3}{(x - 1)^2} \qquad \blacksquare$$

Example 3 Express the fraction $\dfrac{2x^2 + x + 1}{x^3 + x}$ as the sum of fractions with denominators of the smallest possible degree.

Solution Factoring the denominator suggests that this fraction can be expressed as the sum of two fractions, one with a denominator of x and the other with a denominator of $x^2 + 1$.

$$\frac{2x^2 + x + 1}{x(x^2 + 1)} = \frac{}{x} + \frac{}{x^2 + 1}$$

Because the denominator x of the first fraction is of first degree, its numerator must be of degree 0—that is, a constant. Because the denominator $x^2 + 1$ of the second equation is of second degree, the numerator might be a first-degree polynomial or a constant. You can allow for both possibilities by using a numerator of $Bx + C$. If $B = 0$, then $Bx + C$ is a constant. If $B \neq 0$, then $Bx + C$ is a first-degree polynomial. Thus,

$$\frac{2x^2 + x + 1}{x(x^2 + 1)} = \frac{A}{x} + \frac{Bx + C}{x^2 + 1} = \frac{A(x^2 + 1) + (Bx + C)x}{x(x^2 + 1)}$$

$$= \frac{Ax^2 + A + Bx^2 + Cx}{x(x^2 + 1)} = \frac{(A + B)x^2 + Cx + A}{x(x^2 + 1)}$$

Equate the corresponding coefficients of the polynomials $2x^2 + x + 1$ and

$(A + B)x^2 + Cx + A$ to produce the following system of equations:

$$\begin{cases} A + B = 2 \\ \quad\; C = 1 \\ \quad\; A = 1 \end{cases}$$

The solutions are $A = 1$, $B = 1$, $C = 1$. Therefore, the given fraction can be written as a sum:

$$\frac{2x^2 + x + 1}{x^3 + x} = \frac{1}{x} + \frac{x + 1}{x^2 + 1}$$
■

The process illustrated in these examples can be summarized: Let $\frac{P(x)}{Q(x)}$ be the quotient of two polynomials with real coefficients, with the degree of $P(x)$ less than the degree of $Q(x)$. Suppose also that the fraction $\frac{P(x)}{Q(x)}$ has been simplified. The polynomial $Q(x)$ can always be factored as a product of first-degree and irreducible second-degree expressions.

If all the identical factors of $Q(x)$ are collected into single factors of the form $(ax + b)^n$ and the form $(ax^2 + bx + c)^n$, then the partial fractions required for the decomposition of $\frac{P(x)}{Q(x)}$ can be found. Each factor of $Q(x)$ of the form $(ax + b)^n$ generates a sum of n partial fractions of the form

$$\frac{A_1}{ax + b} + \frac{A_2}{(ax + b)^2} + \cdots + \frac{A_n}{(ax + b)^n}$$

where each A_i represents a constant. Each factor of the form $(ax^2 + bx + c)^n$ generates the sum of n fractions of the form

$$\frac{B_1 x + C_1}{ax^2 + bx + c} + \frac{B_2 x + C_2}{(ax^2 + bx + c)^2} + \cdots + \frac{B_n x + C_n}{(ax^2 + bx + c)^n}$$

where each B_i and C_i is a constant. After finding a least common denominator and adding the fractions, we obtain a fractional expression that must be equivalent to $\frac{P(x)}{Q(x)}$. Equating the corresponding coefficients of the numerators gives a system of linear equations that can be solved for the constants A_i, B_i, and C_i.

Example 4 What fractions should be used in the decomposition of the rational expression

$$\frac{3x^7 - 5x^5 + 3x + 2}{x^3(x - 3)(x + 2)^2(2x^2 + x + 3)^2(x^2 + 1)^3}$$

Solution The factor x^3 in the denominator requires three possible fractions in the decomposition:

$$\frac{A}{x} + \frac{B}{x^2} + \frac{C}{x^3}$$

The factor $x - 3$ adds one more to the list:

$$\frac{A}{x} + \frac{B}{x^2} + \frac{C}{x^3} + \frac{D}{x - 3}$$

The factor $(x + 2)^2$ generates two more fractions, each with a constant as numerator:

$$\frac{A}{x} + \frac{B}{x^2} + \frac{C}{x^3} + \frac{D}{x - 3} + \frac{E}{x + 2} + \frac{F}{(x + 2)^2}$$

The factor $(2x^2 + x + 3)^2$ produces two more fractions, each requiring first-degree numerators:

$$\frac{A}{x} + \frac{B}{x^2} + \frac{C}{x^3} + \frac{D}{x - 3} + \frac{E}{x + 2} + \frac{F}{(x + 2)^2} + \frac{Gx + H}{2x^2 + x + 3}$$

$$+ \frac{Jx + K}{(2x^2 + x + 3)^2}$$

Finally, the factor $(x^2 + 1)^3$ requires three more fractions, also with first-degree numerators:

$$\frac{A}{x} + \frac{B}{x^2} + \frac{C}{x^3} + \frac{D}{x - 3} + \frac{E}{x + 2} + \frac{F}{(x + 2)^2} + \frac{Gx + H}{2x^2 + x + 3}$$

$$+ \frac{Jx + K}{(2x^2 + x + 3)^2} + \frac{Lx + M}{x^2 + 1} + \frac{Nx + P}{(x^2 + 1)^2} + \frac{Rx + S}{(x^2 + 1)^3}$$

If you find a common denominator and combine the fractions, equating the corresponding coefficients of the numerators will give 16 equations in 16 variables. These can be solved for the variables $A, B, C, \ldots, S$. ∎

Example 5 Express the fraction $\dfrac{x^2 + 4x + 2}{x^2 + x}$ as the sum of several fractions with denominators of the smallest degree possible.

Solution The method of partial fractions requires that the degree of the numerator be less than the degree of the denominator. Because the degree of both the numerator and denominator are the same in this example, you must perform a long division and express the given fraction in $quotient + \dfrac{remainder}{divisor}$ form:

$$
\begin{array}{r}
1 \\
x^2 + x \overline{)\, x^2 + 4x + 2} \\
\underline{x^2 + x} \\
3x + 2
\end{array}
$$

Hence,

$$\frac{x^2 + 4x + 2}{x^2 + x} = 1 + \frac{3x + 2}{x^2 + x}$$

Because the degree of the numerator of the fraction $\dfrac{3x + 2}{x^2 + x}$ is less than the degree of the denominator, you can find the partial fraction decomposition of this fraction:

$$\frac{3x + 2}{x^2 + x} = \frac{3x + 2}{x(x + 1)}$$

$$= \frac{A}{x} + \frac{B}{x + 1}$$

$$= \frac{A(x + 1) + Bx}{x(x + 1)}$$

$$= \frac{(A + B)x + A}{x(x + 1)}$$

Equate the corresponding coefficients in the numerator, and solve the resulting system of equations:

$$\begin{cases} A + B = 3 \\ A = 2 \end{cases}$$

The solution is $A = 2$, $B = 1$, and the decomposition of the given fraction is

$$\frac{x^2 + 4x + 2}{x^2 + x} = 1 + \frac{2}{x} + \frac{1}{x + 1}$$ ■

Exercise 8.6

In Exercises 1–24, decompose each expression into partial fractions.

1. $\dfrac{3x + 1}{(x + 1)(x - 1)}$

2. $\dfrac{-2x + 11}{x^2 - x - 6}$

3. $\dfrac{-3x^2 + x - 5}{(x + 1)(x^2 + 2)}$

4. $\dfrac{-x^2 - 3x - 5}{x^3 + x^2 + 2x + 2}$

5. $\dfrac{-2x^2 + x - 2}{x^3 - x^2}$

6. $\dfrac{2x^2 - 7x + 2}{x(x - 1)^2}$

7. $\dfrac{2x^2 + 1}{x^4 + x^2}$

8. $\dfrac{x^2 + x + 1}{x^3}$

9. $\dfrac{5x^2 + 2x + 2}{x^3 + x}$

10. $\dfrac{-2x^3 + 7x^2 + 6}{x^2(x^2 + 2)}$

11. $\dfrac{x^3 + 4x^2 + 2x + 1}{x^4 + x^3 + x^2}$

12. $\dfrac{x^3 + 4x^2 + 3x + 6}{(x^2 + 2)(x^2 + x + 2)}$

13. $\dfrac{x^3 + 3x^2 + 6x + 6}{(x^2 + x + 5)(x^2 + 1)}$

14. $\dfrac{x^2 - 2x - 3}{(x - 1)^3}$

15. $\dfrac{x^4 - x^3 + x^2 - x + 1}{x(x^2 + 1)^2}$

16. $\dfrac{x^2 + 2}{x^3 + 3x^2 + 3x + 1}$

17. $\dfrac{x^3 + 3x^2 + 2x + 4}{(x^2 + 1)(x^2 + x + 2)}$

18. $\dfrac{4x^3 + 5x^2 + 3x + 4}{x^2(x^2 + 1)}$

19. $\dfrac{2x^4 + 6x^3 + 20x^2 + 22x + 25}{x(x^2 + 2x + 5)^2}$

20. $\dfrac{3x^3 + 5x^2 + 3x + 1}{x^2(x^2 + x + 1)}$

21. $\dfrac{2x^3 + 6x^2 + 3x + 2}{x^3 + x^2}$

22. $\dfrac{x^3}{x^2 + 3x + 2}$

23. $\dfrac{x^4 + x^3 + x^2 + x + 1}{x^2}$

24. $\dfrac{x^3 + 2x^2 + 3x + 4}{x^3}$

8.7   SYSTEMS OF INEQUALITIES AND LINEAR PROGRAMMING

We now consider the graphs of systems of inequalities.

Example 1   Graph the solution set of the system

$$\begin{cases} x + y \leq 1 \\ 2x - y > 2 \end{cases}$$

Solution   On the same set of coordinate axes, graph each inequality. See Figure 8-3. The graph of the inequality $x + y \leq 1$ includes the line graph of the equation $x + y = 1$ and all points below it. Because the boundary line is included, it is drawn as a solid line. The graph of the inequality $2x - y > 2$ contains only those points below the line graph of the equation $2x - y = 2$. Because the boundary line is not included, it is drawn as a broken line. The area that is shaded twice represents the simultaneous solutions of the given system of inequalities. Any point in the doubly shaded region has coordinates that satisfy both inequalities in the system.

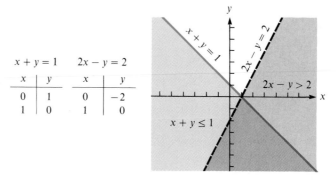

$x + y = 1$		$2x - y = 2$	
x	y	x	y
0	1	0	-2
1	0	1	0

Figure 8-3   ■

Example 2   Graph the solution set of the system

$$\begin{cases} y < x^2 \\ y > \dfrac{x^2}{4} - 2 \end{cases}$$

Solution   The graph of the equation $y = x^2$ is a parabola opening upward with vertex at the origin. See Figure 8-4. The points with coordinates that satisfy the inequality $y < x^2$ are those points below the parabola.

The graph of $y = \frac{x^2}{4} - 2$ is also a parabola opening upward. However, this time the points with coordinates that satisfy the inequality are those points above the parabola. Thus, the graph of the solution set of this system is the shaded area between the two parabolas.

$y = x^2$		$y = \dfrac{x^2}{4} - 2$	
x	y	x	y
0	0	0	-2
1	1	2	-1
-1	1	-2	-1
2	4	4	2
-2	4	-4	2

$y < x^2$

$y > \dfrac{x^2}{4} - 2$

Figure 8-4

Example 3 Graph the solution set of the system

$$\begin{cases} x \geq 1 \\ y \geq x \\ 4x + 5y < 20 \end{cases}$$

Solution The graph of the solution set of the inequality $x \geq 1$ includes those points on the graph of the equation $x = 1$ and to the right. See Figure 8-5**a**. The graph of the solution set of the inequality $y \geq x$ includes those points on the graph of the equation $y = x$ and above it. See Figure 8-5**b**. The graph of the solution set of the inequality $4x + 5y < 20$ includes those points below the line graph of the equation $4x + 5y = 20$. See Figure 8-5**c**.

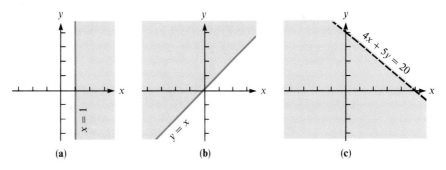

(a)   (b)   (c)

Figure 8-5

If these three graphs are merged onto a single coordinate system, the graph of the original system of inequalities includes those points within the shaded triangle together with the points on the sides of the triangle drawn as solid lines. See Figure 8-6.

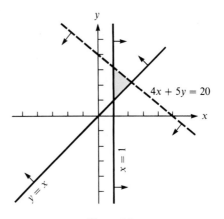

Figure 8-6   ■

Linear Programming

Systems of inequalities provide the basis for an area of applied mathematics known as **linear programming**. Linear programming is used to help answer such questions as "How can a business make as much money as possible?" or "How can I plan a nutritious menu at the least cost?" In such problems, the solution depends on certain **constraints**: The business has limited resources, and the nutritious meal must contain sufficient vitamins, minerals, and so on. Any solution that satisfies the constraints is called a **feasible solution**. In linear programming, the constraints are expressed as a system of linear inequalities, and the quantity that is to be maximized (or minimized) is expressed as a linear function of several variables.

Example 4   Many ordered pairs (x, y) satisfy each inequality in the system

$$\begin{cases} x + y \geq 1 \\ x - y \leq 1 \\ x - y \geq 0 \\ x \leq 2 \end{cases}$$

If $Z = y - 2x$, which of these pairs will produce the greatest value of Z?

Solution   Find the solution of the given system of inequalities, and find the coordinates of each corner of region R, as shown in Figure 8-7. Then rewrite the equation

$$Z = y - 2x$$

in the equivalent form

$$y = 2x + Z$$

This is the equation of a straight line with slope of 2 and y-intercept of Z. Many such lines pass through the region R. To decide which of these provides the greatest value of Z, refer to Figure 8-8 and find the line with the greatest y-intercept. It is line l passing through point P, the left-most corner of R. The coordinates of P are $(\frac{1}{2}, \frac{1}{2})$. Thus, the greatest value of Z possible (subject to the given constraints) is

$$Z = \frac{1}{2} - 2\left(\frac{1}{2}\right) = -\frac{1}{2}$$

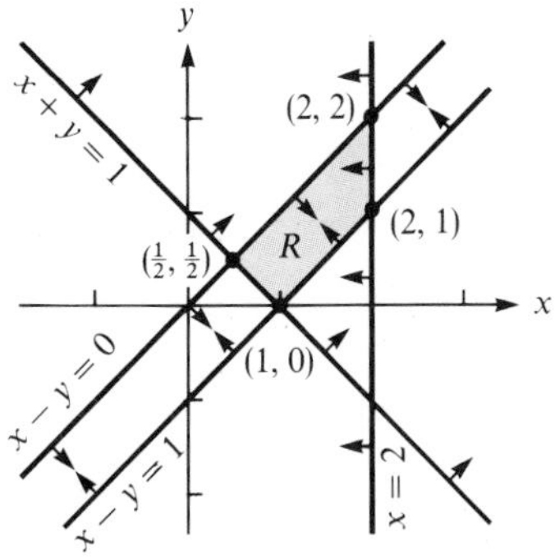

Figure 8-7

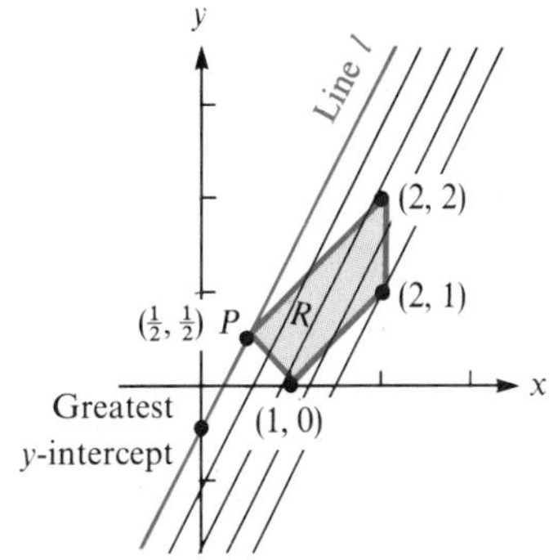

Figure 8-8 ■

Example 4 illustrates this fact:

The maximum (or minimum) value of a linear function that is subject to the constraints of a system of linear inequalities in two variables is always attained at a corner or along an entire edge of the region R that represents the solution of the system.

Example 5   Fred and Donna are in a part-time business manufacturing clock cases. Fred must work 4 hours and Donna 2 hours to complete one case for a grandfather clock. To build one case for a wall clock, Fred must work 3 hours and Donna 4 hours. Neither partner wishes to work more than 20 hours per week. If they receive $80 for each grandfather clock and $64 for each wall clock, how many of each should they build each week to maximize their profit?

Solution   If the partners manufacture cases for x grandfather clocks and y wall clocks each week, their profit P (in dollars) is

$P =$ | The profit on one grandfather clock | $\cdot$ | the number of grandfather clocks | $+$ | the profit on one wall clock | $\cdot$ | the number of wall clocks.

$$P = 80x + 64y$$

The time requirements are summarized in the following chart:

Partner	Time for one grandfather clock	Time for one wall clock
Fred	4 hours	3 hours
Donna	2 hours	4 hours

The profit function is subject to the following constraints:

$$\begin{cases} x \geq 0 \\ y \geq 0 \\ 4x + 3y \leq 20 \\ 2x + 4y \leq 20 \end{cases}$$

The inequalities $x \geq 0$ and $y \geq 0$ state that the number of clock cases to be built cannot be negative. The inequality $4x + 3y \leq 20$ is a constraint on Fred's time because he spends 4 hours on each of the x grandfather clocks and 3 hours on each of the y wall clocks, and his total time cannot exceed 20 hours. Similarly, the inequality $2x + 4y \leq 20$ is a constraint on Donna's time.

Graph each of the constraints to find the **feasibility region R**, as in Figure 8-9. The four corners of region R have coordinates of $(0, 0)$, $(0, 5)$, $(2, 4)$, and $(5, 0)$. Substitute each of these number pairs into the equation $P = 80x + 64y$ to find the maximum profit, P.

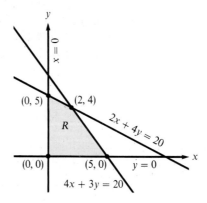

Figure 8-9

Corner	Profit
$(0, 0)$	$P = \$80(0) + \$64(0) = \$\ \ 0$
$(0, 5)$	$P = \$80(0) + \$64(5) = \$320$
$(2, 4)$	$P = \$80(2) + \$64(4) = \$416$
$(5, 0)$	$P = \$80(5) + \$64(0) = \$400$

Donna and Fred will maximize their profits if they build cases for 2 grandfather clocks and 4 wall clocks each week. If they do so, they will earn \$416.   ■

Exercise 8.7

In Exercises 1–20, graph the solution set of each system of inequalities.

1. $\begin{cases} y < 3x + 2 \\ y < -2x + 3 \end{cases}$

2. $\begin{cases} y \le x - 2 \\ y \ge 2x + 1 \end{cases}$

3. $\begin{cases} 3x + 2y > 6 \\ x + 3y \le 2 \end{cases}$

4. $\begin{cases} x + y < 2 \\ x + y \le 1 \end{cases}$

5. $\begin{cases} 3x + y \le 1 \\ -x + 2y \ge 9 \end{cases}$

6. $\begin{cases} x + 2y < 3 \\ 2x - 4y < 8 \end{cases}$

7. $\begin{cases} 2x - y > 4 \\ y < -x^2 + 2 \end{cases}$

8. $\begin{cases} x \le y^2 \\ y \ge x \end{cases}$

9. $\begin{cases} y > x^2 - 4 \\ y < -x^2 + 4 \end{cases}$

10. $\begin{cases} x \ge y^2 \\ y \ge x^2 \end{cases}$

11. $\begin{cases} 2x + 3y \le 5 \\ 3x + y \le 1 \\ x \le 0 \end{cases}$

12. $\begin{cases} 2x + y \le 2 \\ y \ge x \\ x \ge 0 \end{cases}$

13. $\begin{cases} x - y < 4 \\ y \ge 0 \\ xy = 12 \end{cases}$

14. $\begin{cases} xy \le 1 \\ x \ge 0 \\ y \ge 0 \end{cases}$

15. $\begin{cases} x \ge 0 \\ y \ge 0 \\ 9x + 3y \le 18 \\ 3x + 6y \le 18 \end{cases}$

16. $\begin{cases} x + y \ge 1 \\ x - y \le 1 \\ x - y \ge 0 \\ x \le 2 \end{cases}$

17. $\begin{cases} y < \sqrt{x} \\ x \ge 0 \end{cases}$

18. $\begin{cases} y < -\sqrt{x} \\ x \ge 0 \end{cases}$

19. $\begin{cases} |x - 2| > 3 \\ |y| > 1 \end{cases}$

20. $\begin{cases} |x + 2| < 3 \\ |y| > 2 \end{cases}$

In Exercises 21–26, maximize P subject to the given constraints.

21. $P = 2x + 3y$
$\begin{cases} x \ge 0 \\ y \ge 0 \\ x + y \le 4 \end{cases}$

22. $P = 3x + 2y$
$\begin{cases} x \ge 0 \\ y \ge 0 \\ x + y \le 4 \end{cases}$

23. $P = y + \dfrac{1}{2}x$
$\begin{cases} 2y - x \le 1 \\ y - 2x \ge -2 \\ x \ge 0 \\ y \ge 0 \end{cases}$

24. $P = 4y - x$
$\begin{cases} 2y - x \le 1 \\ x \le 2 \\ x + y \ge 1 \\ y \ge 0 \end{cases}$

25. $P = 2x + y$
$\begin{cases} y - x \le 2 \\ 2x + 3y \le 6 \\ 3x + y \le 3 \\ y \ge 0 \end{cases}$

26. $P = 3x - 2y$
$\begin{cases} y - x \le 1 \\ x - y \le 1 \\ x \ge -1 \\ x \le 1 \end{cases}$

27. Sally and Sandra each have 12 hours a week to make furniture. They obtain a \$50 profit from each table and a \$10 profit from each chair. Sally must work 3 hours and Sandra 2 to make a chair. Sally must work 2 hours and Sandra 6 to make a table. How many tables and how many chairs should they make each week to maximize profits?

28. Two machines, *A* and *B*, work 24 hours a day to manufacture chewing gum and bubble gum. There is a profit of \$150 per case of chewing gum and \$100 per case of bubble gum. To make a case of chewing gum, machine *A* must run 2 hours and machine *B* must run 8 hours. To make a case of bubble gum, machine *A* must run 4 hours and machine *B* must run 2 hours. How many cases of each should be produced each week to maximize profits?

29. Sarah and her sister Heidi have decided to plant a garden. Sarah wants to plant strawberries; Heidi wants to plant pumpkins. Neither crop may use more than $\frac{3}{4}$ of the 40 square meters of available space. Each square meter of strawberries will earn the children $4, and each square meter of pumpkins will earn the children $3. The children plan to eat half of the strawberries themselves. How should the planting be divided to maximize the total income?

30. A distributor of hybrid corn has two storehouses, A and B. At A is stored 110 tons of corn; 190 tons are stored at B. Farmer X ordered 60 tons to be delivered to his farm, and farmer Y ordered 80 tons. The shipping costs appear in the following table:

Storehouse	Farmer	Costs per ton
A	X	$5
A	Y	$7
B	X	$8
B	Y	$14

How should the orders be filled to minimize the shipping costs? (*Hint:* Let x represent the number of tons shipped from A to X, and let y represent the number of tons shipped from A to Y.)

31. Bill packs two foods for his camping trip. One ounce of food X costs 35¢ and provides 150 calories and 21 units of vitamins. One ounce of food Y costs 27¢ and provides 60 calories and 42 units of vitamins. Every day, Bill needs at least 3000 calories and at least 1260 units of vitamins, but he does not want to carry more than 60 ounces of food for each day of his trip. How much of each food should Bill pack to minimize the cost?

32. To manufacture two products X and Y, three workers are scheduled as follows:

Worker	Hours required for product X	Hours required for product Y	Time available per week
A	1	4	18
B	2	3	12
C	2	1	6

The profit on product X is $40, and on product Y, $60. How many of each product should be produced to maximize the profit?

REVIEW EXERCISES

In Review Exercises 1–4, solve each system of equations by the method of graphing.

1. $\begin{cases} 2x - y = -1 \\ x + y = 7 \end{cases}$

2. $\begin{cases} 5x + 2y = 1 \\ 2x - y = -5 \end{cases}$

3. $\begin{cases} y = 5x + 7 \\ x = y - 7 \end{cases}$

4. $\begin{cases} x = y + 5 \\ y = -4 + \dfrac{x}{2} \end{cases}$

In Review Exercises 5–8, solve each system of equations by substitution.

5. $\begin{cases} y = 3x + 2 \\ y = 5x \end{cases}$

6. $\begin{cases} 2y + x = 0 \\ x = y + 3 \end{cases}$

7. $\begin{cases} 2x + y = -3 \\ x - y = 3 \end{cases}$

8. $\begin{cases} \dfrac{x + y}{2} + \dfrac{x - y}{3} = 1 \\ y = 3x - 2 \end{cases}$

In Review Exercises 9–12, solve each system of equations by addition.

9. $\begin{cases} x + 5y = 7 \\ 3x + y = -7 \end{cases}$

10. $\begin{cases} 2x + 3y = 11 \\ 3x - 7y = -41 \end{cases}$

11. $\begin{cases} 2(x + y) - x = 0 \\ 3(x + y) + 2y = 1 \end{cases}$

12. $\begin{cases} \dfrac{x + y}{2} + \dfrac{x - y}{3} = \dfrac{7}{2} \\ \dfrac{x + y}{5} + \dfrac{x - y}{2} = \dfrac{5}{2} \end{cases}$

In Review Exercises 13–16, solve each system of equations by any method.

13. $\begin{cases} 3x + 2y - z = 2 \\ x + y - z = 0 \\ 2x + 3y - z = 1 \end{cases}$

14. $\begin{cases} 5x - y + z = 3 \\ 3x + y + 2z = 2 \\ x + y = 2 \end{cases}$

15. $\begin{cases} 2x - y + z = 1 \\ x - y + 2z = 3 \\ x - y + z = 1 \end{cases}$

16. $\begin{cases} x + 2y - z = -6 \\ x + y - z = -4 \\ 3y - 2z = -12 \end{cases}$

In Review Exercises 17–22, solve each system of equations by matrix methods, if possible.

17. $\begin{cases} 2x + 5y = 7 \\ 3x - y = 2 \end{cases}$

18. $\begin{cases} x + 3y - z = 8 \\ 2x + y - 2z = 11 \\ x - y + 5z = -8 \end{cases}$

19. $\begin{cases} x + 3y + z = 7 \\ 2x - y + z = 0 \\ 3x + 2y + 2z = 7 \end{cases}$

20. $\begin{cases} x + y + z = 4 \\ 3x - 2y - 2z = -3 \\ 4x - y - z = 0 \end{cases}$

21. $\begin{cases} w + x - 3y + z = 2 \\ x - y + 2z = 0 \\ w + y - z = 2 \\ w - x + 2z = -3 \end{cases}$

22. $\begin{cases} w + x + z = 3 \\ x - y + z = 3 \\ w + 2x - y + 2z = 6 \\ y - z = -2 \end{cases}$

In Review Exercises 23–32, perform the indicated matrix arithmetic.

23. $\begin{bmatrix} 3 & 2 & 1 \\ 3 & 2 & 1 \end{bmatrix} + \begin{bmatrix} -2 & 1 & 3 \\ 1 & -2 & 1 \end{bmatrix}$

24. $\begin{bmatrix} 2 & 3 & 5 \\ 1 & -2 & 4 \\ 2 & 1 & -2 \end{bmatrix} - \begin{bmatrix} 0 & -2 & 1 \\ 3 & 4 & -2 \\ 6 & -4 & 1 \end{bmatrix}$

25. $\begin{bmatrix} 2 & 3 \\ -1 & 2 \end{bmatrix} \begin{bmatrix} 1 & -2 \\ -3 & 1 \end{bmatrix}$

26. $\begin{bmatrix} -2 & 3 & 5 \\ 1 & -2 & -3 \end{bmatrix} \begin{bmatrix} 2 & 1 \\ -1 & 2 \\ -2 & 3 \end{bmatrix}$

27. $\begin{bmatrix} 1 & -3 & 2 \end{bmatrix} \begin{bmatrix} 2 \\ 1 \\ 3 \end{bmatrix}$

28. $\begin{bmatrix} 1 & -1 & -2 \\ 2 & -1 & 1 \end{bmatrix} \begin{bmatrix} 1 & 3 & -1 \\ 2 & -1 & 5 \\ 1 & -5 & 3 \end{bmatrix}$

29. $\begin{bmatrix} 1 \\ 2 \\ 1 \\ 5 \end{bmatrix} \begin{bmatrix} 2 & -1 & 1 & 3 \end{bmatrix}$

30. $\begin{bmatrix} 1 & -5 & 3 \\ 2 & 1 & -1 \end{bmatrix} \begin{bmatrix} 2 \\ -2 \\ 3 \end{bmatrix} + \begin{bmatrix} 1 & -1 \\ -1 & 3 \end{bmatrix} \begin{bmatrix} 1 \\ -2 \end{bmatrix}$

31. $\begin{bmatrix} 1 & -3 & 2 \end{bmatrix} \begin{bmatrix} 2 \\ 1 \\ -5 \end{bmatrix} + \begin{bmatrix} 1 & -3 \end{bmatrix} \begin{bmatrix} 2 \\ 5 \end{bmatrix} + \begin{bmatrix} 6 \end{bmatrix}$

32. $\left(\begin{bmatrix} 1 & -3 \\ 3 & 1 \end{bmatrix} + \begin{bmatrix} -1 & 3 \\ 1 & 1 \end{bmatrix} \right) \begin{bmatrix} 1 \\ -5 \end{bmatrix}$

In Review Exercises 33–38, find the inverse of each matrix, if possible.

33. $\begin{bmatrix} 1 & 3 \\ -3 & 5 \end{bmatrix}$

34. $\begin{bmatrix} 4 & 7 \\ 5 & 9 \end{bmatrix}$

35. $\begin{bmatrix} 1 & 3 & -5 \\ 0 & 1 & 9 \\ 0 & 0 & 1 \end{bmatrix}$

36. $\begin{bmatrix} 1 & 0 & 0 \\ 2 & 0 & -2 \\ 1 & 2 & 2 \end{bmatrix}$

37. $\begin{bmatrix} 1 & 0 & 8 \\ 3 & 7 & 6 \\ 1 & 2 & 3 \end{bmatrix}$

38. $\begin{bmatrix} -1 & 1 & 0 \\ -2 & 1 & 0 \\ 3 & -1 & -1 \end{bmatrix}$

In Review Exercises 39–40, use the inverse of the coefficient matrix to solve each system of equations.

39.
$$\begin{cases} 4x - y + 2z = 0 \\ x + y + 2z = 1 \\ x \quad + z = 0 \end{cases}$$

40.
$$\begin{cases} w + 3x + y + 3z = 1 \\ w + 4x + y + 3z = 2 \\ x + y \quad = 1 \\ w + 2x - y + 2z = 1 \end{cases}$$

In Review Exercises 41–44, evaluate each determinant.

41. $\begin{vmatrix} 3 & -2 \\ 1 & -3 \end{vmatrix}$

42. $\begin{vmatrix} 1 & 3 & -1 \\ 1 & 2 & 1 \\ 1 & 0 & 2 \end{vmatrix}$

43. $\begin{vmatrix} 1 & -2 & 3 \\ 2 & -1 & 3 \\ 1 & -1 & 0 \end{vmatrix}$

44. $\begin{vmatrix} 1 & 2 & 3 & 4 \\ -1 & 3 & -3 & 2 \\ 0 & 0 & 0 & -1 \\ 3 & 3 & 4 & 3 \end{vmatrix}$

In Review Exercises 45–48, use Cramer's rule to solve each system of equations.

45.
$$\begin{cases} x + 3y = -5 \\ -2x + y = -4 \end{cases}$$

46.
$$\begin{cases} x - y + z = -1 \\ 2x - y + 3z = -4 \\ x - 3y + z = -1 \end{cases}$$

47.
$$\begin{cases} x - 3y + z = 7 \\ x + y - 3z = -9 \\ x + y + z = 3 \end{cases}$$

48.
$$\begin{cases} w + x - y + z = 4 \\ 2w + x \quad + z = 4 \\ x + 2y + z = 0 \\ w \quad + y + z = 2 \end{cases}$$

In Review Exercises 49–52, decompose each fraction into partial fractions.

49. $\dfrac{4x^2 + 4x + 1}{x^3 + x}$

50. $\dfrac{4x^3 + 3x + x^2 + 2}{x^4 + x^2}$

51. $\dfrac{x^2 + 5}{x^3 + x^2 + 5x}$

52. $\dfrac{x^2 + 1}{(x + 1)^3}$

In Review Exercises 53–56, maximize P subject to the given conditions.

53. $P = 2x + y$
$$\begin{cases} x \geq 0 \\ y \geq 0 \\ x + y \leq 3 \end{cases}$$

54. $P = 2x - 3y$
$$\begin{cases} x \geq 0 \\ y \leq 3 \\ x - y \leq 4 \end{cases}$$

55. $P = 3x - y$
$$\begin{cases} y \geq 1 \\ y \leq 2 \\ y \leq 3x + 1 \\ x \leq 1 \end{cases}$$

56. $P = y - 2x$
$$\begin{cases} x + y \geq 1 \\ x \leq 1 \\ y \leq \dfrac{x}{2} + 2 \\ x + y \leq 2 \end{cases}$$

57. A company manufactures two fertilizers, X and Y. Each 50-pound bag of fertilizer requires three ingredients, which are available in the limited quantities shown below:

Ingredient	Number of pounds in fertilizer X	Number of pounds in fertilizer Y	Total number of pounds available
Nitrogen	6	10	20,000
Phosphorus	8	6	16,400
Potash	6	4	12,000

The profit on each bag of fertilizer X is $6, and on each bag of Y, $5. How many bags of each product should be produced to maximize the profit?

9

CONIC SECTIONS AND QUADRATIC SYSTEMS

The graphs of second-degree equations in x and y represent figures that have interested mathematicians since the time of the ancient Greeks. However, the equations of those graphs were not carefully studied until the seventeenth century, when René Descartes (1596–1650) and Blaise Pascal (1623–1662) began investigating them.

Definition. If A, B, C, D, E, and F are real numbers, and if at least one of A, B, and C is not 0, then

$$Ax^2 + Bxy + Cy^2 + Dx + Ey + F = 0$$

is called the **general form of a second-degree equation in x and y**.

Descartes discovered that the graphs of second-degree equations always fall into one of seven categories: a single point, a pair of straight lines, a circle, a parabola, an ellipse, a hyperbola, or no graph at all. These graphs are called **conic sections** because each is the intersection of a plane and a right-circular cone. See Figure 9-1.

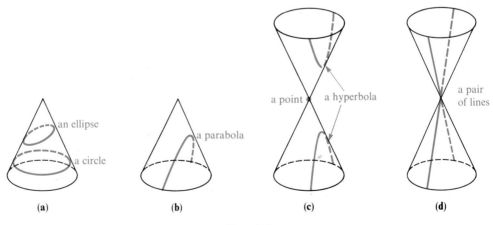

(a) (b) (c) (d)

Figure 9-1

The conic sections have many practical applications. For example, the properties of parabolas are used in building flashlights, satellite antennas, and solar furnaces. The orbits of the planets around the sun are ellipses. Hyperbolas are used in navigation and the design of gears.

9.1 THE CIRCLE

The most familiar of the conic sections is the circle.

Definition.   A **circle** is the set of all points in a plane that are a fixed distance from a point called its **center**. The fixed distance is called the **radius of the circle**.

To find the general equation of a circle with radius r and center at the point $C(h, k)$, we must find all points $P(x, y)$ such that the length of the line segment PC is r. See Figure 9-2. We can use the distance formula to find the length of CP, which is r:

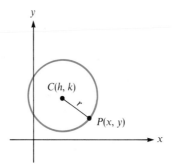

Figure 9-2

$$r = \sqrt{(x - h)^2 + (y - k)^2}$$

We square both sides to get

$$r^2 = (x - h)^2 + (y - k)^2$$

This equation is called the **standard form of the equation of a circle**.

Theorem.   Any equation that can be written in the form

$$(x - h)^2 + (y - k)^2 = r^2$$

has a graph that is a circle with radius r and center at the point (h, k).

If $r = 0$, the circle reduces to a single point called a **point circle**. If $r < 0$, a circle does not exist. If the center of the circle is the origin, then $(h, k) = (0, 0)$, and we have the following result.

Theorem.   Any equation that can be written in the form

$$x^2 + y^2 = r^2$$

has a graph that is a circle with radius r and with center at the origin.

We can use the previous theorems to write the equations of many circles.

Example 1   Find the equation of the circle with radius 5 and center $(3, 2)$. Express the equation in general form.

Solution   Substitute 5 for r, 3 for h, and 2 for k in the standard form of the equation of the circle and simplify:

$$(x - h)^2 + (y - k)^2 = r^2$$
$$(x - 3)^2 + (y - 2)^2 = 5^2$$
$$x^2 - 6x + 9 + y^2 - 4y + 4 = 25$$
$$x^2 + y^2 - 6x - 4y - 12 = 0$$

This final equation is a special case of the general form of a second-degree equation. The coefficient of the xy-term equals 0, and the coefficients of x^2 and y^2 are both 1.   ■

Example 2   Find the equation of the circle with endpoints of its diameter at $(8, -3)$ and $(-4, 13)$.

Solution   First find the center (h, k) of the circle by finding the midpoint of its diameter. Use the midpoint formulas with $(x_1, y_1) = (8, -3)$ and $(x_2, y_2) = (-4, 13)$:

$$h = \frac{x_1 + x_2}{2} \qquad k = \frac{y_1 + y_2}{2}$$

$$h = \frac{8 + (-4)}{2} \qquad k = \frac{-3 + 13}{2}$$

$$= \frac{4}{2} \qquad\qquad = \frac{10}{2}$$

$$= 2 \qquad\qquad\quad = 5$$

Thus, the center of the circle is the point $(h, k) = (2, 5)$.

To find the radius of the circle, use the distance formula to find the distance between the center and one endpoint of the diameter. Because one endpoint is $(8, -3)$, substitute 8 for x_1, -3 for y_1, 2 for x_2, and 5 for y_2 in the distance formula and simplify:

$$r = \sqrt{(x_2 - x_1)^2 + (y_2 - y_1)^2}$$
$$r = \sqrt{(2 - 8)^2 + [5 - (-3)]^2}$$
$$= \sqrt{(-6)^2 + (8)^2}$$
$$= \sqrt{36 + 64}$$
$$= \sqrt{100}$$
$$= 10$$

Thus, the radius of the circle is 10.

To find the equation of the circle with radius 10 and center at the point (2, 5), substitute **2** for *h*, **5** for *k*, and **10** for *r* in the standard form of the equation of the circle and simplify:

$$(x - h)^2 + (y - k)^2 = \quad r^2$$
$$(x - 2)^2 + (y - 5)^2 = \quad 10^2$$
$$x^2 - 4x + 4 + y^2 - 10y + 25 = 100 \qquad \text{Remove parentheses.}$$
$$x^2 + y^2 - 4x - 10y - 71 = \quad 0 \qquad \text{Simplify.} \qquad \blacksquare$$

Example 3   Graph the circle $x^2 + y^2 - 4x + 2y = 20$.

Solution   To find the coordinates of the center and the radius, write the equation in standard form by completing the square on both *x* and *y* and then simplifying:

$$x^2 + y^2 - 4x + 2y = 20$$
$$x^2 - 4x + y^2 + 2y = 20$$
$$x^2 - 4x + 4 + y^2 + 2y + 1 = 20 + 4 + 1 \qquad \text{Add 4 and 1 to both sides to}$$
$$\text{complete the square.}$$
$$(x - 2)^2 + (y + 1)^2 = 25 \qquad \text{Factor } x^2 - 4x + 4 \text{ and}$$
$$y^2 + 2y + 1.$$
$$(x - 2)^2 + [y - (-1)]^2 = 5^2$$

Note that the radius of the circle is 5 and the coordinates of its center are $h = 2$ and $k = -1$. Plot the center of the circle and construct the circle with a radius of 5 units, as shown in Figure 9-3.

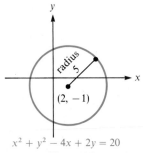

$$x^2 + y^2 - 4x + 2y = 20$$
Figure 9-3 $\blacksquare$

Exercise 9.1

In Exercises 1–26, write an equation for the circle with the given properties.

1. Center at the origin; $r = 1$
2. Center at the origin; $r = 4$
3. Center at $(6, 8)$; $r = 4$
4. Center at $(5, 3)$; $r = 2$
5. Center at $(-5, 3)$ and tangent to the y-axis
6. Center at $(-7, -2)$ and tangent to the x-axis
7. Center at $(3, -4)$; $r = \sqrt{2}$
8. Center at $(-9, 8)$; $r = 2\sqrt{3}$
9. Ends of diameter at $(3, -2)$ and $(3, 8)$
10. Ends of diameter at $(5, 9)$ and $(-5, -9)$
11. Ends of diameter at $(-6, 9)$ and $(-4, -7)$
12. Ends of diameter at $(17, 0)$ and $(-3, -3)$
13. Center at $(-3, 4)$ and circle passing through the origin
14. Center at $(4, 0)$ and circle passing through the origin
15. Center at $(-2, -6)$ and circle passing through the origin
16. Center at $(-19, -13)$ and circle passing through the origin
17. Center at $(0, -3)$ and circle passing through $(6, 8)$
18. Center at $(2, 4)$ and circle passing through $(1, 1)$
19. Center at $(5, 8)$ and circle passing through $(-2, -9)$
20. Center at $(7, -5)$ and circle passing through $(-3, -7)$
21. Center at $(-4, -2)$ and circle passing through $(3, 5)$
22. Center at $(0, -7)$ and circle passing through $(0, 7)$
23. Radius of 6 and center at the intersection of $3x + y = 1$ and $-2x - 3y = 4$
24. Radius of 8 and center at the intersection of $x + 2y = 8$ and $2x - 3y = -5$
25. Radius of $\sqrt{10}$ and center at the intersection of $x - y = 12$ and $3x - y = 12$
26. Radius of $2\sqrt{2}$ and center at the intersection of $6x - 4y = 8$ and $2x + 3y = 7$
27. Can a circle with a radius of 10 have endpoints of its diameter at $(6, 8)$ and $(-2, -2)$?
28. Can a circle with radius 25 have endpoints of its diameter at $(0, 0)$ and $(6, 24)$?

In Exercises 29–38, graph the circle.

29. $x^2 + y^2 - 25 = 0$
30. $x^2 + y^2 - 8 = 0$
31. $(x - 1)^2 + (y + 2)^2 = 4$
32. $(x + 1)^2 + (y - 2)^2 = 9$
33. $x^2 + y^2 + 2x - 26 = 0$
34. $x^2 + y^2 - 4y = 12$
35. $9x^2 + 9y^2 - 12y = 5$
36. $4x^2 + 4y^2 + 4y = 15$
37. $4x^2 + 4y^2 - 4x + 8y + 1 = 0$
38. $9x^2 + 9y^2 - 6x + 18y + 1 = 0$
39. Write the equation of the circle passing through $(0, 8)$, $(5, 3)$, and $(4, 6)$.
40. Write the equation of the circle passing through $(-2, 0)$, $(2, 8)$, and $(5, -1)$.
41. Find the area of the circle $3x^2 + 3y^2 + 6x + 12y = 0$. (*Hint*: $A = \pi r^2$.)
42. Find the circumference of the circle $x^2 + y^2 + 4x - 10y - 20 = 0$. (*Hint*: $C = 2\pi r$.)

9.2 THE PARABOLA

We have encountered parabolas in the discussion of quadratic functions. We now examine their equations in greater detail.

> **Definition.** A **parabola** is the set of all points in a plane such that each point in the set is equidistant from a line l, called the **directrix**, and a fixed point F, called the **focus**. The point on the parabola that is closest to the directrix is called the **vertex**. The line passing through the vertex and the focus is called the **axis**.

We will consider parabolas that open to the left, to the right, upward, and downward and that have a vertex at point (h, k). There is a standard form for the equation of each of these parabolas.

Consider the parabola in Figure 9-4, which opens to the right and has its vertex at the point $V(h, k)$. Let $P(x, y)$ be any point on the parabola. Because each point on the parabola is the same distance from the focus (point F) and from the directrix, we can let $DV = VF = p$, where p is some positive constant. Because of the geometry of the figure,

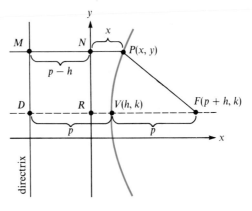

Figure 9-4

$$MP = p - h + x$$

Because of the distance formula,

$$PF = \sqrt{[x - (p + h)]^2 + (y - k)^2}$$

By the definition of the parabola, $MP = PF$. Thus,

$$p - h + x = \sqrt{[x - (p + h)]^2 + (y - k)^2}$$
$$(p - h + x)^2 = [x - (p + h)]^2 + (y - k)^2 \qquad \text{Square both sides.}$$

Finally, we expand the expression on each side of the equation and simplify:

$$p^2 - ph + px - ph + h^2 - hx + px - hx + x^2$$
$$= x^2 - 2px - 2hx + p^2 + 2ph + h^2 + (y - k)^2$$
$$-2ph + 2px = -2px + 2ph + (y - k)^2$$
$$4px - 4ph = (y - k)^2$$
$$4p(x - h) = (y - k)^2$$

The above argument proves the following theorem.

> **Theorem.** The standard form of the equation of a parabola with vertex at point (h, k) and opening to the right is
>
> $$(y - k)^2 = 4p(x - h)$$
>
> where p is the distance from the vertex to the focus.

If the parabola has its vertex at the origin, both h and k are equal to zero, and we have the following theorem.

> **Theorem.** The standard form of the equation of a parabola with vertex at the origin and opening to the right is
>
> $$y^2 = 4px$$
>
> where p is the distance from the vertex to the focus.

Equations of parabolas that open to the right, left, upward, and downward are summarized in Table 9-1. If $p > 0$, then

Table 9-1

Parabola opening	Vertex at origin	Vertex at $V(h, k)$
Right	$y^2 = 4px$	$(y - k)^2 = 4p(x - h)$
Left	$y^2 = -4px$	$(y - k)^2 = -4p(x - h)$
Upward	$x^2 = 4py$	$(x - h)^2 = 4p(y - k)$
Downward	$x^2 = -4py$	$(x - h)^2 = -4p(y - k)$

Example 1 Find the equation of the parabola with vertex at the origin and focus at $(3, 0)$.

Solution Sketch the parabola as in Figure 9-5. Because the focus is to the right of the vertex, the parabola opens to the right. Because the vertex is the origin, the standard form of the equation is $y^2 = 4px$. The distance between the focus and the vertex is 3, which is p. Therefore, the equation of the parabola is $y^2 = 4(3)x$, or

$$y^2 = 12x$$

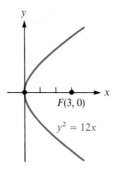

Figure 9-5

Example 2 Find the equation of the parabola that opens upward, has vertex at the point (4, 5), and passes through the point (0, 7).

Solution Because the parabola opens upward, use the standard form $(x - h)^2 = 4p(y - k)$. Because the point (0, 7) is on the curve, substitute **0** for x and **7** for y in the equation. Because the vertex (h, k) is (4, 5), also substitute **4** for h and **5** for k. Then solve the equation to determine p:

$$(x - h)^2 = 4p(y - k)$$
$$(0 - 4)^2 = 4p(7 - 5)$$
$$16 = 8p$$
$$2 = p$$

To find the equation of the parabola, substitute **4** for h, **5** for k, and **2** for p in the standard form of the equation and simplify:

$$(x - h)^2 = 4p(y - k)$$
$$(x - 4)^2 = 4 \cdot 2(y - 5)$$
$$(x - 4)^2 = 8(y - 5)$$

The graph of this equation appears in Figure 9-6.

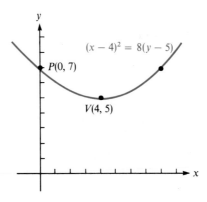

Figure 9-6

Example 3   Find the equations of the two parabolas each having its vertex at (2, 4) and passing through the point (0, 0).

Solution   Sketch the two parabolas as shown in Figure 9-7.

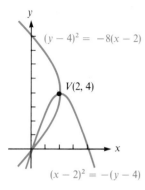

$(y - 4)^2 = -8(x - 2)$

$V(2, 4)$

$(x - 2)^2 = -(y - 4)$

Figure 9-7

Part 1.   To find the parabola that opens to the left, use the standard form of the equation $(y - k)^2 = -4p(x - h)$. Because the curve passes through the point $(x, y) = (0, 0)$ and the vertex is $(h, k) = (2, 4)$, substitute 0 for x, 0 for y, 2 for h, and 4 for k in the equation $(y - k)^2 = -4p(x - h)$ and solve for p:

$$(y - k)^2 = -4p(x - h)$$
$$(0 - 4)^2 = -4p(0 - 2)$$
$$16 = 8p$$
$$2 = p$$

Because $h = 2$, $k = 4$, and $p = 2$ and because the parabola opens to the left, its equation is

$$(y - k)^2 = -4p(x - h)$$
$$(y - 4)^2 = -4(2)(x - 2)$$
$$(y - 4)^2 = -8(x - 2)$$

Part 2.   To find the equation of the parabola that opens downward, use the standard form $(x - h)^2 = -4p(y - k)$. Substitute 2 for h, 4 for k, 0 for x, and 0 for y in the equation and solve for p:

$$(x - h)^2 = -4p(y - k)$$
$$(0 - 2)^2 = -4p(0 - 4)$$
$$4 = 16p$$
$$\frac{1}{4} = p$$

Because $h = 2$, $k = 4$, and $p = \frac{1}{4}$, and the parabola opens downward, its equation is

$$(x - h)^2 = -4p(y - k)$$
$$(x - 2)^2 = -4\left(\frac{1}{4}\right)(y - 4)$$
$$(x - 2)^2 = -(y - 4)$$ ■

Example 4   Find the vertex and y-intercepts of the parabola $y^2 + 8x - 4y = 28$. Then graph the parabola.

Solution   Complete the square on y to write the equation in standard form:

$$y^2 + 8x - 4y = 28$$
$$y^2 - 4y = -8x + 28 \qquad \text{Add } -8x \text{ to both sides.}$$
$$y^2 - 4y + 4 = -8x + 28 + 4 \qquad \text{Add 4 to both sides.}$$
$$(y - 2)^2 = -8(x - 4) \qquad \text{Factor both sides.}$$

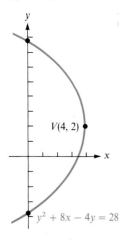

$V(4, 2)$

$y^2 + 8x - 4y = 28$

Figure 9-8

Observe that this equation represents a parabola opening to the left with vertex at (4, 2). To find the points where the graph intersects the y-axis, substitute 0 for x in the equation of the parabola.

$$(y - 2)^2 = -8(x - 4)$$
$$(y - 2)^2 = -8(0 - 4) \qquad \text{Substitute 0 for } x.$$
$$y^2 - 4y + 4 = 32 \qquad \text{Remove parentheses.}$$
$$y^2 - 4y - 28 = 0$$

Use the quadratic formula to determine that the roots of this quadratic equation are $y \approx 7.7$ and $y \approx -3.7$.

The points with coordinates of approximately (0, 7.7) and (0, −3.7) are on the graph of the parabola. Using this information and the knowledge that the graph opens to the left and has a vertex at (4, 2), draw the curve as shown in Figure 9-8.

Example 5 A stone is thrown straight up. The equation $s = 128t - 16t^2$ expresses the height of the stone in feet t seconds after it was thrown. Find the maximum height reached by the stone.

Solution The stone goes straight up and then straight down. The graph of $s = 128t - 16t^2$, expressing the height of the stone t seconds after it was thrown, is a parabola (see Figure 9-9). To find the maximum height reached by the stone, calculate the y-coordinate, k, of the vertex of the parabola. To find k, write the equation of the parabola, $s = 128t - 16t^2$, in standard form. To change this equation into standard form, complete the square on t:

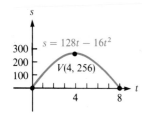

$s = 128t - 16t^2$

$V(4, 256)$

Figure 9-9

$$s = 128t - 16t^2$$
$$16t^2 - 128t = -s \qquad \text{Multiply both sides by } -1.$$
$$t^2 - 8t = \frac{-s}{16} \qquad \text{Divide both sides by 16.}$$
$$t^2 - 8t + 16 = \frac{-s}{16} + 16 \qquad \text{Add 16 to both sides.}$$
$$(t - 4)^2 = \frac{-s + 256}{16} \qquad \text{Factor } t^2 - 8t + 16 \text{ and combine terms.}$$
$$(t - 4)^2 = -\frac{1}{16}(s - 256) \qquad \text{Factor out } -\tfrac{1}{16}.$$

This equation indicates that the maximum height of 256 feet was reached in 4 seconds.

Exercise 9.2

In Exercises 1–16, find the equation of each parabola.

1. Vertex at (0, 0) and focus at (0, 3)

2. Vertex at (0, 0) and focus at (0, −3)

3. Vertex at (0, 0) and focus at (3, 0)

4. Vertex at (0, 0) and focus at (−3, 0)

5. Vertex at (3, 5) and focus at (3, 2)

6. Vertex at (3, 5) and focus at (−3, 5)

7. Vertex at $(3, 5)$ and focus at $(3, -2)$ **8.** Vertex at $(3, 5)$ and focus at $(6, 5)$

9. Vertex at $(2, 2)$ and the parabola passing through $(0, 0)$

10. Vertex at $(-2, -2)$ and the parabola passing through $(0, 0)$

11. Vertex at $(-4, 6)$ and the parabola passing through $(0, 3)$

12. Vertex at $(-2, 3)$ and the parabola passing through $(0, -3)$

13. Vertex at $(6, 8)$ and the parabola passing through $(5, 10)$ and $(5, 6)$

14. Vertex at $(2, 3)$ and the parabola passing through $(1, \frac{13}{4})$ and $(-1, \frac{21}{4})$

15. Vertex at $(3, 1)$ and the parabola passing through $(4, 3)$ and $(2, 3)$

16. Vertex at $(-4, -2)$ and the parabola passing through $(-3, 0)$ and $(\frac{9}{4}, 3)$

In Exercises 17–26, change each equation to standard form and graph each parabola.

17. $y = x^2 + 4x + 5$ **18.** $2x^2 - 12x - 7y = 10$

19. $y^2 + 4x - 6y = -1$ **20.** $x^2 - 2y - 2x = -7$

21. $y^2 + 2x - 2y = 5$ **22.** $y^2 - 4y = -8x + 20$

23. $x^2 - 6y + 22 = -4x$ **24.** $4y^2 - 4y + 16x = 7$

25. $4x^2 - 4x + 32y = 47$ **26.** $4y^2 - 16x + 17 = 20y$

27. A parabolic arch spans 30 meters and has a maximum height of 10 meters. Derive the equation of the arch using the vertex of the arch as the origin.

28. Find the maximum value of y in the parabola $x^2 + 8y - 8x = 8$.

29. A resort owner plans to build and rent n cabins for d dollars per week. The price, d, that she can charge for each cabin depends on the number of cabins she builds, where $d = -45(\frac{n}{32} - \frac{1}{2})$. Find the number of cabins that she should build to maximize her weekly income.

30. A toy rocket is s meters above the earth at the end of t seconds, where $s = -16t^2 + 80\sqrt{3}t$. Find the maximum height of the rocket.

31. An engineer plans to build a tunnel whose arch is in the shape of a parabola. The tunnel will span a two-lane highway that is 8 meters wide. To allow safe passage for most vehicles, the tunnel must be 5 meters high at a distance of 1 meter from the tunnel's edge. What will be the maximum height of the tunnel?

32. The towers of a suspension bridge are 900 feet apart and rise 120 feet above the roadway. The cable between the towers has the shape of a parabola with a vertex 15 feet above the roadway. Find the equation of the parabola with respect to the indicated coordinate system. See Illustration 1.

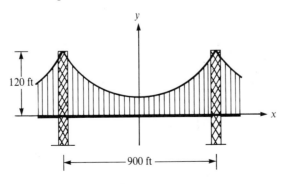

Illustration 1

33. A satellite antenna with a parabolic cross section is a dish 6 feet in diameter and 1 foot deep at its center. How far is the focus from the center of the dish?

34. A stone tossed upward is s meters above the earth after t seconds, where $s = -16t^2 + 128t$. Show that the stone's height x seconds *after* it is thrown is equal to its height x seconds *before* it returns to the ground.

35. Derive the standard form of the equation of a parabola that opens downward and has vertex at the origin.

36. Show that the result in Example 1 is a special case of the general form of the equation of second degree.

In Exercises 37–38, find the equation of the form $y = ax^2 + bx + c$ that determines a parabola passing through the three given points.

37. (1, 8), (−2, −1), and (2, 15)   **38.** (1, −3), (−2, 12), and (−1, 3)

9.3 THE ELLIPSE

A third important conic is the ellipse.

> **Definition.** An **ellipse** is the set of all points P in a plane such that the sum of the distances from P to two other fixed points F and F' is a positive constant.

In the ellipse shown in Figure 9-10, the two fixed points F and F' are called **foci** of the ellipse, the midpoint of the chord FF' is called the **center**, the chord VV' is called the **major axis**, and each endpoint of the major axis is called a **vertex**. The chord BB', perpendicular to the major axis and passing through the center C, is called the **minor axis**.

To keep the algebra manageable, we will derive the equation of the ellipse shown in Figure 9-11, which has its center at $(0, 0)$. Because the origin is the midpoint of the chord FF', we can let $OF = OF' = c$, where $c > 0$. Then the coordinates of point F are $(c, 0)$, and the coordinates of F' are $(-c, 0)$. We also let $P(x, y)$ be any point on the ellipse.

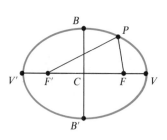

Figure 9-10

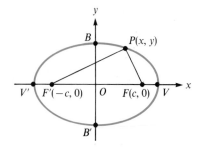

Figure 9-11

The definition of an ellipse requires that the sum of $F'P$ and PF be a positive constant, which we will call $2a$. Thus,

1. $F'P + PF = 2a$

We use the distance formula to compute the lengths of $F'P$ and PF:

$$F'P = \sqrt{[x - (-c)]^2 + y^2}$$
$$PF = \sqrt{(x - c)^2 + y^2}$$

and substitute these values into Equation 1 to obtain

$$\sqrt{[x - (-c)]^2 + y^2} + \sqrt{(x - c)^2 + y^2} = 2a$$

or

$$\sqrt{[x + c]^2 + y^2} = 2a - \sqrt{(x - c)^2 + y^2}$$

We square both sides of this equation and simplify to get

$$(x + c)^2 + y^2 = 4a^2 - 4a\sqrt{(x - c)^2 + y^2} + [(x - c)^2 + y^2]$$
$$x^2 + 2cx + c^2 + y^2 = 4a^2 - 4a\sqrt{(x - c)^2 + y^2} + x^2 - 2cx + c^2 + y^2$$
$$4cx = 4a^2 - 4a\sqrt{(x - c)^2 + y^2}$$
$$cx = a^2 - a\sqrt{(x - c)^2 + y^2}$$
$$cx - a^2 = -a\sqrt{(x - c)^2 + y^2}$$

We square both sides again and simplify to obtain

$$c^2x^2 - 2a^2cx + a^4 = a^2[(x - c)^2 + y^2]$$
$$c^2x^2 - 2a^2cx + a^4 = a^2(x^2 - 2cx + c^2 + y^2)$$
$$c^2x^2 - 2a^2cx + a^4 = a^2x^2 - 2a^2cx + a^2c^2 + a^2y^2$$
$$c^2x^2 + a^4 = a^2x^2 + a^2c^2 + a^2y^2$$
$$a^4 - a^2c^2 = a^2x^2 - c^2x^2 + a^2y^2$$

2. $$a^2(a^2 - c^2) = (a^2 - c^2)x^2 + a^2y^2$$

Because the shortest path between two points is a line segment, $F'P + PF > F'F$. Therefore, $2a > 2c$. This implies that $a > c$ and that $a^2 - c^2$ is a positive number, which we will call b^2. Letting $b^2 = a^2 - c^2$ and substituting into Equation 2, we have

$$a^2b^2 = b^2x^2 + a^2y^2$$

Dividing both sides of this equation by a^2b^2 gives the standard form of the equation for an ellipse with center at the origin and major axis on the x-axis:

$$\frac{x^2}{a^2} + \frac{y^2}{b^2} = 1 \quad \text{where } a > b > 0$$

To find the coordinates of the vertices V and V', we substitute 0 for y and solve for x:

$$\frac{x^2}{a^2} + \frac{y^2}{b^2} = 1$$

$$\frac{x^2}{a^2} + \frac{0^2}{b^2} = 1$$

$$\frac{x^2}{a^2} = 1$$

$$x^2 = a^2$$

$$x = a \qquad \text{or} \qquad x = -a$$

Thus, the coordinates of V are $(a, 0)$, and the coordinates of V' are $(-a, 0)$. In other words, a is the distance between the center of the ellipse, $(0, 0)$, and either of its vertices, and the center of the ellipse is the midpoint of the major axis.

To find the coordinates of B and B', we substitute 0 for x and solve for y:

$$\frac{x^2}{a^2} + \frac{y^2}{b^2} = 1$$

$$\frac{0^2}{a^2} + \frac{y^2}{b^2} = 1$$

$$y^2 = b^2$$

$$y = b \qquad \text{or} \qquad y = -b$$

Thus, the coordinates of B are $(0, b)$, and the coordinates of B' are $(0, -b)$. The distance between the center of the ellipse and either endpoint of the minor axis is b.

Theorem. The standard form of the equation of an ellipse with center at the origin and major axis on the x-axis is

$$\frac{x^2}{a^2} + \frac{y^2}{b^2} = 1 \quad \text{where } a > b > 0$$

If the major axis of an ellipse with center at $(0, 0)$ lies on the y-axis, the standard form of the equation of the ellipse is

$$\frac{y^2}{a^2} + \frac{x^2}{b^2} = 1 \quad \text{where } a > b > 0$$

In either case, the length of the major axis is $2a$, and the length of the minor axis is $2b$.

If we develop the equation of the ellipse with center at (h, k), we obtain the following results.

Theorem.   The standard form of the equation of an ellipse with center at (h, k) and major axis parallel to the x-axis is

$$\frac{(x - h)^2}{a^2} + \frac{(y - k)^2}{b^2} = 1 \quad \text{where } a > b > 0$$

If the major axis of an ellipse with center at (h, k) is parallel to the y-axis, the standard form of the equation of the ellipse is

$$\frac{(y - k)^2}{a^2} + \frac{(x - h)^2}{b^2} = 1 \quad \text{where } a > b > 0$$

In either case, the length of the major axis is $2a$, and the length of the minor axis is $2b$.

Example 1   Find the equation of the ellipse with center at the origin, major axis of length 6 units located on the x-axis, and minor axis of length 4 units.

Solution   Because the center of the ellipse is the origin and the length of the major axis is 6, $a = 3$ and the coordinates of the vertices of the ellipse are $(3, 0)$ and $(-3, 0)$, as shown in Figure 9-12.

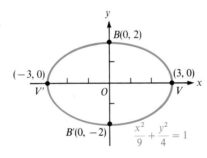

Figure 9-12

Because the length of the minor axis is 4, the value of b is 2 and the coordinates of B and B' are $(0, 2)$ and $(0, -2)$. To find the desired equation, substitute **3** for a and **2** for b in the standard form of the equation of an ellipse with center at the origin and major axis on the x-axis. Then simplify the equation:

$$\frac{x^2}{a^2} + \frac{y^2}{b^2} = 1$$

$$\frac{x^2}{3^2} + \frac{y^2}{2^2} = 1$$

$$\frac{x^2}{9} + \frac{y^2}{4} = 1$$

Example 2   Find the equation of the ellipse with focus (0, 3) and vertices V and V' at (3, 3) and $(-5, 3)$.

Solution   Because the midpoint of the major axis is the center of the ellipse, the coordinates of the center are $(-1, 3)$. Look at Figure 9-13 and note that the major axis is parallel to the x-axis. The standard form of the equation to use is

$$\frac{(x-h)^2}{a^2} + \frac{(y-k)^2}{b^2} = 1 \quad \text{where } a > b > 0$$

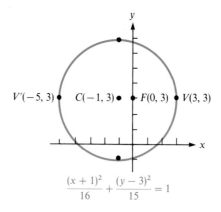

$$\frac{(x+1)^2}{16} + \frac{(y-3)^2}{15} = 1$$

Figure 9-13

The distance between the center of the ellipse and a vertex is $a = 4$; the distance between the focus and the center is $c = 1$. In the ellipse, $b^2 = a^2 - c^2$. From this equation, compute b^2:

$$b^2 = a^2 - c^2$$
$$= 4^2 - 1^2$$
$$= 15$$

To find the equation of the ellipse, substitute -1 for h, 3 for k, 16 for a^2, and 15 for b^2 in the standard form of the equation for an ellipse and simplify:

$$\frac{(x-h)^2}{a^2} + \frac{(y-k)^2}{b^2} = 1$$
$$\frac{[x-(-1)]^2}{16} + \frac{(y-3)^2}{15} = 1$$
$$\frac{(x+1)^2}{16} + \frac{(y-3)^2}{15} = 1$$

∎

Example 3   The orbit of the earth is approximately an ellipse, with the sun at one focus. The ratio of c to a (called the *eccentricity* of the ellipse) is about $\frac{1}{62}$, and the length of the major axis is approximately 186,000,000 miles. How close does the earth get to the sun?

Solution   Assume that this ellipse has its center at the origin and vertices V' and V at $(-93{,}000{,}000, 0)$ and $(93{,}000{,}000, 0)$, as shown in Figure 9-14. This implies that $a = 93{,}000{,}000$.

$$\frac{c}{a} = \frac{1}{62} \quad \text{or} \quad c = \frac{1}{62} a$$

Because $a = 93{,}000{,}000$,

$$c = \frac{1}{62}(93{,}000{,}000)$$

$$= 1{,}500{,}000$$

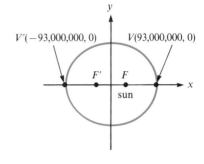

Figure 9-14

FV represents the shortest possible distance between the earth and the sun. (You'll be asked to prove this in the exercises.) Thus,

$$FV = a - c = 93{,}000{,}000 - 1{,}500{,}000 = 91{,}500{,}000 \text{ miles}$$

The earth's point of closest approach to the sun (called the *perigee*) is approximately 91.5 million miles. ∎

Example 4   Graph the ellipse $\dfrac{(x + 2)^2}{4} + \dfrac{(y - 2)^2}{9} = 1$.

Solution   The center of the ellipse is at $(-2, 2)$, and the major axis is parallel to the y-axis. Because $a = 3$, the vertices are 3 units above and below the center at points $(-2, 5)$ and $(-2, -1)$. Because $b = 2$, the endpoints of the minor axis are 2 units to the right and left of the center at points $(0, 2)$ and $(-4, 2)$. Using these four points as guides, sketch the ellipse, as shown in Figure 9-15.

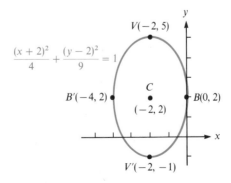

Figure 9-15

Example 5   Graph the equation $4x^2 + 9y^2 - 16x - 18y = 11$.

Solution   Write the equation in standard form by completing the square on x and y as follows:

$$4x^2 + 9y^2 - 16x - 18y = 11$$
$$4x^2 - 16x + 9y^2 - 18y = 11$$
$$4(x^2 - 4x) + 9(y^2 - 2y) = 11$$
$$4(x^2 - 4x + 4) + 9(y^2 - 2y + 1) = 11 + 16 + 9$$
$$4(x - 2)^2 + 9(y - 1)^2 = 36$$
$$\frac{(x - 2)^2}{9} + \frac{(y - 1)^2}{4} = 1$$

You can now see that the graph of the given equation is an ellipse with center at $(2, 1)$ and major axis parallel to the x-axis. Because $a = 3$, the vertices are at $(-1, 1)$ and $(5, 1)$. Because $b = 2$, the endpoints of the minor axis are at $(2, -1)$ and $(2, 3)$. Using these four points as guides, sketch the ellipse, as shown in Figure 9-16.

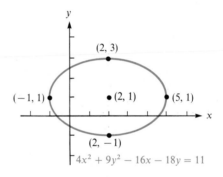

Figure 9-16

Exercise 9.3

In Exercises 1–6, write the equation of the ellipse that has its center at the origin.

1. Focus at $(3, 0)$ and a vertex at $(5, 0)$
2. Focus at $(0, 4)$ and a vertex at $(0, 7)$
3. Focus at $(0, 1)$; $\frac{4}{3}$ is one-half the length of the minor axis.
4. Focus at $(1, 0)$; $\frac{4}{3}$ is one-half the length of the minor axis.
5. Focus at $(0, 3)$ and major axis equal to 8
6. Focus at $(5, 0)$ and major axis equal to 12

In Exercises 7–16, write the equation of each ellipse.

7. Center at $(3, 4)$; $a = 3$, $b = 2$; the major axis is parallel to the y-axis.
8. Center at $(3, 4)$; the curve passes through $(3, 10)$ and $(3, -2)$; $b = 2$.

9. Center at $(3, 4)$; $a = 3$, $b = 2$; the major axis is parallel to the x-axis.

10. Center at $(3, 4)$; the curve passes through $(8, 4)$ and $(-2, 4)$; $b = 2$.

11. Foci at $(-2, 4)$ and $(8, 4)$; $b = 4$

12. Foci at $(-8, 5)$ and $(4, 5)$; $b = 3$

13. Vertex at $(6, 4)$ and foci at $(-4, 4)$ and $(4, 4)$

14. Center at $(-4, 5)$; $\frac{c}{a} = \frac{1}{3}$; vertex at $(-4, -1)$

15. Foci at $(6, 0)$ and $(-6, 0)$; $\frac{c}{a} = \frac{3}{5}$

16. Vertices at $(2, 0)$ and $(-2, 0)$; $\frac{2b^2}{a} = 2$

In Exercises 17–24, graph each ellipse.

17. $\dfrac{x^2}{25} + \dfrac{y^2}{49} = 1$

18. $4x^2 + y^2 = 4$

19. $\dfrac{x^2}{16} + \dfrac{(y + 2)^2}{36} = 1$

20. $(x - 1)^2 + \dfrac{4y^2}{25} = 4$

21. $x^2 + 4y^2 - 4x + 8y + 4 = 0$

22. $x^2 + 4y^2 - 2x - 16y = -13$

23. $16x^2 + 25y^2 - 160x - 200y + 400 = 0$

24. $3x^2 + 2y^2 + 7x - 6y = -1$

25. The moon has an orbit that is an ellipse with the earth at one focus. If the major axis of the orbit is 378,000 miles and the ratio of c to a is approximately $\frac{11}{200}$, how far does the moon get from earth? (This farthest point in an orbit is called the *apogee*.)

26. An arch is a semiellipse 10 meters wide and 5 meters high. Write the equation of the ellipse if the ellipse is centered at the origin.

27. A track is built in the shape of an ellipse and has a maximum length of 100 meters and a maximum width of 60 meters. Write the equation of the ellipse and find its focal width; that is, find the length of a chord that is perpendicular to the major axis and that passes through either focus of the ellipse.

28. An arch has the shape of a semiellipse and has a maximum height of 5 meters. The foci are on the ground with a distance between them of 24 meters. Find the total distance from one focus to any point on the arch and back to the other focus.

29. Consider the ellipse in Illustration 1. If F is a focus of the ellipse and B is an endpoint of the minor axis, use the distance formula to prove that the length of segment FB is a. (*Hint:* Remember that in an ellipse $a^2 - c^2 = b^2$.)

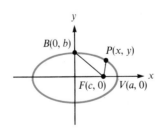

Illustration 1

30. Consider the ellipse in Illustration 1. If F is a focus of the ellipse and P is any point on the ellipse, use the distance formula to show that the length of segment FP is $a - \frac{c}{a}x$. (*Hint:* Remember that in an ellipse $a^2 - c^2 = b^2$.)

31. Consider the ellipse in Illustration 2. Chord AA' passes through the focus F and is perpendicular to the major axis. Show that the length of AA' (called the **focal width** of the ellipse) is $\frac{2b^2}{a}$.

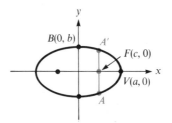

Illustration 2

32. Prove that the segment FV in Example 3 does represent the shortest distance between the earth and the sun. (*Hint:* You might find the result of Exercise 30 helpful.)

33. The ends of a piece of string 6 meters long are attached to two thumbtacks that are 2 meters apart. A pencil catches the loop and draws it tight. As the pencil is moved about the thumbtacks (always keeping the tension), an ellipse is produced with the thumbtacks as foci. Write the equation of the ellipse. (*Hint:* You'll have to establish a coordinate system.)

34. Prove that $a > b$ in the development of the standard form of the equation of an ellipse.

35. Show that the expansion of the standard equation of an ellipse is a special case of the general second-degree equation.

36. The distance between point $P(x, y)$ and the point $(0, 2)$ is $\frac{1}{3}$ of the distance of point P from the line $y = 18$. Find the equation of the curve on which point P lies.

9.4   THE HYPERBOLA

The definition of the hyperbola is similar to the definition of the ellipse except that we demand a constant *difference* of $2a$ instead of a constant sum.

> **Definition.**   A **hyperbola** is the set of all points P in a plane such that the difference of the distances from point P to two other points in the plane, F and F', is a positive constant.

Points F and F' (see Figure 9-17) are called the **foci** of the hyperbola, and the midpoint of chord FF' is called the **center** of the hyperbola. The points V and V', where the hyperbola intersects the line segment FF', are called the **vertices** of the hyperbola, and the line segment VV' is called the **transverse axis**.

As with the ellipse, we will develop the equation of the hyperbola centered at the origin. Because the origin is the midpoint of chord FF', we can let $F'O = OF = c > 0$. Therefore, F is at $(c, 0)$ and F' is at $(-c, 0)$. The definition requires that $|F'P - PF| = 2a$, where $2a$ is a positive constant. Using the distance formula to compute the lengths of $F'P$ and PF gives

$$F'P = \sqrt{[x - (-c)]^2 + y^2}$$
$$PF = \sqrt{(x - c)^2 + y^2}$$

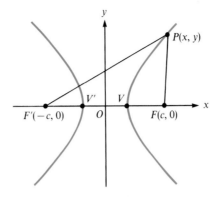

Figure 9-17

Substituting these values into the equation $F'P - PF = 2a$ gives

$$\sqrt{(x + c)^2 + y^2} - \sqrt{(x - c)^2 + y^2} = 2a$$

or

$$\sqrt{(x + c)^2 + y^2} = 2a + \sqrt{(x - c)^2 + y^2}$$

Squaring both sides of this equation and simplifying gives

$$(x + c)^2 + y^2 = 4a^2 + 4a\sqrt{(x - c)^2 + y^2} + (x - c)^2 + y^2$$
$$x^2 + 2cx + c^2 + y^2 = 4a^2 + 4a\sqrt{(x - c)^2 + y^2} + x^2 - 2cx + c^2 + y^2$$
$$4cx = 4a^2 + 4a\sqrt{(x - c)^2 + y^2}$$
$$cx - a^2 = a\sqrt{(x - c)^2 + y^2}$$

Squaring both sides again and simplifying gives

$$c^2x^2 - 2a^2cx + a^4 = a^2(x^2 - 2cx + c^2 + y^2)$$
$$c^2x^2 - 2a^2cx + a^4 = a^2x^2 - 2a^2cx + a^2c^2 + a^2y^2$$
$$c^2x^2 + a^4 = a^2x^2 + a^2c^2 + a^2y^2$$

1. $(c^2 - a^2)x^2 - a^2y^2 = a^2(c^2 - a^2)$

Because $c > a$ (you will be asked to prove this in the exercises), $c^2 - a^2$ is a positive number. Thus, we can let $b^2 = c^2 - a^2$ and substitute b^2 for $c^2 - a^2$ in Equation 1 to get

$$b^2x^2 - a^2y^2 = a^2b^2$$

Dividing both sides of the previous equation by a^2b^2 gives the standard form of the equation for a hyperbola with center at the origin and foci on the x-axis:

$$\frac{x^2}{a^2} - \frac{y^2}{b^2} = 1$$

If $y = 0$, the equation above becomes

$$\frac{x^2}{a^2} = 1 \quad \text{or} \quad x^2 = a^2$$

Solving this equation for x gives

$$x = a \quad \text{or} \quad x = -a$$

This implies that the coordinates of V and V' are $(a, 0)$ and $(-a, 0)$ and that the distance between the center of the hyperbola and either vertex is a. This, in turn, implies that the center of the hyperbola is the midpoint of the segment $V'V$ as well as of the segment FF'.

If $x = 0$, the equation becomes

$$\frac{-y^2}{b^2} = 1 \quad \text{or} \quad y^2 = -b^2$$

Because this equation has no real solutions, the hyperbola cannot intersect the y-axis. These results suggest the following theorem.

Theorem.   The standard form of the equation of a hyperbola with center at the origin and foci on the x-axis is

$$\frac{x^2}{a^2} - \frac{y^2}{b^2} = 1$$

The standard form of the equation of a hyperbola with center at the origin and foci on the y-axis is

$$\frac{y^2}{a^2} - \frac{x^2}{b^2} = 1$$

As with the ellipse, the standard equation of the hyperbola can be developed with center at (h, k). We state the results without proof.

Theorem.   The standard form of the equation of a hyperbola with center at (h, k) and foci on a line parallel to the x-axis is

$$\frac{(x - h)^2}{a^2} - \frac{(y - k)^2}{b^2} = 1$$

The standard form of the equation of a hyperbola with center at (h, k) and foci on a line parallel to the y-axis is

$$\frac{(y - k)^2}{a^2} - \frac{(x - h)^2}{b^2} = 1$$

Example 1   Write the equation of the hyperbola with vertices $(3, -3)$ and $(3, 3)$ and with a focus at $(3, 5)$.

Solution   First, plot the vertices and focus, as shown in Figure 9-18. Note that the foci lie on a vertical line. Therefore, the standard form to use is

$$\frac{(y - k)^2}{a^2} - \frac{(x - h)^2}{b^2} = 1$$

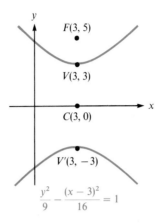

Figure 9-18

Because the center of the hyperbola is midway between the vertices V and V', the center is point $(3, 0)$, so $h = 3$, and $k = 0$. The distance between the vertex and the center of the hyperbola is $a = 3$, and the distance between the focus and the center is $c = 5$. Also, in a hyperbola, $b^2 = c^2 - a^2$. Therefore, $b^2 = 5^2 - 3^2 = 16$. Substituting the values for h, k, a^2, and b^2 into the standard form of the equation gives the desired result:

$$\frac{(y - 0)^2}{9} - \frac{(x - 3)^2}{16} = 1$$

$$\frac{y^2}{9} - \frac{(x - 3)^2}{16} = 1$$

■

Asymptotes of a Hyperbola

The values of a and b play an important role in graphing hyperbolas. To see their significance, we consider the hyperbola

$$\frac{x^2}{a^2} - \frac{y^2}{b^2} = 1$$

The center of this hyperbola is the origin, and the vertices are at $V(a, 0)$ and $V'(-a, 0)$. We plot points V, V', $B(0, b)$, and $B'(0, -b)$ and form rectangle $RSQP$, called the **fundamental rectangle**, as in Figure 9-19. The extended diagonals of this rectangle are asymptotes of the hyperbola. In the exercises, you will be asked

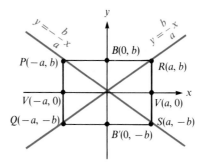

Figure 9-19

to show that the equations of these two lines are

$$y = \frac{b}{a}x \quad \text{and} \quad y = -\frac{b}{a}x$$

To show that the extended diagonals are asymptotes of the hyperbola, we solve the equation

$$\frac{x^2}{a^2} - \frac{y^2}{b^2} = 1$$

for y and modify its form:

$$\frac{x^2}{a^2} - \frac{y^2}{b^2} = 1$$

$$b^2x^2 - a^2y^2 = a^2b^2 \qquad \text{Multiply both sides by } a^2b^2.$$

$$y^2 = \frac{b^2x^2 - a^2b^2}{a^2} \qquad \text{Add } -b^2x^2 \text{ to both sides and divide both sides by } -a^2.$$

$$y^2 = \frac{b^2x^2}{a^2}\left(1 - \frac{a^2}{x^2}\right) \qquad \text{Factor out a } b^2x^2 \text{ from the numerator.}$$

$$y = \pm\frac{bx}{a}\sqrt{1 - \frac{a^2}{x^2}} \qquad \text{Take the square root of both sides.}$$

If $|x|$ grows large without bound, the fraction $\frac{a^2}{x^2}$ in the previous equation approaches 0 and $\sqrt{1 - \frac{a^2}{x^2}}$ approaches 1. Hence, the hyperbola approaches the lines

$$y = \frac{b}{a}x \quad \text{or} \quad y = -\frac{b}{a}x$$

This fact makes it easy to sketch a hyperbola. We convert the equation into standard form, find the coordinates of its vertices, and plot them. Then, we construct the fundamental rectangle and its extended diagonals. Using the vertices and the asymptotes as guides, we make a quick and relatively accurate sketch, as in Figure 9-20. The segment BB' is called the **conjugate axis** of the hyperbola.

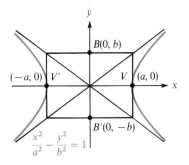

Figure 9-20

Example 2   Graph the hyperbola $x^2 - y^2 - 2x + 4y = 12$.

Solution   First complete the square on x and y to convert the equation into standard form:

$$x^2 - 2x - y^2 + 4y = 12$$
$$x^2 - 2x - (y^2 - 4y) = 12$$
$$x^2 - 2x + 1 - (y^2 - 4y + 4) = 12 - 3$$
$$(x - 1)^2 - (y - 2)^2 = 9$$
$$\frac{(x - 1)^2}{9} - \frac{(y - 2)^2}{9} = 1$$

From the standard form of the equation of a hyperbola, observe that the center is $(1, 2)$, that $a = 3$ and $b = 3$, and that the vertices are on a line segment parallel to the x-axis, as shown in Figure 9-21. Therefore, the vertices V and V' are 3 units to the right and left of the center and have coordinates of $(4, 2)$ and $(-2, 2)$. Points B and B', 3 units above and below the center, have coordinates $(1, 5)$ and $(1, -1)$. After plotting points V, V', B, and B', construct the fundamental rectangle and its extended diagonals. Using the vertices as points on the hyperbola and the extended diagonals as asymptotes, sketch the graph.

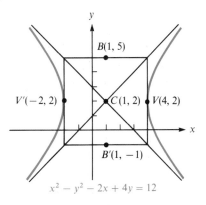

Figure 9-21

This discussion of the hyperbola has considered only cases where the segment that joins the foci is horizontal or vertical. However, there are hyperbolas where this is not true. For example, the graph of the equation $xy = 4$ is a hyperbola with vertices at $(2, 2)$ and $(-2, -2)$, as shown in Figure 9-22.

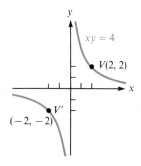

Figure 9-22

Exercise 9.4

In Exercises 1–12, write the equation of each hyperbola.

1. Vertices at $(5, 0)$ and $(-5, 0)$ and focus at $(7, 0)$
2. Focus at $(3, 0)$, vertex at $(2, 0)$, and center at $(0, 0)$
3. Center at $(2, 4)$; $a = 2$, $b = 3$; transverse axis is horizontal
4. Center at $(-1, 3)$, vertex at $(1, 3)$, and focus at $(2, 3)$
5. Center at $(5, 3)$, vertex at $(5, 6)$, hyperbola passes through $(1, 8)$
6. Foci at $(0, 10)$ and $(0, -10)$; $\frac{c}{a} = \frac{5}{4}$
7. Vertices at $(0, 3)$ and $(0, -3)$; $\frac{c}{a} = \frac{5}{3}$
8. Focus at $(4, 0)$, vertex at $(2, 0)$, and center at the origin
9. Center at $(1, -3)$; $a^2 = 4$, $b^2 = 16$
10. Center at $(1, 4)$, focus at $(7, 4)$, and vertex at $(3, 4)$
11. Center at the origin; hyperbola passes through points $(4, 2)$ and $(8, -6)$
12. Center at $(3, -1)$, y-intercept of -1, x-intercept of $3 + \dfrac{3\sqrt{5}}{2}$

In Exercises 13–16, find the area of the fundamental rectangle of each hyperbola.

13. $4(x - 1)^2 - 9(y + 2)^2 = 36$
14. $x^2 - y^2 - 4x - 6y = 6$
15. $x^2 + 6x - y^2 + 2y = -11$
16. $9x^2 - 4y^2 = 18x + 24y + 63$

In Exercises 17–20, write the equation of each hyperbola.

17. Center at $(-2, -4)$; $a = 2$; area of fundamental rectangle is 36 square units
18. Center at $(3, -5)$; $b = 6$; area of fundamental rectangle is 24 square units
19. One vertex at $(6, 0)$, one end of conjugate axis at $(0, \frac{5}{4})$
20. One vertex at $(3, 0)$, one focus at $(-5, 0)$, center at $(0, 0)$

In Exercises 21–29, graph the hyperbola.

21. $\dfrac{x^2}{9} - \dfrac{y^2}{4} = 1$ **22.** $\dfrac{y^2}{4} - \dfrac{x^2}{9} = 1$

23. $4x^2 - 3y^2 = 36$ **24.** $x^2 + 6x - y^2 + 2y = -11$

25. $y^2 - x^2 = 1$ **26.** $x^2 - y^2 - 4x - 6y = 6$

27. $4x^2 - 2y^2 + 8x - 8y = 8$ **28.** $9(y + 2)^2 - 4(x - 1)^2 = 36$

29. $y^2 - 4x^2 + 6y + 32x = 59$

In Exercises 30–32, graph each hyperbola by plotting points.

30. $xy = 9$ **31.** $-xy = 6$ **32.** $-xy = 20$

In Exercises 33–36, find the equation of each curve on which point P lies.

33. The difference of the distances between $P(x, y)$ and the points $(-2, 1)$ and $(8, 1)$ is 6.

34. The difference of the distances between $P(x, y)$ and the points $(3, -1)$ and $(3, 5)$ is 5.

35. The distance between point $P(x, y)$ and the point $(0, 3)$ is $\frac{3}{2}$ of the distance between P and the line $y = -2$.

36. The distance between point $P(x, y)$ and the point $(5, 4)$ is $\frac{5}{3}$ of the distance between P and the line $x = -3$.

37. Prove that $c > a$ for a hyperbola with center at the origin and line segment FF' on the x-axis.

38. Show that the equations of the extended diagonals of the fundamental rectangle of the hyperbola with equation $\frac{x^2}{a^2} - \frac{y^2}{b^2} = 1$ are $y = \frac{b}{a}x$ and $y = -\frac{b}{a}x$.

39. Show that the expansion of the standard form of the equation of a hyperbola is a special case of the general equation of second degree with $B = 0$.

9.5   SOLVING SIMULTANEOUS SECOND-DEGREE EQUATIONS

We now discuss techniques for solving systems of two equations in two variables where at least one of the equations is of second degree.

Example 1   Solve this system of equations by graphing:

$$\begin{cases} x^2 + y^2 = 25 \\ 2x + y = 10 \end{cases}$$

Solution   The graph of the equation $x^2 + y^2 = 25$ is a circle with center at the origin and radius of 5. The graph of the equation $2x + y = 10$ is a straight line. Depending on whether the line is a secant (intersecting the circle at two points) or a tangent (intersecting the circle at one point) or does not intersect the circle at all, there are two, one, or no solutions to the system, respectively. After graphing the circle and the line, as shown in Figure 9-23, note that there are two intersection points, P and P', with the coordinates of $(3, 4)$ and $(5, 0)$. Thus, the solutions to the given system of equations are

$$\begin{cases} x = 3 \\ y = 4 \end{cases} \quad \text{and} \quad \begin{cases} x = 5 \\ y = 0 \end{cases}$$

Verify that these are *exact* solutions.

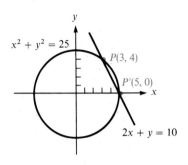

Figure 9-23

Graphical solutions of systems of equations usually give only approximate solutions. A second method, using algebra, can be used to find exact solutions.

Example 2   Solve the following system of equations algebraically.

$$\begin{cases} x^2 + y^2 = 25 \\ 2x + y = 10 \end{cases}$$

Solution   This system contains one equation of second degree and another of first degree. Solve systems of this kind by the substitution method. Solving the linear equation for y gives

$$2x + y = 10$$
$$y = -2x + 10$$

Substitute the expression $-2x + 10$ for y in the second-degree equation, and solve the resulting quadratic equation for x:

$$\begin{aligned}
x^2 + y^2 &= 25 \\
x^2 + (-2x + 10)^2 &= 25 \\
x^2 + 4x^2 - 40x + 100 &= 25 &&\text{Remove parentheses.} \\
5x^2 - 40x + 75 &= 0 &&\text{Combine terms.} \\
x^2 - 8x + 15 &= 0 &&\text{Divide both sides by 5.} \\
(x - 5)(x - 3) &= 0 &&\text{Factor } x^2 - 8x + 15 = 0. \\
x = 5 \quad \text{or} \quad x &= 3
\end{aligned}$$

Because $y = -2x + 10$, if $x = 5$ then $y = 0$, and if $x = 3$ then $y = 4$. The two solutions are

$$\begin{cases} x = 5 \\ y = 0 \end{cases} \quad \text{or} \quad \begin{cases} x = 3 \\ y = 4 \end{cases}$$

Example 3   Solve the following system of equations algebraically:

$$\begin{cases} 4x^2 + 9y^2 = 5 \\ \quad\quad y = x^2 \end{cases}$$

Solution   Solve this system by substitution.

$$4x^2 + 9y^2 = 5$$

$$4y + 9y^2 = 5 \qquad\qquad \text{Substitute } y \text{ for } x^2.$$

$$9y^2 + 4y - 5 = 0 \qquad\qquad \text{Add } -5 \text{ to both sides.}$$

$$(9y - 5)(y + 1) = 0 \qquad\qquad \text{Factor } 9y^2 + 4y^2 - 5.$$

$$9y - 5 = 0 \quad \text{or} \quad y + 1 = 0$$

$$y = \frac{5}{9} \qquad\qquad y = -1$$

Because $y = x^2$, the values of x are found by solving the equations

$$x^2 = \frac{5}{9} \qquad \text{and} \qquad x^2 = -1$$

Because the equation $x^2 = -1$ has no real solutions, this possibility is discarded. The solutions of the equation $x^2 = \frac{5}{9}$ are

$$x = \frac{\sqrt{5}}{3} \qquad \text{or} \qquad x = \frac{-\sqrt{5}}{3}$$

Thus, the solutions of the system are

$$\left(\frac{\sqrt{5}}{3}, \frac{5}{9}\right) \qquad \text{and} \qquad \left(\frac{-\sqrt{5}}{3}, \frac{5}{9}\right) \qquad\qquad\blacksquare$$

Example 4   Solve the following system of equations algebraically:

$$\begin{cases} 3x^2 + 2y^2 = 36 \\ 4x^2 - \quad y^2 = \quad 4 \end{cases}$$

Solution   In this system, both equations are of second degree and in the form $ax^2 + by^2 = c$. Solve systems like this by eliminating one of the variables by addition. Copy the first equation and multiply the second equation by 2 to obtain the equivalent system of equations

$$\begin{cases} 3x^2 + 2y^2 = 36 \\ 8x^2 - 2y^2 = \quad 8 \end{cases}$$

Add the equations to eliminate the variable y, and solve the resulting equation for x:

$$11x^2 = 44$$

$$x^2 = \quad 4$$

$$x = 2 \qquad \text{or} \qquad x = -2$$

To find y, substitute 2 for x and then -2 for x in the first equation and proceed as follows:

For $x = 2$	**For $x = -2$**
$3x^2 + 2y^2 = 36$	$3x^2 + 2y^2 = 36$
$3(2)^2 + 2y^2 = 36$	$3(-2)^2 + 2y^2 = 36$
$12 + 2y^2 = 36$	$12 + 2y^2 = 36$
$2y^2 = 24$	$2y^2 = 24$
$y^2 = 12$	$y^2 = 12$

$$y = +\sqrt{12} \quad \text{or} \quad y = -\sqrt{12} \qquad y = +\sqrt{12} \quad \text{or} \quad y = -\sqrt{12}$$
$$y = 2\sqrt{3} \quad | \quad y = -2\sqrt{3} \qquad y = 2\sqrt{3} \quad | \quad y = -2\sqrt{3}$$

The four solutions of this system are

$$(2, 2\sqrt{3}), \quad (2, -2\sqrt{3}), \quad (-2, 2\sqrt{3}), \quad \text{and} \quad (-2, -2\sqrt{3}) \qquad \blacksquare$$

Exercise 9.5

In Exercises 1–10, solve each system of equations by graphing.

1. $\begin{cases} 8x^2 + 32y^2 = 256 \\ x = 2y \end{cases}$
2. $\begin{cases} x^2 + y^2 = 2 \\ x + y = 2 \end{cases}$
3. $\begin{cases} x^2 + y^2 = 90 \\ y = x^2 \end{cases}$
4. $\begin{cases} x^2 + y^2 = 5 \\ x + y = 3 \end{cases}$

5. $\begin{cases} x^2 + y^2 = 25 \\ 12x^2 + 64y^2 = 768 \end{cases}$
6. $\begin{cases} x^2 + y^2 = 13 \\ y = x^2 - 1 \end{cases}$
7. $\begin{cases} x^2 - 13 = -y^2 \\ y = 2x - 4 \end{cases}$
8. $\begin{cases} x^2 + y^2 = 20 \\ y = x^2 \end{cases}$

9. $\begin{cases} x^2 - 6x - y = -5 \\ x^2 - 6x + y = -5 \end{cases}$
10. $\begin{cases} x^2 - y^2 = -5 \\ 3x^2 + 2y^2 = 30 \end{cases}$

In Exercises 11–36, solve each system of equations algebraically for real values of x and y.

11. $\begin{cases} 25x^2 + 9y^2 = 225 \\ 5x + 3y = 15 \end{cases}$
12. $\begin{cases} x^2 + y^2 = 20 \\ y = x^2 \end{cases}$
13. $\begin{cases} x^2 + y^2 = 2 \\ x + y = 2 \end{cases}$

14. $\begin{cases} x^2 + y^2 = 36 \\ 49x^2 + 36y^2 = 1764 \end{cases}$
15. $\begin{cases} x^2 + y^2 = 5 \\ x + y = 3 \end{cases}$
16. $\begin{cases} x^2 - x - y = 2 \\ 4x - 3y = 0 \end{cases}$

17. $\begin{cases} x^2 + y^2 = 13 \\ y = x^2 - 1 \end{cases}$
18. $\begin{cases} x^2 + y^2 = 25 \\ 2x^2 - 3y^2 = 5 \end{cases}$
19. $\begin{cases} x^2 + y^2 = 30 \\ y = x^2 \end{cases}$

20. $\begin{cases} 9x^2 - 7y^2 = 81 \\ x^2 + y^2 = 9 \end{cases}$
21. $\begin{cases} x^2 + y^2 = 13 \\ x^2 - y^2 = 5 \end{cases}$
22. $\begin{cases} 2x^2 + y^2 = 6 \\ x^2 - y^2 = 3 \end{cases}$

23. $\begin{cases} x^2 + y^2 = 20 \\ x^2 - y^2 = -12 \end{cases}$
24. $\begin{cases} xy = -\dfrac{9}{2} \\ 3x + 2y = 6 \end{cases}$
25. $\begin{cases} y^2 = 40 - x^2 \\ y = x^2 - 10 \end{cases}$

26. $\begin{cases} x^2 - 6x - y = -5 \\ x^2 - 6x + y = -5 \end{cases}$
27. $\begin{cases} y = x^2 - 4 \\ x^2 - y^2 = -16 \end{cases}$
28. $\begin{cases} 6x^2 + 8y^2 = 182 \\ 8x^2 - 3y^2 = 24 \end{cases}$

29. $\begin{cases} x^2 - y^2 = -5 \\ 3x^2 + 2y^2 = 30 \end{cases}$

30. $\begin{cases} \dfrac{1}{x} + \dfrac{1}{y} = 5 \\ \dfrac{1}{x} - \dfrac{1}{y} = -3 \end{cases}$

31. $\begin{cases} \dfrac{1}{x} + \dfrac{2}{y} = 1 \\ \dfrac{2}{x} - \dfrac{1}{y} = \dfrac{1}{3} \end{cases}$

32. $\begin{cases} \dfrac{1}{x} + \dfrac{3}{y} = 4 \\ \dfrac{2}{x} - \dfrac{1}{y} = 7 \end{cases}$

33. $\begin{cases} 3y^2 = xy \\ 2x^2 + xy - 84 = 0 \end{cases}$

34. $\begin{cases} x^2 + y^2 = 10 \\ 2x^2 - 3y^2 = 5 \end{cases}$

35. $\begin{cases} xy = \dfrac{1}{6} \\ y + x = 5xy \end{cases}$

36. $\begin{cases} xy = \dfrac{1}{12} \\ y + x = 7xy \end{cases}$

37. The area of a rectangle is 63 square centimeters, and its perimeter is 32 centimeters. Find the dimensions of the rectangle.

38. The product of two integers is 32 and their sum is 12. Find the integers.

39. The sum of the squares of two numbers is 221, and the sum of the numbers is 212 less. Find the numbers.

40. Grant receives $225 annual income from one investment. Jeff invested $500 more than Grant, but at an annual rate of 1% less. Jeff's annual income is $240. What is the amount and rate of Grant's investment?

41. Carol receives $67.50 annual income from one investment. John invested $150 more than Carol at an annual rate of $1\frac{1}{2}$% more. John's annual income is $94.50. What is the amount and rate of Carol's investment? (*Hint*: There are two answers.)

42. Jim drove 306 miles. Jim's brother made the same trip at a speed 17 miles per hour slower than Jim did and required an extra $1\frac{1}{2}$ hours. What was Jim's rate and time?

9.6   TRANSLATION OF COORDINATE AXES

The graph of the equation

$$(x - 3)^2 + (y - 1)^2 = 4$$

is a circle with radius of 2 and with center at the point (3, 1). See Figure 9-24. If we were to shift the black xy-coordinate system 3 units to the right and 1 unit up, we would establish the colored $x'y'$-coordinate system. With respect to this new $x'y'$-system, the center of the circle is the origin and its equation is

$$x'^2 + y'^2 = 4$$

In this section, we will discuss how to change the equation of a graph by shifting the position of the x- and y-axes. A shift to the left, right, up, or down is called a **translation of the coordinate axes**.

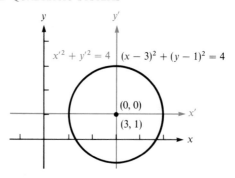

Figure 9-24

Figure 9-25 shows both an xy- and an $x'y'$-coordinate system. The colored x'- and y'-axes are parallel to the black x- and y-axes, respectively, and the unit distance on each is the same. The origin of the $x'y'$-system is the point O' with $x'y'$-coordinates of $(0, 0)$ and with xy-coordinates of (h, k). The $x'y'$-system is called a **translated coordinate system**.

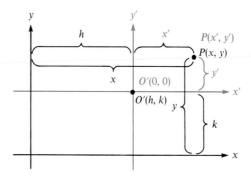

Figure 9-25

Relative to the xy-system in Figure 9-25, the coordinates of point P are (x, y). Relative to the $x'y'$-system, the coordinates of point P are (x', y'). But the geometry of the figure,

1. $\begin{cases} x = x' + h \\ y = y' + k \end{cases}$ or   **2.** $\begin{cases} x' = x - h \\ y' = y - k \end{cases}$

Equations 1 and Equations 2 are called the **translation-of-axes formulas**. They enable us to determine the coordinates of any point with respect to any translated coordinate system.

Example 1   The $x'y'$-coordinates of point P in Figure 9-26 are $(-3, -2)$, and the xy-coordinates of point O' are $(2, 1)$. Find the xy-coordinates of point P.

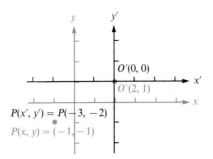

Figure 9-26

Solution   Because you want to find xy-coordinates, use Equations 1 with $(h, k) = (2, 1)$ and $(x', y') = (-3, -2)$:

$$x = x' + h \qquad y = y' + k$$
$$x = -3 + 2 \qquad y = -2 + 1$$
$$= -1 \qquad\qquad = -1$$

The xy-coordinates of point P are $(-1, -1)$. ■

Example 2   The xy-coordinates of point O' in Figure 9-27 are $(-3, -5)$, and the xy-coordinates of point Q are $(1, -2)$. Find the $x'y'$-coordinates of point Q.

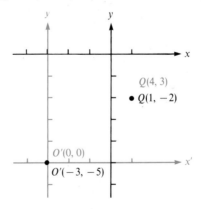

Figure 9-27

Solution   Because you want to find $x'y'$-coordinates, use Equations 2 with $(h, k) = (-3, -5)$ and $(x, y) = (1, -2)$:

$$x' = x - h \qquad y' = y - k$$
$$x' = 1 - (-3) \qquad y' = -2 - (-5)$$
$$= 4 \qquad\qquad\quad = 3$$

The $x'y'$-coordinates of point Q are $(4, 3)$. ■

Example 3   The xy-coordinates of point O' in Figure 9-28 are $(1, 5)$. Find the equation of the line determined by $y = 2x + 3$ in the variables x' and y'.

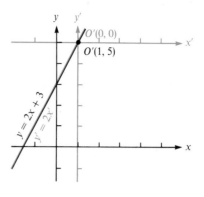

Figure 9-28

Solution   Because the xy-coordinates of point O' are $(h, k) = (1, 5)$, substitute 1 for h and 5 for k in Equations 1 to obtain

$$x = x' + h \qquad y = y' + k$$
$$x = x' + 1 \qquad y = y' + 5$$

Then substitute $x' + 1$ for x and $y' + 5$ for y in the equation $y = 2x + 3$ and simplify:

$$y = 2x + 3$$
$$y' + 5 = 2(x' + 1) + 3$$
$$y' + 5 = 2x' + 2 + 3$$
$$y' + 5 = 2x' + 5$$
$$y' = 2x'$$

In the variables x' and y', the equation of the line is $y' = 2x'$. Note that graphing $y = 2x + 3$ with respect to the black xy-coordinate system gives the same line as graphing $y' = 2x'$ with respect to the colored $x'y'$-coordinate system.   ■

Example 4   Find the equation of the parabola $y + 3 = (x - 2)^2$ with respect to a translated coordinate system with origin at $O'(2, -3)$. Graph the resulting equation with respect to the translated coordinate system.

Solution   Because the origin is translated to the point $(h, k) = (2, -3)$, substitute 2 for h and -3 for k in the translation-of-axes formulas:

$$x = x' + h \qquad y = y' + k$$
$$x = x' + 2 \qquad y = y' + (-3)$$
$$y = y' - 3$$

To obtain the equation of the same parabola with respect to the translated axes, substitute $x' + 2$ for x and $y' - 3$ for y in the given equation:

$$y + 3 = (x - 2)^2$$
$$(y' - 3) + 3 = [(x' + 2) - 2]^2$$
$$y' = x'^2 \qquad \text{Simplify.}$$

The graph of the equation $y' = x'^2$ is a parabola with vertex at the origin of the $x'y'$-system. See Figure 9-29.

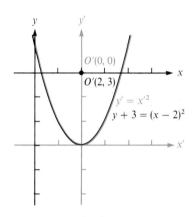

$O'(0, 0)$
$O'(2, 3)$
$y' = x'^2$
$y + 3 = (x - 2)^2$

Figure 9-29

■

Example 5 To what point should the origin of the xy-coordinate system be translated to remove the first-degree terms in the equation

$$4x^2 + y^2 + 8x - 6y + 9 = 0$$

Graph the resulting equation with respect to the translated coordinate system.

Solution Begin by completing the square in both x and y and simplifying:

$$4x^2 + y^2 + 8x - 6y + 9 = 0$$
$$4x^2 + 8x + y^2 - 6y = -9 \qquad \text{Rearrange terms and add } -9 \text{ to both sides.}$$
$$4(x^2 + 2x) + y^2 - 6y = -9 \qquad \text{Factor 4 from } 4x^2 + 8x.$$
$$4(x^2 + 2x + 1) + y^2 - 6y + 9 = -9 + 4 + 9 \qquad \text{Add 4 and 9 to both sides.}$$
$$4(x + 1)^2 + (y - 3)^2 = 4 \qquad \text{Factor both trinomials.}$$
$$\frac{(x + 1)^2}{1} + \frac{(y - 3)^2}{4} = 1 \qquad \text{Divide both sides by 4.}$$

Let the origin of the $x'y'$-coordinate system be the point $O'(h, k) = O'(-1, 3)$. Then, by the translation-of-axes formulas, you have $x' = x + 1$ and $y' = y - 3$.

Substitute x' for $x + 1$ and y' for $y - 3$ to obtain the equation

$$\frac{(x + 1)^2}{1} + \frac{(y - 3)^2}{4} = 1$$

$$\frac{x'^2}{1} + \frac{y'^2}{4} = 1$$

Note that this final equation contains no first-degree terms. Its graph is an ellipse centered at the origin of an $x'y'$-coordinate system whose origin has been translated to the point $(-1, 3)$ of the xy-system. See Figure 9-30.

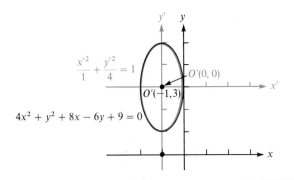

Figure 9-30   ■

Example 6   Show that the slope of a nonvertical line is not affected by a translation of axes.

Solution   In the xy-system, the slope-intercept form of the equation of a nonvertical line with slope m is

$$y = mx + b$$

Use the translation-of-axes formulas to find the equation of the *same* line with respect to an $x'y'$-system whose origin has been translated to the point (h, k) of the xy-system. Substitute $x' + h$ for x and $y' + k$ for y and simplify. Proceed as follows:

$$\begin{aligned}
y &= mx + b \\
y' + k &= m(x' + h) + b \\
y' + k &= mx' + mh + b \qquad \text{Remove parentheses.} \\
y' &= mx' + mh + b - k \qquad \text{Add } -k \text{ to both sides.}
\end{aligned}$$

This final equation is in slope-intercept form and represents a line with y'-intercept of $mh + b - k$ and a slope of m. Thus, the translation does not change the slope of the line.   ■

Exercise 9.6

In Exercises 1–4, the origin of the $x'y'$-system is at the point $(1, -3)$ of the xy-system. Find the xy-coordinates of the point whose $x'y'$-coordinates are given.

1. $P(2, 4)$ **2.** $Q(-1, 3)$ **3.** $R(0, 0)$ **4.** $S(1, -3)$

In Exercises 5–8, the origin of the $x'y'$-system is at the point $(-2, 4)$ of the xy-system. Find the $x'y'$-coordinates of the point whose xy-coordinates are given.

5. $P(2, -4)$ **6.** $Q(-2, 4)$ **7.** $R(0, 0)$ **8.** $S(4, 2)$

In Exercises 9–12, the origin of the $x'y'$-system is at the point $(0, -5)$ of the xy-system. Express each equation in terms of the variables x' and y'. Draw a sketch that shows the relation of the graphs of the equations to both coordinate systems.

9. $y = 3x - 2$ **10.** $y = -3x - 4$ **11.** $y = x^2 - 5$ **12.** $x = y^2 + 10y + 25$

In Exercises 13–16, the origin of the $x'y'$-system is at the point $(3, -2)$ of the xy-system. Express each equation in terms of the variables x' and y'. Draw a sketch that shows the relation of the graphs of the equations to both coordinate systems.

13. $x^2 + y^2 = 9$ **14.** $2x^2 + 3y^2 = 12$ **15.** $x^2 - 6x + 7 = y$ **16.** $y^2 + 4y + 7 = x$

In Exercises 17–22, determine the point to which the origin of the $x'y'$-system should be translated to eliminate the first-degree terms of the given equation. Draw a sketch that shows the relation of the graphs of the equations to both coordinate systems.

17. $x^2 + y^2 + 4x - 10y - 6 = 0$ **18.** $x^2 + y^2 - 8x - 2y + 1 = 0$

19. $2x^2 + y^2 + 4x + 2y - 1 = 0$ **20.** $4x^2 + 9y^2 + 24x - 18y + 9 = 0$

21. $x^2 - y^2 - 6x - 4y + 4 = 0$ **22.** $4x^2 - 9y^2 + 8x - 36y = 68$

23. Show that the radius of a circle is not affected by a translation of axes.

24. Show that no translation of axes will remove the xy-term of the equation $xy = 1$.

In Exercises 25–26, suppose that the equation $Ax^2 + Bxy + Cy^2 + Dx + Ey + F = 0$ is changed into $A'x'^2 + B'x'y' + C'y'^2 + D'x' + E'y' + F' = 0$ by substituting $x' + h$ for x and $y' + k$ for k and then simplifying.

25. Show that $A + C = A' + C'$.   **26.** Show that $B^2 - 4AC = B'^2 - 4A'C'$.

9.7 ROTATION OF COORDINATE AXES

In this section, we discuss how to change the equation of a graph by rotating the xy-axes through some angle θ. This process is called **rotation of the coordinate axes**.

Figure 9-31 shows both an xy- and an $x'y'$-coordinate system with a common origin O, and the same unit of measure on all axes. The colored x'- and y'-axes are rotated through some angle θ with respect to the black x- and y-axes. The $x'y'$-system is called a **rotated coordinate system**.

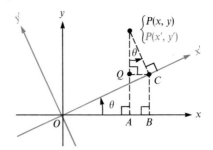

Figure 9-31

Suppose the xy-coordinates of point P are (x, y) and its $x'y'$-coordinates are (x', y'). From point P, we draw PC perpendicular to the x'-axis and PA perpendicular to the x-axis. Segments CQ and CB are perpendicular to PA and the x-axis, respectively, as indicated in the figure. This forms similar triangles OBC and PQC, in which

$$\sin \theta = \frac{BC}{OC} = \frac{BC}{x'} \quad \text{or} \quad BC = x' \sin \theta$$

$$\cos \theta = \frac{OB}{OC} = \frac{OB}{x'} \quad \text{or} \quad OB = x' \cos \theta$$

$$\sin \theta = \frac{QC}{PC} = \frac{AB}{y'} \quad \text{or} \quad AB = y' \sin \theta$$

and

$$\cos \theta = \frac{QP}{PC} = \frac{QP}{y'} \quad \text{or} \quad QP = y' \cos \theta$$

Now the variables x and y can be expressed in terms of x', y', and θ:

$$x = OA \qquad\qquad\qquad y = AP$$
$$\quad = OB - AB \qquad\qquad = BC + QP$$
$$\quad = x' \cos \theta - y' \sin \theta \qquad = x' \sin \theta + y' \cos \theta$$

The equations relating the xy-coordinates of point P to its coordinates in the $x'y'$-system are as follows:

The Equations of Rotation.

$$x = x' \cos \theta - y' \sin \theta$$
$$y = x' \sin \theta + y' \cos \theta$$

Example 1   The xy-coordinate system is rotated 30° to form the $x'y'$-coordinate system as in Figure 9-32. The $x'y'$-coordinates of P are $(\sqrt{3} + 1, \sqrt{3} - 1)$. What are the xy-coordinates of P?

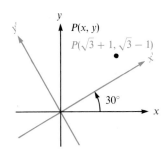

Figure 9-32

Solution   Use the equations

$$x = x' \cos \theta - y' \sin \theta$$
$$y = x' \sin \theta + y' \cos \theta$$

with $x' = \sqrt{3} + 1$, $y' = \sqrt{3} - 1$, and $\theta = 30°$.

$$x = x' \cos \theta - y' \sin \theta$$

$$x = (\sqrt{3} + 1) \frac{\sqrt{3}}{2} - (\sqrt{3} - 1) \frac{1}{2} \qquad \text{Substitute } \sqrt{3} + 1 \text{ for } x', \sqrt{3} - 1 \text{ for } y',$$
$$\qquad\qquad\qquad\qquad\qquad\qquad \tfrac{\sqrt{3}}{2} \text{ for } \cos 30°, \text{ and } \tfrac{1}{2} \text{ for } \sin 30°.$$

$$= \frac{3}{2} + \frac{\sqrt{3}}{2} - \frac{\sqrt{3}}{2} + \frac{1}{2}$$

$$= 2$$

$$y = x' \sin \theta + y' \cos \theta$$

$$= (\sqrt{3} + 1) \frac{1}{2} + (\sqrt{3} - 1) \frac{\sqrt{3}}{2} \qquad \text{Substitute } \sqrt{3} + 1 \text{ for } x', \sqrt{3} - 1 \text{ for } y',$$
$$\qquad\qquad\qquad\qquad\qquad\qquad \tfrac{1}{2} \text{ for } \sin 30°, \text{ and } \tfrac{\sqrt{3}}{2} \text{ for } \cos 30°.$$

$$= \frac{\sqrt{3}}{2} + \frac{1}{2} + \frac{3}{2} - \frac{\sqrt{3}}{2}$$

$$= 2$$

With respect to the xy-system, the coordinates of point P are $(2, 2)$.   ∎

Example 2   Through what angle θ should the $x'y'$-system be rotated to transform the equation $xy = 1$ into an equation with no term involving the product $x'y'$? Graph the resulting equation.

Solution Use the rotation equations to transform the equation $xy = 1$ into a new equation with variables x' and y':

$$xy = 1$$
$$(x' \cos \theta - y' \sin \theta)(x' \sin \theta + y' \cos \theta) = 1$$
$$x'^2 \cos \theta \sin \theta + x'y' \cos^2 \theta - x'y' \sin^2 \theta - y'^2 \sin \theta \cos \theta = 1$$

1. $$x'^2 \cos \theta \sin \theta + x'y'(\cos^2 \theta - \sin^2 \theta) - y'^2 \sin \theta \cos \theta = 1$$

Because Equation 1 is to be free of any term involving the product $x'y'$, set the coefficient of $x'y'$ equal to zero and solve for θ:

$$\cos^2 \theta - \sin^2 \theta = 0$$
$$\cos 2\theta = 0 \qquad \text{Use the identity } \cos 2\theta = \cos^2 \theta - \sin^2 \theta.$$
$$2\theta = 90°$$
$$\theta = 45°$$

Thus, the xy-coordinate system should be rotated through an angle of 45°. When this is done Equation 1 becomes

$$x'^2 \cos 45° \sin 45° + x'y' \cdot 0 - y'^2 \sin 45° \cos 45° = 1 \qquad \begin{array}{l}\text{Substitute 45° for} \\ \theta \text{ in Equation 1.}\end{array}$$

$$x'^2 \frac{\sqrt{2}}{2} \cdot \frac{\sqrt{2}}{2} - y'^2 \frac{\sqrt{2}}{2} \cdot \frac{\sqrt{2}}{2} = 1 \qquad \begin{array}{l}\text{Substitute } \frac{\sqrt{2}}{2} \text{ for} \\ \cos 45° \text{ and for} \\ \sin 45°.\end{array}$$

$$\frac{x'^2}{2} - \frac{y'^2}{2} = 1 \qquad \text{Simplify.}$$

The graph of this equation is the hyperbola shown in Figure 9-33.

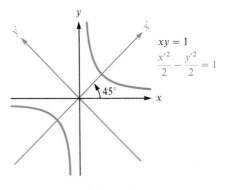

Figure 9-33

A rotation of the coordinate axes can transform the general second-degree equation

$$Ax^2 + Bxy + Cy^2 + Dx + Ey + F = 0$$

into another equation with no xy-term. To determine the appropriate angle θ, use the following result. The proof is omitted.

The Angle of Rotation Formula.   If the coordinate axes are rotated through an angle θ, $0° < \theta < 90°$, where

$$\cot 2\theta = \frac{A - C}{B}$$

then the equation

$$Ax^2 + Bxy + Cy^2 + Dx + Ey + F = 0$$

will be transformed into an equation with no xy-term.

To determine the required rotation equations, we must use $\cot 2\theta$ to determine the values of $\sin \theta$ and $\cos \theta$.

Example 3   Transform the equation

$$17x^2 - 48xy + 31y^2 + 49 = 0$$

into an equation with no $x'y'$-term. Graph the resulting equation.

Solution   Rotate the coordinate system through an acute angle θ, where $\cot 2\theta = \frac{A-C}{B}$. To do so, first evaluate $\cot 2\theta$.

$$\cot 2\theta = \frac{A - C}{B}$$

$$= \frac{17 - 31}{-48} \qquad \text{Substitute 17 for } A, 31 \text{ for } C, \text{ and } -48 \text{ for } B.$$

$$= \frac{7}{24}$$

To find the values of $\sin \theta$ and $\cos \theta$ required by the rotation equations, evaluate $\cos 2\theta$ and then use the identities

$$\sin \theta = \sqrt{\frac{1 - \cos 2\theta}{2}}$$

and

$$\cos \theta = \sqrt{\frac{1 + \cos 2\theta}{2}}$$

To evaluate $\cos 2\theta$, sketch a right triangle such as the one in Figure 9-34 and use the fact that $\cot 2\theta = \frac{7}{24}$ and the Pythagorean Theorem to determine the sides of the triangle. You can then see that $\cos 2\theta = \frac{7}{25}$.

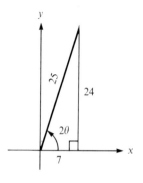

Figure 9-34

Thus,

$$\sin \theta = \sqrt{\frac{1 - \cos 2\theta}{2}} \qquad \text{and} \qquad \cos \theta = \sqrt{\frac{1 + \cos 2\theta}{2}}$$

$$\sin \theta = \sqrt{\frac{1 - \frac{7}{25}}{2}} \qquad\qquad \cos \theta = \sqrt{\frac{1 + \frac{7}{25}}{2}}$$

$$= \sqrt{\frac{9}{25}} \qquad\qquad\qquad = \sqrt{\frac{16}{25}}$$

$$= \frac{3}{5} \qquad\qquad\qquad\qquad = \frac{4}{5}$$

The required rotation is given by the equations

$$x = x' \cos \theta - y' \sin \theta$$
$$y = x' \sin \theta + y' \cos \theta$$

where $\frac{3}{5} = \sin \theta$ and $\frac{4}{5} = \cos \theta$:

$$x = x'\left(\frac{4}{5}\right) - y'\left(\frac{3}{5}\right)$$

$$y = x'\left(\frac{3}{5}\right) - y'\left(\frac{4}{5}\right)$$

To transform the given equation $17x^2 - 48xy + 31y^2 + 49 = 0$ into an equation with no xy-term, substitute $\frac{4}{5}x' - \frac{3}{5}y'$ for x and $\frac{3}{5}x' - \frac{4}{5}y'$ for y in the given equation:

$$17x^2 - 48xy + 31y^2 + 49 = 0$$

$$17\left(\frac{4}{5}x' - \frac{3}{5}y'\right)^2 - 48\left(\frac{4}{5}x' - \frac{3}{5}y'\right)\left(\frac{3}{5}x' - \frac{4}{5}y'\right) + 31\left(\frac{3}{5}x' - \frac{4}{5}y'\right)^2 + 49 = 0$$

Perform the indicated operations to obtain

$$17\left(\frac{16}{25}x'^2 - 2 \cdot \frac{12}{25}x'y' + \frac{9}{25}y'^2\right) - 48\left(\frac{12}{25}x'^2 + \frac{16}{25}x'y' - \frac{9}{25}x'y' - \frac{12}{25}y'^2\right)$$

$$+ 31\left(\frac{9}{25}x'^2 + 2 \cdot \frac{12}{25}x'y' + \frac{16}{25}y'^2\right) + 49 = 0$$

Then remove parentheses and combine terms to obtain

$$\frac{-25}{25}x'^2 + 0x'y' + \frac{1225}{25}y'^2 + 49 = 0$$

or

$$\frac{x'^2}{49} - y'^2 = 1$$

The graph of this transformed equation is the hyperbola that appears in Figure 9-35.

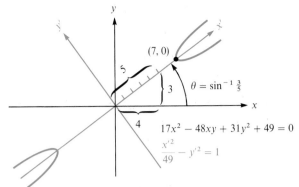

$(7, 0)$

$\theta = \sin^{-1} \frac{3}{5}$

$17x^2 - 48xy + 31y^2 + 49 = 0$

$$\frac{x'^2}{49} - y'^2 = 1$$

Figure 9-35

Exercise 9.7

In Exercises 1–6, an xy-coordinate system is rotated through an angle θ to form an $x'y'$-coordinate system. Find the xy-coordinates of the point whose $x'y'$-coordinates are given.

1. $P(2\sqrt{3}, 0); \theta = 30°$
2. $Q(\sqrt{3} - 1, \sqrt{3} + 1); \theta = 30°$
3. $R(2, 6); \theta = 45°$
4. $S(\sqrt{2} - 1, \sqrt{2} + 1); \theta = 45°$
5. $T(7, 4); \theta = 60°$
6. $U(0, -8); \theta = 60°$

In Exercises 7–10, determine the angle θ through which the axes must be rotated to remove the xy-term. Do not transform the equation.

7. $3x^2 - \sqrt{3}xy + 2y^2 - 3x + y - 10 = 0$
8. $5x^2 - 8xy + 5y^2 + y + 3 = 0$
9. $5\sqrt{3}x^2 + 8xy - 3\sqrt{3}y^2 - 9 = 0$
10. $x^2 - 6xy - 5y^2 - 27 = 0$

In Exercises 11–20, perform a rotation of axes to remove the xy-term. Graph the equation.

11. $xy = 2$
12. $xy = 18$
13. $x^2 + 3xy + y^2 = 2$
14. $5x^2 + 26xy + 5y^2 - 72 = 0$
15. $x^2 + 2xy + y^2 - 2\sqrt{2}x + 2\sqrt{2}y = 0$
16. $x^2 - 2\sqrt{3}xy - y^2 - 2 = 0$
17. $2x^2 - 4xy + 5y^2 - 6 = 0$
18. $7x^2 + 2\sqrt{3}xy + 5y^2 = 8$
19. $6x^2 + 24xy - y^2 - 30 = 0$
20. $5x^2 - 4xy + 8y^2 - 36 = 0$

21. Suppose that the equation

$$Ax^2 + Bxy + Cy^2 + Dx + Ey + F = 0$$

is changed into

$$A'x'^2 + B'x'y' + C'y'^2 + D'x' + E'y' + F' = 0$$

by rotating the axes through an angle θ. Show that

$$A + C = A' + C'$$

22. Show that the equation $x^2 + y^2 = 1$ is not affected by any rotation of axes.

1. Write the equation of the circle with center at the origin and curve passing through point (5, 5).

2. Write the equation of the circle with center at the origin and curve passing through point (6, 8).

3. Write the equation of a circle with endpoints of its diameter at $(-2, 4)$ and (12, 16).

4. Write the equation of a circle with endpoints of its diameter at $(-3, -6)$ and (7, 10).

5. Write in standard form the equation of the circle $x^2 + y^2 - 6x + 4y = 3$ and graph the circle.

6. Write in standard form the equation of the circle $x^2 + 4x + y^2 - 10y = -13$ and graph the circle.

7. Write the equation of the parabola with vertex at the origin and curve passing through $(-8, 4)$ and $(-8, -4)$.

8. Write the equation of the parabola with vertex at the origin and curve passing through $(-8, 4)$ and (8, 4).

9. Write the equation of the parabola $y = ax^2 + bx + c$ in standard form to show that the x-coordinate of the vertex of the parabola is $-\frac{b}{2a}$.

10. Find the equation of the parabola with vertex at $(-2, 3)$, curve passing through point $(-4, -8)$, and opening downward.

11. Graph the parabola $x^2 - 4y - 2x + 9 = 0$.

12. Graph the parabola $y^2 - 6y = 4x - 13$.

13. Graph $y^2 - 4x - 2y + 13 = 0$.

14. Write the equation of the ellipse with center at the origin, major axis that is horizontal and 12 units long, and minor axis 8 units long.

15. Write the equation of the ellipse with center at point $(-2, 3)$ and curve passing through points $(-2, 0)$ and (2, 3).

16. Graph the ellipse $4x^2 + y^2 - 16x + 2y = -13$.

17. Graph the curve $x^2 + 9y^2 - 6x - 18y + 9 = 0$.

18. Write the equation of the hyperbola with vertices at points $(-3, 3)$ and (3, 3) and a focus at point (5, 3).

19. Graph the hyperbola $9x^2 - 4y^2 - 16y - 18x = 43$.

20. Graph the hyperbola $-2xy = 9$.

21. Solve the following system of equations by graphing:
$$\begin{cases} 3x^2 + y^2 = 52 \\ x^2 - y^2 = 12 \end{cases}$$

22. Solve the system in Review Exercise 21 algebraically.

23. Solve the following system of equations by graphing:
$$\begin{cases} x^2 + y^2 = 16 \\ -\sqrt{3}\, y + 4\sqrt{3} = 3x \end{cases}$$

24. Solve the system in Review Exercise 23 algebraically.

25. Solve the following system of equations by graphing:

$$\begin{cases} \dfrac{x^2}{16} + \dfrac{y^2}{12} = 1 \\ \dfrac{x^2}{1} - \dfrac{y^2}{3} = 1 \end{cases}$$

26. Solve the system in Review Exercise 25 algebraically.

In Review Exercises 27–30, the origin of the x′y′-system is at the point (2, −3) of the xy-system. Express each equation in terms of the variables x′ and y′. Draw a sketch that shows the relation of the graphs to both coordinate systems.

27. $x^2 + y^2 = 25$

28. $x^2 - y^2 = 4$

29. $4x^2 + 9y^2 = 36$

30. $x^2 - 4x - 3y = 5$

In Review Exercises 31–32, determine the point at which the origin of the x′y′-system should be located to eliminate the first-degree terms of the given equation. Draw a sketch that shows the relation of the graphs of the equations to both coordinate systems.

31. $x^2 + y^2 - 6x - 4y + 12 = 0$

32. $4x^2 - y^2 - 8x + 4y = 4$

33. Determine the rotation of axes needed to remove the xy-term of the equation

$$13x^2 + 6\sqrt{3}xy + 7y^2 - 16 = 0$$

Draw a sketch that shows the relation of the graphs of the equations to both coordinate systems.

10

COMPLEX NUMBERS AND POLAR COORDINATES

All of the numbers used thus far have been real numbers. However, some situations require a new set of numbers called the set of complex numbers.

10.1 COMPLEX NUMBERS

If we solve the equation $x^2 = 9$ for x, we find that it has two solutions:

$$x^2 = 9$$
$$x = \sqrt{9} \quad \text{or} \quad x = -\sqrt{9}$$
$$= 3 \quad \quad = -3$$

If we solved the equation $x^2 = -1$ in a similar way, we would expect it to have two solutions also:

$$x^2 = -1$$
$$x = \sqrt{-1} \quad \text{or} \quad x = -\sqrt{-1}$$

Each proposed solution of the equation $x^2 = -1$ involves the symbol $\sqrt{-1}$. The symbol $\sqrt{-1}$ cannot represent a real number, because no real number has a square that is equal to -1. For years, mathematicians believed that square roots of negative numbers, denoted by symbols such as $\sqrt{-3}$, $\sqrt{-4}$, and $\sqrt{-5}$, were nonsense. In the seventeenth century, these symbols were termed **imaginary numbers** by René Descartes (1596–1650). Mathematicians no longer think of imaginary numbers as impossible. In fact, imaginary numbers have important uses, as in describing the behavior of alternating current in electronics.

The imaginary number $\sqrt{-1}$ occurs often enough to warrant a special symbol: the letter i is used to denote $\sqrt{-1}$. Because i is used to denote $\sqrt{-1}$, it follows that

$$i^2 = -1$$

The powers of i with natural number exponents produce an interesting pattern:

$$i = \sqrt{-1} = i \qquad\qquad i^5 = i^4 \cdot i = 1 \cdot i = i$$
$$i^2 = \sqrt{-1}\sqrt{-1} = -1 \qquad i^6 = i^4 \cdot i^2 = 1(-1) = -1$$
$$i^3 = i^2 \cdot i = -1 \cdot i = -i \qquad i^7 = i^4 \cdot i^3 = 1(-i) = -i$$
$$i^4 = i^2 \cdot i^2 = (-1)(-1) = 1 \qquad i^8 = i^4 \cdot i^4 = 1(1) = 1$$

The pattern continues $i, -1, -i, 1$.

We can simplify powers of i involving negative integral exponents by rationalizing denominators:

$$i^{-1} = \frac{1}{i} = \frac{1}{i} \cdot \frac{i}{i} = \frac{i}{-1} = -i$$

$$i^{-2} = \frac{1}{i^2} = \frac{1}{-1} = -1$$

$$i^{-3} = \frac{1}{i^3} = \frac{1}{i^3} \cdot \frac{i}{i} = \frac{i}{1} = i$$

$$i^{-4} = \frac{1}{i^4} = \frac{1}{1} = 1$$

If we define i^0 to be 1, then the familiar pattern for the powers of i carries over to integral exponents:

$$\vdots$$
$$i^{-3} = i$$
$$i^{-2} = -1$$
$$i^{-1} = -i$$
$$i^0 = 1$$
$$i^1 = i$$
$$i^2 = -1$$
$$i^3 = -i$$
$$i^4 = 1$$
$$\vdots$$

The sequence $i, -1, -i, 1$ repeats endlessly. Because the powers of i form a repeating sequence, it is easy to compute large powers of i.

Example 1   Simplify i^{365}.

Solution
$$i^{365} = (i^4)^{91} \cdot i^1$$
$$= 1^{91} \cdot i$$
$$= i$$

Note that the number 365 leaves a remainder of 1 when it is divided by 4. ∎

Example 2   Suppose that i is raised to the nth power, where n is a natural number. Discuss the possible values for i^n.

Solution   If the exponent n of i^n is a multiple of 4, then $i^n = 1$.
If n leaves a remainder of 1 when divided by 4, then $i^n = i$.
If n leaves a remainder of 2 when divided by 4, then $i^n = -1$.
If n leaves a remainder of 3 when divided by 4, then $i^n = -i$. ∎

Recall that if at least one of two numbers x and y is not negative, then

$$\sqrt{xy} = \sqrt{x}\sqrt{y}$$

This allows us to write the square root of any negative number in the form bi, where b is a real number and $i = \sqrt{-1}$. For example,

$$\sqrt{-9} = \sqrt{9(-1)} = \sqrt{9}\sqrt{-1} = 3i$$

and

$$\sqrt{-2} = \sqrt{2(-1)} = \sqrt{2}\sqrt{-1} = \sqrt{2}\,i$$

Because a square root of a number is one of two equal factors of that number, we know that $\sqrt{-3}\sqrt{-3} = -3$. It is incorrect to apply the rule $\sqrt{xy} = \sqrt{x}\sqrt{y}$ in this case:

$$\sqrt{-3}\sqrt{-3} \neq \sqrt{(-3)(-3)} = \sqrt{9} = 3$$

This example illustrates that if x and y are both negative, then the rule $\sqrt{xy} = \sqrt{x}\sqrt{y}$ is not valid.

Numbers such as $3 + 4i$ and $-1 + 9i$ that indicate the sum of a real number and an imaginary number are called **complex numbers**.

Definition. A **complex number** is any number that can be expressed in the form $a + bi$, where a and b are real numbers and $i = \sqrt{-1}$.

The number a is called the **real part** and the number b is called the **imaginary part** of the complex number $a + bi$.

If $b = 0$, the complex number $a + bi$ is the real number a. Thus, any real number is a complex number with a zero imaginary part. If $a = 0$ and $b \neq 0$, the complex number $a + bi$ is the imaginary number bi. Thus, any imaginary number is a complex number with a zero real part. It follows that the real number set and the imaginary number set are both subsets of the complex number set.

We must accept some definitions before doing any arithmetic with complex numbers.

Equality of Complex Numbers. Two complex numbers are equal if and only if their real parts are equal and their imaginary parts are equal. Thus, if $a + bi$ and $c + di$ are two complex numbers, then

$$a + bi = c + di \quad \text{if and only if} \quad a = c \text{ and } b = d$$

Example 3   For what real number values of x and y is $3x + 4i = (4y + x) + xi$?

Solution   Because the imaginary parts of these two complex numbers must be equal, $x = 4$. Because the real parts of these two complex numbers must be equal, $3x = 4y + x$. Solve the system of equations

$$\begin{cases} x = 4 \\ 3x = 4y + x \end{cases}$$

by substituting 4 for x in the equation $3x = 4y + x$ and solving for y. You will find that $y = 2$. Hence, the solution is $x = 4$ and $y = 2$.   ■

Complex numbers can be added and multiplied as if they were binomials.

Addition of Complex Numbers.   Two complex numbers such as $a + bi$ and $c + di$ are added as if they were algebraic binomials:

$$(a + bi) + (c + di) = (a + c) + (b + d)i$$

Note that the previous definition implies that the sum of two complex numbers is another complex number.

Example 4   Find the sum of $3 + 4i$ and $2 + 7i$.

Solution
$$\begin{aligned}
(3 + 4i) + (2 + 7i) &= 3 + 4i + 2 + 7i \\
&= 3 + 2 + 4i + 7i \\
&= 5 + 11i
\end{aligned}$$
   ■

Multiplication of Complex Numbers.   Two complex numbers such as $a + bi$ and $c + di$ are multiplied as if they were algebraic binomials, with $i^2 = -1$:

$$(a + bi)(c + di) = (ac - bd) + (ad + bc)i$$

Note that the previous definition implies that the product of two complex numbers is another complex number.

Example 5   Find the product of $3 + 4i$ and $2 + 7i$.

Solution
$$\begin{aligned}
(3 + 4i)(2 + 7i) &= 6 + 21i + 8i + 28i^2 \\
&= 6 + 21i + 8i - 28 \\
&= -22 + 29i
\end{aligned}$$
   ■

It is a good idea to express complex numbers in $a + bi$ form before attempting any algebraic manipulations. This will help you avoid making errors in determining the sign of the result.

Example 6   Find the product of $-2 + \sqrt{-16}$ and $4 - \sqrt{-9}$.

Solution   First change each complex number to $a + bi$ form:

$$-2 + \sqrt{-16} = -2 + \sqrt{16}\sqrt{-1}$$
$$= -2 + 4i$$
$$4 - \sqrt{-9} = 4 - \sqrt{9}\sqrt{-1}$$
$$= 4 - 3i$$

Then, find the product of $-2 + 4i$ and $4 - 3i$:

$$(-2 + 4i)(4 - 3i) = -8 + 6i + 16i - 12i^2$$
$$= -8 + 6i + 16i + 12$$
$$= 4 + 22i$$ ∎

The following definition is useful in finding the quotient of two complex numbers.

Definition.   The complex numbers $a + bi$ and $a - bi$ are called **conjugates** of each other.

The conjugate of $3 + 4i$ is $3 - 4i$, and the conjugate of $-\frac{1}{2} - 4i$ is $-\frac{1}{2} + 4i$.

Example 7   Write the number $\dfrac{3}{2 + i}$ in $a + bi$ form.

Solution   Multiply both the numerator and the denominator by the conjugate of the denominator and simplify:

$$\frac{3}{2 + i} = \frac{3}{2 + i} \cdot \frac{2 - i}{2 - i}$$
$$= \frac{6 - 3i}{4 - 2i + 2i - i^2}$$
$$= \frac{6 - 3i}{4 + 1}$$
$$= \frac{6}{5} - \frac{3}{5}i$$
$$= \frac{6}{5} + \left(-\frac{3}{5}\right)i$$

It is common practice to accept $\frac{6}{5} - \frac{3}{5}i$ as a substitute for $\frac{6}{5} + (-\frac{3}{5})i$. ∎

Example 8 Write the number $\dfrac{2 - \sqrt{-64}}{3 + \sqrt{-1}}$ in $a + bi$ form.

Solution $\dfrac{2 - \sqrt{-64}}{3 + \sqrt{-1}} = \dfrac{2 - 8i}{3 + i}$ Change each complex number to $a + bi$ form.

$$= \frac{2 - 8i}{3 + i} \cdot \frac{3 - i}{3 - i}$$

$$= \frac{6 - 2i - 24i + 8i^2}{9 - 3i + 3i - i^2}$$

$$= \frac{-2 - 26i}{9 + 1}$$

$$= \frac{2(-1 - 13i)}{10}$$

$$= -\frac{1}{5} - \frac{13}{5}i$$ ■

A generalization of the process involved in Examples 7 and 8 shows that the quotient of two complex numbers is another complex number.

The solutions of certain quadratic equations are complex numbers. The following example shows how the quadratic formula can be used to find these solutions.

Example 9 Use the quadratic formula to solve the quadratic equation $x^2 - 4x + 5 = 0$.

Solution Substitute 1 for a, -4 for b, and 5 for c in the quadratic formula, and simplify:

$$x = \frac{-b \pm \sqrt{b^2 - 4ac}}{2a}$$

$$= \frac{-(-4) \pm \sqrt{(-4)^2 - 4(1)(5)}}{2(1)}$$

$$= \frac{4 \pm \sqrt{16 - 20}}{2}$$

$$= \frac{4 \pm \sqrt{-4}}{2}$$

$$= \frac{4 \pm 2i}{2}$$

$$= 2 \pm i$$

Thus,

$$x = 2 + i \quad \text{or} \quad x = 2 - i$$ ■

Exercise 10.1

In Exercises 1–12, simplify each expression.

1. i^9

2. i^{27}

3. i^{38}

4. i^{99}

5. $\dfrac{1}{i^5}$

6. $\dfrac{1}{i^{10}}$

7. $\dfrac{1}{i^{99}}$

8. $\dfrac{1}{i^{1776}}$

9. i^{-1984}

10. i^{-1492}

11. i^{-1111}

12. i^{-2001}

In Exercises 13–18, solve for the real numbers x and y. You may have to solve a system of equations after using the definition of equality of complex numbers.

13. $x + (x + y)i = y - i$

14. $x + 3y + 2xi + yi = 5x + 2 + 2yi$

15. $x + iy = y + 2xi$

16. $x + y + (x - y)i = x + iy$

17. $3x - 2yi = 2 + (x + y)i$

18. $\begin{cases} 2 + (x + y)i = 2 - i \\ x + 3i = 2 + 3i \end{cases}$

In Exercises 19–46, perform any indicated operations, and express the final answer in a + bi form.

19. $(2 - 7i) + (3 + i)$

20. $(-7 + 2i) + (2 - 8i)$

21. $(5 - 6i) - (7 + 4i)$

22. $(11 + 2i) - (13 - 5i)$

23. $(14i + 2) + (2 - 4i)$

24. $(5 + 8i) - (23i - 32)$

25. $(2 + \sqrt{-9})(-3 - \sqrt{-16})$

26. $(-2 - \sqrt{-16})(4 + \sqrt{-25})$

27. $(-11 + \sqrt{-25})(-2 - \sqrt{-36})$

28. $(6 + \sqrt{-49})(6 - \sqrt{-49})$

29. $(\sqrt{-16} + 3)(2 + \sqrt{-9})$

30. $(12 - \sqrt{-4})(\sqrt{-25} + 7)$

31. $(2 + 3i)^2$

32. $(3 - 4i)^2$

33. $(2 - i)^3$

34. $(2 + 3i)^3$

35. $\dfrac{1}{2 + i}$

36. $\dfrac{-2}{3 - i}$

37. $\dfrac{3}{4i^2}$

38. $\dfrac{-11}{3i^3}$

39. $\dfrac{2i}{7 + i}$

40. $\dfrac{-3i}{2 + 5i}$

41. $\dfrac{2 + i}{3 - i}$

42. $\dfrac{4 - \sqrt{-25}}{2 + \sqrt{-9}}$

43. $\dfrac{-\sqrt{-16} + \sqrt{5}}{-8 + \sqrt{-4}}$

44. $\dfrac{34 + 2i}{\sqrt{2} - 4i}$

45. $\dfrac{2 + i\sqrt{3}}{3 + i}$

46. $\dfrac{3}{4 - i\sqrt{2}}$

47. Verify that $1 - i$ is a square root of $-2i$ by showing that $(1 - i)^2 = -2i$.

48. Verify that $2 - i$ is a square root of $3 - 4i$ by showing that $(2 - i)^2 = 3 - 4i$.

49. Verify that $1 + 2i$ is a square root of $-3 + 4i$ by showing that $(1 + 2i)^2 = -3 + 4i$.

50. Verify that $2 + i$ is a square root of $3 + 4i$ by showing that $(2 + i)^2 = 3 + 4i$.

51. Verify that $-\dfrac{1}{2} + \dfrac{\sqrt{3}}{2}i$ and $-\dfrac{1}{2} - \dfrac{\sqrt{3}}{2}i$ are roots of $x^2 + x + 1 = 0$.

52. Verify that $\dfrac{1}{2} + \dfrac{\sqrt{5}}{2}i$ and $\dfrac{1}{2} - \dfrac{\sqrt{5}}{2}i$ are not roots of $x^2 - x - 1 = 0$.

In Exercises 53–58, use the quadratic formula to solve each equation.

53. $x^2 + 2x + 2 = 0$

54. $x^2 + 4x + 8 = 0$

55. $x^2 + 4x + 5 = 0$

56. $x^2 + 2x + 5 = 0$

57. $9x^2 - 6x + 2 = 0$

58. $4x^2 - 4x + 5 = 0$

In Exercises 59–62, use a calculator to write each expression in $a + bi$ form.

59. $(2.3 + 4.5i)(8.9 - 3.2i)$

60. $(6.73 - 3.25i)^2 + (1.75 + 2.21i)$

61. $\dfrac{4.7 + 11.2i}{5.2 - 3.7i} - (6.5 - 7.4i)$

62. $\dfrac{29.8 - 45.3i}{-7.4 + 27.3i}$

63. Show that the product of the complex number $a + bi$ and its conjugate $a - bi$ is the real number $a^2 + b^2$.

64. Show that the quotient $\dfrac{a + bi}{c + di}$ is a complex number.

10.2   GRAPHING COMPLEX NUMBERS

The complex numbers share many of the properties of real numbers. However, there is no way of ordering the complex numbers that is consistent with the ordering established for the real numbers. For example, it makes no sense to say that $3i$ is larger than 5 or that $5 + 5i$ is greater than $-5 + i$.

Because the complex numbers are not linearly ordered, it is impossible to graph them on a number line in any way that preserves the ordering of the real numbers. However, there *is* a method for graphing the complex numbers. To graph complex numbers, we construct two perpendicular axes and consider one axis to be the axis of real numbers and the other to be the axis of imaginary numbers. (See Figure 10-1.) These axes determine what is called the **complex plane**. Although these axes resemble the x-axis and y-axis encountered in previous chapters, instead of plotting x- and y-values, we plot ordered pairs of real numbers (a, b), where a and b are the real and imaginary parts of the complex number $a + bi$. The axis of reals is used for plotting a and the axis of imaginaries is used for plotting b.

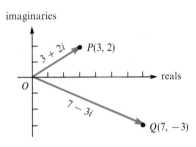

Figure 10-1

Example 1   Graph the complex number $3 + 2i$.

Solution   To graph the complex number $3 + 2i$, plot the point $P(3, 2)$ as in Figure 10-1. The vector drawn from the origin, point O, to point P is the graph of the complex number $3 + 2i$.   ■

Example 2   Graph the complex number $7 - 3i$.

Solution   To graph the complex number $7 - 3i$, plot the point $Q(7, -3)$ as in Figure 10-1. The vector **OQ** is the graph of the complex number $7 - 3i$.   ■

The length of the vector that represents a complex number is considered to be the **absolute value** of that complex number.

Absolute Value of a Complex Number.   If $a + bi$ is a complex number, then

$$|a + bi| = \sqrt{a^2 + b^2}$$

See Figure 10-2 and note that $\sqrt{a^2 + b^2}$ is the length of the hypotenuse of a right triangle with sides of length $|a|$ and $|b|$. On the complex plane, $|a + bi|$ is the distance between the point (a, b) and the origin $(0, 0)$. This is consistent with the definition of the absolute value of a real number. The expression $|x|$ represents the distance on the number line between the point with coordinate x and the origin O.

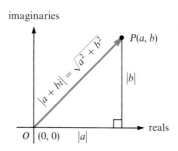

Figure 10-2

Example 3   Find the value of $|3 + 4i|$.

Solution $|3 + 4i| = \sqrt{3^2 + 4^2} = \sqrt{25} = 5$ ■

Example 4   Evaluate $|2 - 5i|$.

Solution $|2 - 5i| = \sqrt{2^2 + (-5)^2} = \sqrt{29}$ ■

Exercise 10.2

In Exercises 1–12, graph the given complex number.

1. $2 + 3i$

2. $2 - 3i$

3. $-2 + 3i$

4. $-2 - 3i$

5. $(4 + i)^2$

6. $(4 - i)^3$

7. $\dfrac{2 + i}{3 - i}$

8. $\dfrac{-2 + 2i}{-3 - i}$

9. 6

10. -8

11. $7i$

12. $-5i$

In Exercises 13–24, compute each absolute value.

13. $|2 + 3i|$

14. $|-5 - i|$

15. $|-7 + 7i|$

16. $\left|\dfrac{1}{2} - \dfrac{1}{4}i\right|$

17. $|6|$

18. $|-6|$

19. $|5i|$

20. $|-4i|$

21. $\left|\dfrac{-3i}{2 + i}\right|$

22. $\left|\dfrac{5i}{i - 2}\right|$

23. $\left|\dfrac{4 - i}{4 + i}\right|$

24. $|i^{365}|$

In Exercises 25–30, let $z = a + bi$. Assume that $\bar{z}$ represents the conjugate of $\bar{z}$, that is, $\bar{z} = a - bi$.

25. Show that $|z| = |\bar{z}|$.

26. Show that $|z| + |\bar{z}| = 2|z|$.

27. Show that $|z||\bar{z}| = |z|^2$.

28. Show that $|z + \bar{z}| = 2|a|$.

29. Show that $\sqrt{z\bar{z}} = |z|$.

30. Show that $z = \bar{z}$ if and only if z is a real number.

31. Show that the addition of two complex numbers is commutative. Do this by adding the complex numbers $a + bi$ and $c + di$ in both orders and observing that the sums are equal.

32. Show that the multiplication of two complex numbers is commutative. Do this by multiplying the complex numbers $a + bi$ and $c + di$ in both orders and observing that the products are equal.

33. Show that the addition of complex numbers is associative.

34. Find three examples of complex numbers that are reciprocals of their own conjugates.

In Exercises 35–36, let $x = a + bi$ and $y = c + di$. The bar over each symbol is read as "the conjugate of." For example, $\overline{x + y}$ means "the conjugate of the sum of x and y."

35. Show that $\overline{x + y} = \bar{x} + \bar{y}$.

36. Show that $\overline{xy} = \bar{x}\,\bar{y}$.

10.3   TRIGONOMETRIC FORM OF A COMPLEX NUMBER

If a complex number is written in the form $a + bi$, it is said to be written in **rectangular form**. It is also possible to write a complex number in *trigonometric form*. To do so, we refer to Figure 10-3, note that $a = r \cos \theta$ and $b = r \sin \theta$, and proceed as follows:

$$a + bi = r \cos \theta + r \sin \theta i \qquad \text{Substitute } r \cos \theta \text{ for } a \text{ and } r \sin \theta \text{ for } b.$$
$$= r(\cos \theta + i \sin \theta) \qquad \text{Factor out } r.$$

The expression $r(\cos \theta + i \sin \theta)$ is called the **trigonometric form** of the complex number $a + bi$.

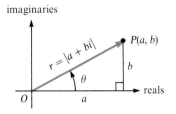

Figure 10-3

If $r = |a + bi|$ and θ is any angle for which $\sin \theta = \frac{b}{r}$ and $\cos \theta = \frac{a}{r}$, then

$$a + bi = r(\cos \theta + i \sin \theta)$$

If a complex number is written in $r(\cos \theta + i \sin \theta)$ form, the number r is called the **modulus**, and the angle θ is called the **argument**. The notation $r(\cos \theta + i \sin \theta)$ is often abbreviated as r cis θ. Thus

$$r \text{ cis } \theta = r(\cos \theta + i \sin \theta)$$

Example 1 Write the complex number $3 + 4i$ in trigonometric form.

Solution To write $3 + 4i$ in the form $r(\cos \theta + i \sin \theta)$, you must find values for r and θ. To do this, graph the complex number as in Figure 10-4. Because triangle OAP is a right triangle, you have $r^2 = 3^2 + 4^2$ or $r = 5$. From the figure, it follows that $\tan \theta = \frac{4}{3}$, so $\theta \approx 53.1°$. Substituting these values for r and θ gives

$$3 + 4i \approx 5(\cos 53.1° + i \sin 53.1°)$$

Note that $5(\cos 53.1° + i \sin 53.1°)$ can be written as 5 cis $53.1°$.

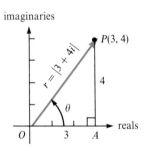

Figure 10-4

Example 2 Write the complex number $-2 - 7i$ in trigonometric form.

Solution To write $-2 - 7i$ in the form $r(\cos \theta + i \sin \theta)$, the values of r and θ must be found. Graph the number $-2 - 7i$ as in Figure 10-5 and calculate r:

$$r = |-2 - 7i| = \sqrt{(-2)^2 + (-7)^2} = \sqrt{4 + 49} = \sqrt{53}$$

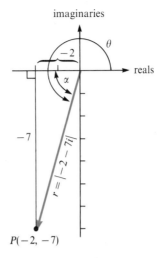

Figure 10-5

Since $\tan \alpha = \frac{7}{2}$, it follows that $\alpha \approx 74.1°$. Find θ by noting that θ is a third-quadrant angle, so that

$$\theta = 180° + \alpha \approx 180° + 74.1° = 254.1°$$

Thus,

$$-2 - 7i \approx \sqrt{53}(\cos 254.1° + i \sin 254.1°)$$ ∎

Example 3   Change the complex number $10(\cos 60° + i \sin 60°)$ to $a + bi$ form.

Solution
$$10(\cos 60° + i \sin 60°) = 10\left(\frac{1}{2} + i\frac{\sqrt{3}}{2}\right)$$
$$= 5 + 5\sqrt{3}\,i$$ ∎

It is easy to add two complex numbers in $a + bi$ form—simply add their real parts and add their imaginary parts. However, it is difficult to multiply and divide complex numbers written in $a + bi$ form. Fortunately, it is easy to multiply and divide complex numbers if they are written in trigonometric form.

Suppose that $N_1 = r_1(\cos \theta_1 + i \sin \theta_1)$ and $N_2 = r_2(\cos \theta_2 + i \sin \theta_2)$. The product N_1N_2 can be found as follows:

$$N_1N_2 = r_1(\cos \theta_1 + i \sin \theta_1) \cdot r_2(\cos \theta_2 + i \sin \theta_2)$$
$$= r_1r_2(\cos \theta_1 \cos \theta_2 + i \cos \theta_1 \sin \theta_2 + i \sin \theta_1 \cos \theta_2 - \sin \theta_1 \sin \theta_2)$$
$$= r_1r_2[(\cos \theta_1 \cos \theta_2 - \sin \theta_1 \sin \theta_2) + i(\cos \theta_1 \sin \theta_2 + \sin \theta_1 \cos \theta_2)]$$
$$= r_1r_2[\cos(\theta_1 + \theta_2) + i \sin(\theta_1 + \theta_2)]$$

This result can be stated more compactly by using r cis θ notation:

$$r_1 \text{ cis } \theta_1 \cdot r_2 \text{ cis } \theta_2 = r_1r_2 \text{ cis}(\theta_1 + \theta_2)$$

Thus, to form a product of two complex numbers written in trigonometric form, multiply the moduli r_1 and r_2 and add the arguments θ_1 and θ_2. This property generalizes to products containing any number of factors:

$$r_1(\cos \theta_1 + i \sin \theta_1) \cdot r_2(\cos \theta_2 + i \sin \theta_2) \cdots \cdot r_n(\cos \theta_n + i \sin \theta_n)$$

$$= r_1 r_2 \cdots \cdot r_n [\cos(\theta_1 + \theta_2 + \cdots + \theta_n) + i \sin(\theta_1 + \theta_2 + \cdots + \theta_n)]$$

The previous equation can be written more compactly as

$$r_1 \text{ cis } \theta_1 \cdot r_2 \text{ cis } \theta_2 \cdots \cdot r_n \text{ cis } \theta_n = r_1 r_2 \cdots \cdot r_n \text{ cis}(\theta_1 + \theta_2 + \cdots + \theta_n)$$

Example 4   Find the product of $2(\cos 40° + i \sin 40°)$ and $3(\cos 30° + i \sin 30°)$.

Solution   To find the product, multiply the moduli and add the arguments:

$$2(\cos 40° + i \sin 40°) \cdot 3(\cos 30° + i \sin 30°)$$
$$= 2 \cdot 3[\cos(40° + 30°) + i \sin(40° + 30°)]$$
$$= 6(\cos 70° + i \sin 70°)$$ ∎

Example 5   Find the product of $2 \text{ cis } 10°$, $4 \text{ cis } 20°$, and $9 \text{ cis } 50°$.

Solution   To find the product, multiply the moduli and add the arguments:

$$(2 \text{ cis } 10°)(4 \text{ cis } 20°)(9 \text{ cis } 50°) = 2 \cdot 4 \cdot 9 \text{ cis}(10° + 20° + 50°)$$
$$= 72 \text{ cis } 80°$$ ∎

If two complex numbers are written in trigonometric form, their quotient is found by dividing their moduli and subtracting their arguments. To show that this is true, we let $N_1 = r_1(\cos \theta_1 + i \sin \theta_1)$ and $N_2 = r_2(\cos \theta_2 + i \sin \theta_2)$. Then we have

$$\frac{N_1}{N_2} = \frac{r_1(\cos \theta_1 + i \sin \theta_1)}{r_2(\cos \theta_2 + i \sin \theta_2)}$$

$$= \frac{r_1}{r_2} \cdot \frac{\cos \theta_1 + i \sin \theta_1}{\cos \theta_2 + i \sin \theta_2} \cdot \frac{\cos \theta_2 - i \sin \theta_2}{\cos \theta_2 - i \sin \theta_2}$$

$$= \frac{r_1}{r_2} \cdot \frac{\cos \theta_1 \cos \theta_2 - i \cos \theta_1 \sin \theta_2 + i \sin \theta_1 \cos \theta_2 - i^2 \sin \theta_1 \sin \theta_2}{\cos^2 \theta_2 - i^2 \sin^2 \theta_2}$$

$$= \frac{r_1}{r_2} \cdot \frac{\cos \theta_1 \cos \theta_2 + \sin \theta_1 \sin \theta_2 + i(\sin \theta_1 \cos \theta_2 - \cos \theta_1 \sin \theta_2)}{\cos^2 \theta_2 + \sin^2 \theta_2}$$

$$= \frac{r_1}{r_2} \cdot \frac{\cos(\theta_1 - \theta_2) + i \sin(\theta_1 - \theta_2)}{1}$$

$$= \frac{r_1}{r_2} [\cos(\theta_1 - \theta_2) + i \sin(\theta_1 - \theta_2)]$$

Thus, to divide complex numbers written in trigonometric form, we use the following rule:

$$\frac{r_1(\cos \theta_1 + i \sin \theta_1)}{r_2(\cos \theta_2 + i \sin \theta_2)} = \frac{r_1}{r_2} [\cos(\theta_1 - \theta_2) + i \sin(\theta_1 - \theta_2)]$$

or in r cis θ notation,

$$\frac{r_1 \text{ cis } \theta_1}{r_2 \text{ cis } \theta_2} = \frac{r_1}{r_2} \text{cis}(\theta_1 - \theta_2)$$

Example 6   Divide 8 cis 110° by 4 cis 50°.

Solution   Use the division rule:

$$\frac{8 \text{ cis } 110°}{4 \text{ cis } 50°} = \frac{8}{4} \text{cis}(110° - 50°)$$
$$= 2 \text{ cis } 60°$$

Note that 2 cis 60° can be written in the form 2(cos 60° + i sin 60°).   ∎

Exercise 10.3

In Exercises 1–12, write the given complex number in trigonometric form.

1. $6 + 0i$	**2.** $-7 + 0i$	**3.** $0 - 3i$	**4.** $0 + 4i$
5. $-1 - i$	**6.** $-1 + i$	**7.** $3 + 3i\sqrt{3}$	**8.** $7 - 7i$
9. $-1 - \sqrt{3}\,i$	**10.** $-3\sqrt{3} + 3i$	**11.** $-\sqrt{3} - i$	**12.** $1 + i\sqrt{3}$

In Exercises 13–24, write the given complex number in $a + bi$ form.

13. $2(\cos 30° + i \sin 30°)$

14. $5(\cos 45° + i \sin 45°)$

15. $7(\cos 90° + i \sin 90°)$

16. $12(\cos 0° + i \sin 0°)$

17. $-2\left(\cos \dfrac{2\pi}{3} + i \sin \dfrac{2\pi}{3}\right)$

18. $3\left(\cos \dfrac{4\pi}{3} + i \sin \dfrac{4\pi}{3}\right)$

19. $\dfrac{1}{2}(\cos \pi + i \sin \pi)$

20. $-\dfrac{2}{3}(\cos 2\pi + i \sin 2\pi)$

21. -3 cis $225°$

22. 3 cis $300°$

23. 11 cis $\dfrac{11\pi}{6}$

24. 9 cis 3

In Exercises 25–36, find each product.

25. $[4(\cos 30° + i \sin 30°)][2(\cos 60° + i \sin 60°)]$

26. $[3(\cos 45° + i \sin 45°)][2(\cos 120° + i \sin 120°)]$

27. $(\cos 300° + i \sin 300°)(\cos 0° + i \sin 0°)$

28. $[5(\cos 85° + i \sin 85°)][2(\cos 65° + i \sin 65°)]$

29. $[2(\cos \pi + i \sin \pi)][3(\cos \pi + i \sin \pi)]$

30. $\left(\cos \dfrac{\pi}{2} + i \sin \dfrac{\pi}{2}\right)\left(\cos \dfrac{3\pi}{2} + i \sin \dfrac{3\pi}{2}\right)$

31. $\left[2\left(\cos \dfrac{\pi}{3} + i \sin \dfrac{\pi}{3}\right)\right]\left[3\left(\cos \dfrac{\pi}{6} + i \sin \dfrac{\pi}{6}\right)\right]$

32. $\left[3\left(\cos\dfrac{5\pi}{6} + i\sin\dfrac{5\pi}{6}\right)\right]\left[4\left(\cos\dfrac{7\pi}{6} + i\sin\dfrac{7\pi}{6}\right)\right]$

33. $(3 \text{ cis } 12°)(2 \text{ cis } 22°)(5 \text{ cis } 82°)$

34. $(2 \text{ cis } 50°)(3 \text{ cis } 100°)(6 \text{ cis } 2°)$

35. $\left(3 \text{ cis }\dfrac{\pi}{2}\right)\left(4 \text{ cis }\dfrac{\pi}{3}\right)\left(3 \text{ cis }\dfrac{\pi}{4}\right)$

36. $\left(4 \text{ cis }\dfrac{\pi}{6}\right)\left(2 \text{ cis }\dfrac{2\pi}{3}\right)\left(\text{cis }\dfrac{\pi}{4}\right)$

In Exercises 37–44, find each quotient.

37. $\dfrac{12(\cos 60° + i\sin 60°)}{2(\cos 30° + i\sin 30°)}$

38. $\dfrac{24(\cos 150° + i\sin 150°)}{48(\cos 50° + i\sin 50°)}$

39. $\dfrac{18(\cos \pi + i\sin \pi)}{12\left(\cos \dfrac{\pi}{2} + i\sin \dfrac{\pi}{2}\right)}$

40. $\dfrac{15(\cos 2\pi + i\sin 2\pi)}{45(\cos \pi + i\sin \pi)}$

41. $\dfrac{12 \text{ cis } 250°}{5 \text{ cis } 120°}$

42. $\dfrac{365 \text{ cis } 370°}{20 \text{ cis } 255°}$

43. $\dfrac{\text{cis }\dfrac{2\pi}{3}}{2 \text{ cis }\dfrac{\pi}{6}}$

44. $\dfrac{250 \text{ cis }\dfrac{7\pi}{16}}{50 \text{ cis }\dfrac{\pi}{3}}$

In Exercises 45–48, simplify each expression.

45. $\dfrac{(2 \text{ cis } 60°)(3 \text{ cis } 20°)}{6 \text{ cis } 40°}$

46. $\dfrac{36 \text{ cis } 200°}{(2 \text{ cis } 40°)(9 \text{ cis } 10°)}$

47. $\dfrac{48 \text{ cis }\dfrac{11\pi}{6}}{\left(3 \text{ cis }\dfrac{\pi}{3}\right)\left(4 \text{ cis }\dfrac{2\pi}{3}\right)}$

48. $\dfrac{(96 \text{ cis } \pi)(12 \text{ cis } 2\pi)}{\left(48 \text{ cis }\dfrac{\pi}{2}\right)\left(3 \text{ cis }\dfrac{3\pi}{2}\right)}$

10.4 DE MOIVRE'S THEOREM

If two complex numbers are written in trigonometric form, their product is found by multiplying their moduli and adding their arguments. This fact makes it easy to find powers of complex numbers that are expressed in trigonometric form. For example, to find the cube of $3(\cos 40° + i\sin 40°)$, we proceed as follows:

$$[3(\cos 40° + i\sin 40°)]^3$$
$$= [3(\cos 40° + i\sin 40°)][3(\cos 40° + i\sin 40°)][3(\cos 40° + i\sin 40°)]$$
$$= 3^3[\cos(40° + 40° + 40°) + i\sin(40° + 40° + 40°)]$$
$$= 27[\cos 3(40°) + i\sin 3(40°)]$$
$$= 27(\cos 120° + i\sin 120°)$$

The generalization of the previous example is called **De Moivre's theorem.**

> **De Moivre's Theorem.**   If n is a real number and $r (\cos \theta + i \sin \theta)$ is a complex number in trigonometric form, then
>
> $$[r(\cos \theta + i \sin \theta)]^n = r^n[\cos n\theta + i \sin n\theta]$$
>
> or
>
> $$(r \text{ cis } \theta)^n = r^n \text{ cis } n\theta$$

This theorem was first developed about 1730 by the French mathematician Abraham De Moivre.

Example 1   Find $[2(\cos 15° + i \sin 15°)]^4$.

Solution   Use De Moivre's theorem:

$$[2(\cos 15° + i \sin 15°)]^4 = 2^4(\cos 4 \cdot 15° + i \sin 4 \cdot 15°)$$
$$= 16(\cos 60° + i \sin 60°)$$

This result can be changed to $a + bi$ form if desired:

$$16(\cos 60° + i \sin 60°) = 16\left(\frac{1}{2} + i\frac{\sqrt{3}}{2}\right) = 8 + 8\sqrt{3}\,i \qquad \blacksquare$$

Example 2   Find $[\sqrt{2}(\cos 10° + i \sin 10°)]^{10}$.

Solution   Use De Moivre's theorem:

$$[\sqrt{2}(\cos 10° + i \sin 10°]^{10} = (\sqrt{2})^{10}(\cos 10 \cdot 10° + i \sin 10 \cdot 10°)$$
$$= 32(\cos 100° + i \sin 100°) \qquad \blacksquare$$

De Moivre's theorem can be used to find all of the nth roots of any number. Because both real and complex numbers can be written in the form $a + bi$, they can be written in trigonometric form as well. Thus, one nth root of $a + bi$ is

$$\sqrt[n]{a + bi} = (a + bi)^{\frac{1}{n}}$$
$$= [r(\cos \theta + i \sin \theta)]^{\frac{1}{n}}$$
$$= \sqrt[n]{r}\left(\cos \frac{\theta}{n} + i \sin \frac{\theta}{n}\right)$$

Recall that the equation $x^2 = 9$ has two distinct roots, 3 and -3, and each qualifies as a square root of 9. In like manner, the equation $x^n = a + bi$ has n distinct roots, and each qualifies as an nth root of the complex number $a + bi$. It follows that there are n distinct nth roots of any complex number.

Because $\sin \theta = \sin(\theta + k \cdot 360°)$ and $\cos \theta = \cos(\theta + k \cdot 360°)$ for all integers k, De Moivre's theorem implies that

$$[r(\cos \theta + i \sin \theta)]^{\frac{1}{n}} = \{r[\cos(\theta + k \cdot 360°) + i \sin(\theta + k \cdot 360°)]\}^{\frac{1}{n}}$$

$$= r^{\frac{1}{n}}\left(\cos \frac{\theta + k \cdot 360°}{n} + i \sin \frac{\theta + k \cdot 360°}{n}\right)$$

Substituting the numbers $0, 1, 2, \ldots, (n-1)$ for k yields the n nth roots of the given complex number.

Example 3   Find the three cube roots of 8.

Solution   Because 8 can be expressed as $8 + 0i$, graph the complex number $8 + 0i$ as in Figure 10-6 to see that $r = 8$ and $\theta = 0°$. Write $8 + 0i$ in trigonometric form and use the equation

$$[r(\cos \theta + i \sin \theta)]^{\frac{1}{n}} = r^{\frac{1}{n}}\left[\cos \frac{\theta + k \cdot 360°}{n} + i \sin \frac{\theta + k \cdot 360°}{n}\right]$$

imaginaries

reals

$P(8, 0°)$

Figure 10-6

Substituting the values for n, r, and θ gives

$$[8(\cos 0° + i \sin 0°)]^{\frac{1}{3}} = 8^{\frac{1}{3}}\left[\cos \frac{0° + k \cdot 360°}{3} + i \sin \frac{0° + k \cdot 360°}{3}\right]$$

Substituting 0 for k and replacing $8^{\frac{1}{3}}$ with 2 gives

$$2\left(\cos \frac{0°}{3} + i \sin \frac{0°}{3}\right) = 2(\cos 0° + i \sin 0°)$$

$$= 2(1 + 0i)$$

$$= 2$$

Now substitute 1 for k:

$$2\left(\cos \frac{0° + 360°}{3} + i \sin \frac{0° + 360°}{3}\right) = 2(\cos 120° + i \sin 120°)$$

$$= 2\left(-\frac{1}{2} + i \frac{\sqrt{3}}{2}\right)$$

$$= -1 + i\sqrt{3}$$

Finally, substitute 2 for k:

$$2\left(\cos\frac{0° + 720°}{3} + i\sin\frac{0° + 720°}{3}\right) = 2(\cos 240° + i\sin 240°)$$

$$= 2\left(-\frac{1}{2} + i\frac{-\sqrt{3}}{2}\right)$$

$$= -1 - i\sqrt{3}$$

The numbers 2, $-1 + i\sqrt{3}$, and $-1 - i\sqrt{3}$ are the three cube roots of 8. If these three cube roots of 8 are graphed in the complex plane, they are equally spaced around a circle of radius 2. If the endpoints of these vectors are joined by straight line segments, an equilateral triangle is formed, as in Figure 10-7.

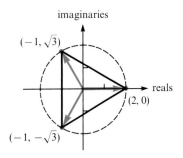

Figure 10-7

Example 4 Find the four fourth roots of $-16i$.

Solution Express $-16i$ as $0 - 16i$ and graph it to determine that $r = 16$ and $\theta = 270°$. Hence,

$$\sqrt[4]{-16i} = [16(\cos 270° + i\sin 270°)]^{\frac{1}{4}}$$

$$= 16^{\frac{1}{4}}\left(\cos\frac{270° + k\cdot 360°}{4} + i\sin\frac{270° + k\cdot 360°}{4}\right)$$

Substitute 0, 1, 2, and 3 for k to get the four fourth roots of $-16i$:

$$2(\cos\ 67.5° + i\sin\ 67.5°) \approx\ \ \ 0.77 + 1.85i$$
$$2(\cos 157.5° + i\sin 157.5°) \approx -1.85 + 0.77i$$
$$2(\cos 247.5° + i\sin 247.5°) \approx -0.77 - 1.85i$$
$$2(\cos 337.5° + i\sin 337.5°) \approx\ \ \ 1.85 - 0.77i$$

Graphs of these four fourth roots are equally spaced about a circle of radius 2, and the endpoints of these vectors are the vertices of a square. See Figure 10-8.

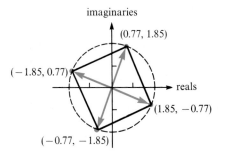

Figure 10-8

Example 5   Find the five fifth roots of $-4 - 4i$.

Solution   Express $-4 - 4i$ in trigonometric form by determining that $r = 4\sqrt{2}$ and $\theta = 225°$. Then,

$$[4\sqrt{2}(\cos 225° + i \sin 225°)]^{\frac{1}{5}}$$
$$= (4\sqrt{2})^{\frac{1}{5}}\left(\cos\frac{225° + k \cdot 360°}{5} + i \sin\frac{225° + k \cdot 360°}{5}\right)$$
$$= \sqrt{2}\left(\cos\frac{225° + k \cdot 360°}{5} + i \sin\frac{225° + k \cdot 360°}{5}\right)$$

Substituting 0, 1, 2, 3, and 4 for k generates the five fifth roots of $-4 - 4i$:

$$\sqrt{2}(\cos\ \ 45° + i \sin\ \ 45°)$$
$$\sqrt{2}(\cos 117° + i \sin 117°)$$
$$\sqrt{2}(\cos 189° + i \sin 189°)$$
$$\sqrt{2}(\cos 261° + i \sin 261°)$$
$$\sqrt{2}(\cos 333° + i \sin 333°)$$

If these five fifth roots are graphed, they are equally spaced about a circle with radius $\sqrt{2}$, and the endpoints of these vectors are the vertices of a regular pentagon. See Figure 10-9.

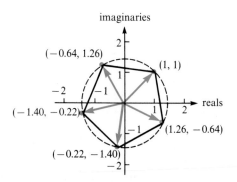

Figure 10-9

If $n > 2$ and the n nth roots of a complex number are graphed on the complex plane, the endpoints of the vectors that represent each root will always be at the vertices of a regular polygon.   ∎

Exercise 10.4

In Exercises 1–12, find the indicated power. Leave all answers in trigonometric form.

1. $[3(\cos 30° + i \sin 30°)]^3$

2. $[4(\cos 15° + i \sin 15°)]^6$

3. $(\cos 15° + i \sin 15°)^{12}$

4. $[2(\cos 120° + i \sin 120°)]^6$

5. $(5 \text{ cis } 2°)^5$

6. $(0.5 \text{ cis } 100°)^3$

7. $\left[3\left(\cos \dfrac{\pi}{4} + i \sin \dfrac{\pi}{4} \right) \right]^4$

8. $\left[2\left(\cos \dfrac{3\pi}{2} + i \sin \dfrac{3\pi}{2} \right) \right]^6$

9. $[4(\cos 3 + i \sin 3)]^4$

10. $[2(\cos 5 + i \sin 5)]^{20}$

11. $\left(\dfrac{1}{3} \text{ cis } \dfrac{\pi}{2} \right)^3$

12. $\left(\dfrac{1}{2} \text{ cis } \dfrac{\pi}{6} \right)^5$

In Exercises 13–18, find the indicated nth root of each expression in $a + bi$ form.

13. A cube root of $8(\cos 180° + i \sin 180°)$

14. A fifth root of $32(\cos 150° + i \sin 150°)$

15. A fifth root of $(\cos 300° + i \sin 300°)$

16. A fourth root of $64(\cos \pi + i \sin \pi)$

17. A sixth root of $64(\cos 2\pi + i \sin 2\pi)$

18. A sixth root of $3^6(\cos \pi + i \sin \pi)$

In Exercises 19–26, find and graph the indicated roots of each complex number. Change answers to $a + bi$ form only if that answer is exact. Otherwise, leave the answers in trigonometric form.

19. The three cube roots of -8

20. The four fourth roots of 16

21. The two square roots of i

22. The three cube roots of i

23. The five fifth roots of $-i$

24. The three cube roots of $\dfrac{\sqrt{2}}{2} + \dfrac{\sqrt{2}}{2} i$

25. The four fourth roots of $-8 + 8\sqrt{3}i$

26. The six sixth roots of $-i$

In Exercises 27–30, substitute the given value of n into De Moivre's theorem with $r = 1$, and raise the binomial on the left to the nth power. Follow the additional directions.

27. $n = 2$. Set the real parts of the complex numbers equal to each other and thereby show that $\cos 2\theta = \cos^2 \theta - \sin^2 \theta$.

28. $n = 2$. Set the imaginary parts of the complex numbers equal to each other and thereby show that $\sin 2\theta = 2 \cos \theta \sin \theta$.

29. $n = 3$. Set the imaginary parts of the complex numbers equal to each other and thereby show that $\sin 3\theta = 3 \cos^2 \theta \sin \theta - \sin^3 \theta$.

30. $n = 3$. Set the real parts of the complex numbers equal to each other and thereby show that $\cos 3\theta = \cos^2 \theta - 3 \cos \theta \sin^2 \theta$.

31. Use the right side of the identity in Exercise 29 to show that $\sin 3\theta = 3 \sin \theta - 4 \sin^3 \theta$.

32. Use the right side of the identity in Exercise 30 to show that $\cos 3\theta = 4 \cos^3 \theta - 3 \cos \theta$.

10.5 POLAR COORDINATES

Some equations such as $(x^2 + y^2)^{\frac{3}{2}} = x$ are difficult to graph using the x- and y-coordinates of the Cartesian rectangular coordinate system. However, these equations can often be written in a form using the variables r (a radius) and θ (an angle). These coordinates enable us to graph such equations on an alternative coordinate system called the **polar coordinate system**.

The polar coordinate system is based on a ray called the **polar axis** and its source called the **pole**. In Figure 10-10, ray OA is the polar axis and point O is the pole. Any point $P(r, \theta)$ in the plane can be located if the length of a radius and an angle in standard position are known. For example, the polar coordinates $(10, 30°)$ determine the position of point R. If θ is in radians, the coordinates $(5, \frac{2\pi}{3})$ determine the point Q. In Figure 10-11, point L is determined by the coordinates $(7, 225°)$ and point M by the coordinates $(6, \frac{5\pi}{3})$. Note that point M is also determined by many other sets of polar coordinates, such as $(6, 660°)$ or $(6, -60°)$. Any set of polar coordinates locates a single point. However, any point has infinitely many sets of polar coordinates.

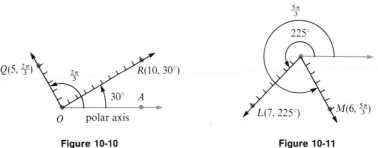

Figure 10-10   Figure 10-11

To plot point P with coordinates (r, θ) when r is positive, we draw angle θ in standard position and count r units along the terminal side of θ. See Figure 10-12. To plot point Q with coordinates (r, θ) when r is negative, we draw angle θ in standard position and count $|r|$ units along the extension through the origin of the terminal side of θ. For example, to graph the point $P(-2, \frac{\pi}{6})$, we first draw an angle of $\frac{\pi}{6}$ in standard position, as in Figure 10-13. We then draw the extension of ray OC in the opposite direction to obtain ray OB and count 2 units along ray OB to find point P. The graphs of the three points $R(5, \pi)$, $Q(-6, 100°)$, and $P(5, -30°)$ are shown in Figure 10-14.

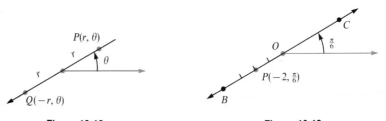

Figure 10-12   Figure 10-13

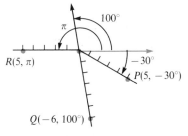

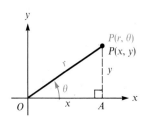

Figure 10-14   Figure 10-15

There is a relationship between the rectangular coordinates (x, y) and the polar coordinates (r, θ) of a point. Suppose point P in Figure 10-15 has rectangular coordinates (x, y) and polar coordinates (r, θ). Draw PA perpendicular to the x-axis to form right triangle OAP with $OA = x$, $AP = y$, and $OP = r$. Because angle θ is in standard position, the hypotenuse OP is the terminal side of angle θ. Thus,

$$\cos \theta = \frac{x}{r} \quad \text{and} \quad \sin \theta = \frac{y}{r}$$

or

$$x = r \cos \theta \quad \text{and} \quad y = r \sin \theta$$

These formulas can be used to convert from polar coordinates to rectangular coordinates.

Formulas to Convert from Polar to Rectangular Coordinates.

$$x = r \cos \theta$$
$$y = r \sin \theta$$

Example 1 If the polar coordinates of point P are $(10, 150°)$, find the rectangular coordinates.

Solution
$$x = r \cos \theta = 10 \cos(150°) = 10\left(\frac{-\sqrt{3}}{2}\right) = -5\sqrt{3}$$

$$y = r \sin \theta = 10 \sin(150°) = 10\left(\frac{1}{2}\right) = 5$$

The rectangular coordinates of point P are $(-5\sqrt{3}, 5)$. ■

To find polar coordinates from the rectangular coordinates of point P, again refer to Figure 10-15. From the right triangle OAP, it follows that

$$r^2 = x^2 + y^2 \quad \text{and} \quad \tan \theta = \frac{y}{x}$$

To find possible polar coordinates for point P, let $r = \sqrt{x^2 + y^2}$. Then, find an angle θ $\left(\text{equal to } \tan^{-1} \dfrac{y}{x} \right)$ whose terminal side passes through the point (x, y). That is, if x is negative, for example, and y is positive, choose θ to be a second-quadrant angle. If x and y are both negative, choose θ to be a third-quadrant angle.

Formulas to Convert from Rectangular to Polar Coordinates.

$$r = \sqrt{x^2 + y^2}$$

$$\theta = \tan^{-1} \frac{y}{x}$$

where the terminal side of θ passes through the point (x, y). If $x = 0$, choose θ to be $90°$ (if $y > 0$) or $270°$ (if $y < 0$).

Example 2   If the rectangular coordinates of point P are $(4, 3)$, find polar coordinates for P.

Solution   Refer to Figure 10-16. First, find the r-coordinate of point P as follows:

$$r = \sqrt{x^2 + y^2}$$
$$= \sqrt{4^2 + 3^2}$$
$$= 5$$

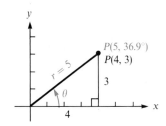

Figure 10-16

Second, determine the θ-coordinate of point P. Because x and y are both positive, θ is a first-quadrant angle:

$$\theta = \tan^{-1} \frac{y}{x}$$

$$= \tan^{-1} \frac{3}{4}$$

$$\approx 36.9°$$

One possible choice of polar coordinates for point P is $(5, 36.9°)$, where θ is rounded to the nearest tenth of a degree.   ■

Example 3 Change the rectangular coordinates $(-1, -4)$ to polar coordinates.

Solution First, find the r-coordinate of point P:

$$r = \sqrt{x^2 + y^2}$$
$$= \sqrt{(-1)^2 + (-4)^2}$$
$$= \sqrt{17}$$

Then, determine the θ-coordinate of point P. Because x and y are both negative, θ is a third-quadrant angle:

$$\theta = \tan^{-1}\frac{y}{x}$$
$$= \tan^{-1}\left(\frac{-4}{-1}\right)$$
$$= \tan^{-1} 4$$
$$\approx 256.0° \qquad \text{Note that } 256.0° = 76.0° + 180°.$$

One possible choice of polar coordinates for point P is $(\sqrt{17}, 256.0°)$, where θ is rounded to the nearest tenth of a degree.

There are other possible choices for the polar coordinates of P. One is $(-\sqrt{17}, 76.0°)$, which has a *negative* value of r. See Figure 10-17.

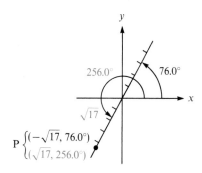

Figure 10-17

Equations involving the variables r and θ can be graphed in a polar coordinate system.

Example 4 Graph the polar equation $r = \theta$.

Solution Make a table of values, plot the points, and join them with a smooth curve as in Figure 10-18. As θ increases, r increases; the graph is a spiral called an **Archimedean spiral**.

$r = \theta$

θ	r
0	0
$\frac{\pi}{6}$	0.52
$\frac{\pi}{3}$	1.05
$\frac{\pi}{2}$	1.57
$\frac{2\pi}{3}$	2.09
$\frac{5\pi}{6}$	2.62
π	3.14
$\frac{3\pi}{2}$	4.71

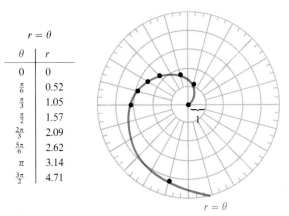

$r = \theta$

Figure 10-18

Example 5   Change the rectangular equation $x^2 + y^2 = y$ to polar coordinates and graph the curve.

Solution   Substitute r^2 for $x^2 + y^2$ and $r \sin \theta$ for y in the original equation:

$$x^2 + y^2 = y$$
$$r^2 = r \sin \theta$$

If $r = 0$, the graph is the point at the pole for all θ. If $r \neq 0$, you can divide both sides of the equation by r to obtain the equation

$$r = \sin \theta$$

Make a table of values, plot the points, and graph as in Figure 10-19. The graph is a circle with center at $(\frac{1}{2}, \frac{\pi}{2})$ and with radius of $\frac{1}{2}$. You will be asked in an exercise to graph $x^2 + y^2 = y$ on a set of rectangular coordinate axes to verify that the graph is the described circle.

$r = \sin \theta$

θ	r
0	0
$\frac{\pi}{6}$	0.5
$\frac{\pi}{3}$	0.87
$\frac{\pi}{2}$	1
$\frac{2\pi}{3}$	0.87
$\frac{5\pi}{6}$	0.5
π	0
$\frac{7\pi}{6}$	-0.5
$\frac{3\pi}{2}$	-1

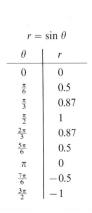
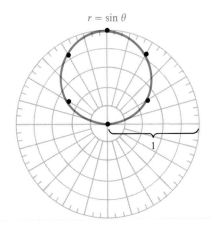

$r = \sin \theta$

Figure 10-19

Example 6 Change the rectangular equation $(x^2 + y^2)^{\frac{3}{2}} = x$ to an equation having variables of r and θ. Then graph the curve using polar coordinates.

Solution Because $x^2 + y^2 = r^2$ and $x = r \cos \theta$, you have

$$(x^2 + y^2)^{\frac{3}{2}} = x$$
$$r^3 = r \cos \theta$$

If $r = 0$, the graph is the pole for all θ. If $r \neq 0$, you can divide both sides by r and obtain

$$r^2 = \cos \theta$$
$$r = \pm\sqrt{\cos \theta}$$

Make a table of values and plot the points, as in Figure 10-20. Because r^2 is positive, $\cos \theta$ must be positive. Therefore, θ is in QI or QIV.

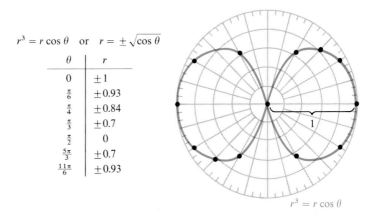

$r^3 = r \cos \theta$ or $r = \pm\sqrt{\cos \theta}$

θ	r
0	± 1
$\frac{\pi}{6}$	± 0.93
$\frac{\pi}{4}$	± 0.84
$\frac{\pi}{3}$	± 0.7
$\frac{\pi}{2}$	0
$\frac{5\pi}{3}$	± 0.7
$\frac{11\pi}{6}$	± 0.93

$r^3 = r \cos \theta$

Figure 10-20 ◼

Example 7 Change the polar equation $r(3 \cos \theta + 2 \sin \theta) = 7$ to an equation with rectangular coordinates.

Solution Use the distributive law to remove parentheses:

$$r(3 \cos \theta + 2 \sin \theta) = 7$$
$$3r \cos \theta + 2r \sin \theta = 7$$

Because $r \cos \theta = x$ and $r \sin \theta = y$, you can write this equation as

$$3x + 2y = 7$$

This is the equation of a line that can be graphed easily. ◼

Example 8 Change the polar equation $r^2 = 9 \cos 2\theta$ to an equation with rectangular coordinates.

Solution   Recall that $\cos 2\theta = \cos^2 \theta - \sin^2 \theta$, and substitute $\cos^2 \theta - \sin^2 \theta$ for $\cos 2\theta$ in the original equation:

$$r^2 = 9 \cos 2\theta$$
$$= 9(\cos^2 \theta - \sin^2 \theta)$$
$$= 9 \cos^2 \theta - 9 \sin^2 \theta$$

Multiply both sides by r^2 to obtain

$$r^4 = 9r^2 \cos^2 \theta - 9r^2 \sin^2 \theta$$
$$= 9(r \cos \theta)^2 - 9(r \sin \theta)^2$$
$$(x^2 + y^2)^2 = 9x^2 - 9y^2 \qquad \blacksquare$$

Example 9   Change the rectangular equation $(x^2 + y^2)^3 = 8x^2y^2$ to an equation with polar coordinates.

Solution   Substitute $r \cos \theta$ for x, $r \sin \theta$ for y, and r^2 for $x^2 + y^2$ in the rectangular equation and simplify:

$$(x^2 + y^2)^3 = 8x^2y^2$$
$$(r^2)^3 = 8r^2 \cos^2 \theta \, r^2 \sin^2 \theta$$
$$r^6 = 8r^4 \cos^2 \theta \sin^2 \theta$$

If $r = 0$, the graph is the point at the pole for all θ. If $r \neq 0$, you can divide both sides by r^4 to obtain

$$r^2 = 8 \cos^2 \theta \sin^2 \theta$$
$$= 2(2 \cos \theta \sin \theta)(2 \cos \theta \sin \theta)$$
$$= 2(\sin 2\theta)(\sin 2\theta)$$
$$r^2 = 2 \sin^2 2\theta \qquad \blacksquare$$

Exercise 10.5

In Exercises 1–20, the polar coordinates of point P are given. Find the rectangular coordinates of point P.

1. $(2, 30°)$ **2.** $(5, 135°)$ **3.** $(7, 300°)$ **4.** $(20, 225°)$

5. $(-3, 60°)$ **6.** $(-7, 210°)$ **7.** $\left(2, \dfrac{\pi}{2}\right)$ **8.** $\left(4, \dfrac{3\pi}{2}\right)$

9. $\left(-2, \dfrac{13\pi}{6}\right)$ **10.** $(-5, 3\pi)$ **11.** $\left(5, \dfrac{13\pi}{4}\right)$ **12.** $\left(3, \dfrac{17\pi}{6}\right)$

13. $(2, -30°)$ **14.** $(6, -225°)$ **15.** $(-10, -90°)$ **16.** $(-15, -45°)$

17. $(0, 39°)$ **18.** $(0, 0°)$ **19.** $(6, 1230°)$ **20.** $(35, 11.5\pi)$

In Exercises 21–36, the rectangular coordinates of point P are given. Find polar coordinates for point P.

21. $(1, 1)$ **22.** $(1, \sqrt{3})$ **23.** $(2\sqrt{3}, -2)$ **24.** $(-2, -2\sqrt{3})$

25. $(-\sqrt{3}, -1)$ **26.** $(-1, \sqrt{3})$ **27.** $(-\sqrt{3}, 1)$ **28.** $(0, 3)$

29. $(0, 0)$ **30.** $(7, 0)$ **31.** $(-5, 0)$ **32.** $(7, 7)$

33. $(3, -3)$ **34.** (π, π) **35.** $(7, 7\sqrt{3})$ **36.** $(-\sqrt{2}, -\sqrt{6})$

In Exercises 37–48, each equation contains rectangular coordinates. Change each equation to an equation containing polar coordinates.

37. $x = 3$

38. $y = -7$

39. $3x + 2y = 3$

40. $2x - y = 7$

41. $x^2 + y^2 = 9x$

42. $yx = 12$

43. $(x^2 + y^2)^3 = 4x^2y^2$

44. $x^2 + y^2 = 9$

45. $x^2 = 2x - x^2$

46. $(x^2 + y^2)^2 = x^2 - y^2$

47. $x^2 = 2y + 1$

48. $y^2 = 2x + 1$

In Exercises 49–60, each equation contains polar coordinates. Change each equation to an equation containing rectangular coordinates.

49. $r = 3$

50. $r \sin \theta = 4$

51. $\cos \theta = \dfrac{5}{r}$

52. $3r \cos \theta + 2r \sin \theta = 2$

53. $r = \dfrac{1}{1 + \sin \theta}$

54. $r = \dfrac{1}{1 - \cos \theta}$

55. $r^2 = \sin 2\theta$

56. $r^2 = \cos 2\theta$

57. $\theta = \pi$

58. $\theta = 90°$

59. $r(2 - \cos \theta) = 2$

60. $r = 3 \csc \theta + 2 \sec \theta$

In Exercises 61–72, graph the given equation.

61. $r = -\theta; \theta \geq 0$

62. $r \sin \theta = 3$

63. $r \cos \theta = -3$

64. $r \cos \theta + r \sin \theta = 1$

65. $r = \cos 2\theta$

66. $r = \sin 2\theta$

67. $r = \sin 3\theta$

68. $r = 3 \cos 3\theta$

69. $r = 2(1 + \sin \theta)$

70. $r = \sqrt{2 \cos \theta}$

71. $r = 2 + \cos \theta$

72. $r = 3(1 - \cos \theta)$

73. Graph $x^2 + y^2 = y$ on a Cartesian coordinate system and show that its graph is identical to the graph in Figure 10-19.

10.6   MORE ON POLAR COORDINATES

In the previous section, you graphed several equations in polar coordinates by making an extensive table of values and plotting many points. A different approach will be used in this section. Although some specific points will be plotted, the behavior of functions of θ will be considered as θ increases from 0 to 2π.

Example 1   Graph $r = \sin 2\theta$.

Solution   Note that $\sin 2\theta$, and hence r as well, is 0 when $\theta = 0, \frac{\pi}{2}, \pi$, and $\frac{3\pi}{2}$. For these values, the curve passes through the pole. When $\theta = \frac{\pi}{4}$ or $\frac{5\pi}{4}$, the value of $\sin 2\theta = 1$. When θ is $\frac{3\pi}{4}$ or $\frac{7\pi}{4}$, the value of $\sin 2\theta = -1$. Therefore, the curve passes through the points $(1, \frac{\pi}{4})$, $(-1, \frac{3\pi}{4})$, $(1, \frac{5\pi}{4})$, and $(-1, \frac{7\pi}{4})$ and several times through the pole: $(0, 0)$, $(0, \frac{\pi}{2})$, $(0, \pi)$, and $(0, \frac{3\pi}{2})$. Note that these last four ordered pairs represent four different sets of polar coordinates for the pole. See Figure 10-21**a**.

As θ increases from 0 to $\frac{\pi}{4}$, the value of r increases from 0 to 1; draw that portion of the curve as in Figure 10-21**b**. As θ continues to increase from $\frac{\pi}{4}$ to $\frac{\pi}{2}$, r decreases, and the curve returns to the pole as in Figure 10-21**c**. When θ continues to increase from $\frac{\pi}{2}$ to $\frac{3\pi}{4}$, the value of r goes from 0 to -1 and the points (r, θ) trace the additional portion of the curve shown in Figure 10-21**d**. As θ increases from $\frac{3\pi}{4}$ to π, r goes from -1 back to 0 and the curve continues,

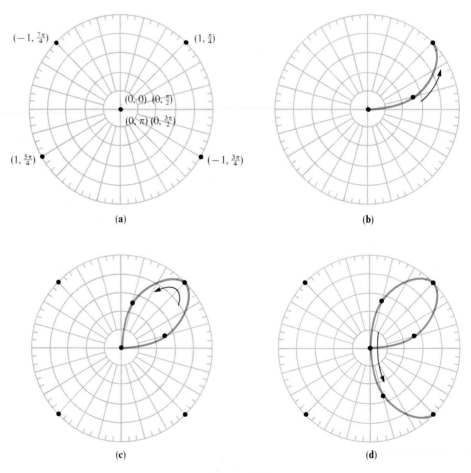

(a)

(b)

(c)

(d)

Figure 10-21

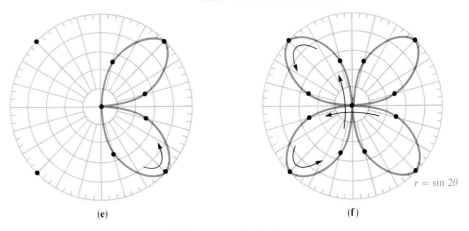

(e)   (f)

Figure 10-21 (continued)

to develop as in Figure 10-21e. The complete curve, called a **four-leaved rose**, is shown in Figure 10-21f. ∎

Example 2 Graph $r = \cos 3\theta$.

Solution When $\theta = \frac{\pi}{6}, \frac{\pi}{2}, \frac{5\pi}{6}, \frac{7\pi}{6}, \frac{3\pi}{2}$, or $\frac{11\pi}{6}$, the corresponding value of $\cos 3\theta$, and hence of r, is 0. When $\theta = 0, \frac{2\pi}{3}$, or $\frac{4\pi}{3}$, r reaches its maximum value of 1. When $\theta = \frac{\pi}{3}$, π, or $\frac{5\pi}{3}$, r reaches its minimum value of -1. Therefore, the curve passes through the points $(1, 0)$, $(1, \frac{2\pi}{3})$, $(1, \frac{4\pi}{3})$, $(-1, \frac{\pi}{3})$, $(-1, \pi)$, and $(-1, \frac{5\pi}{3})$. The curve also passes through the pole several times at $(0, \frac{\pi}{6})$, $(0, \frac{\pi}{2})$, $(0, \frac{5\pi}{6})$, $(0, \frac{7\pi}{6})$, $(0, \frac{3\pi}{2})$, and $(0, \frac{11\pi}{6})$. Note that these last six pairs of coordinates represent six different pairs of coordinates for the pole. Note also that the first six pairs of coordinates represent only three distinct points, as indicated in Figure 10-22a.

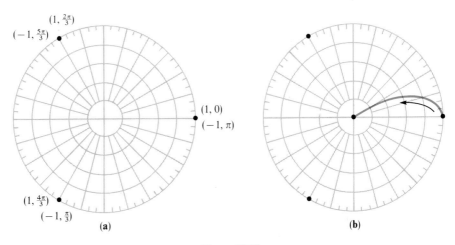

(a)   (b)

Figure 10-22

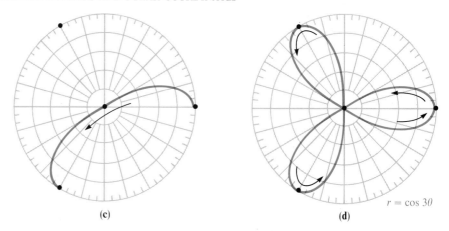

(c)   (d)

Figure 10.22 (continued)

As θ increases from 0 to $\frac{\pi}{6}$, cos 3θ decreases from 1 to 0, and that portion of the curve is traced in Figure 10-22**b**. As θ continues to increase from $\frac{\pi}{6}$ to $\frac{\pi}{3}$, the value of cos 3θ continues to decrease to its minimum value of -1; the curve continues to develop as in Figure 10-22**c**. The complete graph of $r = \cos 3\theta$, called a **three-leaved rose**, appears in Figure 10-22**d**. ■

All of the rose curves fit into two categories.

Theorem.   If n is an odd integer greater than 1, then $r = \cos n\theta$ or $r = \sin n\theta$ represents an n-leaved rose.

If n is even, then $r = \cos n\theta$ or $r = \sin n\theta$ represents a $2n$-leaved rose.

Example 3   Graph the curve $r = 1 - \sin \theta$.

Solution   The easiest values of r to compute are those associated with the quadrantal angle values of θ. When $\theta = 0$ or π, then $r = 1$. When $\theta = \frac{\pi}{2}$, then $r = 0$. When $\theta = \frac{3\pi}{2}$, then $r = 2$. These four points—(1, 0), (0, $\frac{\pi}{2}$), (1, π), and (2, $\frac{3\pi}{2}$)—are the intercepts of the curve with the polar axis and the perpendicular line to the polar axis at the pole. See Figure 10-23**a**. As θ increases from 0 to π, the value of $1 - \sin \theta$, and hence the value of r, decreases from 1 to 0 and then increases back to 1. This accounts for the two bumps in the curve in Figure 10-23**b**. As θ increases from π to $\frac{3\pi}{2}$, the third-quadrant loop of Figure 10-23**c** is formed. The complete curve, called a **cardioid**, is shown in Figure 10-23**d**.

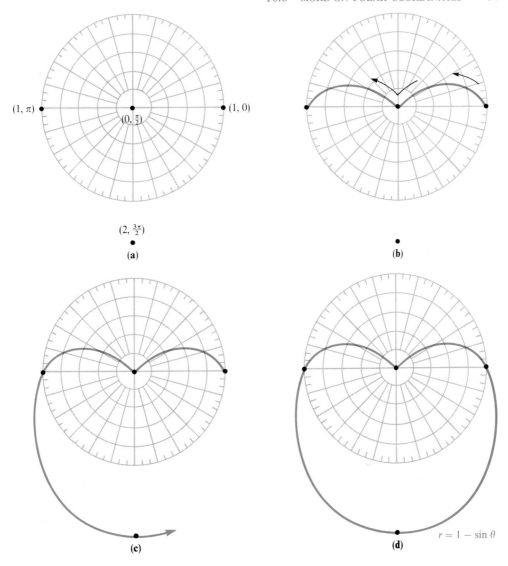

$(1, \pi)$ $(0, \frac{\pi}{2})$ $(1, 0)$

$(2, \frac{3\pi}{2})$

(a)

(b)

(c)

(d) $r = 1 - \sin \theta$

Figure 10-23

Example 4   Graph the curve $r\theta = \pi$.

Solution   Because the product of variables r and θ is a constant, r and θ are inversely proportional—as θ increases, r decreases. Write the equation as $r = \frac{\pi}{\theta}$, and calculate the intercepts of the curve with the polar axis and a line perpendicular to the axis through the pole: At $\theta = \frac{\pi}{2}, r = \frac{\pi}{\pi/2} = 2$; at $\theta = \pi, r = 1$; and at $\theta = \frac{3\pi}{2}$, $r = \frac{2}{3}$.

These points and a portion of the curve are plotted in Figure 10-24**a**; the curve is a spiral. The difficulty is in determining the shape of the curve as θ approaches

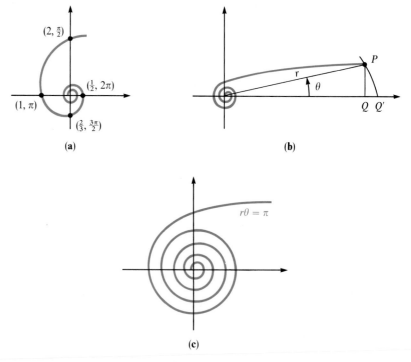

(a)

(b)

(c)

Figure 10-24

0 (and r approaches infinity). If θ is close to 0, and r is consequently very large, the distance PQ of Figure 10-24**b** is very close to the length of arc PQ', centered at the pole. Because this arc length is $r\theta$ or π, the distance PQ is approximately π. The horizontal line that is π units above the polar axis is an asymptote of this curve. The graph of $r\theta = \pi$ is called a **hyperbolic spiral** and its graph appears in Figure 10-24**c**. The tail heading off to the right approaches the horizontal line that is parallel to the polar axis and π units above it. Note that unlike the Archimedean spiral of the previous section, the hyperbolic spiral does not pass through the pole. ∎

The following chart shows the graphs and the general equations for the most important polar curves.

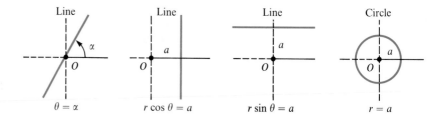

Line	Line	Line	Circle
$\theta = \alpha$	$r\cos\theta = a$	$r\sin\theta = a$	$r = a$

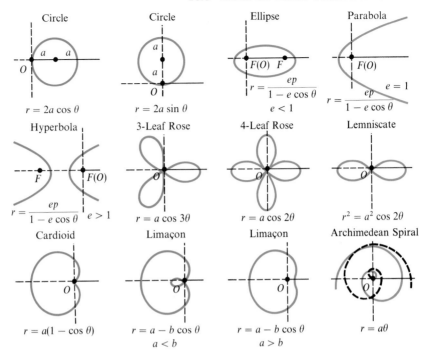

Exercise 10.6

Each of the following curves is important enough to have a name. The equation and the name of the curve are given. Graph each polar equation.

1. $r = 2 \cos 2\theta$; four-leaved rose

2. $r = 3 \sin 2\theta$; four-leaved rose

3. $r = \sin 3\theta$; three-leaved rose

4. $r = 2 \cos 3\theta$; three-leaved rose

5. $r = 2(1 + \cos \theta)$; cardioid

6. $r = 3(1 + \sin \theta)$; cardioid

7. $r = 2 + \cos \theta$; limaçon

8. $r = \dfrac{1}{2} + \cos \theta$; limaçon

9. $r^2 = 2 \sin 2\theta$; lemniscate

10. $r = \sin \theta \cos^2 \theta$; bifolium

11. $r^2 \theta = \pi, r > 0$; lituus   (*Hint*: This curve has a horizontal asymptote.)

12. $r = \tan \theta$; kappa curve $\Big\}$

13. $r = \sin \theta \tan \theta$; cissoid   (*Hint*: These curves have vertical asymptotes.)

14. $r = 2a \cos \theta$; circle

15. $r = \dfrac{3}{2 - \cos \theta}$; ellipse

16. $r = \dfrac{4}{2 - 3 \cos \theta}$; hyperbola

17. $r = \dfrac{4}{1 + \sin \theta}$; parabola

18. $r = \dfrac{6}{4 - 3 \cos \theta}$; ellipse

In Review Exercises 1–4, simplify each expression.

1. i^{11} **2.** i^{5003} **3.** i^{-33} **4.** i^{-1812}

5. Solve $x + (x + y)i = 2y + 2i$ for x and y.

6. Solve $3x + (x - y)i = 2y + 3 + 7i$.

In Review Exercises 7–18, perform any indicated operations and express the final answer in $a + bi$ form.

7. $(3 + 2i) + (-7 - i)$ **8.** $(-2 - i) - (3 - 2i)$

9. $(2 + i)(-3 - i)$ **10.** $(3 + 2i)(5 - 3i)$

11. $\dfrac{1}{5i}$ **12.** $\dfrac{13}{-6i}$

13. $\dfrac{2}{4 + i}$ **14.** $\dfrac{-5}{3 - i}$

15. $\dfrac{1 + \sqrt{-1}}{1 - \sqrt{-1}}$ **16.** $\dfrac{2 + \sqrt{-1}}{3 + \sqrt{-16}}$

17. $\dfrac{2 + 3i}{1 - \sqrt{2}i}$ **18.** $\dfrac{3 - i}{1 - \sqrt{3}i}$

In Review Exercises 19–22, graph each complex number.

19. $4 - 5i$ **20.** $-7 + 2i$ **21.** 6 **22.** $3i$

In Review Exercises 23–26, compute each absolute value.

23. $|8 + 3i|$ **24.** $|10 - 10i|$ **25.** $\left|\dfrac{3i}{i + 3}\right|$ **26.** $\left|\dfrac{4 - 3i}{4 + 3i}\right|$

In Review Exercises 27–30, write the given complex number in trigonometric form.

27. $-2 + 2i$ **28.** $5 - 5i$ **29.** $3 + 3i\sqrt{3}$ **30.** 4

In Review Exercises 31–34, write the given complex number in $a + bi$ form.

31. $3(\cos 60° + i \sin 60°)$ **32.** $2(\cos 330° + i \sin 330°)$

33. $3\left(\cos \dfrac{4\pi}{3} + i \sin \dfrac{4\pi}{3}\right)$ **34.** $7\left(\cos \dfrac{5}{6}\pi + i \sin \dfrac{5}{6}\pi\right)$

In Review Exercises 35–40, perform the indicated operation. Simplify, but leave your answer in trigonometric form.

35. $(\text{cis } 60°)(\text{cis } 50°)$ **36.** $(2 \text{ cis } 330°)(3 \text{ cis } 240°)$

37. $\left[3\left(\cos \dfrac{\pi}{12} + i \sin \dfrac{\pi}{12}\right)\right]\left[2\left(\cos \dfrac{\pi}{6} + i \sin \dfrac{\pi}{6}\right)\right]$

38. $\left[7\left(\cos \dfrac{\pi}{5} + i \sin \dfrac{\pi}{5}\right)\right]\left[3\left(\cos \dfrac{4\pi}{5} + i \sin \dfrac{4\pi}{5}\right)\right]$

39. $\dfrac{10 \text{ cis } 60°}{5 \text{ cis } 10°}$ **40.** $\dfrac{20(\cos 50° + i \sin 50°)}{30(\cos 40° + i \sin 40°)}$

41. Find one cube root of $\cos 60° + i \sin 60°$.

42. Find one fourth root of $7\sqrt{2} + 7\sqrt{2}\,i$.

43. Find the three cube roots of 125.

44. Find the four fourth roots of 81.

In Review Exercises 45–48, change the polar coordinates to rectangular coordinates.

45. $(5, 60°)$ **46.** $(-2, 390°)$ **47.** $\left(-1, \dfrac{7\pi}{6}\right)$ **48.** $\left(10, -\dfrac{5\pi}{4}\right)$

In Review Exercises 49–52, change the rectangular coordinates to polar coordinates.

49. $(-\sqrt{2}, \sqrt{2})$ **50.** $(-\sqrt{3}, 1)$ **51.** $(1, 0)$ **52.** $(1, -\sqrt{3})$

In Review Exercises 53–56, change the rectangular equation to a polar equation.

53. $4xy = 4$ **54.** $x + 2y = 2$

55. $x^2 = 3y$ **56.** $(x^2 + y^2)^2 = 4xy$

In Review Exercises 57–60, change the polar equation to a rectangular equation.

57. $r^2 = 9 \cos 2\theta$ **58.** $r = 5 \sin \theta$

59. $r = \dfrac{1}{4 + \sin \theta}$ **60.** $r = \dfrac{2}{1 - \cos \theta}$

In Review Exercises 61–64, graph each polar equation.

61. $r = \dfrac{6}{1 + \sin \theta}$ **62.** $r = \dfrac{2}{1 - \sin \theta}$

63. $r = 4(1 + \cos \theta)$ **64.** $r = 8 - 4 \cos \theta$

11

THEORY OF EQUATIONS

There are some equations of degree greater than 2 that can be solved by factoring. For example, the equation

$$x^3 - 3x^2 + 2x = 0$$

can be solved as follows:

$$x^3 - 3x^2 + 2x = 0$$
$$x(x^2 - 3x + 2) = 0 \qquad \text{Factor out } x.$$
$$x(x - 1)(x - 2) = 0 \qquad \text{Factor } x^2 - 3x + 2.$$
$$x = 0 \quad \text{or} \quad x - 1 = 0 \quad \text{or} \quad x - 2 = 0 \qquad \text{Set each factor equal to 0.}$$
$$x = 1 \qquad \qquad x = 2$$

The solution set of the given equation is $\{0, 1, 2\}$.

In this chapter we will discuss how to solve more complicated polynomial equations.

11.1 THE FACTOR AND REMAINDER THEOREMS

A **polynomial equation** is an equation that can be written in the form $P(x) = 0$, where

$$P(x) = a_n x^n + a_{n-1} x^{n-1} + \cdots + a_1 x + a_0$$

is a polynomial of degree n (n is a natural number). A **zero of the polynomial** $P(x)$ is any number r for which $P(r) = 0$. It follows that a zero of the polynomial $P(x)$ is a solution or root of the equation $P(x) = 0$. For example, 2 is a zero of the polynomial

$$P(x) = x^2 - 3x + 2$$

because

$$P(2) = 2^2 - 3(2) + 2$$
$$= 4 - 6 + 2$$
$$= 0$$

Note that 2 is a root of the polynomial equation

$$x^2 - 3x + 2 = 0$$

Before attempting to find zeros of more complicated polynomials, we need to know whether a given polynomial even has a zero. This question was answered by Karl Friedrich Gauss (1777–1855) when he proved the **fundamental theorem of algebra**.

> **The Fundamental Theorem of Algebra.**   If $P(x)$ is a polynomial with positive degree, then $P(x)$ has at least one zero.

The fundamental theorem points out that polynomials such as

$$2x + 3 \quad \text{or} \quad 32.75x^{1984} + ix^3 - (2 + i)x - 5$$

all have zeros. It may be difficult to find the zeros, and we may have to settle for approximations of the zeros, but the zeros do exist.

There is a relationship between a zero r of a polynomial $P(x)$ and the results of a long division of $P(x)$ by the binomial $x - r$. This relationship can be illustrated by an example.

Leonhard Euler (1707–1783) Euler was one of the most prolific mathematicians of all time. He did much of his work after he became blind.

Example 1   Let $P(x) = 3x^3 - 5x^2 + 3x - 10$.

a.   Find $P(1)$ and $P(-2)$.

b.   Divide $P(x)$ by $x - 1$ and by $x + 2$.

Solution   **a.** $P(1) = 3(1)^3 - 5(1)^2 + 3(1) - 10 \qquad P(-2) = 3(-2)^3 - 5(-2)^2 + 3(-2) - 10$

$\qquad\qquad = 3 - 5 + 3 - 10 \qquad\qquad\qquad\qquad = 3(-8) - 5(4) + 3(-2) - 10$

$\qquad\qquad = -9 \qquad\qquad\qquad\qquad\qquad\qquad\quad = -60$

b.

$$
\begin{array}{r}
3x^2 - 2x + 1 \\
x - 1 \overline{)\ 3x^3 - 5x^2 + 3x - 10} \\
\underline{3x^3 - 3x^2} \\
-2x^2 + 3x \\
\underline{-2x^2 + 2x} \\
+ x - 10 \\
\underline{x - 1} \\
-9
\end{array}
\qquad
\begin{array}{r}
3x^2 - 11x + 25 \\
x + 2 \overline{)\ 3x^3 - 5x^2 + 3x - 10} \\
\underline{3x^3 + 6x^2} \\
-11x^2 + 3x \\
\underline{-11x^2 - 22x} \\
25x - 10 \\
\underline{25x + 50} \\
-60
\end{array}
$$

Note the results of Example 1. When $P(x)$ was divided by $x - 1$, the remainder was $P(1)$, or -9. When $P(x)$ was divided by $x - (-2)$, or $x + 2$, the remainder was $P(-2)$, or -60. These results are not coincidental. The following theorem, called the **remainder theorem**, asserts that the division of any polynomial $P(x)$ by the binomial $x - r$ yields $P(r)$ as the remainder.

> **The Remainder Theorem.**   If $P(x)$ is a polynomial, r is a real or complex number, and $P(x)$ is divided by $x - r$, then the remainder is $P(r)$.

Proof   To divide $P(x)$ by $x - r$, we must find a quotient $Q(x)$ and a remainder $R(x)$ such that

$$\textbf{dividend} = \textbf{divisor} \cdot \textbf{quotient} + \textbf{remainder}$$
$$\downarrow \qquad \downarrow \qquad \downarrow \qquad \qquad \downarrow$$
$$P(x) \quad = (x - r) \cdot \quad Q(x) \quad + \quad R(x)$$

Furthermore, the degree of the remainder $R(x)$ must be less than the degree of the divisor $x - r$. Because the divisor is of degree 1, the remainder must be a constant R. The expression $P(x) = (x - r)Q(x) + R$ indicates that the polynomial on the left side of the equation is the same as the polynomial on the right. In particular, the values that these polynomials assume for any replacement of the variable x must be equal. Replacing x with the number r, we have $P(r) = (r - r)Q(r) + R$. Because $(r - r) = 0$, it follows that $P(r) = R$; that is, the value of the polynomial $P(x)$ attained at $x = r$ is the remainder produced by dividing $P(x)$ by $x - r$. The proof is complete. $\square$

The **factor theorem**, a corollary to the remainder theorem, applies when the remainder R is 0.

> **The Factor Theorem.**   Let $P(x)$ be any polynomial and let r be a real or complex number. Then, $P(r) = 0$ if and only if $x - r$ is a factor of $P(x)$.

Proof   First, assume that $P(r) = 0$ and prove that $x - r$ is a factor of $P(x)$. Divide $P(x)$ by $x - r$. The remainder theorem asserts that the remainder must be $P(r)$. But $P(r)$, by assumption, is 0. Hence, $x - r$ is a factor of $P(x)$.

Conversely, assume that $x - r$ is a factor of $P(x)$ and prove that $P(r) = 0$. Because $x - r$ is a factor of $P(x)$, dividing $P(x)$ by $x - r$ gives a remainder of 0. The remainder theorem asserts that this remainder is $P(r)$. Hence, $P(r) = 0$.
$\square$

Example 2   Let $P(x) = 3x^3 - 5x^2 + 3x - 10$. Show that $P(2) = 0$, and use the factor theorem to factor $P(x)$.

Solution   Calculate $P(2)$:

$$P(x) = 3x^3 - 5x^2 + 3x - 10$$
$$P(2) = 3(2)^3 - 5(2)^2 + 3(2) - 10 \qquad \text{Substitute 2 for } x.$$
$$= 3 \cdot 8 - 5 \cdot 4 + 6 - 10$$
$$= 0$$

Because $P(2) = 0$, it follows (by the factor theorem) that $x - 2$ is a factor of $P(x)$. To determine the other factor of $P(x)$, divide $P(x)$ by $x - 2$:

$$
\begin{array}{r}
3x^2 + x + 5 \\
x - 2 \overline{)\, 3x^3 - 5x^2 + 3x - 10} \\
\underline{3x^3 - 6x^2} \\
x^2 + 3x \\
\underline{x^2 - 2x} \\
5x - 10 \\
\underline{5x - 10} \\
0
\end{array}
$$

Thus, $P(x)$ factors as

$$P(x) = (x - 2)(3x^2 + x + 5)$$ ■

Example 3   Solve the equation $3x^3 - 5x^2 + 3x - 10 = 0$.

Solution   This equation is related to the polynomial of Example 2, so you can use the work already done there:

$$3x^3 - 5x^2 + 3x - 10 = 0$$
$$(x - 2)(3x^2 + x + 5) = 0$$

To solve for x, set each factor equal to 0 and apply the quadratic formula to the equation $3x^2 + x + 5 = 0$. The complete solution set is

$$\left\{ 2, -\frac{1}{6} + \frac{\sqrt{59}}{6}i, -\frac{1}{6} - \frac{\sqrt{59}}{6}i \right\}$$ ■

Example 4   Find a polynomial $P(x)$ that has zeros of 2, 3, and -5.

Solution   By the factor theorem, if $2, 3$, and -5 are zeros of $P(x)$, then $x - 2$, $x - 3$, and $x - (-5)$ are all factors of $P(x)$. Hence,

$$
\begin{aligned}
P(x) &= (x - 2)(x - 3)(x + 5) \\
&= (x^2 - 5x + 6)(x + 5) \\
&= x^3 - 19x + 30
\end{aligned}
$$

The polynomial $P(x) = x^3 - 19x + 30$ has zeros of 2, 3, and -5.   ■

Example 5   Is $x + 2$ a factor of the polynomial $P(x) = x^4 - 7x^2 - 6x$?

Solution   By the factor theorem, $x + 2$ will be a factor of $P(x)$ if -2 is a zero of $P(x)$. So, evaluate $P(-2)$ and see if it is a zero of $x^4 - 7x^2 - 6x$:

$$
\begin{aligned}
P(x) &= x^4 - 7x^2 - 6x \\
P(-2) &= (-2)^4 - 7(-2)^2 - 6(-2) \\
&= 16 - 28 + 12 \\
&= 0
\end{aligned}
$$

Because -2 is a zero of the polynomial $P(x)$, then $x - (-2)$, or $x + 2$, is a factor of $P(x) = x^4 - 7x^2 - 6x$. ∎

Example 6   Find the three cube roots of -1.

Solution   The three cube roots of -1 are solutions of the equation $x^3 = -1$, or $x^3 + 1 = 0$. One of the cube roots of -1 is the value -1. Because of the factor theorem, $x - (-1)$ must be a factor of $x^3 + 1$. Hence, $x + 1$ divides $x^3 + 1$. Obtain the other factor of $x^3 + 1$ by long division.

$$
\begin{array}{r}
x^2 - x + 1 \\
x + 1 \overline{)\, x^3 \qquad\qquad + 1} \\
\underline{x^3 + x^2} \\
-x^2 \\
\underline{-x^2 - x} \\
x + 1 \\
\underline{x + 1} \\
0
\end{array}
$$

Thus, the equation $x^3 + 1 = 0$ factors as

$$(x + 1)(x^2 - x + 1) = 0$$

Setting each factor equal to 0 and solving for x gives the three cube roots of -1:

$$x = -1, \quad x = \frac{1}{2} + \frac{\sqrt{3}}{2}i, \quad x = \frac{1}{2} - \frac{\sqrt{3}}{2}i$$ ∎

Exercise 11.1

In Exercises 1–6, let $P(x) = 2x^4 - 2x^3 + 5x^2 - 1$. Evaluate the polynomial by substituting the given value of x into the polynomial and simplifying. Then evaluate the polynomial by using the remainder theorem.

1. $P(2)$ 2. $P(-1)$ 3. $P(0)$ 4. $P(1)$ 5. $P(-4)$ 6. $P(4)$
7. Let $P(x) = x^5 - 1$. Find $P(2)$ using the remainder theorem.
8. Let $P(x) = x^5 + 4x^2 - 1$. Find $P(-3)$ using the remainder theorem.

In Exercises 9–16, use the factor theorem to decide whether each statement is true. If not, so indicate.

9. $x - 1$ is a factor of $x^7 - 1$.
10. $x - 2$ is a factor of $x^3 - x^2 + 2x - 8$.
11. $x - 1$ is a factor of $3x^5 + 4x^2 - 7$.
12. $x + 1$ is a factor of $3x^5 + 4x^2 - 7$.
13. $x + 3$ is a factor of $2x^3 - 2x^2 + 1$.
14. $x - 3$ is a factor of $3x^5 - 3x^4 + 5x^2 - 13x - 6$.
15. $x - 1$ is a factor of $x^{1984} - x^{1776} + x^{1492} - x^{1066}$.
16. $x + 1$ is a factor of $x^{1984} + x^{1776} - x^{1492} - x^{1066}$.
17. Completely solve $x^3 + 3x^2 - 13x - 15 = 0$, given that $x = -1$ is a root.
18. Completely solve $x^4 + 4x^3 - 10x^2 - 28x - 15 = 0$, given that $x = -1$ is a double root.

In Exercises 19–28, find a polynomial of lowest degree that has the indicated zeros.

19. $1, 1, 1$ **20.** $1, 0, -1$

21. $2, 4, 5$ **22.** $7, 6, 3$

23. $-1, 1, -\sqrt{2}, \sqrt{2}$ **24.** $0, 0, 0, \sqrt{3}, -\sqrt{3}$

25. $\sqrt{2}, i, -i$ **26.** $i, i, 2$

27. $1 + i, 1 - i, 0$ **28.** $2 + i, 2 - i, i$

In Exercises 29–32, find the three cube roots of each number.

29. 1 **30.** 64 **31.** -125 **32.** -216

33. Completely solve $x^4 - 5x^3 + 7x^2 - 5x + 6 = 0$, given that $x = 3$ and $x = 2$ are roots.

34. Completely solve $x^4 + 2x^3 - 3x^2 - 4x + 4 = 0$, given that $x = 1$ and $x = -2$ are roots.

35. Completely solve $x^4 - 2x^3 - 9x^2 + 2x + 8 = 0$, given that $x = 4$ and $x = -1$ are roots.

36. If 0 is a zero of $P(x) = a_n x^n + a_{n-1} x^{n-1} + \cdots + a_1 x + a_0$, what is a_0?

37. If 0 occurs as a zero twice in the polynomial $P(x) = a_n x^n + a_{n-1} x^{n-1} + \cdots + a_1 x + a_0$, what is a_1?

38. Explain why the fundamental theorem of algebra guarantees that every polynomial equation has at least one root.

39. Explain why the fundamental theorem of algebra and the factor theorem guarantee that an nth-degree polynomial equation has n roots.

40. The fundamental theorem of algebra demands that a polynomial be of positive degree. Would the theorem still be true if the polynomial were of degree 0? Explain.

11.2 SYNTHETIC DIVISION

There is a shortcut, called **synthetic division**, that can be used to divide a higher degree polynomial written in descending powers of x by a binomial of the form $x - r$. To see how this method works, we consider the long division of $2x^3 + 4x^2 - 3x + 10$ by $x - 3$:

$$
\begin{array}{r}
2x^2 + 10x + 27 \\
x - 3 \overline{)\, 2x^3 + 4x^2 - 3x + 10} \\
\underline{2x^3 - 6x^2} \\
10x^2 - 3x \\
\underline{10x^2 - 30x} \\
27x + 10 \\
\underline{27x - 81} \\
\text{(remainder)} \quad 91
\end{array}
\qquad
\begin{array}{r}
2 + 10 + 27 \\
1 - 3 \overline{)\, 2 + 4 - 3 + 10} \\
\underline{2 - 6} \\
10 - 3 \\
\underline{10 - 30} \\
27 + 10 \\
\underline{27 - 81} \\
\text{(remainder)} \quad 91
\end{array}
$$

On the left is the complete long division. On the right is a modified version of the long division in which the variables have been removed. We can shorten the work on the right even further by omitting the numbers printed in color:

$$
\begin{array}{r}
2 + 10 + 27 \\
-3\,\overline{)\,2 + 4 - 3 + 10} \\
-6 \\
\overline{10} \\
-30 \\
\overline{27} \\
-81 \\
\overline{\phantom{(\text{remainder})}\,91}
\end{array}
$$

(remainder)   91

We can then compress the work vertically to obtain

$$
\begin{array}{r}
2 + 10 + 27 \\
-3\,\overline{)\,2 + 4 - 3 + 10} \\
-6 - 30 - 81 \\
\overline{10 \quad 27 \quad 91}
\end{array}
$$

There is no reason why the quotient, represented by the numbers 2, 10, and 27, must appear above the long division symbol. If we write the 2 on the bottom line, the bottom line gives both the coefficients of the quotient and the remainder. The top line can be eliminated, and the division now appears as

$$
\begin{array}{r}
-3\,\underline{}\;2 \quad +4 \quad -3 \quad +10 \\
-6 \quad -30 \quad -81 \\
\overline{2 \quad 10 \quad 27 \quad 91}
\end{array}
$$

The bottom line was obtained by subtracting the middle line from the top line. If we replace the -3 in the divisor with a 3, the signs of every entry in the middle line are reversed in the division process. Then, the bottom line can be obtained by addition. Thus, we have this final form of the synthetic division:

$$
\begin{array}{r}
3\,\underline{}\;2 \quad 4 \quad -3 \quad 10 \\
6 \quad 30 \quad 81 \\
\overline{2 \quad 10 \quad 27 \,|\, 91}
\end{array}
$$

The coefficients of the dividend.

The coefficients of the quotient and the remainder.

Thus,

$$
\frac{2x^3 + 4x^2 - 3x + 10}{x - 3} = 2x^2 + 10x + 27 + \frac{91}{x - 3}
$$

Example 1   Use synthetic division to divide $3x^4 - 8x^3 + 10x + 3$ by $x - 2$.

Solution   Begin by writing the coefficients of the dividend and the 2 from the divisor in the following form:

$$
2\,\underline{}\;3 \quad -8 \quad 0 \quad 10 \quad 3
$$

Write 0 for the coefficient of the missing term of x^2.

Then follow these steps:

$$2 \,\rvert\, 3 \quad -8 \quad 0 \quad 10 \quad 3$$
$$ \downarrow$$
$$ \overline{3}$$

Begin by bringing down the 3.

$$2 \,\rvert\, 3 \quad -8 \quad 0 \quad 10 \quad 3$$
$$ 6$$
$$ \overline{3 \quad -2}$$

Multiply 2 and 3 and add the product to -8 to get -2.

$$2 \,\rvert\, 3 \quad -8 \quad 0 \quad 10 \quad 3$$
$$ 6 \quad -4$$
$$ \overline{3 \quad -2 \quad -4}$$

Multiply 2 and $--2$ and add the product to 0 to get -4.

$$2 \,\rvert\, 3 \quad -8 \quad 0 \quad 10 \quad 3$$
$$ 6 \quad -4 \quad -8$$
$$ \overline{3 \quad -2 \quad -4 \quad 2}$$

Multiply 2 and -4 and add the product to 10 to get 2.

$$2 \,\rvert\, 3 \quad -8 \quad 0 \quad 10 \quad 3$$
$$ 6 \quad -4 \quad -8 \quad 4$$
$$ \overline{3 \quad -2 \quad -4 \quad 2 \,\rvert\, 7}$$

Multiply 2 and 2 and add the product to 3 to get 7.

Thus,

$$\frac{3x^4 - 8x^3 + 10x + 3}{x - 2} = 3x^3 - 2x^2 - 4x + 2 + \frac{7}{x - 2}$$

∎

Example 2 Use synthetic division to find $P(-2)$ if $P(x) = 5x^3 + 3x^2 - 21x - 1$.

Solution Because of the remainder theorem, $P(-2)$ will be the remainder when $P(x)$ is divided by $x - (-2)$. Perform the division as follows:

$$-2 \,\rvert\, 5 \quad 3 \quad -21 \quad -1 \qquad\qquad -2 \,\rvert\, 5 \quad 3 \quad -21 \quad -1$$
$$ -10 \qquad\qquad\qquad\qquad\quad -10 \quad 14$$
$$ \overline{5 \quad -7} \qquad\qquad\qquad\qquad \overline{5 \quad -7 \quad -7}$$

$$-2 \,\rvert\, 5 \quad 3 \quad -21 \quad -1$$
$$ -10 \quad 14 \quad 14$$
$$ \overline{5 \quad -7 \quad -7 \,\rvert\, 13}$$

Because the remainder is 13, $P(-2) = 13$.

∎

Example 3 Find $P(i)$, where $i = \sqrt{-1}$ and $P(x) = x^3 - x^2 + x - 1$.

Solution Use synthetic division.

$$i \,\rvert\, 1 \quad -1 \quad +1 \quad -1$$
$$ i \quad -1-i \quad +1$$
$$ \overline{1 \quad i-1 \quad -i \,\rvert\, 0}$$

Because the remainder is 0, $P(i) = 0$, and i is a zero of $P(x)$.

∎

Example 4   Graph $y = x^3 + x^2 - 2x$.

Solution   Let $x = 0$ and find the y-intercept; if $x = 0$, then $y = 0$. Then use synthetic division to help find coordinates of other points on the graph. For example, if $x = 1$, then

$$
\begin{array}{r|rrrr}
1 & 1 & 1 & -2 & 0 \\
  &   & 1 & 2 & 0 \\
\hline
  & 1 & 2 & 0 & 0 \\
\end{array}
$$

The point with coordinates $(1, 0)$ lies on the graph. As another example, if $x = -1$, then

$$
\begin{array}{r|rrrr}
-1 & 1 & 1 & -2 & 0 \\
   &   & -1 & 0 & 2 \\
\hline
   & 1 & 0 & -2 & 2 \\
\end{array}
$$

The point with coordinates $(-1, 2)$ lies on the graph. These coordinates and others that satisfy the equation appear in the graph in Figure 11-1.

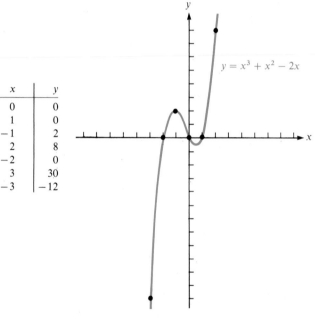

x	y
0	0
1	0
-1	2
2	8
-2	0
3	30
-3	-12

$y = x^3 + x^2 - 2x$

Figure 11-1

Example 5   Graph $y = x^4 - 5x^2 + 4$.

Solution   Because x appears only with even powers, the graph is symmetric about the y-axis. Because $y = 4$ when $x = 0$, the graph passes through the point $(0, 4)$. Use synthetic division to help find coordinates of other points on the graph. For example, if $x = -3$, then

$$
\begin{array}{r|rrrrr}
-3 & 1 & 0 & -5 & 0 & 4 \\
& & -3 & 9 & -12 & 36 \\
\hline
& 1 & -3 & 4 & -12 & 40
\end{array}
$$

The graph is shown in Figure 11-2.

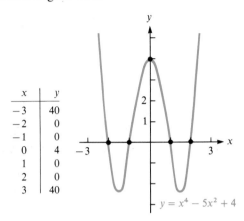

x	y
-3	40
-2	0
-1	0
0	4
1	0
2	0
3	40

$y = x^4 - 5x^2 + 4$

Figure 11-2

Exercise 11.2

In Exercises 1–8, assume that $P(x) = 5x^3 + 2x^2 - x + 1$. Use synthetic division to find each value of $P(x)$.

1. $P(2)$ **2.** $P(-2)$ **3.** $P(-5)$ **4.** $P(3)$

5. $P(0)$ **6.** $P(5)$ **7.** $P(-i)$ **8.** $P(i)$

In Exercises 9–16, assume that $P(x) = 2x^4 - x^2 + 2$. Use synthetic division to find each value of $P(x)$.

9. $P(1)$ **10.** $P(-1)$ **11.** $P(-2)$ **12.** $P(3)$

13. $P(\frac{1}{2})$ **14.** $P(\frac{1}{3})$ **15.** $P(i)$ **16.** $P(-i)$

In Exercises 17–24, assume that $P(x) = x^4 - 8x^3 + 8x + 14x^2 - 15$. Write the terms of $P(x)$ in descending powers of x and use synthetic division to find each value of $P(x)$.

17. $P(1)$ **18.** $P(0)$ **19.** $P(-3)$ **20.** $P(-1)$

21. $P(3)$ **22.** $P(5)$ **23.** $P(2)$ **24.** $P(-5)$

In Exercises 25–32, assume that $P(x) = 8 - 8x^2 + x^5 - x^3$. Write the terms of $P(x)$ in descending powers of x and use synthetic division to find each value of $P(x)$.

25. $P(0)$ **26.** $P(1)$ **27.** $P(2)$ **28.** $P(-2)$

29. $P(i)$ **30.** $P(-i)$ **31.** $P(-2i)$ **32.** $P(2i)$

In Exercises 33–38, use synthetic division to express $P(x) = 3x^3 - 2x^2 - 6x - 4$ in the form (divisor)(quotient) + remainder for each of the given divisors.

33. $x + 1$ **34.** $x - 1$ **35.** $x - 2$ **36.** $x + 2$ **37.** x **38.** $x - 7$

In Exercises 39–44, use synthetic division to perform each division.

39. $(7x^3 - 3x^2 - 5x + 1) \div (x + 1)$

40. $(2x^3 + 4x^2 - 3x + 8) \div (x - 3)$

41. $\dfrac{4x^4 - 3x^3 - x + 5}{x - 3}$

42. $\dfrac{x^4 + 5x^3 - 2x^2 + x - 1}{x + 1}$

43. $\dfrac{3x^5 - 768x}{x - 4}$

44. $\dfrac{x^5 - 4x^2 + 4x + 4}{x + 3}$

In Exercises 45–48, use synthetic division to find each power. (Hint: $y = 3^5$ is equivalent to $y = x^5$ when $x = 3$.)

45. 3^5

46. 4^5

47. 2^7

48. 3^6

In Exercises 49–56, use synthetic division to help graph each equation.

49. $y = x^3 - 4x$

50. $y = -x^3 + 4x$

51. $y = x^3 - x^2 - 2x$

52. $y = x^3 - x^2 - 6x$

53. $y = -x^4 + x^2$

54. $y = x^5 - 2x^3$

55. $y = x^5 - 3x^4 - 5x^3 + 15x^2 + 4x - 12$

56. $y = x^4 + 2x^3 - 5x^2 - 6x$

57. Let $P(x) = 5x^3 - 10x^2 + kx - 9$. For what value of k will the number 3 be a zero of $P(x)$?

58. Let $P(x) = 3x^3 + 8x^2 + kx - 6$. For what values of k will the number -2 be a zero of $P(x)$?

59. Use a calculator and synthetic division to find $P(1.3)$ where $P(x) = 2.5x^3 - 0.78x^2 - 2.7x + 4.3$.

60. Use a calculator and synthetic division to find $P(0.13)$ where $P(x) = 2.1x^3 - 1.2x^2 + 3.5x - 1.8$.

11.3 DESCARTES' RULE OF SIGNS AND BOUNDS ON ROOTS

The remainder theorem and synthetic division provide a way of verifying that a particular number is a root of a polynomial equation, but they do not provide the solutions. Selecting numbers at random, checking to see if they work, and hoping for the best is not an efficient technique! Some guidelines are needed to indicate how many solutions to expect, the kind of solutions to expect, and where they can be found. This section presents several theorems that provide such guidelines. The first theorem tells how many solutions to expect when solving a polynomial equation.

> **Theorem.** If multiple roots are counted individually, the polynomial equation $P(x) = 0$ with complex coefficients (which include real coefficients) and degree $n > 0$ has exactly n roots among the complex numbers.

Proof Let $P(x)$ be a polynomial of degree n. The fundamental theorem of algebra asserts that $P(x)$ has a zero, r_1. Therefore, the equation $P(x) = 0$ has r_1 as a root. The factor theorem guarantees that $x - r_1$ is a factor of $P(x)$. Thus,

$$P(x) = (x - r_1)Q_1(x)$$

If the lead coefficient of the nth-degree polynomial $P(x)$ is a_n, then $Q_1(x)$ is a polynomial of degree $n - 1$ whose lead coefficient is also a_n. The fundamental theorem of algebra asserts that $Q_1(x)$ also has a zero, which we call r_2. According to the factor theorem, $x - r_2$ is a factor of $Q_1(x)$, and

$$P(x) = (x - r_1)(x - r_2)Q_2(x)$$

where $Q_2(x)$ is a polynomial of degree $n - 2$ with lead coefficient a_n. This process can continue only to n factors of the form $x - r_i$; the final quotient $Q_n(x)$ is a polynomial of degree $n - n$, or degree 0. A polynomial of degree 0 with lead coefficient a_n is simply the constant a_n. The original polynomial $P(x)$ factors completely as

$$P(x) = a_n(x - r_1)(x - r_2)(x - r_3) \cdots (x - r_n)$$

Each of the n values r_i is a zero of $P(x)$ and a root of the equation $P(x) = 0$. There are no other roots, because no single factor in this product becomes 0 for any value of x not included in the list $r_1, r_2, r_3, \ldots, r_n$.

The theorem is proved.   □

The r_i in the above proof need not be different. Any root that occurs k times is called a **root of multiplicity k**.

The next theorem points out a pattern in the complex roots of real polynomial equations.

Theorem.   If a polynomial equation $P(x) = 0$ with real coefficients has a complex root $a + bi$, with $b \neq 0$, then the conjugate $a - bi$ is a root also. (This theorem is often stated as "complex roots of real polynomial equations occur in conjugate pairs.")

The proof of this theorem is omitted.

Example 1   Form a fourth-degree equation with real coefficients and a double root of i.

Solution   Because i is a root of multiplicity two and a fourth-degree equation must have four roots, two more roots are needed. According to the previous theorem, the missing roots must be the conjugates of the given roots. Thus, the complete solution set is

$$\{+i, +i, -i, -i\}$$

The equation is

$$(x - i)(x + i)(x - i)(x + i) = 0$$
$$(x^2 + 1)(x^2 + 1) = 0$$
$$x^4 + 2x^2 + 1 = 0$$

■

Example 2   Can a quadratic equation have a double root of i? If so, find such an equation.

Solution   Complex roots of a quadratic equation with *real* coefficients will be conjugates. A quadratic equation can have two nonreal, nonconjugate roots if the equation has coefficients that are not real. If i is a solution of multiplicity two, the equation is

$$(x - i)(x - i) = 0$$
$$x^2 - 2ix - 1 = 0$$ ∎

Descartes' Rule of Signs

René Descartes is credited with a theorem known as **Descartes' rule of signs** that enables us to look at a polynomial equation and estimate the number of positive, negative, and nonreal roots.

If a polynomial is written in descending powers of x and we scan it from left to right, a "variation in sign" occurs whenever successive terms have opposite signs. For example,

$$P(x) = 3x^5 - 2x^4 - 5x^3 + x^2 - x - 9$$

has three variations in sign, and

$$P(-x) = 3(-x)^5 - 2(-x)^4 - 5(-x)^3 + (-x)^2 - (-x) - 9$$
$$= -3x^5 - 2x^4 + 5x^3 + x^2 + x - 9$$

has two variations in sign.

> **Descartes' Rule of Signs.**   If $P(x)$ is a polynomial with real coefficients, the number of positive roots of the polynomial equation $P(x) = 0$ is either equal to the number of variations in sign of $P(x)$ or less than that by an even number.
>
> The number of negative roots of $P(x) = 0$ is either equal to the number of variations in sign of $P(-x)$ or less than that by an even number.

The proof of this theorem is omitted.

Example 3   What possible roots can be expected for the polynomial equation

$$P(x) = x^8 + x^6 + x^4 + x^2 + 1 = 0$$

Solution   Because the equation is an eighth-degree polynomial equation, it must have eight roots. Because there are no variations in sign for $P(x)$, none of the roots can be positive. Because there are no variations in sign for $P(-x)$, none of the roots can be negative. Because $P(0) = 1$, zero is not a root. Thus, all eight roots are complex numbers with nonzero imaginary parts, and they occur in four conjugate pairs.   ∎

Example 4 Discuss the possibilities for the roots of $3x^3 - 2x^2 + x - 5 = 0$.

Solution Let $P(x) = 3x^3 - 2x^2 + x - 5$. There are three variations in sign $P(x)$, so there could be three positive solutions or only one (because 1 is less than 3 by an even number). Because $P(-x) = -3x^3 - 2x^2 - x - 5$ has no variations in sign, there are no negative roots. Furthermore, 0 is not a root.

If there are three positive roots, all the roots are accounted for; there would be no other possibilities. If there is only one positive root, the remaining two roots must be nonreal. The following list indicates these two possibilities.

Number of positive roots	Number of negative roots	Number of nonreal roots
3	0	0
1	0	2

The number of nonreal roots is the number needed to bring the total number of roots up to three. ∎

Example 5 Discuss the possibilities for the roots of $5x^5 - 3x^3 - 2x^2 + x - 1 = 0$.

Solution $P(x)$ has three variations in sign; there are either three positive solutions or only one. Because $P(-x) = -5x^5 + 3x^3 - 2x^2 - x - 1$ has two variations in sign, there are either two negative roots or none. Each line of the following list indicates a possible combination of positive, negative, and nonreal roots.

Number of positive roots	Number of negative roots	Number of nonreal roots
1	0	4
3	0	2
1	2	2
3	2	0

Note that in each case the number of nonreal roots is even. This is expected, because this polynomial has real coefficients and its complex roots must occur in conjugate pairs. ∎

Bounds for Roots

A final theorem provides a way of finding **bounds** for the roots, enabling us to concentrate our efforts on those regions where roots can be found. This theorem is also presented without proof.

Theorem. Let the lead coefficient of the polynomial $P(x)$ with real co-efficients be positive, and do a synthetic division of the coefficients of $P(x)$ by the positive number c. If each term in the last row is nonnegative, then no number greater than c can be a root of $P(x) = 0$. (c is an *upper bound* of the real roots.)

If $P(x)$ is divided synthetically by a negative number d, and the signs in the last row alternate,* then no value less than d can be a root of $P(x) = 0$. (d is a *lower bound* of the real roots.)

Example 6   Establish best integer bounds for the roots of $18x^3 - 3x^2 - 37x + 12 = 0$.

Solution   Try several synthetic divisions, looking for a nonnegative last row (if you synthetically divide by a positive number) or the alternating-sign last row (if you synthetically divide by a negative number). Trying 1 first gives

$$
\begin{array}{r|rrrr}
1 & 18 & -3 & -37 & 12 \\
  &   & 18 & 15 & -22 \\
\hline
  & +18 & +15 & -22 & -10
\end{array}
$$

Because some of the signs in the last row are negative, 1 is not an upper bound of the roots of the equation. Now try 2.

$$
\begin{array}{r|rrrr}
2 & 18 & -3 & -37 & 12 \\
  &   & 36 & 66 & 58 \\
\hline
  & +18 & +33 & +29 & +70
\end{array}
$$

Because the last row is entirely positive, no number greater than 2 can be a root. Thus, the smallest or best integer upper bound of the roots is 2.

Now try a negative divisor such as -2.

$$
\begin{array}{r|rrrr}
-2 & 18 & -3 & -37 & 12 \\
   &   & -36 & 78 & -82 \\
\hline
   & +18 & -39 & +41 & -70
\end{array}
$$

The alternating signs in the last row indicate that no number less than -2 can be a root. Try -1 next.

$$
\begin{array}{r|rrrr}
-1 & 18 & -3 & -37 & 12 \\
   &   & -18 & 21 & 16 \\
\hline
   & +18 & -21 & -16 & +28
\end{array}
$$

Because the signs in the last row do not alternate, -1 is not a lower bound. Thus, the smallest or best integer lower bound is -2.

All of the real roots of this equation lie between -2 and 2. ∎

---

* If a 0 appears in the third row, that 0 can be assigned either a plus or a minus sign to help the signs alternate.

Example 7   Find bounds for the roots of the equation $x^5 - x^4 - 3x^3 + 3x^2 - 10x + 10 = 0$.

Solution   Perform several divisions and watch for the desired nonnegative last row (if synthetically dividing by a positive number) or the alternating-sign last row (if synthetically dividing by a negative number):

$$
\begin{array}{r|rrrrrr}
-3 & 1 & -1 & -3 & 3 & -10 & 10 \\
 & & -3 & 12 & -27 & 72 & -186 \\
\hline
 & +1 & -4 & +9 & -24 & +62 & -176 \\
\end{array}
$$

$$
\begin{array}{r|rrrrrr}
-2 & 1 & -1 & -3 & 3 & -10 & 10 \\
 & & -2 & 6 & -6 & 6 & 8 \\
\hline
 & +1 & -3 & +3 & -3 & -4 & +18 \\
\end{array}
$$

Note the alternation of signs in the synthetic division by -3. You can conclude that -3 is a lower bound; that is, the equation has no roots less than -3. That claim cannot be made for -2, however, because the signs do not alternate.

Now look for an upper bound by dividing synthetically by various positive values:

$$
\begin{array}{r|rrrrrr}
3 & 1 & -1 & -3 & 3 & -10 & 10 \\
 & & 3 & 6 & 9 & 36 & 78 \\
\hline
 & +1 & +2 & +3 & +12 & +26 & +88 \\
\end{array}
$$

$$
\begin{array}{r|rrrrrr}
2 & 1 & -1 & -3 & 3 & -10 & 10 \\
 & & 2 & 2 & -2 & 2 & -16 \\
\hline
 & +1 & +1 & -1 & +1 & -8 & -6 \\
\end{array}
$$

When you divide synthetically by 3, you obtain a completely positive last row, which indicates that 3 is an upper bound. You cannot say the same thing for 2 because its last row contains negative numbers.   ■

Exercise 11.3

1. How many roots does $x^{10} = 1$ have?

2. How many roots does $x^{40} = 1$ have?

3. One root of $x(3x^4 - 2) = 12x$ is 0. How many other roots are there?

4. One root of $3x^2(x^7 - 14x + 3) = 0$ is 0. How many other roots are there?

In Exercises 5–18, use Descartes' rule of signs to find the number of possible positive, negative, and nonreal roots of each equation. **Do not attempt to find the roots.**

5. $3x^3 + 5x^2 - 4x + 3 = 0$

6. $3x^3 - 5x^2 - 4x - 3 = 0$

7. $2x^3 + 7x^2 + 5x + 4 = 0$

8. $-2x^3 - 7x^2 - 5x - 4 = 0$

9. $8x^4 = -5$

10. $-3x^3 = -5$

11. $x^4 + 8x^2 - 5x = 10$

12. $5x^7 + 3x^6 - 2x^5 + 3x^4 + 9x^3 + x^2 + x + 1 = 0$

13. $-x^{10} - x^8 - x^6 - x^4 - x^2 - 1 = 0$

14. $x^{10} + x^8 + x^6 + x^4 + x^2 + 1 = 0$

15. $x^9 + x^7 + x^5 + x^3 + x = 0$ (Is 0 a root?)

16. $-x^9 - x^7 - x^5 - x^3 = 0$ (Is 0 a root?)

17. $-2x^4 - 3x^2 + 2x + 3 = 0$

18. $-7x^5 - 6x^4 + 3x^3 - 2x^2 + 7x - 4 = 0$

In Exercises 19–26, find the best integer bounds for the roots of each equation.

19. $x^2 - 5x - 6 = 0$

20. $6x^2 + x - 1 = 0$

21. $6x^2 - 13x - 110 = 0$

22. $3x^2 + 12x + 24 = 0$

23. $x^5 + x^4 - 8x^3 - 8x^2 + 15x + 15 = 0$

24. $12x^3 + 20x^2 - x - 6 = 0$

25. $2x^3 + 9x^2 - 5x = 41$

26. $x^4 - 34x^2 = -225$

27. Prove that any odd-degree polynomial equation with real coefficients must have at least one real root.

28. If a, b, c, and d are positive numbers, prove that $ax^4 + bx^2 + cx - d = 0$ has exactly two nonreal roots.

11.4 RATIONAL ROOTS OF POLYNOMIAL EQUATIONS

This section considers a method for actually finding the rational roots of polynomial equations with integral coefficients.

Theorem. If the polynomial equation

$$P(x) = a_nx^n + a_{n-1}x^{n-1} + a_{n-2}x^{n-2} + \cdots + a_1x + a_0 = 0$$

has integral coefficients and the rational number p/q (in lowest terms) is a root of the equation, then p is a factor of the constant term a_0, and q is a factor of the lead coefficient a_n.

Proof Let p/q be a rational root in lowest terms of the equation $P(x) = 0$. The equation is satisfied by p/q:

$$a_n\left(\frac{p}{q}\right)^n + a_{n-1}\left(\frac{p}{q}\right)^{n-1} + a_{n-2}\left(\frac{p}{q}\right)^{n-2} + \cdots + a_1\left(\frac{p}{q}\right) + a_0 = 0$$

By multiplying both sides of the equation by the lowest common denominator q^n, we clear the equation of fractions. (Remember that p, q, and each of the a_i are integers.)

1. $a_np^n + {}_{n-1}p^{n-1}q + a_{n-2}p^{n-2}q^2 + \cdots + a_1pq^{n-1} + a_0q^n = 0$

Note that all the terms but the last share a common factor of p. We rewrite the equation in the form

$$p(a_np^{n-1} + a_{n-1}p^{n-2}q + a_{n-2}p^{n-3}q^2 + \cdots + a_1q^{n-1}) = -a_0q^n$$

Because p is a factor of the left side, it must also be a factor of the right side. It cannot be a factor of q^n, because the fraction p/q is in lowest terms, and p and q share no common factor. Therefore, p and q^n share no common factors either. It follows that p must be a factor of a_0.

We return to Equation 1, note that all terms but the first share a common factor of q, and rewrite Equation 1 as

$$q(a_{n-1}p^{n-1} + a_{n-2}p^{n-2}q + a_{n-3}p^{n-3}q^2 + \cdots + a_0q^{n-1}) = -a_np^n$$

Now q is a factor of the left side and must therefore be a factor of the right side as well. Because q cannot be a factor of p^n, it must be a factor of a_n.

The theorem is proved.   □

Example 1   What are the only possible rational roots of the equation

$$\frac{1}{2}x^4 + \frac{2}{3}x^3 + 3x^2 - \frac{3}{2}x + 3 = 0$$

Solution   The previous theorem applies to polynomial equations with *integral* coefficients. To clear this equation of its fractional coefficients, multiply both sides by 6 to get

$$3x^4 + 4x^3 + 18x^2 - 9x + 18 = 0$$

The only possible numerators available for a rational root are the factors of the constant term 18: ± 1, ± 2, ± 3, ± 6, ± 9, and ± 18. The only possible denominators are the factors of the lead coefficient 3: ± 1 and ± 3. You can form the list of possible rational solutions by listing all the combinations of values from these two sets.

$$\pm\frac{1}{1},\ \pm\frac{2}{1},\ \pm\frac{3}{1},\ \pm\frac{6}{1},\ \pm\frac{9}{1},\ \pm\frac{18}{1},\ \pm\frac{1}{3},\ \pm\frac{2}{3},\ \pm\frac{3}{3},\ \pm\frac{6}{3},\ \pm\frac{9}{3},\ \pm\frac{18}{3}$$

Several of these are duplicates, so you can condense the list to obtain

Possible Rational Roots

$$\pm 1,\ \pm 2,\ \pm 3,\ \pm 6,\ \pm 9,\ \pm 18,\ \pm\frac{1}{3},\ \pm\frac{2}{3}$$ ■

Example 2   Prove that $\sqrt{2}$ is irrational.

Solution $\sqrt{2}$ is a real root of the polynomial equation $x^2 - 2 = 0$. Any rational solution of the equation must have a numerator of either ± 1 or ± 2 and a denominator of ± 1. The only possible rational solutions, therefore, are ± 1 and ± 2, but none of these satisfies the equation. Therefore, the solution $\sqrt{2}$ must be irrational.   ■

Example 3   Solve the polynomial equation $P(x) = x^3 - 7x + 6 = 0$.

Solution   Because the equation is of third degree, it must have three roots. According to Descartes' rule of signs, there are two possible combinations of positive, negative, and nonreal roots. They are summarized as follows.

Number of positive roots	Number of negative roots	Number of nonreal roots
2	1	0
0	1	2

The only possible rational roots are

$$\frac{\pm 6}{\pm 1}, \frac{\pm 3}{\pm 1}, \frac{\pm 2}{\pm 1}, \frac{\pm 1}{\pm 1}$$

or

$$-6, \quad -3, \quad -2, \quad -1, \quad 1, \quad 2, \quad 3, \quad 6$$

Check each one, crossing out those that do not satisfy the equation. Start, for example, with $x = 3$:

$$
\begin{array}{r|rrrr}
3 & 1 & 0 & -7 & 6 \\
  &   & 3 & 9 & 6 \\
\hline
  & 1 & 3 & 2 & \!\mid 12 \\
\end{array}
$$

Because the remainder is not 0, the number 3 is not a root and can be crossed off the list. Because the last row in the synthetic division is entirely positive, 3 is an upper bound. Thus, 6 cannot be a root either, and you can cross it off the list as well:

$$-6, \quad -3, \quad -2, \quad -1, \quad 1, \quad 2, \quad 3, \quad \cancel{6}$$

Now try $x = 2$:

$$
\begin{array}{r|rrrr}
2 & 1 & 0 & -7 & 6 \\
  &   & 2 & 4 & -6 \\
\hline
  & 1 & 2 & -3 & \!\mid 0 \\
\end{array}
$$

Because the remainder is 0, the number 2 is a root.

Thus, the binomial $x - 2$ is a factor of $P(x)$, and any remaining roots must be supplied by the remaining factor, which is the quotient $1x^2 + 2x - 3$. The other roots can be found by solving the equation

$$x^2 + 2x - 3 = 0$$

This equation, called the **depressed equation**, is a quadratic equation and can be solved by factoring:

$$x^2 + 2x - 3 = 0$$
$$(x - 1)(x + 3) = 0$$
$$x - 1 = 0 \quad \text{or} \quad x + 3 = 0$$
$$x = 1 \qquad \qquad x = -3$$

The solution set is $\{2, 1, -3\}$. ■

Example 4   Solve the equation $x^7 - 2x^6 - 5x^5 + 6x^4 - x^3 + 2x^2 + 5x - 6 = 0$.

Solution   Being of seventh degree, the equation must have seven roots. According to Descartes' rule of signs, there are six possible combinations of positive, negative, and nonreal roots for this equation.

Number of positive roots	Number of negative roots	Number of nonreal roots
5	2	0
3	2	2
1	2	4
5	0	2
3	0	4
1	0	6

The only possible rational roots are

$$-6, \; -3, \; -2, \; -1, \; 1, \; 2, \; 3, \; 6$$

Check each one, crossing off those that do not satisfy the equation. Begin with -3:

$$
\begin{array}{r|rrrrrrrr}
-3 & 1 & -2 & -5 & 6 & -1 & 2 & 5 & -6 \\
   &   & -3 & 15 & -30 & 72 & -213 & 633 & -1914 \\
\hline
   & 1 & -5 & 10 & -24 & 71 & -211 & 638 & -1920
\end{array}
$$

Because the last number in the synthetic division is not 0, -3 is not a root and can be crossed off the list. Because the last row is alternately positive and negative, -3 is a lower bound. Thus, you can cross off -6 as well:

$$-\!\!\!\!/\,6, \; -\!\!\!\!/\,3, \; -2, \; -1, \; 1, \; 2, \; 3, \; 6$$

Now try -2:

$$
\begin{array}{r|rrrrrrrr}
-2 & 1 & -2 & -5 & 6 & -1 & 2 & 5 & -6 \\
   &   & -2 & 8 & -6 & 0 & 2 & -8 & 6 \\
\hline
   & 1 & -4 & 3 & 0 & -1 & 4 & -3 & 0
\end{array}
$$

Because the remainder is 0, -2 is a root.

This root is negative, so you can revise the chart of positive/negative/nonreal possibilities. (Until now, 0 negative roots was a possibility.)

Number of positive roots	Number of negative roots	Number of nonreal roots
5	2	0
3	2	2
1	2	4

Because -2 is a root, the factor theorem asserts that $x - (-2)$, or $x + 2$, is a factor of $P(x)$. Any remaining roots can be found by solving the depressed equation

$$x^6 - 4x^5 + 3x^4 - x^2 + 4x - 3 = 0$$

Because the constant term of this equation is different from the constant term of the original equation, you can cross off some other possible rational roots. The numbers 2 and 6 must go because neither is a factor of the constant, 3. The number -2 cannot be a root a second time because -2 is not a factor of 3. The list of candidates is now

$$-\cancel{6}, \; -\cancel{3}, \; -2, \; -1, \; 1, \cancel{2}, \; 3, \cancel{6}$$

The only solution found thus far is -2. Try -1 next because you know there must be one more negative root. The coefficients are those of the depressed equation. (Don't forget the missing x^3.)

$$
\begin{array}{r|rrrrrrr}
-1 & 1 & -4 & 3 & 0 & -1 & 4 & -3 \\
   &   & -1 & 5 & -8 & 8 & -7 & 3 \\
\hline
   & 1 & -5 & 8 & -8 & 7 & -3 & 0
\end{array}
$$

Because the remainder is 0, the number -1 is another root. The roots found so far are -1 and -2. The root -1 cannot appear again because there can be only two negative roots, and you have found them both. The current list of candidates is now

$$-\cancel{6}, \; -\cancel{3}, \; -\cancel{2}, \; -\cancel{1}, \; 1, \cancel{2}, \; 3, \cancel{6}$$

Other roots can be found by solving the depressed equation

$$x^5 - 5x^4 + 8x^3 - 8x^2 + 7x - 3 = 0$$

Try 1 next:

$$
\begin{array}{r|rrrrrr}
1 & 1 & -5 & 8 & -8 & 7 & -3 \\
  &   & 1 & -4 & 4 & -4 & 3 \\
\hline
  & 1 & -4 & 4 & -4 & 3 & 0
\end{array}
$$

The solution 1 joins the growing list of solutions. To see if 1 is a multiple solution, try it again in the depressed equation.

$$
\begin{array}{r|rrrrr}
1 & 1 & -4 & 4 & -4 & 3 \\
  &   & 1 & -3 & 1 & -3 \\
\hline
  & 1 & -3 & 1 & -3 & 0
\end{array}
$$

Again, 1 is a root. Will it work a third time?

$$
\begin{array}{r|rrrr}
1 & 1 & -3 & 1 & -3 \\
  &   & 1 & -2 & -1 \\
\hline
  & 1 & -2 & -1 & -4
\end{array}
$$

No, 1 is only a double root.

Solutions found thus far are -2, -1, 1, and 1. Now try 3:

$$
\begin{array}{r|rrrr}
3 & 1 & -3 & 1 & -3 \\
  &   & 3 & 0 & 3 \\
\hline
  & 1 & 0 & 1 & 0
\end{array}
$$

The solution 3 is added to the list of roots. So far the roots are $-2, -1, 1, 1,$ and 3.

The depressed equation is now the quadratic equation $x^2 + 1 = 0$, which can be solved as follows:

$$x^2 + 1 = 0$$
$$x^2 = -1$$
$$x = i \quad \text{or} \quad x = -i$$

Because i and $-i$ are solutions, the complete solution set of the original seventh-degree equation is

$$\{-2, \quad -1, \quad 1, \quad 1, \quad 3, \quad i, \quad -i\}$$

Note that this list contains three positives, two negatives, and two conjugate complex numbers. This combination was one of the predicted possibilities.

■

Exercise 11.4

In Exercises 1–22, find all roots for each equation.

1. $x^3 - 2x^2 - x + 2 = 0$
2. $x^3 + 2x^2 - x - 2 = 0$
3. $x^4 - 10x^3 + 35x^2 - 50x + 24 = 0$
4. $x^4 + 4x^3 + 6x^2 + 4x + 1 = 0$
5. $x^5 - 2x^4 - 2x^3 + 4x^2 + x - 2 = 0$
6. $x^5 - x^3 - 8x^2 + 8 = 0$
7. $x^4 + 3x^3 - 13x^2 - 9x + 30 = 0$
8. $x^4 - 8x^3 + 14x^2 + 8x - 15 = 0$
9. $x^7 - 12x^5 + 48x^3 - 64x = 0$
10. $x^7 + 7x^6 + 21x^5 + 35x^4 + 35x^3 + 21x^2 + 7x + 1 = 0$
11. $6x^5 - 7x^4 - 48x^3 + 81x^2 - 4x - 12 = 0$
12. $x^6 - 3x^5 - x^4 + 9x^3 - 10x^2 + 12x - 8 = 0$
13. $4x^5 - 12x^4 + 15x^3 - 45x^2 - 4x + 12 = 0$
14. $12x^4 + 20x^3 - 41x^2 + 20x - 3 = 0$
15. $3x^4 - 14x^3 + 11x^2 + 16x - 12 = 0$
16. $2x^4 - x^3 - 2x^2 - 4x - 40 = 0$
17. $4x^4 - 8x^3 - x^2 + 8x - 3 = 0$
18. $3x^3 - 2x^2 + 12x - 8 = 0$
19. $30x^3 - 47x^2 - 9x + 18 = 0$
20. $x^{-5} - 8x^{-4} + 25x^{-3} - 38x^{-2} + 28x^{-1} - 8 = 0$
21. $1 - x^{-1} - x^{-2} - 2x^{-3} = 0$
22. $x^3 - \frac{19}{6}x^2 + \frac{1}{6}x + 1 = 0$
23. If n is an even positive integer and c is a positive constant, prove that the equation $x^n + c = 0$ has no real roots.
24. If n is an even positive integer and c is a positive constant, prove that the equation $x^n - c = 0$ has exactly two real roots.

11.5   IRRATIONAL ROOTS OF POLYNOMIAL EQUATIONS

First-degree equations are easy to solve, and quadratic equations can be solved by the quadratic formula. There are also formulas for solving general third- and fourth-degree polynomial equations, although these formulas are complicated.

However, there are no explicit algebraic formulas for solving polynomial equations of degree 5 or greater. This fact was proven for fifth-degree equations by the Norwegian mathematician Niels Henrik Abel (1802–1829) and for equations of degree greater than 5 by the French mathematician Évariste Galois (1811–1832). To solve a polynomial equation of high degree with integer coefficients, we could use the methods of Section 11.4 to find the rational roots. Once they were found, however, the remaining depressed equation would have to be a first- or second-degree equation or we could not finish. The purpose of this section is to provide techniques for approximating the real roots of equations of high degree, even though the exact roots cannot be found. The following theorem provides one method for locating a root.

Theorem.   Let $P(x)$ be a polynomial with real coefficients. If $P(a)$ and $P(b)$ have opposite signs, there is at least one number r between a and b for which $P(r) = 0$.

Justification.   A proof of this theorem requires the use of calculus. The theorem becomes plausible if we consider the graph of the polynomial $y = P(x)$. Graphs of polynomials are *continuous* curves, a technical term that means, roughly, that they can be drawn without lifting the pencil from the paper. If $P(a)$ and $P(b)$ have opposite signs, the points $A(a, P(a))$ and $B(b, P(b))$ on the graph of $y = P(x)$ lie on opposite sides of the x-axis, which separates them like a fence. The continuous curve joining A and B has no gaps; thus, it must cross the x-axis at least once. The point of crossing, $x = r$, is a zero of $P(x)$, and a solution of the equation $P(x) = 0$.   □

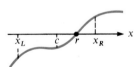

Figure 11-3

The previous theorem provides a method for finding roots of $P(x) = 0$ to any degree of accuracy desired. Suppose we find, by trial and error, the numbers x_L and x_R (for left and right) that straddle a root; that is, $x_L < x_R$ and $P(x_L)$ and $P(x_R)$ have opposite signs. For purposes of discussion, $P(x_L)$ will be negative and $P(x_R)$ will be positive. We compute a number c that is halfway between x_L and x_R (c is the average of x_L and x_R), and then evaluate $P(c)$. If $P(c)$ is 0, we've found a root. More likely, however, $P(c)$ will not be 0.

If $P(c)$ is negative, the root, r, lies between c and x_R, as shown in Figure 11-3. In such a case, let c become the *new* x_L, and repeat the procedure.

If $P(c)$ is positive, however, the root lies between x_L and c, as shown in Figure 11-4. In this case, let c become the new x_R, and repeat the process.

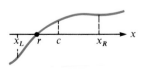

Figure 11-4

At any stage in this procedure, the root is contained between the current numbers x_L and x_R. If the original bounds were, say, 1 unit apart, after 10 repetitions of this procedure, the root would be contained between fences that were 2^{-10}, or about 0.001, units apart. After 20 repetitions the bounds are only 0.000001 units apart. The actual zero of $P(x)$ is within 0.000001 of either x_L or x_R. This procedure, called **binary chopping**, is well suited for solving equations by computer.

Example 1 Find $\sqrt{2}$ to two decimal places.

Solution $\sqrt{2}$ is a root of the polynomial equation $P(x) = x^2 - 2 = 0$. Note that the values $P(1) = 1^2 - 2 = -1$ and $P(2) = 2^2 - 2 = 2$ have opposite signs. Set x_L equal to 1 and x_R equal to 2 and compute the midpoint, $c = \dfrac{1+2}{2} = 1.5$.

Because $P(c) = P(1.5) = 0.25$ is a positive number, c becomes the new x_R and we calculate a new c. Tabulating the information in a chart helps keep things straight.

Step	x_L	c	x_R	$P(x_L)$	$P(c)$	$P(x_R)$
0	1	1.5	2	negative	positive	positive
1	1	1.25	1.5	negative	negative	positive

At this point, $P(c)$ and $P(x_R)$ are of opposite signs, so c becomes the new x_L. The process continues, but without a hand calculator it can be difficult.

Step	x_L	c	x_R	$P(x_L)$	$P(c)$	$P(x_R)$
0	1	1.5	2	negative	positive	positive
1	1	1.25	1.5	negative	negative	positive
2	1.25	1.375	1.5	negative	negative	positive
3	1.375	1.4375	1.5	negative	positive	positive
4	1.375	1.40625	1.4375	negative	negative	positive
5	1.40625	1.421875	1.4375	negative	positive	positive
6	1.40625	1.4140625	1.421875	negative	negative	positive
7	1.4140625	1.41796875	1.421875	negative	positive	positive
8	1.4140625	1.416015625	1.41796875	negative	positive	positive

In Step 7, the bounds x_R and x_L agree to only one decimal place; our approximation of $\sqrt{2}$ is 1.4. In Step 8, x_R and x_L agree to two decimal places. Thus, $\sqrt{2} \approx 1.41$. ■

Exercise 11.5

In Exercises 1–10, show that each equation has at least one real root between the numbers specified.

1. $2x^2 + x - 3 = 0$; $-2, -1$

2. $2x^3 + 17x^2 + 31x - 20 = 0$; $-1, 2$

3. $3x^3 - 11x^2 - 14x = 0$; 4, 5

4. $2x^3 - 3x^2 + 2x - 3 = 0$; 1, 2

5. $x^4 - 8x^2 + 15 = 0$; 1, 2

6. $x^4 - 8x^2 + 15 = 0$; 2, 3

7. $30x^3 + 10 = 61x^2 + 39x$; 2, 3

8. $30x^3 + 10 = 61x^2 + 39x$; $-1, 0$

9. $30x^3 + 10 = 61x^2 + 39x$; 0, 1

10. $5x^3 - 9x^2 - 4x + 9 = 0$; $-1, 2$

11. Use binary chopping to evaluate $\sqrt{3}$ to two decimal places.

12. Use binary chopping to evaluate $\sqrt[3]{53}$ to two decimal places.

13. Use binary chopping to evaluate $\sqrt{5}$ to three decimal places.

14. Use binary chopping to evaluate $\sqrt[3]{102}$ to three decimal places.

15. Use binary chopping to find a root of $2x^3 - x^2 + 2x - 1 = 0$ to two decimal places.

16. Use binary chopping to find a root of $35x^3 + 12x^2 + 8x + 1 = 0$ to one decimal place.

REVIEW EXERCISES

In Review Exercises 1–4, let $P(x) = 4x^4 + 2x^3 - 3x - 2$. Use synthetic division to evaluate the polynomial for the given value.

1. $P(0)$ **2.** $P(2)$ **3.** $P(-3)$ **4.** $P(\frac{1}{2})$

In Review Exercises 5–8, use the factor theorem to decide whether each statement is true. If not, so indicate.

5. $x - 2$ is a factor of $x^3 + 4x^2 - 2x + 4$.

6. $x + 3$ is a factor of $2x^4 + 10x^3 + 4x^2 + 7x + 21$.

7. $x - 5$ is a factor of $x^5 - 3125$.

8. $x - 6$ is a factor of $x^5 - 6x^4 - 4x + 24$.

9. Find the three cube roots of -64.

10. Find the three cube roots of 343.

11. Find the polynomial of lowest degree with zeros of -1, 2, and $\frac{3}{2}$.

12. Find the polynomial equation of lowest degree with roots of 1, -3, and $\frac{1}{2}$.

13. Use synthetic division to find the quotient when the polynomial $3x^4 + 2x^2 + 3x + 7$ is divided by $x - 3$.

14. Use synthetic division to find the quotient when the polynomial $5x^5 - 4x^4 + 3x^3 - 2x^2 + x - 1$ is divided by $x + 2$.

15. How many roots does the polynomial equation $3x^6 - 4x^5 + 3x + 2 = 0$ have?

16. How many roots does the equation $x^{1984} - 1 = 0$ have?

In Review Exercises 17–20, use Descartes' rule of signs to find the number of possible positive, negative, and nonreal roots.

17. $3x^4 + 2x^3 - 4x + 2 = 0$ **18.** $4x^5 + 3x^4 + 2x^3 + x^2 + x = 7$

19. $x^4 + x^2 + 24{,}567 = 0$ **20.** $-x^7 - 5 = 0$

21. Find all roots of the equation $2x^3 + 17x^2 + 41x + 30 = 0$.

22. Find all roots of the equation $3x^3 + 2x^2 + 2x = 1$.

23. Show that $5x^3 + 37x^2 + 59x + 18 = 0$ has a root between $x = 0$ and $x = -1$.

24. Show that $6x^3 - x^2 - 10x - 3 = 0$ has a root between $x = 1$ and $x = 2$.

25. Use binary chopping to find $\sqrt{7}$ to the nearest hundredth.

26. Use binary chopping to find an approximation of the root of the equation $0 = 3x - 1$. What is the exact root?

12 NATURAL NUMBER FUNCTIONS AND PROBABILITY

Karl Friedrich Gauss (1777–1855) Many people consider Gauss to be the greatest mathematician of all time. He often is called the "prince of the mathematicians."

The brilliant German mathematician Karl Friedrich Gauss (1777–1855) was once a student in the class of a very strict teacher. To keep the class busy one day, the teacher asked the students to add together all of the natural numbers from 1 through 100. Gauss immediately wrote the sum on his slate and put it on the teacher's desk.

Gauss's solution was relatively simple. He recognized that in the sum $1 + 2 + 3 + \cdots + 98 + 99 + 100$, the first number (1) added to the last number (100) was 101. Similarly, the second number (2) added to the second from the last number (99) was 101. The third number (3) added to the third from the last number (98) was 101 also. Gauss realized that this pattern continued, and because there were fifty pairs of numbers there were fifty sums of 101. He multiplied 101 by 50 and obtained the correct answer of 5050.

This story illustrates a group of problems involving long strings of numbers called *sequences*. We begin a discussion of sequences by considering a method of proof that is used to prove sequence formulas. This method of proof, called **mathematical induction**, was first used extensively by Giuseppe Peano (1858–1932).

12.1 MATHEMATICAL INDUCTION

Suppose we ask a theatergoer whether everyone in line for a movie gained admittance. The theatergoer answers that "the first person in line was admitted." Does this response answer our question? Certainly more than one person was admitted, but this does not mean that everyone was admitted. As a matter of fact, we know little more now than before we asked the question.

Meeting a second theatergoer, we ask the question again. This time the answer is "they promised that, if anyone was admitted, the person next in line would also be admitted." On the basis of this response we know that, if anyone was admitted, the person behind was admitted also. But a promise that begins with "if someone is admitted" does not guarantee that anyone actually was admitted.

However, when we consider both the first and second answers, we know that everyone in line gained admittance. The first theatergoer said that the first person got in; the second theatergoer said that, if anyone got in, the next person in line got in also. Because the first person was admitted, the second person was admitted also. And, if the second person was admitted, then so was the third. This pattern would have continued until everyone was admitted to the theater.

This situation is very similar to a children's game played with dominoes. The dominoes are placed on end fairly close together in a row. When the first domino is pushed over, it falls against the second, knocking it down. The second domino, in turn, knocks down the third, which topples the fourth, and so on until all of the dominoes fall. Two things must happen to guarantee that all of the dominoes fall: (1) The first domino must be knocked over, and (2) every domino that falls must topple the next one. When both of these conditions are met, it is certain that all of the dominoes will fall.

The preceding examples illustrate the basic idea that underlies the principle of mathematical induction.

The Axiom of Mathematical Induction. If a statement involving the natural number n has the two properties that

1. the statement is true for $n = 1$, and
2. the statement is true for $n = k + 1$ whenever it is true for $n = k$,

then the statement is true for all natural numbers.

The axiom of mathematical induction provides a method for proving many theorems. Note that any such proof by induction involves two parts: We must first show that the formula is true for the natural number 1, and then show that, *if* the formula is true for any natural number k, then it also is true for the natural number $k + 1$. A proof by induction is complete only if both of the required properties are established.

Let us return to Gauss's problem of finding the sum of the first 100 natural numbers. There is a formula for finding the sum of the first n natural numbers:

$$1 + 2 + 3 + \cdots + n = \frac{n(n + 1)}{2}$$

To show that this formula is correct, we will prove it by using the axiom of mathematical induction.

Example 1   Use mathematical induction to prove that the formula

$$1 + 2 + 3 + \cdots + n = \frac{n(n + 1)}{2}$$

is true for every natural number n.

Solution   The proof has two parts.

Part 1.   Verify that the formula is true for the value $n = 1$. Substituting $n = 1$ into the term n on the left side of the equation yields a single term, the number 1. Substituting the number 1 for n on the right side, the formula becomes

$$1 = \frac{n(n + 1)}{2}$$

$$1 = \frac{(1)(1 + 1)}{2}$$

$$1 = 1$$

Thus, the formula is true for $n = 1$, and Part 1 of the proof is complete.

Part 2. Assume that the given formula is true when n is replaced by *some* natural number k. By this assumption, called the **induction hypothesis**, you accept that

1. $1 + 2 + 3 + \cdots + k = \dfrac{k(k + 1)}{2}$

is a true statement. The plan is to show that the given formula is true for the next natural number, $k + 1$. Do this by verifying the equation

2. $1 + 2 + 3 + \cdots + k + (k + 1) = \dfrac{(k + 1)[(k + 1) + 1]}{2}$

obtained from the given formula by replacing n with $k + 1$.

Compare the left sides of Equations 1 and 2 and note that the left side of Equation 2 contains an extra term of $k + 1$. Hence, add $k + 1$ to both sides of Equation 1 (which was assumed to be true) to obtain the equation

$$1 + 2 + 3 + \cdots + k + (k + 1) = \frac{k(k + 1)}{2} + (k + 1)$$

Because both terms on the right side of this equation have a common factor of $k + 1$, the right side factors and the equation can be rewritten as follows:

$$1 + 2 + 3 + \cdots + k + (k + 1) = (k + 1)\left(\frac{k}{2} + 1\right)$$

$$= \frac{(k + 1)(k + 2)}{2}$$

$$= \frac{(k + 1)[(k + 1) + 1]}{2}$$

This final result is Equation 2. Because the truth of Equation 1 implies the truth of Equation 2, Part 2 of the proof is complete. Parts 1 and 2 together establish that the formula

$$1 + 2 + 3 + \cdots + n = \frac{n(n + 1)}{2}$$

is true for any natural number n. ■

Here is a brief overview of Example 1.

1. Did the first domino fall? That is, is the formula

$$1 + 2 + 3 + \cdots + n = \frac{n(n + 1)}{2}$$

true for $n = 1$? Yes, Part 1 verified this.

2. Will toppling any domino knock over the next domino? If the given formula is true for the value $n = k$, is it also true for the value $n = k + 1$? Yes, Part 2 of the proof verified this.

Because both of the induction requirements were verified, the formula is true for all natural numbers n.

Example 2   Use mathematical induction to prove the formula

$$1 + 5 + 9 + \cdots + (4n - 3) = n(2n - 1)$$

for all natural numbers n.

Solution   The proof has two parts.

Part 1.   First verify the formula for the value $n = 1$. Substituting the value $n = 1$ into the term $4n - 3$ on the left side of the formula gives the single term 1. After substituting the same value into the right side, the equation becomes

$$1 = 1[2(1) - 1]$$
$$1 = 1$$

Thus, the formula is true for $n = 1$, and Part 1 of the proof is complete.

Part 2.   The induction hypothesis is the assumption that the formula is true for $n = k$. Hence, you assume that

$$1 + 5 + 9 + \cdots + (4k - 3) = k(2k - 1)$$

is a true statement. Because the truth of this assumption must guarantee the truth of the formula for $k + 1$ terms, add the $(k + 1)$th term to both sides of the induction hypothesis formula. In this example, the terms on the left side increase by 4, so the $(k + 1)$th term is $(4k - 3) + 4$, or $4k + 1$. Adding $4k + 1$ to both sides of the induction hypothesis formula gives

$$1 + 5 + 9 + \cdots + (4k - 3) + (4k + 1) = k(2k - 1) + (4k + 1)$$

Simplify the right side and rewrite the equation as follows:

$$1 + 5 + 9 + \cdots + (4k - 3) + [4(k + 1) - 3] = 2k^2 + 3k + 1$$
$$= (k + 1)(2k + 1)$$
$$= (k + 1)[2(k + 1) - 1]$$

Because the above equation has the same form as the given formula, except that $k + 1$ appears in place of n, the truth of the formula for $n = k$ implies the truth of the formula for $n = k + 1$. Part 2 of the proof is complete.

Because both of the induction requirements have been verified, the given formula is proved for all natural numbers.   ∎

Example 3   Prove that $\dfrac{1}{2} + \dfrac{1}{4} + \dfrac{1}{8} + \cdots + \dfrac{1}{2^n} < 1$.

Solution   The proof is by induction.

Part 1.   Verify the formula for $n = 1$. Substituting 1 for n on the left side of the inequality gives $\frac{1}{2} < 1$. Thus, the formula is true for $n = 1$, and Part 1 of the proof is complete.

Part 2.   The induction hypothesis is the assumption that the inequality is true for $n = k$. Thus, assume that

$$\frac{1}{2} + \frac{1}{4} + \frac{1}{8} + \cdots + \frac{1}{2^k} < 1$$

Multiply both sides of the above inequality by $\frac{1}{2}$ to obtain the inequality

$$\frac{1}{2}\left(\frac{1}{2} + \frac{1}{4} + \frac{1}{8} + \cdots + \frac{1}{2^k}\right) < 1\left(\frac{1}{2}\right)$$

or

$$\frac{1}{4} + \frac{1}{8} + \frac{1}{16} + \cdots + \frac{1}{2^{k+1}} < \frac{1}{2}$$

Now add $\frac{1}{2}$ to both sides of this inequality to obtain

$$\frac{1}{2} + \frac{1}{4} + \frac{1}{8} + \frac{1}{16} + \cdots + \frac{1}{2^{k+1}} < \frac{1}{2} + \frac{1}{2}$$

or

$$\frac{1}{2} + \frac{1}{4} + \frac{1}{8} + \frac{1}{16} + \cdots + \frac{1}{2^{k+1}} < 1$$

The resulting inequality is the same as the original except that $k + 1$ appears in place of n. The truth of the inequality for $n = k$ implies the truth of the inequality for $n = k + 1$. Part 2 of the proof is complete.

Because both of the induction requirements have been verified, this inequality is true for all natural numbers.   ∎

There are statements that are not true when $n = 1$, but that are true for all natural numbers equal to or greater than some given natural number, say q. In these cases, verify the given statements for $n = q$ in Part 1 of the induction proof. After establishing Part 2 of the induction proof, the given statement is proved for all natural numbers that are greater than or equal to q.

Exercise 12.1

In Exercises 1–4, verify each given formula for n = 1, 2, 3, and 4.

1. $5 + 10 + 15 + \cdots + 5n = \dfrac{5n(n + 1)}{2}$

2. $1^2 + 2^2 + 3^2 + \cdots + n^2 = \dfrac{n(n + 1)(2n + 1)}{6}$

3. $7 + 10 + 13 + \cdots + (3n + 4) = \dfrac{n(3n + 11)}{2}$

4. $1(3) + 2(4) + 3(5) + \cdots + n(n + 2) = \dfrac{n}{6}(n + 1)(2n + 7)$

In Exercises 5–20, prove each of the following formulas by mathematical induction, if possible.

5. $2 + 4 + 6 + \cdots + 2n = n(n + 1)$

6. $1 + 3 + 5 + \cdots + (2n - 1) = n^2$

7. $3 + 7 + 11 + \cdots + (4n - 1) = n(2n + 1)$

8. $4 + 8 + 12 + \cdots + 4n = 2n(n + 1)$

9. $10 + 6 + 2 + \cdots + (14 - 4n) = 12n - 2n^2$

10. $8 + 6 + 4 + \cdots + (10 - 2n) = 9n - n^2$

11. $2 + 5 + 8 + \cdots + (3n - 1) = \dfrac{n(3n + 1)}{2}$

12. $3 + 6 + 9 + \cdots + 3n = \dfrac{3n(n + 1)}{2}$

13. $1^2 + 2^2 + 3^2 + \cdots + n^2 = \dfrac{n(n + 1)(2n + 1)}{6}$

14. $1 + 2 + 3 + \cdots + (n - 1) + n + (n - 1) + \cdots + 3 + 2 + 1 = n^2$

15. $\dfrac{1}{3} + 2 + \dfrac{11}{3} + \cdots + \left(\dfrac{5}{3}n - \dfrac{4}{3}\right) = n\left(\dfrac{5}{6}n - \dfrac{1}{2}\right)$

16. $\dfrac{1}{1 \cdot 2} + \dfrac{1}{2 \cdot 3} + \dfrac{1}{3 \cdot 4} + \cdots + \dfrac{1}{n(n + 1)} = \dfrac{n}{n + 1}$

17. $\dfrac{1}{2} + \dfrac{1}{4} + \dfrac{1}{8} + \cdots + \left(\dfrac{1}{2}\right)^n = 1 - \left(\dfrac{1}{2}\right)^n$

18. $\dfrac{1}{3} + \dfrac{2}{9} + \dfrac{4}{27} + \cdots + \dfrac{1}{3}\left(\dfrac{2}{3}\right)^{n-1} = 1 - \left(\dfrac{2}{3}\right)^n$

19. $2^0 + 2^1 + 2^2 + 2^3 + \cdots + 2^{n-1} = 2^n - 1$

20. $1^3 + 2^3 + 3^3 + \cdots + n^3 = \left[\dfrac{n(n + 1)}{2}\right]^2$

21. Prove that $x - y$ is a factor of $x^n - y^n$. (*Hint:* Consider subtracting and adding xy^k to the binomial $x^{k+1} - y^{k+1}$.)

22. Prove that $n < 2^n$.

23. There are $180°$ in the sum of the angles of any triangle. Prove by induction that $(n - 2)180°$ gives the sum of the angles of any simple polygon when n is the number of sides of that polygon. (*Hint:* If a polygon has $k + 1$ sides, it has $k - 2$ sides plus three more sides.)

24. Consider the equation $1 + 3 + 5 + \cdots + 2n - 1 = 3n - 2$.
   a. Is the equation true for $n = 1$?
   b. Is the equation true for $n = 2$?
   c. Is the equation true for all natural numbers n?

25. If $1 + 2 + 3 + \cdots + n = \dfrac{n}{2}(n + 1) + 1$ were true for $n = k$, show that it would be true for $n = k + 1$.

   Is it true for $n = 1$?

26. Prove that $n + 1 = 1 + n$ for each natural number n.

27. If n is any natural number, prove that $7^n - 1$ is divisible by 6.

28. Prove that $1 + 2n < 3^n$ for $n > 1$.

29. Prove that, if r is a real number where $r \neq 1$, then $1 + r + r^2 + \cdots + r^n = \dfrac{1 - r^{n+1}}{1 - r}$.

30. The expression a^m where m is a natural number was defined in Section 1.3. An alternative definition of a^m, useful in proofs by induction, is (Part 1) $a^1 = a$ and (Part 2) $a^{m+1} = a^m \cdot a$. Use mathematical induction on n to prove the familiar law of exponents, $a^m a^n = a^{m+n}$.

12.2   SEQUENCES, SERIES, AND SUMMATION NOTATION

We now formally define a **sequence**.

> **Definition.**   A **sequence** is a function whose domain is the set of natural numbers.

Since a sequence is a function whose domain is the set of natural numbers, we can write its values as a list of numbers. For example, if n is a natural number, the function defined by $f(n) = 2n - 1$ generates the list

$$1, 3, 5, \ldots, 2n - 1, \ldots$$

It is common to call such a list, as well as the function, a sequence. The number 1 is the first term of this sequence, the number 3 is the second term, and the expression $2n - 1$ represents the **general**, or **nth**, **term** of the sequence. Likewise, if n is a natural number, then the function defined by $f(n) = 3n^2 + 1$ generates the list

$$4, 13, 28, \ldots, 3n^2 + 1, \ldots$$

The number 4 is the first term, 13 is the second term, 28 is the third term, and $3n^2 + 1$ is the general term.

Because the domain of any sequence is the infinite set of natural numbers, the sequence itself is an unending list of numbers. Note that a constant function such as $g(n) = 1$ is a sequence also; it generates the list $1, 1, 1, \ldots$.

Many times, sequences do not lend themselves to functional notation because it is difficult or even impossible to write the general term—the expression that shows how the terms are constructed. In such cases, if there is a pattern that is assumed to be continued, it is acceptable simply to list several terms of the sequence. Some examples of sequences follow:

$$1^2, 2^2, 3^2, \ldots, n^2, \ldots$$
$$3, 9, 19, 33, \ldots, 2n^2 + 1, \ldots$$

$$1, 3, 6, 10, 15, 21, \ldots, \frac{n(n+1)}{2}, \ldots$$

$1, 1, 2, 3, 5, 8, 13, 21, \ldots$ (Fibonacci sequence)

$2, 3, 5, 7, 11, 13, 17, 19, 23, \ldots$ (prime numbers)

The fourth example listed is called the Fibonacci sequence, after the twelfth-century mathematician Leonardo of Pisa—known to his friends as Fibonacci. After the two 1's in the Fibonacci sequence, each term is the sum of the two terms that immediately precede it. The Fibonacci sequence occurs in the study of botany, for example, in the growth patterns of certain plants.

To add the terms of a sequence, we replace each comma between the terms with a plus sign, forming what is called a **series**. Because each sequence is infinite, the number of terms in the series associated with it is infinite also. Two examples of infinite series are

$$1^2 + 2^2 + 3^2 + \cdots + n^2 + \cdots$$

and

$$1 + 2 + 3 + 5 + 8 + 13 + 21 + \cdots$$

There is a shorthand method of indicating the sum of the first n terms, or the **nth partial sum** of a sequence. This method, called **summation notation**, involves the symbol $\sum$, which is capital sigma in the Greek alphabet. The expression

$$\sum_{n=1}^{3} (2n^2 + 1)$$

designates the sum of the three terms obtained if we successively substitute the natural numbers 1, 2, and 3 for n in the expression $2n^2 + 1$. Hence,

$$\sum_{n=1}^{3} (2n^2 + 1) = [2(1)^2 + 1] + [2(2)^2 + 1] + [2(3)^2 + 1]$$

$$= 3 + 9 + 19$$

$$= 31$$

Example 1   Evaluate $\displaystyle\sum_{n=1}^{4} (n^2 - 1)$.

Solution   In this example, n is said to run from 1 to 4. Hence, substitute 1, 2, 3, and 4 for n in the expression $n^2 - 1$, and find the sum of the resulting values:

$$\sum_{n=1}^{4} (n^2 - 1) = (1^2 - 1) + (2^2 - 1) + (3^2 - 1) + (4^2 - 1)$$

$$= 0 + 3 + 8 + 15$$

$$= 26$$ ■

Example 2   Evaluate $\displaystyle\sum_{n=3}^{5} (3n + 2)$.

Solution   In this example, n runs from 3 to 5. Hence, substitute 3, 4, and 5 for n in the expression $3n + 2$, find the sum of the resulting values:

$$\sum_{n=3}^{5} (3n + 2) = [3(3) + 2] + [3(4) + 2] + [3(5) + 2]$$

$$= 11 + 14 + 17$$

$$= 42$$ ∎

The following theorems give three properties of summations.

Theorem.   If c is a constant, then $\displaystyle\sum_{k=1}^{n} c = nc$.

Proof   Because c is a constant, each term is c for each value of k as k runs from 1 to n.

$$\sum_{k=1}^{n} c = \overbrace{c + c + c + c + \cdots + c}^{n \text{ number of } c\text{'s}} = nc$$ □

In words, this theorem states that the summation of a constant as k runs from 1 to n is n times that constant.

Example 3   Evaluate $\displaystyle\sum_{n=1}^{5} 13$.

Solution $$\sum_{n=1}^{5} 13 = 13 + 13 + 13 + 13 + 13$$

$$= 5(13)$$

$$= 65$$ ∎

Theorem.   If c is a constant, then $\displaystyle\sum_{k=1}^{n} cf(k) = c \sum_{k=1}^{n} f(k)$.

Proof $$\sum_{k=1}^{n} cf(k) = cf(1) + cf(2) + cf(3) + \cdots + cf(n)$$

$$= c[f(1) + f(2) + f(3) + \cdots + f(n)] \qquad \text{Factor out } c.$$

$$= c \sum_{k=1}^{n} f(k)$$ □

In words, this theorem states that a constant factor may be brought outside a summation sign.

Example 4   Show that $\displaystyle\sum_{k=1}^{3} 5k^2 = 5 \sum_{k=1}^{3} k^2$.

Solution
$$\sum_{k=1}^{3} 5k^2 = 5(1)^2 + 5(2)^2 + 5(3)^2$$
$$= 5 + 20 + 45$$
$$= 70$$

$$5 \sum_{k=1}^{3} k^2 = 5[(1)^2 + (2)^2 + (3)^2]$$
$$= 5[1 + 4 + 9]$$
$$= 5(14)$$
$$= 70 \qquad \blacksquare$$

Theorem. $\displaystyle\sum_{k=1}^{n} [f(k) + g(k)] = \sum_{k=1}^{n} f(k) + \sum_{k=1}^{n} g(k)$

Proof
$$\sum_{k=1}^{n} [f(k) + g(k)] = [f(1) + g(1)] + [f(2) + g(2)]$$
$$+ [f(3) + g(3)] + \cdots + [f(n) + g(n)]$$
$$= [f(1) + f(2) + f(3) + \cdots + f(n)]$$
$$+ [g(1) + g(2) + g(3) + \cdots + g(n)]$$
$$= \sum_{k=1}^{n} f(k) + \sum_{k=1}^{n} g(k) \qquad \square$$

In words, this theorem states that the summation of a sum is equal to the sum of the summations.

Example 5   Show that $\displaystyle\sum_{k=1}^{3} (k + k^2) = \sum_{k=1}^{3} k + \sum_{k=1}^{3} k^2$.

Solution
$$\sum_{k=1}^{3} (k + k^2) = (1 + 1^2) + (2 + 2^2) + (3 + 3^2)$$
$$= 2 + 6 + 12$$
$$= 20$$

$$\sum_{k=1}^{3} k + \sum_{k=1}^{3} k^2 = (1 + 2 + 3) + (1^2 + 2^2 + 3^2)$$
$$= 6 + 14$$
$$= 20 \qquad \blacksquare$$

Example 6   Evaluate $\displaystyle\sum_{k=1}^{5} (2k - 1)^2$ directly. Then expand the binomial, apply the previous theorems, and evaluate the expression again.

Solution   *Part 1.* $\displaystyle\sum_{k=1}^{5} (2k - 1)^2 = 1 + 9 + 25 + 49 + 81 = 165$

Part 2. $\displaystyle\sum_{k=1}^{5} (2k - 1)^2 = \sum_{k=1}^{5} (4k^2 - 4k + 1)$

$\displaystyle = \sum_{k=1}^{5} 4k^2 + \sum_{k=1}^{5} (-4k) + \sum_{k=1}^{5} 1$ The summation of a sum is the sum of the summations.

$\displaystyle = 4 \sum_{k=1}^{5} k^2 - 4 \sum_{k=1}^{5} k + \sum_{k=1}^{5} 1$ Bring the constant factors outside the summation sign.

$\displaystyle = 4 \sum_{k=1}^{5} k^2 - 4 \sum_{k=1}^{5} k + 5$ The summation of a constant as k runs from 1 to 5 is 5 times that constant.

$= 4(1 + 4 + 9 + 16 + 25) - 4(1 + 2 + 3 + 4 + 5) + 5$

$= 4(55) - 4(15) + 5 = 220 - 60 + 5$

$= 165$

Note that the sum is 165, regardless of the method used.   ■

Exercise 12.2

1. Write the first eight terms of the sequence defined by the function $f(n) = 5n(n - 1)$.

2. Write the first six terms of the sequence defined by the function $f(n) = n\left(\dfrac{n - 1}{2}\right)\left(\dfrac{n - 2}{3}\right)$.

In Exercises 3–8, write the fifth term in each of the given sequences.

3. 1, 6, 11, 16, . . .

4. 1, 8, 27, 64, . . .

5. $a, a + d, a + 2d, a + 3d, \ldots$

6. $a, ar, ar^2, ar^3, \ldots$

7. 1, 3, 6, 10, . . .

8. 20, 17, 13, 8, . . .

In Exercises 9–16, find the sum of the first five terms of the sequence with the given general term.

9. n

10. $2k$

11. 3

12. $4k^0$

13. $2\left(\dfrac{1}{3}\right)^n$

14. $(-1)^n$

15. $3n - 2$

16. $2k + 1$

In Exercises 17–28, evaluate each sum.

17. $\displaystyle\sum_{k=1}^{5} 2k$

18. $\displaystyle\sum_{k=3}^{6} 3k$

19. $\displaystyle\sum_{k=3}^{4} (-2k^2)$

20. $\displaystyle\sum_{k=1}^{100} 5$

21. $\displaystyle\sum_{k=1}^{5} (3k - 1)$

22. $\displaystyle\sum_{n=2}^{5} (n^2 + 3n)$

23. $\displaystyle\sum_{k=1}^{1000} \dfrac{1}{2}$

24. $\displaystyle\sum_{x=4}^{5} \dfrac{2}{x}$

25. $\displaystyle\sum_{x=3}^{4} \dfrac{1}{x}$

26. $\displaystyle\sum_{x=2}^{6} (3x^2 + 2x) - 3 \sum_{x=2}^{6} x^2$

27. $\displaystyle\sum_{x=1}^{4} (4x + 1)^2 - \sum_{x=1}^{4} (4x - 1)^2$

28. $\displaystyle\sum_{x=0}^{10} (2x - 1)^2 + 4 \sum_{x=0}^{10} x(1 - x)$

In Exercises 29–31, use mathematical induction to prove each statement.

29. $\displaystyle\sum_{k=1}^{n} (4k - 3) = n(2n - 1)$

30. $\displaystyle\sum_{k=1}^{n} (5k - 3) = \frac{n(5n - 1)}{2}$

31. $\displaystyle\sum_{k=1}^{n} (6k + 4) = n(3n + 7)$

32. Construct an example to disprove the proposition that the summation of a product is the product of the summations. In other words, prove that

$$\sum_{k=1}^{n} f(k)g(k) \quad \text{is not always equal to} \quad \sum_{k=1}^{n} f(k) \sum_{k=1}^{n} g(k)$$

12.3 ARITHMETIC AND GEOMETRIC PROGRESSIONS

Some important sequences are called **progressions**. One of these is the arithmetic progression.

Definition.   An **arithmetic progression** is a sequence of the form

$$a, a + d, a + 2d, a + 3d, \ldots, a + (n - 1)d, \ldots$$

where a is the first term, $a + (n - 1)d$ is the nth term, and d is the common difference.

In this definition, note that the second term of the progression has an addend of d, the third term has an addend of $2d$, the fourth term has an addend of $3d$, and so on. This is why the nth term has an addend of $(n - 1)d$.

Example 1   For an arithmetic progression that has a first term of 7 and a common difference of 5, write the first six terms and the 21st term of the progression.

Solution   Because the first term, a, is 7 and the common difference, d, is 5, the first six terms are

$$7, \quad 7 + 5, \quad 7 + 2(5), \quad 7 + 3(5), \quad 7 + 4(5), \quad 7 + 5(5)$$

or

$$7, \quad 12, \quad 17, \quad 22, \quad 27, \quad 32$$

The nth term is $a + (n - 1)d$, and, because you are looking for the 21st term, $n = 21$

$$n\text{th term} = a + (n - 1)d$$
$$21\text{st term} = 7 + (21 - 1)5$$
$$= 7 + (20)5$$
$$= 107$$

The 21st term is 107.   ∎

Example 2   For an arithmetic progression with the first three terms 2, 6, and 10, find the 98th term.

Solution   In this example, $a = 2$, $n = 98$, and $d = 6 - 2 = 10 - 6 = 4$. The nth term is given by the formula $a + (n - 1)d$. Therefore, the 98th term is

$$n\text{th term} = a + (n - 1)d$$
$$98\text{th term} = 2 + (98 - 1)4$$
$$= 2 + (97)4$$
$$= 390$$
■

   Numbers inserted between a first and last term to form a segment of an arithmetic progression are called **arithmetic means**. In this type of problem, the last term, l, is considered the nth term:

$$l = a + (n - 1)d$$

Example 3   Insert three arithmetic means between the numbers -3 and 12.

Solution   Begin by finding the common difference d. In this example the first term is -3 and the last term is 12. Because you are inserting three terms, the total number of terms is five. Thus, $a = -3$, $l = 12$, and $n = 5$. The formula for the nth (or last) term is

$$l = a + (n - 1)d$$

Substituting 12 for l, -3 for a, and 5 for n in the formula and solving for d gives

$$12 = -3 + (5 - 1)d$$
$$15 = 4d$$
$$\frac{15}{4} = d$$

Once the common difference has been found, the arithmetic means are the second, third, and fourth terms of the arithmetic progression with a first term of -3 and a fifth term of 12:

$$a + d = -3 + \frac{15}{4} = \frac{3}{4}$$
$$a + 2d = -3 + \frac{30}{4} = 4\frac{1}{2}$$
$$a + 3d = -3 + \frac{45}{4} = 8\frac{1}{4}$$

The three arithmetic means are $\frac{3}{4}$, $4\frac{1}{2}$, and $8\frac{1}{4}$.   ■

   The formula stated in the following theorem gives the sum of the first n terms of an arithmetic progression.

Theorem. The formula

$$S_n = \frac{n(a + l)}{2}$$

gives the sum of the first n terms of an arithmetic progression. In this formula, a is the first term, l is the last (or nth) term, and n is the number of terms.

Proof We write the first n terms of an arithmetic progression letting S_n represent their sum, rewrite the same sum in reverse order, and add the equations together term by term:

$$S_n = \qquad a \qquad + \qquad (a + d) \qquad + \cdots + \; [a + (n - 2)d] + [a + (n - 1)d]$$
$$S_n = [a + (n - 1)d] + [a + (n - 2)d] + \cdots + \qquad (a + d) \qquad + \qquad a$$
$$2S_n = [2a + (n - 1)d] + [2a + (n - 1)d] + \cdots + [2a + (n - 1)d] + [2a + (n - 1)d]$$

Because there are n equal terms on the right side of the previous equation,

$$2S_n = n[2a + (n - 1)d]$$

or

$$2S_n = n\{a + [a + (n - 1)d]\}$$

Because $a + (n - 1)d = l$, we make that substitution in the right side of the above equation and divide both sides by 2 to obtain

$$S_n = \frac{n(a + l)}{2}$$

The theorem is proved. (Exercise 45 will ask you to prove this theorem again using mathematical induction.) □

Example 4 Find the sum of the first 30 terms of the arithmetic progression $5, 8, 11, \ldots$.

Solution In this example, $a = 5$, $n = 30$, $d = 3$, and $l = 5 + 29(3) = 92$. Substituting these values into the formula $S_n = \dfrac{n(a + l)}{2}$ and simplifying gives

$$S_{30} = \frac{30(5 + 92)}{2} = 15(97) = 1455$$

The sum of the first 30 terms is 1455. ■

Another important sequence is the **geometric progression**.

> **Definition.** A **geometric progression** is a sequence of the form
>
> $$a, ar, ar^2, ar^3, \ldots, ar^{n-1}, \ldots$$
>
> where a is the first term, ar^{n-1} is the nth term, and r is the common ratio.

In this definition, note that the second term of the progression has a factor of r^1, the third term has a factor of r^2, the fourth term has a factor of r^3, and so on. This explains why the nth term has a factor of r^{n-1}.

Example 5   For a geometric progression with a first term of 3 and a common ratio of 2, write the first six terms and the 15th term of the progression.

Solution   Write the first six terms of the geometric progression:

$$3, \quad 3(2), \quad 3(2)^2, \quad 3(2)^3, \quad 3(2)^4, \quad 3(2)^5$$

or

$$3, \quad 6, \quad 12, \quad 24, \quad 48, \quad 96$$

To obtain the 15th term, substitute 15 for n, 3 for a, and 2 for r in the formula for the nth term:

$$n\text{th term} = ar^{n-1}$$
$$15\text{th term} = 3(2)^{15-1}$$
$$= 3(2)^{14}$$
$$= 3(16,384)$$
$$= 49,152 \qquad \blacksquare$$

Example 6   For a geometric progression with the first three terms 9, 3, and 1, find the eighth term.

Solution   In this example, $a = 9$, $r = \frac{1}{3}$, $n = 8$, and the nth term is ar^{n-1}. To obtain the eighth term, substitute these values into the expression for the nth term:

$$n\text{th term} = ar^{n-1}$$
$$8\text{th term} = 9\left(\frac{1}{3}\right)^{8-1}$$
$$= 9\left(\frac{1}{3}\right)^7$$
$$= \frac{1}{243} \qquad \blacksquare$$

As with arithmetic progressions, numbers may be inserted between a first and last term to form a segment of a geometric progression. The numbers inserted are called **geometric means**. In this type of problem, the last term, l, is considered to be the nth term: $l = ar^{n-1}$.

Example 7   Insert two geometric means between the numbers 4 and 256.

Solution   Begin by finding the common ratio. The first term, a, is 4. Because 256 is to be the fourth term, $n = 4$ and $l = 256$. Substituting these values into the formula for the nth term of a geometric progression, and solving for r gives

$$ar^{n-1} = l$$
$$4r^{4-1} = 256$$
$$r^3 = 64$$
$$r = 4$$

The common ratio is 4. The two geometric means are the second and third terms of the geometric progression:

$$ar = 4 \cdot 4 = 16$$
$$ar^2 = 4 \cdot 4^2 = 4 \cdot 16 = 64$$

The first four terms of the geometric progression are 4, 16, 64, and 256; 16 and 64 are geometric means between 4 and 256.   ∎

There is a formula that gives the sum of the first n terms of a geometric progression.

Theorem.   The formula

$$S_n = \frac{a - ar^n}{1 - r} \quad (r \neq 1)$$

gives the sum of the first n terms of a geometric progression. In this formula, S_n is the sum, a is the first term, r is the common ratio, and n is the number of terms.

Proof   We write out the sum of the first n terms of the geometric progression:

1. $S_n = a + ar + ar^2 + \cdots + ar^{n-3} + ar^{n-2} + ar^{n-1}$

Multiplying both sides of this equation by r gives

2. $S_n r = \quad ar + ar^2 + \quad \cdots \quad + ar^{n-2} + ar^{n-1} + ar^n$

We now subtract Equation 2 from Equation 1 and solve for S_n:

$$S_n - S_n r = a - ar^n$$
$$S_n(1 - r) = a - ar^n$$
$$S_n = \frac{a - ar^n}{1 - r}$$

The theorem is proved. (Exercise 46 will ask you to prove this theorem using mathematical induction.)   □

Example 8   Find the sum of the first six terms of the geometric progression 8, 4, 2, ...

Solution   In this example, $a = 8$, $n = 6$, and $r = \frac{1}{2}$. Substituting these values into the formula for the sum of the first n terms of a geometric progression gives

$$S_n = \frac{a - ar^n}{1 - r}$$

$$S_6 = \frac{8 - 8\left(\frac{1}{2}\right)^6}{1 - \frac{1}{2}}$$

$$= 2\left(\frac{63}{8}\right)$$

$$= \frac{63}{4}$$

The sum of the first six terms is $\frac{63}{4}$.   ■

Under certain conditions, it is possible to find the sum of all the terms in an infinite geometric progression. To define this sum, we consider the geometric progression

$$a_1, a_2, a_3, \ldots$$

The first partial sum, S_1, of the progression is a_1. Hence,

$$S_1 = a_1$$

The second partial sum, S_2, of this progression is $a_1 + a_2$. Hence,

$$S_2 = a_1 + a_2$$

In general, the nth partial sum, S_n, of this progression is

$$S_n = a_1 + a_2 + a_3 + \cdots + a_n$$

If the nth partial sum, S_n, of an infinite geometric progression approaches some number S as n becomes large without bound, then S is called the **sum of the infinite geometric progression**. The symbol $\sum_{n=1}^{\infty} a_n$, where ∞ is the symbol for infinity, denotes the sum, S, of the infinite geometric progression; $S = \sum_{n=1}^{\infty} a_n$, provided that sum exists.

To develop a formula for finding the sum of all the terms in an infinite geometric progression, we consider the formula

$$S_n = \frac{a - ar^n}{1 - r} \quad (r \neq 1)$$

If $|r| < 1$ and a is a constant, then the term ar^n, or $a(r^n)$, approaches 0 as n becomes large without bound. Hence, when n is very large, the value of ar^n is extremely small, and the term ar^n in the formula can be ignored. This argument leads to the following theorem.

Theorem.   If $|r| < 1$, then the sum of the terms of an infinite geometric progression is given by the formula

$$S = \frac{a}{1 - r}$$

where a is the first term and r is the common ratio.

Example 9   Change $0.444\ldots$ to a common fraction.

Solution   Write the decimal as an infinite geometric series and find its sum:

$$S = \frac{4}{10} + \frac{4}{100} + \frac{4}{1000} + \frac{4}{10{,}000} + \cdots$$

$$S = \frac{4}{10} + \frac{4}{10}\left(\frac{1}{10}\right) + \frac{4}{10}\left(\frac{1}{10}\right)^2 + \frac{4}{10}\left(\frac{1}{10}\right)^3 + \cdots$$

Because the common ratio is $\frac{1}{10}$ and $\left|\frac{1}{10}\right| < 1$, use the formula for the sum of an infinite geometric series:

$$S = \frac{a}{1 - r} = \frac{\dfrac{4}{10}}{1 - \dfrac{1}{10}} = \frac{\dfrac{4}{10}}{\dfrac{9}{10}} = \frac{4}{9}$$

Long division will verify that changing $\frac{4}{9}$ to a decimal fraction gives $0.444\ldots$. ■

Exercise 12.3

In Exercises 1–6, write the first six terms of the arithmetic progressions with the given properties.

1. $a = 1$ and $d = 2$
2. $a = -12$ and $d = -5$
3. $a = 5$ and the third term is 2.
4. $a = 4$ and the fifth term is 12.
5. The seventh term is 24, and the common difference is $\frac{5}{2}$.
6. The 20th term is -49, and the common difference is -3.

In Exercises 7–10, find the sum of the first n terms of each arithmetic progression.

7. $5 + 7 + 9 + \cdots$ (to 15 terms)
8. $\displaystyle\sum_{n=1}^{10} (-n - 2)$

9. $\displaystyle\sum_{n=1}^{20}\left(\frac{3}{2}n + 12\right)$ $\qquad\qquad$ **10.** $\displaystyle\sum_{n=1}^{10}\left(\frac{2}{3}n + \frac{1}{3}\right)$

11. In an arithmetic progression, the 25th term is 10 and the common difference is $\frac{1}{2}$. Find the sum of the first 30 terms.

12. In an arithmetic progression, the 15th term is 86 and the first term is 2. Find the sum of the first 100 terms.

13. If the fifth term of an arithmetic progression is 14 and the second term is 5, find the 15th term.

14. Can an arithmetic progression have a first term of 4, a 25th term of 126, and a common difference of $4\frac{1}{4}$? If not, explain why.

15. Insert three arithmetic means between 10 and 20.

16. Insert five arithmetic means between 5 and 15.

17. Insert four arithmetic means between -7 and $\frac{2}{3}$.

18. Insert three arithmetic means between -11 and -2.

In Exercises 19–26, write the first four terms of each geometric progression with the given properties.

19. $a = 10$ and $r = 2$ $\qquad\qquad$ **20.** $a = -3$ and $r = 2$

21. $a = -2$ and $r = 3$ $\qquad\qquad$ **22.** $a = 64$ and $r = \frac{1}{2}$

23. $a = 3$ and $r = \sqrt{2}$ $\qquad\qquad$ **24.** $a = 2$ and $r = \sqrt{3}$

25. $a = 2$, and the fourth term is 54. $\qquad\qquad$ **26.** The third term is 4, and $r = \frac{1}{2}$.

In Exercises 27–32, find the sum of the indicated terms of each geometric progression.

27. $4, 8, 16, \ldots$ (to 5 terms) $\qquad\qquad$ **28.** $9, 27, 81, \ldots$ (to 6 terms)

29. $2, -6, 18, \ldots$ (to 10 terms) $\qquad\qquad$ **30.** $\dfrac{1}{8}, \dfrac{1}{4}, \dfrac{1}{2}, \ldots$ (to 12 terms)

31. $\displaystyle\sum_{n=1}^{6} 3\left(\frac{3}{2}\right)^{n-1}$ $\qquad\qquad$ **32.** $\displaystyle\sum_{n=1}^{6} 12\left(-\frac{1}{2}\right)^{n-1}$

In Exercises 33–36, find the sum of each infinite geometric progression.

33. $6 + 4 + \dfrac{8}{3} + \cdots$ $\qquad\qquad$ **34.** $8 + 4 + 2 + 1 + \cdots$

35. $\displaystyle\sum_{n=1}^{\infty} 12\left(-\frac{1}{2}\right)^{n-1}$ $\qquad\qquad$ **36.** $\displaystyle\sum_{n=1}^{\infty} 1\left(\frac{1}{3}\right)^{n-1}$

37. Insert three geometric means between 10 and 20.

38. Insert five geometric means between -5 and 5, if possible.

39. Insert four geometric means between 2 and 2048.

40. Insert three geometric means between 162 and 2.

In Exercises 41–44, change each decimal to a common fraction.

41. $0.555\ldots$ $\qquad$ **42.** $0.666\ldots$ $\qquad$ **43.** $0.252525\ldots$ $\qquad$ **44.** $0.373737\ldots$

45. Use mathematical induction to prove the formula for finding the sum of the first n terms of an arithmetic progression.

46. Use mathematical induction to prove the formula for finding the sum of the first n terms of a geometric progression.

47. If Justin earns 1¢ on the first day of May, 2¢ on the second day, 4¢ on the third day, and the pay continues to double each day throughout the month, what will his total earnings be for the month?

48. A single arithmetic mean between two numbers is called *the* arithmetic mean of the two numbers. Similarly, a single geometric mean between two numbers is called *the* geometric mean of the two numbers. Find the arithmetic mean and the geometric mean between the numbers 4 and 64. Which is larger, the arithmetic or the geometric mean?

49. Use the definitions in Exercise 48 to compute the arithmetic mean and the geometric mean between $\frac{1}{2}$ and $\frac{7}{8}$. Which is larger, the arithmetic or geometric mean?

50. If a and b are positive numbers and $a \neq b$, prove that their arithmetic mean is greater than their geometric mean. (*Hint:* See Exercises 48 and 49.)

51. Find the indicated sum: $\displaystyle\sum_{n=1}^{100} \frac{1}{n(n+1)}$. (*Hint:* Use partial fractions first.)

52. Find the indicated sum: $\displaystyle\sum_{k=1}^{100} \ln\left(\frac{k}{k+1}\right)$.

12.4 APPLICATIONS OF PROGRESSIONS

The following examples illustrate some applications of arithmetic and geometric progressions.

Example 1 A town with a population of 3500 people has a predicted growth rate of 6% over the preceding year for the next 20 years. How many people are expected to live in the town 20 years from now?

Solution Let p_0 be the initial population of the town. After 1 year, there will be a different population, p_1. The initial population (p_0) plus the growth (the product of p_0 and the rate of growth, r) will equal the new population after 1 year (p_1):

$$p_1 = p_0 + p_0 r = p_0(1 + r)$$

The population of the town at the end of 2 years will be p_2, and

$$p_2 = p_1 + p_1 r$$
$$p_2 = p_1(1 + r)$$
$$p_2 = p_0(1 + r)(1 + r)$$
$$p_2 = p_0(1 + r)^2$$

The population at the end of the third year will be $p_3 = p_0(1 + r)^3$. Writing the terms in a sequence yields

$$p_0, \quad p_0(1 + r), \quad p_0(1 + r)^2, \quad p_0(1 + r)^3, \quad p_0(1 + r)^4, \ldots$$

This is a geometric progression with p_0 as the first term and $1 + r$ as the common ratio. Recall that the nth term is given by the formula $l = ar^{n-1}$. In this example $p_0 = 3500$, $1 + r = 1.06$, and, because the population after 20 years will be the value of the 21st term of the geometric progression, $n = 21$. The population after 20 years is $p = 3500(1.06)^{20}$. Use a calculator to find that $p \approx 11{,}225$. ■

Example 2   A woman deposits \$2500 in a bank at 9% annual interest compounded daily. If the investment is left untouched for 6 years, how much money will be in the account?

Solution   This problem is similar to that of Example 1. Let the initial amount in the account be A_0. At the end of the first day, the amount in the account is

$$A_1 = A_0 + A_0\left(\frac{r}{365}\right) = A_0\left(1 + \frac{r}{365}\right)$$

The amount in the bank after the second day is

$$A_2 = A_1 + A_1\left(\frac{r}{365}\right) = A_1\left(1 + \frac{r}{365}\right) = A_0\left(1 + \frac{r}{365}\right)^2$$

Just as in Example 1, the amounts in the account each day form a geometric progression.

$$A_0, \quad A_0\left(1 + \frac{r}{365}\right), \quad A_0\left(1 + \frac{r}{365}\right)^2, \quad A_0\left(1 + \frac{r}{365}\right)^3, \ldots$$

where A_0 is the initial deposit and r is the annual rate of interest.

Because the interest is compounded daily for 6 years, the amount in the bank at the end of 6 years will be the 2191th term ($6 \cdot 365 + 1$) of the progression. The amount in the account at the end of 6 years is

$$A_{2191} = 2500\left(1 + \frac{0.09}{365}\right)^{2190}$$

Use a calculator to find that $A_{2191} \approx \$4289.73$. ■

Example 3   The equation $S = 16t^2$ represents the distance in feet, S, that an object will fall in t seconds. After 1 second, the object has fallen 16 feet. After 2 seconds, the object has fallen 64 feet. After 3 seconds the object has fallen 144 feet. In other words, the object fell 16 feet during the first second, 48 feet during the next second, and 80 feet during the third second. Thus, the sequence 16, 48, 80, ... represents the distance an object will fall during the first second, second second, third second, and so forth. Find the distance the object falls during the 12th second.

Solution   The sequence 16, 48, 80, ... is an arithmetic progression with $a = 16$ and $d = 32$. To find the 12th term, substitute these values into the formula $l = a + (n - 1)d$

and simplify:

$$l = a + (n - 1)d$$
$$l = 16 + 11(32)$$
$$l = 16 + 352$$
$$l = 368$$

During the 12th second, the object falls 368 feet.   ■

Example 4   A pump can remove 20% of the gas in a container with each stroke. What percent of the gas will remain in the container after six strokes?

Solution   Let V represent the volume of the container. Because each stroke of the pump removes 20% of the gas, 80% of the gas remains after each stroke, and you have the sequence

$$V, \quad 0.80 \, V, \quad 0.80(0.80 \, V), \quad 0.80[0.80(0.80 \, V)], \ldots$$

This can be written as the geometric progression

$$V, \quad 0.8 \, V, \quad (0.8)^2 \, V, \quad (0.8)^3 \, V, \quad (0.8)^4 \, V, \ldots$$

You wish to know the amount of gas remaining after six strokes. This amount is the seventh term, l, of the progression:

$$l = ar^{n-1}$$
$$l = V(0.8)^6$$

Use a calculator to find that approximately 26% of the gas remains after six strokes of the pump.   ■

Exercise 12.4

Decide whether each of the following exercises involves an arithmetic or geometric progression and then solve each problem. You may use a calculator.

1. The number of students studying college algebra this year at State College is 623. If a trend has been established that the following year's enrollment is always 10% higher than the preceding year, how many professors will be needed in 8 years to teach college algebra if one professor can handle 60 students?

2. If Amelia borrows $5500 interest free from her mother to buy a new car and agrees to pay her mother back at the rate of $105 per month, how much does she still owe after four years?

3. A Super Ball can always rebound to 95% of the height from which it was dropped. How high will the ball rise after the 13th bounce if it was dropped from a height of 10 meters?

4. If Philip invests $1000 in a 1-year certificate of deposit at $6\frac{3}{4}\%$ annual interest compounded daily, how much interest will be earned that year?

5. If a single cell divides into two cells every 30 minutes, how many cells will there be at the end of ten hours?

6. If a lawn tractor, which cost c dollars when new, depreciates 20% of its previous year's value each year, how much is the lawn tractor worth after 5 years?

7. Maria can invest $1000 at $7\frac{1}{2}\%$ compounded annually or at $7\frac{1}{4}\%$ compounded daily. If she invests the money for a year, which is the best investment?

8. Find how many feet a brick will travel during the 10th second of its fall.

9. If the population of the world were to double every 30 years, approximately how many people would be on Earth in the year 3000? (Consider the population in 1980 to be 4 billion, and use 1980 as the base year.)

10. If Linda deposits $1300 in a bank at 7% interest compounded annually, how much will be in the bank 17 years later? (Assume that there are no other transactions on the account.)

11. If a house purchased for $50,000 in 1978 appreciates in value by 6% each year, how much will the house be worth in the year 2000?

12. Calculate the value of $1000 left on deposit for 10 years at an annual rate of 7% compounded annually.

13. Calculate the value of $1000 left on deposit for 10 years at an annual rate of 7% compounded quarterly.

14. Calculate the value of $1000 left on deposit for 10 years at an annual rate of 7% compounded monthly.

15. Calculate the value of $1000 left on deposit for 10 years at an annual rate of 7% compounded daily.

16. Calculate the value of $1000 left on deposit for 10 years at an annual rate of 7% compounded hourly.

17. When John was 20 years old, he opened an individual retirement account by investing $2000 that will earn 11% interest compounded quarterly. How much will his investment be worth when John is 65 years old?

18. One lone bacterium divides to form two bacteria every 5 minutes. If two bacteria multiply enough to fill a petri dish completely in 2 hours, how long will it take one bacterium to fill the dish?

19. A legend tells of an ancient king who was grateful to the inventor of the game of chess and offered to grant him any request. The man was shrewd, and he said, "My request is modest, Your Majesty. Simply place one grain of wheat on the first square on the chessboard, two grains on the second, four on the third, and so on, with each square holding double that of the square before. Do this until the board is full." The king, thinking he'd gotten off lightly, readily agreed. How many grains did the king need to fill the chessboard?

20. Estimate the size of the wheat pile in Exercise 19. (*Hint*: There are about one-half million grains of wheat in a bushel.)

21. Does $0.999999 = 1$? Explain.   22. Does $0.999\ldots = 1$? Explain.

12.5 THE BINOMIAL THEOREM

In this section we discuss a method to raise binomials to positive integral powers. To this end, we consider the following binomial expansions:

$$(a + b)^0 = 1$$
$$(a + b)^1 = a + b$$
$$(a + b)^2 = a^2 + 2ab + b^2$$

$$(a + b)^3 = a^3 + 3a^2b + 3ab^2 + b^3$$
$$(a + b)^4 = a^4 + 4a^3b + 6a^2b^2 + 4ab^3 + b^4$$
$$(a + b)^5 = a^5 + 5a^4b + 10a^3b^2 + 10a^2b^3 + 5ab^4 + b^5$$
$$(a + b)^6 = a^6 + 6a^5b + 15a^4b^2 + 20a^3b^3 + 15a^2b^4 + 6ab^5 + b^6$$

Four patterns are apparent in the above expansions.

1. Each expansion has one more term than the power of the binomial.
2. The degree of each term in each expansion equals the exponent of the binomial.
3. The first term in each expansion is a raised to the power of the binomial.
4. The exponents of a decrease by one in each successive term, and the exponents of b, beginning with b^0 in the first term, increase by one in each successive term.

To make another pattern apparent, we write the coefficients of each of the binomial expansions in a triangular array:

Blaise Pascal (1623–1662)
Pascal made several contributions to the field of probability.

$(a + b)^0$	1
$(a + b)^1$	1 1
$(a + b)^2$	1 2 1
$(a + b)^3$	1 3 3 1
$(a + b)^4$	1 4 6 4 1
$(a + b)^5$	1 5 10 10 5 1
$(a + b)^6$	1 6 15 20 15 6 1

In this triangular array, each entry other than the 1's is the sum of the closest pair of numbers in the line immediately above it. For example, the 6 in the bottom row is the sum of the 1 and the 5 above it, and the 20 is the sum of the two 10's above it.

The triangular array, named after the French mathematician Blaise Pascal (1623–1662), continues with the same pattern forever. The next two lines are shown below.

$(a + b)^7$	1 7 21 35 35 21 7 1
$(a + b)^8$	1 8 28 56 70 56 28 8 1

Example 1   Expand $(x + y)^6$.

Solution   The first term in the expansion is x^6, and the exponents of x decrease by one in each successive term. The y will appear in the second term, and the exponents of y will increase in each successive term, concluding when the term y^6 is reached. The variables in the expansion are

$$x^6 \qquad x^5y \qquad x^4y^2 \qquad x^3y^3 \qquad x^2y^4 \qquad xy^5 \qquad y^6$$

Using Pascal's triangle, you can find the coefficients of these variables. Because the binomial is raised to the 6th power, choose the row in Pascal's triangle whose second entry is 6. The coefficients of the variables are the numbers in that row:

1   6   15   20   15   6   1

Putting these two pieces of information together, the expansion is

$$(x + y)^6 = x^6 + 6x^5y + 15x^4y^2 + 20x^3y^3 + 15x^2y^4 + 6xy^5 + y^6$$ ∎

Example 2   Expand $(x - y)^6$.

Solution   To expand $(x - y)^6$, rewrite the binomial in the form

$$[x + (-y)]^6$$

The expansion is

$$[x + (-y)]^6 = x^6 + 6x^5(-y) + 15x^4(-y)^2 + 20x^3(-y)^3 + 15x^2(-y)^4$$
$$+ 6x(-y)^5 + (-y)^6$$
$$= x^6 - 6x^5y + 15x^4y^2 - 20x^3y^3 + 15x^2y^4 - 6xy^5 + y^6$$

In general, in the binomial expansion of $(x - y)^n$, the sign of the first term, x^n, is $+$, the sign of the second term is $-$, and the signs continue to alternate.   ∎

Another method for expanding a binomial, called the **binomial theorem**, uses **factorial notation**.

Definition.   The symbol $n!$ (read either as "n factorial" or as "factorial n") is defined as

$$n! = n(n - 1)(n - 2)(n - 3) \cdots (3)(2)(1)$$

where n is a natural number.

Example 3   Evaluate **a.** 3!, **b.** 6!, and **c.** 10!.

Solution   **a.** $3! = 3 \cdot 2 \cdot 1 = 6$

b. $6! = 6 \cdot 5 \cdot 4 \cdot 3 \cdot 2 \cdot 1 = 720$

c. $10! = 10 \cdot 9 \cdot 8 \cdot 7 \cdot 6 \cdot 5 \cdot 4 \cdot 3 \cdot 2 \cdot 1 = 3,628,800$ ∎

There are two fundamental properties of factorials.

Property 1.   By definition, $0! = 1$.
Property 2. $n \cdot (n - 1)! = n!$

Example 4   Show that **a.** $6 \cdot 5! = 6!$ and that **b.** $8 \cdot 7! = 8!$

Solution   **a.** $6 \cdot 5! = 6(5 \cdot 4 \cdot 3 \cdot 2 \cdot 1) = 6 \cdot 5 \cdot 4 \cdot 3 \cdot 2 \cdot 1 = 6!$

b. $8 \cdot 7! = 8(7 \cdot 6 \cdot 5 \cdot 4 \cdot 3 \cdot 2 \cdot 1) = 8 \cdot 7 \cdot 6 \cdot 5 \cdot 4 \cdot 3 \cdot 2 \cdot 1 = 8!$ ∎

We now state the binomial theorem.

The Binomial Theorem.   If n is any positive integer, then

$$(a + b)^n = a^n + \frac{n!}{1!(n-1)!} a^{n-1}b + \frac{n!}{2!(n-2)!} a^{n-2}b^2$$

$$+ \frac{n!}{3!(n-3)!} a^{n-3}b^3 + \cdots + \frac{n!}{r!(n-r)!} a^{n-r}b^r + \cdots + b^n$$

A proof of the binomial theorem appears in Appendix I.

In the binomial theorem, the exponents of the variables in each term on the right side follow familiar patterns: The sum of the exponents of a and b in each term is n, the exponents of a decrease, and the exponents of b increase. Only the method of finding the coefficients is different. Except for the first and last term, $n!$ is the numerator of each coefficient. If the exponent of b is r in a particular term, the two factors $r!$ and $(n-r)!$ form the denominator of the fractional coefficient.

Example 5   Use the binomial theorem to expand $(a + b)^5$.

Solution   Substituting directly into the binomial theorem gives

$$(a + b)^5 = a^5 + \frac{5!}{1!(5-1)!} a^4b + \frac{5!}{2!(5-2)!} a^3b^2 + \frac{5!}{3!(5-3)!} a^2b^3$$

$$+ \frac{5!}{4!(5-4)!} ab^4 + b^5$$

$$= a^5 + \frac{5 \cdot 4!}{1 \cdot 4!} a^4b + \frac{5 \cdot 4 \cdot 3!}{2 \cdot 1 \cdot 3!} a^3b^2 + \frac{5 \cdot 4 \cdot 3!}{3! \cdot 2 \cdot 1} a^2b^3$$

$$+ \frac{5 \cdot 4!}{4! \cdot 1} ab^4 + b^5$$

$$= a^5 + 5a^4b + 10a^3b^2 + 10a^2b^3 + 5ab^4 + b^5$$

Note that the coefficients in this example are the same numbers that appear in the sixth row of Pascal's triangle (the row whose second entry is 5).   ∎

Example 6   Find the expansion of $(2x - 3y)^4$.

Solution   Note that $(2x - 3y)^4 = (2x + [-3y])^4$. To find the expansion of $(2x + [-3y])^4$, let $a = 2x$ and $b = -3y$. Then find the expansion of $(a + b)^4$. Substituting 4 for *n* in the binomial theorem gives

$$(a + b)^4 = a^4 + \frac{4!}{1!(4 - 1)!} a^3 b + \frac{4!}{2!(4 - 2)!} a^2 b^2 + \frac{4!}{3!(4 - 3)!} ab^3 + b^4$$

$$= a^4 + \frac{4 \cdot 3!}{3!} a^3 b + \frac{4 \cdot 3 \cdot 2!}{2 \cdot 1 \cdot 2!} a^2 b^2 + \frac{4 \cdot 3!}{3!} ab^3 + b^4$$

$$= a^4 + 4a^3 b + 6a^2 b^2 + 4ab^3 + b^4$$

In this expansion, substitute $2x$ for a and $-3y$ for b, and simplify to obtain

$$(2x - 3y)^4 = (2x)^4 + 4(2x)^3(-3y) + 6(2x)^2(-3y)^2 + 4(2x)(-3y)^3 + (-3y)^4$$

$$= 16x^4 - 96x^3 y + 216x^2 y^2 - 216xy^3 + 81y^4 \qquad \blacksquare$$

Suppose we wish to find the fifth term of the expansion of $(a + b)^{11}$. It is possible to raise the binomial $a + b$ to the 11th power and then look at the fifth term, but that would be tedious. However, this task is easy if we use the binomial theorem.

Example 7   Find the fifth term of the expansion of $(a + b)^{11}$.

Solution   The exponent of b in the fifth term of this expansion is 4, because the exponent for b is always 1 less than the number of the term. Because the exponent of b added to the exponent of a must equal 11, the exponent of a must be 7. The variables of the fifth term appear as $a^7 b^4$.

Because of the binomial theorem, the number in the numerator of the coefficient is $n!$, which in this case is 11!. The factors in the denominator are 4! and $(11 - 4)!$. Thus, the complete fifth term of the expansion of $(a + b)^{11}$ is

$$\frac{11!}{4!7!} a^7 b^4 = \frac{11 \cdot 10 \cdot 9 \cdot 8 \cdot 7!}{4 \cdot 3 \cdot 2 \cdot 1 \cdot 7!} a^7 b^4 = 330a^7 b^4 \qquad \blacksquare$$

Example 8   Find the sixth term of the expansion of $(a + b)^9$.

Solution   The exponent of b is 5, and the exponent of a is $9 - 5$ or 4 in the sixth term. The factors in the denominator of the coefficient are 5! and $(9 - 5)!$, and 9! is the numerator. Thus, the sixth term of the expansion of $(a + b)^9$ is

$$\frac{9!}{5!(9 - 5)!} a^4 b^5 = \frac{9 \cdot 8 \cdot 7 \cdot 6 \cdot 5!}{5!4!} a^4 b^5 = \frac{9 \cdot 8 \cdot 7 \cdot 6}{4 \cdot 3 \cdot 2 \cdot 1} a^4 b^5 = 126a^4 b^5 \qquad \blacksquare$$

Example 9   Find the third term of the expansion of $(3x - 2y)^6$.

Solution   Let $a = 3x$ and $b = -2y$ and use the binomial theorem to find the third term in the expansion of $(a + b)^6$:

$$\frac{6!}{2!(6 - 2)!} a^4 b^2 = \frac{6 \cdot 5 \cdot 4!}{2 \cdot 1 \cdot 4!} a^4 b^2 = 15a^4 b^2$$

Replacing a with $3x$ and b with $-2y$ in the term $15a^4b^2$ gives the third term of the expansion of $(3x - 2y)^6$.

$$15a^4b^2 = 15(3x)^4(-2y)^2$$
$$= 15(3)^4(-2)^2x^4y^2$$
$$= 4860x^4y^2 \qquad \blacksquare$$

Exercise 12.5

In Exercises 1–10, evaluate each expression.

1. $4!$

2. $-5!$

3. $3! \cdot 5!$

4. $0! \cdot 7!$

5. $6! + 6!$

6. $5! - 2!$

7. $\dfrac{9!}{12!}$

8. $\dfrac{8!}{5!}$

9. $\dfrac{18!}{6!(18 - 6)!}$

10. $\dfrac{15!}{9!(15 - 9)!}$

In Exercises 11–22, use the binomial theorem to expand each binomial.

11. $(a + b)^4$

12. $(a + b)^3$

13. $(a - b)^5$

14. $(x - y)^6$

15. $(2x - y)^3$

16. $(x + 2y)^5$

17. $(2x + y)^4$

18. $(2x - y)^4$

19. $(4x - 3y)^4$

20. $(5x + 2y)^5$

21. $(6x - 3y)^2$

22. $\left(\dfrac{x}{2} + \dfrac{y}{3}\right)^4$

In Exercises 23–38, find the required term in the expansion of the given expression.

23. $(a + b)^4$; third term

24. $(a - b)^4$; second term

25. $(a + b)^7$; fifth term

26. $(a + b)^5$; fourth term

27. $(a - b)^5$; sixth term

28. $(a + b)^{12}$; twelfth term

29. $(x - y)^8$; seventh term

30. $(2x - y)^4$; second term

31. $(\sqrt{2}\,x + y)^5$; third term

32. $(\sqrt{2}\,x - 3y)^5$; third term

33. $(x + 2y)^9$; eighth term

34. $(3x - 5y)^6$; fourth term

35. $(a + b)^r$; fourth term

36. $(a - b)^r$; fifth term

37. $(a + b)^n$; rth term

38. $(a + b)^n$; $(r + 1)$th term

39. Find the sum of the numbers in each row of the first 10 rows of Pascal's triangle. What is the pattern?

40. Show that the sum of the coefficients in the binomial expansion of $(x + y)^n$ is 2^n. (*Hint*: Let $x = y = 1$.)

41. Find the constant term in the expansion of $\left(a - \dfrac{1}{a}\right)^{10}$.

42. Find the coefficient of x^5 in the expansion of $\left(x + \dfrac{1}{x}\right)^9$.

43. Find the coefficient of x^8 in the expansion of $\left(\sqrt{x} + \dfrac{1}{2x}\right)^{25}$.

44. Find the constant term in the expansion of $\left(\dfrac{1}{a} + a\right)^8$.

12.6 PERMUTATIONS AND COMBINATIONS

Lydia plans to go to dinner and then attend a movie. If she has a choice of four restaurants and three movies, in how many ways can Lydia spend her evening? There are four choices of restaurants and, for any one of these options, there are three choices of movies. The choices are shown in the tree diagram in Figure 12-1.

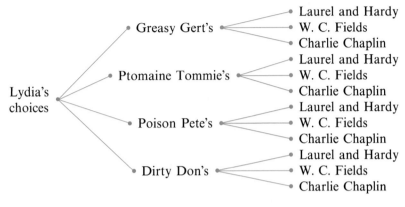

Figure 12-1

This diagram shows that Lydia has 12 ways to spend her evening. One possibility is to visit Ptomaine Tommie's and see W. C. Fields. Another is to dine at Dirty Don's and enjoy Laurel and Hardy.

Any situation that can have several outcomes is called an **event**. Lydia's first event (choosing a restaurant) can occur in 4 ways. Her second event (choosing a movie) can occur in 3 ways. Lydia has 4 times 3, or 12, ways to spend her evening. This example illustrates the **multiplication principle for events**.

> **The Multiplication Principle for Events.** Let E_1 and E_2 be two events. If E_1 can be done in a_1 ways, and if—after E_1 has occurred—E_2 can be done in a_2 ways, then the event "E_1 followed by E_2" can be done in $a_1 \cdot a_2$ ways.

The multiplication principle can be extended to n events.

Example 1   If Frank has four ways to travel from New York to Chicago, three ways to travel from Chicago to Denver, and six ways to travel from Denver to San Francisco, in how many ways can he go from New York to San Francisco?

Solution   Let E_1 be the event "going from New York to Chicago," let E_2 be the event "going from Chicago to Denver," and let E_3 be the event "going from Denver to San Francisco." Because there are 4 ways to accomplish E_1, 3 ways to accomplish E_2, and 6 ways to accomplish E_3, the number of routes that Frank can follow is

$$4 \cdot 3 \cdot 6 = 72$$ ■

Permutations

Suppose we wish to arrange 7 books on a shelf. We can fill the first space with any of the 7 books, the second space with any of the remaining 6 books, the third space with any of the remaining 5 books, and so on, until there is only one space to fill with the last book. According to the multiplication principle, the number of ways that we can arrange the books is

$$7 \cdot 6 \cdot 5 \cdot 4 \cdot 3 \cdot 2 \cdot 1 = 5040$$

When computing the number of possible arrangements of the elements in a set such as books on a shelf, we are determining the number of **permutations** of the elements in that set. The number of permutations of 7 books, using all the books, is 5040. The symbol $_nP_r$, which is used in expressing permutation problems, is read as "the number of permutations of n things r at a time."

Example 2   If there are 7 flags of 7 different colors to hang on a flagpole, how many different signals can be sent if only 3 flags are used?

Solution   You are asked to find $_7P_3$, the number of permutations of 7 things using only 3 of them. Any one of the 7 flags can hang in the top position on the flagpole. In the middle position, any one of the 6 remaining flags can hang, and in the bottom position can hang any one of the remaining 5 flags. Therefore, according to the multiplication principle,

$$_7P_3 = 7 \cdot 6 \cdot 5 = 210$$

It is possible to send 210 different signals.   ■

Although it is acceptable to write $_7P_3 = 7 \cdot 6 \cdot 5$, there is an advantage in changing the form of this answer to obtain a convenient formula. To get this formula, we will multiply both the numerator and the denominator of $\dfrac{7 \cdot 6 \cdot 5}{1}$ by 4!.

$$_7P_3 = 7 \cdot 6 \cdot 5 = \frac{7 \cdot 6 \cdot 5 \cdot 4 \cdot 3 \cdot 2 \cdot 1}{4 \cdot 3 \cdot 2 \cdot 1} = \frac{7!}{4!} = \frac{7!}{(7-3)!}$$

This idea is generalized in the following theorem.

> **Theorem.**   The formula for computing the number of permutations of n things r at a time is
>
> $$_nP_r = \frac{n!}{(n-r)!}$$
>
> If $r = n$, then
>
> $$_nP_n = n!$$

The second part of the previous theorem is true because

$$_nP_n = \frac{n!}{(n-n)!} = \frac{n!}{0!} = n! \qquad \text{Remember } 0! = 1.$$

Example 3   In how many ways can a baseball manager arrange the batting order if there are 25 players on the team?

Solution   To find the number of permutations of 25 things 9 at a time, $_{25}P_9$, use the formula $_nP_r = \frac{n!}{(n-r)!}$.

$$_{25}P_9 = \frac{25!}{(25-9)!}$$

$$= \frac{25!}{16!}$$

$$= \frac{25 \cdot 24 \cdot 23 \cdot 22 \cdot 21 \cdot 20 \cdot 19 \cdot 18 \cdot 17 \cdot \mathbf{16!}}{\mathbf{16!}}$$

$$\approx 741,354,768,000$$

The number of permutations is approximately 741,354,768,000.   ■

Example 4   In how many ways can 5 people stand in a line if 2 people refuse to stand next to each other?

Solution   The total number of ways that 5 people can stand in line is

$$_5P_5 = 5! = 5 \cdot 4 \cdot 3 \cdot 2 \cdot 1 = 120$$

Now find the number of ways that five people can stand in line if two people *insist* on standing together by considering those two people as a single person. Then, there are 4 people to stand in line, and this can be done in $_4P_4 = 4! = 24$ ways. However, there are two arrangements for the pair that insist on standing together, because either could be first. Hence, there are $2 \cdot 4!$ ways that the 5 people can stand in line if 2 people insist on standing together.

The number of ways that 5 people can stand in line if two people *refuse* to stand together is 5! = 120 (the total number of ways to arrange 5 people) minus 2 · 4! = 48 (the number of ways to arrange five people if two must stand together):

$$120 - 48 = 72$$

There are 72 ways to stand 5 people in a line if 2 people refuse to stand next to each other.   ■

Example 5   In how many ways can 5 people be seated at a round table?

Solution   If you were to seat 5 people in a row, there would be 5! possible arrangements. However, the situation is different when seating people at a round table. At a round table, each person has a neighbor to the left and to the right. If each person moves one place to the left, everyone will still have the same neighbors as before. This same situation applies if everyone moves two, three, four, or five places. Hence, you must divide 5! by 5 to get rid of these duplications. The number of ways that 5 people can be seated at a round table is

$$\frac{5!}{5} = 4! = 4 \cdot 3 \cdot 2 \cdot 1 = 24$$ ■

The results of Example 5 can be generalized into the following theorem.

> **Theorem.**   There are $(n - 1)!$ ways to place n things in a circle.

Combinations

Suppose a class of 12 students selects a committee of 3 to plan a party. A possible committee is John, Sally, and Joe. In this situation, the order of the 3 is not important, because the committee of John, Sally, and Joe is the same as the committee of Sally, Joe, and John. For the moment, however, we assume that order is important and compute the number of permutations of 12 things 3 at a time:

$$_{12}P_3 = \frac{12!}{(12 - 3)!} = \frac{12 \cdot 11 \cdot 10 \cdot 9!}{9!} = 1320$$

This result indicates the number of ways of arranging 3 people if there are 12 people to choose from. However, in this situation we do not care about the order of the trio. Because there are six ways (3!) of ordering the committee of 3 students, the calculation of $_{12}P_3 = 1320$ provides an answer that is 6 times too big. Actually, the number of possible committees that could plan the party is the number of permutations of 12 things taken 3 at a time divided by 6:

$$\frac{_{12}P_3}{6} = \frac{1320}{6} = 220$$

When choosing committee members and in other cases of selection where order is not important, we are interested in **combinations**, not permutations. The symbols $_nC_r$ and $\binom{n}{r}$ both mean the number of combinations of n things r at a time. If a committee of r people is chosen from a total of n people, the number of committees is $_nC_r$, and there will be $r!$ arrangements of each committee. If we consider the committee as an ordered grouping, the number of orderings of r people selected from a group of n people is $_nP_r$. Therefore, the number of *combinations* of n things r at a time multiplied by $r!$ is equal to the number of *permutations* of n things r at a time. This relationship is shown by the equation

$$r!_nC_r = {}_nP_r$$

We divide both sides of this equation by $r!$ to obtain the formula for computing $_nC_r$, or $\binom{n}{r}$.

$$_nC_r = \binom{n}{r} = \frac{{}_nP_r}{r!} = \frac{n!}{r!(n-r)!}$$

This reasoning leads to the following theorem.

Theorem. The formula for computing the number of combinations of n things r at a time is

$$_nC_r = \binom{n}{r} = \frac{n!}{r!(n-r)!}$$

In the exercises, you will be asked to prove the following theorem.

Theorem. If n is a whole number, then

$$_nC_n = 1 \quad \text{and} \quad {}_nC_0 = 1$$

Example 6 If Carla must read 4 books from a reading list of 10 books, how many choices does she have?

Solution Because the order in which the books are read is unimportant, calculate the number of combinations of 10 things 4 at a time:

$$_{10}C_4 = \frac{10!}{4!(10-4)!} = \frac{10 \cdot 9 \cdot 8 \cdot 7 \cdot 6!}{4 \cdot 3 \cdot 2 \cdot 1 \cdot 6!}$$
$$= \frac{10 \cdot 9 \cdot 8 \cdot 7}{4 \cdot 3 \cdot 2}$$
$$= 210$$

Carla has 210 options. ∎

Example 7   A class consists of 15 boys and 8 girls. In how many ways can a debate team be chosen that will have 3 boys and 3 girls on the team?

Solution   There are $_{15}C_3$ ways of choosing the three boys and $_8C_3$ ways of choosing the three girls. By the multiplication principle, there are $_{15}C_3 \cdot {_8C_3}$ ways of choosing members of the debate team:

$$_{15}C_3 \cdot {_8C_3} = \frac{15!}{3!(15-3)!} \cdot \frac{8!}{3!(8-3)!}$$

$$= \frac{15 \cdot 14 \cdot 13 \cdot 12!}{6 \cdot 12!} \cdot \frac{8 \cdot 7 \cdot 6 \cdot 5!}{6 \cdot 5!}$$

$$= \frac{15 \cdot 14 \cdot 13}{6} \cdot \frac{8 \cdot 7 \cdot 6}{6}$$

$$= 25,480$$

There are 25,480 ways to place 3 boys and 3 girls on the debate team.   ■

Note that the formula $_nC_r = n!/[r!(n-r)!]$ gives the coefficient of the $(r+1)$th term of the binomial expansion of $(a+b)^n$. This implies that the coefficients of a binomial expansion can be used to solve problems involving combinations. The binomial theorem is restated below, this time using combination notation.

The Binomial Theorem.   If n is any positive integer, then

$$(a+b)^n = \binom{n}{0}a^n + \binom{n}{1}a^{n-1}b + \binom{n}{2}a^{n-2}b^2 + \cdots$$

$$+ \binom{n}{r}a^{n-r}b^r + \cdots + \binom{n}{n}b^n$$

Example 8   Use Pascal's triangle to compute $_7C_5$.

Solution   Consider the eighth row of Pascal's triangle and the corresponding combinations:

$$
\begin{array}{cccccccc}
1 & 7 & 21 & 35 & 35 & 21 & 7 & 1 \\
\binom{7}{0} & \binom{7}{1} & \binom{7}{2} & \binom{7}{3} & \binom{7}{4} & \binom{7}{5} & \binom{7}{6} & \binom{7}{7}
\end{array}
$$

$$_7C_5 = \binom{7}{5} = 21$$ ■

When discussing permutations and combinations, a "word" is a distinguishable arrangement of letters. For example, 6 words can be formed with the letters a, b, and c if all 3 letters are used exactly once. The six words are *abc*, *acb, bac, bca, cab,* and *cba.* If there are *n* distinct letters and each letter is used once, the number of distinct words that can be formed is $n! = {}_nP_n$. It is more complicated to compute the number of distinguishable words that can be formed with *n* letters if some of the letters are duplicates.

Example 9   Find the number of "words" that can be formed if each of the 6 letters of the word *little* is used once.

Solution   For the moment, pretend that all of the letters of the word *little* are distinguishable: "LitTle." The number of words that can be formed using each letter once is $6! = {}_6P_6$. However, in reality you cannot tell the *l*'s or the *t*'s apart. Therefore, divide by an appropriate number to get rid of these duplications; because there are 2! orderings of the two *l*'s and 2! orderings of the two *t*'s, divide by $2! \cdot 2!$. The number of words that can be formed using each letter of the word *little* is

$$\frac{{}_6P_6}{2!2!} = \frac{6!}{2!2!} = \frac{6 \cdot 5 \cdot 4 \cdot 3 \cdot 2 \cdot 1}{2 \cdot 1 \cdot 2 \cdot 1} = 180$$ ■

Example 9 illustrates the following general principle.

> **Theorem.**   If a word with *n* letters has *a* of one letter, *b* of another letter, and so on, then the number of distinguishable words that can be formed using each letter of the *n*-lettered word exactly once is
>
> $$\frac{n!}{a!b! \cdots}$$

Exercise 12.6

1. A lunch room has a machine with eight kinds of sandwiches, a machine with four kinds of soda, a machine with both white and chocolate milk, and a machine with three kinds of ice cream. How many different lunches can be chosen? (Consider a lunch to be one sandwich, one drink, and one ice cream.)

2. How many six-digit license plates can be manufactured if no license plate number begins with 0?

3. How many different seven-digit phone numbers can be used in one area code if no phone number begins with 0 or 1?

4. In how many ways can the letters of the word *number* be arranged?

5. In how many ways can the letters of the word *number* be arranged if the *e* and *r* must remain next to each other?

6. In how many ways can the letters of the word *number* be arranged if the *e* and *r* cannot be side by side?

7. How many ways can five Scrabble tiles bearing the letters *F, F, F, L,* and *U* be arranged to spell the word *fluff*?

8. How many ways can six Scrabble tiles bearing the letters *B, E, E, E, F,* and *L* be arranged to spell the word *feeble*?

In Exercises 9–24, evaluate each expression.

9. $_7P_4$

10. $_8P_3$

11. $_7C_4$

12. $_8C_3$

13. $_5P_5$

14. $_5P_0$

15. $\binom{5}{4}$

16. $\binom{8}{4}$

17. $\binom{5}{0}$

18. $\binom{5}{5}$

19. $_5P_4 \cdot {_5C_3}$

20. $_3P_2 \cdot {_4C_3}$

21. $\binom{5}{3}\binom{4}{3}\binom{3}{3}$

22. $\binom{5}{5}\binom{6}{6}\binom{7}{7}\binom{8}{8}$

23. $\binom{68}{66}$

24. $\binom{100}{99}$

25. In how many arrangements can 8 girls be placed in a line?

26. In how many arrangements can 5 girls and 5 boys be placed in a line if the girls and boys alternate?

27. In how many arrangements can 5 girls and 5 boys be placed in a line if all the boys line up first?

28. In how many arrangements can 5 girls and 5 boys be placed in a line if all the girls line up first?

29. How many permutations does a combination lock have if each combination has 3 numbers, no two numbers of the combination are the same, and the lock dial has 30 notches? Wouldn't it be better to call these locks "permutation locks"? Explain.

30. How many permutations does a combination lock have if each combination has 3 numbers, no two numbers of the combination are the same, and the lock has 100 notches?

31. In how many ways can 8 people be seated at a round table?

32. In how many ways can 7 people be seated at a round table?

33. In how many ways can 6 people be seated at a round table if 2 of the people insist on sitting together?

34. In how many ways can 6 people be seated at a round table if 2 of the people refuse to sit together?

35. In how many ways can 7 children be arranged in a circle if Sally and John must be kept apart and Martha and Peter want to be together?

36. In how many ways can 8 children be arranged in a circle if Laura and Scott want to be together but Billy and Paula don't?

37. In how many groupings can 4 candy bars be selected from 10 different candy bars?

38. How many hands of 5 cards can be selected from a deck of 52 cards?

39. How many possible bridge hands are there? (*Hint:* There are 13 cards in a bridge hand and 52 cards in a deck.)

40. How many words can be formed from the letters of the word *igloo* if each of the 5 letters is to be used once?

41. How many words can be formed from the letters of the word *parallel* if each letter is to be used once?

42. How many words can be formed from the letters of the word *banana* if each letter is to be used once?

43. How many license plates can be made using two different letters followed by four different digits if the first digit cannot be 0 and the letter *O* is not used?

44. If there are seven class periods in a school day and a typical student takes 5 classes, how many different time patterns are possible for the student?

45. From a bucket containing 6 red and 8 white golf balls, in how many ways can we draw 6 golf balls of which 3 are red and 3 are white?

46. In how many ways can you select a committee of 3 Republicans and 3 Democrats from a group containing 18 Democrats and 11 Republicans?

47. In how many ways can you select a committee of 4 Democrats and 3 Republicans from a group containing 12 Democrats and 10 Republicans?

48. In how many ways can you select a group of 5 red cards and 2 black cards from a deck containing 10 red cards and 8 black cards?

49. In how many ways can a husband and wife choose 2 different dinners from a menu of 17 dinners?

50. In how many ways can 7 people stand in a row if 2 of the people refuse to stand together?

51. How many lines are determined by 8 points if no 3 points lie on a straight line?

52. How many lines are determined by 10 points if no 3 points lie on a straight line?

53. Use Pascal's triangle to find $_8C_5$.

54. Use Pascal's triangle to find $_{10}C_8$.

55. How many teams can a baseball manager put on the field if the entire squad consists of 25 players? (There are 9 players on the field at a time. Assume that all players can play all positions.)

56. Prove that $_nC_n = 1$ and that $_nC_0 = 1$.

57. Prove that $\binom{n}{r} = \binom{n}{n-r}$.

58. Show that the binomial theorem can be expressed in the form

$$(a + b)^n = \sum_{k=0}^{n} \binom{n}{k} a^{n-k} b^k$$

12.7 PROBABILITY

Definition. If an experiment can have n distinct and equally likely outcomes and if E is an event that can occur in f of these ways, then the probability of E is

$$P(E) = \frac{f}{n}$$

Because $0 \le f \le n$, it follows that $0 \le f/n \le 1$ and that all probabilities must have values from 0 to 1. An event that cannot happen has probability 0, and an event that is certain to happen has probability 1.

Saying that the probability of tossing heads with a single toss of a coin is $\frac{1}{2}$ means that, if a fair coin is tossed a very large number of times, the ratio of the number of heads to the total number of tosses is very nearly $\frac{1}{2}$. As the number of tosses approaches infinity, this ratio approaches $\frac{1}{2}$ more and more closely. To say that the probability of rolling a 5 with a single roll of a die is $\frac{1}{6}$ is to say that, as the number of rolls approaches infinity, the ratio of the number of favorable outcomes (rolling a 5) to the total number of outcomes approaches $\frac{1}{6}$.

In order to calculate probabilities, we sometimes need to exhibit the total list of possible outcomes. Such a listing is called a **sample space**.

Example 1   Exhibit the sample space of the event "rolling two dice a single time."

Solution   The sample space is the listing of all possible outcomes. Use ordered-pair notation, and let the first number of the pair be the result on the first die and the second number the result on the second die:

$$
\begin{array}{cccccc}
(1, 1) & (1, 2) & (1, 3) & (1, 4) & (1, 5) & (1, 6) \\
(2, 1) & (2, 2) & (2, 3) & (2, 4) & (2, 5) & (2, 6) \\
(3, 1) & (3, 2) & (3, 3) & (3, 4) & (3, 5) & (3, 6) \\
(4, 1) & (4, 2) & (4, 3) & (4, 4) & (4, 5) & (4, 6) \\
(5, 1) & (5, 2) & (5, 3) & (5, 4) & (5, 5) & (5, 6) \\
(6, 1) & (6, 2) & (6, 3) & (6, 4) & (6, 5) & (6, 6)
\end{array}
$$

This sample space contains 36 ordered pairs. Because there are 6 possible outcomes with the first die and 6 possible outcomes with the second die, the multiplication principle for events tells you to expect $6 \cdot 6$, or 36 total possible outcomes. ■

Example 2   On a single toss of two dice, what is the probability of tossing a sum of seven?

Solution   The sample space for this event is listed in Example 1. The favorable outcomes are the ones that give a sum of 7: (1, 6), (2, 5), (3, 4), (4, 3), (5, 2), and (6, 1). Because there are 6 favorable outcomes among the 36 possible outcomes.

$$
P(\text{tossing a 7}) = \frac{6}{36} = \frac{1}{6}
$$

■

Example 3   What is the probability of being dealt 5 cards, all hearts, from an ordinary deck of cards?

Solution   The number of ways to draw 5 hearts from the 13 hearts is $_{13}C_5$. The number of ways of drawing 5 cards from the complete deck is $_{52}C_5$. The desired probability is the ratio of the number of favorable outcomes to the number of possible outcomes.

$$
P(\text{5 hearts}) = \frac{_{13}C_5}{_{52}C_5}
$$

$$P(5 \text{ hearts}) = \frac{\dfrac{13!}{5!8!}}{\dfrac{52!}{5!47!}}$$

$$= \frac{13!}{5!8!} \cdot \frac{5!47!}{52!}$$

$$= \frac{13 \cdot 12 \cdot 11 \cdot 10 \cdot 9 \cdot 8!}{8!} \cdot \frac{47!}{52 \cdot 51 \cdot 50 \cdot 49 \cdot 48 \cdot 47!}$$

$$= \frac{13 \cdot 12 \cdot 11 \cdot 10 \cdot 9}{52 \cdot 51 \cdot 50 \cdot 49 \cdot 48}$$

$$= \frac{33}{66,640} \qquad \blacksquare$$

There is a multiplication property for probabilities that is similar to the multiplication principle for events that occur in succession.

> **Multiplication Property of Probabilities.** If $P(A)$ represents the probability of event A, and $P(B|A)$ represents the probability that event B will occur after event A, then $P(A \text{ and } B) = P(A) \cdot P(B|A)$.

Example 4 A box contains 40 cubes of the same size. Of these cubes, 17 are red, 13 are blue, and the rest are yellow. If 2 cubes are drawn at random, without replacement, what is the probability that 2 yellow cubes will be drawn?

Solution Of the 40 cubes in the box, 10 are yellow. Thus, the probability of getting a yellow cube on the first draw is

$$P(\text{yellow cube on first draw}) = \frac{10}{40} = \frac{1}{4}$$

Because there is no replacement after the first draw, 39 cubes remain in the box, and 9 of these are yellow. The probability of getting a yellow cube on the second draw is

$$P(\text{yellow cube on the second draw}) = \frac{9}{39} = \frac{3}{13}$$

The probability of drawing 2 yellow cubes in succession is the product of the probability of drawing a yellow cube on the first draw and the probability of drawing a yellow cube on the second draw.

$$P(\text{drawing two yellow cubes}) = \frac{1}{4} \cdot \frac{3}{13} = \frac{3}{52} \qquad \blacksquare$$

Example 5   Repeat Example 3 using the multiplication property of probabilities.

Solution   The probability of drawing a heart on the first draw is $\frac{13}{52}$, on the second draw $\frac{12}{51}$, on the third draw $\frac{11}{50}$, on the fourth draw $\frac{10}{49}$, and on the fifth draw $\frac{9}{48}$. By the multiplication property of probabilities,

$$P(5 \text{ hearts in a row}) = \frac{13}{52} \cdot \frac{12}{51} \cdot \frac{11}{50} \cdot \frac{10}{49} \cdot \frac{9}{48}$$

$$= \frac{33}{66,640}$$

Exercise 12.7

In Exercises 1–4, an ordinary die is tossed. Find the probability of each event.

1.   Tossing a 2

2.   Tossing a number greater than 4

3.   Tossing a number larger than 1 but less than 6

4.   Tossing an odd number

In Exercises 5–8, balls numbered from 1 to 42 are placed in a container and stirred. If one is drawn at random, find the probability of each event.

5.   The number is less than 20.

6.   The number is less than 50.

7.   The number is a prime number.

8.   The number is less than 10 or greater than 40.

In Exercises 9–12, refer to the spinner in Illustration 1. If the spinner is spun, find the probability of each event. Assume that the spinner never stops on a line.

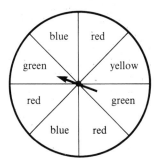

Illustration 1

9.   The spinner stops on red.

10.   The spinner stops on green.

11.   The spinner stops on orange.

12.   The spinner stops on yellow.

In Exercises 13–30, find the probability of each given event.

13.   Rolling a sum of 4 on one roll of two dice

14.   Drawing a diamond on one draw from a card deck

15.   Drawing two aces in succession from a card deck if the card is replaced and the deck shuffled after the first draw

16. Drawing two aces from a card deck without replacing the card after the first draw

17. Drawing a red egg from a basket containing 5 red eggs and 7 azure eggs

18. Getting 2 red eggs in a single scoop from a bucket containing 5 red eggs and 7 cyan eggs

19. Drawing a bridge hand of 13 cards, all of one suit

20. Drawing 6 diamonds from a card deck without replacing the cards after each draw

21. Drawing 5 aces from a card deck without replacing the cards after each draw

22. Drawing 5 clubs from the black cards in a card deck

23. Drawing a face card from a card deck

24. Drawing 6 face cards in a row from a card deck without replacing the cards after each draw

25. Drawing 5 fuchsia cubes from a bowl containing 5 fuchsia cubes and 1 beige cube

26. Rolling a sum of 4 with one roll of three dice

27. Rolling a sum of 11 with one roll of three dice

28. Picking, at random, 5 Republicans from a group containing 8 Republicans and 10 Democrats

29. Tossing 3 heads in 5 tosses of a fair coin   30. Tossing 5 heads in 5 tosses of a fair coin

In Exercises 31–37, assume that the probability that an airplane engine will fail during a torture test is $\frac{1}{2}$ and that the aircraft in question has 4 engines.

31. Construct a sample space showing each of the possible outcomes after the torture test.

32. Find the probability that all of the engines will survive the test.

33. Find the probability that exactly 1 engine will survive.

34. Find the probability that exactly 2 engines will survive.

35. Find the probability that exactly 3 engines will survive.

36. Find the probability that no engines will survive.

37. Find the sum of the probabilities in Exercises 32 through 36.

In Exercises 38–40, assume that a survey of 282 people is taken to determine the opinions of blacks, whites, and Orientals on a proposed piece of legislation, with the following results:

	Number that favor	Number that oppose	Number with no opinion	Total
Black	70	32	17	119
White	83	24	10	117
Oriental	23	15	8	46
Total	176	71	35	282

A person is chosen at random from those surveyed. Refer to the chart to find each probability.

38. What is the probability that the person favors the legislation?

39. If a black is chosen, what is the probability that he/she opposes the legislation?

40. If the person opposes the legislation, what is the probability that the person is Oriental?

12.8 COMPUTATION OF COMPOUND PROBABILITIES

Sometimes we need to compute the probability of one event *or* another, or one event *and* another. Such events are called **compound events**.

Suppose we wish to find the probability of drawing an ace *or* a heart from an ordinary card deck. The probability of drawing an ace is $\frac{4}{52}$, and the probability of drawing a heart is $\frac{13}{52}$. However, the probability of drawing an ace *or* a heart is not the sum of these two probabilities. Because the ace of hearts was counted twice, once as an ace and once as a heart, and the probability of drawing the ace of hearts is $\frac{1}{52}$, we must subtract $\frac{1}{52}$ from the sum of $\frac{4}{52}$ and $\frac{13}{52}$. Thus,

$$P(\text{ace } or \text{ heart}) = P(\text{ace}) + P(\text{heart}) - P(\text{ace of hearts})$$

$$= \frac{4}{52} + \frac{13}{52} - \frac{1}{52}$$

$$= \frac{16}{52}$$

$$= \frac{4}{13}$$

This example suggests the following theorem.

Theorem.   If A and B are two events, then

$$P(A \text{ or } B) = P(A) + P(B) - P(A \text{ and } B)$$

If event A and event B are **mutually exclusive** (that is, if one event occurs, the other cannot), then $P(A \text{ and } B) = 0$ and we have another theorem.

Theorem.   If A and B cannot occur simultaneously, then

$$P(A \text{ or } B) = P(A) + P(B)$$

Because the events A and $\bar{A}$ (read as "not A") are mutually exclusive,

$$P(A \text{ or } \bar{A}) = P(A) + P(\bar{A})$$

Because either event A or event $\bar{A}$ must happen, $P(A \text{ or } \bar{A}) = 1$. Thus,

$$P(A \text{ or } \bar{A}) = 1$$
$$P(A) + P(\bar{A}) = 1$$
$$P(\bar{A}) = 1 - P(A) \qquad \text{Add } -P(A) \text{ to both sides.}$$

This result gives a third theorem about compound probabilities.

> **Theorem.**   If A is any event, then
> $$P(\bar{A}) = 1 - P(A)$$

Example 1   A guidance counselor tells a student that his probability of earning a grade of D in algebra is $\frac{1}{5}$, and his probability of failing is $\frac{1}{25}$. What is the probability that the student earns a C or better?

Solution   The probability of earning a D or F is given by

$$P(D \text{ or } F) = P(D) + P(F) \qquad \text{Note that } P(D \text{ and } F) = 0.$$
$$= \frac{1}{5} + \frac{1}{25}$$
$$= \frac{6}{25}$$

The probability that the student will receive a C or better is given by

$$P(C \text{ or better}) = 1 - P(D \text{ or } F)$$
$$= 1 - \frac{6}{25}$$
$$= \frac{19}{25}$$

The student's probability of earning a C or better is $\frac{19}{25}$.   ■

If two events do not influence each other, they are called **independent events**.

> **Definition.**   The events A and B are said to be **independent events** if and only if $P(B) = P(B|A)$.

Substituting $P(B)$ for $P(B|A)$ in the multiplication property for probabilities gives a formula for computing probabilities of compound independent events.

> **Theorem.**   If A and B are independent events, then
> $$P(A \text{ and } B) = P(A) \cdot P(B)$$

For example, P(drawing an ace and tossing a head) = P(drawing an ace) $\cdot$ P(tossing a head). This is true because neither event influences the other. Consequently,

$$P(\text{drawing an ace and tossing a head}) = \frac{4}{52} \cdot \frac{1}{2} = \frac{1}{26}$$

Example 2   The probability that a baseball player can get a hit is $\frac{1}{3}$. What is the probability that she will bat safely three times in a row?

Solution   Assume that the three events, each time at bat, are independent; one time at bat does not influence the player's chances of getting a hit on another turn at bat. Because $P(E_1) = \frac{1}{3}$, $P(E_2) = \frac{1}{3}$, and $P(E_3) = \frac{1}{3}$,

$$P(E_1 \text{ and } E_2 \text{ and } E_3) = \frac{1}{3} \cdot \frac{1}{3} \cdot \frac{1}{3} = \frac{1}{27}$$

The probability that the baseball player bats safely three times in a row is $\frac{1}{27}$.

■

Example 3   A die is tossed three times. What is the probability that the outcome is a six on the first toss, an even number on the second toss, and an odd prime number on the third toss?

Solution   The probability of a six on any toss is $P(\text{six}) = \frac{1}{6}$. Because there are three even integers represented on the faces of a die, the probability of tossing an even number is $P(\text{even number}) = \frac{3}{6} = \frac{1}{2}$. The numbers 3 and 5 are the only odd prime numbers on the faces of a die, so the probability of tossing an odd prime is $P(\text{odd prime}) = \frac{2}{6} = \frac{1}{3}$. Because these three events are independent, the probability of the three events happening in succession is the product of the probabilities:

$$P(\text{six and even number and odd prime}) = \frac{1}{6} \cdot \frac{1}{2} \cdot \frac{1}{3} = \frac{1}{36}$$

■

Example 4   The probability that the drug Flake Off will cure dandruff is $\frac{1}{8}$. If the drug is used, the probability that the patient will have side effects is $\frac{1}{6}$. What is the probability that a patient who uses the drug will be cured and will suffer no side effects?

Solution   The probability that Flake Off will work is $P(\text{works}) = \frac{1}{8}$, and the probability that the patient will have no side effects is $P(\text{no side effects}) = 1 - P(\text{side effects}) = 1 - \frac{1}{6} = \frac{5}{6}$. These events are independent, so $P(\text{cure and no side effects}) = \frac{1}{8} \cdot \frac{5}{6} = \frac{5}{48}$.

■

Exercise 12.8

In Exercises 1–4, assume that you draw one card from a card deck. Find the probability of the given event.

1. Drawing a black card
2. Drawing a jack
3. Drawing a black card or an ace
4. Drawing a red card or a face card

In Exercises 5–8, assume that you draw two cards from a card deck without replacement. Find the probability of the given event.

5. Drawing two aces
6. Drawing three aces
7. Drawing a club and then drawing another black card
8. Drawing a heart and then drawing a spade

In Exercises 9–12, assume that you roll two dice once.

9. What is the probability of rolling a sum of 7 or 6?
10. What is the probability of rolling a sum of 5 or an even sum?
11. What is the probability of rolling a sum of 10 or an odd sum?
12. What is the probability of rolling a sum of 12 or 1?

In Exercises 13–16, assume that you are dealing with a bucket that contains 7 beige capsules, 3 cyan capsules, and 6 magenta capsules. You make a single draw from the bucket, taking one capsule.

13. What is the probability of drawing a beige or a cyan capsule?
14. What is the probability of drawing a magenta capsule?
15. What is the probability of not drawing a cyan capsule?
16. What is the probability of not drawing either a beige or a cyan capsule?

In Exercises 17–19, assume that you are using the same bucket of capsules as in Exercises 13–16.

17. On two draws from the bucket, what is the probability of drawing a beige capsule followed by a magenta capsule? (Assume that the capsule is returned to the bucket after the first draw.)
18. On two draws from the bucket, what is the probability of drawing one cyan and one magenta capsule? (Assume that the capsule is not returned to the bucket after the first draw.)
19. On three successive draws from the bucket (without replacement), what is the probability of failing to draw a beige capsule?
20. Jeff rolls a die and draws one card from a card deck. What is the probability of his rolling a four and drawing a four?
21. Three people are in an elevator together. What is the probability that all three were born on the same day of the week?
22. Three people are on a bus together. What is the probability that at least one was born on a different day of the week than the others?
23. Five people are in a room together. What is the probability that all five were born on a different day of the year?

24. Five people are on a bus together. What is the probability that at least two of them were born on the same day of the year?

25. If the probability that Rick will solve a problem is $\frac{1}{4}$ and the probability that Dinah will solve it is $\frac{2}{5}$, what is the probability that at least one of them will solve the problem?

26. A certain bugle call is based on four pitches and is five notes long. If a child can play these four pitches on a bugle, what is the probability that the first five notes that the child plays will be the bugle call? (Assume that the child is equally likely to play any of the four pitches each time a note is blown.)

27. Valerie visits her cabin in Canada. The probability that her lawn mower will start is $\frac{1}{2}$, the probability that her gas power saw will start is $\frac{1}{3}$, and the probability that her outboard motor will start is $\frac{3}{4}$. What is the probability that all three will start and that Valerie will have a nice vacation? What is the probability that none will start? That exactly one will start? That exactly two will start? What is the sum of your answers?

28. Three children will leave Thailand to start a new life in either the United States or France. The probability that May Xao will go to France is $\frac{1}{3}$, that Tou Lia will go to France is $\frac{1}{2}$, and that May Moua will go to France is $\frac{1}{6}$. What is the probability that exactly two of them will end up in the United States?

12.9 ODDS AND MATHEMATICAL EXPECTATION

There is a concept known as **mathematical odds** that is related to probability.

> **Definition.**   The **odds for an event** is the probability of a favorable outcome divided by the probability of an unfavorable outcome.

> **Definition.**   The **odds against an event** is the probability of an unfavorable outcome divided by the probability of a favorable outcome.

Example 1   The probability that a horse will win a race is $\frac{1}{4}$. What are the odds for and the odds against the horse?

Solution   Because the probability that the horse will win is $\frac{1}{4}$, the probability that the horse will not win is $\frac{3}{4}$. Therefore, the odds for the horse are

$$\frac{\dfrac{1}{4}}{\dfrac{3}{4}} = \frac{1}{3}$$

or 1 to 3, and the odds against the horse are

$$\frac{\dfrac{3}{4}}{\dfrac{1}{4}} = 3$$

or 3 to 1. (Note that the odds for an event is the reciprocal of the odds against the event.) ∎

Suppose we have a chance to play a simple game with the following rules:

1. Roll a single die once.
2. If a six appears, win $3.
3. If a five appears, win $1.
4. If any other number appears, win 50¢.
5. The cost to play (one roll of the die) is $1.

Whether it would be advisable to play this game depends on what we can expect to win. In this case the probability of any of the six outcomes—rolling a 1, 2, 3, 4, 5, or 6—is $\frac{1}{6}$, and the winnings are $3, $1, and 50¢. The expected winnings can be found by using the following equation and simplifying the right side:

$$E = \frac{1}{6}(3) + \frac{1}{6}(1) + \frac{1}{6}(0.50) + \frac{1}{6}(0.50) + \frac{1}{6}(0.50) + \frac{1}{6}(0.50)$$

$$= \frac{1}{6}(6)$$

$$= 1$$

Over the long run, we could expect to win $1 with every play of the game. However, it does cost $1 to play the game, so the expected net gain or loss of playing the game is 0. Because the expected winnings equal the admission price, the game is said to be *fair*.

Definition. If a certain event has n different outcomes with probabilities $p_1, p_2, p_3, \ldots, p_n$ and the winnings assigned to each outcome are $x_1, x_2, x_3, \ldots, x_n$, the expected winnings, or **mathematical expectation**, $E(x)$, is given by

$$E(x) = p_1 x_1 + p_2 x_2 + p_3 x_3 + \cdots + p_n x_n$$

Example 2 It costs $1 to play the following game: Roll two dice, collect $5 if you roll a sum of 7, and collect $2 if you roll a sum of 11. All other numbers pay nothing. Is it wise to play this game?

Solution   The probability of rolling a 7 on a single roll of two dice is $\frac{6}{36}$, the probability of rolling an 11 is $\frac{2}{36}$, and the probability. of rolling something else is $\frac{28}{36}$. The mathematical expectation is

$$E(x) = \frac{6}{36}(5) + \frac{2}{36}(2) + \frac{28}{36}(0)$$

$$= \frac{17}{18}$$

$$\approx \$0.944$$

By playing the game over and over for a long period of time, you can expect to retrieve a little less than 95¢ for every dollar spent. For the fun of playing, the cost is about 5¢ a game. If the game is enjoyable, it might be worth the expected loss of a nickel. However, the game is slightly unfair.   ∎

Exercise 12.9

In Exercises 1–6, assume a single roll of a die.

1.   What is the probability of rolling a 6?
2.   What are the odds in favor of rolling a 6?
3.   What are the odds against rolling a 6?
4.   What is the probability of rolling an even number?
5.   What are the odds in favor of rolling an even number?
6.   What are the odds against rolling an even number?

In Exercises 7–12, assume a single roll of two dice.

7.   What is the probability of rolling a sum of 6?
8.   What are the odds in favor of rolling a sum of 6?
9.   What are the odds against rolling a sum of 6?
10.   What is the probability of rolling an even sum?
11.   What are the odds in favor of rolling an even sum?
12.   What are the odds against rolling an even sum?

In Exercises 13–16, assume that you are drawing one card from a card deck.

13.   What are the odds in favor of drawing a queen?
14.   What are the odds against drawing a black card?
15.   What are the odds in favor of drawing a face card?
16.   What are the odds against drawing a diamond?
17.   If the odds in favor of victory are 5 to 2, what is the probability of victory?
18.   If the odds in favor of victory are 5 to 2, what are the odds against victory?
19.   If the odds against winning are 90 to 1, what are the odds in favor of winning?
20.   What are the odds in favor of rolling a 7 on a single toss of two dice?

21. What are the odds against tossing four heads in a row with a fair nickel?

22. What are the odds in favor of a couple having four girl babies in succession? (Assume $P(\text{boy}) = \frac{1}{2}$.)

23. The odds against a horse are 8 to 1. What is the probability that the horse will win?

24. The odds against a horse are 1 to 1. What is the probability that the horse will lose?

25. It costs $2 to play the following game:
  a. Draw one card from a card deck.   **b.** Collect $5 if an ace is drawn.
  c. Collect $4 if a king is drawn.   **d.** Collect nothing for all other cards drawn.
  Is it wise to play this game? Explain.

26. One thousand tickets are sold for a lottery with two grand prizes of $800. What is a "fair" price for the tickets?

27. What are the odds against a couple having three baby boys in a row? (Assume $P(\text{boy}) = \frac{1}{2}$.)

28. What are the odds in favor of tossing at least three heads in five tosses of a fair coin?

29. Suppose you toss a coin five times and collect $5 if you toss five heads, $4 if you toss four heads, $3 if you toss three heads, and no money for any other combinations. How much should you pay to play the game if the game is to be fair?

30. If you toss two dice one time and collect $10 for double sixes and $1 for double ones, what is a fair price for playing the game?

31. Counting an ace as 1, a face card as 10, and all others at their numerical value, what is the expected value if you draw one card from a card deck?

32. What is the expected sum of one roll of two dice?

33. A multiple-choice test of eight questions gives five possible answers for each question. Only one of the answers for each question is right. What is the probability of getting seven right answers by simple guessing?

34. In the situation described in Exercise 33, what are the odds in favor of getting seven answers right?

REVIEW EXERCISES

1. Verify the following formula for $n = 1$, $n = 2$, $n = 3$, and $n = 4$, and then prove the formula by mathematical induction:

$$1^3 + 2^3 + 3^3 + \cdots + n^3 = \frac{n^2(n+1)^2}{4}$$

2. Evaluate $\sum_{k=1}^{4} 3k^2$.

3. Evaluate $\sum_{k=5}^{8} (k^3 + 3k^2)$.

4. Evaluate $\sum_{k=1}^{30} \left(\frac{3}{2} k - 12 \right) - \frac{3}{2} \sum_{k=1}^{30} k$

In Review Exercises 5–8, find the required term of each arithmetic progression.

5. $5, 9, 13, \ldots$; 29th term

6. $8, 15, 22, \ldots$; 40th term

7. $6, -1, -8, \ldots$; 15th term

8. $\frac{1}{2}, -\frac{3}{2}, -\frac{7}{2}, \ldots$; 35th term

In Review Exercises 9–12, find the required term of each geometric progression.

9. $81, 27, 9, \ldots$; 11th term

10. $2, 6, 18, \ldots$; 9th term

11. $9, \dfrac{9}{2}, \dfrac{9}{4}, \ldots$; 15th term

12. $8, -\dfrac{8}{5}, \dfrac{8}{25}, \ldots$; 7th term

In Review Exercises 13–16, find the sum of the first 40 terms in each progression.

13. $5, 9, 13, \ldots$

14. $8, 15, 22, \ldots$

15. $6, -1, -8, \ldots$

16. $\dfrac{1}{2}, -\dfrac{3}{2}, -\dfrac{7}{2}, \ldots$

In Review Exercises 17–20, find the sum of the first eight terms in each progression.

17. $81, 27, 9, \ldots$

18. $2, 6, 18, \ldots$

19. $9, \dfrac{9}{2}, \dfrac{9}{4}, \ldots$

20. $8, -\dfrac{8}{5}, \dfrac{8}{25}, \ldots$

In Review Exercises 21–24, find the sum of each infinite progression, if possible.

21. $\dfrac{1}{3}, \dfrac{1}{6}, \dfrac{1}{12}, \ldots$

22. $\dfrac{1}{5}, -\dfrac{2}{15}, \dfrac{4}{45}, \ldots$

23. $1, \dfrac{3}{2}, \dfrac{9}{4}, \ldots$

24. $0.5, 0.25, 0.125, \ldots$

In Review Exercises 25–28, use the formula for the sum of the terms of an infinite geometric progression to change each decimal into a common fraction.

25. $0.333\ldots$

26. $0.999\ldots$

27. $0.171717\ldots$

28. $0.454545\ldots$

29. Insert three arithmetic means between 2 and 8.

30. Insert five arithmetic means between 10 and 100.

31. Insert three geometric means between 2 and 8.

32. Insert four geometric means between -2 and 64.

33. Find the sum of the first 8 terms of the progression $\frac{1}{3}, 1, 3, \ldots$.

34. Find the seventh term of the progression $2\sqrt{2}, 4, 4\sqrt{2}, \ldots$.

35. Find the single geometric mean between 4 and 64.

36. If Leonard invests $3000 in a 6-year certificate of deposit at the annual rate of 7.75% compounded daily, how much money will be in the account when it matures?

37. The enrollment at Hometown College is growing at the rate of 5% over each previous year's enrollment. If the enrollment is currently 4000 students, what will the enrollment be 10 years from now? What was it 5 years ago?

38. A house trailer that originally cost $10,000 depreciates in value at the rate of 10% per year. How much will the trailer be worth after 10 years?

In Review Exercises 39–42, use the binomial theorem to find the expansion of each expression.

39. $(x - y)^3$

40. $(u + 2v)^3$

41. $(4a - 5b)^5$

42. $(\sqrt{7}\,r + \sqrt{3}\,s)^4$

In Review Exercises 43–46, find the required term of each expansion.

43. $(a + b)^8$; fourth term

44. $(2x - y)^5$; third term

45. $(x - y)^9$; seventh term

46. $(4x + 7)^6$; fourth term

In Review Exercises 47–58, evaluate each expression.

47. $_6P_6$

48. $\binom{7}{4}$

49. $0!$

50. $_{10}P_2 \cdot {}_{10}C_2$

51. $_8P_6 \cdot {}_8C_6$

52. $\binom{8}{5}\binom{6}{2}$

53. $_7C_5 \cdot {}_4P_0$

54. $_{12}C_0 \cdot {}_{11}C_0$

55. $\dfrac{_8P_5}{_8C_5}$

56. $\dfrac{_8C_5}{_{13}C_5}$

57. $\dfrac{_6C_3}{_{10}C_3}$

58. $\dfrac{_{13}C_5}{_{52}C_5}$

59. Make a tree diagram to illustrate the possible results of tossing a coin four times.

60. State the multiplication principle for events.

61. State the multiplication property for probabilities.

62. In how many ways can you draw a five-card poker hand of 3 aces and 2 kings?

63. What is the probability of drawing the hand described in Review Exercise 62?

64. What is the probability of *not* drawing the hand described in Review Exercise 62?

65. In how many ways can 10 teenagers be seated at a round table if 2 girls wish to sit with their boyfriends?

66. How many distinguishable words can be formed from the letters of the word *casserole* if each letter is used exactly once?

67. What is the probability of having a 13-card bridge hand consisting of 4 aces, 4 kings, 4 queens, and 1 jack?

68. Find the probability of choosing a committee of 3 boys and 2 girls from a group of 8 boys and 6 girls.

69. Find the probability of drawing a club or a spade on one draw from a card deck.

70. Find the probability of drawing a black card or a king on one draw from a card deck.

71. What is the probability of getting an ace-high royal flush (ace, king, queen, jack, and ten of hearts) in poker?

72. What is the probability of being dealt 13 cards of one suit in a bridge hand?

73. What is the probability of getting 3 heads or fewer on 4 tosses of a fair coin?

74. What are the odds against a horse if the probability that the horse will win is $\frac{7}{8}$?

75. What are the odds in favor of a couple having 4 baby girls in a row?

76. What are the expected earnings if you collect $1 for every head you get when you toss a fair coin 4 times?

77. If the probability that Joe will marry is $\frac{5}{6}$ and the probability that John will marry is $\frac{3}{4}$, what are the odds against either one becoming a husband?

78. If the odds against Priscilla's graduation from college are $\frac{10}{11}$, what is the probability that she will graduate?

79. If the probability that a drug cures a certain disease is 0.83 and we give the drug to 800 people with the disease, what is the number of people that we can expect to be cured?

80. If the total number of subsets that a set with n elements can have is 2^n, explain why $\binom{n}{0} + \binom{n}{1} + \binom{n}{2} + \cdots + \binom{n}{n} = 2^n$.

A PROOF OF THE BINOMIAL THEOREM

The binomial theorem can be proved for positive integral exponents by using mathematical induction.

The Binomial Theorem. If n is a positive integer, then

$$(a + b)^n = a^n + \frac{n!}{1!(n-1)!}\, a^{n-1}b + \frac{n!}{2!(n-2)!}\, a^{n-2}b^2 + \cdots$$

$$+ \frac{n!}{r!(n-r)!}\, a^{n-r}b^r + \cdots + b^n$$

Proof As in all induction proofs, there are two parts.

Part 1. Substituting the number 1 for n on both sides of the equation, we have

$$(a + b)^1 = a^1 + \frac{1!}{1!(1-1)!}\, a^{1-1}b^1$$

$$a + b = a + a^0 b$$

$$a + b = a + b$$

and the theorem is true when $n = 1$. Part 1 is complete.

Part 2. We write expressions for two general terms in the statement of the induction hypothesis. We assume that the theorem is true for $n = k$:

$$(a + b)^k = a^k + \frac{k!}{1!(k-1)!}\, a^{k-1}b + \frac{k!}{2!(k-2)!}\, a^{k-2}b^2 + \cdots$$

$$+ \frac{k!}{(r-1)!(k-r+1)!}\, a^{k-r+1}b^{r-1}$$

$$+ \frac{k!}{r!(k-r)!}\, a^{k-r}b^r + \cdots + b^k$$

We multiply both sides of the equation above by $a + b$ and hope to obtain a similar equation in which the quantity $k + 1$ replaces all of the n values in the binomial theorem:

$$(a + b)^k(a + b)$$

$$= (a + b)\left[a^k + \frac{k!}{1!(k-1)!} a^{k-1}b + \frac{k!}{2!(k-2)!} a^{k-2}b^2 + \cdots \right.$$

$$\left. + \frac{k!}{(r-1)!(k-r+1)!} a^{k-r+1}b^{r-1} + \frac{k!}{r!(k-r)!} a^{k-r}b^r + \cdots + b^k \right]$$

We distribute the multiplication first by a and then by b:

$$(a + b)^{k+1} = \left[a^{k+1} + \frac{k!}{1!(k-1)!} a^k b + \frac{k!}{2!(k-2)!} a^{k-1}b^2 + \cdots \right.$$

$$\left. + \frac{k!}{(r-1)!(k-r+1)!} a^{k-r+2}b^{r-1} + \frac{k!}{r!(k-r)!} a^{k-r+1}b^r + \cdots + ab^k \right]$$

$$+ \left[a^k b + \frac{k!}{1!(k-1)!} a^{k-1}b^2 + \frac{k!}{2!(k-2)!} a^{k-2}b^3 + \cdots \right.$$

$$\left. + \frac{k!}{(r-1)!(k-r+1)!} a^{k-r+1}b^r + \frac{k!}{r!(k-r)!} a^{k-r}b^{r+1} + \cdots + b^{k+1} \right]$$

Combining like terms, we have

$$(a + b)^{k+1} = a^{k+1} + \left[\frac{k!}{1!(k-1)!} + 1 \right] a^k b$$

$$+ \left[\frac{k!}{2!(k-2)!} + \frac{k!}{1!(k-1)!} \right] a^{k-1}b^2 + \cdots$$

$$+ \left[\frac{k!}{r!(k-r)!} + \frac{k!}{(r-1)!(k-r+1)!} \right] a^{k-r+1}b^r + \cdots + b^{k+1}$$

These results may be written as

$$(a + b)^{k+1} = a^{k+1} + \frac{(k+1)!}{1!(k+1-1)!} a^{(k+1)-1}b + \frac{(k+1)!}{2!(k+1-2)!} a^{(k+1)-2}b^2$$

$$+ \cdots + \frac{(k+1)!}{r!(k+1-r)!} a^{(k+1)-r}b^r + \cdots + b^{k+1}$$

This formula has precisely the same form as the binomial theorem with the quantity $k + 1$ replacing all of the original n values. Therefore, the truth of the theorem for $n = k$ implies the truth of the theorem for $n = k + 1$. Because both parts of the axiom of mathematical induction are verified, the theorem is proved.

$\square$

ANSWERS TO SELECTED EXERCISES

Exercise 1.1 (page 5)

1. 53   **3.**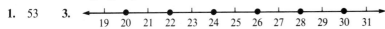

5. false; it can be divided by 3 or 9   **7.** true

9. false; the product of two primes such as 3 and 5 is a composite: $3 \cdot 5 = 15$

11. false; the square of a prime number such as 7 is a composite: $7^2 = 49$ **13.** true   **15.** true

17. false; the even integer 2 is a prime number   **19.** true   **21.** true

23. false; if two of the integers are odd and one is even, their sum is even: $3 + 5 + 10 = 18$

25. false; 0 can be written in the required form: $\frac{0}{5}$ **27.** true   **29.** 0.25   **31.** 0.222...

33. $-0.41666...$ **35.** $\frac{3}{10}$ **37.** $-\frac{25}{33}$ **39.** $\frac{41}{333}$ **41.** $\frac{1151}{3330}$ **43.** $-\frac{853}{99}$ **45.** $\frac{1601}{990}$

49. no; $0.999 = \frac{999}{1000}$, which is not equal to 1

Exercise 1.2 (page 13)

1. true   **3.** false; all integers are rational numbers   **5.** false; $\mathbf{R} \cup \mathbf{H} = \mathscr{R}$ **7.** true

9. (graph, open circles at 3 and 10)   **11.** (graph at -4, 0)   **13.** (graph at 3, 7)   **15.** (graph at -2, 10)

17. No graph; the set is empty.   **19.** (graph at -2, 3)   **21.** (graph at 4, 8)

23. (graph at -2, 0, 1, 2)   **25.** (graph at -3, 0, 2)   **27.** 7   **29.** 1

31. $\sqrt{2} - 1$ **33.** 10   **35.** -30 **37.** 9   **39.** 3 units   **41.** Transitive

43. reflexive and transitive   **45.** reflexive, symmetric, and transitive   **47.** reflexive, symmetric, and transitive

49. commutative property of addition   **51.** closure property   **53.** transitive property of equality

55. reflexive property of equality   **57.** closure property   **59.** commutative property of addition

61. symmetric property of equality   **63.** associative property of multiplication

Exercise 1.3 (page 21)

1. 13   **3.** 25   **5.** 0   **7.** $\frac{1}{6}$ **9.** x^5 **11.** $\frac{1}{y^5}$ **13.** z^6 **15.** $27x^3$ **17.** x^6y^3 **19.** $\frac{x^6}{y^3}$

21. $-243x^5y^{10}$ **23.** $\frac{27x^6}{y^9z^3}$ **25.** $\frac{x^{16}}{y^8}$ **27.** $\frac{9}{25}x^{10}$ **29.** 1   **31.** $-x^{10}y^{5x}$ **33.** $-x^{8xy}y^{12xy}$

35. x^4 **37.** x^{2n^2+n} **39.** x^5 **41.** $\frac{64z^7}{25y^6}$ **43.** $\frac{1}{m^{14}n^{16}p^{12}}$ **45.** 4   **47.** -4 **49.** 8

51. 18   **53.** 20   **55.** 0   **57.** $\frac{1}{24}$ **59.** 0   **61.** $\frac{27}{4}$ **63.** 3.72×10^5 **65.** 1.77×10^8

67. 7×10^{-3} **69.** 6.93×10^{-7} **71.** 1×10^{12} **73.** 1×10^{-12} **75.** 9.97×10^{-3} **77.** 937,000

79. 0.0000221   **81.** 3.2   **83.** 1.17×10^4 **85.** 7×10^4 **87.** 1.986×10^4 meters per minute

89. 1.67248×10^{-15} gram

Exercise 1.4   (page 30)

1. 7   **3.** -8 **5.** 2   **7.** -4 **9.** 3   **11.** $\frac{2}{3}$ **13.** 3   **15.** 8   **17.** -100 **19.** $\frac{1}{8}$

21. $\frac{1}{512}$ **23.** $-\frac{1}{27}$ **25.** $\frac{32}{243}$ **27.** $\frac{9}{16}$ **29.** $7x$ **31.** xy^2 **33.** $2y$ **35.** x^6y^3 **37.** $\frac{1}{x^6y^8}$

39. $\frac{a^2c^6}{b^5}$ **41.** $2|a|$ **43.** $|r|s^2$ **45.** $3x$ **47.** $\frac{1}{4y^4z^2}$ **49.** $\sqrt{2}$ **51.** $17x\sqrt{2}$ **53.** $2y^2\sqrt{3y}$

55. $12\sqrt[3]{3}$ **57.** $6z\sqrt[4]{3z}$ **59.** 45   **61.** $\sqrt{3}$ **63.** $\frac{2\sqrt{x}}{x}$ **65.** $\sqrt[3]{4}$ **67.** $\sqrt[3]{5a^2}$ **69.** $\frac{2b\sqrt[4]{27a^2}}{3a}$

71. $\frac{\sqrt{2xy}}{2y}$ **73.** $\frac{u\sqrt[3]{6uv^2}}{3v}$ **75.** $\frac{2\sqrt{3}}{9}$ **77.** $\frac{\sqrt{2x}}{4x} - \frac{\sqrt{2y}}{2y^2}$ **79.** $\sqrt{3}$ **81.** $x\sqrt[3]{25x}$ **83.** $\frac{x\sqrt{y}}{3}$

85. $\frac{3y}{x}$ **87.** $2y$ **89.** $x\sqrt{2y}$ **91.** $y\sqrt[3]{2xy}$ **93.** $\frac{x\sqrt{2}}{4}$ **95.** $\sqrt{3}$ **97.** $\sqrt[5]{4x^3}$ **99.** $\sqrt[6]{32}$

101. $\sqrt[4]{3}$ **103.** $x \geq 0$ **105.** $x \leq 0$

Exercise 1.5   (page 40)

1. a polynomial, degree of 2, trinomial   **3.** not a polynomial   **5.** a polynomial, degree of 3, binomial
7. a polynomial, degree of 0, monomial   **9.** a polynomial, undefined degree, monomial
11. not a polynomial   **13.** $6x^3 - 3x^2 - 8x$ **15.** $4x^3 + 14$ **17.** $-x^2 + 14$ **19.** $-28x + 96$
21. $-4x^2 + x$ **23.** $8x^3y^7$ **25.** $-4r^3s - 4rs^3$ **27.** $x^2 - 25$ **29.** $3x^2 + 4x - 4$
31. $2x^2 - 5x - 12$ **33.** $10x^2 + 13x - 3$ **35.** $6 - 16x + 8x^2$ **37.** $-6x^2 + 7x + 3$
39. $9x^3 - x^2 - 27x + 3$ **41.** $5x^3 + 2x^2 - 5x - 2$ **43.** $9x^2 + 54x + 81$ **45.** $27x^3 - 27x^2 + 9x - 1$
47. $x^3y^3 + 2xy^2 - x^2y - 2$ **49.** $6x^3 + 14x^2 - 5x - 3$ **51.** $x^4 + 2x^3 + x^2 - 1$ **53.** $xy + x^{\frac{3}{2}}y^{\frac{1}{2}}$
55. $a - b$ **57.** $\sqrt{3} + 1$ **59.** $x(\sqrt{7} - 2)$ **61.** $2(\sqrt{5} + \sqrt{2})$ **63.** $\frac{x(x + \sqrt{3})}{x^2 - 3}$ **65.** $\frac{(y + \sqrt{2})^2}{y^2 - 2}$
67. $9 - 4\sqrt{5}$ **69.** $\frac{\sqrt{3} + 3 - \sqrt{2} - \sqrt{6}}{2}$ **71.** $\frac{1}{\sqrt{x + 3} + \sqrt{x}}$ **73.** $3x + 2 + \frac{3}{x + 3}$ **75.** $x - 7$
77. $x - 3$ **79.** $x^3 - 3 + \frac{3}{x^2 - 2}$ **81.** $x^4 + 2x^3 + 4x^2 + 8x + 16$ **83.** $6x^2 + x - 12$
85. $3x^2 - x + 2$ **87.** x^{5a-1} **89.** $b^{3a^2 + 3a}$ **91.** a^{4x-1}

Exercise 1.6   (page 47)

1. $3(x - 2)$ **3.** $4x^2(2 + x)$ **5.** $7xy(y + 2x)$ **7.** $3abc(a + 2b + 3c)$ **9.** $(x + y)(b - a)$
11. $(4a + b)(1 - 3a)$ **13.** $(x + 1)(3x^2 - 1)$ **15.** $t(y + c)(2x - 3)$ **17.** $(a + b)(x + y + z)$
19. $(2x + y)(3c + d)$ **21.** $(2x + 3)(2x - 3)$ **23.** $(2 + 3r)(2 - 3r)$ **25.** $(x + z + 5)(x + z - 5)$
27. not factorable over the integers   **29.** $(x + y - z)(x - y + z)$ **31.** $-4xy$
33. $(x^2 + y^2)(x + y)(x - y)$ **35.** $(x^4 + 8z^2)(x^4 - 8z^2)$ **37.** $3(x + 2)(x - 2)$
39. $2x(3y + 2)(3y - 2)$ **41.** $(x + 7)(x + 3)$ **43.** $(x - 6)(x + 2)$ **45.** not factorable over the integers
47. $(4x - 3)(3x + 2)$ **49.** $(6y - 5)(4y - 3)$ **51.** $(6a + 5)(4a - 3)$ **53.** $(3x + 7)(2x + 5)$
55. $(2x - 5)(3x + 7)$ **57.** $2(6y - 35)(y + 1)$ **59.** $x(6x + 7)(x - 5)$ **61.** $(2x - 7)(3x + 5)$
63. $(6x - 5)(x - 7)$ **65.** $(x^2 + 5)(x^2 - 3)$ **67.** $2(x^2y^2z^2 - 8xyz + 40)$ **69.** $(2z - 3)(4z^2 + 6z + 9)$
71. $2(x + 10)(x^2 - 10x + 100)$ **73.** $(x + y - 4)(x^2 + 2xy + y^2 + 4x + 4y + 16)$ **75.** $-x(x^2 + 3x + 3)$
77. $(2a + y)(2a - y)(4a^2 - 2ay + y^2)(4a^2 + 2ay + y^2)$ **79.** $x(2 + 3y)(2x + 3y)$
81. $(4x^2 + y^2)(16x^4 - 4x^2y^2 + y^4)$ **83.** $(x - 3 + 12y)(x - 3 - 12y)$ **85.** $(a + b - 5)(a + b + 2)$
87. $(3u + 3v + 4)(2u + 2v + 1)$ **89.** $(x + 2)(x^2 - 2x + 4)(x - 1)(x^2 + x + 1)$ **91.** $(a + b)(c + d + 1)$
93. $(x^2 + x + 2)(x^2 - x + 2)$ **95.** $(x^2 + x + 4)(x^2 - x + 4)$ **97.** $(2a^2 - a + 1)(2a^2 + a + 1)$

99. $2(x^2 + 2x + 2)(x^2 - 2x + 2)$ **101.** $2(\frac{3}{2}x + 1)$ **103.** $x^{\frac{1}{2}}(x^{\frac{1}{2}} + 1)$ **105.** $ab(b^{\frac{1}{2}} - a^{\frac{1}{2}})$ **107.** $\dfrac{3 - \sqrt{5}}{2}$

Exercise 1.7   (page 56)

1. $\dfrac{x + 4}{x - 4}$ **3.** $\dfrac{3x - 4}{2x - 1}$ **5.** $\dfrac{x(x - 1)}{x + 1}$ **7.** $\frac{1}{2}$ **9.** $\dfrac{z - 4}{(z + 2)(z - 2)}$ **11.** $\dfrac{x - 1}{x}$ **13.** $\dfrac{x(x + 1)(x + 1)}{x + 2}$

15. 1 **17.** $\dfrac{(x + 12)(2x - 1)}{(2x + 1)(2x - 3)}$ **19.** $\dfrac{x(x + 3)}{x + 1}$ **21.** $\dfrac{x + 5}{x + 3}$ **23.** $\dfrac{-3x^2 + 7x + 4}{(x - 1)(x + 1)}$ **25.** $\dfrac{2a - 4}{(a + 4)(a - 4)}$

27. $\dfrac{-2}{(3a + 4)(a + 1)}$ **29.** $\dfrac{1}{x + 2}$ **31.** $\dfrac{2x - 5}{2x(x - 2)}$ **33.** $\dfrac{2x^2 + 19x + 1}{(x - 4)(x + 4)}$ **35.** 0

37. $\dfrac{-x^4 + 3x^3 - 18x^2 - 58x + 72}{(x + 5)(x - 5)(x - 4)(x + 4)}$ **39.** $\dfrac{b}{2c}$ **41.** $81a$ **43.** -1 **45.** $\dfrac{y + x}{x^2 y^2}$ **47.** $\dfrac{y + x}{y - x}$

49. $\dfrac{a^2(3x - 4ab)}{ax + b}$ **51.** $\dfrac{x - 2}{x + 2}$ **53.** $\dfrac{3x^2 y^2}{xy - 1}$ **55.** $\dfrac{3x^2}{x^2 + 1}$ **57.** $\dfrac{x^2 - 3x - 4}{x^2 + 5x - 3}$ **59.** $\dfrac{x}{x + 1}$

61. $\dfrac{5x + 1}{x - 1}$ **63.** $\dfrac{1}{1 + x}$

REVIEW EXERCISES   (page 58)

1.
11 13   17 19

2. $0.925925\ldots$ **3.** $\dfrac{853}{990}$

4. no; because it is a terminating decimal, it represents a rational number   **5.** ◄───────○────────●───►
3    $\frac{1}{3}$

6. ◄──●───────○──►
$\frac{2}{3}$ $\frac{5}{3}$ **7.** ◄──────○───────●──►
-1 0 **8.** ◄───○──────────○──►
0 4 **9.** 6 **10.** -1

11. associative property of addition   **12.** commutative property of addition   **13.** additive identity property
14. associative property of multiplication   **15.** multiplicative inverse property

16. distributive property or the multiplicative identity property   **17.** $\dfrac{x^9}{y^6}$ **18.** $-\dfrac{x^6}{y^4}$ **19.** $\dfrac{8x^6}{y^6}$

20. $\dfrac{27}{x^3 y^3}$ **21.** $\dfrac{y^8}{9}$ **22.** $\dfrac{y^4}{9x^4}$ **23.** 0 **24.** $-\frac{1}{3}$ **25.** $\frac{1}{4}$ **26.** $\frac{1}{4}$ **27.** 2 **28.** 1 **29.** 6

30. -7 **31.** -2 **32.** 4 **33.** -4 **34.** -32 **35.** xy^2 **36.** x **37.** $x^6 y$ **38.** $\dfrac{y^2}{x^7}$

39. $|x|y^2$ **40.** x **41.** $x^6|y|$ **42.** $\dfrac{y^2}{|x^7|}$ **43.** $\dfrac{2\sqrt{5}}{5}$ **44.** $2\sqrt{2}$ **45.** $\dfrac{\sqrt[3]{4}}{2}$ **46.** $\dfrac{2\sqrt[3]{5}}{5}$

47. $\sqrt{3} + 1$ **48.** $2(\sqrt{3} + \sqrt{2})$ **49.** $7\sqrt{2}$ **50.** 0 **51.** $5 + 2\sqrt{6}$
52. -1 **53.** $\sqrt{6} + \sqrt{2} + \sqrt{3} + 1$ **54.** -5 **55.** a polynomial, degree of 3, binomial
56. a polynomial, degree of 2, trinomial   **57.** a polynomial, degree of 2, monomial   **58.** not a polynomial

59. $x^2 + 3x - 2$ **60.** $x^2 + x + 1$ **61.** $3x - 3 + \dfrac{4}{x + 1}$ **62.** $x - 3 + \dfrac{8x + 5}{x^2 + 3x}$

63. $3x(x + 1)(x - 1)$ **64.** $5(x - 1)(x^2 + x + 1)$ **65.** $(3x + 8)(2x - 3)$ **66.** $(3a + x)(a - 1)$
67. $(2x - 5)(4x^2 + 10x + 25)$ **68.** $2(3x + 2)(x - 4)$ **69.** $(x + 3 + 2t)(x + 3 - 2t)$ **70.** prime
71. $(2z + 7)(4z^2 - 14z + 49)$ **72.** $(7b + 1)(7b + 1)$ **73.** $(11z - 2)(11z - 2)$

74. $8(2y - 5)(4y^2 + 10y + 25)$ **75.** $(y - 2z)(2x - w)$ **76.** $(x^2 - x + 1)(x^2 + x + 1)(x^4 - x^2 + 1)$

77. $(x - 2)(x + 3)$ **78.** $\dfrac{2x - 5}{x - 2}$ **79.** $\dfrac{x + 1}{5}$ **80.** $\dfrac{x^2(x + 4)}{(x - 3)(x^2 + 8x + 4)}$ **81.** $\dfrac{(x - 2)(x + 3)(x - 3)}{(x - 1)(x + 2)^2}$

82. 1 **83.** $\dfrac{3x^2 - 10x + 10}{(x - 4)(x + 5)}$ **84.** $\dfrac{2x^2 + 20x + 2}{(x - 2)(x + 3)}$ **85.** $\dfrac{3x^3 - 12x^2 + 11x}{(x - 1)(x - 2)(x - 3)}$

86. $\dfrac{-5x - 6}{(x + 1)(x + 2)}$ **87.** $\dfrac{-x^3 + x^2 - 12x - 15}{x^2(x + 1)}$ **88.** $\dfrac{3x}{x + 1}$ **89.** $\dfrac{20}{3x}$ **90.** $\dfrac{y}{2}$

91. $\dfrac{y + x}{xy(x - y)}$ **92.** $\dfrac{y + x}{x - y}$

Exercise 2.1   (page 67)

1. all real numbers   **3.** all real numbers except 3 and -3 **5.** all real numbers greater than or equal to -5
7. all real numbers except 3, -3, and -4 **9.** $x = 5$, conditional equation   **11.** no solution
13. $x = 7$, conditional equation   **15.** no solution   **17.** identity
19. $b = 6$, conditional equation   **21.** identity   **23.** $x = 1$ **25.** $z = 9$ **27.** $z = 10$ **29.** $x = -3$
31. $x = 6$ **33.** $x = -2$ **35.** $x = \frac{5}{2}$ **37.** $x = 1$ **39.** $x = -\frac{14}{11}$ **41.** identity   **43.** $y = 2$
45. $s = 4$ **47.** $x = -4$ **49.** no solution   **51.** $n = 17$ **53.** $x = -\frac{2}{5}$ **55.** $x = \frac{2}{3}$ **57.** $n = 3$
59. $y = 5$ **61.** no solutions

Exercise 2.2   (page 73)

1. 7   **3.** 17 and 37   **5.** \$10,000   **7.** 327   **9.** 8 of each kind   **11.** $\frac{4}{3}$ hours   **13.** $\frac{190}{9}$ hours
15. 1 liter   **17.** 600 cubic centimeters   **19.** about 45.5%   **21.** 84   **23.** 12 miles per hour
25. 300 miles   **27.** 600 pounds of barley, 1637 pounds of oats, 163 pounds of soybean meal

Exercise 2.3   (page 81)

1. $3, -2$ **3.** $12, -12$ **5.** $2, -\frac{5}{2}$ **7.** $2, \frac{3}{5}$ **9.** $\frac{3}{5}, -\frac{5}{3}$ **11.** $\frac{3}{2}, \frac{1}{2}$ **13.** $3, -3$
15. $5\sqrt{2}, -5\sqrt{2}$ **17.** $3, -1$ **19.** $2, -4$ **21.** $5, 3$ **23.** $2, -3$ **25.** $0, 25$ **27.** $\frac{2}{3}, -2$

29. $\dfrac{-5 + \sqrt{5}}{2}, \dfrac{-5 - \sqrt{5}}{2}$ **31.** $\dfrac{-2 + \sqrt{7}}{3}, \dfrac{-2 - \sqrt{7}}{3}$ **33.** $2\sqrt{3}, -2\sqrt{3}$ **35.** $3, -\frac{5}{2}$

37. $2, -\frac{1}{5}$ **39.** $1, -2$ **41.** $\dfrac{-5 + \sqrt{13}}{6}, \dfrac{-5 - \sqrt{13}}{6}$ **43.** $\dfrac{-1 + \sqrt{61}}{10}, \dfrac{-1 - \sqrt{61}}{10}$

45. $3, -4$ **47.** $\frac{3}{2}, -\frac{1}{4}$ **49.** $\frac{1}{2}, -\frac{4}{3}$ **51.** $\frac{5}{6}, -\frac{2}{5}$ **53.** $4 + 2\sqrt{2}, 4 - 2\sqrt{2}$

Exercise 2.4   (page 87)

1. rational and equal   **3.** not real numbers   **5.** rational and unequal   **7.** $2, 10$ **9.** yes
11. $2, -2, 3, -3$ **13.** $2, -2, 5, -5$ **15.** $\sqrt{5}, -\sqrt{5}, 3\sqrt{2}, -3\sqrt{2}$ **17.** $\frac{3}{2}, -\frac{3}{2}$
19. $1, 144$ **21.** $\frac{1}{64}$ **23.** $-1, -2, 3, 4$ **25.** $1, -1$ **27.** $-\frac{1}{2}, 5$ **29.** -2 **31.** 27 **33.** 2
35. $3, 4$ **37.** $3, 5$ **39.** 5 **41.** no solutions   **43.** no solutions   **45.** $\frac{1}{3}, -\frac{1}{3}, \frac{1}{2}, -\frac{1}{2}$
47. $x = \dfrac{2c}{-b \pm \sqrt{b^2 - 4ca}}$ **49.** When you divide both sides by $x - 1$, you divide by 0.
51. 10, 11, 12, 13, 14 or $-2, -1, 0, 1, 2$

Exercise 2.5   (page 91)

1. 6 and 8   **3.** 4 feet by 8 feet   **5.** 9 centimeters
7. 20 miles per hour going, 10 miles per hour returning   **9.** 7 hours   **11.** 25 seconds
13. about 9.5 seconds   **15.** 4 hours   **17.** about 9.5 hours   **19.** no
21. Matilda at 8%, Maude at 7%   **23.** 10

Exercise 2.6   (page 94)

1. $p = \dfrac{k}{2.2}$ **3.** $b_2 = \dfrac{2A}{h} - b_1$ **5.** $r = \sqrt{\dfrac{3V}{\pi h}}$ **7.** $s = \dfrac{f(P_n - L)}{i}$ **9.** $r = \dfrac{r_1 r_2}{r_1 + r_2}$

11. $n = \dfrac{l - a + d}{d}$ **13.** $N = \dfrac{\sum x^2}{\sigma^2 + \mu^2}$ **15.** $y = \dfrac{-3x^2 + 1}{3x - 7}$ **17.** $a = s(1 - r) + lr$

19. $r_3 = \dfrac{-Rr_1 r_2}{Rr_2 + Rr_1 - r_1 r_2}$ **21.** $y = \dfrac{-x \pm \sqrt{x^2 + 4ax}}{2}$ **23.** $r = \dfrac{3V + \pi h^3}{3\pi h^2}$ **25.** $h = \dfrac{6V}{B + B' + 4M}$

27. $x = \dfrac{r - y}{1 + ry}$ **29.** $y = \dfrac{y_2 - y_1}{x_2 - x_1}(x - x_1) + y_1$ **31.** $y_1 = \dfrac{y_2 x - y_2 x_1 + yx_1 - yx_2}{x - x_2}$

Exercise 2.7   (page 100)

1. ⊶ (open circle at 1)   **3.** ⊶ (closed at 1)   **5.** ⊶ (open at 1)   **7.** ⊶ (closed at 1)

9. (open at $-\frac{10}{3}$)   **11.** (closed at 3)   **13.** (closed at 5)   **15.** (closed at 14)

17. (open at 6, closed at 9)   **19.** (open at 8, closed at 22)   **21.** (closed at -11 and 4)   **23.** (closed at 5 and 21)

25. (open at 0)   **27.** (open at 0)   **29.** (open at 3 and $\frac{14}{3}$)   **31.** (open at 2)

33. (open at -4 and $\frac{5}{6}$)   **35.** (open at -2)   **37.** all x such that $x > 1$

39. all x such that $x < -4$ **41.** all x such that $x > -4$ **43.** all x such that $x < 3$
45. from 5.25 to 6.25 miles   **47.** between $16\frac{2}{3}$ and 20 centimeters   **49.** $40 + 2w < P < 60 + 2w$

Exercise 2.8   (page 106)

1. 7   **3.** 0   **5.** 2   **7.** $\pi - 2$ **9.** $0, -4$ **11.** $2, -\frac{4}{3}$ **13.** $\frac{14}{3}, -2$ **15.** no values of x

17. $x \geq 0$ **19.** $-\frac{3}{2}$ **21.** $0, -6$ **23.** 0   **25.** $3, \frac{3}{5}$ **27.** (open at -3 and 9)

29. (closed at -3 and 9)   **31.** (closed at -7 and 3)   **33.** (closed at -5 and $\frac{5}{3}$)   **35.** (open at -3)

37. (open at -1 and $\frac{1}{5}$)   **39.** (open at -8, closed at $-\frac{4}{3}$)   **41.** (open at $-2, -\frac{1}{2}, 1$)   **43.** (open at $-\frac{7}{3}, -\frac{2}{3}, \frac{4}{3}, 3$)

45. (open at $-7, -1, 11, 17$)   **47.** (open at -18, closed -6, 10, open 22)   **49.** $|x - 5| < 2$ **51.** $|x - 2| \leq \frac{5}{2}$

53. $0 < |x - 2| < 2$ **55.** $0 < |x| < 5$ **57.** $[-\frac{1}{2}, \infty)$ **59.** $(-\infty, 0)$

Exercise 2.9 (page 110)

1. $-4\ -3$

3. $2\ \ \ 3$

5. $-3\ -2$

7. $-\frac{1}{2}\ -\frac{1}{3}$

9. $\frac{1}{3}\ \ \frac{1}{2}$

11. $-\frac{3}{2}\ \ 1$

13. $-3\ \ \ 2$

15. $-1\ \ 0\ \ 1$

17. $-3\ -2\ \ \ 2$

19. $-6\ -5\ -3\ -1$

21. $-9\ -4\ -2$

23. $0\ \ \frac{3}{2}$

25. $0\ \ \frac{3}{2}$

27. $2\ \ \frac{13}{5}$

29. $-2\ -\frac{5}{4}$

31. $-\sqrt{7}\ -1\ \ \ 1\ \ \sqrt{7}$

33. $-3\ \ \ 0\ \ \ 1$

35. 0

37. $0\ \ \ 1\ \ \ 2$

39. $-1\ \ \ 0\ \ \ 1$

41. $2\ \ \frac{14}{3}$

REVIEW EXERCISES (page 111)

1. all real numbers   **2.** all real numbers except 0   **3.** all real numbers   **4.** all real numbers
5. all x such that $x \ge 0$ **6.** all real numbers except 2 and 3   **7.** $\frac{16}{3}$, conditional equation
8. -14, conditional equation   **9.** 4, conditional equation   **10.** no solution
11. 7, conditional equation   **12.** identity   **13.** 7, conditional equation   **14.** $\frac{1}{3}$, conditional equation
15. $2, -\frac{3}{2}$ **16.** $\frac{1}{4}, -\frac{4}{3}$ **17.** $0, \frac{8}{5}$ **18.** $\frac{2}{3}, \frac{4}{9}$ **19.** 3, 5   **20.** $-2, -4$
21. $\dfrac{1+\sqrt{21}}{10}, \dfrac{1-\sqrt{21}}{10}$ **22.** $0, \frac{1}{5}$ **23.** $2, -7$ **24.** $9, -\frac{2}{3}$ **25.** $\dfrac{-1+\sqrt{21}}{10}, \dfrac{-1-\sqrt{21}}{10}$
26. $\dfrac{-4+\sqrt{6}}{2}, \dfrac{-4-\sqrt{6}}{2}$ **27.** $2, -3$ **28.** $4, -2$ **29.** $1, 1, -1, -1$ **30.** $1, -1, 6, -6$
31. 5   **32.** $4, -4$ **33.** 0   **34.** no solutions   **35.** $\frac{1}{3}$ **36.** 2, 10   **37.** 1.5 liters
38. $33\frac{27}{31}$ hours   **39.** $5\frac{1}{7}$ hours   **40.** $3\frac{1}{3}$ ounces   **41.** \$4500 at 11%, \$5500 at 14%
42. George can paint the house in 36 hours; his son can paint the house in 45 hours.
43. either 95 by 110 yards or 55 by 190 yards
44. The jet flies at 440 miles per hour and the propeller plane flies at 320 miles per hour.
45. $f_1 = \dfrac{ff_2}{f_2 - f}$ **46.** $C = \dfrac{C_1 C_2}{C_2 + C_1}$ **47.** $l = \dfrac{a + Sr - S}{r}$ **48.** $b = s - \dfrac{A^2}{s(s-a)(s-c)}$
49. $y = \dfrac{x^2}{2 - x}$ **50.** $x = \dfrac{-y \pm \sqrt{y^2 + 8y}}{2}$

51. 7

52. $-\frac{1}{5}$

53. $\frac{35}{3}$

54. -15

55. $\frac{5}{3}$

56. 2

57. 7

58. $-3\ \ \ 5$

59. $-\frac{9}{2}\ \ \frac{15}{2}$

60. 0 **61.** 8

62. -2 **63.** -9 **64.** -1 **65.** $-6\ \ \ 0$ **66.** $2\ \ \frac{8}{3}$ **67.** $5, -7$

68. 0   **69.** $-5\ \ \ 1$ **70.** $-\frac{7}{2}\ -2\ -1\ \ \frac{1}{2}$ **71.** $-1\ \ \ 3$

72. $-\frac{3}{2}\ \ 1$ **73.** $-7\ \ \frac{3}{5}$ **74.** $-1\ -\frac{1}{7}$ **75.** $-\frac{5}{2}\ -\frac{2}{3}$

76. $-\frac{5}{2}\ \ \ 3$ **77.** $-2\ \ \ 1\ \ \ 3$ **78.** $-7\ -1\ \ \ 1$ **79.** $-1\ -\frac{1}{2}\ \ \ 3$

80. 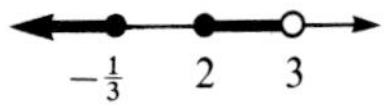

81. 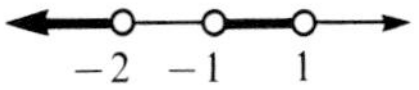

82. 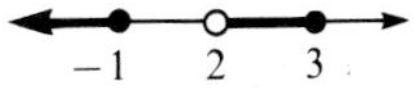

Exercise 3.1 (page 122)

1. I **3.** III **5.** I **7.** positive x-axis **9.** positive y-axis **11.** negative x-axis

13. 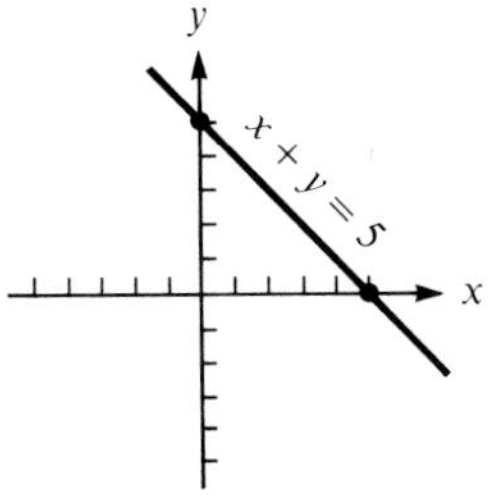

15. 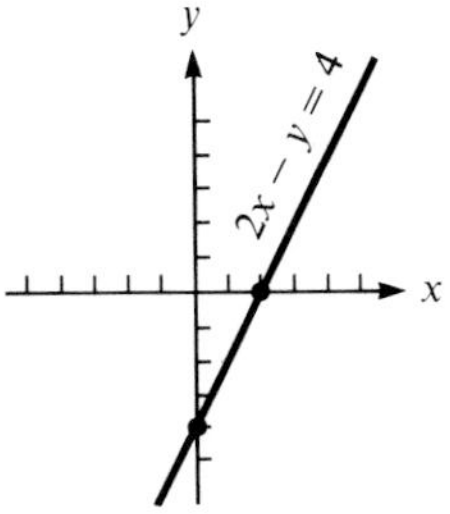

17. 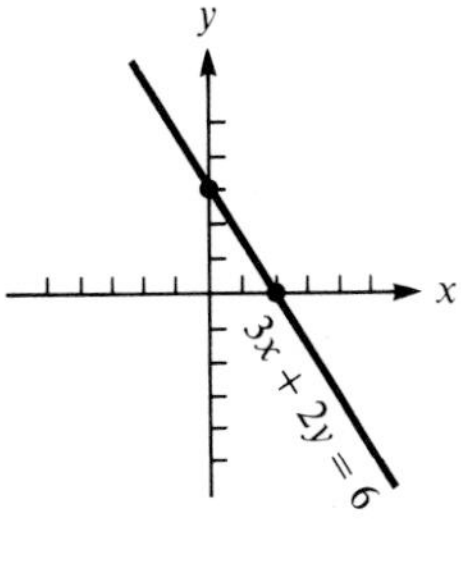

19. 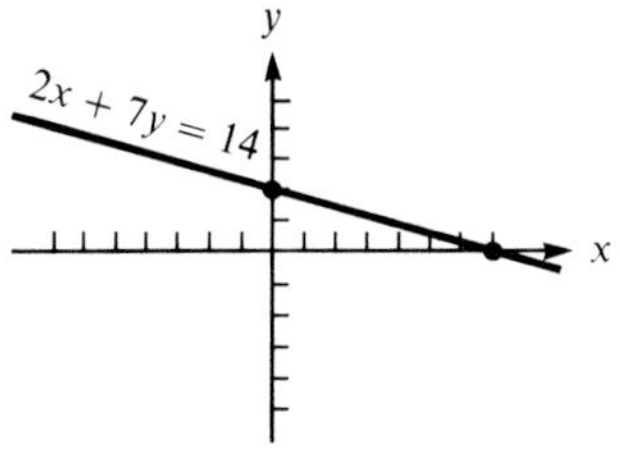

21. 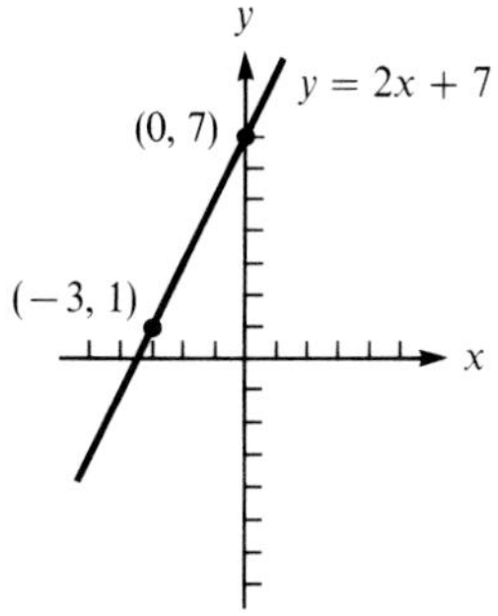

23. 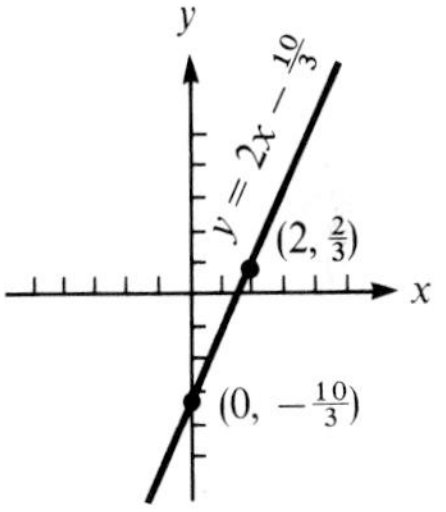

25. 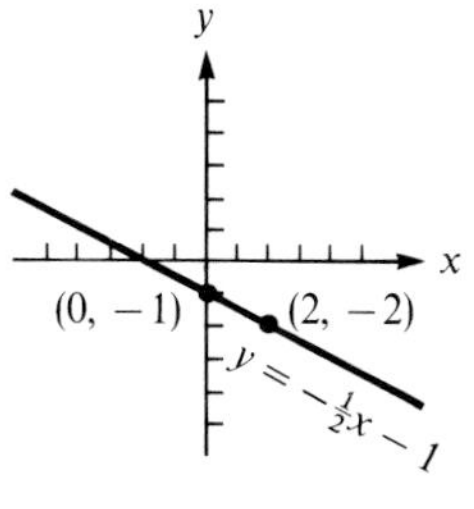

27. 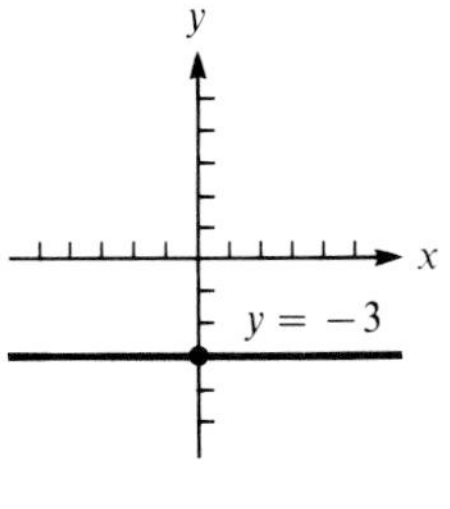

29. 5 **31.** $\sqrt{13}$ **33.** $\sqrt{a^2 + b^2}$

35. 5 **37.** 10 **39.** $2\sqrt{2}$ **41.** 2 **43.** 3 **45.** $\sqrt{x^2 + y^2}$ **47.** (4, 6) **49.** (0, 1)

51. $\left(\frac{\sqrt{5}}{2}, \frac{\sqrt{5}}{2}\right)$ **53.** $\left(\frac{a+2}{2}, \frac{2+a}{2}\right)$ **55.** (5, 6) **57.** (5, 15) **63.** $(-4, -3)$

65. (5, 0), (0, 5), $(-5, 0)$, $(0, -5)$ **67.** $\sqrt{2}$

Exercise 3.2 (page 129)

1. 5 **3.** $-\frac{7}{4}$ **5.** -2 **7.** no defined slope **9.** -1 **11.** 3 **13.** $\frac{1}{2}$ **15.** $\frac{2}{3}$ **17.** 0
19. perpendicular **21.** parallel **23.** perpendicular **25.** parallel **27.** perpendicular
29. neither **31.** They do not lie on a straight line. **33.** None are perpendicular.
35. *PQ* and *PR* are perpendicular.

Exercise 3.3   (page 137)

1. $y = 2x$ **3.** $y = 2x + \frac{7}{2}$ **5.** $y = -5x + 20$ **7.** $y = 3x - 2$ **9.** $y = \sqrt{2}\,x + \sqrt{2}$
11. $y = ax + \frac{1}{a}$ **13.** $3x - 2y = 0$ **15.** $3x + y = -4$ **17.** $\sqrt{2}\,x - y = -\sqrt{2}$ **19.** $x + y = 5$
21. $6x + y = 16$ **23.** $13x + 9y = 0$ **25.** $m = \frac{42}{5};\ b = \frac{21}{5}$ **27.** no defined slope; no y-intercept
29. $m = -1;\ b = 2$ **31.** $m = 3;\ b = 11$ **33.** $m = \frac{1}{3};\ b = \frac{7}{3}$ **35.** $x - 2y = -5$ **37.** $8x + 7y = 12$
39. $11x + 2y = 0$ **41.** $x - y = -7$ **43.** $3x - 4y = -12$ **45.** $2x + 3y = 30$ **47.** $5x - y = 20$
49. $x + 3y = -15$ **51.** $5x - 3y = 15$ **53.** $y = \frac{2}{3}x + \frac{20}{3}$ **55.** $y = -\frac{1}{4}x$ **61.** $y = \frac{1}{4}x$

Exercise 3.4   (page 145)

1. function   **3.** not a function   **5.** function   **7.** not a function   **9.** not a function
11. domain is the set of all real numbers, range is the set of real numbers, $f(3) = 14$
13. domain is the set of all real numbers, range is the set of nonnegative numbers, $f(3) = 9$
15. domain is the set of all real numbers except -1, range is the set of all real numbers except 0, $f(3) = \frac{3}{4}$
17. domain is the set of all x such that $|x| \le 4$, range is the set of all y such that $0 \le y \le 4$, $f(3) = \sqrt{7}$
19. domain is the set of all real numbers except 0 and 2, $f(-2) = \frac{1}{4}$, $f(0)$ is undefined, $f(a) = \dfrac{2}{a^2 - 2a}$
21. domain is the set of all x such that $|x| \ge 1$, $f(-2) = \sqrt{3}$, $f(0)$ is undefined, $f(a) = \sqrt{a^2 - 1}$
23. domain is the set of all real numbers except -3, $f(-2) = -2$, $f(0) = 0$, $f(a) = \dfrac{a}{a + 3}$
25. domain is the set of all real numbers $f(-2) = -\frac{1}{3}$, $f(0) = 0$, $f(a) = \dfrac{a}{\sqrt{a^2 + 32}}$ **27.** function
29. function   **31.** not a function   **33.**

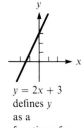

$y = 2x + 3$
defines y
as a
function of x

35.

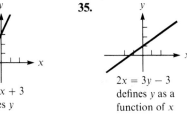

$2x = 3y - 3$
defines y as a
function of x

37.

$y = -x^2$
defines y
as a function
of x

39.

$y = \sqrt{x}$ defines
y as a function of x

41.

$y = x^2 + x$ defines y
as a function of x

43.

The phrase "y
is the greatest
integer less than
or equal to x"
defines y as a
function of x

45.

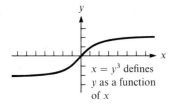

$x = y^3$ defines
y as a function
of x

47.

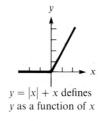

$y = |x| + x$ defines
y as a function of x

49.

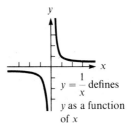

$y = \dfrac{1}{x}$ defines
y as a function
of x

51.

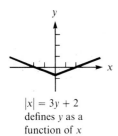

$|x| = 3y + 2$
defines y as a
function of x

53.

$|x| = 3y^2$ does
not define a function

55. $F = \frac{9}{5}C + 32$ **57.** $C = \frac{3}{2500}n + \frac{7}{2}$

59. $r = \dfrac{1}{2\pi}C$ **61.** $p = 4\sqrt{A}$ **63.** $V = A^{\frac{3}{2}}$

Exercise 3.5   (page 156)

1.

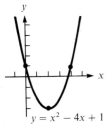

$y = x^2 - 4x + 1$

3.

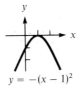

$y = -(x - 1)^2$

5.

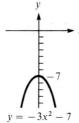

$y = -3x^2 - 7$

7.

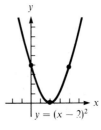

$y = (x - 2)^2$

9.

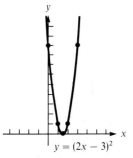

$y = (2x - 3)^2$

11. $(2, 0)$ **13.** $(-3, -12)$ **15.** $(3, 1)$

17. symmetric about the origin; odd   **19.** symmetric about the x-axis; neither   **21.** no symmetry; neither

23. symmetric about the y-axis; even   **25.** symmetric about the y-axis; even

27.

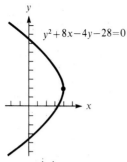

$y^2 + 8x - 4y - 28 = 0$

a parabola,
but not a function

29. 20 units   **31.** $1\frac{1}{2}$ seconds   **33.** 36 units   **35.**

$y = x^3$

37.

$y = x^3 + x^2$

39.

$y = x^5 - x^3$

41.

$y = x^4 - 2x^2 + 1$

43.

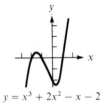

$y = x^3 + 2x^2 - x - 2$

45.

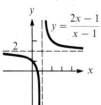

$y = x^{10}$

47. 25 by 25 feet   **49.** Both numbers are 3.   **51.** $\frac{1}{2}$ **53.** 6 by $4\frac{1}{2}$
55. when $x \le 0$ or $x \ge 1$ **57.** always increasing

Exercise 3.6   (page 164)

1.

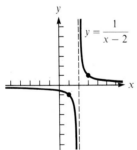

$y = \dfrac{1}{x - 2}$

3.

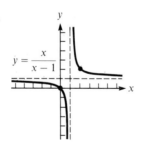

$y = \dfrac{x}{x - 1}$

5.

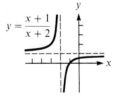

$y = \dfrac{x + 1}{x + 2}$

7.

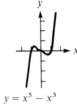

$y = \dfrac{2x - 1}{x - 1}$

2

9.

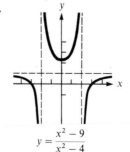

$y = \dfrac{x^2 - 9}{x^2 - 4}$

11.

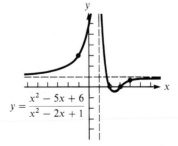

$y = \dfrac{x^2 - 5x + 6}{x^2 - 2x + 1}$

13.

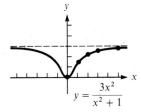

$$y = \frac{3x^2}{x^2 + 1}$$

15.

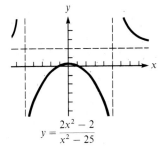

$$y = \frac{2x^2 - 2}{x^2 - 25}$$

17.

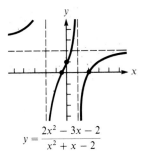

$$y = \frac{2x^2 - 3x - 2}{x^2 + x - 2}$$

19.

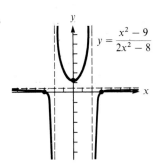

$$y = \frac{x^2 - 9}{2x^2 - 8}$$

21.

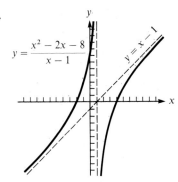

$$y = \frac{x^2 - 2x - 8}{x - 1}$$

$$y = x - 1$$

23.

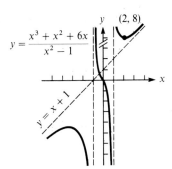

$$y = \frac{x^3 + x^2 + 6x}{x^2 - 1}$$

$(2, 8)$

$y = x + 1$

25. no

Exercise 3.7   (page 170)

1. $(f + g)(x) = 5x - 1$; all real numbers   **3.** $(f \cdot g)(x) = 6x^2 - x - 2$; all real numbers

5. $(f - g)(x) = x + 1$; all real numbers

7. $(f/g)(x) = \dfrac{x^2 + x}{x^2 - 1} = \dfrac{x(x + 1)}{(x + 1)(x - 1)} = \dfrac{x}{x - 1}$; all real numbers except 1 and -1 **9.** 7   **11.** 1   **13.** 12

15. no value   **17.** $f(x) = 3x^2, g(x) = 2x$ **19.** $f(x) = 3x^2, g(x) = x^2 - 1$ **21.** $f(x) = 3x^3, g(x) = -x$

23. $f(x) = x + 9, g(x) = x - 2$ **25.** 11   **27.** -17 **29.** 190   **31.** 145

33. $(f \circ g)(x) = 3x + 3$; all real numbers   **35.** $(f \circ f)(x) = 9x$; all real numbers

37. $(g \circ f)(x) = 2x^2$; all real numbers   **39.** $(f \circ f)(x) = x^4$; all real numbers

41. $(f \circ g)(x) = \sqrt{x^2 + 1}$; all real numbers   **43.** $(f \circ f)(x) = \sqrt{\sqrt{x}}$ or $\sqrt[4]{x}$; all nonnegative numbers

45. $(g \circ f)(x) = x$; all $x \geq -1$ **47.** $(g \circ g)(x) = x^4 - 2x^2$; all real numbers   **49.** $f(x) = 3x - 2, g(x) = x$

51. $f(x) = x - 2, g(x) = x^2$ **53.** $f(x) = x^2, g(x) = x - 2$ **55.** $f(x) = \sqrt{x}, g(x) = x + 2$

57. $f(x) = x + 2, g(x) = \sqrt{x}$ **59.** $f(x) = x, g(x) = x$ **63.** $(f \circ f)(x) = -\dfrac{1}{x}$

Exercise 3.8   (page 177)

1. one-to-one   **3.** not one-to-one

5. not one-to-one (there are at least 3 values of $x(1, -1, 0)$ such that $y = 0$)   **7.** not one-to-one

9. not one-to-one   **11.** one-to-one   **13.** not one-to-one

15. $(f \circ g)(x) = f(g(x)) = 5(\frac{1}{5}x) = x$

$(g \circ f)(x) = g(f(x)) = \frac{1}{5}(5x) = x$

17. $(f \circ g)(x) = f(g(x)) = \dfrac{\dfrac{1}{x-1} + 1}{\dfrac{1}{x-1}} = \dfrac{\dfrac{x}{x-1}}{\dfrac{1}{x-1}} = x$

$(g \circ f)(x) = g(f(x)) = \dfrac{1}{\dfrac{x+1}{x} - 1} = \dfrac{1}{\dfrac{1}{x}} = x$

19. $y = \frac{1}{3}x$ **21.** $y = \dfrac{x-2}{3}$ **23.** $y = \dfrac{1-3x}{x}$ **25.** $y = \dfrac{1}{2x}$

27.

29.

31.

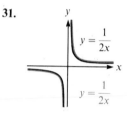

33.

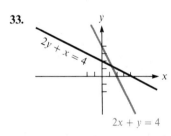

35.

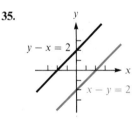

37. $f^{-1}(x) = -\sqrt{x+3}$ $(y \le 0)$

39. $f^{-1}(x) = \sqrt[4]{x+8}$ $(y \ge 0)$ **41.** $f^{-1}(x) = \sqrt{4-x^2}$ $(0 \le y \le 2)$

43. domain is the set of all real numbers except 1, range is the set of all real numbers except 1

45. domain is the set of all real numbers except 0, range is the set of all real numbers except -2

Exercise 3.9   (page 182)

1. $x = 14$ **3.** $x = -3$ or $x = 2$ **5.** 18 girls   **7.** $\frac{1}{2}$ **9.** 1000   **11.** 1   **13.** $\frac{21}{4}$

15. -8 **17.** $\frac{160}{11}$ cubic feet   **19.** 3 seconds   **21.** 20 volts   **23.** 432 hertz   **25.** $\dfrac{\sqrt{3}}{4}$

Exercise 3.10 (page 186)

1.

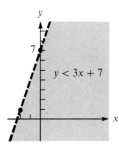

$y < 3x + 7$

3.

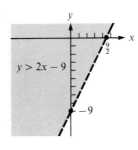

$y > 2x - 9$

5.

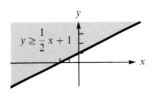

$y \geq \dfrac{1}{2} x + 1$

7.

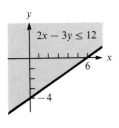

$2x - 3y \leq 12$

9.

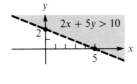

$2x + 5y > 10$

11.

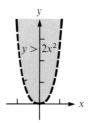

$y > 2x^2$

13.

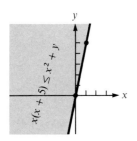

$x(x + 5) \leq x^2 + y$

15.

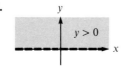

$y > 0$

17.

$x \leq 2$

19.

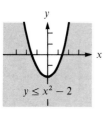

$y \leq x^2 - 2$

21.

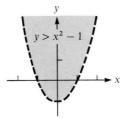

$y > x^2 - 1$

23.

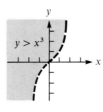

$y > x^3$

25.

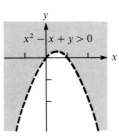

$x^2 - x + y > 0$

27.

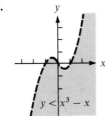

$y < x^3 - x$

29.

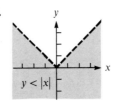

$y < |x|$

31.
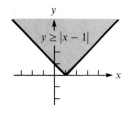

33.
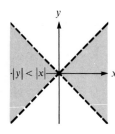

35.
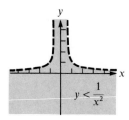

37.
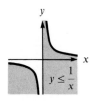

39.
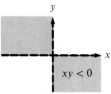

REVIEW EXERCISES   (page 187)

1. $10; (0, 3)$ **2.** $13; (-6, \frac{15}{2})$ **3.** $16; (-\sqrt{3}, 1)$ **4.** $2a\sqrt{2}; (0, 0)$ **5.** -6 **6.** 2 **7.** -1
8. 1 **9.** $7x + 5y = 0$ **10.** $4x + y = -7$ **11.** $2x + y = 9$ **12.** $2x - 3y = -9$ **13.** $y = 17$
14. $x = -5$ **15.** $7x + y = 54$ **16.** $x - 7y = 22$ **17.** $3x - 4y = 6$ **18.** $3x - y = 21$
19. a function, both domain and range are the set of real numbers

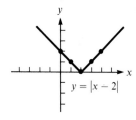

20. a function; domain is the set of real numbers, range is the set of nonnegative real numbers

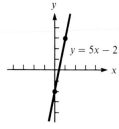

21. a function; domain is the set of all x such that $x \geq 1$, range is the set of nonnegative real numbers

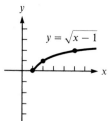

22. not a function

23. a function; domain is the set of all real numbers, range is the set of all real numbers greater than or equal to $\frac{3}{4}$

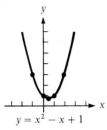

$y = x^2 - x + 1$

24. a function; domain is the set of all real numbers except 0, range is the set $\{-1, 1\}$

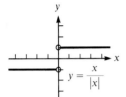

$y = \dfrac{x}{|x|}$

25.

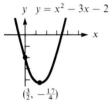

$y = x^2 - 3x - 2$

$\left(\frac{3}{2}, -\frac{17}{4}\right)$

26.

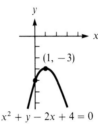

$(1, -3)$

$x^2 + y - 2x + 4 = 0$

27.

$y = x^4 + 2x^2 + 1$

28.

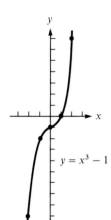

$y = x^3 - 1$

29.

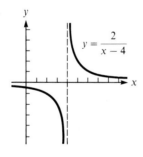

$y = \dfrac{2}{x - 4}$

30.

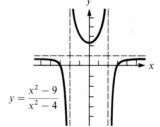

$y = \dfrac{x^2 - 9}{x^2 - 4}$

31.

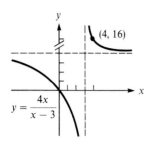

$$y = \frac{4x}{x - 3}$$

32.

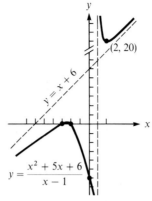

$$y = \frac{x^2 + 5x + 6}{x - 1}$$

33.

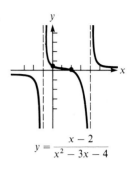

$$y = \frac{x - 2}{x^2 - 3x - 4}$$

34.

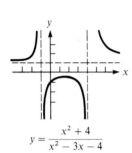

$$y = \frac{x^2 + 4}{x^2 - 3x - 4}$$

35. $f(2) = 6$ **36.** $g(-3) = 0$ **37.** $(f + g)(2) = 11$

38. $(f - g)(-1) = 1$ **39.** $(f \cdot g)(0) = 0$ **40.** $(f/g)(2) = \frac{6}{5}$

41. $(f \circ g)(-2) = 1$ **42.** $(g \circ f)(4) = 31$

43. $(f + g)(x) = \sqrt{x + 1} + x^2 - 1$ **44.** $(f \cdot g)(x) = \sqrt{x + 1}(x^2 - 1)$

45. $(f \circ g)(x) = \sqrt{x^2} = |x|$ **46.** $(g \circ f)(x) = x$

47. all x such that $x > -1$ and $x \neq 1$ **48.** all real numbers x

49. one-to-one, $f^{-1}(x) = \frac{1}{7}x$ **50.** not one-to-one   **51.** one-to-one, $f^{-1}(x) = \dfrac{1 + x}{x}$

52. one-to-one, $f^{-1}(x) = \sqrt[3]{\dfrac{3}{x}}$ **53.** one-to-one, $f^{-1}(x) = \dfrac{2 + 3x}{x - 1}$ **54.** not one-to-one

55. The domain is the set of real numbers except $\frac{1}{2}$; the range is the set of real numbers except 1.
56. The domain is the set of real numbers except 0; the range is the set of real numbers except $\frac{1}{2}$.
57. $\frac{9}{5}$ pounds   **58.** $\frac{1000}{3}$ cubic centimeters   **59.** about 2.8   **60.** about 117 ohms

61.

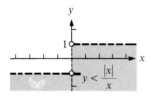

$y \leq 3x + 7$

62.

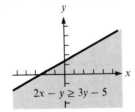

$2x - y \geq 3y - 5$

63.

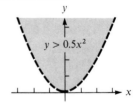

$y > 0.5x^2$

64.

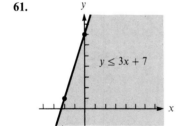

$y < \dfrac{|x|}{x}$

65.

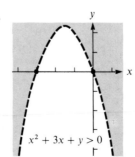

$x^2 + 3x + y > 0$

66.

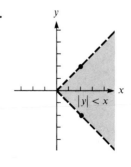

$|y| < x$

Exercise 4.1   (page 195)

1.
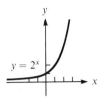
$y = 2^x$

3.

$y = (\frac{1}{3})^x$

5.

$y = 2.5^x$

7.
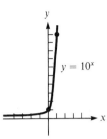
$y = 10^x$

9.
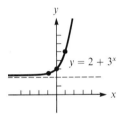
$y = 5(2^x)$

11.
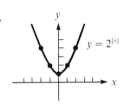
$y = 3(2^x)$

13.
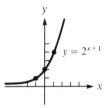
$y = 2^{x+1}$

15.
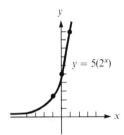
$y = 2 + 3^x$

17.
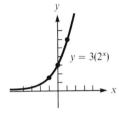
$y = 2^{|x|}$

19. $b = 5$ **21.** no value of b
23. $b = \frac{1}{2}$ **25.** $\frac{32}{243}A_0$
27. $A_0 2^{-\frac{3000}{5700}} \approx 0.6943 A_0$
29. \$1342.53   **31.** \$2,273,996.13
33. 1.68×10^8 **35.** 2.83

Exercise 4.2   (page 200)

1.

$y = -e^x$

3.
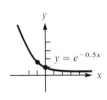
$y = e^{-0.5x}$

5.
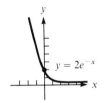
$y = 2e^{-x}$

7.
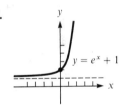
$y = e^x + 1$

9. yes   **11.** no   **13.** no   **15.** no   **17.** \$19,171.75   **19.** \$4500   **21.** \$8753.36
23. 9.44×10^5 **25.** 2.6   **27.** 202

Exercise 4.3   (page 211)

1. 3   **3.** -3 **5.** 2   **7.** 2   **9.** 7   **11.** 4   **13.** $-\frac{3}{2}$ **15.** $\frac{2}{3}$ **17.** 5   **19.** $\frac{3}{2}$
21. $\frac{1}{9}$ **23.** 8   **25.**

$y = \log_5 x$

27.
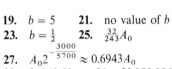
$y = \log_{\frac{1}{3}} x$

29.
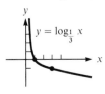
$y = \ln x^2$
$= 2 \ln x$

31.

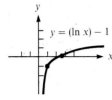

$y = (\ln x) - 1$

33.

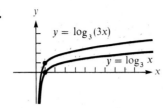

$y = \log_3 (3x)$

$y = \log_3 x$

35.

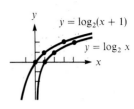

$y = \log_2(x + 1)$

$y = \log_2 x$

37.

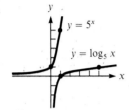

$y = 5^x$

$y = \log_5 x$

39. no value of b　　**41.** 3　　**43.** no value of b　　**45.** yes　　**47.** no

49. true　**51.** false　**53.** true　**55.** false　**57.** true　**59.** false　**61.** false　**63.** true
65. true　**67.** false　**69.** true　**71.** 1.4472　**73.** 0.3521　**75.** 1.1972　**77.** 2.4014
79. 2.0493　**81.** 0.5119　**83.** 69.4079　**85.** 129.0719　**87.** 4.77　**89.** from 0.00126 to 0.000501

91. 0.71 volt　　**93.** $10 \log \dfrac{P_O}{P_I} = 10 \log \dfrac{kE_O{}^2}{kE_I{}^2} = 10 \log \left(\dfrac{E_O}{E_I}\right)^2 = 20 \log \dfrac{E_O}{E_I}$　　**95.** 4.4

97. 2500 micrometers　　**99.** 19.9 hours　　**101.** $L = L_0 + k \ln 2$ where $L_0 = k \ln I$
103. $L = L_0 + k \ln 3$ where $L_0 = k \ln I$

Exercise 4.4 (page 218)

1. $x = \dfrac{\log 5}{\log 4} \approx 1.16$　　**3.** $x = \dfrac{\log 2}{\log 13} + 1 \approx 1.27$　　**5.** $x = \dfrac{\log 2}{\log 3 - \log 2} \approx 1.71$　　**7.** $x = 0$

9. $x = \sqrt{\dfrac{1}{\log 7}} \approx 1.09$　　**11.** $x = 0$ or $x = \dfrac{\log 9}{\log 8} \approx 1.06$　　**13.** 7　　**15.** 4　　**17.** $10, -10$　　**19.** 50

21. 20　　**23.** 10　　**25.** 3, 4　　**27.** 1, 100　　**29.** 9　　**31.** 1.771　　**33.** 2.322
35. about 5.146 years　　**37.** about 42.7 days　　**39.** about 4200 years old　　**41.** 4.03 years, 3.92 years
43. approximately 6.96%　　**45.** about 3.15 days

REVIEW EXERCISES (page 220)

1.

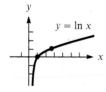

$y = \left(\dfrac{6}{5}\right)^x$

2.

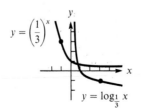

$y = \left(\dfrac{3}{4}\right)^x$

3.

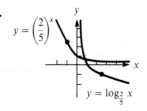

$y = \log_{10} x$

4.

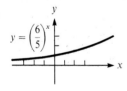

$y = \ln x$

5.

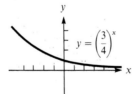

$y = \left(\dfrac{1}{3}\right)^x$

$y = \log_{\frac{1}{3}} x$

6.

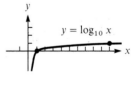

$y = \left(\dfrac{2}{5}\right)^x$

$y = \log_{\frac{2}{5}} x$

7. 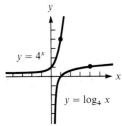   **8.** 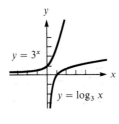   **9.** 8   **10.** $\frac{1}{9}$ **11.** 3   **12.** 2

13. 1   **14.** $\frac{1}{2}$ **15.** $\frac{1}{6}$ **16.** 2   **17.** -2 **18.** 0   **19.** 27   **20.** $\frac{1}{5}$ **21.** 32

22. 9   **23.** 27   **24.** -1 **25.** $\frac{1}{8}$ **26.** 2   **27.** 4   **28.** 2   **29.** 10   **30.** $\frac{1}{25}$

31. $x = 2.8665$ **32.** no value   **33.** $z = -2.0149$ **34.** $M = 209.5579$

35. $x = \dfrac{\log 7}{\log 3}$ **36.** $x = \dfrac{\log 1.2}{5.6 \log 3.4}$ **37.** $x = \dfrac{\log 3}{\log 3 - \log 2}$ **38.** $x = 4$ or $x = 25$

39. $x = 4$ **40.** $x = 2$ **41.** $x = \dfrac{\ln 9}{\ln 2}$ **42.** no value   **43.** $x = \dfrac{e}{e-1}$ **44.** $x = 1$

45. approximately 9034 years old   **46.** 7.94×10^{-4}

47. $\text{pH} = \log_{10} \dfrac{1}{[\text{H}^+]} = \log_{10} [\text{H}^+]^{-1} = -\log_{10} [\text{H}^+]$ **48.** 34.19 years

Exercise 5.1   (page 227)

1. yes; positive   **3.** no; positive   **5.** no; negative   **7.** yes; positive   **9.** QII   **11.** QIV

13. QII   **15.** QIII   **17.** QIV   **19.** QIII   **21.** yes   **23.** no   **25.** yes   **27.** yes

29. 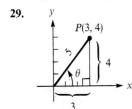 $\begin{cases} \sin\theta = \frac{4}{5}, \cos\theta = \frac{3}{5}, \tan\theta = \frac{4}{3}, \\ \csc\theta = \frac{5}{4}, \sec\theta = \frac{5}{3}, \cot\theta = \frac{3}{4} \end{cases}$

31. 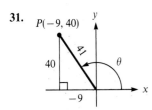 $\begin{cases} \sin\theta = \frac{40}{41}, \cos\theta = \frac{-9}{41}, \tan\theta = \frac{40}{-9}, \\ \csc\theta = \frac{41}{40}, \sec\theta = \frac{41}{-9}, \cot\theta = \frac{-9}{40} \end{cases}$

33. 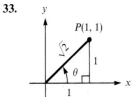 $\begin{cases} \sin\theta = \frac{\sqrt{2}}{2}, \cos\theta = \frac{\sqrt{2}}{2}, \tan\theta = 1, \\ \cos\theta = \sqrt{2}, \sec\theta = \sqrt{2}, \cot\theta = 1 \end{cases}$

35.

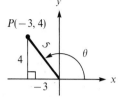

$$\begin{cases} \sin \theta = \frac{4}{5}, \cos \theta = \frac{-3}{5}, \tan \theta = -\frac{4}{3} \\ \csc \theta = \frac{5}{4}, \sec \theta = -\frac{5}{3}, \cot \theta = \frac{-3}{4} \end{cases}$$

37.

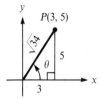

$$\begin{cases} \sin \theta = \frac{5\sqrt{34}}{34}, \cos \theta = \frac{3\sqrt{34}}{34}, \tan \theta = \frac{5}{3} \\ \csc \theta = \frac{\sqrt{34}}{5}, \sec \theta = \frac{\sqrt{34}}{3}, \cot \theta = \frac{3}{5} \end{cases}$$

39.

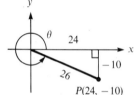

$$\begin{cases} \sin \theta = \frac{-5}{13}, \cos \theta = \frac{12}{13}, \tan \theta = \frac{-5}{12} \\ \csc \theta = -\frac{13}{5}, \sec \theta = \frac{13}{12}, \cot \theta = -\frac{12}{5} \end{cases}$$

41. $\cos \theta = \frac{4}{5}, \tan \theta = \frac{3}{4}, \csc \theta = \frac{5}{3}, \sec \theta = \frac{5}{4}, \cot \theta = \frac{4}{3}$ **43.** $\sin \theta = \frac{-12}{13}, \tan \theta = \frac{12}{5}, \csc \theta = \frac{13}{-12}, \sec \theta = \frac{13}{-5}$

45. $\cos \theta = \frac{40}{41}, \tan \theta = \frac{-9}{40}, \csc \theta = \frac{41}{-9}, \sec \theta = \frac{41}{40}, \cot \theta = \frac{40}{-9}$

47. $\sin \theta = \frac{4}{5}, \cos \theta = \frac{-3}{5}, \tan \theta = \frac{-4}{3}, \cot \theta = \frac{-3}{4}$ **49.** $\sin \theta = \frac{-40}{41}, \csc \theta = \frac{41}{-40}, \sec \theta = \frac{41}{9}, \cot \theta = \frac{9}{-40}$

Exercise 5.2 (page 233)

1. $\dfrac{\cos \theta}{\sin \theta} = \dfrac{\frac{x}{r}}{\frac{y}{r}} = \dfrac{x}{r} \cdot \dfrac{r}{y} = \dfrac{x}{y} = \cot \theta$ **3.** $\tan \theta = \dfrac{y}{x} = \dfrac{1}{\frac{x}{y}} = \dfrac{1}{\frac{x}{r} \cdot \frac{r}{y}} = \dfrac{1}{\cos \theta \csc \theta}$

5. $\sin^2 \theta + \sin^2 \theta \cot^2 \theta = \dfrac{y^2}{r^2} + \dfrac{y^2}{r^2} \cdot \dfrac{x^2}{y^2} = \dfrac{y^2}{r^2} + \dfrac{x^2}{r^2} = \dfrac{y^2 + x^2}{r^2} = \dfrac{r^2}{r^2} = 1$

7. $\cot^2 \theta + \sin^2 \theta = \dfrac{x^2}{y^2} + \dfrac{y^2}{r^2} = \dfrac{r^2 - y^2}{y^2} + \dfrac{r^2 - x^2}{r^2} = \dfrac{r^2}{y^2} - 1 + 1 - \dfrac{x^2}{r^2} = \dfrac{r^2}{y^2} - \dfrac{x^2}{r^2} = \csc^2 \theta - \cos^2 \theta$

9. $\dfrac{\cos \theta}{\sin \theta} = \cot \theta = \dfrac{1}{\tan \theta}$ **11.** $\tan \theta = \dfrac{\sin \theta}{\cos \theta} = \dfrac{1}{\cos \theta \csc \theta}$

13. $\sin^2 \theta + \sin^2 \theta \cot^2 \theta = \sin^2 \theta (1 + \cot^2 \theta) = \sin^2 \theta \csc^2 \theta = 1$

15. $\cot^2 \theta + \sin^2 \theta = \csc^2 \theta - 1 + 1 - \cos^2 \theta = \csc^2 \theta - \cos^2 \theta$

17. $\cos \theta = \frac{3}{5}, \tan \theta = \frac{4}{3}, \csc \theta = \frac{5}{4}, \sec \theta = \frac{5}{3}, \cot \theta = \frac{3}{4}$

19. $\sin \theta = \frac{12}{13}, \tan \theta = \frac{-12}{5}, \csc \theta = \frac{13}{12}, \sec \theta = \frac{-13}{5}, \cot \theta = \frac{-5}{12}$

21. $\sin \theta = \frac{4}{5}, \cos \theta = \frac{-3}{5}, \csc \theta = \frac{5}{4}, \sec \theta = \frac{-5}{3}, \cot \theta = \frac{-3}{4}$

23. $\sin \theta = \frac{40}{41}, \cos \theta = \frac{9}{41}, \tan \theta = \frac{40}{9}, \csc \theta = \frac{41}{40}, \sec \theta = \frac{41}{9}$

25. $\sin \theta = \frac{-12}{13}, \cos \theta = \frac{-5}{13}, \tan \theta = \frac{12}{5}, \csc \theta = \frac{-13}{12}, \cot \theta = \frac{5}{12}$

27. 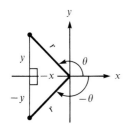   **29.** 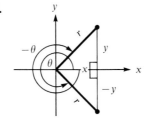   **31.** negative   **33.** positive

35. odd   **37.** even   **39.** odd

Exercise 5.3   (page 243)

1. $\sin 135° = \dfrac{\sqrt{2}}{2}$; $\cos 135° = -\dfrac{\sqrt{2}}{2}$; $\tan 135° = -1$

3. $\sin 450° = \sin 90° = 1$; $\cos 450° = \cos 90° = 0$; $\tan 450° = \tan 90°$ is undefined

5. $\sin(-240°) = -\sin 240° = \dfrac{\sqrt{3}}{2}$; $\cos(-240°) = \cos 240° = -\dfrac{1}{2}$; $\tan(-240°) = -\tan 240° = -\sqrt{3}$

7. $\sin 540° = \sin 180° = 0$; $\cos 540° = \cos 180° = -1$; $\tan 540° = \tan 180° = 0$

9. $\csc 225° = -\sqrt{2}$; $\sec 225° = -\sqrt{2}$; $\cot 225° = 1$

11. $\csc 1080° = \csc 0°$ is undefined; $\sec 1080° = \sec 0° = 1$; $\cot 1080° = \cot 0°$ is undefined

13. $\csc(-210°) = -\csc 210° = 2$; $\sec(-210°) = \sec 210° = -\dfrac{2\sqrt{3}}{3}$; $\cot(-210°) = -\cot 210° = -\sqrt{3}$

15. $\csc 585° = \csc 225° = -\sqrt{2}$; $\sec 585° = \sec 225° = -\sqrt{2}$; $\cot 585° = \cot 225° = 1$ **17.** 1   **19.** 0

21. 2   **23.** $-\dfrac{23}{4}$ **25.** $\dfrac{\sqrt{6}}{4} + 1$ **27.** $\dfrac{8\sqrt{3} - \sqrt{6}}{3}$ **29.** $\theta = 30°$ **31.** $\theta = 300°$ **33.** $\theta = 210°$

35. $\theta = 315°$ **37.** $\theta = 135°$ **39.** $\theta = 90°, 270°$ **41.** $\theta = 330°$ **43.** 30°, 150°   **45.** 30°, 210°

47. 120°, 240°   **49.** impossible   **51.** 210°   **53.** $\sin 17° = 0.2924$; $\cos 17° = 0.9563$; $\tan 17° = 0.3057$

55. $\sin 73° = 0.9563$; $\cos 73° = 0.2924$; $\tan 73° = 3.271$ **57.** $\sin 89° = 0.9998$; $\cos 89° = 0.0175$; $\tan 89° = 57.29$

59. $\sin 90° = 1$; $\cos 90° = 0$; $\tan 90°$ is undefined; $\csc 90° = 1$; $\sec 90°$ is undefined; $\cot 90° = 0$

61. $\sin 23.1° = 0.3923$ **63.** $\cos 133.7° = -0.6909$ **65.** $\tan 223.5° = 0.9490$ **67.** $\csc 312.4° = -1.3542$

69. $\sec(-47.4°) = \sec 47.4° = 1.4774$ **71.** $\cot 640.6° = -0.1871$ **73.** 14.0°   **75.** 110.0°

77. 207.0°   **79.** 280.0°   **81.** 49.0°   **83.** 220.0°

Exercise 5.4   (page 250)

1. $\dfrac{1}{12}\pi$ **3.** $\dfrac{2}{3}\pi$ **5.** $\dfrac{7}{6}\pi$ **7.** $\dfrac{5}{3}\pi$ **9.** $\dfrac{13}{3}\pi$ **11.** $-\dfrac{26}{9}\pi$ **13.** 135°   **15.** 450°   **17.** 240°

19. $\dfrac{1080°}{\pi} \approx 343.77°$ **21.** $-\dfrac{1800°}{\pi} \approx -572.96°$ **23.** $\dfrac{2250°}{\pi} \approx 716.20°$ **25.** 39 centimeters

27. 2970 miles   **29.** 38.6° N   **31.** 104.72 square units   **33.** $\dfrac{15\pi}{4} \approx 11.8$ centimeters

35. 11.2 million miles   **37.** $2\pi \frac{\text{rad}}{\text{h}}$ **39.** $\dfrac{\pi}{1800} \frac{\text{rad}}{\text{sec}}$ **41.** $\dfrac{4\pi}{59} \frac{\text{rad}}{\text{day}}$ **43.** $\dfrac{176}{3} \frac{\text{rad}}{\text{sec}}$ **45.** $1000\pi \frac{\text{ft}}{\text{min}}$

47. $12\pi \frac{\text{in}}{\text{sec}}$ **49.** approximately $14.9 \frac{\text{in}}{\text{sec}}$ **51.** $\dfrac{40}{3}$ rpm   **53.** approximately 933 mph

Exercise 5.5   (page 258)

1. $\dfrac{1}{2}$ **3.** $-\dfrac{\sqrt{3}}{2}$ **5.** 1   **7.** 2   **9.** $-\dfrac{\sqrt{3}}{2}$ **11.** -1 **13.** 0.9093   **15.** -0.1455

17. 0.8163   **19.** 3.5253   **21.** 0.6421   **23.** -0.1411 **25.** $(0, -1)$ **27.** $(-1, 0)$ **29.** $(-1, 0)$

31. $(0, 1)$ **33.** $\left(\dfrac{\sqrt{2}}{2}, \dfrac{\sqrt{2}}{2}\right)$ **35.** $\left(-\dfrac{\sqrt{2}}{2}, -\dfrac{\sqrt{2}}{2}\right)$ **37.** $\left(\dfrac{1}{2}, \dfrac{\sqrt{3}}{2}\right)$ **39.** $\left(\dfrac{-1}{2}, \dfrac{-\sqrt{3}}{2}\right)$

41. $\left(\dfrac{-1}{2}, \dfrac{-\sqrt{3}}{2}\right)$ **43.** $\left(\dfrac{1}{2}, \dfrac{\sqrt{3}}{2}\right)$ **45.** $\left(\dfrac{\sqrt{3}}{2}, \dfrac{-1}{2}\right)$ **47.** $\left(\dfrac{\sqrt{3}}{2}, \dfrac{-1}{2}\right)$ **51.** 4π seconds

53. $48 \dfrac{\text{newtons}}{\text{meter}}$ **55.** approximately 560 pounds

Exercise 5.6 (page 265)

1. $2; 2\pi$ **3.** $1; \dfrac{2\pi}{9}$ **5.** $1; 6\pi$ **7.** $1; 10\pi$ **9.** $3; 4\pi$ **11.** $\frac{1}{2}; 2$ **13.** $3; 1$ **15.** $\dfrac{1}{3}; \dfrac{2\pi^2}{3}$

17.

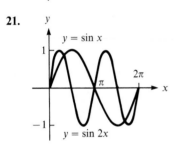

19.

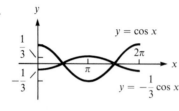

21.

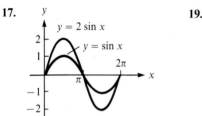

23.

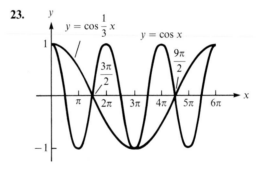

25.

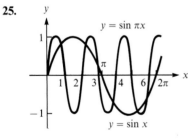

27.

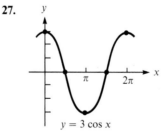

29.

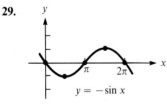

31.

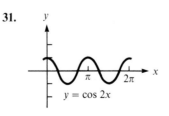

33.

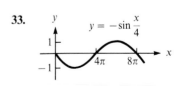

35.

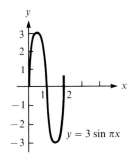

$y = 3 \sin \pi x$

37.

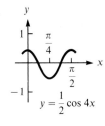

$y = \frac{1}{2} \cos 4x$

39.

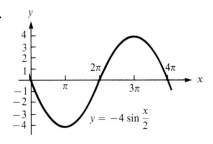

$y = -4 \sin \frac{x}{2}$

Exercise 5.7   (page 269)

1. π **3.** 2π **5.** $\dfrac{\pi}{3}$ **7.** 2   **9.** 6π **11.** $\frac{3}{2}$ **13.** 4   **15.** $2\pi^2$ **17.** $\frac{2}{3}\pi^2$

19.

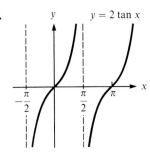

$y = 2 \tan x$

21.

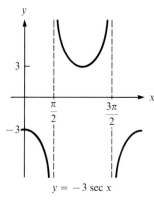

$y = -3 \sec x$

23.

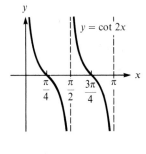

$y = \cot 2x$

25.

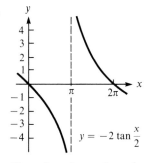

$y = -2 \tan \dfrac{x}{2}$

27.

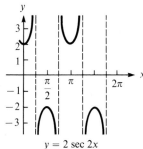

$y = 2 \sec 2x$

29.

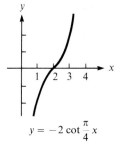

$y = -2 \cot \dfrac{\pi}{4} x$

31.   The value of y can be as large as you want.   **33.** $\ldots, -\pi, 0, \pi, 2\pi, 3\pi, \ldots$

Exercise 5.8   (page 276)

1. 2 units up; 2π **3.** 1 unit down; π **5.** 7 units up; $\dfrac{2\pi}{5}$ **7.** 3 units up; 2π **9.** 5 units down; π

11. 6 units up; 1   **13.** $2\pi; \dfrac{\pi}{3}$ to the right   **15.** $2\pi; \dfrac{\pi}{6}$ to the left   **17.** 1; no phase shift

19. $\pi; \pi$ to the right   **21.** $2\pi; \dfrac{\pi}{4}$ to the left   **23.** $\pi; \dfrac{\pi}{2}$ to the left   **25.** $2; \frac{1}{2}$ to the left

27. 6π; 18π to the right   **29.** $\frac{\pi}{7}$; $\frac{3}{2}\pi$ to the right   **31.**

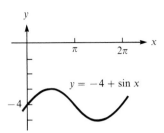

$y = -4 + \sin x$

33.

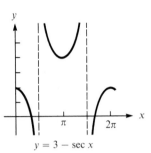

$y = 3 - \sec x$

35.

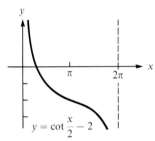

$y = \cot \dfrac{x}{2} - 2$

37.

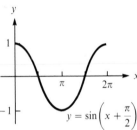

$y = \sin\left(x + \dfrac{\pi}{2}\right)$

39.

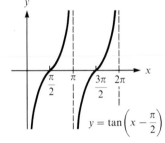

$y = \tan\left(x - \dfrac{\pi}{2}\right)$

41.

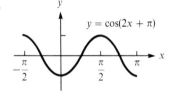

$y = \cos(2x + \pi)$

43.

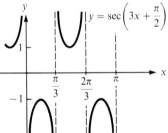

$y = \sec\left(3x + \dfrac{\pi}{2}\right)$

45.

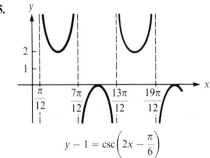

$y - 1 = \csc\left(2x - \dfrac{\pi}{6}\right)$

47.

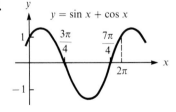

$y = \sin x + \cos x$

49.

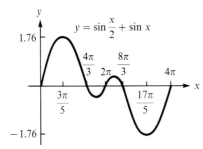

$y = \sin \dfrac{x}{2} + \sin x$

51.

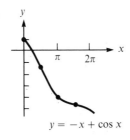

$y = -x + \cos x$

REVIEW EXERCISES (page 277)

1. yes   **2.** no   **3.** no   **4.** yes   **5.** no   **6.** no

7. $\cos \theta = -\dfrac{\sqrt{51}}{10}$, $\tan \theta = \dfrac{7\sqrt{51}}{51}$, $\csc \theta = -\dfrac{10}{7}$, $\sec \theta = -\dfrac{10\sqrt{51}}{51}$, $\cot \theta = \dfrac{\sqrt{51}}{7}$

8. $\sin \theta = -\dfrac{7\sqrt{130}}{130}$, $\cos \theta = -\dfrac{9\sqrt{130}}{130}$, $\csc \theta = -\dfrac{\sqrt{130}}{7}$, $\sec \theta = -\dfrac{\sqrt{130}}{9}$, $\cot \theta = \dfrac{9}{7}$

9. $\sin \theta = \dfrac{\sqrt{51}}{10}$, $\tan \theta = -\dfrac{\sqrt{51}}{7}$, $\csc \theta = \dfrac{10\sqrt{51}}{51}$, $\sec \theta = -\dfrac{10}{7}$, $\cot \theta = -\dfrac{7\sqrt{51}}{51}$

10. $\sin \theta = \dfrac{-8\sqrt{145}}{145}$, $\cos \theta = \dfrac{9\sqrt{145}}{145}$, $\tan \theta = \dfrac{-8}{9}$, $\csc \theta = -\dfrac{\sqrt{145}}{8}$, $\sec \theta = \dfrac{\sqrt{145}}{9}$

11. $\dfrac{1}{\sec \theta} = \cos \theta = \cos \theta \dfrac{\sin \theta}{\sin \theta} = \sin \theta \cot \theta$ **12.** $\cos \theta \csc \theta = \cos \theta \dfrac{1}{\sin \theta} = \dfrac{\cos \theta}{\sin \theta} = \cot \theta$ **13.** $\dfrac{\sqrt{6}}{4}$

14. $\frac{1}{2}$ **15.** $\frac{9}{16}$ **16.** 0   **17.** $\sin 930° = \frac{-1}{2}$, $\cos 930° = \dfrac{-\sqrt{3}}{2}$, $\tan 930° = \dfrac{\sqrt{3}}{3}$

18. $\sin 1380° = -\dfrac{\sqrt{3}}{2}$, $\cos 1380° = \dfrac{1}{2}$, $\tan 1380° = -\sqrt{3}$

19. $\sin(-300°) = \dfrac{\sqrt{3}}{2}$, $\cos(-300°) = \dfrac{1}{2}$, $\tan(-300°) = \sqrt{3}$

20. $\sin(-585°) = \dfrac{\sqrt{2}}{2}$, $\cos(-585°) = \dfrac{-\sqrt{2}}{2}$, $\tan(-585°) = -1$

21. $\sin 15° \approx 0.2588$, $\cos 15° \approx 0.9659$, $\tan 15° \approx 0.2679$
22. $\sin 160° \approx 0.3420$, $\cos 160° \approx -0.9397$, $\tan 160° \approx -0.3640$
23. $\sin 265° \approx -0.9962$, $\cos 265° \approx -0.0872$, $\tan 265° \approx 11.430$
24. $\sin 340° \approx -0.3420$, $\cos 340° \approx 0.9397$, $\tan 340° \approx -0.3640$
25. $\sin(-160°) \approx -0.3420$, $\cos(-160°) \approx -0.9397$, $\tan(-160°) \approx 0.3640$
26. $\sin(-340°) \approx 0.3420$, $\cos(-340°) \approx 0.9397$, $\tan(-340°) \approx 0.3640$ **27.** 119°   **28.** 211°   **29.** 317°
30. 57.7°   **31.** 100°   **32.** 287°   **33.** $\frac{7}{12}\pi$ **34.** $\frac{65}{36}\pi$ **35.** $\frac{53}{30}\pi$ **36.** $\frac{-7}{12}\pi$ **37.** 570°
38. $-150°$ **39.** 1260°   **40.** 458.366°   **41.** $\frac{1}{2}$ **42.** $\dfrac{\sqrt{3}}{2}$ **43.** $-\sqrt{3}$ **44.** 2

45. 2750 miles   **46.** 41.5° N   **47.** 19 square centimeters   **48.** $\dfrac{\pi}{43,200} \dfrac{\text{rad}}{\text{sec}}$ **49.** $\dfrac{1125}{\pi}$ rpm

50. $\dfrac{1000\pi}{3} \dfrac{\text{ft}}{\text{min}}$ **51.** $\left(\dfrac{-\sqrt{3}}{2}, \dfrac{-1}{2} \right)$ **52.** $\left(-\dfrac{\sqrt{2}}{2}, -\dfrac{\sqrt{2}}{2} \right)$ **53.** 0.7539   **54.** -2.1850

55. undefined   **56.** -0.3983 **57.** $4; \frac{2\pi}{3}$ **58.** $\frac{1}{8}; \frac{\pi}{2}$ **59.** $\frac{1}{3}; 6\pi$ **60.** $0.875; 8\pi$ **61.** 2 units up

62. $\frac{\pi}{6}$ to the left   **63.** 4 units up; $\frac{21}{2}$ to the left   **64.** 1 unit down; $\frac{5}{2}\pi$ to the right

65.

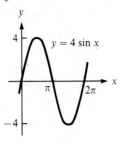

66.

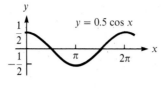

67.

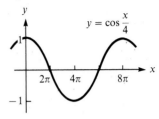

68.

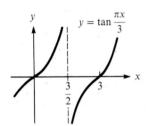

69.

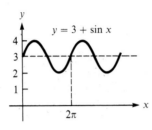

70.

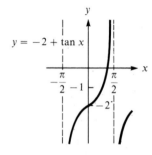

71.

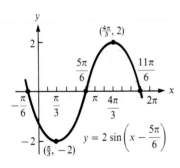

72.

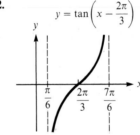

73.

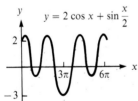

74.

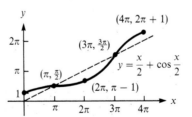

Exercise 6.1   (page 288)

1. angle $B = 53°$, $a = 12$, $b = 16$ **3.** angle $B = 21.3°$, $a = 206$, $c = 221$ **5.** 15 feet   **7.** 721 feet
9. $4.0°$ **11.** 29 feet   **13.** 319 miles   **15.** S 62.9° W   **17.** 5.8 miles   **19.** 685 mph
21. 60.7 feet   **23.** 9180 feet; 1.74 miles   **25.** 1450 feet   **27.** $48.3°$

Exercise 6.2   (page 295)

1. 3 mph; 7 mph   **3.** 63°   **5.** 352 mph   **7.** N 11.6° W   **9.** 323 pounds   **11.** 384 pounds
13. 1160 pounds   **15.** 13.7°   **17.** 224 pounds; 48.1°   **19.** 13°
21. $(F_1)_y = 80$ pounds; $(F_1)_x = 95$ pounds   **23.** N 79.9° E   **25.** 58 pounds

Exercise 6.3   (page 302)

1. 61.0 centimeters   **3.** 1410 kilometers   **5.** 65.9 centimeters   **7.** 54°   **9.** 90°   **11.** 37.85°
13. 210 pounds   **15.** 5° if 210 pounds is used for the resultant force   **17.** 1090 pounds
19. 36.7 nautical miles   **21.** 131 meters   **23.** 69.3°, 64.4°, 46.3°   **27.** 86°   **29.** 85.2°   **31.** 90.0°
33. 1 hour

Exercise 6.4   (page 313)

1. 67 kilometers   **3.** 256 meters   **5.** 2.55 meters   **7.** 305 meters   **9.** 49.2 centimeters
11. 1.0 miles   **13.** 218   **15.** 3.97   **17.** 180 yards   **19.** 420 feet   **21.** 7.1°
23. 2.5 nautical miles   **25.** 31.5°   **27.** 61.6°   **29.** 12 or 2.1   **31.** not a triangle
33. 2900 feet or 780 feet   **35.** 957 feet

Exercise 6.5   (page 319)

1. 190 square feet   **3.** 301 square centimeters   **5.** 6.5 square units   **7.** 6 square units
9. 42 square units   **11.** 960 square units   **13.** 0.001 square millimeter
15. 195,000 square kilometers   **17.** 31,400 square feet   **19.** 72 square meters
21. $\sqrt{2295}$ square centimeters   **23.** 126,000 square feet
25. $\dfrac{s(s-a)}{bc} = \dfrac{(a+b+c)(b+c-a)}{4bc} = \dfrac{1}{2}\left(\dfrac{(a+b+c)(b+c-a)}{2bc}\right) = \frac{1}{2}(1 + \cos A) = \cos^2 \dfrac{A}{2}$
27. Area of $\triangle ABC = \frac{1}{2}(AC)\frac{1}{2}(DB)\sin\alpha = \frac{1}{4}(AC)(DB)\sin\alpha$ **29.** $A = \frac{1}{4}b^2 \cot \dfrac{\alpha}{2} = \dfrac{b^2 \cos^2(\frac{\alpha}{2})}{2 \sin\alpha}$ **31.** $\frac{1}{4}b^2\sqrt{15}$
$\underline{\;\;+ \text{Area of } \triangle ACD = \frac{1}{2}(AC)\frac{1}{2}(DB)\sin\alpha = \frac{1}{4}(AC)(DB)\sin\alpha\;\;}$
$\frac{1}{2}(AC)(DB)\sin\alpha$

Exercise 6.6   (page 327)

1. $\langle 7, -5\rangle$ **3.** $\langle 6, -9\rangle$ **5.** $\langle 9, -8\rangle$ **7.** $\sqrt{13}$ **9.** $\sqrt{5}$ **11.** $\sqrt{13} + \sqrt{2}$ **13.** $8i + 8j$
15. $5\sqrt{3}\,i + 5j$ **17.** $18.6i + 14.1j$ **19.** 9   **21.** 7   **23.** 0   **25.** 45°   **27.** 150°   **29.** 36.9°
31. perpendicular   **33.** not perpendicular   **35.** perpendicular   **37.** $\frac{63}{13}$ **39.** 0

REVIEW EXERCISES   (page 328)

1. 17.3°   **2.** 59.7 feet   **3.** 55 miles   **4.** 150 miles   **5.** 10 pounds   **6.** 11.5°   **7.** 7.6
8. 32   **9.** 0.6   **10.** 11.1   **11.** 25°   **12.** 37°, 143°   **13.** 18.6°, 161.4°   **14.** 51.31°, 128.69°
15. 65.7   **16.** 70.8   **17.** 14°, 82°   **18.** 42.4°   **19.** 640 square units   **20.** 1200 square units
21. 39 square units   **22.** 110 square units   **23.** 1.4 square units   **24.** 12,000 square units
25. 2400 square units   **26.** 67.3 square units   **27.** 558 miles   **28.** about 27.2°, 48.8°, and 104.0°
29. 180 feet   **30.** 611 feet   **31.** 280 square meters   **32.** 24 square units   **33.** $\langle 0, 29\rangle$
34. $3\sqrt{58} - \sqrt{29}$ **35.** $\langle 25, 10\rangle$ **36.** $\sqrt{145}$ **37.** 90°   **38.** 0°   **39.** 75°   **40.** 120°

Exercise 7.1   (page 335)

1. identity   **3.** not an identity   **5.** identity   **7.** not an identity   **9.** identity

Exercise 7.2   (page 343)

1. $\sin 195° = \sin(45° + 150°) = -\frac{\sqrt{6}+\sqrt{2}}{4}$ **3.** $\tan 195° = \tan(225° - 30°) = \frac{3-\sqrt{3}}{3+\sqrt{3}}$

5. $\cos\frac{11\pi}{12} = \cos(\frac{\pi}{6} + \frac{3\pi}{4}) = \frac{-\sqrt{6}-\sqrt{2}}{4}$ **7.** $\cos\frac{19\pi}{12} = \cos(\frac{11\pi}{6} - \frac{\pi}{4}) = \frac{\sqrt{6}-\sqrt{2}}{4}$

9. $\sin 255° = \sin(210° + 45°) = \frac{-\sqrt{2}+\sqrt{6}}{4}$ **11.** $\tan 105° = \tan(60° + 45°) = \frac{\sqrt{3}+1}{1-\sqrt{3}}$

13. $\cos\frac{\pi}{12} = \cos(\frac{\pi}{3} - \frac{\pi}{4}) = \frac{\sqrt{2}+\sqrt{6}}{4}$ **15.** $\sin\frac{5\pi}{12} = \sin(\frac{2\pi}{3} - \frac{\pi}{4}) = \frac{\sqrt{6}+\sqrt{2}}{4}$

17. $\sin(60° + \theta) = \sin 60° \cos\theta + \cos 60° \sin\theta = \frac{\sqrt{3}}{2}\cos\theta + \frac{1}{2}\sin\theta$

19. $\tan(\pi + x) = \dfrac{\tan\pi + \tan x}{1 - \tan\pi\tan x} = \dfrac{\tan x}{1} = \tan x$ **21.** $\cos(\pi - x) = \cos\pi\cos x + \sin\pi\sin x = -\cos x$

23. $\sin(10° + 30°) = \sin 40°$ **25.** $\tan(75° + 40°) = \tan 115°$ **27.** $\cos(120° - 40°) = \cos 80°$

29. $\sin(x + 2x) = \sin 3x$ **31.** $\sin(\alpha + \beta) = -\frac{56}{65}; \cos(\alpha - \beta) = \frac{63}{65}$

33. $\tan(\alpha + \beta) = \frac{140}{171}; \tan(\alpha - \beta) = -\frac{220}{21}$ **35.** $\sin\alpha = \frac{416}{425}; \cos\alpha = \frac{87}{425}$

Exercise 7.3   (page 348)

1. $\sin 2\alpha$ **3.** $\sin 6\theta$ **5.** $\cos 2\beta$ **7.** $\cos B$ **9.** $2\sin 2\theta$ **11.** $(\sin 4\theta)^2 = \sin^2 4\theta$ **13.** $\frac{1}{2}\cos 2\alpha$

15. $\cos 18\theta$ **17.** $\sin^2 10\theta$ **19.** $\tan 8C$ **21.** $\tan A$ **23.** $\cos 8x$ **25.** $-\cos 10x$ **27.** $\frac{\sqrt{3}}{2}$

29. undefined   **31.** -0.5 **33.** 0 **35.** $\frac{\sqrt{3}}{2}$ **37.** 0.5

39. $\sin 2\theta = \frac{120}{169}; \cos 2\theta = -\frac{119}{169}; \tan 2\theta = -\frac{120}{119}$ **41.** $\sin 2\theta = \frac{120}{169}; \cos 2\theta = -\frac{119}{169}; \tan 2\theta = -\frac{120}{119}$

43. $\sin 2\theta = \frac{24}{25}; \cos 2\theta = \frac{7}{25}; \tan 2\theta = \frac{24}{7}$ **45.** $\sin 2\theta = -\frac{336}{625}; \cos 2\theta = -\frac{527}{625}, \tan 2\theta = \frac{336}{527}$

47. $\sin 2\theta = \frac{720}{1681}; \cos 2\theta = -\frac{1519}{1681}; \tan 2\theta = -\frac{720}{1519}$ **49.** $\sin 2\theta = -\frac{720}{1681}; \cos 2\theta = \frac{1519}{1681}; \tan 2\theta = -\frac{720}{1519}$

Exercise 7.4   (page 355)

1. $\dfrac{\sqrt{2+\sqrt{3}}}{2}$ **3.** $-2-\sqrt{3}$ **5.** $\dfrac{\sqrt{2-\sqrt{2}}}{2}$ **7.** $\dfrac{\sqrt{2+\sqrt{3}}}{2}$ **9.** $\sqrt{3}-2$ **11.** 1

13. $\sin\frac{\theta}{2} = \frac{\sqrt{10}}{10}; \cos\frac{\theta}{2} = \frac{3\sqrt{10}}{10}; \tan\frac{\theta}{2} = \frac{1}{3}$ **15.** $\sin\frac{\theta}{2} = \frac{2\sqrt{5}}{5}; \cos\frac{\theta}{2} = -\frac{\sqrt{5}}{5}; \tan\frac{\theta}{2} = -2$

17. $\sin\frac{\theta}{2} = \frac{3\sqrt{34}}{34}; \cos\frac{\theta}{2} = -\frac{5\sqrt{34}}{34}; \tan\frac{\theta}{2} = -\frac{3}{5}$ **19.** $\sin\frac{\theta}{2} = \frac{\sqrt{82}}{82}; \cos\frac{\theta}{2} = \frac{9\sqrt{82}}{82}; \tan\frac{\theta}{2} = \frac{1}{9}$

21. $\sin\frac{\theta}{2} = \frac{4\sqrt{17}}{17}; \cos\frac{\theta}{2} = \frac{\sqrt{17}}{17}; \tan\frac{\theta}{2} = 4$ **23.** $\sin\frac{\theta}{2} = \frac{\sqrt{6}}{6}; \cos\frac{\theta}{2} = \frac{-\sqrt{30}}{6}; \tan\frac{\theta}{2} = \frac{-\sqrt{5}}{5}$ **25.** $\cos 15°$

27. $\tan 100°$ **29.** $\tan 40°$ **31.** $\tan\pi = 0$ **33.** $\tan\frac{x}{4}$ **35.** $\tan 5A$

Exercise 7.5   (page 363)

1. $\frac{1}{4}$ **3.** $\frac{1}{4}$ **5.** $\frac{\sqrt{2}}{4}$ **7.** $\frac{1}{4}$ **9.** $\frac{\sqrt{3}-2}{4}$ **11.** $\frac{1}{2}(1 + \frac{\sqrt{3}}{2})$ **13.** $\frac{\sqrt{6}}{4}$ **15.** $-\frac{\sqrt{2}}{2}$ **17.** $\frac{-\sqrt{2}}{2}$

19. $\frac{\sqrt{6}}{2}$ **21.** 0 **23.** $\frac{\sqrt{2}}{2}$ **25.** $10\sin(x + 53.1°)$ **27.** $10\sin(x - 53.1°)$ **29.** $\sqrt{5}\sin(x + 26.6°)$

31. $\sqrt{2}\sin(x + 45°)$ **33.** $\sqrt{26}\sin(x + 101.3°)$ or $-\sqrt{26}\sin(x - 78.7°)$ **35.** $2\sqrt{3}\sin(x - 60°)$

37.   **39.**

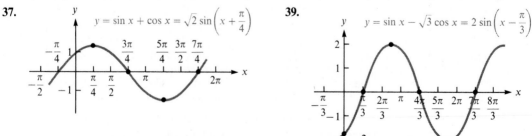

$y = \sin x + \cos x = \sqrt{2}\sin\left(x + \frac{\pi}{4}\right)$

$y = \sin x - \sqrt{3}\cos x = 2\sin\left(x - \frac{\pi}{3}\right)$

41.

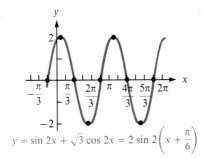

$$y = \sin 2x + \sqrt{3} \cos 2x = 2 \sin 2\left(x + \frac{\pi}{6}\right)$$

Exercise 7.6 (page 369)

1. $\ldots, -300°, -240°, 60°, 120°, 420°, 480°, \ldots$ **3.** $\ldots, -180°, -90°, 0°, 90°, 180°, 270°, 360°, 450°, \ldots$
5. $120°$ **7.** $0°, 90°, 180°, 270°$ **9.** $90°, 270°$ **11.** $0°, 180°$ **13.** $45°, 135°, 225°, 315°$
15. $0°, 120°, 240°$ **17.** $210°, 330°$ **19.** $90°, 210°, 270°, 330°$ **21.** $45°, 225°$ **23.** $0°, 240°$
25. $0°, 120°, 240°$ **27.** $0°, 180°$ **29.** $30°, 150°, 210°, 330°$ **31.** $45°$ **33.** $0°, 30°, 150°, 180°$
35. $90°, 120°, 240°, 270°$ **37.** $0°, 30°, 90°, 150°, 180°, 210°, 270°, 330°$ **39.** $0°, 240°$
41. $60°, 120°, 240°, 300°$ **43.** $0°, 180°$ **45.** $45°, 135°, 315°$ **47.** $0°, 120°$ **49.** $30°$
51. $0°, 240°$ **53.** $\frac{\pi}{4}, \frac{5\pi}{4}$ **55.** $0, \frac{\pi}{2}, \pi, \frac{3\pi}{2}$ **57.** $0, \frac{\pi}{4}, \frac{\pi}{2}, \frac{3\pi}{4}, \pi, \frac{5\pi}{4}, \frac{3\pi}{2}, \frac{7\pi}{4}$ **59.** $0, \frac{\pi}{3}, \frac{\pi}{2}, \frac{2\pi}{3}, \pi, \frac{4\pi}{3}, \frac{3\pi}{2}, \frac{5\pi}{3}$
61. $\frac{\pi}{12}, \frac{\pi}{6}, \frac{\pi}{4}, \frac{5\pi}{12}, \frac{\pi}{2}, \frac{7\pi}{12}, \frac{3\pi}{4}, \frac{5\pi}{6}, \frac{11\pi}{12}, \frac{13\pi}{12}, \frac{7\pi}{6}, \frac{5\pi}{4}, \frac{17\pi}{12}, \frac{3\pi}{2}, \frac{19\pi}{12}, \frac{7\pi}{4}, \frac{11\pi}{6}, \frac{23\pi}{12}$ **63.** $\frac{\pi}{4}, \frac{3\pi}{4}, \frac{5\pi}{4}, \frac{7\pi}{4}$
65. $\frac{\pi}{6}, \frac{\pi}{2}, \frac{5\pi}{6}, \frac{7\pi}{6}, \frac{3\pi}{2}, \frac{11\pi}{6}$ **67.** $\frac{\pi}{2}, \frac{3\pi}{2}$ **69.** $\frac{7\pi}{6}, \frac{11\pi}{6}$
71. $\frac{\pi}{20}, \frac{3\pi}{20}, \frac{\pi}{4}, \frac{7\pi}{20}, \frac{9\pi}{20}, \frac{11\pi}{20}, \frac{13\pi}{20}, \frac{3\pi}{4}, \frac{17\pi}{20}, \frac{19\pi}{20}, \frac{21\pi}{20}, \frac{23\pi}{20}, \frac{5\pi}{4}, \frac{27\pi}{20}, \frac{29\pi}{20}, \frac{31\pi}{20}, \frac{33\pi}{20}, \frac{7\pi}{4}, \frac{37\pi}{20}, \frac{39\pi}{20}$
73. $0, \frac{\pi}{4}, \pi, \frac{5\pi}{4}$ **75.** $\frac{\pi}{6}, \frac{5\pi}{6}, \frac{7\pi}{6}, \frac{11\pi}{6}$

Exercise 7.7 (page 377)

1. $\frac{\pi}{6}$ **3.** $\frac{\pi}{2}$ **5.** $\frac{\pi}{4}$ **7.** no value   **9.** $\frac{3\pi}{4}$ **11.** $\frac{\pi}{3}$ **13.** $\sin \frac{\pi}{6} = \frac{1}{2}; \cos \frac{\pi}{6} = \frac{\sqrt{3}}{2}; \tan \frac{\pi}{6} = \frac{\sqrt{3}}{3}$
15. $\sin 0 = 0; \cos 0 = 1; \tan 0 = 0$ **17.** $\sin \frac{5\pi}{6} = \frac{1}{2}; \cos \frac{5\pi}{6} = -\frac{\sqrt{3}}{2}; \tan \frac{5\pi}{6} = -\frac{\sqrt{3}}{3}$
19. $\sin \frac{\pi}{2} = 1; \cos \frac{\pi}{2} = 0; \tan \frac{\pi}{2} =$ is undefined   **21.** $\sin \pi = 0; \cos \pi = -1; \tan \pi = 0$
23. $\sin \frac{\pi}{4} = \frac{\sqrt{2}}{2}; \cos \frac{\pi}{4} = \frac{\sqrt{2}}{2}; \tan \frac{\pi}{4} = 1$ **25.** $\frac{1}{2}$ **27.** 1 **29.** 1 **31.** $\frac{1}{2}$ **33.** $-\sqrt{3}$ **35.** $\frac{\sqrt{2}}{2}$
37. $\frac{3}{5}$ **39.** $\frac{12}{13}$ **41.** $\frac{-4}{3}$ **43.** $\frac{12}{5}$ **45.** $\frac{12}{13}$ **47.** $\frac{40}{41}$ **49.** $\sin(\frac{\pi}{6} + \frac{\pi}{3}) = 1$ **51.** $\frac{24}{25}$
53. $\sin 2(\frac{\pi}{4}) = 1$ **55.** $\tan 2(\frac{\pi}{4})$ is undefined   **57.** $\sin \frac{1}{2}(\frac{\pi}{3}) = \frac{1}{2}$ **59.** $\frac{1}{4}$ **61.** $\dfrac{x}{\sqrt{1 + x^2}}$ **63.** $\dfrac{x}{\sqrt{1 - x^2}}$
65. $\sqrt{1 - x^2}$ **67.** $2x\sqrt{1 - x^2}$ **69.** $\dfrac{2x}{1 - x^2}$ **71.** $1 - 2x^2$ **73.** $-\frac{\pi}{6}$ **75.** $\frac{\pi}{4}$ **77.** $\frac{\pi}{6}$
79. π **81.** $-\frac{\pi}{2}$

REVIEW EXERCISES (page 379)

5. $\sin 71°$ **6.** $\tan(-31°) = -\tan 31°$ **7.** $\cos \frac{4\pi}{11}$ **8.** $\cos \frac{2\pi}{3} = -\frac{1}{2}$
13. $\sin 2\theta = -\frac{120}{169}; \cos 2\theta = \frac{119}{169}; \tan 2\theta = -\frac{120}{119}$ **14.** $\sin 2\theta = \frac{120}{169}; \cos 2\theta = \frac{119}{169}; \tan 2\theta = \frac{120}{119}$
15. $\sin 2\theta = -\frac{840}{841}; \cos 2\theta = \frac{41}{841}; \tan 2\theta = -\frac{840}{41}$ **16.** $\sin 2\theta = \frac{840}{841}; \cos 2\theta = \frac{41}{841}; \tan 2\theta = \frac{840}{41}$
17. $\sin 2\theta = \frac{24}{25}; \cos 2\theta = \frac{-7}{25}; \tan 2\theta = \frac{-24}{7}$ **18.** $\sin 2\theta = -\frac{24}{25}; \cos 2\theta = -\frac{7}{25}; \tan 2\theta = \frac{24}{7}$
19. $\sin \frac{\theta}{2} = \frac{\sqrt{26}}{26}; \cos \frac{\theta}{2} = \frac{5\sqrt{26}}{26}; \tan \frac{\theta}{2} = \frac{1}{5}$ **20.** $\sin \frac{\theta}{2} = \frac{\sqrt{26}}{26}; \cos \frac{\theta}{2} = -\frac{5\sqrt{26}}{26}; \tan \frac{\theta}{2} = -\frac{1}{5}$
21. $\sin \frac{\theta}{2} = \frac{\sqrt{26}}{26}; \cos \frac{\theta}{2} = -\frac{5\sqrt{26}}{26}; \tan \frac{\theta}{2} = -\frac{1}{5}$ **22.** $\sin \frac{\theta}{2} = \frac{2\sqrt{5}}{5}; \cos \frac{\theta}{2} = \frac{\sqrt{5}}{5}; \tan \frac{\theta}{2} = 2$
23. $\sin \frac{\theta}{2} = \frac{2\sqrt{5}}{5}; \cos \frac{\theta}{2} = \frac{\sqrt{5}}{5}; \tan \frac{\theta}{2} = 2$ **24.** $\sin \frac{\theta}{2} = \frac{2\sqrt{5}}{5}; \cos \frac{\theta}{2} = -\frac{\sqrt{5}}{5}; \tan \frac{\theta}{2} = -2$ **25.** $-\frac{2+\sqrt{3}}{4}$

26. $\frac{1}{2}(-\frac{\sqrt{3}}{2} + 1)$ **27.** $\frac{1-\sqrt{3}}{4}$ **28.** $\frac{1}{2}(-\frac{1}{2} - \frac{\sqrt{3}}{2})$ **29.** $2\sin 6° \cos 1°$ **30.** $2\cos 226° \sin 86°$

31. $-2\sin\frac{2\pi}{5}\sin\frac{\pi}{5}$ **32.** $2\cos\frac{5\pi}{14}\cos\frac{\pi}{14}$ **33.** $-\frac{\sqrt{2}}{2}$ **34.** $\frac{\sqrt{6}}{2}$ **35.** $-\frac{\sqrt{2}}{2}$ **36.** $\frac{\sqrt{6}}{2}$

37. $y = \sqrt{5}\sin(x + 63.4°)$ **38.** $y = -\sqrt{2}\sin(x + 315°)$ **39.** $45°, 225°$ **40.** $0°, 45°, 225°$

41. $45°, 135°, 225°, 315°$ **42.** no solutions   **43.** $\frac{\pi}{3}$ **44.** $\frac{2\pi}{3}$ **45.** $-\frac{\pi}{3}$ **46.** no value   **47.** $\frac{\pi}{2}$

48. no value   **49.** $-\frac{\pi}{6}$ **50.** $\frac{\pi}{6}$ **51.** $\frac{\pi}{2}$ **52.** π **53.** 0 **54.** 0 **55.** 1 **56.** $\frac{\sqrt{3}}{2}$

57. $\frac{\sqrt{3}}{2}$ **58.** 0 **59.** $\sqrt{1 - u^2}$ **60.** $\frac{2u}{1 - u^2}$

Exercise 8.1   (page 388)

1.

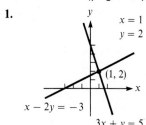

3.

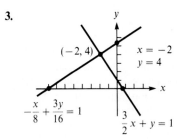

5. $x = 0, y = 0$
7. $x = 3, y = -2$
9. $x = 2, y = 3$
11. no solution
13. $x = 3, y = 1$
15. $x = 3, y = 2$
17. $x = -3, y = 0$
19. $x = 1, y = -\frac{1}{2}$

21. all (x, y) such that $x = 2y$ **23.** $x = 2, y = 2$ **25.** $x = 1, y = 2, z = 0$ **27.** $x = 0, y = -\frac{1}{3}, z = -\frac{1}{3}$
29. $x = 1, y = 2, z = -1$ **31.** $x = 1, y = 0, z = 5$ **33.** no solution
35. $x = 2 - y, y =$ any number, $z = 2 - y$ **37.** $x = 1, y = 1, z = 2$ **39.** $x = 2 - y, y =$ any number, $z = 1$
41. 50 units from A, the rest from B and C in any combination
43. 15 hours cooking hamburgers, 10 hours pumping gas, and 5 hours as a janitor
45. 14 nickels, 12 dimes, 6 quarters

Exercise 8.2   (page 398)

1. $x = 2, y = -1$ **3.** $x = -2, y = 0$ **5.** $x = 3, y = 1$ **7.** no solution   **9.** $x = 1, y = 0, z = 2$
11. $x = 2, y = -2, z = 1$ **13.** $x = 1, y = 1, z = 2$ **15.** $x = -1, y = 3, z = 1$ **17.** $x = 1, y = -3$
19. $x = \frac{8}{7} + \frac{1}{7}z, y = \frac{10}{7} - \frac{4}{7}z, z =$ any number   **21.** $w = y, x = 1, y =$ any number, $z = 1$ **23.** no solution
25. $x = 1, y = 2, z = 1, t = 1$ **27.** $x = 1, y = 2, z = 0, t = 1$ **29.** $x = \pm2, y = \pm1, z = \pm3$
31. $x = 16, y = 1, z = 0$

Exercise 8.3   (page 408)

1. $x = 2, y = 5$ **3.** no values   **5.** $x = 1, y = 2$ **7.** $x = 2, y = 2$ **9.** $\begin{bmatrix} -1 & 2 & 1 \\ -6 & 0 & 0 \end{bmatrix}$

11. $\begin{bmatrix} 5 & -4 & 3 \\ -7 & -4 & -2 \\ 0 & 4 & -8 \end{bmatrix}$ **13.** not possible   **15.** $[5 \quad 7 \quad 9]$ **17.** $\begin{bmatrix} 3 & 2 & -7 \\ 9 & 19 & 15 \\ 0 & -4 & -1 \end{bmatrix}$

19. $\begin{bmatrix} 2 & -2 \\ 3 & 10 \end{bmatrix}$ **21.** $\begin{bmatrix} -22 & -22 \\ -105 & 126 \end{bmatrix}$ **23.** $\begin{bmatrix} 4 & 2 & 10 \\ 5 & -2 & 4 \\ 2 & -2 & 1 \end{bmatrix}$ **25.** $[32]$ **27.** not possible

29. $[30 \quad 36 \quad 42]$ **31.** $\begin{bmatrix} 2 & 0 & 3 \\ 3 & 0 & 5 \end{bmatrix}$ **33.** $\begin{bmatrix} 12 & -4 & 9 \\ 16 & -2 & 2 \\ 10 & -2 & 1 \end{bmatrix}$ **35.** $\begin{bmatrix} 7 \\ 6 \end{bmatrix}$

37. $\begin{bmatrix} 4 & 5 \\ -7 & -1 \end{bmatrix}$ **39.** not possible   **41.** $\begin{bmatrix} 24 & 16 \\ 39 & 26 \end{bmatrix}$ **43.** $\begin{bmatrix} 64 & 64 \\ 64 & 64 \end{bmatrix}$

45. One example is $\begin{bmatrix} 1 & 0 \\ 0 & 0 \end{bmatrix}$.

Exercise 8.4 (page 413)

1. $\begin{bmatrix} 3 & 4 \\ 2 & 3 \end{bmatrix}$ **3.** $\begin{bmatrix} 5 & -7 \\ -2 & 3 \end{bmatrix}$ **5.** $\begin{bmatrix} -40 & 16 & 9 \\ 13 & -5 & -3 \\ 5 & -2 & -1 \end{bmatrix}$ **7.** $\begin{bmatrix} 4 & 1 & -3 \\ -5 & -1 & 4 \\ -1 & -1 & 1 \end{bmatrix}$

9. no inverse   **11.** $\begin{bmatrix} 1 & -2 & 1 \\ 0 & 1 & -2 \\ 0 & 0 & 1 \end{bmatrix}$ **13.** no inverse   **15.** $\begin{bmatrix} 1 & -2 & 1 & 0 \\ 0 & 1 & -2 & 1 \\ 0 & 0 & 1 & -2 \\ 0 & 0 & 0 & 1 \end{bmatrix}$

17. $x = 23, y = 17$ **19.** $x = 70, y = -30$ **21.** $x = -10, y = 4, z = 1$ **23.** $x = 7, y = -9, z = -1$
25. $x = 4, y = -6, z = 3$

29. $A^2 = \begin{bmatrix} 1 & 0 \\ 0 & 1 \end{bmatrix}, A^3 = \begin{bmatrix} 0 & 1 \\ 1 & 0 \end{bmatrix}, A^4 = \begin{bmatrix} 1 & 0 \\ 0 & 1 \end{bmatrix}. A^n = \begin{bmatrix} 0 & 1 \\ 1 & 0 \end{bmatrix}$ if n is odd, $A^n = \begin{bmatrix} 1 & 0 \\ 0 & 1 \end{bmatrix}$ if n is even.

33. $A^2 = \begin{bmatrix} 1 & 0 \\ 2 & 1 \end{bmatrix}, A^3 = \begin{bmatrix} 1 & 0 \\ 3 & 1 \end{bmatrix}, A^n = \begin{bmatrix} 1 & 0 \\ n & 1 \end{bmatrix}$

35. $x = \pm 4$ **39.** $E = \begin{bmatrix} 1 & 0 & 0 \\ 0 & 1 & 0 \\ 3 & 0 & 1 \end{bmatrix}, E^{-1} = \begin{bmatrix} 1 & 0 & 0 \\ 0 & 1 & 0 \\ -3 & 0 & 1 \end{bmatrix}$

Exercise 8.5 (page 423)

1. 8   **3.** 1   **5.** -54 **7.** -7 **9.** 86   **11.** -2 **13.** 12   **15.** 1   **17.** $x = 1, y = 2$
19. $x = 3, y = 0$ **21.** $x = 1, y = 0, z = 1$ **23.** $x = 1, y = -1, z = 2$ **25.** $x = 6, y = 6, z = 12$
27. $p = 1, q = 1, r = 1, s = 1$

31. For example, $\begin{vmatrix} 1 & 0 \\ 0 & 1 \end{vmatrix} + \begin{vmatrix} 1 & 0 \\ 0 & 1 \end{vmatrix} = 1 + 1 = 2$, however $\begin{vmatrix} \begin{bmatrix} 1 & 0 \\ 0 & 1 \end{bmatrix} + \begin{bmatrix} 1 & 0 \\ 0 & 1 \end{bmatrix} \end{vmatrix} = \begin{vmatrix} 2 & 0 \\ 0 & 2 \end{vmatrix} = 4$

33. $\begin{vmatrix} a & b & c \\ 0 & d & e \\ 0 & 0 & f \end{vmatrix} = f \begin{vmatrix} a & b \\ 0 & d \end{vmatrix} = adf$ **39.** $x = 8$ **41.** $x = -1$

43. domain is the set of $n \times n$ matrices, range is the set of real numbers
45. Yes, because $|AB| = |A||B|$, and $|AB| = 0$ implies that $|A||B| = 0$, and therefore $|A| = 0$ or $|B| = 0$.

Exercise 8.6 (page 429)

1. $\dfrac{1}{x + 1} + \dfrac{2}{x - 1}$ **3.** $\dfrac{1}{x^2 + 2} - \dfrac{3}{x + 1}$ **5.** $\dfrac{1}{x} + \dfrac{2}{x^2} - \dfrac{3}{x - 1}$ **7.** $\dfrac{1}{x^2} + \dfrac{1}{x^2 + 1}$ **9.** $\dfrac{2}{x} + \dfrac{3x + 2}{x^2 + 1}$

11. $\dfrac{1}{x} + \dfrac{1}{x^2} + \dfrac{2}{x^2 + x + 1}$ **13.** $\dfrac{1}{x^2 + x + 5} + \dfrac{x + 1}{x^2 + 1}$ **15.** $\dfrac{-1}{x^2 + 1} - \dfrac{x}{(x^2 + 1)^2} + \dfrac{1}{x}$

17. $\dfrac{1}{x^2 + 1} + \dfrac{x + 2}{x^2 + x + 2}$ **19.** $\dfrac{1}{x} + \dfrac{x}{x^2 + 2x + 5} + \dfrac{x + 2}{(x^2 + 2x + 5)^2}$ **21.** $2 + \dfrac{1}{x} + \dfrac{2}{x^2} + \dfrac{3}{x + 1}$

23. $x^2 + x + 1 + \dfrac{1}{x} + \dfrac{1}{x^2}$

Exercise 8.7   (page 435)

1. $y = -2x + 3$
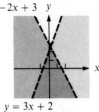
$y = 3x + 2$

3.
$3x + 2y = 6$
$x + 3y = 2$
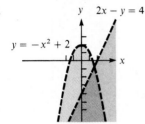

5.
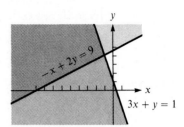
$-x + 2y = 9$
$3x + y = 1$

7. $2x - y = 4$
$y = -x^2 + 2$
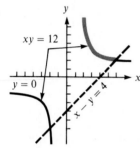

9.
$y = x^2 - 4$
$y = -x^2 + 4$
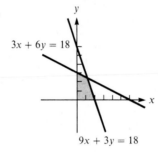

11.
$x = 0$
$2x + 3y = 5$
$3x + y = 1$
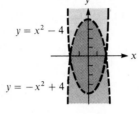

13.
$xy = 12$
$y = 0$
$x - y = 4$
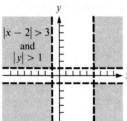

15.
$3x + 6y = 18$
$9x + 3y = 18$
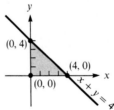

17.
$y < \sqrt{x}$
and
$x \geq 0$
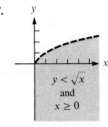

19.
$|x - 2| > 3$
and
$|y| > 1$
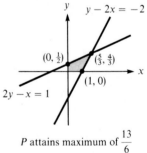

21.
$(0, 4)$
$(0, 0)$
$(4, 0)$
$x + y = 4$

P attains maximum of
12 at $(0, 4)$
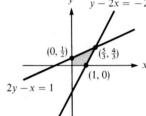

23.
$y - 2x = -2$
$(0, \frac{1}{2})$
$(\frac{5}{3}, \frac{4}{3})$
$(1, 0)$
$2y - x = 1$

P attains maximum of $\dfrac{13}{6}$

at $\left(\dfrac{5}{3}, \dfrac{4}{3}\right)$

25.
$3x + y = 3$
$y - x = 2$
$2x + 3y = 6$
$(0, 3)$
$(0, 2)$
$(\frac{3}{7}, \frac{12}{7})$
$(-2, 0)$
$(3, 0)$
$(0, 0)$
$(1, 0)$
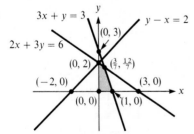

P attains maximum value

of $\dfrac{18}{7}$ at $\left(\dfrac{3}{7}, \dfrac{12}{7}\right)$

27. 2 tables, no chairs   **29.** 10 square meters of strawberries, 30 square meters of pumpkins
31. 10 ounces of X and 25 ounces of Y per day

REVIEW EXERCISES   (page 436)

1.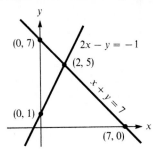

2.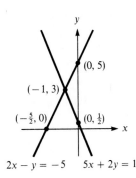

3.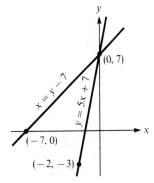

4.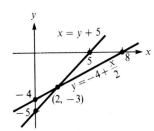

5. $x = 1, y = 5$ **6.** $x = 2, y = -1$ **7.** $x = 0, y = -3$

8. $x = 1, y = 1$ **9.** $x = -3, y = 2$ **10.** $x = -2, y = 5$ **11.** $x = 2, y = -1$ **12.** $x = 4, y = 1$
13. $x = 1, y = 0, z = 1$ **14.** $x = 1, y = 1, z = -1$ **15.** $x = 0, y = 1, z = 2$ **16.** $x = 1, y = -2, z = 3$
17. $x = 1, y = 1$ **18.** $x = 3, y = 1, z = -2$ **19.** $x = 1 - \frac{4}{7}z, y = 2 - \frac{1}{7}z, z = $ any number
20. no solution   **21.** $w = 1, x = 2, y = 0, z = -1$ **22.** $w = 2 - z, x = 1, y = -2 + z, z = $ any number
23. $\begin{bmatrix} 1 & 3 & 4 \\ 4 & 0 & 2 \end{bmatrix}$ **24** $\begin{bmatrix} 2 & 5 & 4 \\ -2 & -6 & 6 \\ -4 & 5 & -3 \end{bmatrix}$ **25.** $\begin{bmatrix} -7 & -1 \\ -7 & -4 \end{bmatrix}$ **26.** $\begin{bmatrix} -17 & 19 \\ 10 & -12 \end{bmatrix}$

27. $[5]$ **28.** $\begin{bmatrix} -3 & 14 & -12 \\ 1 & 2 & -4 \end{bmatrix}$ **29.** $\begin{bmatrix} 2 & -1 & 1 & 3 \\ 4 & -2 & 2 & 6 \\ 2 & -1 & 1 & 3 \\ 10 & -5 & 5 & 15 \end{bmatrix}$ **30.** $\begin{bmatrix} 24 \\ -8 \end{bmatrix}$ **31.** $[-18]$

32. $\begin{bmatrix} 0 \\ -6 \end{bmatrix}$ **33.** $\begin{bmatrix} \frac{5}{14} & -\frac{3}{14} \\ \frac{3}{14} & \frac{1}{14} \end{bmatrix}$ **34.** $\begin{bmatrix} 9 & -7 \\ -5 & 4 \end{bmatrix}$ **35.** $\begin{bmatrix} 1 & -3 & 32 \\ 0 & 1 & -9 \\ 0 & 0 & 1 \end{bmatrix}$ **36.** $\begin{bmatrix} 1 & 0 & 0 \\ -\frac{3}{2} & \frac{1}{2} & \frac{1}{2} \\ 1 & -\frac{1}{2} & 0 \end{bmatrix}$

37. $\begin{bmatrix} 9 & 16 & -56 \\ -3 & -5 & 18 \\ -1 & -2 & 7 \end{bmatrix}$ **38.** $\begin{bmatrix} 1 & -1 & 0 \\ 2 & -1 & 0 \\ 1 & -2 & -1 \end{bmatrix}$ **39.** $x = 1, y = 2, z = -1$

40. $w = 1, x = 1, y = 0, z = -1$ **41.** -7 **42.** 3 **43.** -6 **44.** -25 **45.** $x = 1, y = -2$
46. $x = 1, y = 0, z = -2$ **47.** $x = 1, y = -1, z = 3$ **48.** $w = 1, x = 0, y = -1, z = 2$
49. $\dfrac{1}{x} + \dfrac{3x + 4}{x^2 + 1}$ **50.** $\dfrac{3}{x} + \dfrac{2}{x^2} + \dfrac{x - 1}{x^2 + 1}$ **51.** $\dfrac{1}{x} + \dfrac{-1}{x^2 + x + 5}$ **52.** $\dfrac{1}{x + 1} - \dfrac{2}{(x + 1)^2} + \dfrac{2}{(x + 1)^3}$

53.

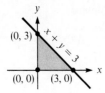

P attains maximum
of 6 at (3, 0)

54.

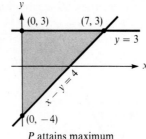

P attains maximum
of 12 at (0, −4)

55.

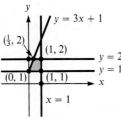

P attains maximum
of 2 at (1, 1)

56.

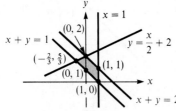

P attains maximum

of 3 at $\left(-\dfrac{2}{3}, \dfrac{5}{3}\right)$

57.   1000 bags of X, 1400 bags of Y

Exercise 9.1   (page 445)

1. $x^2 + y^2 = 1$ **3.** $(x - 6)^2 + (y - 8)^2 = 16$ **5.** $(x + 5)^2 + (y - 3)^2 = 25$ **7.** $(x - 3)^2 + (y + 4)^2 = 2$

9. $(x - 3)^2 + (y - 3)^2 = 25$ **11.** $(x + 5)^2 + (y - 1)^2 = 65$ **13.** $(x + 3)^2 + (y - 4)^2 = 25$

15. $(x + 2)^2 + (y + 6)^2 = 40$ **17.** $x^2 + (y + 3)^2 = 157$ **19.** $(x - 5)^2 + (y - 8)^2 = 338$

21. $(x + 4)^2 + (y + 2)^2 = 98$ **23.** $(x - 1)^2 + (y + 2)^2 = 36$ **25.** $x^2 + (y + 12)^2 = 10$ **27.** no

29.

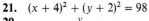

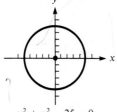

$x^2 + y^2 - 25 = 0$

31.

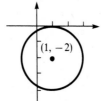

$(x - 1)^2 + (y + 2)^2 = 4$

33.

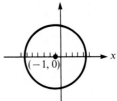

$x^2 + y^2 + 2x - 26 = 0$

35.

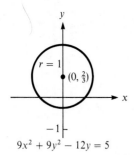

$9x^2 + 9y^2 - 12y = 5$

37.

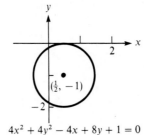

$4x^2 + 4y^2 - 4x + 8y + 1 = 0$

39. $x^2 + (y - 3)^2 = 25$
41. $A = 5\pi$

Exercise 9.2 (page 450)

1. $x^2 = 12y$ **3.** $y^2 = 12x$ **5.** $(x - 3)^2 = -12(y - 5)$ **7.** $(x - 3)^2 = -28(y - 5)$

9. $(x - 2)^2 = -2(y - 2)$ or $(y - 2)^2 = -2(x - 2)$ **11.** $(x + 4)^2 = -\frac{16}{3}(y - 6)$ or $(y - 6)^2 = \frac{9}{4}(x + 4)$

13. $(y - 8)^2 = -4(x - 6)$ **15.** $(x - 3)^2 = \frac{1}{2}(y - 1)$ **17.** **19.**

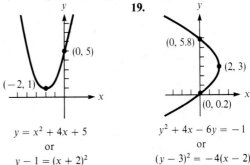

$y = x^2 + 4x + 5$
or
$y - 1 = (x + 2)^2$

$y^2 + 4x - 6y = -1$
or
$(y - 3)^2 = -4(x - 2)$

21. **23.** **25.**

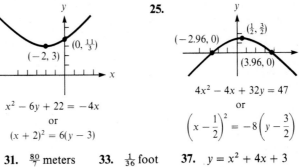

$y^2 + 2x - 2y = 5$
or
$(y - 1)^2 = -2(x - 3)$

$x^2 - 6y + 22 = -4x$
or
$(x + 2)^2 = 6(y - 3)$

$4x^2 - 4x + 32y = 47$
or
$\left(x - \frac{1}{2}\right)^2 = -8\left(y - \frac{3}{2}\right)$

27. $x^2 = -\frac{45}{2}y$ **29.** 8 cabins **31.** $\frac{80}{7}$ meters **33.** $\frac{1}{36}$ foot **37.** $y = x^2 + 4x + 3$

Exercise 9.3 (page 458)

1. $\dfrac{x^2}{25} + \dfrac{y^2}{16} = 1$ **3.** $\dfrac{9x^2}{16} + \dfrac{9y^2}{25} = 1$ **5.** $\dfrac{x^2}{7} + \dfrac{y^2}{16} = 1$ **7.** $\dfrac{(x - 3)^2}{4} + \dfrac{(y - 4)^2}{9} = 1$

9. $\dfrac{(x - 3)^2}{9} + \dfrac{(y - 4)^2}{4} = 1$ **11.** $\dfrac{(x - 3)^2}{41} + \dfrac{(y - 4)^2}{16} = 1$ **13.** $\dfrac{x^2}{36} + \dfrac{(y - 4)^2}{20} = 1$ **15.** $\dfrac{x^2}{100} + \dfrac{y^2}{64} = 1$

17. **19.** **21.**

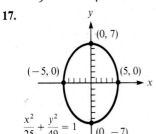

$\dfrac{x^2}{25} + \dfrac{y^2}{49} = 1$

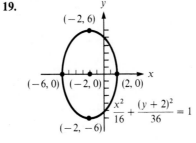

$\dfrac{x^2}{16} + \dfrac{(y + 2)^2}{36} = 1$

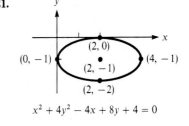

$x^2 + 4y^2 - 4x + 8y + 4 = 0$

23.

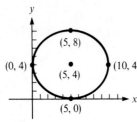

$(5, 8)$
$(0, 4)$ $\quad$ $(5, 4)$ $\quad$ $(10, 4)$
$(5, 0)$

$16x^2 + 25y^2 - 160x - 200y + 400 = 0$

25. 199,395 miles

27. $\dfrac{x^2}{2500} + \dfrac{y^2}{900} = 1$; 36 meters

29. In the ellipse, $b^2 = a^2 - c^2$, or $a = \sqrt{b^2 + c^2}$. Using the distance formula, you have $d(FB) = \sqrt{b^2 + c^2}$.

31. Substitute c for x in the equation $\dfrac{x^2}{a^2} + \dfrac{y^2}{b^2} = 1$. Use the fact that $c^2 = a^2 - b^2$, and solve for y to determine FA'. The focal width is $2y$.

33. The thumbtacks are the foci at $(\pm 1, 0)$. Hence, $c = 1$. The string is 6 meters, so $2a = 6$, or $a^2 = 9$. Because $b^2 = a^2 - c^2$, you have $b^2 = 8$. The equation is $\dfrac{x^2}{9} + \dfrac{y^2}{8} = 1$.

Exercise 9.4 (page 466)

1. $\dfrac{x^2}{25} - \dfrac{y^2}{24} = 1$ $\quad$ **3.** $\dfrac{(x-2)^2}{4} - \dfrac{(y-4)^2}{9} = 1$ $\quad$ **5.** $\dfrac{(y-3)^2}{9} - \dfrac{(x-5)^2}{9} = 1$ $\quad$ **7.** $\dfrac{y^2}{9} - \dfrac{x^2}{16} = 1$

9. $\dfrac{(x-1)^2}{4} - \dfrac{(y+3)^2}{16} = 1$ or $\dfrac{(y+3)^2}{4} - \dfrac{(x-1)^2}{16} = 1$ $\quad$ **11.** $\dfrac{x^2}{10} - \dfrac{3y^2}{20} = 1$ $\quad$ **13.** 24 square units

15. 12 square units $\quad$ **17.** $\dfrac{(x+2)^2}{4} - \dfrac{4(y+4)^2}{81} = 1$ $\quad$ **19.** $\dfrac{x^2}{36} - \dfrac{16y^2}{25} = 1$

21.

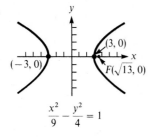

$(3, 0)$
$(-3, 0)$
$F(\sqrt{13}, 0)$
$\dfrac{x^2}{9} - \dfrac{y^2}{4} = 1$

23.

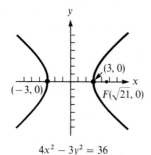

$(3, 0)$
$(-3, 0)$
$F(\sqrt{21}, 0)$
$4x^2 - 3y^2 = 36$

25.

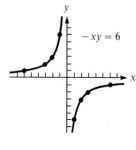

$y^2 - x^2 = 1$

27.

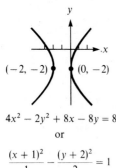

$(-2, -2)$ $\quad$ $(0, -2)$

$4x^2 - 2y^2 + 8x - 8y = 8$
or
$\dfrac{(x+1)^2}{1} - \dfrac{(y+2)^2}{2} = 1$

29.

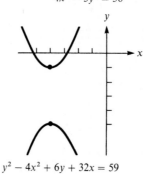

$y^2 - 4x^2 + 6y + 32x = 59$

31.

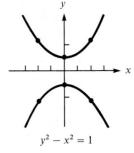

$-xy = 6$

33. $\dfrac{(x-3)^2}{9} - \dfrac{(y-1)^2}{16} = 1$ **35.** $4x^2 - 5y^2 - 60y = 0$

37. From the geometry of Figure 9-17, $PF + F'F > F'P$. This is equivalent to $F'P - PF < F'F$ or $2a < 2c$, which implies that $c > a$.

Exercise 9.5   (page 470)

1.

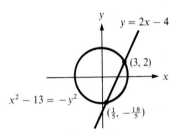

3.

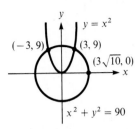

5.

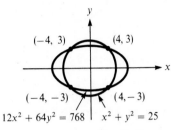

7.

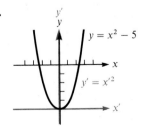

9.

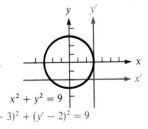

11. $(3, 0), (0, 5)$ **13.** $(1, 1)$

15. $(1, 2), (2, 1)$ **17.** $(-2, 3), (2, 3)$ **19.** $(\sqrt{5}, 5), (-\sqrt{5}, 5)$ **21.** $(3, 2), (3, -2), (-3, 2), (-3, -2)$
23. $(2, 4), (2, -4), (-2, 4), (-2, -4)$ **25.** $(-\sqrt{15}, 5), (\sqrt{15}, 5), (-2, -6), (2, -6)$
27. $(0, -4), (-3, 5), (3, 5)$ **29.** $(-2, 3), (2, 3), (-2, -3), (2, -3)$ **31.** $(3, 3)$
33. $(6, 2), (-6, -2), (\sqrt{42}, 0), (-\sqrt{42}, 0)$ **35.** $(\frac{1}{2}, \frac{1}{3}), (\frac{1}{3}, \frac{1}{2})$ **37.** 7 by 9 centimeters   **39.** 14 and -5
41. Either \$750 at 9% or \$900 at 7.5%

Exercise 9.6   (page 477)

1. $P(3, 1)$ **3.** $R(1, -3)$ **5.** $P(4, -8)$ **7.** $R(2, -4)$ **9.**

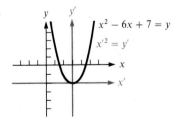

11.

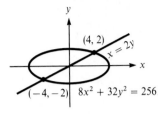

13.

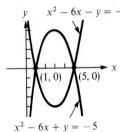

15.

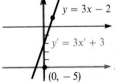

17.

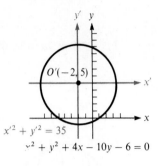

$x'^2 + y'^2 = 35$

$x^2 + y^2 + 4x - 10y - 6 = 0$

19.

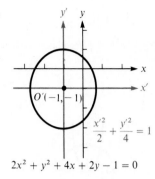

$\dfrac{x'^2}{2} + \dfrac{y'^2}{4} = 1$

$2x^2 + y^2 + 4x + 2y - 1 = 0$

21.

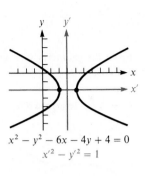

$x^2 - y^2 - 6x - 4y + 4 = 0$

$x'^2 - y'^2 = 1$

Exercise 9.7   (page 483)

1. $(3, \sqrt{3})$ **3.** $(\sqrt{3} - 3, 1 + 3\sqrt{3})$ **5.** $\left(\dfrac{7 - 2\sqrt{3}}{2}, \dfrac{7\sqrt{3} + 4}{2}\right)$ **7.** $60°$ **9.** $15°$

11.

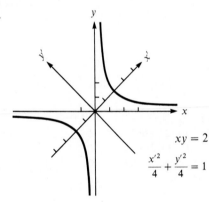

$xy = 2$

$\dfrac{x'^2}{4} + \dfrac{y'^2}{4} = 1$

13.

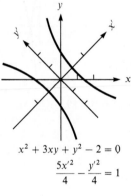

$x^2 + 3xy + y^2 - 2 = 0$

$\dfrac{5x'^2}{4} - \dfrac{y'^2}{4} = 1$

15.

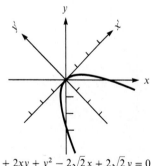

$x^2 + 2xy + y^2 - 2\sqrt{2}\,x + 2\sqrt{2}\,y = 0$

$x'^2 + 2y' = 0$

17.

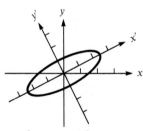

$2x^2 - 4xy + 5y^2 - 6 = 0$

$x'^2 + 6y'^2 = 6$

19.

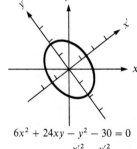

$6x^2 + 24xy - y^2 - 30 = 0$

$\dfrac{x'^2}{2} - \dfrac{y'^2}{3} = 1$

REVIEW EXERCISES (page 484)

1. $x^2 + y^2 = 50$ **2.** $x^2 + y^2 = 100$ **3.** $(x - 5)^2 + (y - 10)^2 = 85$ **4.** $(x - 2)^2 + (y - 2)^2 = 89$

5.

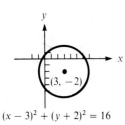

$(x - 3)^2 + (y + 2)^2 = 16$

6.

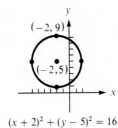

$(x + 2)^2 + (y - 5)^2 = 16$

7. $y^2 = -2x$ **8.** $x^2 = 16y$

9. $\left(x + \dfrac{b}{2a}\right)^2 = \dfrac{1}{a}\left(y + \dfrac{b^2 - 4ac}{4a}\right)$

10. $(x + 2)^2 = -\frac{4}{11}(y - 3)$

11.

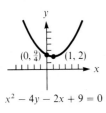

$x^2 - 4y - 2x + 9 = 0$

12.

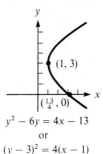

$y^2 - 6y = 4x - 13$
or
$(y - 3)^2 = 4(x - 1)$

13.

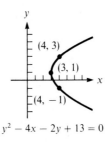

$y^2 - 4x - 2y + 13 = 0$

14. $\dfrac{x^2}{36} + \dfrac{y^2}{16} = 1$ **15.** $\dfrac{(x + 2)^2}{16} + \dfrac{(y - 3)^2}{9} = 1$ **16.**

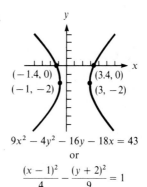

$4x^2 + y^2 - 16x + 2y = -13$
or
$\dfrac{(x - 2)^2}{1} + \dfrac{(y + 1)^2}{4} = 1$

17.

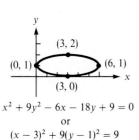

$x^2 + 9y^2 - 6x - 18y + 9 = 0$
or
$(x - 3)^2 + 9(y - 1)^2 = 9$

18. $\dfrac{x^2}{9} - \dfrac{(y - 3)^2}{16} = 1$ **19.**

$9x^2 - 4y^2 - 16y - 18x = 43$
or
$\dfrac{(x - 1)^2}{4} - \dfrac{(y + 2)^2}{9} = 1$

20.

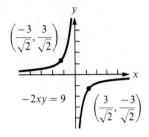

$\left(\dfrac{-3}{\sqrt{2}}, \dfrac{3}{\sqrt{2}}\right)$

$-2xy = 9$

$\left(\dfrac{3}{\sqrt{2}}, \dfrac{-3}{\sqrt{2}}\right)$

21.

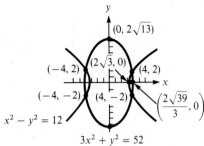

$(0, 2\sqrt{13})$

$(-4, 2)$ $(2\sqrt{3}, 0)$ $(4, 2)$

$(-4, -2)$ $(4, -2)$

$\left(\dfrac{2\sqrt{39}}{3}, 0\right)$

$x^2 - y^2 = 12$

$3x^2 + y^2 = 52$

22. $(4, 2), (4, -2), (-4, 2), (-4, -2)$

23.

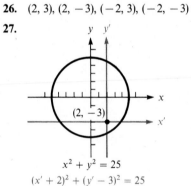

$x^2 + y^2 = 16$

$(0, 4)$

$(4, 0)$

$(-4, 0)$

$(3.46, -2)$

$(0, -4)$ $-\sqrt{3}y + 4\sqrt{3} = 3x$

24. $(0, 4), (2\sqrt{3}, -2)$

25.

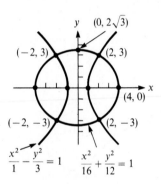

$(0, 2\sqrt{3})$

$(-2, 3)$ $(2, 3)$

$(4, 0)$

$(-2, -3)$ $(2, -3)$

$\dfrac{x^2}{1} - \dfrac{y^2}{3} = 1$ $\dfrac{x^2}{16} + \dfrac{y^2}{12} = 1$

26. $(2, 3), (2, -3), (-2, 3), (-2, -3)$

27.

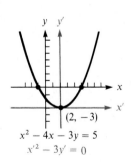

$(2, -3)$

$x^2 + y^2 = 25$

$(x' + 2)^2 + (y' - 3)^2 = 25$

28.

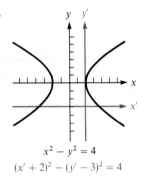

$x^2 - y^2 = 4$

$(x' + 2)^2 - (y' - 3)^2 = 4$

29.

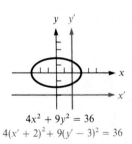

$4x^2 + 9y^2 = 36$

$4(x' + 2)^2 + 9(y' - 3)^2 = 36$

30.

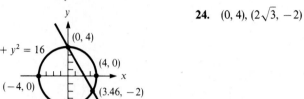

$(2, -3)$

$x^2 - 4x - 3y = 5$

$x'^2 - 3y' = 0$

31.

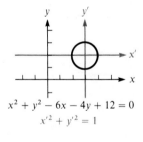

$x^2 + y^2 - 6x - 4y + 12 = 0$

$x'^2 + y'^2 = 1$

32.

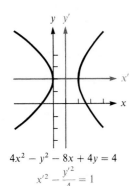

$$4x^2 - y^2 - 8x + 4y = 4$$

$$x'^2 - \frac{y'^2}{4} = 1$$

33.

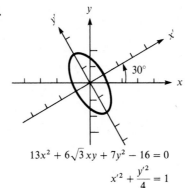

30°

$$13x^2 + 6\sqrt{3}\,xy + 7y^2 - 16 = 0$$

$$x'^2 + \frac{y'^2}{4} = 1$$

Exercise 10.1   (page 492)

1. i **3.** -1 **5.** $-i$ **7.** i **9.** 1 **11.** i **13.** $x = -\frac{1}{2}; y = -\frac{1}{2}$ **15.** $x = 0; y = 0$

17. $x = \frac{2}{3}; y = -\frac{2}{9}$ **19.** $5 - 6i$ **21.** $-2 - 10i$ **23.** $4 + 10i$ **25.** $6 - 17i$ **27.** $52 + 56i$

29. $-6 + 17i$ **31.** $-5 + 12i$ **33.** $2 - 11i$ **35.** $\frac{2}{5} - \frac{1}{5}i$ **37.** $-\frac{3}{4} + 0i$ **39.** $\frac{1}{25} + \frac{7}{25}i$

41. $\frac{1}{2} + \frac{1}{2}i$ **43.** $\dfrac{-2 - 2\sqrt{5}}{17} + \dfrac{16 - \sqrt{5}}{34}i$ **45.** $\dfrac{6 + \sqrt{3}}{10} + \dfrac{3\sqrt{3} - 2}{10}i$

47. $(1 - i)(1 - i) = 1 - i - i - 1 = -2i$ **49.** $(1 + 2i)(1 + 2i) = 1 + 2i + 2i - 4 = -3 + 4i$

53. $-1 + i, -1 - i$ **55.** $-2 + i, -2 - i$ **57.** $\frac{1}{3} + \frac{1}{3}i, \frac{1}{3} - \frac{1}{3}i$ **59.** $34.87 + 32.69i$

61. approximately $-6.92 + 9.26i$

Exercise 10.2   (page 495)

1. imaginaries

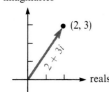

(2, 3)

$2 + 3i$

reals

3. imaginaries

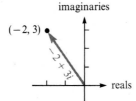

$(-2, 3)$

$-2 + 3i$

reals

5. imaginaries

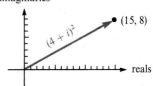

(15, 8)

$(4 + i)^2$

reals

7. imaginaries

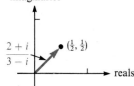

$\dfrac{2 + i}{3 - i}$

$(\frac{1}{2}, \frac{1}{2})$

reals

9. imaginaries

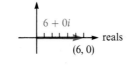

$6 + 0i$

reals

(6, 0)

11. imaginaries

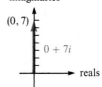

(0, 7)

$0 + 7i$

reals

13. $\sqrt{13}$ **15.** $7\sqrt{2}$ **17.** 6 **19.** 5 **21.** $\frac{3\sqrt{5}}{5}$ **23.** 1

Exercise 10.3   (page 499)

1. $6(\cos 0° + i \sin 0°)$ **3.** $3(\cos 270° + i \sin 270°)$ **5.** $\sqrt{2}(\cos 225° + i \sin 225°)$

7. $6(\cos 60° + i \sin 60°)$ **9.** $2(\cos 240° + i \sin 240°)$ **11.** $2(\cos 210° + i \sin 210°)$ **13.** $\sqrt{3} + i$

15. $0 + 7i$ **17.** $1 - \sqrt{3}\,i$ **19.** $-\frac{1}{2} + 0i$ **21.** $\dfrac{3\sqrt{2}}{2} + \dfrac{3\sqrt{2}}{2}\,i$ **23.** $\dfrac{11\sqrt{3}}{2} - \dfrac{11}{2}\,i$

25. $8(\cos 90° + i \sin 90°)$ **27.** $\cos 300° + i \sin 300°$ **29.** $6(\cos 2\pi + i \sin 2\pi)$ **31.** $6(\cos \frac{\pi}{2} + i \sin \frac{\pi}{2})$

33. $30 \operatorname{cis} 116°$ **35.** $36 \operatorname{cis} \frac{13\pi}{12}$ **37.** $6(\cos 30° + i \sin 30°)$ **39.** $\frac{3}{2}(\cos \frac{\pi}{2} + i \sin \frac{\pi}{2})$ **41.** $\frac{12}{5} \operatorname{cis} 130°$

43. $\frac{1}{2} \operatorname{cis} \frac{\pi}{2}$ **45.** $\operatorname{cis} 40°$ **47.** $4 \operatorname{cis} \frac{5\pi}{6}$

Exercise 10.4   (page 505)

1. $27(\cos 90° + i \sin 90°)$ **3.** $\cos 180° + i \sin 180°$ **5.** $3125 \operatorname{cis} 10°$ **7.** $81(\cos \pi + i \sin \pi)$

9. $256(\cos 12 + i \sin 12)$ **11.** $\frac{1}{27} \operatorname{cis} \frac{3\pi}{2}$ **13.** $1 + \sqrt{3}\,i$ **15.** $\frac{1}{2} + \frac{\sqrt{3}}{2}i$ **17.** $1 + \sqrt{3}\,i$

19. $1 + \sqrt{3}\,i, -2, 1 - \sqrt{3}\,i$ **21.** $\frac{\sqrt{2}}{2} + \frac{\sqrt{2}}{2}i, -\frac{\sqrt{2}}{2} - \frac{\sqrt{2}}{2}i$

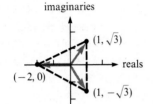

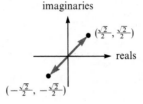

23. $\cos 54° + i \sin 54°, \cos 126° + i \sin 126°, \cos 198° + i \sin 198°, \cos 270° + i \sin 270°, \cos 342° + i \sin 342°$

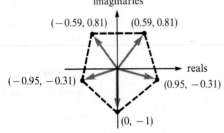

25. $\sqrt{3} + i, -1 + i\sqrt{3}, -\sqrt{3} - i, 1 - i\sqrt{3}$

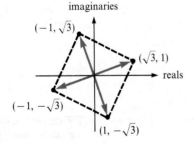

Exercise 10.5   (page 512)

1. $(\sqrt{3}, 1)$ **3.** $(\frac{7}{2}, -\frac{7\sqrt{3}}{2})$ **5.** $(\frac{-3}{2}, \frac{-3\sqrt{3}}{2})$ **7.** $(0, 2)$ **9.** $(-\sqrt{3}, -1)$ **11.** $(\frac{-5\sqrt{2}}{2}, \frac{-5\sqrt{2}}{2})$

13. $(\sqrt{3}, -1)$ **15.** $(0, 10)$ **17.** $(0, 0)$ **19.** $(-3\sqrt{3}, 3)$ **21.** $(\sqrt{2}, 45°)$ **23.** $(4, 330°)$

25. $(2, 210°)$ **27.** $(2, 150°)$ **29.** $(0, \theta°)$ **31.** $(5, 180°)$ **33.** $(3\sqrt{2}, 315°)$ **35.** $(14, 60°)$

37. $r \cos \theta = 3$ **39.** $r(3 \cos \theta + 2 \sin \theta) = 3$ **41.** $r = 9 \cos \theta$ **43.** $r^2 = 4 \cos^2 \theta \sin^2 \theta$

45. $\theta = \frac{\pi}{2}$ or $r = \sec \theta$ **47.** $r^2 \cos^2 \theta - 2r \sin \theta = 1$ **49.** $x^2 + y^2 = 9$ **51.** $x = 5$

53. $\sqrt{x^2 + y^2} + y = 1$ **55.** $(x^2 + y^2)^2 = 2xy$ **57.** $y = 0$ **59.** $2\sqrt{x^2 + y^2} - x = 2$

61.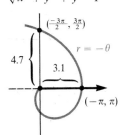

63.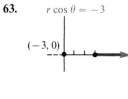

65.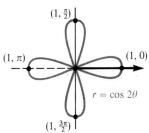

67.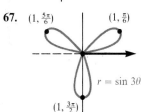

69.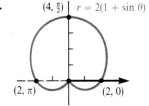

71.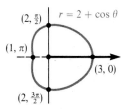

73.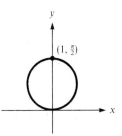

Exercise 10.6   (page 519)

1.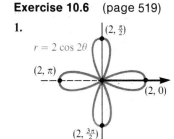

3.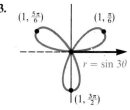

5.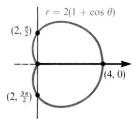

7.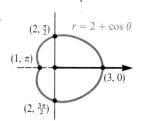

9.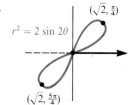

11.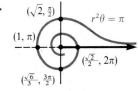

13.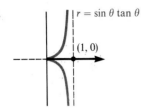

15.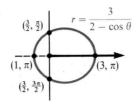

17.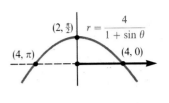

REVIEW EXERCISES (page 520)

1. $-i$ **2.** $-i$ **3.** $-i$ **4.** $\frac{1}{i^{1812}} = 1$ **5.** $(\frac{4}{3}, \frac{2}{3})$ **6.** $(-11, -18)$ **7.** $-4 + i$ **8.** $-5 + i$

9. $-5 - 5i$ **10.** $21 + i$ **11.** $0 - \frac{1}{5}i$ **12.** $0 + \frac{13}{6}i$ **13.** $\frac{8}{17} - \frac{2}{17}i$ **14.** $-\frac{3}{2} - \frac{1}{2}i$ **15.** $0 + i$

16. $\frac{2}{5} - \frac{1}{5}i$ **17.** $\frac{2 - 3\sqrt{2}}{3} + \frac{3 + 2\sqrt{2}}{3}i$ **18.** $\frac{3 + \sqrt{3}}{4} + \frac{3\sqrt{3} - 1}{4}i$ **19.** imaginaries

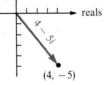

20.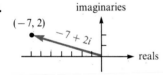

21.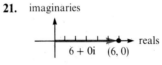

22.

23. $\sqrt{73}$ **24.** $10\sqrt{2}$ **25.** $\frac{3\sqrt{10}}{10}$ **26.** 1 **27.** $2\sqrt{2}(\cos 135° + i \sin 135°)$

28. $5\sqrt{2}(\cos 315° + i \sin 315°)$ **29.** $6(\cos 60° + i \sin 60°)$ **30.** $4(\cos 0° + i \sin 0°)$ **31.** $\frac{3}{2} + \frac{3\sqrt{3}}{2}i$

32. $\sqrt{3} - i$ **33.** $-\frac{3}{2} - \frac{3\sqrt{3}}{2}i$ **34.** $-\frac{7\sqrt{3}}{2} + \frac{7}{2}i$ **35.** cis 110°   **36.** 6 cis 570°   **37.** $6(\cos \frac{\pi}{4} + i \sin \frac{\pi}{4})$

38. $21(\cos \pi + i \sin \pi)$ **39.** 2 cis 50°   **40.** $\frac{2}{3}(\cos 10° + i \sin 10°)$ **41.** $\cos 20° + i \sin 20°$

42. $\sqrt[4]{14}(\cos \frac{45°}{4} + i \sin \frac{45°}{4})$ **43.** $5, \frac{-5 + 5\sqrt{3}i}{2}, \frac{-5 - 5\sqrt{3}i}{2}$ **44.** $3, 3i, -3, -3i$ **45.** $(\frac{5}{2}, \frac{5\sqrt{3}}{2})$

46. $(-\sqrt{3}, -1)$ **47.** $(\frac{\sqrt{3}}{2}, \frac{1}{2})$ **48.** $(-5\sqrt{2}, 5\sqrt{2})$ **49.** $(2, 135°)$ **50.** $(2, 150°)$ **51.** $(1, 0°)$

52. $(2, 300°)$ **53.** $r^2 \cos \theta \sin \theta = 1$ **54.** $r(\cos \theta + 2 \sin \theta) = 2$ **55.** $r \cos^2 \theta = 3 \sin \theta$

56. $r^2 = 4 \cos \theta \sin \theta$ **57.** $(x^2 + y^2)^2 = 9x^2 - 9y^2$ **58.** $x^2 + y^2 = 5y$ **59.** $4\sqrt{x^2 + y^2} + y = 1$

60. $\sqrt{x^2 + y^2} - x = 2$ **61.**

62.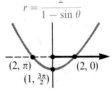

63.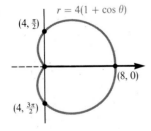

64.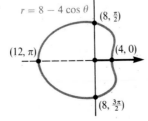

Exercise 11.1   (page 526)

1. $P(2) = 35$ **3.** $P(0) = -1$ **5.** $P(-4) = 719$ **7.** $P(2) = 31$ **9.** true   **11.** true   **13.** false
15. true   **17.** $3, -1, -5$ **19.** $x^3 - 3x^2 + 3x - 1$ **21.** $x^3 - 11x^2 + 38x - 40$ **23.** $x^4 - 3x^2 + 2$
25. $x^3 - \sqrt{2}x^2 + x - \sqrt{2}$ **27.** $x^3 - 2x^2 + 2x$ **29.** $1, \dfrac{-1 + \sqrt{3}i}{2}, \dfrac{-1 - \sqrt{3}i}{2}$
31. $-5, \dfrac{5 + 5\sqrt{3}i}{2}, \dfrac{5 - \sqrt{3}i}{2}$ **33.** $3, 2, i, -i$ **35.** $4, -1, 1, -2$ **37.** $a_1 = 0$

Exercise 11.2   (page 531)

1. $P(2) = 47$ **3.** $P(-5) = -569$ **5.** $P(0) = 1$ **7.** $P(-i) = -1 + 6i$ **9.** $P(1) = 3$
11. $P(-2) = 30$ **13.** $P(\frac{1}{2}) = \frac{15}{8}$ **15.** $P(i) = 5$ **17.** $P(1) = 0$ **19.** $P(-3) = 384$ **21.** $P(3) = 0$
23. $P(2) = 9$ **25.** $P(0) = 8$ **27.** $P(2) = 0$ **29.** $P(i) = 16 + 2i$ **31.** $P(-2i) = 40 - 40i$
33. $(x + 1)(3x^2 - 5x - 1) - 3$ **35.** $(x - 2)(3x^2 + 4x + 2) + 0$ **37.** $x(3x^2 - 2x - 6) - 4$
39. $7x^2 - 10x + 5 + \dfrac{-4}{x + 1}$ **41.** $4x^3 + 9x^2 + 27x + 80 + \dfrac{245}{x - 3}$ **43.** $3x^4 + 12x^3 + 48x^2 + 192x$
45. $3^5 = 243$ **47.** $2^7 = 128$

49.   **51.**   **53.**

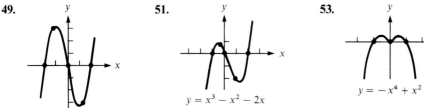

$y = x^3 - 4x$ $y = x^3 - x^2 - 2x$ $y = -x^4 + x^2$

55.

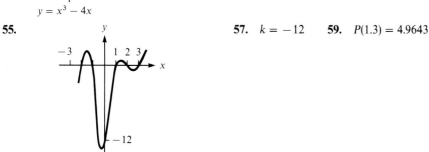

$y = x^5 - 3x^4 - 5x^3 + 15x^2 + 4x - 12$

57. $k = -12$ **59.** $P(1.3) = 4.9643$

Exercise 11.3   (page 537)

1. 10   **3.** 4   **5.** 0 or 2 positive; 1 negative; 0 or 2 nonreal   **7.** 0 positive; 1 or 3 negative;
0 or 2 nonreal   **9.** 0 positive; 0 negative; 4 nonreal   **11.** 1 positive; 1 negative; 2 nonreal
13. 0 positive; 0 negative; 10 nonreal   **15.** 0 positive, 0 negative, 8 nonreal, 1 root of 0
17. 1 positive, 1 negative, 2 nonreal   **19.** 2, 3   **21.** $-4, 6$ **23.** $-4, 3$ **25.** $-5, 2$
27. An odd-degree polynomial equation must have an odd number of roots. Since complex roots occur in
conjugate pairs, one root must be left over, and it is real.

Exercise 11.4   (page 543)

1. $-1, 1, 2$ **3.** $1, 2, 3, 4$ **5.** $-1, -1, 1, 1, 2$ **7.** $-5, 2, \sqrt{3}, -\sqrt{3}$ **9.** $0, -2, -2, -2, 2, 2, 2$
11. $-3, -\frac{1}{3}, \frac{1}{2}, 2, 2$ **13.** $-\frac{1}{2}, \frac{1}{2}, 3, 2i, -2i$ **15.** $-1, \frac{2}{3}, 2, 3$ **17.** $-1, \frac{1}{2}, 1, \frac{3}{2}$

19. $-\frac{3}{5}, \frac{2}{3}, \frac{3}{2}$ **21.** $2, -\frac{1}{2}+\frac{\sqrt{3}}{2}i, -\frac{1}{2}-\frac{\sqrt{3}}{2}i$

23. There can be no positive or negative roots. 0 is not a root. Hence, all roots are nonreal.

Exercise 11.5 (page 545)

1. $P(-2) = 3; P(-1) = -2$ **3.** $P(4) = -40; P(5) = 30$ **5.** $P(1) = 8; P(2) = -1$
7. $P(2) = -72; P(3) = 154$ **9.** $P(0) = 10; P(1) = -60$ **11.** $\sqrt{3} \approx 1.73$ **13.** $\sqrt{5} \approx 2.236$
15. $x \approx 0.50$

REVIEW EXERCISES (page 546)

1. $P(0) = -2$ **2.** $P(2) = 72$ **3.** $P(-3) = 277$ **4.** $P(\frac{1}{2}) = -3$ **5.** false   **6.** false

7. true   **8.** true   **9.** $-4, 2+2\sqrt{3}i, 2-2\sqrt{3}i$ **10.** $7, \dfrac{-7+7\sqrt{3}i}{2}, \dfrac{-7-7\sqrt{3}i}{2}$

11. $2x^3 - 5x^2 - x + 6$ **12.** $2x^3 + 3x^2 - 8x + 3$ **13.** $3x^3 + 9x^2 + 29x + 90$ with a remainder of 277
14. $5x^4 - 14x^3 + 31x^2 - 64x + 129$ with a remainder of -259 **15.** 6   **16.** 1984
17. 2 or 0 positive; 2 or 0 negative; 0, 2, or 4 nonreal   **18.** 1 positive; 0, 2, or 4 negative; 4, 2, or 0 nonreal
19. 0 positive; 0 negative; 4 nonreal   **20.** 0 positive; 1 negative; 6 nonreal   **21.** $-5, -\frac{3}{2}, -2$

22. $\dfrac{1}{3}, \dfrac{-1+\sqrt{3}i}{2}, \dfrac{-1-\sqrt{3}i}{2}$ **23.** $P(0) = 18; P(-1) = -9$ **24.** $P(1) = -8; P(2) = 21$

25. $\sqrt{7} \approx 2.65$ **26.** $0.33; \frac{1}{3}$

Exercise 12.1 (page 552)

25. no

Exercise 12.2 (page 557)

1. 0, 10, 30, 60, 100, 150, 210, 280   **3.** 21   **5.** $a + 4d$ **7.** 15   **9.** 15   **11.** 15   **13.** $\frac{242}{243}$
15. 35   **17.** 30   **19.** -50 **21.** 40   **23.** 500   **25.** $\frac{7}{12}$ **27.** 160

Exercise 12.3 (page 564)

1. 1, 3, 5, 7, 9, 11   **3.** $5, \frac{7}{2}, 2, \frac{1}{2}, -1, -\frac{5}{2}$ **5.** $9, \frac{23}{2}, 14, \frac{33}{2}, 19, \frac{43}{2}$ **7.** 285   **9.** 555   **11.** $157\frac{1}{2}$
13. 44   **15.** $\frac{25}{2}, 15, \frac{35}{2}$ **17.** $-\frac{82}{15}, -\frac{59}{15}, -\frac{36}{15}, -\frac{13}{15}$ **19.** 10, 20, 40, 80   **21.** $-2, -6, -18, -54$
23. $3, 3\sqrt{2}, 6, 6\sqrt{2}$ **25.** 2, 6, 18, 54   **27.** 124   **29.** $-29,524$ **31.** $\frac{1995}{32}$ **33.** 18   **35.** 8
37. $10\sqrt[4]{2}, 10\sqrt{2}, 10\sqrt[4]{8}$ **39.** 8, 32, 128, 512   **41.** $\frac{5}{9}$ **43.** $\frac{25}{99}$ **47.** $\$21,474,836.47$
49. Arithmetic mean is $\frac{11}{16}$; geometric mean is $\sqrt{7}/4$; the arithmetic mean is larger.   **51.** $1 - \frac{1}{101} = \frac{100}{101}$

Exercise 12.4 (page 568)

1. 23   **3.** 5.13 meters   **5.** 1,048,576   **7.** She will earn $0.19 more on the $7\frac{1}{4}\%$ investment.
9. approximately 6.87×10^{19} **11.** $\$180,176.87$ **13.** $\$2001.60$ **15.** $\$2013.62$ **17.** $\$264,094.58$

19. 9.2234×10^{18} grains   **21.** No, because $0.999999 = \dfrac{999999}{1000000}$, which does not equal 1.

Exercise 12.5 (page 574)

1. 24   **3.** 720   **5.** 1440   **7.** $\frac{1}{1320}$ **9.** 18,564   **11.** $a^4 + 4a^3b + 6a^2b^2 + 4ab^3 + b^4$
13. $a^5 - 5a^4b + 10a^3b^2 - 10a^2b^3 + 5ab^4 - b^5$ **15.** $8x^3 - 12x^2y + 6xy^2 - y^3$

17. $16x^4 + 32x^3y + 24x^2y^2 + 8xy^3 + y^4$ **19.** $256x^4 - 768x^3y + 864x^2y^2 - 432xy^3 + 81y^4$
21. $36x^2 - 36xy + 9y^2$ **23.** $6a^2b^2$ **25.** $35a^3b^4$ **27.** $-b^5$ **29.** $28x^3y^6$ **31.** $20\sqrt{2}x^3y^2$
33. $4608x^2y^7$ **35.** $\dfrac{r!}{3!(r-3)!}a^{r-3}b^3$ **37.** $\dfrac{n!}{(r-1)!(n-r+1)!}a^{n-r+1}b^{r-1}$

39. 1, 2, 4, 8, 16, 32, 64, 128, 256, 512. They are sequential powers of 2.

41. The sixth term is $\dfrac{10!}{6!(10-6)!}a^5\left(\dfrac{1}{a}\right)^5$, or 210.

43. $(x^{\frac{1}{2}})^{25-r}\left(\dfrac{1}{x}\right)^r$ must be x^8. Hence, $r = 3$. Coefficient is $\dfrac{25!}{3!(25-3)!}\cdot\dfrac{1}{2^3} = \dfrac{575}{2}$.

Exercise 12.6   (page 581)

1. 144   **3.** 8,000,000   **5.** 240   **7.** 6   **9.** 840   **11.** 35   **13.** 120   **15.** 5   **17.** 1
19. 1200   **21.** 40   **23.** 2278   **25.** 40,320   **27.** 14,400   **29.** 24,360   **31.** 5040   **33.** 48
35. 144   **37.** 210   **39.** approximately 6.35×10^{11} **41.** 3360   **43.** 2,721,600   **45.** 1120
47. 59,400   **49.** 272   **51.** 28   **53.** 56   **55.** 2,042,975 (ignoring the order of the players)

Exercise 12.7   (page 586)

1. $\frac{1}{6}$ **3.** $\frac{2}{3}$ **5.** $\frac{19}{42}$ **7.** $\frac{13}{42}$ **9.** $\frac{3}{8}$ **11.** 0   **13.** $\frac{1}{12}$ **15.** $\frac{1}{169}$ **17.** $\frac{5}{12}$
19. approximately 6.3×10^{-12} **21.** 0   **23.** $\frac{3}{13}$ **25.** $\frac{1}{6}$ **27.** $\frac{1}{8}$ **29.** $\frac{5}{16}$
31. (S-survive; F-fail) SSSS, SSSF, SSFS, SFSS, FSSS, SSFF, SFSF, FSSF, SFFS, FSFS, FFSS, SFFF, FSFF, FFSF, FFFS, FFFF   **33.** $\frac{1}{4}$ **35.** $\frac{1}{4}$
37. 1; Exercises 32 through 36 exhaust all possibilities and are mutually exclusive.   **39.** $\frac{32}{119}$

Exercise 12.8   (page 591)

1. $\frac{1}{2}$ **3.** $\frac{7}{13}$ **5.** $\frac{1}{221}$ **7.** $\frac{25}{204}$ **9.** $\frac{11}{36}$ **11.** $\frac{7}{12}$ **13.** $\frac{5}{8}$ **15.** $\frac{13}{16}$ **17.** $\frac{21}{128}$ **19.** $\frac{3}{20}$
21. $\frac{1}{49}$ **23.** 0.973   **25.** $\frac{11}{20}$ **27.** $\frac{1}{8}, \frac{1}{12}, \frac{3}{8}, \frac{5}{12}, 1$

Exercise 12.9   (page 594)

1. $\frac{1}{6}$ **3.** 5 to 1   **5.** 1 to 1   **7.** $\frac{5}{36}$ **9.** 31 to 5   **11.** 1 to 1   **13.** 1 to 12   **15.** 3 to 10
17. $\frac{5}{7}$ **19.** 1 to 90   **21.** 15 to 1   **23.** $\frac{1}{9}$ **25.** No, the expected winnings are $\$\frac{9}{13}$.   **27.** 7 to 1
29. $1.72   **31.** 6.54   **33.** $\frac{32}{390,625}$

REVIEW EXERCISES   (page 595)

2. 90   **3.** 1718   **4.** -360 **5.** 117   **6.** 281   **7.** -92 **8.** -67.5 **9.** $\dfrac{1}{3^6}$

10. 13,122   **11.** $\dfrac{9}{2^{14}}$ **12.** $\dfrac{8}{15,625}$ **13.** 3320   **14.** 5780   **15.** -5220 **16.** -1540

17. $\dfrac{3280}{27}$ **18.** 6560   **19.** $\dfrac{2295}{128}$ **20.** $\dfrac{520,832}{78,125}$ **21.** $\frac{2}{3}$ **22.** $\frac{3}{25}$ **23.** no sum   **24.** 1

25. $\frac{1}{3}$ **26.** 1   **27.** $\frac{17}{99}$ **28.** $\frac{5}{11}$ **29.** $\frac{7}{2}, 5, \frac{13}{2}$ **30.** 25, 40, 55, 70, 85   **31.** $2\sqrt{2}, 4, 4\sqrt{2}$

32. $-2\sqrt[5]{-5}, -2\sqrt[5]{25}, -2\sqrt[5]{-125}, -2\sqrt[5]{625}$ **33.** $\dfrac{3280}{3}$ **34.** $16\sqrt{2}$ **35.** 16   **36.** $4775.80

37. 6516, 3134   **38.** $3487   **39.** $x^3 - 3x^2y + 3xy^2 - y^3$ **40.** $u^3 + 6u^2v + 12uv^2 + 8v^3$
41. $1024a^5 - 6400a^4b + 16000a^3b^2 - 20000a^2b^3 + 12500ab^4 - 3125b^5$

42. $49r^4 + 28\sqrt{21}r^3s + 126r^2s^2 + 12\sqrt{21}\,rs^3 + 9s^4$ **43.** $56a^5b^3$ **44.** $80x^3y^2$ **45.** $84x^3y^6$

46. $439,040x^3$ **47.** 720   **48.** 35   **49.** 1   **50.** 4050   **51.** 564,480   **52.** 840   **53.** 21

54. 1   **55.** 120   **56.** $\dfrac{56}{1287}$ **57.** $\frac{1}{6}$ **58.** $\dfrac{429}{866,320}$ **62.** 24   **63.** 0.00000923

64. $\dfrac{2,598,936}{2,598,960} = \dfrac{108,289}{108,290}$ **65.** 20,160   **66.** 90,720   **67.** approximately 6.30×10^{-12} **68.** 0.4196

69. $\frac{1}{2}$ **70.** $\frac{7}{13}$ **71.** $\dfrac{1}{2,598,960}$ **72.** approximately 6.2992×10^{-12} **73.** $\frac{15}{16}$ **74.** 1 to 7

75. $\frac{1}{15}$ **76.** \$2.00   **77.** 23 to 1   **78.** $\frac{11}{21}$ **79.** 664

TABLES

Table A Values of the Trigonometric Functions

Radians	Degrees	Sin	Cos	Tan	Cot		
.0000	.0°	.0000	1.0000	.0000	—	90.0°	1.5708
.0017	.1°	.0017	1.0000	.0017	573.0	89.9°	1.5691
.0035	.2°	.0035	1.0000	.0035	286.5	89.8°	1.5673
.0052	.3°	.0052	1.0000	.0052	191.0	89.7°	1.5656
.0070	.4°	.0070	1.0000	.0070	143.2	89.6°	1.5638
.0087	.5°	.0087	1.0000	.0087	114.6	89.5°	1.5621
.0105	.6°	.0105	.9999	.0105	95.49	89.4°	1.5603
.0122	.7°	.0122	.9999	.0122	81.85	89.3°	1.5586
.0140	.8°	.0140	.9999	.0140	71.62	89.2°	1.5568
.0157	.9°	.0157	.9999	.0157	63.66	89.1°	1.5551
.0175	1.0°	.0175	.9998	.0175	57.29	89.0°	1.5533
.0192	1.1°	.0192	.9998	.0192	52.08	88.9°	1.5516
.0209	1.2°	.0209	.9998	.0209	47.74	88.8°	1.5499
.0227	1.3°	.0227	.9997	.0227	44.07	88.7°	1.5481
.0244	1.4°	.0244	.9997	.0244	40.92	88.6°	1.5464
.0262	1.5°	.0262	.9997	.0262	38.19	88.5°	1.5446
.0279	1.6°	.0279	.9996	.0279	35.80	88.4°	1.5429
.0297	1.7°	.0297	.9996	.0297	33.69	88.3°	1.5411
.0314	1.8°	.0314	.9995	.0314	31.82	88.2°	1.5394
.0332	1.9°	.0332	.9995	.0332	30.14	88.1°	1.5376
.0349	2.0°	.0349	.9994	.0349	28.64	88.0°	1.5359
.0367	2.1°	.0366	.9993	.0367	27.27	87.9°	1.5341
.0384	2.2°	.0384	.9993	.0384	26.03	87.8°	1.5324
.0401	2.3°	.0401	.9992	.0402	24.90	87.7°	1.5307
.0419	2.4°	.0419	.9991	.0419	23.86	87.6°	1.5289
.0436	2.5°	.0436	.9990	.0437	22.90	87.5°	1.5272
.0454	2.6°	.0454	.9990	.0454	22.02	87.4°	1.5254
.0471	2.7°	.0471	.9989	.0472	21.20	87.3°	1.5237
.0489	2.8°	.0488	.9988	.0489	20.45	87.2°	1.5219
.0506	2.9°	.0506	.9987	.0507	19.74	87.1°	1.5202
.0524	3.0°	.0523	.9986	.0524	19.08	87.0°	1.5184
.0541	3.1°	.0541	.9985	.0542	18.46	86.9°	1.5167
.0559	3.2°	.0558	.9984	.0559	17.89	86.8°	1.5149
.0576	3.3°	.0576	.9983	.0577	17.34	86.7°	1.5132
.0593	3.4°	.0593	.9982	.0594	16.83	86.6°	1.5115
.0611	3.5°	.0610	.9981	.0612	16.35	86.5°	1.5097
.0628	3.6°	.0628	.9980	.0629	15.89	86.4°	1.5080
.0646	3.7°	.0645	.9979	.0647	15.46	86.3°	1.5062
.0663	3.8°	.0663	.9978	.0664	15.06	86.2°	1.5045
.0681	3.9°	.0680	.9977	.0682	14.67	86.1°	1.5027
.0698	4.0°	.0698	.9976	.0699	14.30	86.0°	1.5010
.0716	4.1°	.0715	.9974	.0717	13.95	85.9°	1.4992
.0733	4.2°	.0732	.9973	.0734	13.62	85.8°	1.4975
.0750	4.3°	.0750	.9972	.0752	13.30	85.7°	1.4957
.0768	4.4°	.0767	.9971	.0769	13.00	85.6°	1.4940
.0785	4.5°	.0785	.9969	.0787	12.71	85.5°	1.4923
.0803	4.6°	.0802	.9968	.0805	12.43	85.4°	1.4905
.0820	4.7°	.0819	.9966	.0822	12.16	85.3°	1.4888
.0838	4.8°	.0837	.9965	.0840	11.91	85.2°	1.4870
.0855	4.9°	.0854	.9963	.0857	11.66	85.1°	1.4853
		Cos	Sin	Cot	Tan	Degrees	Radians

Table A *(Continued)*

Radians	Degrees	Sin	Cos	Tan	Cot		
.0873	5.0°	.0872	.9962	.0875	11.43	85.0°	1.4835
.0890	5.1°	.0889	.9960	.0892	11.20	84.9°	1.4818
.0908	5.2°	.0906	.9959	.0910	10.99	84.8°	1.4800
.0925	5.3°	.0924	.9957	.0928	10.78	84.7°	1.4783
.0942	5.4°	.0941	.9956	.0945	10.58	84.6°	1.4765
.0960	5.5°	.0958	.9954	.0963	10.39	84.5°	1.4748
.0977	5.6°	.0976	.9952	.0981	10.20	84.4°	1.4731
.0995	5.7°	.0993	.9951	.0998	10.02	84.3°	1.4713
.1012	5.8°	.1011	.9949	.1016	9.845	84.2°	1.4696
.1030	5.9°	.1028	.9947	.1033	9.677	84.1°	1.4678
.1047	6.0°	.1045	.9945	.1051	9.514	84.0°	1.4661
.1065	6.1°	.1063	.9943	.1069	9.357	83.9°	1.4643
.1082	6.2°	.1080	.9942	.1086	9.205	83.8°	1.4626
.1100	6.3°	.1097	.9940	.1104	9.058	83.7°	1.4608
.1117	6.4°	.1115	.9938	.1122	8.915	83.6°	1.4591
.1134	6.5°	.1132	.9936	.1139	8.777	83.5°	1.4573
.1152	6.6°	.1149	.9934	.1157	8.643	83.4°	1.4556
.1169	6.7°	.1167	.9932	.1175	8.513	83.3°	1.4539
.1187	6.8°	.1184	.9930	.1192	8.386	83.2°	1.4521
.1204	6.9°	.1201	.9928	.1210	8.264	83.1°	1.4504
.1222	7.0°	.1219	.9925	.1228	8.144	83.0°	1.4486
.1239	7.1°	.1236	.9923	.1246	8.028	82.9°	1.4469
.1257	7.2°	.1253	.9921	.1263	7.916	82.8°	1.4451
.1274	7.3°	.1271	.9919	.1281	7.806	82.7°	1.4434
.1292	7.4°	.1288	.9917	.1299	7.700	82.6°	1.4416
.1309	7.5°	.1305	.9914	.1317	7.596	82.5°	1.4399
.1326	7.6°	.1323	.9912	.1334	7.495	82.4°	1.4382
.1344	7.7°	.1340	.9910	.1352	7.396	82.3°	1.4364
.1361	7.8°	.1357	.9907	.1370	7.300	82.2°	1.4347
.1379	7.9°	.1374	.9905	.1388	7.207	82.1°	1.4329
.1396	8.0°	.1392	.9903	.1405	7.115	82.0°	1.4312
.1414	8.1°	.1409	.9900	.1423	7.026	81.9°	1.4294
.1431	8.2°	.1426	.9898	.1441	6.940	81.8°	1.4277
.1449	8.3°	.1444	.9895	.1459	6.855	81.7°	1.4259
.1466	8.4°	.1461	.9893	.1477	6.772	81.6°	1.4242
.1484	8.5°	.1478	.9890	.1495	6.691	81.5°	1.4224
.1501	8.6°	.1495	.9888	.1512	6.612	81.4°	1.4207
.1518	8.7°	.1513	.9885	.1530	6.535	81.3°	1.4190
.1536	8.8°	.1530	.9882	.1548	6.460	81.2°	1.4172
.1553	8.9°	.1547	.9880	.1566	6.386	81.1°	1.4155
.1571	9.0°	.1564	.9877	.1584	6.314	81.0°	1.4137
.1588	9.1°	.1582	.9874	.1602	6.243	80.9°	1.4120
.1606	9.2°	.1599	.9871	.1620	6.174	80.8°	1.4102
.1623	9.3°	.1616	.9869	.1638	6.107	80.7°	1.4085
.1641	9.4°	.1633	.9866	.1655	6.041	80.6°	1.4067
.1658	9.5°	.1650	.9863	.1673	5.976	80.5°	1.4050
.1676	9.6°	.1668	.9860	.1691	5.912	80.4°	1.4032
.1693	9.7°	.1685	.9857	.1709	5.850	80.3°	1.4015
.1710	9.8°	.1702	.9854	.1727	5.789	80.2°	1.3998
.1728	9.9°	.1719	.9851	.1745	5.730	80.1°	1.3980
		Cos	Sin	Cot	Tan	Degrees	Radians

Table A *(Continued)*

Radians	Degrees	Sin	Cos	Tan	Cot		
.1745	10.0°	.1736	.9848	.1763	5.671	80.0°	1.3963
.1763	10.1°	.1754	.9845	.1781	5.614	79.9°	1.3945
.1780	10.2°	.1771	.9842	.1799	5.558	79.8°	1.3928
.1798	10.3°	.1788	.9839	.1817	5.503	79.7°	1.3910
.1815	10.4°	.1805	.9836	.1835	5.449	79.6°	1.3893
.1833	10.5°	.1822	.9833	.1853	5.396	79.5°	1.3875
.1850	10.6°	.1840	.9829	.1871	5.343	79.4°	1.3858
.1868	10.7°	.1857	.9826	.1890	5.292	79.3°	1.3840
.1885	10.8°	.1874	.9823	.1908	5.242	79.2°	1.3823
.1902	10.9°	.1891	.9820	.1926	5.193	79.1°	1.3806
.1920	11.0°	.1908	.9816	.1944	5.145	79.0°	1.3788
.1937	11.1°	.1925	.9813	.1962	5.097	78.9°	1.3771
.1955	11.2°	.1942	.9810	.1980	5.050	78.8°	1.3753
.1972	11.3°	.1959	.9806	.1998	5.005	78.7°	1.3736
.1990	11.4°	.1977	.9803	.2016	4.959	78.6°	1.3718
.2007	11.5°	.1994	.9799	.2035	4.915	78.5°	1.3701
.2025	11.6°	.2011	.9796	.2053	4.872	78.4°	1.3683
.2042	11.7°	.2028	.9792	.2071	4.829	78.3°	1.3666
.2059	11.8°	.2045	.9789	.2089	4.787	78.2°	1.3648
.2077	11.9°	.2062	.9785	.2107	4.745	78.1°	1.3631
.2094	12.0°	.2079	.9781	.2126	4.705	78.0°	1.3614
.2112	12.1°	.2096	.9778	.2144	4.665	77.9°	1.3596
.2129	12.2°	.2113	.9774	.2162	4.625	77.8°	1.3579
.2147	12.3°	.2130	.9770	.2180	4.586	77.7°	1.3561
.2164	12.4°	.2147	.9767	.2199	4.548	77.6°	1.3544
.2182	12.5°	.2164	.9763	.2217	4.511	77.5°	1.3526
.2199	12.6°	.2181	.9759	.2235	4.474	77.4°	1.3509
.2217	12.7°	.2198	.9755	.2254	4.437	77.3°	1.3491
.2234	12.8°	.2215	.9751	.2272	4.402	77.2°	1.3474
.2251	12.9°	.2233	.9748	.2290	4.366	77.1°	1.3456
.2269	13.0°	.2250	.9744	.2309	4.331	77.0°	1.3439
.2286	13.1°	.2267	.9740	.2327	4.297	76.9°	1.3422
.2304	13.2°	.2284	.9736	.2345	4.264	76.8°	1.3404
.2321	13.3°	.2300	.9732	.2364	4.230	76.7°	1.3387
.2339	13.4°	.2317	.9728	.2382	4.198	76.6°	1.3369
.2356	13.5°	.2334	.9724	.2401	4.165	76.5°	1.3352
.2374	13.6°	.2351	.9720	.2419	4.134	76.4°	1.3334
.2391	13.7°	.2368	.9715	.2438	4.102	76.3°	1.3317
.2409	13.8°	.2385	.9711	.2456	4.071	76.2°	1.3299
.2426	13.9°	.2402	.9707	.2475	4.041	76.1°	1.3282
.2443	14.0°	.2419	.9703	.2493	4.011	76.0°	1.3265
.2461	14.1°	.2436	.9699	.2512	3.981	75.9°	1.3247
.2478	14.2°	.2453	.9694	.2530	3.952	75.8°	1.3230
.2496	14.3°	.2470	.9690	.2549	3.923	75.7°	1.3212
.2513	14.4°	.2487	.9686	.2568	3.895	75.6°	1.3195
.2531	14.5°	.2504	.9681	.2586	3.867	75.5°	1.3177
.2548	14.6°	.2521	.9677	.2605	3.839	75.4°	1.3160
.2566	14.7°	.2538	.9673	.2623	3.812	75.3°	1.3142
.2583	14.8°	.2554	.9668	.2642	3.785	75.2°	1.3125
.2601	14.9°	.2571	.9664	.2661	3.758	75.1°	1.3107
		Cos	**Sin**	**Cot**	**Tan**	**Degrees**	**Radians**

Table A *(Continued)*

Radians	Degrees	Sin	Cos	Tan	Cot		
.2618	15.0°	.2588	.9659	.2679	3.732	75.0°	1.3090
.2635	15.1°	.2605	.9655	.2698	3.706	74.9°	1.3073
.2653	15.2°	.2622	.9650	.2717	3.681	74.8°	1.3055
.2670	15.3°	.2639	.9646	.2736	3.655	74.7°	1.3038
.2688	15.4°	.2656	.9641	.2754	3.630	74.6°	1.3020
.2705	15.5°	.2672	.9636	.2773	3.606	74.5°	1.3003
.2723	15.6°	.2689	.9632	.2792	3.582	74.4°	1.2985
.2740	15.7°	.2706	.9627	.2811	3.558	74.3°	1.2968
.2758	15.8°	.2723	.9622	.2830	3.534	74.2°	1.2950
.2775	15.9°	.2740	.9617	.2849	3.511	74.1°	1.2933
.2793	16.0°	.2756	.9613	.2867	3.487	74.0°	1.2915
.2810	16.1°	.2773	.9608	.2886	3.465	73.9°	1.2898
.2827	16.2°	.2790	.9603	.2905	3.442	73.8°	1.2881
.2845	16.3°	.2807	.9598	.2924	3.420	73.7°	1.2863
.2862	16.4°	.2823	.9593	.2943	3.398	73.6°	1.2846
.2880	16.5°	.2840	.9588	.2962	3.376	73.5°	1.2828
.2897	16.6°	.2857	.9583	.2981	3.354	73.4°	1.2811
.2915	16.7°	.2874	.9578	.3000	3.333	73.3°	1.2793
.2932	16.8°	.2890	.9573	.3019	3.312	73.2°	1.2776
.2950	16.9°	.2907	.9568	.3038	3.291	73.1°	1.2758
.2967	17.0°	.2924	.9563	.3057	3.271	73.0°	1.2741
.2985	17.1°	.2940	.9558	.3076	3.251	72.9°	1.2723
.3002	17.2°	.2957	.9553	.3096	3.230	72.8°	1.2706
.3019	17.3°	.2974	.9548	.3115	3.211	72.7°	1.2689
.3037	17.4°	.2990	.9542	.3134	3.191	72.6°	1.2671
.3054	17.5°	.3007	.9537	.3153	3.172	72.5°	1.2654
.3072	17.6°	.3024	.9532	.3172	3.152	72.4°	1.2636
.3089	17.7°	.3040	.9527	.3191	3.133	72.3°	1.2619
.3107	17.8°	.3057	.9521	.3211	3.115	72.2°	1.2601
.3124	17.9°	.3074	.9516	.3230	3.096	72.1°	1.2584
.3142	18.0°	.3090	.9511	.3249	3.078	72.0°	1.2566
.3159	18.1°	.3107	.9505	.3269	3.060	71.9°	1.2549
.3176	18.2°	.3123	.9500	.3288	3.042	71.8°	1.2531
.3194	18.3°	.3140	.9494	.3307	3.024	71.7°	1.2514
.3211	18.4°	.3156	.9489	.3327	3.006	71.6°	1.2497
.3229	18.5°	.3173	.9483	.3346	2.989	71.5°	1.2479
.3246	18.6°	.3190	.9478	.3365	2.971	71.4°	1.2462
.3264	18.7°	.3206	.9472	.3385	2.954	71.3°	1.2444
.3281	18.8°	.3223	.9466	.3404	2.937	71.2°	1.2427
.3299	18.9°	.3239	.9461	.3424	2.921	71.1°	1.2409
.3316	19.0°	.3256	.9455	.3443	2.904	71.0°	1.2392
.3334	19.1°	.3272	.9449	.3463	2.888	70.9°	1.2374
.3351	19.2°	.3289	.9444	.3482	2.872	70.8°	1.2357
.3368	19.3°	.3305	.9438	.3502	2.856	70.7°	1.2339
.3386	19.4°	.3322	.9432	.3522	2.840	70.6°	1.2322
.3403	19.5°	.3338	.9426	.3541	2.824	70.5°	1.2305
.3421	19.6°	.3355	.9421	.3561	2.808	70.4°	1.2287
.3438	19.7°	.3371	.9415	.3581	2.793	70.3°	1.2270
.3456	19.8°	.3387	.9409	.3600	2.778	70.2°	1.2252
.3473	19.9°	.3404	.9403	.3620	2.762	70.1°	1.2235
		Cos	**Sin**	**Cot**	**Tan**	**Degrees**	**Radians**

Table A *(Continued)*

Radians	Degrees	Sin	Cos	Tan	Cot		
.3491	20.0°	.3420	.9397	.3640	2.747	70.0°	1.2217
.3508	20.1°	.3437	.9391	.3659	2.733	69.9°	1.2200
.3526	20.2°	.3453	.9385	.3679	2.718	69.8°	1.2182
.3543	20.3°	.3469	.9379	.3699	2.703	69.7°	1.2165
.3560	20.4°	.3486	.9373	.3719	2.689	69.6°	1.2147
.3578	20.5°	.3502	.9367	.3739	2.675	69.5°	1.2130
.3595	20.6°	.3518	.9361	.3759	2.660	69.4°	1.2113
.3613	20.7°	.3535	.9354	.3779	2.646	69.3°	1.2095
.3630	20.8°	.3551	.9348	.3799	2.633	69.2°	1.2078
.3648	20.9°	.3567	.9342	.3819	2.619	69.1°	1.2060
.3665	21.0°	.3584	.9336	.3839	2.605	69.0°	1.2043
.3683	21.1°	.3600	.9330	.3859	2.592	68.9°	1.2025
.3700	21.2°	.3616	.9323	.3879	2.578	68.8°	1.2008
.3718	21.3°	.3633	.9317	.3899	2.565	68.7°	1.1990
.3735	21.4°	.3649	.9311	.3919	2.552	68.6°	1.1973
.3752	21.5°	.3665	.9304	.3939	2.539	68.5°	1.1956
.3770	21.6°	.3681	.9298	.3959	2.526	68.4°	1.1938
.3787	21.7°	.3697	.9291	.3979	2.513	68.3°	1.1921
.3805	21.8°	.3714	.9285	.4000	2.500	68.2°	1.1903
.3822	21.9°	.3730	.9278	.4020	2.488	68.1°	1.1886
.3840	22.0°	.3746	.9272	.4040	2.475	68.0°	1.1868
.3857	22.1°	.3762	.9265	.4061	2.463	67.9°	1.1851
.3875	22.2°	.3778	.9259	.4081	2.450	67.8°	1.1833
.3892	22.3°	.3795	.9252	.4101	2.438	67.7°	1.1816
.3910	22.4°	.3811	.9245	.4122	2.426	67.6°	1.1798
.3927	22.5°	.3827	.9239	.4142	2.414	67.5°	1.1781
.3944	22.6°	.3843	.9232	.4163	2.402	67.4°	1.1764
.3962	22.7°	.3859	.9225	.4183	2.391	67.3°	1.1746
.3979	22.8°	.3875	.9219	.4204	2.379	67.2°	1.1729
.3997	22.9°	.3891	.9212	.4224	2.367	67.1°	1.1711
.4014	23.0°	.3907	.9205	.4245	2.356	67.0°	1.1694
.4032	23.1°	.3923	.9198	.4265	2.344	66.9°	1.1676
.4049	23.2°	.3939	.9191	.4286	2.333	66.8°	1.1659
.4067	23.3°	.3955	.9184	.4307	2.322	66.7°	1.1641
.4084	23.4°	.3971	.9178	.4327	2.311	66.6°	1.1624
.4102	23.5°	.3987	.9171	.4348	2.300	66.5°	1.1606
.4119	23.6°	.4003	.9164	.4369	2.289	66.4°	1.1589
.4136	23.7°	.4019	.9157	.4390	2.278	66.3°	1.1572
.4154	23.8°	.4035	.9150	.4411	2.267	66.2°	1.1554
.4171	23.9°	.4051	.9143	.4431	2.257	66.1°	1.1537
.4189	24.0°	.4067	.9135	.4452	2.246	66.0°	1.1519
.4206	24.1°	.4083	.9128	.4473	2.236	65.9°	1.1502
.4224	24.2°	.4099	.9121	.4494	2.225	65.8°	1.1484
.4241	24.3°	.4115	.9114	.4515	2.215	65.7°	1.1467
.4259	24.4°	.4131	.9107	.4536	2.204	65.6°	1.1449
.4276	24.5°	.4147	.9100	.4557	2.194	65.5°	1.1432
.4294	24.6°	.4163	.9092	.4578	2.184	65.4°	1.1414
.4311	24.7°	.4179	.9085	.4599	2.174	65.3°	1.1397
.4328	24.8°	.4195	.9078	.4621	2.164	65.2°	1.1380
.4346	24.9°	.4210	.9070	.4642	2.154	65.1°	1.1362
		Cos	Sin	Cot	Tan	Degrees	Radians

Table A *(Continued)*

Radians	Degrees	Sin	Cos	Tan	Cot		
.4363	25.0°	.4226	.9063	.4663	2.145	65.0°	1.1345
.4381	25.1°	.4242	.9056	.4684	2.135	64.9°	1.1327
.4398	25.2°	.4258	.9048	.4706	2.125	64.8°	1.1310
.4416	25.3°	.4274	.9041	.4727	2.116	64.7°	1.1292
.4433	25.4°	.4289	.9033	.4748	2.106	64.6°	1.1275
.4451	25.5°	.4305	.9026	.4770	2.097	64.5°	1.1257
.4468	25.6°	.4321	.9018	.4791	2.087	64.4°	1.1240
.4485	25.7°	.4337	.9011	.4813	2.078	64.3°	1.1222
.4503	25.8°	.4352	.9003	.4834	2.069	64.2°	1.1205
.4520	25.9°	.4368	.8996	.4856	2.059	64.1°	1.1188
.4538	26.0°	.4384	.8988	.4877	2.050	64.0°	1.1170
.4555	26.1°	.4399	.8980	.4899	2.041	63.9°	1.1153
.4573	26.2°	.4415	.8973	.4921	2.032	63.8°	1.1135
.4590	26.3°	.4431	.8965	.4942	2.023	63.7°	1.1118
.4608	26.4°	.4446	.8957	.4964	2.014	63.6°	1.1100
.4625	26.5°	.4462	.8949	.4986	2.006	63.5°	1.1083
.4643	26.6°	.4478	.8942	.5008	1.997	63.4°	1.1065
.4660	26.7°	.4493	.8934	.5029	1.988	63.3°	1.1048
.4677	26.8°	.4509	.8926	.5051	1.980	63.2°	1.1030
.4695	26.9°	.4524	.8918	.5073	1.971	63.1°	1.1013
.4712	27.0°	.4540	.8910	.5095	1.963	63.0°	1.0996
.4730	27.1°	.4555	.8902	.5117	1.954	62.9°	1.0978
.4747	27.2°	.4571	.8894	.5139	1.946	62.8°	1.0961
.4765	27.3°	.4586	.8886	.5161	1.937	62.7°	1.0943
.4782	27.4°	.4602	.8878	.5184	1.929	62.6°	1.0926
.4800	27.5°	.4617	.8870	.5206	1.921	62.5°	1.0908
.4817	27.6°	.4633	.8862	.5228	1.913	62.4°	1.0891
.4835	27.7°	.4648	.8854	.5250	1.905	62.3°	1.0873
.4852	27.8°	.4664	.8846	.5272	1.897	62.2°	1.0856
.4869	27.9°	.4679	.8838	.5295	1.889	62.1°	1.0838
.4887	28.0°	.4695	.8829	.5317	1.881	62.0°	1.0821
.4904	28.1°	.4710	.8821	.5340	1.873	61.9°	1.0804
.4922	28.2°	.4726	.8813	.5362	1.865	61.8°	1.0786
.4939	28.3°	.4741	.8805	.5384	1.857	61.7°	1.0769
.4957	28.4°	.4756	.8796	.5407	1.849	61.6°	1.0751
.4974	28.5°	.4772	.8788	.5430	1.842	61.5°	1.0734
.4992	28.6°	.4787	.8780	.5452	1.834	61.4°	1.0716
.5009	28.7°	.4802	.8771	.5475	1.827	61.3°	1.0699
.5027	28.8°	.4818	.8763	.5498	1.819	61.2°	1.0681
.5044	28.9°	.4833	.8755	.5520	1.811	61.1°	1.0664
.5061	29.0°	.4848	.8746	.5543	1.804	61.0°	1.0647
.5079	29.1°	.4863	.8738	.5566	1.797	60.9°	1.0629
.5096	29.2°	.4879	.8729	.5589	1.789	60.8°	1.0612
.5114	29.3°	.4894	.8721	.5612	1.782	60.7°	1.0594
.5131	29.4°	.4909	.8712	.5635	1.775	60.6°	1.0577
.5149	29.5°	.4924	.8704	.5658	1.767	60.5°	1.0559
.5166	29.6°	.4939	.8695	.5681	1.760	60.4°	1.0542
.5184	29.7°	.4955	.8686	.5704	1.753	60.3°	1.0524
.5201	29.8°	.4970	.8678	.5727	1.746	60.2°	1.0507
.5219	29.9°	.4985	.8669	.5750	1.739	60.1°	1.0489
		Cos	Sin	Cot	Tan	Degrees	Radians

Table A *(Continued)*

Radians	Degrees	Sin	Cos	Tan	Cot		
.5236	30.0°	.5000	.8660	.5774	1.732	60.0°	1.0472
.5253	30.1°	.5015	.8652	.5797	1.725	59.9°	1.0455
.5271	30.2°	.5030	.8643	.5820	1.718	59.8°	1.0437
.5288	30.3°	.5045	.8634	.5844	1.711	59.7°	1.0420
.5306	30.4°	.5060	.8625	.5867	1.704	59.6°	1.0402
.5323	30.5°	.5075	.8616	.5890	1.698	59.5°	1.0385
.5341	30.6°	.5090	.8607	.5914	1.691	59.4°	1.0367
.5358	30.7°	.5105	.8599	.5938	1.684	59.3°	1.0350
.5376	30.8°	.5120	.8590	.5961	1.678	59.2°	1.0332
.5393	30.9°	.5135	.8581	.5985	1.671	59.1°	1.0315
.5411	31.0°	.5150	.8572	.6009	1.664	59.0°	1.0297
.5428	31.1°	.5165	.8563	.6032	1.658	58.9°	1.0280
.5445	31.2°	.5180	.8554	.6056	1.651	58.8°	1.0263
.5463	31.3°	.5195	.8545	.6080	1.645	58.7°	1.0245
.5480	31.4°	.5210	.8536	.6104	1.638	58.6°	1.0228
.5498	31.5°	.5225	.8526	.6128	1.632	58.5°	1.0210
.5515	31.6°	.5240	.8517	.6152	1.625	58.4°	1.0193
.5533	31.7°	.5255	.8508	.6176	1.619	58.3°	1.0175
.5550	31.8°	.5270	.8499	.6200	1.613	58.2°	1.0158
.5568	31.9°	.5284	.8490	.6224	1.607	58.1°	1.0140
.5585	32.0°	.5299	.8480	.6249	1.600	58.0°	1.0123
.5603	32.1°	.5314	.8471	.6273	1.594	57.9°	1.0105
.5620	32.2°	.5329	.8462	.6297	1.588	57.8°	1.0088
.5637	32.3°	.5344	.8453	.6322	1.582	57.7°	1.0071
.5655	32.4°	.5358	.8443	.6346	1.576	57.6°	1.0053
.5672	32.5°	.5373	.8434	.6371	1.570	57.5°	1.0036
.5690	32.6°	.5388	.8425	.6395	1.564	57.4°	1.0018
.5707	32.7°	.5402	.8415	.6420	1.558	57.3°	1.0001
.5725	32.8°	.5417	.8406	.6445	1.552	57.2°	.9983
.5742	32.9°	.5432	.8396	.6469	1.546	57.1°	.9966
.5760	33.0°	.5446	.8387	.6494	1.540	57.0°	.9948
.5777	33.1°	.5461	.8377	.6519	1.534	56.9°	.9931
.5794	33.2°	.5476	.8368	.6544	1.528	56.8°	.9913
.5812	33.3°	.5490	.8358	.6569	1.522	56.7°	.9896
.5829	33.4°	.5505	.8348	.6594	1.517	56.6°	.9879
.5847	33.5°	.5519	.8339	.6619	1.511	56.5°	.9861
.5864	33.6°	.5534	.8329	.6644	1.505	56.4°	.9844
.5882	33.7°	.5548	.8320	.6669	1.499	56.3°	.9826
.5899	33.8°	.5563	.8310	.6694	1.494	56.2°	.9809
.5917	33.9°	.5577	.8300	.6720	1.488	56.1°	.9791
.5934	34.0°	.5592	.8290	.6745	1.483	56.0°	.9774
.5952	34.1°	.5606	.8281	.6771	1.477	55.9°	.9756
.5969	34.2°	.5621	.8271	.6796	1.471	55.8°	.9739
.5986	34.3°	.5635	.8261	.6822	1.466	55.7°	.9721
.6004	34.4°	.5650	.8251	.6847	1.460	55.6°	.9704
.6021	34.5°	.5664	.8241	.6873	1.455	55.5°	.9687
.6039	34.6°	.5678	.8231	.6899	1.450	55.4°	.9669
.6056	34.7°	.5693	.8221	.6924	1.444	55.3°	.9652
.6074	34.8°	.5707	.8211	.6950	1.439	55.2°	.9634
.6091	34.9°	.5721	.8202	.6976	1.433	55.1°	.9617
		Cos	Sin	Cot	Tan	Degrees	Radians

Table A *(Continued)*

Radians	Degrees	Sin	Cos	Tan	Cot		
.6109	35.0°	.5736	.8192	.7002	1.428	55.0°	.9599
.6126	35.1°	.5750	.8181	.7028	1.423	54.9°	.9582
.6144	35.2°	.5764	.8171	.7054	1.418	54.8°	.9564
.6161	35.3°	.5779	.8161	.7080	1.412	54.7°	.9547
.6178	35.4°	.5793	.8151	.7107	1.407	54.6°	.9530
.6196	35.5°	.5807	.8141	.7133	1.402	54.5°	.9512
.6213	35.6°	.5821	.8131	.7159	1.397	54.4°	.9495
.6231	35.7°	.5835	.8121	.7186	1.392	54.3°	.9477
.6248	35.8°	.5850	.8111	.7212	1.387	54.2°	.9460
.6266	35.9°	.5864	.8100	.7239	1.381	54.1°	.9442
.6283	36.0°	.5878	.8090	.7265	1.376	54.0°	.9425
.6301	36.1°	.5892	.8080	.7292	1.371	53.9°	.9407
.6318	36.2°	.5906	.8070	.7319	1.366	53.8°	.9390
.6336	36.3°	.5920	.8059	.7346	1.361	53.7°	.9372
.6353	36.4°	.5934	.8049	.7373	1.356	53.6°	.9355
.6370	36.5°	.5948	.8039	.7400	1.351	53.5°	.9338
.6388	36.6°	.5962	.8028	.7427	1.347	53.4°	.9320
.6405	36.7°	.5976	.8018	.7454	1.342	53.3°	.9303
.6423	36.8°	.5990	.8007	.7481	1.337	53.2°	.9285
.6440	36.9°	.6004	.7997	.7508	1.332	53.1°	.9268
.6458	37.0°	.6018	.7986	.7536	1.327	53.0°	.9250
.6475	37.1°	.6032	.7976	.7563	1.322	52.9°	.9233
.6493	37.2°	.6046	.7965	.7590	1.317	52.8°	.9215
.6510	37.3°	.6060	.7955	.7618	1.313	52.7°	.9198
.6528	37.4°	.6074	.7944	.7646	1.308	52.6°	.9180
.6545	37.5°	.6088	.7934	.7673	1.303	52.5°	.9163
.6562	37.6°	.6101	.7923	.7701	1.299	52.4°	.9146
.6580	37.7°	.6115	.7912	.7729	1.294	52.3°	.9128
.6597	37.8°	.6129	.7902	.7757	1.289	52.2°	.9111
.6615	37.9°	.6143	.7891	.7785	1.285	52.1°	.9093
.6632	38.0°	.6157	.7880	.7813	1.280	52.0°	.9076
.6650	38.1°	.6170	.7869	.7841	1.275	51.9°	.9058
.6667	38.2°	.6184	.7859	.7869	1.271	51.8°	.9041
.6685	38.3°	.6198	.7848	.7898	1.266	51.7°	.9023
.6702	38.4°	.6211	.7837	.7926	1.262	51.6°	.9006
.6720	38.5°	.6225	.7826	.7954	1.257	51.5°	.8988
.6737	38.6°	.6239	.7815	.7983	1.253	51.4°	.8971
.6754	38.7°	.6252	.7804	.8012	1.248	51.3°	.8954
.6772	38.8°	.6266	.7793	.8040	1.244	51.2°	.8936
.6789	38.9°	.6280	.7782	.8069	1.239	51.1°	.8919
.6807	39.0°	.6293	.7771	.8098	1.235	51.0°	.8901
.6824	39.1°	.6307	.7760	.8127	1.230	50.9°	.8884
.6842	39.2°	.6320	.7749	.8156	1.226	50.8°	.8866
.6859	39.3°	.6334	.7738	.8185	1.222	50.7°	.8849
.6877	39.4°	.6347	.7727	.8214	1.217	50.6°	.8831
.6894	39.5°	.6361	.7716	.8243	1.213	50.5°	.8814
.6912	39.6°	.6374	.7705	.8273	1.209	50.4°	.8796
.6929	39.7°	.6388	.7694	.8302	1.205	50.3°	.8779
.6946	39.8°	.6401	.7683	.8332	1.200	50.2°	.8762
.6964	39.9°	.6414	.7672	.8361	1.196	50.1°	.8744
		Cos	**Sin**	**Cot**	**Tan**	**Degrees**	**Radians**

Table A *(Continued)*

Radians	Degrees	Sin	Cos	Tan	Cot		
.6981	40.0°	.6428	.7660	.8391	1.192	50.0°	.8727
.6999	40.1°	.6441	.7649	.8421	1.188	49.9°	.8709
.7016	40.2°	.6455	.7638	.8451	1.183	49.8°	.8692
.7034	40.3°	.6468	.7627	.8481	1.179	49.7°	.8674
.7051	40.4°	.6481	.7615	.8511	1.175	49.6°	.8657
.7069	40.5°	.6494	.7604	.8541	1.171	49.5°	.8639
.7086	40.6°	.6508	.7593	.8571	1.167	49.4°	.8622
.7103	40.7°	.6521	.7581	.8601	1.163	49.3°	.8604
.7121	40.8°	.6534	.7570	.8632	1.159	49.2°	.8587
.7138	40.9°	.6547	.7559	.8662	1.154	49.1°	.8570
.7156	41.0°	.6561	.7547	.8693	1.150	49.0°	.8552
.7173	41.1°	.6574	.7536	.8724	1.146	48.9°	.8535
.7191	41.2°	.6587	.7524	.8754	1.142	48.8°	.8517
.7208	41.3°	.6600	.7513	.8785	1.138	48.7°	.8500
.7226	41.4°	.6613	.7501	.8816	1.134	48.6°	.8482
.7243	41.5°	.6626	.7490	.8847	1.130	48.5°	.8465
.7261	41.6°	.6639	.7478	.8878	1.126	48.4°	.8447
.7278	41.7°	.6652	.7466	.8910	1.122	48.3°	.8430
.7295	41.8°	.6665	.7455	.8941	1.118	48.2°	.8412
.7313	41.9°	.6678	.7443	.8972	1.115	48.1°	.8395
.7330	42.0°	.6691	.7431	.9004	1.111	48.0°	.8378
.7348	42.1°	.6704	.7420	.9036	1.107	47.9°	.8360
.7365	42.2°	.6717	.7408	.9067	1.103	47.8°	.8343
.7383	42.3°	.6730	.7396	.9099	1.099	47.7°	.8325
.7400	42.4°	.6743	.7385	.9131	1.095	47.6°	.8308
.7418	42.5°	.6756	.7373	.9163	1.091	47.5°	.8290
.7435	42.6°	.6769	.7361	.9195	1.087	47.4°	.8273
.7453	42.7°	.6782	.7349	.9228	1.084	47.3°	.8255
.7470	42.8°	.6794	.7337	.9260	1.080	47.2°	.8238
.7487	42.9°	.6807	.7325	.9293	1.076	47.1°	.8221
.7505	43.0°	.6820	.7314	.9325	1.072	47.0°	.8203
.7522	43.1°	.6833	.7302	.9358	1.069	46.9°	.8186
.7540	43.2°	.6845	.7290	.9391	1.065	46.8°	.8168
.7557	43.3°	.6858	.7278	.9424	1.061	46.7°	.8151
.7575	43.4°	.6871	.7266	.9457	1.057	46.6°	.8133
.7592	43.5°	.6884	.7254	.9490	1.054	46.5°	.8116
.7610	43.6°	.6896	.7242	.9523	1.050	46.4°	.8098
.7627	43.7°	.6909	.7230	.9556	1.046	46.3°	.8081
.7645	43.8°	.6921	.7218	.9590	1.043	46.2°	.8063
.7662	43.9°	.6934	.7206	.9623	1.039	46.1°	.8046
.7679	44.0°	.6947	.7193	.9657	1.036	46.0°	.8029
.7697	44.1°	.6959	.7181	.9691	1.032	45.9°	.8011
.7714	44.2°	.6972	.7169	.9725	1.028	45.8°	.7994
.7732	44.3°	.6984	.7157	.9759	1.025	45.7°	.7976
.7749	44.4°	.6997	.7145	.9793	1.021	45.6°	.7959
.7767	44.5°	.7009	.7133	.9827	1.018	45.5°	.7941
.7784	44.6°	.7022	.7120	.9861	1.014	45.4°	.7924
.7802	44.7°	.7034	.7108	.9896	1.011	45.3°	.7906
.7819	44.8°	.7046	.7096	.9930	1.007	45.2°	.7889
.7837	44.9°	.7059	.7083	.9965	1.003	45.1°	.7871
.7854	45.0°	.7071	.7071	1.0000	1.000	45.0°	.7854
		Cos	Sin	Cot	Tan	Degrees	Radians

Table B Base-10 Logarithms

N	0	1	2	3	4	5	6	7	8	9
1.0	.0000	.0043	.0086	.0128	.0170	.0212	.0253	.0294	.0334	.0374
1.1	.0414	.0453	.0492	.0531	.0569	.0607	.0645	.0682	.0719	.0755
1.2	.0792	.0828	.0864	.0899	.0934	.0969	.1004	.1038	.1072	.1106
1.3	.1139	.1173	.1206	.1239	.1271	.1303	.1335	.1367	.1399	.1430
1.4	.1461	.1492	.1523	.1553	.1584	.1614	.1644	.1673	.1703	.1732
1.5	.1761	.1790	.1818	.1847	.1875	.1903	.1931	.1959	.1987	.2014
1.6	.2041	.2068	.2095	.2122	.2148	.2175	.2201	.2227	.2253	.2279
1.7	.2304	.2330	.2355	.2380	.2405	.2430	.2455	.2480	.2504	.2529
1.8	.2553	.2577	.2601	.2625	.2648	.2672	.2695	.2718	.2742	.2765
1.9	.2788	.2810	.2833	.2856	.2878	.2900	.2923	.2945	.2967	.2989
2.0	.3010	.3032	.3054	.3075	.3096	.3118	.3139	.3160	.3181	.3201
2.1	.3222	.3243	.3263	.3284	.3304	.3324	.3345	.3365	.3385	.3404
2.2	.3424	.3444	.3464	.3483	.3502	.3522	.3541	.3560	.3579	.3598
2.3	.3617	.3636	.3655	.3674	.3692	.3711	.3729	.3747	.3766	.3784
2.4	.3802	.3820	.3838	.3856	.3874	.3892	.3909	.3927	.3945	.3962
2.5	.3979	.3997	.4014	.4031	.4048	.4065	.4082	.4099	.4116	.4133
2.6	.4150	.4166	.4183	.4200	.4216	.4232	.4249	.4265	.4281	.4298
2.7	.4314	.4330	.4346	.4362	.4378	.4393	.4409	.4425	.4440	.4456
2.8	.4472	.4487	.4502	.4518	.4533	.4548	.4564	.4579	.4594	.4609
2.9	.4624	.4639	.4654	.4669	.4683	.4698	.4713	.4728	.4742	.4757
3.0	.4771	.4786	.4800	.4814	.4829	.4843	.4857	.4871	.4886	.4900
3.1	.4914	.4928	.4942	.4955	.4969	.4983	.4997	.5011	.5024	.5038
3.2	.5051	.5065	.5079	.5092	.5105	.5119	.5132	.5145	.5159	.5172
3.3	.5185	.5198	.5211	.5224	.5237	.5250	.5263	.5276	.5289	.5302
3.4	.5315	.5328	.5340	.5353	.5366	.5378	.5391	.5403	.5416	.5428
3.5	.5441	.5453	.5465	.5478	.5490	.5502	.5514	.5527	.5539	.5551
3.6	.5563	.5575	.5587	.5599	.5611	.5623	.5635	.5647	.5658	.5670
3.7	.5682	.5694	.5705	.5717	.5729	.5740	.5752	.5763	.5775	.5786
3.8	.5798	.5809	.5821	.5832	.5843	.5855	.5866	.5877	.5888	.5899
3.9	.5911	.5922	.5933	.5944	.5955	.5966	.5977	.5988	.5999	.6010
4.0	.6021	.6031	.6042	.6053	.6064	.6075	.6085	.6096	.6107	.6117
4.1	.6128	.6138	.6149	.6160	.6170	.6180	.6191	.6201	.6212	.6222
4.2	.6232	.6243	.6253	.6263	.6274	.6284	.6294	.6304	.6314	.6325
4.3	.6335	.6345	.6355	.6365	.6375	.6385	.6395	.6405	.6415	.6425
4.4	.6435	.6444	.6454	.6464	.6474	.6484	.6493	.6503	.6513	.6522
4.5	.6532	.6542	.6551	.6561	.6571	.6580	.6590	.6599	.6609	.6618
4.6	.6628	.6637	.6646	.6656	.6665	.6675	.6684	.6693	.6702	.6712
4.7	.6721	.6730	.6739	.6749	.6758	.6767	.6776	.6785	.6794	.6803
4.8	.6812	.6821	.6830	.6839	.6848	.6857	.6866	.6875	.6884	.6893
4.9	.6902	.6911	.6920	.6928	.6937	.6946	.6955	.6964	.6972	.6981
5.0	.6990	.6998	.7007	.7016	.7024	.7033	.7042	.7050	.7059	.7067
5.1	.7076	.7084	.7093	.7101	.7110	.7118	.7126	.7135	.7143	.7152
5.2	.7160	.7168	.7177	.7185	.7193	.7202	.7210	.7218	.7226	.7235
5.3	.7243	.7251	.7259	.7267	.7275	.7284	.7292	.7300	.7308	.7316
5.4	.7324	.7332	.7340	.7348	.7356	.7364	.7372	.7380	.7388	.7396

Table B *(Continued)*

N	0	1	2	3	4	5	6	7	8	9
5.5	.7404	.7412	.7419	.7427	.7435	.7443	.7451	.7459	.7466	.7474
5.6	.7482	.7490	.7497	.7505	.7513	.7520	.7528	.7536	.7543	.7551
5.7	.7559	.7566	.7574	.7582	.7589	.7597	.7604	.7162	.7619	.7627
5.8	.7634	.7642	.7649	.7657	.7664	.7672	.7679	.7686	.7694	.7701
5.9	.7709	.7716	.7723	.7731	.7738	.7745	.7752	.7760	.7767	.7774
6.0	.7782	.7789	.7796	.7803	.7810	.7818	.7825	.7832	.7839	.7846
6.1	.7853	.7860	.7868	.7875	.7882	.7889	.7896	.7903	.7910	.7917
6.2	.7924	.7931	.7938	.7945	.7952	.7959	.7966	.7973	.7980	.7987
6.3	.7993	.8000	.8007	.8014	.8021	.8028	.8035	.8041	.8048	.8055
6.4	.8062	.8069	.8075	.8082	.8089	.8096	.8102	.8109	.8116	.8122
6.5	.8129	.8136	.8142	.8149	.8156	.8162	.8169	.8176	.8182	.8189
6.6	.8195	.8202	.8209	.8215	.8222	.8228	.8235	.8241	.8248	.8254
6.7	.8261	.8267	.8274	.8280	.8287	.8293	.8299	.8306	.8312	.8319
6.8	.8325	.8331	.8338	.8344	.8351	.8357	.8363	.8370	.8376	.8382
6.9	.8388	.8395	.8401	.8407	.8414	.8420	.8426	.8432	.8439	.8445
7.0	.8451	.8457	.8463	.8470	.8476	.8482	.8488	.8494	.8500	.8506
7.1	.8513	.8519	.8525	.8531	.8537	.8543	.8549	.8555	.8561	.8567
7.2	.8573	.8579	.8585	.8591	.8597	.8603	.8609	.8615	.8621	.8627
7.3	.8633	.8639	.8645	.8651	.8657	.8663	.8669	.8675	.8681	.8686
7.4	.8692	.8698	.8704	.8710	.8716	.8722	.8727	.8733	.8739	.8745
7.5	.8751	.8756	.8762	.8768	.8774	.8779	.8785	.8791	.8797	.8802
7.6	.8808	.8814	.8820	.8825	.8831	.8837	.8842	.8848	.8854	.8859
7.7	.8865	.8871	.8876	.8882	.8887	.8893	.8899	.8904	.8910	.8915
7.8	.8921	.8927	.8932	.8938	.8943	.8949	.8954	.8960	.8965	.8971
7.9	.8976	.8982	.8987	.8993	.8998	.9004	.9009	.9015	.9020	.9025
8.0	.9031	.9036	.9042	.9047	.9053	.9058	.9063	.9069	.9074	.9079
8.1	.9085	.9090	.9096	.9101	.9106	.9112	.9117	.9122	.9128	.9133
8.2	.9138	.9143	.9149	.9154	.9159	.9165	.9170	.9175	.9180	.9186
8.3	.9191	.9196	.9201	.9206	.9212	.9217	.9222	.9227	.9232	.9238
8.4	.9243	.9248	.9253	.9258	.9263	.9269	.9274	.9279	.9284	.9289
8.5	.9294	.9299	.9304	.9309	.9315	.9320	.9325	.9330	.9335	.9330
8.6	.9345	.9350	.9355	.9360	.9365	.9370	.9375	.9380	.9385	.9390
8.7	.9395	.9400	.9405	.9410	.9415	.9420	.9425	.9430	.9435	.9440
8.8	.9445	.9450	.9455	.9460	.9465	.9469	.9474	.9479	.9484	.9489
8.9	.9494	.9499	.9504	.9509	.9513	.9518	.9523	.9528	.9533	.9538
9.0	.9542	.9547	.9552	.9557	.9562	.9566	.9571	.9576	.9581	.9586
9.1	.9590	.9595	.9600	.9605	.9609	.9614	.9619	.9624	.9628	.9633
9.2	.9638	.9643	.9647	.9652	.9657	.9661	.9666	.9671	.9675	.9680
9.3	.9685	.9689	.9694	.9699	.9703	.9708	.9713	.9717	.9722	.9727
9.4	.9731	.9736	.9741	.9745	.9750	.9754	.9759	.9763	.9768	.9773
9.5	.9777	.9782	.9786	.9791	.9795	.9800	.9805	.9809	.9814	.9818
9.6	.9823	.9827	.9832	.9836	.9841	.9845	.9850	.9854	.9859	.9863
9.7	.9868	.9872	.9877	.9881	.9886	.9890	.9894	.9899	.9903	.9908
9.8	.9912	.9917	.9921	.9926	.9930	.9934	.9939	.9943	.9948	.9952
9.9	.9956	.9961	.9965	.9969	.9974	.9978	.9983	.9987	.9991	.9996

Table C Base-*e* Logarithms

N	0	1	2	3	4	5	6	7	8	9
1.0	.0000	.0100	.0198	.0296	.0392	.0488	.0583	.0677	.0770	.0862
1.1	.0953	.1044	.1133	.1222	.1310	.1398	.1484	.1570	.1655	.1740
1.2	.1823	.1906	.1989	.2070	.2151	.2231	.2311	.2390	.2469	.2546
1.3	.2624	.2700	.2776	.2852	.2927	.3001	.3075	.3148	.3221	.3293
1.4	.3365	.3436	.3507	.3577	.3646	.3716	.3784	.3853	.3920	.3988
1.5	.4055	.4121	.4187	.4253	.4318	.4383	.4447	.4511	.4574	.4637
1.6	.4700	.4762	.4824	.4886	.4947	.5008	.5068	.5128	.5188	.5247
1.7	.5306	.5365	.5423	.5481	.5539	.5596	.5653	.5710	.5766	.5822
1.8	.5878	.5933	.5988	.6043	.6098	.6152	.6206	.6259	.6313	.6366
1.9	.6419	.6471	.6523	.6575	.6627	.6678	.6729	.6780	.6831	.6881
2.0	.6931	.6981	.7031	.7080	.7129	.7178	.7227	.7275	.7324	.7372
2.1	.7419	.7467	.7514	.7561	.7608	.7655	.7701	.7747	.7793	.7839
2.2	.7885	.7930	.7975	.8020	.8065	.8109	.8154	.8198	.8242	.8286
2.3	.8329	.8372	.8416	.8459	.8502	.8544	.8587	.8629	.8671	.8713
2.4	.8755	.8796	.8838	.8879	.8920	.8961	.9002	.9042	.9083	.9123
2.5	.9163	.9203	.9243	.9282	.9322	.9361	.9400	.9439	.9478	.9517
2.6	.9555	.9594	.9632	.9670	.9708	.9746	.9783	.9821	.9858	.9895
2.7	.9933	.9969	1.0006	.0043	.0080	.0116	.0152	.0188	.0225	.0260
2.8	1.0296	.0332	.0367	.0403	.0438	.0473	.0508	.0543	.0578	.0613
2.9	.0647	.0682	.0716	.0750	.0784	.0818	.0852	.0886	.0919	.0953
3.0	1.0986	.1019	.1053	.1086	.1119	.1151	.1184	.1217	.1249	.1282
3.1	.1314	.1346	.1378	.1410	.1442	.1474	.1506	.1537	.1569	.1600
3.2	.1632	.1663	.1694	.1725	.1756	.1787	.1817	.1848	.1878	.1909
3.3	.1939	.1969	.2000	.2030	.2060	.2090	.2119	.2149	.2179	.2208
3.4	.2238	.2267	.2296	.2326	.2355	.2384	.2413	.2442	.2470	.2499
3.5	1.2528	.2556	.2585	.2613	.2641	.2669	.2698	.2726	.2754	.2782
3.6	.2809	.2837	.2865	.2892	.2920	.2947	.2975	.3002	.3029	.3056
3.7	.3083	.3110	.3137	.3164	.3191	.3218	.3244	.3271	.3297	.3324
3.8	.3350	.3376	.3403	.3429	.3455	.3481	.3507	.3533	.3558	.3584
3.9	.3610	.3635	.3661	.3686	.3712	.3737	.3762	.3788	.3813	.3838
4.0	1.3863	.3888	.3913	.3938	.3962	.3987	.4012	.4036	.4061	.4085
4.1	.4110	.4134	.4159	.4183	.4207	.4231	.4255	.4279	.4303	.4327
4.2	.4351	.4375	.4398	.4422	.4446	.4469	.4493	.4516	.4540	.4563
4.3	.4586	.4609	.4633	.4656	.4679	.4702	.4725	.4748	.4770	.4793
4.4	.4816	.4839	.4861	.4884	.4907	.4929	.4951	.4974	.4996	.5019
4.5	1.5041	.5063	.5085	.5107	.5129	.5151	.5173	.5195	.5217	.5239
4.6	.5261	.5282	.5304	.5326	.5347	.5369	.5390	.5412	.5433	.5454
4.7	.5476	.5497	.5518	.5539	.5560	.5581	.5602	.5623	.5644	.5665
4.8	.5686	.5707	.5728	.5748	.5769	.5790	.5810	.5831	.5851	.5872
4.9	.5892	.5913	.5933	.5953	.5974	.5994	.6014	.6034	.6054	.6074
5.0	1.6094	.6114	.6134	.6154	.6174	.6194	.6214	.6233	.6253	.6273
5.1	.6292	.6312	.6332	.6351	.6371	.6390	.6409	.6429	.6448	.6467
5.2	.6487	.6506	.6525	.6544	.6563	.6582	.6601	.6620	.6639	.6658
5.3	.6677	.6696	.6715	.6734	.6752	.6771	.6790	.6808	.6827	.6845
5.4	.6864	.6882	.6901	.6919	.6938	.6956	.6974	.6993	.7011	7029

Use the properties of logarithms and ln $10 \approx 2.3026$ to find logarithms of numbers less than 1 or greater than 10.

TABLE C BASE-*e* LOGARITHMS   661

Table C *(Continued)*

N	0	1	2	3	4	5	6	7	8	9
5.5	1.7047	.7066	.7084	.7102	.7120	.7138	.7156	.7174	.7192	.7210
5.6	.7228	.7246	.7263	.7281	.7299	.7317	.7334	.7352	.7370	.7387
5.7	.7405	.7422	.7440	.7457	.7475	.7492	.7509	.7527	.7544	.7561
5.8	.7579	.7596	.7613	.7630	.7647	.7664	.7681	.7699	.7716	.7733
5.9	.7750	.7766	.7783	.7800	.7817	.7834	.7851	.7867	.7884	.7901
6.0	1.7918	.7934	.7951	.7967	.7984	.8001	.8017	.8034	.8050	.8066
6.1	.8083	.8099	.8116	.8132	.8148	.8165	.8181	.8197	.8213	.8229
6.2	.8245	.8262	.8278	.8294	.8310	.8326	.8342	.8358	.8374	.8390
6.3	.8405	.8421	.8437	.8453	.8469	.8485	.8500	.8516	.8532	.8547
6.4	.8563	.8579	.8594	.8610	.8625	.8641	.8656	.8672	.8687	.8703
6.5	1.8718	.8733	.8749	.8764	.8779	.8795	.8810	.8825	.8840	.8856
6.6	.8871	.8886	.8901	.8916	.8931	.8946	.8961	.8976	.8991	.9006
6.7	.9021	.9036	.9051	.9066	.9081	.9095	.9110	.9125	.9140	.9155
6.8	.9169	.9184	.9199	.9213	.9228	.9242	.9257	.9272	.9286	.9301
6.9	.9315	.9330	.9344	.9359	.9373	.9387	.9402	.9416	.9430	.9445
7.0	1.9459	.9473	.9488	.9502	.9516	.9530	.9544	.9559	.9573	.9587
7.1	.9601	.9615	.9629	.9643	.9657	.9671	.9685	.9699	.9713	.9727
7.2	.9741	.9755	.9769	.9782	.9796	.9810	.9824	.9838	.9851	.9865
7.3	.9879	.9892	.9906	.9920	.9933	.9947	.9961	.9974	.9988	2.0001
7.4	2.0015	.0028	.0042	.0055	.0069	.0082	.0096	.0109	.0122	.0136
7.5	2.0149	.0162	.0176	.0189	.0202	.0215	.0229	.0242	.0255	.0268
7.6	.0281	.0295	.0308	.0321	.0334	.0347	.0360	.0373	.0386	.0399
7.7	.0412	.0425	.0438	.0451	.0464	.0477	.0490	.0503	.0516	.0528
7.8	.0541	.0554	.0567	.0580	.0592	.0605	.0618	.0631	.0643	.0656
7.9	.0669	.0681	.0694	.0707	.0719	.0732	.0744	.0757	.0769	.0782
8.0	2.0794	.0807	.0819	.0832	.0844	.0857	.0869	.0882	.0894	.0906
8.1	.0919	.0931	.0943	.0956	.0968	.0980	.0992	.1005	.1017	.1029
8.2	.1041	.1054	.1066	.1078	.1090	.1102	.1114	.1126	.1138	.1150
8.3	.1163	.1175	.1187	.1199	.1211	.1223	.1235	.1247	.1258	.1270
8.4	.1282	.1294	.1306	.1318	.1330	.1342	.1353	.1365	.1377	.1389
8.5	2.1401	.1412	.1424	.1436	.1448	.1459	.1471	.1483	.1494	.1506
8.6	.1518	.1529	.1541	.1552	.1564	.1576	.1587	.1599	.1610	.1622
8.7	.1633	.1645	.1656	.1668	.1679	.1691	.1702	.1713	.1725	.1736
8.8	.1748	.1759	.1770	.1782	.1793	.1804	.1815	.1827	.1838	.1849
8.9	.1861	.1872	.1883	.1894	.1905	.1917	.1928	.1939	.1950	.1961
9.0	2.1972	.1983	.1994	.2006	.2017	.2028	.2039	.2050	.2061	.2072
9.1	.2083	.2094	.2105	.2116	.2127	.2138	.2148	.2159	.2170	.2181
9.2	.2192	.2203	.2214	.2225	.2235	.2246	.2257	.2268	.2279	.2289
9.3	.2300	.2311	.2322	.2332	.2343	.2354	.2364	.2375	.2386	.2396
9.4	.2407	.2418	.2428	.2439	.2450	.2460	.2471	.2481	.2492	.2502
9.5	2.2513	.2523	.2534	.2544	.2555	.2565	.2576	.2586	.2597	.2607
9.6	.2618	.2658	.2638	.2649	.2659	.2670	.2680	.2690	.2701	.2711
9.7	.2721	.2732	.2742	.2752	.2762	.2773	.2783	.2793	.2803	.2814
9.8	.2824	.2834	.2844	.2854	.2865	.2875	.2885	.2895	.2905	.2915
9.9	.2925	.2935	.2946	.2956	.2966	.2976	.2986	.2996	.3006	.3016

Use the properties of logarithms and in $10 \approx 2.3026$ to find logarithms of numbers less than 1 or greater than 10.

Table D Powers and Roots

n	n^2	$\sqrt{n}$	n^3	$\sqrt[3]{n}$	n	n^2	$\sqrt{n}$	n^3	$\sqrt[3]{n}$
1	1	1.000	1	1.000	51	2,601	7.141	132,651	3.708
2	4	1.414	8	1.260	52	2,704	7.211	140,608	3.733
3	9	1.732	27	1.442	53	2,809	7.280	148,877	3.756
4	16	2.000	64	1.587	54	2,916	7.348	157,464	3.780
5	25	2.236	125	1.710	55	3,025	7.416	166,375	3.803
6	36	2.449	216	1.817	56	3,136	7.483	175,616	3.826
7	49	2.646	343	1.913	57	3,249	7.550	185,193	3.849
8	64	2.828	512	2.000	58	3,364	7.616	195,112	3.871
9	81	3.000	729	2.080	59	3,481	7.681	205,379	3.893
10	100	3.162	1,000	2.154	60	3,600	7.746	216,000	3.915
11	121	3.317	1,331	2.224	61	3,721	7.810	226,981	3.936
12	144	3.464	1,728	2.289	62	3,844	7.874	238,328	3.958
13	169	3.606	2,197	2.351	63	3,969	7.937	250,047	3.979
14	196	3.742	2,744	2.410	64	4,096	8.000	262,144	4.000
15	225	3.873	3,375	2.466	65	4,225	8.062	274,625	4.021
16	256	4.000	4,096	2.520	66	4,356	8.124	287,496	4.041
17	289	4.123	4,913	2.571	67	4,489	8.185	300,763	4.062
18	324	4.243	5,832	2.621	68	4,624	8.246	314,432	4.082
19	361	4.359	6,859	2.668	69	4,761	8.307	328,509	4.102
20	400	4.472	8,000	2.714	70	4,900	8.367	343,000	4.121
21	441	4.583	9,261	2.759	71	5,041	8.426	357,911	4.141
22	484	4.690	10,648	2.802	72	5,184	8.485	373,248	4.160
23	529	4.796	12,167	2.844	73	5,329	8.544	389,017	4.179
24	576	4.899	13,824	2.884	74	5,476	8.602	405,224	4.198
25	625	5.000	15,625	2.924	75	5,625	8.660	421,875	4.217
26	676	5.099	17,576	2.962	76	5,776	8.718	438,976	4.236
27	729	5.196	19,683	3.000	77	5,929	8.775	456,533	4.254
28	784	5.292	21,952	3.037	78	6,084	8.832	474,552	4.273
29	841	5.385	24,389	3.072	79	6,241	8.888	493,039	4.291
30	900	5.477	27,000	3.107	80	6,400	8.944	512,000	4.309
31	961	5.568	29,791	3.141	81	6,561	9.000	531,441	4.327
32	1,024	5.657	32,768	3.175	82	6,724	9.055	551,368	4.344
33	1,089	5.745	35,937	3.208	83	6,889	9.110	571,787	4.362
34	1,156	5.831	39,304	3.240	84	7,056	9.165	592,704	4.380
35	1,225	5.916	42,875	3.271	85	7,225	9.220	614,125	4.397
36	1,296	6.000	46,656	3.302	86	7,396	9.274	636,056	4.414
37	1,369	6.083	50,653	3.332	87	7,569	9.327	658,503	4.431
38	1,444	6.164	54,872	3.362	88	7,744	9.381	681,472	4.448
39	1,521	6.245	59,319	3.391	89	7,921	9.434	704,969	4.465
40	1,600	6.325	64,000	3.420	90	8,100	9.487	729,000	4.481
41	1,681	6.403	68,921	3.448	91	8,281	9.539	753,571	4.498
42	1,764	6.481	74,088	3.476	92	8,464	9.592	778,688	4.514
43	1,849	6.557	79,507	3.503	93	8,649	9.644	804,357	4.531
44	1,936	6.633	85,184	3.530	94	8,836	9.695	830,584	4.547
45	2,025	6.708	91,125	3.557	95	9,025	9.747	857,375	4.563
46	2,116	6.782	97,336	3.583	96	9,216	9.798	884,736	4.579
47	2,209	6.856	103,823	3.609	97	9,409	9.849	912,673	4.595
48	2,304	6.928	110,592	3.634	98	9,604	9.899	941,192	4.610
49	2,401	7.000	117,649	3.659	99	9,801	9.950	970,299	4.626
50	2,500	7.071	125,000	3.684	100	10,000	10.000	1,000,000	4.642

INDEX

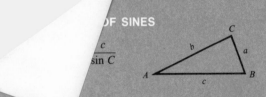

$$\frac{c}{\sin C}$$

6.5 AREAS OF TRIANGLES

$$\text{Area} = k = \tfrac{1}{2}ab \sin C$$
$$\phantom{\text{Area} = k} = \tfrac{1}{2}ac \sin B \left.\right\} \text{SAS}$$
$$\phantom{\text{Area} = k} = \tfrac{1}{2}bc \sin A$$

$$k = \frac{c^2 \sin A \sin B}{2 \sin C}$$
$$ = \frac{a^2 \sin B \sin C}{2 \sin A} \left.\right\} \begin{array}{l} \text{AAS} \\ \text{and} \\ \text{ASA} \end{array}$$
$$ = \frac{b^2 \sin C \sin A}{2 \sin B}$$

$$k = \sqrt{s(s-a)(s-b)(s-c)} \left.\right\} \text{SSS}$$
$$\text{where } s = \tfrac{1}{2}(a + b + c)$$

6.6 DOT PRODUCT OF TWO VECTORS

$$\mathbf{V} \cdot \mathbf{W} = |\mathbf{V}|\,|\mathbf{W}| \cos \theta$$

If $\mathbf{V} = a\mathbf{i} + b\mathbf{j}$ and $\mathbf{W} = c\mathbf{i} + d\mathbf{j}$, then

$$\mathbf{V} \cdot \mathbf{W} = ac + bd$$

7.2 FUNCTIONS OF TWO ANGLES

$$\cos(A + B) = \cos A \cos B - \sin A \sin B$$
$$\cos(A - B) = \cos A \cos B + \sin A \sin B$$
$$\sin(A + B) = \sin A \cos B + \cos A \sin B$$
$$\sin(A - B) = \sin A \cos B - \cos A \sin B$$
$$\tan(A + B) = \frac{\tan A + \tan B}{1 - \tan A \tan B}$$
$$\tan(A - B) = \frac{\tan A - \tan B}{1 + \tan A \tan B}$$

7.3 THE DOUBLE-ANGLE IDENTITIES

$$\sin 2A = 2 \sin A \cos A$$
$$\cos 2A = \cos^2 A - \sin^2 A \qquad \tan 2A = \frac{2 \tan A}{1 - \tan^2 A}$$
$$ = 2 \cos^2 A - 1$$
$$ = 1 - 2 \sin^2 A$$

7.4 THE HALF-ANGLE IDENTITIES

$$\sin \frac{A}{2} = \pm \sqrt{\frac{1 - \cos A}{2}} \qquad \tan \frac{A}{2} = \frac{\sin A}{1 + \cos A}$$
$$\cos \frac{A}{2} = \pm \sqrt{\frac{1 + \cos A}{2}} \qquad \phantom{\tan \frac{A}{2}} = \frac{1 - \cos A}{\sin A}$$

7.5 PRODUCT-TO-SUM AND SUM-TO-PRODUCT FORMULAS

$$\sin A \cos B = \tfrac{1}{2}[\sin(A + B) + \sin(A - B)]$$
$$\cos A \sin B = \tfrac{1}{2}[\sin(A + B) - \sin(A - B)]$$
$$\sin A \sin B = \tfrac{1}{2}[\cos(A - B) - \cos(A + B)]$$
$$\cos A \cos B = \tfrac{1}{2}[\cos(A + B) + \cos(A - B)]$$

$$\sin A + \sin B = 2 \sin \frac{A + B}{2} \cos \frac{A - B}{2}$$

$$\sin A - \sin B = 2 \cos \frac{A + B}{2} \sin \frac{A - B}{2}$$

$$\cos A + \cos B = 2 \cos \frac{A + B}{2} \cos \frac{A - B}{2}$$

$$\cos A - \cos B = -2 \sin \frac{A + B}{2} \sin \frac{A - B}{2}$$

7.5 SUMS OF THE FORM $A \sin x + B \cos x$

$$A \sin x + B \cos x = \sqrt{A^2 + B^2} \sin(x + \phi)$$

$$\text{where } \sin \phi = \frac{B}{\sqrt{A^2 + B^2}} \quad \text{and} \quad \cos \phi = \frac{A}{\sqrt{A^2 + B^2}}$$

7.7 THE GRAPHS OF THE INVERSE TRIGONOMETRIC FUNCTIONS

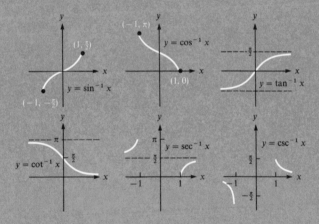